E-PLANE INTEGRATED CIRCUITS

The Artech House Microwave Library

Introduction to Microwaves by Fred E. Gardiol
Lossy Transmission Lines by Fred E. Gardiol
Microwaves Made Simple: Principles and Applications by W. Stephen Cheung, Frederic H. Levien, et al.
Microwaves Made Simple — the Workbook by Salvatore Algeri, W. Stephen Cheung, and Lawrence A. Stark
Microwave Tubes by A. S. Gilmour, Jr.
Electric Filters by Martin Hasler and Jacques Neirynck
Nonlinear Circuits by Martin Hasler and Jacques Neirynck
Microwave Technology by Erich Pehl
Receiving Systems Design by Stephen J. Erst
Microwave Mixers by Stephen A. Maas
Feedback Maximization by B.J. Lurie
High-Power Microwave Sources, Victor L. Granatstein and Igor Alexeff, eds.
Applications of GaAs MESFETs by Robert Soares, et al.
GaAs Processing Techniques by Ralph E. Williams
GaAs FET Principles and Technology by James V. DiLorenzo and Deen D. Khandelwal
E-Plane Integrated Circuits by P. Pramanick and P. Bhartia
Evanescent Mode Microwave Components by George F. Craven and Richard F. Skedd
Filters with Helical and Folded Helical Resonators by Peter Vizmuller
Design Guide for Monopulse Elements on Circular Ground Planes by Melvin M. Weiner et al.
Dielectric Resonators, Darko Kajfez and Pierre Guillon, eds.
Modern Spectrum Analyzer Theory and Applications by Morris Engelson
Microwave Materials and Fabrication Techniques by Thomas S. Laverghetta
Handbook of Microwave Testing by Thomas S. Laverghetta
Microwave Measurements and Techniques by Thomas S. Laverghetta
Linear Active Circuits: Design and Analysis by William Rynone, Jr.
The Design of Impedance-Matching Networks for Radio-Frequency and Microwave Amplifiers by Pieter L.D. Abrie
Microwave Filters, Impedance Matching Networks, and Coupling Structures by G.L. Matthaei, Leo Young, and E.M.T. Jones
Analysis, Design, and Applications of Fin Lines by Bharathi Bhat and Shiban Koul
Microwave Engineer's Handbook, 2 vol., Theodore Saad, ed.
Handbook of Microwave Integrated Circuits by R.K. Hoffmann
Microwave Integrated Circuits, Jeffrey Frey and Kul Bhasin, eds.
Computer-Aided Design of Microwave Circuits by K.C. Gupta, Ramesh Garg, and Rakesh Chadha
Microstrip Lines and Slotlines by K.C. Gupta, R. Garg, and I.J. Bahl
Advanced Mathematics for Practicing Engineers by Kurt Arbenz and Alfred Wohlhauser
Microwave Circuit Design Using Programmable Calculators by J. Lamar Allen and Max Medley, Jr.
Stripline Circuit Design by Harlan Howe, Jr.
Microwave Transmission Line Filters by J.A.G. Malherbe
Microwave Diode Control Devices by Robert V. Garver
Tables for Active Filter Design by Mario Biey and Amedeo Premoli
Design Tables for Discrete Time Normalized Lowpass Filters by Arild Lacroix and Karl-Heinz Witte
Active Filter Design by Arthur B. Williams
Laser Applications by W.V. Smith
Microwave Remote Sensing, 3 vol., by F.T. Ulaby, R.K. Moore, and A.K. Fung

E-PLANE INTEGRATED CIRCUITS

PRAKASH BHARTIA
PROTAP PRAMANICK

Artech House

ARTECH HOUSE, INC.
685 Canton Street
Norwood, MA 02062

International Standard Book Number: 0-89006-242-0
Library of Congress Catalog Card Number: 87-71738

10 9 8 7 6 5 4 3 2 1

Contents

Preface

In 1974, the *IEEE Transactions on Microwave Theory and Techniques* published Paul Meier's article, "Integrated Fin-Line Millimeter Components," which showed the potential of E-plane integration for *millimeter-wave* applications (see Sec. 5.1.1). Since then, a constant and steady stream of publications on fin lines and related E-plane circuits have appeared in the literature.

In the early 1980s, E-plane technology was in transition between the laboratory and commercial world. Today its applications are extensive, especially in military and industrial systems. However, a significant amount of research is still conducted on E-plane circuits and systems. Most of the advances are relatively significant and the whole subject is nearing its maturity. We, therefore, have felt that it is time to review the entire field. Perhaps, a comprehensive treatise would have been more appropriate in view of the current needs. However, these technological advances run the risk of becoming obsolete in a very short time due to the rapid and significant changes in the technology. As a result, we have presented this review as a reprint collection, so that over time, we may delete obsolete articles and replace them with newer articles in subsequent editions of this book.

We express our deep gratitude to the authors and publishers of the articles in this collection for their permission to reprint, and their timely responses in making available reproducible copies of the papers.

The Editors

PRAKASH BHARTIA, Ph.D.
Director, Research and Development
Communications and Space
Department of National Defense
Ottawa, Ontario

PROTAP PRAMANICK, Ph.D.
Senior Member Technical Staff
COM DEV Ltd.
Cambridge, Ontario

Chapter 1
Millimeter Waves and Fin Lines

The term *fin line* first appeared in connection with the launching of two orthogonal modes in a circular waveguide [1,2]. Over the past thirteen years, the activities in millimeter-wave technology have grown significantly in the search for suitable transmission media. Figure 1 shows a general survey of all such planar and quasi-planar transmission media.

TRANSMISSION LINE STRUCTURES	CHARACTERISTIC IMPEDANCE IN Ω	FREQUENCY OF OPERATION IN GHz	LOSS	DISPERSION	DEVICE/COMPONENT COMPATIBILITY INTEGRATION LEVEL	
					IN SERIES	IN PARALLEL
MICROSTRIP	25-95	UP TO 100	MEDIUM	LOW	EASY	DIFFICULT
COPLANAR LINE	30-150	UP TO 60	HIGH	HIGH	EASY	EASY
SLOT LINE	60-200	——	HIGH	NON-TEM MODE	DIFFICULT	EASY
SUSPENDED STRIPLINE	40-150	>100	LOW	HIGH	EASY	DIFFICULT
FIN LINE	10-400	30-100	LOW	LOW	EASY	EASY

Fig. 1

Conventional metal waveguides offer minimum loss, but they are not amenable to integrated circuit fabrication. Microstrips, coplanar waveguides, and slotlines are used extensively in integrated circuits [3]. However, these transmission media suffer from tolerancing problems because they require either very narrow strip widths or very narrow slot widths at millimeter wavelengths, thus making them impractical. They are often incompatible with hybrid devices and have relatively high ohmic losses. Therefore, E-plane technology depends on the fin line as the main transmission medium. Figure 2 shows some of the most commonly used fin-line structures. Such structures can be considered as shielded slotlines suspended in the E-plane of a rectangular waveguide — hence, the origin of the term "E-plane." The suspended fins concentrate the field energy in the fin-gap region, as shown in Figure 3, leading to the capacitive loading of dominant HE modes of the slab-loaded waveguide. This effectively lowers the dominant mode's cut-off frequency to a considerable extent, and that of the next higher-order mode by a much smaller extent, thus leading to a larger bandwidth of the dominant mode.

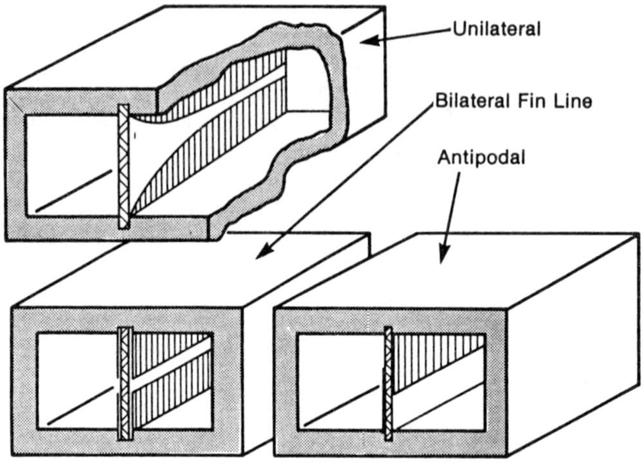

Fig. 2

Any device that is connected across the fin gap is subjected to larger power densities, which results in efficient matching. However, the field concentration results in larger conduction and dielectric losses due to the presence of the dielectric slab and higher current densities at the sharp fin edges. The typical loss in a fin line is on the order of 0.1 dB per wavelength. Consequently, it is unsuitable for long-distance power transmission.

Among all fin lines, the *unilateral fin line* is the simplest and best suited for the fabrication of millimeter-wave fin-line components [4]. The fins etched on both sides of the bilateral fin line produce lower transmission loss and greater flexibility in biasing active devices, as in *pin* diode attenuators and mixers with antiparallel, switched diodes. The bilateral fin line offers a wave impedance as low as 100 Ω, whereas the *antipodal fin line* offers an impedance level on the order of 10 Ω.

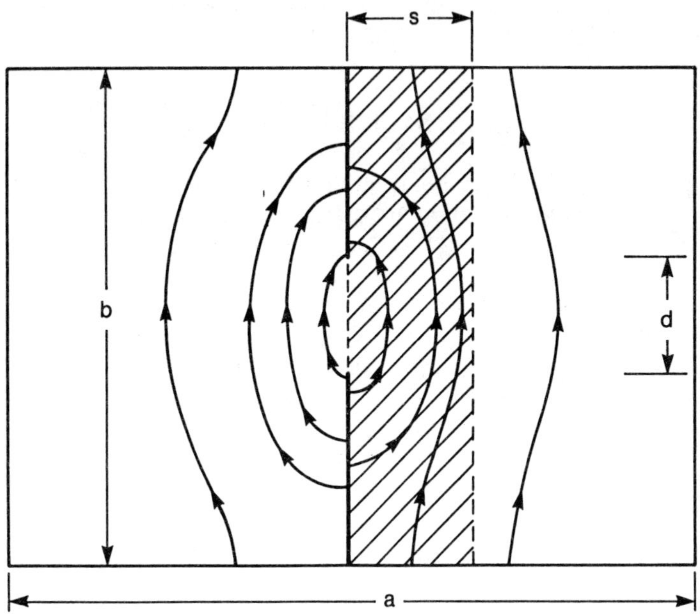

Fig. 3

Millimeter-wave integrated circuits have greater intensity, and fin lines can concentrate fields on tiny diodes. Thus, millimeter-wave integrated circuits provide greater flexibility than monolithic devices, and make the fabrication of multiband devices possible. High-level integration, suitability for mass production, and low-cost circuit design make E-plane technology an "economy-plane" technology, which results in the realization of virtually all important circuit components, including RF front ends.

Although E-plane technology has been proven in the laboratory from A-band through N-band, the physical dimensions of the beam-lead devices and required printing accuracy (on the order of 10 mm) prevent these circuits from becoming a standard technology above 100 GHz.

Although many important papers have appeared in the literature, much remains to be done. For instance, nothing has yet been published on the experimental research being done on the fin line's power-handling capabilities. Likewise, we need to develop simplified design techniques for building E-plane filters rather than using complicated numerical techniques. Another area requiring further study is that of the properties of fin lines on semiconductor substrates, which may lead to the development of millimeter-wave monolithic integrated circuits.

In summary, E-plane technology is adequately developed for the commercial manufacturing of millimeter-wave components up to 100 GHz, and has crossed the transition between laboratory and industry to becoming a standard technology.

REFERENCES

1. S.D. Robertson, "The Ultra-Bandwidth Finline Coupler," *IRE Trans. Microwave Theory and Techniques*, Vol. 3, No. 1, 1955, pp. 45–48.
2. S.D. Robertson, "Recent Advances in Finline Circuits," *IRE Trans. Microwave Theory and Techniques*, Vol. 4, 1956, pp. 263–267.
3. J. Frey and K. Bhasin, eds., *Microwave Integrated Circuits,* 2nd Ed., Artech House, 1985.
4. H. Hofmann, H. Meinel, and B. Addseck, "New Integrated Millimeter-Wave Components Using Finlines," *IEEE Symp. Digest,* 1978, pp. 21–23.

BIBLIOGRAPHY

1. K. Solbach, "Status of Printed Millimeter-Wave E-Plane Circuits," *IEEE Trans. Microwave Theory and Techniques*, Vol. MTT-31, No. 2, Feb. 1983, pp. 107–121.
2. B. Adelseck, H. Callsen, H. Meinel, W. Menzel, and K. Solbach, "A Survey of Planar Integrated Millimeter-Wave Components," *The Radio and Electronic Engineer*, Vol. 52, No. 1, 1982, pp. 46–50.
3. R.N. Bates, M.D. Coleman, S.J. Nightingale, and R. Davis, "E-plane Drops Millimeter Cost," *Microwave Systems News,* Dec. 1980, pp. 74–80.
4. P.J. Meier, "Millimeter-Wave Integrated Circuits Suspended in the E-plane of Rectangular Waveguide," *IEEE Trans. Microwave Theory and Techniques*, Vol. MTT-26, No. 10, Oct. 1978, pp. 726–733

Chapter 2
Analysis

Most E-plane circuits are not amenable to post-production tuning. Therefore, any successful design requires a sound theoretical basis. As in any other transmission line used in distributed circuits, the most important parameters in a fin line are the propagation constant and the characteristic impedance.

The propagation constant is defined uniquely. Because the fin line is a non-TEM line, it supports a hybrid mode of propagation. As a result, the characteristic impedance cannot be defined uniquely. It depends on the choice of definition, but all definitions merge for small fin gaps.

Papers in this chapter will address the analysis techniques of fin lines for theoretical predictions of the propagation constant and characteristic impedance. As we shall see, most researchers favor the *power-voltage* definition of the characteristic impedance.

2.1 NUMERICAL TECHNIQUES

2.1.1 "Analysis of Millimetre-Wave Integrated Fin Line" by Simons

This paper is an extension of the transverse resonance method used for analyzing slotlines by S. Cohn (Reference 9 in the paper). The method is simple to implement on a desk-top computer. The technique assumes zero strip thickness, but this may not be the case for W- and D-band fin lines. The transverse resonance method is sufficiently accurate only for small fin gaps. The paper presents computed results for fin gaps up to 0.5 mm in Ka-band housing. Results are given for insulated, unilateral, bilateral, single-ridge, and coupled-unilateral fin lines.

Because this method is not suitable for large fin gaps, it cannot be used in designing fin-line tapers to match a full-height waveguide to a small fin gap. However, it is a simple and accurate method, and this paper supplies useful design data for many applications.

2.1.2 "Dispersion of Planar Waveguides for Millimeter-Wave Application" by Hofmann

This was the first paper on the theoretical analysis of fin lines. The numerical data presented in the paper have been used as a reference by many subsequent researchers.

Hofmann solves the associated boundary value problem in the space domain using Galerkin's procedure. He also analyzes other planar and quasi-planar transmission lines such as suspended microstrips, slotlines, coplanar strip lines, *et cetera*, enclosed in a metallic housing. Hofmann reasonably addresses the relative convergence problem of the solution for the propagation constant. The characteristic impedance results, which are based on the voltage-current definition, are adequately accurate for small slots because the computation neglects the currents on the walls of the housing.

Before this paper was published, fin-line propagation constant had been evaluated through experiments only.

2.1.3 "Dispersion Characteristics in Fin Lines with One and Two Dielectric Layers Arbitrarily Located in the Waveguide" by Chaves Fernandes and Giarola

This paper presents the most generalized technique for analyzing all kinds of fin-lines and quasi-planar transmission lines. This technique solves the associated eigenvalue problem using the spectral-domain technique in conjunction with Galerkin's procedure. The analysis assumes zero metalization thickness of infinite conductivity. Therefore, the suitability of the method is questionable at higher frequencies. However, this is an important paper for understanding the spectral-domain immittance matrix approach for a full-wave solution of not only any fin line, but also of any kind of planar and quasi-planar transmission line.

2.1.4 "Millimeter-Wave Fin-Line Characteristics" by Knorr and Shayda

This was the first paper on the simple but adequately accurate analysis of unilateral fin lines using the spectral-domain method. In its simplest form, this method neglects the longitudinal slot fields, yet it yields results for the propagation constant within one percent of experimental results. Subsequent and more accurate computations have shown that the inaccuracy may be within two percent near the lower edge of the useful frequency band.

The characteristic impedance results, based on the power-voltage definition, differ by six percent with the results obtained using other numerical techniques. However, this paper presents a wide range of design data, from the most commonly used RT Duroid 5880 substrate, up to W-band.

2.1.5 "A Modified Mode-Matching Technique and Its Application to a Class of Quasi-Planar Transmission Lines" by Vahldieck and Bornemann

This paper presents the first rigorous hybrid-mode analysis to determine the propagation constant in a class of quasi-planar transmission line structures. This is the most generalized work in the sense that it considers the most realistic fin-line structure, which includes the finite metalization thickness, the substrate-supporting grooves, and an arbitrary number of substrate dielectric subregions. The method is based on the mode-matching technique. The method, however, fails for zero fin or strip thickness; i.e., for a negligible fin thickness, the inaccuracy in the computation may be unacceptable. This stems from the assumption of an equal number of modes all over the cross section. In this paper, we see the effects of finite-strip thickness and substrate-supporting grooves on the performance of a fin line.

2.1.6 "Formulation of Singular Integral Equation Technique for Planar Transmission Lines" by Omar and Schunemann

This paper, based on *singular integral equation techniques*, shows that complex- and backward-wave modes, which exist in rectangular and circular waveguides with dielectric inserts, can exist in fin lines. This versatile method computes the first 30 modes accurately. This paper discusses the mode behavior above and below cut-off, in detail. In fact, this is the only paper that presents a complete picture of mode behavior in fin line.

2.1.7 "The Transmission-Line Matrix Method—Theory and Applications" by Hoefer

While all other methods obtain the harmonic solution of Maxwell's equations, the *transmission-line matrix* (TLM) analysis is a true exception. It simulates wave propagation in the time domain.

This paper is an outcome of the author's long involvement with the TLM method for analyzing various kinds of fin lines and related structures.

This excellent paper gives historical background of the method, reviews the theoretical foundations, derives the basic algorithms for simulating the propagation of waves in bounded and unbounded media, and shows how the eigenvalues and the field configuration of resonant structures can be obtained with the Fourier transform. It also discusses various sources of error and points out the possible pitfalls. The end of the paper supplies an exhaustive list of applications of the TLM method, including the fin-line application.

Although we have extensively researched the transmission-line matrix method (the references for this paper span over four decades), this method is superior to only the finite-element method in fin-line analysis.

2.1.8 "Analysis of Fin Line with Finite Metallization Thickness" by Kitazawa and Mittra

This paper describes the hybrid-mode formulation, in which the determinant equation for the propagation constant is formulated in the space domain by the network analytical technique, i.e., by conventional circuit theory. The authors originally used the method in analyzing the unilateral fin line with zero fin thickness. The paper gives computed results on unilateral and bilateral fin lines, and shows the effects of finite metalization thickness on the propagation constant and characteristic impedance. However, unlike the mode-matching method, this hybrid-mode method does not allow for the important effect of substrate-supporting grooves.

2.1.9 "Efficient Analysis of Fin Line with Finite Metallization Thickness" by Piotrowski

The method presented in this paper is the most convenient one for analyzing fin-line structures, in terms of analytical rigor, generality, and simplicity. In fact, we can program this method on a programmable calculator. The propagation constants obtained by this method are as accurate as those obtained by the singular integral equation method. However, the characteristic impedance results differ by six percent from those of Kitazawa and Mittra (Sec. 2.1.8).

2.1.10 "Systematic Characterization of the Spectrum of Unilateral Finline" by Olley and Rozzi

This paper focuses on a systematic characterization of the complete mode family of a unilateral fin line. The authors use the space-domain approach together with transverse equivalent network considerations, and call the method *transverse resonance diffraction.* When the edge conditions are properly satisfied, the method yields a high degree of accuracy. Although the paper is sufficiently thorough, it pays no attention to characteristic impedance behavior.

2.2 CLOSED-FORM SOLUTION

2.2.1 "Equivalent Relative Permittivity and Unloaded Q Factor of Integrated Finline" by Meier

This paper defines the propagation constant of fin lines in terms of the ridged waveguide, and presents an equivalence between a dielectric slab-loaded, ridged waveguide and fin line. It also compares the measured losses of fin lines with other transmission lines.

2.2.2 "Accurate Analysis Equations and Synthesis Technique for Unilateral Fin-lines" by Pramanick and Bhartia

2.2.3 "Simple Formulae for Dispersion in Bilateral Fin-Lines" by Pramanick and Bhartia

2.2.4 "Accurate Analysis and Synthesis Equations for Insulated Fin-Lines" by Pramanick and Bhartia

E-plane circuit design is cumbersome for several reasons. First, most of the numerical techniques in the preceding section require complicated computer programming and a lot of computer time. Moreover, solutions for propagation constants may suffer from the relative convergence problem. To simplify the circuit designer's task, these three papers present closed-form design equations for the fundamental modes in various fin lines. The equations explicitly determine the propagation constants and the characteristic impedances based on the ridged-waveguide model. These equations are vital to computer-aided design of E-plane circuits.

2.2.5 "Frequency Dependence of Equivalent Dielectric Constant in Finlines" by Pramanick and Bhartia

This paper develops a simple LSE model to accurately predict the frequency dependence of the equivalent dielectric constant in fin line. The equations of this paper, together with those in the previous three papers, predict the phase velocity in fin line very accurately.

2.2.6 "Sensitivity Analysis of the Fin-Line Housing" by El Hennawy and Schunemann

This paper analyzes the sensitivity of fin-line characteristics to the housing. The results are derived from the partial derivatives of the equations, which we obtain from transverse resonance analysis. The authors conclude that there are no tight tolerance limits for the housing dimensions.

BIBLIOGRAPHY

Transverse Resonance Analysis

1. A.M.K. Saad and G. Begemann, "Electrical Performance of Finlines of Various Configurations," *IEE Proc. Microwaves, Optics, and Antennas*, Part H, Vol. 1, Jan. 1977, pp. 81–88.
2. J.K. Piotrowski, "Accurate and Simple Formulas for Dispersion in Finlines," *IEEE MTT-S Symposium Digest*, 1984, pp. 333–336.

3. J.K. Piotrowski and W. Wojtasiak, "A Waveguide-to-Microstrip Transition," *Proc. European Microwave Conf.*, 1986, pp. 505–910.

Spectral-Domain Method

1. L.P. Schmidt and T. Itoh, "Dominant and Higher-Order Modes in Finlines," *IEEE Trans. Microwave Theory and Techniques*, Vol. MTT-28, No. 9, Sept. 1980, pp. 981–985.
2. L.P. Schmidt, T. Itoh, and H. Hofmann, "Characteristics of Unilateral Finline Structures with Arbitrarily Located Slots," *IEEE Trans. Microwave Theory and Techniques*, Vol. MTT-29, No. 4, April 1981, pp. 352–355.
3. D.M. Syahkal and J.B. Davis, "An Accurate Unified Solution to Various Finline Structures, of Constant Phase, Characteristic Impedance and Attenuation," *IEEE Trans. Microwave Theory and Techniques*, Vol. MTT-30, No. 11, Nov. 1982, pp. 1854–1861. (A qualitative estimate of losses in unilateral, coupled, antipodal and bilateral finlines can be obtained from this paper)
4. L.P. Schmidt and T. Itoh, "Characteristics of Generalized Finline for Millimeter-Wave Application," *International J. Infrared and Millimeter-Waves*, Vol. 2, No. 3, 1981, pp. 427–436.

Space-Domain Method

1. H. Hofmann, "Finline Dispersion," *Electronics Letters*, Vol. 12, No. 17, Aug. 1972, pp. 428–429.
2. C.A. Olley and T.E. Rozzi, "Characterization of Unilateral Finline Mode Spectrum Including Loss," *Proc. European Microwave Conf.*, 1986, pp. 511–516.

Mode-Matching Method

1. A. Beyer, "Analysis of Characteristics of an Earthed Finline," *IEEE Trans. Microwave Theory and Techniques*, Vol. MTT-29, No. 7, July 1981, pp. 676–680.
2. A.M.K. Saad and K. Schunemann, "Efficient Eigenmode Analysis for Planar Transmission Lines," *IEEE Trans. Microwave Theory and Techniques*, Vol. MTT-30, No. 12, Dec. 1982, pp. 2125–2132.
3. R. Vahldieck, "Accurate Hybrid-Mode Analysis of Various Finline Configurations Including Multilayered Dielectrics, Finite Metallization Thickness and Substrate - Supporting Grooves," *IEEE Trans. Microwave Theory and Techniques*, Vol. MTT-32, No. 11, Nov. 1984, pp. 1454–1460.
4. J. Bornemann, "Rigorous Field Theory Analysis of Quasiplanar Lines," *IEE Proc.*, Vol. 132, Pt. H, Feb. 1985, pp. 1–6.
5. R. Vahldieck and W.J.R. Hoefer, "The Influence of Metallization Thickness and

Mounting Grooves on the Characteristics of Finlines," *IEEE MTT-S Symposium Digest*, 1985, pp. 182–184.

6. J. Siegel, "Phasenkonstante und Wellenwiderstand einer Schlilzleitung mit Rechteckigem Schirm und Endlicher Metallisierumgsdicke," *Frequenz*, Vol. 31, July 1977, pp. 216–220.

Transmission-Line Matrix Method

1. W.J.R. Hoefer and A. Ros, "Finline Parameters Calculated with the TLM Method," *IEEE MTT-S Symposium Digest*, May 1979, pp. 341–343.
2. Y.C. Shih, "Cutoff Frequencies in Finlines Calculated with Two-Dimensional TLM-Program," *IEEE MTT-S Symposium Digest*, 1980, pp. 261–263.
3. W.J.R. Hoefer and Y.C. Shih, "Field Configuration of Fundamental and Higher-Order Modes in Finlines Obtained with the TLM Method," paper presented at URSI and IEEE AP-S Symposium, Quebec, June 2–6, 1980.
4. Y.C. Shih and W.J.R. Hoefer, "Dominant and Second-Order Mode Cut-Off Frequencies in Finlines Calculated with a Two-Dimensional TLM Program," *IEEE Trans. Microwave Theory and Techniques*, Vol. MTT-28, No. 12, pp. 1443–1448, Dec. 1980.
5. Y.C. Shih and W.J.R. Hoefer, "The Accuracy of TLM Analysis of Finned Rectangular Waveguides," *IEEE Trans. Microwave Theory and Techniques*, Vol. MTT-28, No. 7, July 1980, pp. 743–746.

Network Analytical Technique in the Space Domain

1. Y. Hayashi, E. Farr, S. Wilson, and R. Mittra, "Analysis of Dominant and Higher-Order Modes in Unilateral Finlines," *Archiv fur Elek Ubertra*, Vol. 37, No. 3/4, March–April 1983, pp. 117–122.

Closed-Form Solutions

1. P. Pramanick, R.R. Monsour, and R.H. Macphie, "Computer-Aided Design Models for Unilateral Finlines with Finite Metallization Thickness and Arbitrarily Located Slots," *IEEE MTT-S Symposium Digest*, Las Vegas, 1987.
2. P. Pramanick and P. Bhartia, "Computer-Aided Design Models for Millimeter-Wave Finlines and Suspended Substrate Microstrip Lines," *IEEE Trans. Microwave Theory and Techniques*, Vol. MTT-35, No. 12, Dec. 1985, pp. 1429–1435.
3. A.K. Sharma and W.J.R. Hoefer, "Empirical Expressions for Finline Design," *IEEE Trans. Microwave Theory and Techniques*, Vol. MTT-31, No. 4, April 1983, pp. 350–356.
4. M.N. Burton and W.J.R. Hoefer, "Closed-Form Expressions for the Parameters of Finned and Ridged Waveguides," *IEEE Trans. Microwave Theory and Techniques*, Vol. MTT-30, No. 12, Dec. 1982, pp. 2190–2194.

2.1.1

Analysis of millimetre-wave integrated fin line

Rainee N. Simons, B.E., M. Tech., Ph.D., Mem. I.E.E.E.

Indexing terms: *Microwave components, Microwave circuits and networks*

Abstract: The paper presents an analytical technique for determining the dispersion and also the characteristic impedance on integrated fin-line structures. The computed dispersion and also the characteristic impedance are graphically illustrated for the various structures. The method of analysis is accurate for small slot widths.

1 Introduction

The integrated fin line is a very useful transmission line for millimetre-wave integrated circuits (e.g. see Reference 1). The literature reports several techniques for determining the characteristics of fin lines [2–8]. The purpose of this study is to present an analytical technique for determining the dispersion and also the characteristic impedance on integrated fin-line structures. The analysis takes into consideration a rectangular waveguide housing of arbitrary dimensions with an E-plane printed-circuit insert. The printed-circuit insert carries a pair of slots and is sandwiched between several layers of dielectrics. The dielectric layers are assumed to be isotropic and homogeneous, and are of arbitrary thickness and relative permittivity. The rectangular housing and the zero-thickness metalisation on the printed-circuit insert are assumed to have infinite conductivity. The analysis is carried out using Cohn's technique for slot line on a dielectric substrate [9, 10]. The computed results are graphically illustrated for the various structures.

2 Analysis

Consider a general fin-line structure with double slot and multilayer dielectric substrate as shown in Fig. 1. This structure can support two independent modes of propagation; namely, the even mode and the odd mode. For the even excitation, an electric wall is placed along the plane of symmetry $y = 0$, and the upper half of the structure is isolated. For the odd excitation, the electric wall is replaced by a magnetic wall. The formula for the total susceptance B_{te} at the plane of the iris is as follows, for the even excitation:

$$\eta B_{te} = \frac{au_1}{2b}\tan\left[\frac{\pi u_1 d_1}{ap} - \tan^{-1}\left\{\frac{v}{u_1}\coth(vkh_1)\right\}\right] + \frac{au_2}{2b}\tan\left[\frac{\pi u_2 d_2}{ap} + \tan^{-1}\left\{\frac{u_3}{u_2}\tan\left(\frac{\pi u_3 d_3}{ap} - \tan^{-1}\left\{\frac{v}{u_3}\coth(vkh_2)\right\}\right)\right\}\right] + \frac{1}{p}\left\{(\epsilon_{r1} + \epsilon_{r2} - 2p^2)I_1 + \sum_{n=1,2,3,\ldots}^{\infty}(M_{n1} + M_{n2}) \times \frac{\sin^2(n\pi\delta/2)}{n(n\pi\delta/2)^2}\cos^2(n\pi\bar{\delta}/2)\right\} \tag{1}$$

where $\eta = 376.7\,\Omega$, $\delta = w/b$, $k = 2\pi/\lambda$, $\bar{\delta} = (w/b) + (s/b)$, $v = (p^2 - 1)^{1/2}$ and

$$F_n = \left[1 + \left(\frac{bv}{anp}\right)^2\right]^{1/2} \tag{2}$$

$$I_1 = \frac{1}{(\pi\delta)^2}\left\{\frac{(\pi\bar{\delta})^2}{2}\ln(\pi\bar{\delta}) - \frac{(\pi\delta)^2}{2}\ln(\pi\delta) - \frac{[\pi(\bar{\delta}+\delta)]^2}{4}\ln[\pi(\bar{\delta}+\delta)] - \frac{[\pi(\bar{\delta}-\delta)]^2}{4}\ln[\pi(\bar{\delta}-\delta)] + \frac{3}{2}(\pi\delta)^2 + \frac{(\pi\delta)^4}{144} + \frac{(\pi\delta)^2(\pi\bar{\delta})^2}{48}\right\} \tag{3}$$

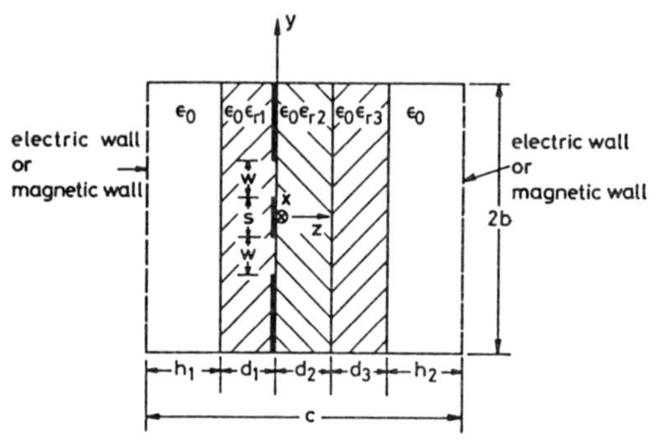

Fig. 1 *An arbitrary fin-line structure with two slots and several layers of dielectric*

For $i = 1,2$, or 3

$$u_i = (\epsilon_{ri} - p^2)^{1/2} \tag{4}$$

$$F_{ni} = \left[1 - \frac{bu_i}{anp}^2\right]^{1/2} \tag{5}$$

For F_{ni} real, $i = 1$ or 2,

$$M_{ni} = \frac{\epsilon_{ri}\tanh r_{ni} - p^2 F_{ni}^2 \coth q_{ni}}{\left[1 + \left(\frac{b}{an}\right)^2\right]F_{ni}} - u_i^2 \tag{6}$$

For F_{ni} imaginary replace F_{ni} by $j|F_{ni}|$

$$r_{ni} = \frac{\pi n d_i F_{ni}}{b} + \tanh^{-1}\left(\frac{\epsilon_{ri+1}F_{ni}}{\epsilon_{ri}F_{ni+1}}\right)\tanh\left\{\frac{\pi n d_{i+1}F_{ni+1}}{b} + \tanh^{-1}\left[\frac{F_{ni+1}}{\epsilon_{ri+1}}F_n\coth\left(\frac{\pi_n F_n h i}{b}\right)\right]\right\} \tag{7}$$

Paper 2359H, received 27th July 1982
The author is with the Centre for Applied Research in Electronics, Indian Institute of Technology, Hauz Khas, New Delhi-110016, India.

R.N. Simons, *IEE Proceedings*, Vol. 130, Pt. H, No. 2, March 1983, pp. 166–169.

$$q_{ni} = \frac{\pi n d_i F_{ni}}{b} + \coth^{-1}\left(\frac{F_{ni+1}}{F_{ni}}\right) \coth\left\{\frac{\pi n d_{i+1} F_{ni+1}}{b} + \coth^{-1}\left[\frac{F_n}{F_{ni+1}} \coth\left(\frac{\pi_n F_n h_i}{b}\right)\right]\right\} \quad (8)$$

For the case when the air region is terminated by a magnetic wall, replace coth (vkh_i) by tanh (vkh_i) and coth $(\pi n F_n h_i/b)$ by tanh $(\pi n F_n h_i/b)$. The formula for the total susceptance B_{to} at the plane of the iris is as follows for the odd excitation:

$$\eta B_{to} = \frac{1}{p}\left\{(\epsilon_{r1} + \epsilon_{r2} - 2p^2) I_2 + \sum_{n=0,\,1,\,2,\ldots\ldots}^{\infty} (M_{n1} + M_{n2}) \frac{\sin^2\left(\frac{2n+1}{2}\right)\frac{\pi\delta}{2}}{\left(\frac{2n+1}{2}\right)\left(\frac{2n+1}{2}\right)\frac{\pi\delta}{2}\Big|^2} \sin^2\left(\frac{2n+1}{2}\right)\frac{\pi\bar{\delta}}{2}\right\} \quad (9)$$

where

$$I_2 = \frac{1}{(\pi\delta)^2}\left\{-\frac{(\pi\bar{\delta})^2}{2} \ln\frac{\pi\bar{\delta}}{4} - \frac{(\pi\delta)^2}{2} \ln\frac{\pi\delta}{4} + \frac{[\pi(\bar{\delta}+\delta)]^2}{4} \ln\frac{[\pi(\bar{\delta}+\delta)]}{4} + \frac{[\pi(\bar{\delta}-\delta)]^2}{4} \ln\frac{[\pi(\bar{\delta}-\delta)]}{4} + \frac{(\pi\bar{\delta})^2(\pi\delta)^2}{96}\right\} \quad (10)$$

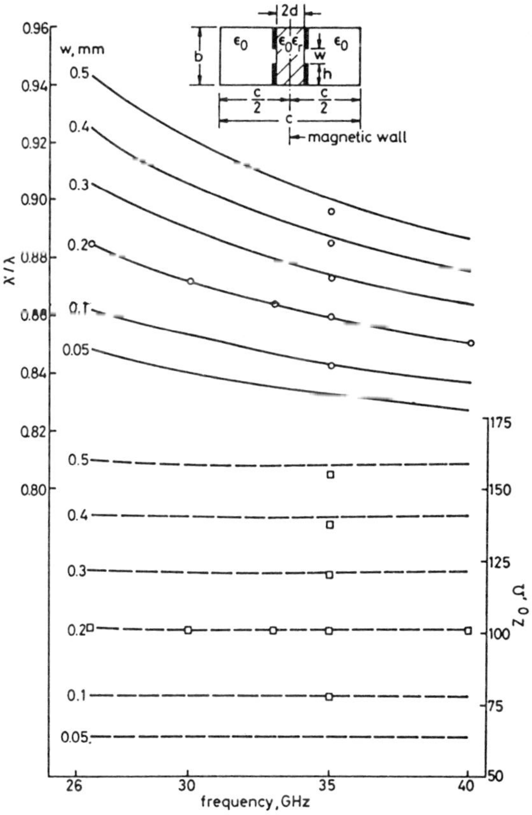

Fig. 2 *Dispersion and characteristic impedance of bilateral fin line*

$\epsilon_r = 2.22$, $d = 0.254$ mm, $b = 3.556$ mm,
$c = 7.112$ mm, $h = s/2 = (b - w)/2$
——— λ'/λ
– – – – Z_0
○ ○ ○ λ'/λ, Schmidt
□ □ □ Z_0, Schmidt

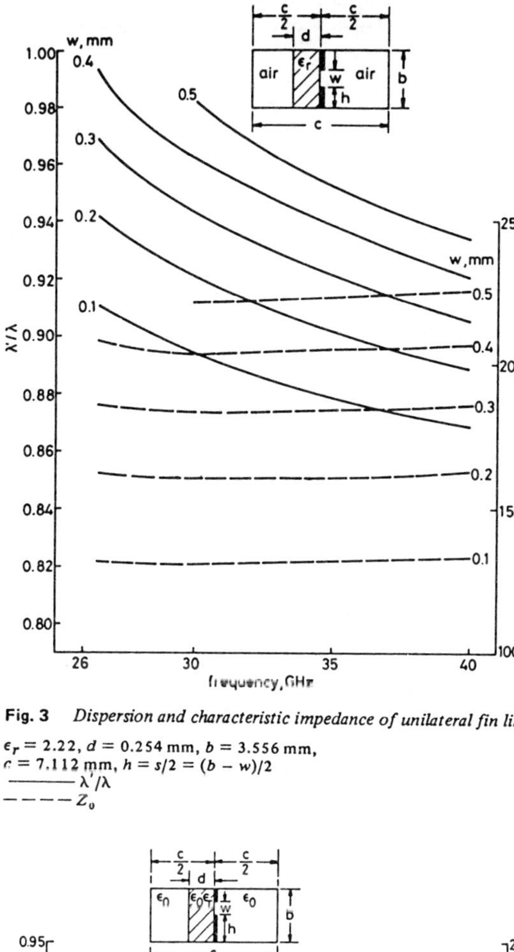

Fig. 3 *Dispersion and characteristic impedance of unilateral fin line*

$\epsilon_r = 2.22$, $d = 0.254$ mm, $b = 3.556$ mm,
$c = 7.112$ mm, $h = s/2 = (b - w)/2$
——— λ'/λ
– – – – Z_0

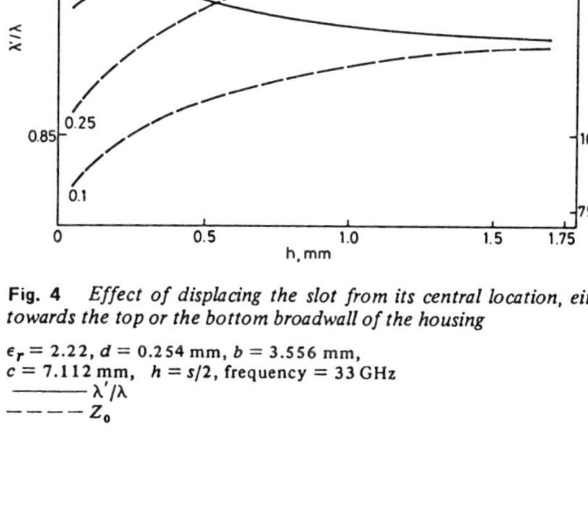

Fig. 4 *Effect of displacing the slot from its central location, either towards the top or the bottom broadwall of the housing*

$\epsilon_r = 2.22$, $d = 0.254$ mm, $b = 3.556$ mm,
$c = 7.112$ mm, $h = s/2$, frequency = 33 GHz
——— λ'/λ
– – – – Z_0

The corresponding expressions for F_n, F_{nl}, M_{nl}, r_{nl} and q_{nl} are obtained by replacing the factor n by $(2n + 1)/2$ in eqns. 2, 5, 6, 7 and 8. The above method of analysis is accurate for $w \leqslant d$, $w/b \leqslant 0.15$ and $w \leqslant \lambda/4\sqrt{\epsilon_r}$, where λ is the free-space wavelength, d is the substrate thickness and ϵ_r is the relative permitivity of the substrate [9]. This range of slot width is adequate for most fin-line applications (e.g. see Reference 11).

The procedure for computing the normalised guide wavelength λ'/λ and also the characteristic impedance Z_0 are explained in Reference 9.

3 Numerical results

3.1 Bilateral fin line

The right half of the structure is isolated for the purpose of analysis by placing a magnetic wall along the place of symmetry (insert in Fig. 2). The expression for the total susceptance B_t at the plane of the iris is obtained by setting $d_2 = d_3 = 0$, $h_1 = 0$ and $d_1 + h_2 = c/2$ in eqn. 1. The slot is located in the centre for $h = s/2 = (b - w)/2$. The computed λ'/λ and z_0 are graphically illustrated in Fig. 2. The λ'/λ are also z_0 computed at a few points, using the rigorous and accurate spectral-domain technique* [4], are also plotted in Fig. 2 for comparison. It is observed that, for small slot widths, the results obtained using the above technique are in agreement with those obtained using the spectral-domain technique.

3.2 Unilateral fin line

The expression for the total susceptance B_t is obtained by setting $d_2 = d_3 = 0$, and $d_1 + h_1 = h_2 = c/2$ in eqn. 1. The computed λ'/λ and also z_0 are illustrated in Fig. 3. The effect of displacing the slot from its central location, either towards the top or the bottom broadwall of the housing on λ'/λ and also z_0, is illustrated in Fig. 4.

3.3 Asymmetrical fin line

Besides, if $h = (s/2)$ is set equal to zero, the resulting structure is an asymmetrical fin line. The asymmetrical fin-line characteristics, namely, λ'/λ and also z_0, are illustrated in Fig. 5.

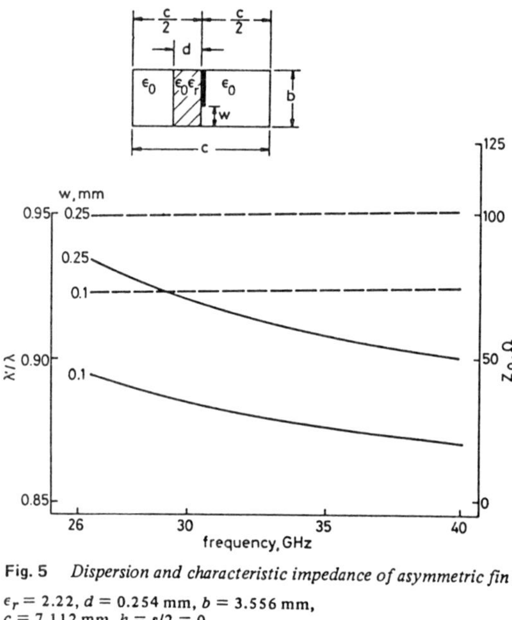

Fig. 5 *Dispersion and characteristic impedance of asymmetric fin line*

$\epsilon_r = 2.22$, $d = 0.254$ mm, $b = 3.556$ mm,
$c = 7.112$ mm, $h = s/2 = 0$
——— λ'/λ
– – – – Z_0

*SCHMIDT, L.P.: Private communication

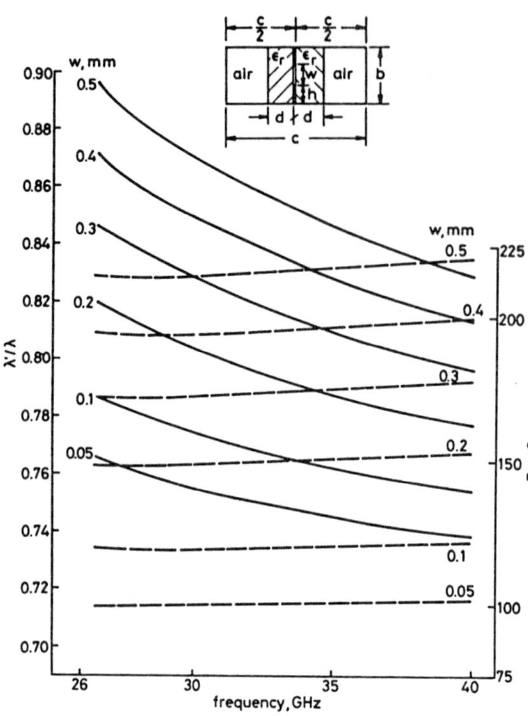

Fig. 6 *Dispersion and characteristic impedance of insulated fin line*

$\epsilon_r = 2.22$, $d = 0.254$ mm, $b = 3.556$ mm,
$c = 7.112$ mm, $h = s/2 = (b - w)/2$
——— λ'/λ
– – – – Z_0

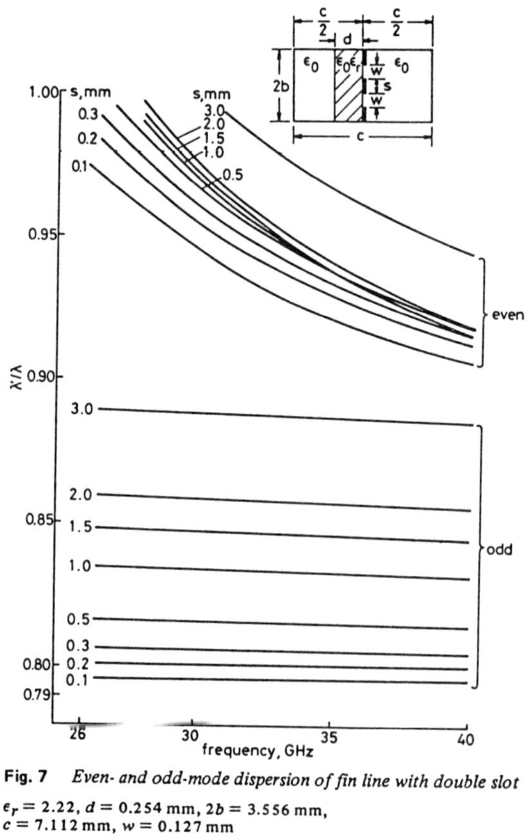

Fig. 7 *Even- and odd-mode dispersion of fin line with double slot*

$\epsilon_r = 2.22$, $d = 0.254$ mm, $2b = 3.556$ mm,
$c = 7.112$ mm, $w = 0.127$ mm

3.4 *Insulated fin line*

The expression for the total susceptance B_t is obtained by setting $d_3 = 0$ and $d_1 + h_1 = d_2 + h_2 = c/2$ in eqn. 1. The computed λ'/λ and also Z_0 are illustrated in Fig. 6.

3.5 *Fin line with double slot*

The expressions for the total susceptance B_{te} and B_{to} are obtained by setting $d_2 = d_3 = 0$ and $d_1 + h_1 = h_2 = c/2$ in eqns. 1 and 10. The computed even and odd mode λ'/λ and also the even and odd mode Z_0 are illsturated in Figs. 7 and 8, respectively.

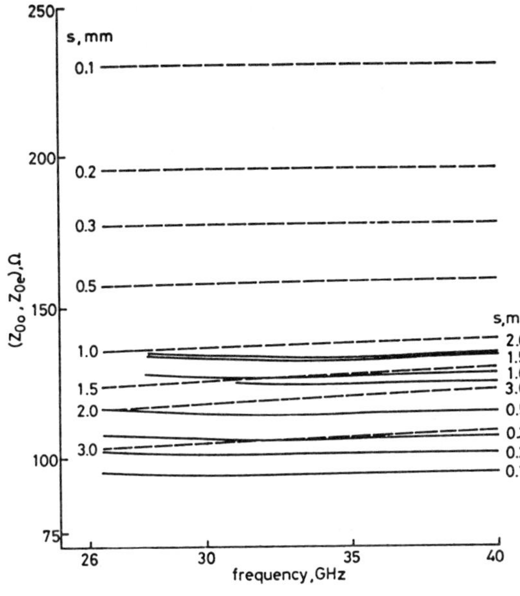

Fig. 8 *Even- and odd-mode characteristic impedance of fin line with double slot*

– – – – Z_{0o}
——— Z_{0e}

4 Conclusion

Briefly, the paper presents an analytical technique, which is accurate for small slot widths, for analysing fin-line structures. The computed dispersion and also the characteristic impedance are graphically illustrated. The results should find extensive applications in the design of millimetre-wave integrated-circuit components, such as balanced mixers, PIN switches and attenuators, detectors, modulators, oscillators and directional couplers.

5 Acknowledgment

The author is extremely grateful to Dr. L.P. Schmidt of AEG Telefunken, West Germany, for several helpful discussions and also for providing computed data on integrated fin lines, obtained via the spectral-domain technique.

6 References

1 MEIER, P.J.: 'Millimeter integrated circuit suspended in the E-plane of rectangular waveguide', *IEEE Trans.*, 1978, **MTT-26**, pp. 726–733

2 SAAD, A.M.K., and BEGEMANN, G.: 'Electrical performance of fin-lines of various configurations', *IEE J. Microwaves, Opt. & Acoust*, 1977, **1**, pp. 81–88

3 SIMONS, R.N., and ARORA, R.K.: 'Double-slot fin-line structure for millimeter-wave applications', *Proc. IEEE*, 1979, **67**, pp. 1159–1160

4 SCHMIDT, L.P., ITOH, T., and HOFMANN, H.: 'Characteristics of unilateral fin-line structure with arbitrarily located slots', *IEEE Trans.*, 1981, **MTT-29**, pp. 352–355

5 KNORR, J.B., and SHAYDA, P.M.: 'Millimeter-Wave fin-line characteristics', *ibid.*, 1980, **MTT-28**, pp. 737–743

6 SHIH, Y.C., and HOEFER, W.J.R.: 'Dominant and second order mode cutoff frequencies in fin-lines calculated with a two-dimensional TLM-program', *ibid.*, 1980, pp. 1443–1448

7 SIMONS, R.N.: 'Fin-line with variable characteristics', *Proc. IEEE*, 1980, **68**, pp. 170–171

8 BEYER, A.: 'Analysis of the characteristics of an earthed fin-line', *IEEE Trans.*, 1981, **MTT-29**, pp. 676–680

9 COHN, S.B.: 'Slot-line on a dielectric substrate', *ibid.*, 1969, **MTT-17**, pp. 768–778

10 COHN, S.B.: 'Sandwich slot-line', *ibid.*, 1971, **MTT-19**, pp. 773–774

11 HENNAWY, H.E., and SCHUNEMANN, K.: 'Computer-aided design of fin-line detectors, modulators and switches', *Arch. Elektron. & Uebertragungstech.*, 1982, **36**, pp. 49–56

2.1.2

Dispersion of Planar Waveguides for Millimeter-Wave Application

by Holger Hofmann *

Report from AEG-Telefunken, Fachgebiet Hochfrequenztechnik, Ulm (Donau)

The dispersion of the phase coefficient and the wave impedance of some planar and quasi-planar structures such as slot-line, coplanar-strip-line, suspended substrate-line, and different fine-line geometries are calculated applying Galerkin's method. The formulation is adapted to the various structures under the aspect of minimizing computational time in the numerical evaluation.

Dispersion von planaren Wellenleitern für Millimeterwellen-Anwendungen

Die Dispersion des Phasenkoeffizienten und des Wellenwiderstandes einiger planarer und quasi-planarer Strukturen, wie Schlitzleitung, Coplanarleitung, ,,Suspended Substrate"-Leitung und verschiedene ,,Fin"-Leitungen, wird mit der Methode von Galerkin berechnet. Die Formulierung wird den verschiedenen Strukturen angepaßt unter dem Gesichtspunkt, minimale Rechenzeit bei der numerischen Auswertung zu erhalten.

1. Introduction

Different planar waveguide structures have been proposed for millimeter-wave applications. The slot-line and the coplanar-strip-line have been treated sufficiently (e.g. [1], [2]). Fin-lines have been applied only recently by Meyer (e.g. [3], [4]) but no theoretical approach is available to date. For the suspended substrate line only quasi-static results have been published (e.g. [5], [6]). However, thinking of the surrounding rectangular waveguide of the fin-line as an electric shield, it is possible to calculate fin-line, suspended substrate-line, slot-line and coplanar-strip-line, in one treatment, if the shield dimensions are allowed to become very large. Six

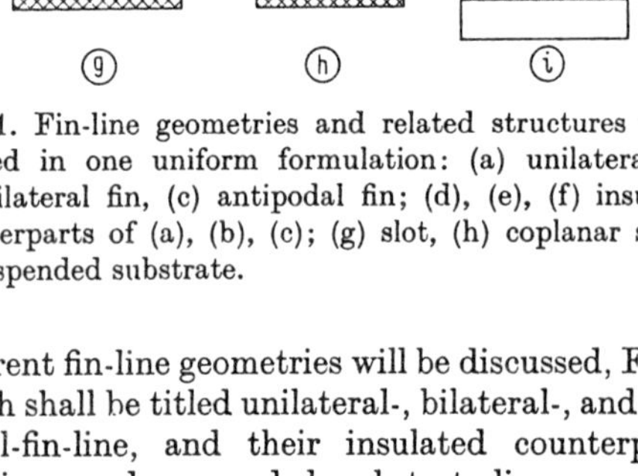

Fig. 1. Fin-line geometries and related structures to be treated in one uniform formulation: (a) unilateral fin, (b) bilateral fin, (c) antipodal fin; (d), (e), (f) insulated counterparts of (a), (b), (c); (g) slot, (h) coplanar strips, (i) suspended substrate.

different fin-line geometries will be discussed, Fig. 1, which shall be titled unilateral-, bilateral-, and antipodal-fin-line, and their insulated counterparts. Fin-lines and suspended substrate-lines are considered quasi-planar since components are modelled by strip structures, keeping the hollow waveguide unchanged.

2. Fourier Representation of the Fields

Using the reduced field quantities

$$\begin{aligned} \boldsymbol{E} &= \sqrt{\varepsilon_0}\,\boldsymbol{E}', & \boldsymbol{D} &= 1/\sqrt{\varepsilon_0}\,\boldsymbol{D}', \\ \boldsymbol{H} &= -\mathrm{j}\sqrt{\mu_0}\,\boldsymbol{H}', & \boldsymbol{B} &= 1/\mathrm{j}\sqrt{\mu_0}\,\boldsymbol{B}', \end{aligned} \tag{1}$$

the primes indicating the classical notation, the fields shall be calculated via the E_x- and H_x-components only:

$$\left(\frac{\partial^2}{\partial x^2} + \frac{\partial^2}{\partial y^2} + (k_0^2 \varepsilon_r - \beta^2)\right)\begin{pmatrix} E_x \\ H_x \end{pmatrix} = 0\,, \tag{2}$$

$$\begin{pmatrix} \boldsymbol{E}_t \\ \boldsymbol{H}_t \end{pmatrix} = \frac{1}{k_0^2 \varepsilon_r - k_x^2}\begin{pmatrix} \partial/\partial x & -k_0\,\boldsymbol{e}_x \times \\ -k_0\,\varepsilon_r\,\boldsymbol{e}_x \times & \partial/\partial x \end{pmatrix} \nabla_t \begin{pmatrix} E_x \\ H_x \end{pmatrix},$$

$$k_x^2 + k_y^2 + \beta^2 = k_0^2\,\varepsilon_r\,, \qquad k_0 = \omega\sqrt{\varepsilon_0\,\mu_0}\,,$$

$$\nabla_t = \partial/\partial y\,\boldsymbol{e}_y - \mathrm{j}\,\beta\,\boldsymbol{e}_z\,,$$

understanding wave propagation of the form $\exp\{\mathrm{j}(\omega t - \beta z)\}$.

The relevant cross-sections are divided into subregions as Fig. 2 shows. The electric and magnetic walls account for the symmetries of the fundamental mode. In Fig. 2c the field space of the antipodal-fin line can be further reduced to $x \geqq 0$ by postulating

$$\begin{aligned} E_x(-x, 2b - y) &= E_x(x, y)\,, \\ H_x(-x, 2b - y) &= H_x(x, y)\,. \end{aligned} \tag{3}$$

In any case the metallic fins are considered to be infinitely thin and the dielectric and metal loss-free.

A complete solution to eq. (2) in the case of the bilateral-fin line, holding for the two subregions is

$$\begin{aligned} E_x^{(1)} &= \sum_{n=1}^{\infty} A_n^{(1)} \cos\tilde{\alpha}_n (x - a) \sin\alpha_n y\,, \\ E_x^{(2)} &= \sum_{n=1}^{\infty} A_n^{(2)} \sin\tilde{\alpha}_n\, x \qquad \sin\alpha_n y\,, \end{aligned} \tag{4}$$

* Dr.-Ing. H. Hofmann, c/o AEG-Telefunken, Postfach 1730, D-7900 Ulm (Donau).

H. Hofmann, *Archiv fur Elek Ubertra*, Vol. 31, No. 1, Jan. 1977, pp. 40–44, 213–216.

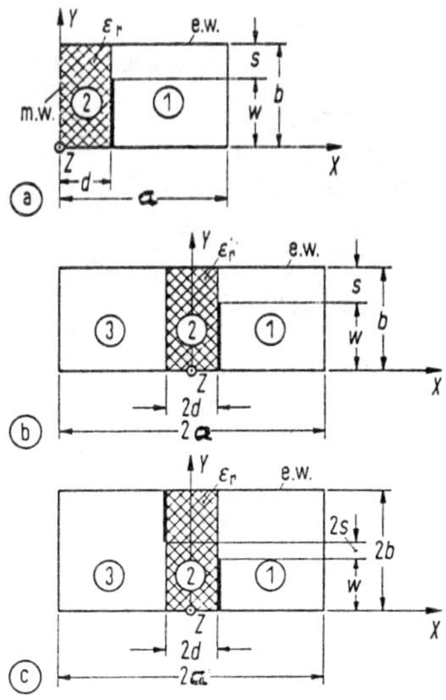

Fig. 2. Division of the field space into subregions for the different geometries; e.w.: electric wall, m.w.: magnetic wall; (a) bilateral fin, (b) unilateral fin, (c) antipodal fin.

with the transversal phase coefficients

$$\alpha_n = n\pi/b\,,\quad \bar{\alpha}_n = \sqrt{k_0^2 - \beta^2 - \alpha_n^2}\,,\quad \tilde{\alpha}_n = \sqrt{k_0^2\varepsilon_r - \beta^2 - \alpha_n^2}\,. \tag{5}$$

The components H_x are obtained by replacing sin → cos, cos → sin and naming the Fourier coefficients B_n, starting the series at $n = 0$. The solution for the unilateral-fin line is achieved by taking in subregion ① the same $E_x^{(1)}$ as above, in subregion ③ by replacing $(x - a) \to (x + a)$ of $E_x^{(1)}$ and by writing in subregion ②

$$E_x^{(2)} = \sum_{n=1}^{\infty} A_n^{(2)} \sin\tilde{\alpha}_n x \sin\alpha_n y + \bar{A}_n^{(2)} \cos\tilde{\alpha}_n x \sin\alpha_n y\,. \tag{6}$$

Finally the series for the antipodal-fin case will be explained. Taking $E_x^{(1)}$ as in eq. (4) with α_n replaced by

$$\alpha_n = n\pi/2b\,, \tag{7}$$

and choosing

$$A_n^{(3)} = -(-1)^n A_n^{(1)}\,,\qquad B_n^{(3)} = -(-1)^n B_n^{(1)} \tag{8}$$

renders the complete solutions in subregions ① and ③, accounting for the symmetry condition of eq. (3). In the same way

$$E_x^{(2)} = \sum_{n=1}^{\infty} A_n^{(2)}(\delta_e \cos\tilde{\alpha}_n x \sin\alpha_n y + \delta_o \sin\tilde{\alpha}_n x \sin\alpha_n y)\,, \tag{9}$$

$$\delta_e = \begin{cases} 0 & \text{for } n \text{ even} \\ 1 & \text{for } n \text{ odd}\,, \end{cases}\qquad \delta_o = \begin{cases} 0 & \text{for } n \text{ odd} \\ 1 & \text{for } n \text{ even} \end{cases}$$

supplies the solution in subregion ②.

Common to all Fourier-series representations of above is, that any of the boundary conditions on the electric and magnetic walls are satisfied.

3. Continuity Conditions

Applying all continuity conditions

$$\boldsymbol{E}_{\tan}^{(i)} - \boldsymbol{E}_{\tan}^{(j)} = 0\,,\qquad (\boldsymbol{H}_{\tan}^{(i)} - \boldsymbol{H}_{\tan}^{(j)}) \times \boldsymbol{e}_x = \boldsymbol{I}\,, \tag{10}$$

where $\boldsymbol{I}$ is a surface current density in the interface, allows all Fourier coefficients to be expressed in terms of $\boldsymbol{I}$, that is

$$A_n^{(i)}, B_n^{(i)} = f\{L_{1n}(I_z), L_{2n}(I_y)\}\,, \tag{11}$$

where the linear operator L applied to any function in its domain is given by

$$L_{1n}(\xi(\tau)) = \int_0^{\hat{y}} \xi(\tau) \sin\alpha_n \tau\, d\tau\,,\qquad L_{2n}(\xi(\tau)) = \int_0^{\hat{y}} \xi(\tau) \cos\alpha_n \tau\, d\tau\,, \tag{12}$$

with $\hat{y} = \begin{cases} b & \text{if } \alpha_n \text{ of eq. (5) is taken} \\ 2b & \text{if } \alpha_n \text{ of eq. (7) is taken.} \end{cases}$

Now the remaining two boundary conditions

$$E_{\tan}^{(1)} = 0\,,\quad x = d\,,\quad 0 \leqq y \leqq w\,, \tag{13}$$

with eq. (11) substituted for the $A_n^{(i)}$, $B_n^{(i)}$ deliver two operator equations in I_y, I_z

$$\sum_n \Gamma_{11} L_{2n}(I_y) \cos\alpha_n y + \sum_n \Gamma_{12} L_{1n}(I_z) \cos\alpha_n y = 0\,,$$
$$\sum_n \Gamma_{21} L_{2n}(I_y) \sin\alpha_n y + \sum_n \Gamma_{22} L_{1n}(I_z) \sin\alpha_n y = 0\,, \tag{14}$$

where the Γ_{ij} are well defined functions rendered by the continuity conditions.

Eqs. (14) are solved applying Galerkin's method: the current densities are represented by appropriate complete Fourier series

$$I_z = \begin{cases} \sum_m C_m i_m(y), & \text{for } 0 \leqq y \leqq w \\ 0 & \text{for } w \leqq y \leqq \hat{y}, \end{cases}\qquad I_y = \begin{cases} \sum_m D_m j_m(y) & \text{for } 0 \leqq y \leqq w \\ 0 & \text{for } w \leqq y \leqq \hat{y}\,. \end{cases} \tag{15}$$

Then eqs. (14) are tested with $i_m(y)$ and $j_m(y)$ respectively, supplying a doubly-infinite system of homogeneous equations in C_m and D_m. Vanishing of the system's determinant is condition for the determination of the phase coefficient β.

For the numerical evaluation the field series are truncated after N_f and the current series after N_i terms. The matrix size depends on N_i only, while each matrix coefficient is a sum over N_f terms. On account of the edge-condition [7] the indices have to be chosen as

$$N_i/N_f = w/\hat{y}\,. \tag{16}$$

On approximating the slot-line by choosing $b \gg b - w$, N_i will be equal to N_f, thus delivering very poor convergence as shall be shown later. In this case it is advantageous to formulate the problem in the slot-fields only instead of in the current densities [2].

4. Slot-field Solution

Instead of taking boundary condition (13) write

$$\begin{aligned} E_y^{(1)} &= f(y), \\ E_z^{(1)} &= g(y), \end{aligned} \qquad x = d,\ 0 \leqq y \leqq \hat{y}, \tag{17}$$

where f, g represent the fields in the slot, and are zero elsewhere. Eq. (17) allows the Fourier coefficients of region ① to be expressed as

$$A_n^{(1)},\, B_n^{(1)} = f\{L_{2n}(f),\, L_{1n}(g)\}\,. \tag{18}$$

On eliminating $A_n^{(1)}$, $B_n^{(1)}$ eqs. (11) and (18) can be rearranged to

$$\begin{aligned} \Lambda_{11} L_{2n}(f) + \Lambda_{12} L_{1n}(g) &= L_{2n}(I_y)\,, \\ \Lambda_{21} L_{2n}(f) + \Lambda_{22} L_{1n}(g) &= L_{1n}(I_z)\,, \end{aligned} \tag{19}$$

where Λ_{ij} are again well defined functions. Now sum eqs. (19) as follows:

$$\begin{aligned} \sum_n 1/\sigma_n L_{2n}(I_y) \cos\alpha_n y &= 1/2\,\hat{y}\, I_y(y)\,, \\ \sum_n L_{1n}(I_z) \sin\alpha_n y &= 1/2\,\hat{y}\, I_z(y) \end{aligned} \tag{20}$$

with $\sigma_n = \begin{cases} 2 & \text{for } n = 0 \\ 1 & \text{for } n \neq 0 \end{cases}$

or in other words

$$\begin{aligned} &\sum_n 1/\sigma_n \Lambda_{11} L_{2n}(f) \cos\alpha_n y + \\ &\quad + \sum_n \Lambda_{12} L_{1n}(g) \cos\alpha_n y = 1/2\,\hat{y}\, I_y(y)\,, \\ &\sum_n \Lambda_{21} L_{2n}(f) \sin\alpha_n y + \\ &\quad + \sum_n \Lambda_{22} L_{1n}(g) \sin\alpha_n y = 1/2\,\hat{y}\, I_z(y)\,. \end{aligned} \tag{21}$$

Galerkin's method is applied to eqs. (21) in writing

$$\begin{aligned} f(y) &= \begin{cases} \sum_m C_m f_m(y) & \text{for } w \leqq y \leqq \hat{y} \\ 0 & \text{for } 0 \leqq y \leqq w\,, \end{cases} \\ g(y) &= \begin{cases} \sum_m D_m g_m(y) & \text{for } w \leqq y \leqq \hat{y} \\ 0 & \text{for } 0 \leqq y \leqq w \end{cases} \end{aligned} \tag{22}$$

and testing with f_m and g_m respectively. The right hand sides will vanish when testing:

$$\int_0^{\hat{y}} I_y(y) f_n(y)\, dy = \int_0^{\hat{y}} I_z(y) g_n(y)\, dy = 0\,, \tag{23}$$

since I_y, I_z and f_n, g_n are defined in complementary field spaces. Thus again a doubly-infinite system of homogeneous equations is sustained.

If now N_i is the truncation index of the interface-fields the ratio

$$N_i/N_f = (\hat{y} - w)/\hat{y} \tag{24}$$

is small resulting in a much faster convergence of the solution, as shall be shown beyond.

For completeness the used Fourier components are stated following. For the earthed fins

$$\begin{aligned} i_m &= \sin\eta_m y\,, \quad j_m = \cos\eta_m y\,, \\ \eta_m &= (m - 1/2)\,\pi/w \end{aligned} \tag{25}$$

is used, while in the insulated fin case

$$\begin{aligned} i_m &= \cos\eta_m(y - y_0)\,, \quad j_m = \sin\eta_m(y - y_0)\,, \\ \eta_m &= m\,\pi/w \end{aligned} \tag{26}$$

is applied, where y_0 provides for a finite spacing of the fin from the waveguide, thus the fin extending from y_0 to $y_0 + w$. The slot fields in turn are represented by

$$\begin{aligned} f_m &= \cos\eta_m(y - w), \quad g_m = \sin\eta_m(y - w), \\ \eta_m &= m\,\pi/(\hat{y} - w)\,. \end{aligned} \tag{27}$$

It should be mentioned that any of the examined problems presents us with a very similar final system of equations (as does, as a matter of fact, the microstrip problem too), so that one FORTRAN program with only little modifications can be used for all cases. The coefficient matrix of the system is all real and symmetric.

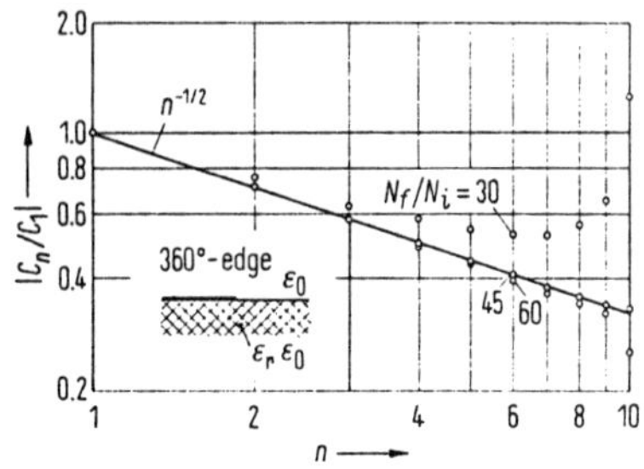

Fig. 3. Asymptotic behaviour of the Fourier coefficients of the transverse electric field in the interface; $N_i = 10$, cover-to-slot ratio $b/s = 30$, $d/\lambda = 0.01$.

5. Relative Convergence and Edge Condition

Though wellknown, there should be some remarks made upon the problem of relative convergence, which will of course emerge in the above calculations. It has been proved [7] that the edge condition supplies the correct ratio N_i/N_f of truncation indices as stated in eqs. (16) and (24). This means, only this ratio will render the correct behaviour of the fields at the edge as N_i, N_f tend to infinity. But the edge condition has also been shown to be approximated much better for a different ratio N_i/N_f if N_i, N_f are finite [8]. Convergence can thus be speeded up extremely.

In Fig. 3 the approximation of the edge condition is shown for the slot-field solution. The Fourier coefficients C_n have to be proportional to $n^{-1/2}$ for large n if the edge condition is to be fulfilled. Best approximation is achieved with

$$N_f/N_i = 1.5\,\hat{y}/(\hat{y} - w)\,. \tag{28}$$

This ratio also gives fastest convergence as is proved in Fig. 4. With $N_i = 1$ for the slot-field solution the relative error of $\varepsilon_{\text{eff}} = (\beta/k_0)^2$ is about $1^0/_{00}$. Fig. 4 also reveals the fact that for large shielding the current formulation is not appropriate, since it gives only very poor convergence.

The distribution of the slot-fields and the current densities is plotted in Fig. 5. Good approximation of the edge behaviour can be recognized in this picture.

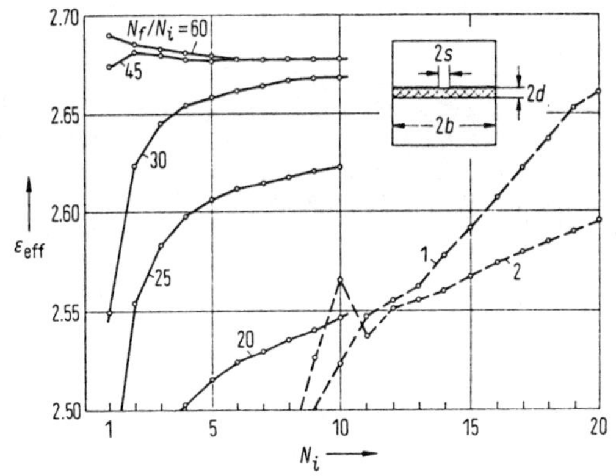

Fig. 4. Relative convergence for different truncation-index ratios;
—— calculation via the slot-fields,
---- calculation via the current densities; cover-to-slot ratio $b/s = 30$, $d/\lambda = 0.01$, $s/d = 1$.

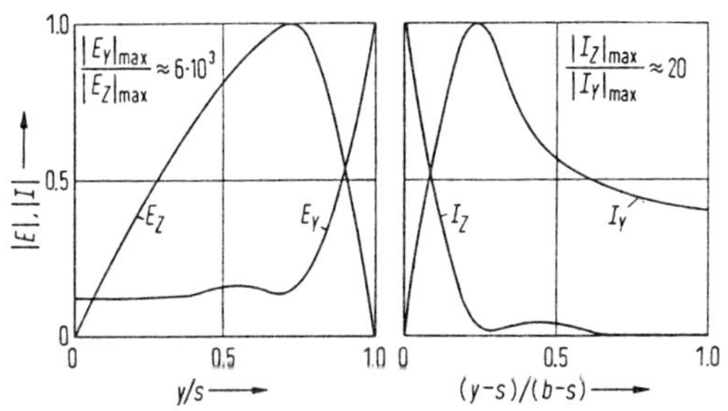

Fig. 5. Slot-fields in the interface (E) and current densities on the conductor (I) versus transverse coordinate; $2s$: slot width, $2b$: cover width.

6. Slot-line and Coplanar-strip-line Solutions

To check the validity of the calculation comparisons with published data of slot-line [1] and coplanar-strip-line [2] are depicted in Fig. 6. Very good agreement is evident. It should be stated that a cover to slot ratio or cover to strip ratio respectively of about 30 is sufficient for not too small values of d/λ. In contrast to the slot-line the configuration of Fig. 1a has a low frequency cut-off so that the shield has to be made very large for d/λ approaching zero if the open structure is to be approximated.

In the case of the coplanar strips the deviation of this theory to the one in [2] grows with decreasing s/d ratio. It is believed that the herein presented results are more accurate since the quasi-static current approximation of [2] is inferior to the complete Fourier series of eq. (26).

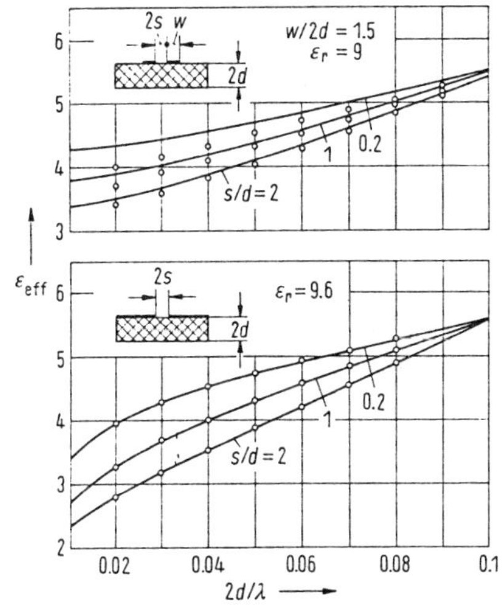

Fig. 6. Dispersion of coplanar-strip line and slot-line: —— this theory, o theory of [2] and [1], respectively.

7. Fin-line Solutions

In Fig. 7 dispersion curves of the bilateral and antipodal fin-lines are presented. Dispersion of the unilateral fin-line is within the plotting accuracy identical to that of the bilateral fin-line. (The latter has only halve the attenuation though.) A Kapton-

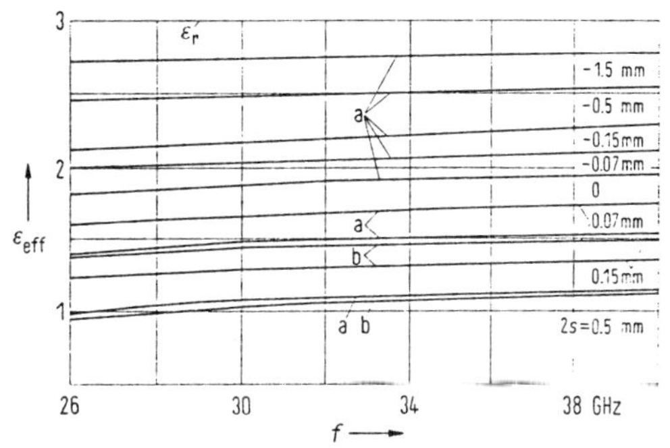

Fig. 7. Dispersion of the bilateral (curves b) and the antipodal (curves a) fin-line. Substrate: Kapton ($\varepsilon_r = 3$, thickness 0.125 mm) within WR 28 waveguide (compare Fig. 8). Negative fin spacing $2s$ implies overlapping of the antipodal fins.

substrate within WR 28 waveguide is taken as an example. Dispersion of the fin-lines is almost linear in the range of Ka-band frequencies, the antipodal fin-line rendering higher ε_{eff} than the other types.

Though very arbitrary, a wave impedance has been defined as

$$Z_L = U/I, \quad (29)$$

where I is the integral over the longitudinal current density and U is the voltage sustained by integrating the E-field along the interface from fin to fin.

Values of this wave impedance compare favourably to those of [1], ranging up to 40% higher. Cohn stated that his results are about 30% high compared to measurement. In Fig. 8 the wave impedance is plotted versus the fin-spacing. Here the advantage of the newly presented antipodal fin-line emerges. While the low limit of the wave impedances of the other two fine-line geometries is given by a minimum etchable slot width, the antipodal fine-line can easily be realized in the very wide range of about 10 to 300 Ω wave impedance by letting the fins overlap. In the vicinity of zero spacing a transition

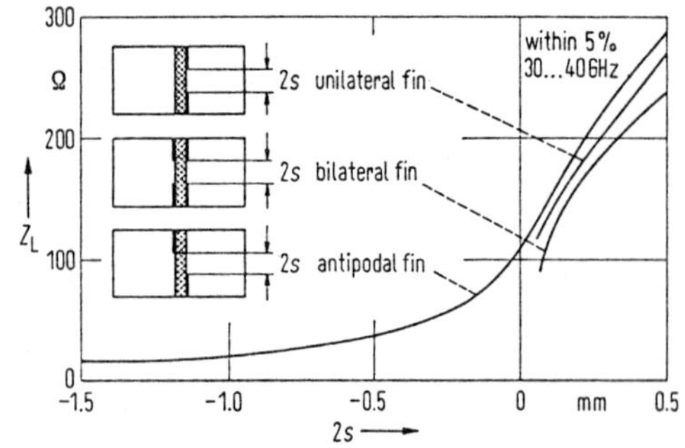

Fig. 8. Wave impedance Z_L versus fin-spacing $2s$. Substrate: Kapton ($\varepsilon_r = 3$, thickness 0.125 mm) within WR 28 waveguide. Negative fin spacing $2s$ implies overlapping of the antipodal fins.

of fine-line behaviour to parallel-plate-line behaviour takes place. For large negative spacing (i.e. overlapping) the wave impedance of this latter line is approached (e.g. $2s = -1.5$ mm renders 18 Ω as compared to 16 Ω with the fin-line). Dispersion of

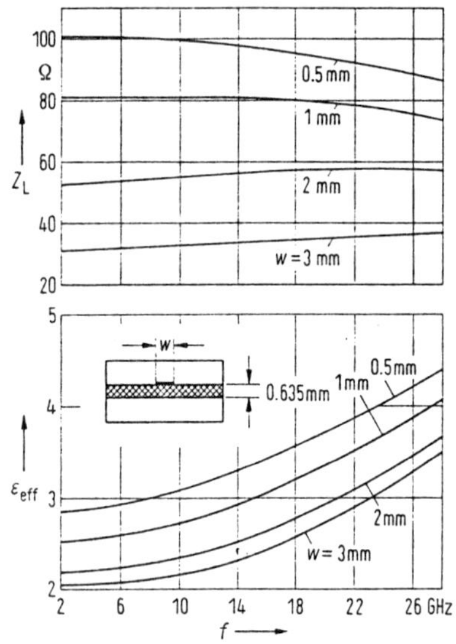

Fig. 9. Wave impedance Z_L and effective dielectric constant ε_{eff} of the suspended substrate line; substrate: $\varepsilon_r = 9.6$, shield: WR 28 waveguide.

the wave impedance is so little that the results are true within 5% in the range of 30 to 40 GHz.

If in the case of the insulated fins a large spacing from the waveguide walls is allowed, the suspended-substrate line can be calculated. Fig. 9 is a plot giving the dispersion of ε_{eff} and Z_L for this line. Alumina substrate inside a WR 28 waveguide was employed as an example.

8. Conclusion

A theory has been presented to calculate the dispersion of the phase coefficient and the wave impedance for a variety of planar structures suitable for mm-wave applications. Included are slot-line, coplanar-strip-line, suspended-substrate line, and unilateral, bilateral, and antipodal fin-lines. Particularly the antipodal fin-line promises interesting applications since its wave impedance can easily be varied in a wide range.

(Received July 14th, 1976.)

References

[1] Mariani, E. A., Heinzmann, C. P., Agrios, J. P., and Cohn, S. B., Slot line characteristics. Transact. Inst. Elect. Electron. Engrs. MTT-17 [1969], 1091–1096.

[2] Knorr, J. B. and Kuchler, K.-D., Analysis of coupled slots and coplanar strips on dielectric substrate. Transact. Inst. Elect. Electron. Engrs. MTT-23 [1975], 541–548.

[3] Meier, P. J., Equivalent relative permittivity and unloaded Q-factor of integrated finline. Electron. Letters 9 [1973], 162–163.

[4] Meier, P. J., New developments with integrated fin-line and related printed millimeter circuits. Inst. Elect. Electron. Engrs. Microwave Symp. Digest [1975], 143–145.

[5] Brenner, H. E., Use a computer to design suspended-substrate ICs. Microwaves 7 [1968], No. 9, 38–45.

[6] Gish, D. L. and Graham, O., Characteristic impedance and phase velocity of a dielectric-supported air strip transmission line with side walls. Transact. Inst. Elect. Electron. Engrs. MTT-18 [1970], 131–148.

[7] Mittra, R. and Lee, S. W., Analytical techniques in the theory of guided waves. MacMillan, New York 1971, pp. 4–14, 38–41.

[8] Jansen, R., Zur numerischen Berechnung geschirmter Streifenleiterstrukturen. AEÜ 29 [1975], 241–247.

2.1.3

DISPERSION CHARACTERISTICS IN FIN LINES WITH ONE AND TWO DIELECTRIC LAYERS ARBITRARILY LOCATED IN THE WAVEGUIDE

Humberto César Chaves Fernandes* and Attílio José Giarola+

ABSTRACT

Using an efficient method of analysis the dispersion characteristics in new unilateral fin-line structures have been obtained. It consists of using an equivalent transmission line theory in the spectral domain of the Fourier transform with the moment method. The analysis is general and may be applied to a variety of structures. Numerical results are obtained for unilateral fin lines with one and two dielectric layers located symmetrically or asymmetrically in the E-plane of the waveguide.

1. INTRODUCTION

Fin-line structures have been under intensive development in the last few years and they are becoming very important in the fabrication of millimeter wave integrated devices and circuits [1]-[3]. Various fin-line configurations have been studied, such as, unilateral, antipodal and bilateral [4]-[7].

With the objective of extending the possible applications of these structures with an increase in integration flexibility, an analysis was developed for the calculation of the dispersion characteristics in unilateral fin lines with one and two dielectric layers, located symmetrically or asymmetrically with respect to the E-plane of the waveguide, as shown in Fig. 1.

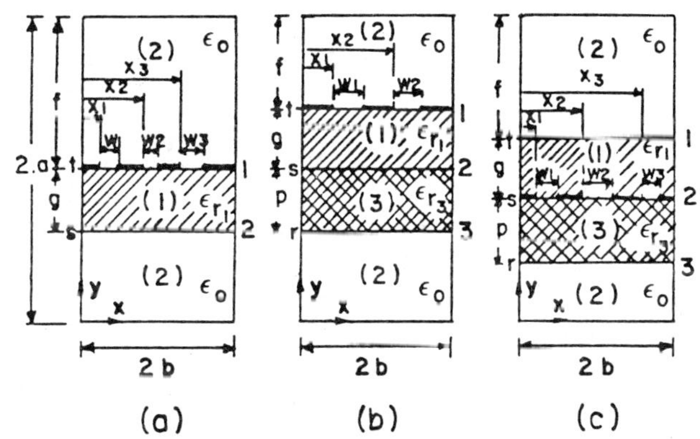

Figure 1 . Unilateral fin-line structures: a) with a single dielectric layer, b) with two adjacent layers, c) and with two dielectric layers oppositely located with respect to the fin. The unilateral fin-lines are asymmetrically located in the E-plane of the waveguide and with arbitrary slots.

The technique chosen for the dispersion analysis consists in using the concept of equivalent transmission line in the spectral domain of the Fourier transform [6]. In this procedure the admittance matrix in the transverse direction of the structure is developed and the method of moments is used to obtain the

*Electrical Engineering Department, Universidade Federal do Rio Grande do Norte (UFRN), 59000 Natal, RN, Brazil. Current address: Electrical Engineering Department, Universidade Estadual de Campinas (UNICAMP), C.P. 1170, 13100 Campinas, SP., Brazil.

+Electrical Engineering Department, Universidade Estadual de Campinas (UNICAMP), C.P. 1170, 13100 Campinas, SP., Brazil.

H.C. Chaves Fernandes and A.J. Giarola, *Proceedings of 13th European Microwave Conference*, 1983, pp. 272–277.

propagation constant. Numerical results are presented for the propagation constant as a function of: a) frequency, b) slot width, c) and distance of the central slots from one of the side walls of the waveguide.

2. THE EIGENVALUES OF THE PROPAGATION CONSTANT

The concept of equivalent transmission line in the spectral domain of the Fourier transform is as excellent method for obtaining the matrix eigenvalues for planar waveguide structures [6]. It has therefore been chosen for the analysis of the fin line structures of Fig. 1.

For the structure of Fig. 1b, for example, only the dielectric layers without metallization in the inner dielectric interfaces will be first considered. The mode spectrum consists only of TE modes in y and TM modes in y, and the modes of the structure will result as the superposition of the fields from those modes.

The Fourier transform for the field components along the y-direction and its corresponding inverse transform are respectively given by:

$$\tilde{E}_y(\alpha,y) = \int_0^{2b} E_y(x,y)\,e^{j\alpha x}dx \qquad (1)$$

and

$$E_y(x,y)e^{-j\beta z} = \frac{1}{4b}\sum_{n=-\infty}^{\infty} \tilde{E}_y(\alpha,y)e^{-j(\alpha x + \beta z)} \qquad (2)$$

where $\alpha = (n\pi/2b)$, $n = 0, \pm1, \pm2,\dots$. The magnetic field components have similar expressions. It is observed from (2) that the fields are expressed as a superposition of plane waves propagating along the x and z directions, with phase constants α and β, respectively. It is equivalent to say that these waves propagate along the u-direction, such that, $\hat{u}=\sin\theta\,\hat{x} + \cos\theta\,\hat{z}$, with $\cos\theta = N_z = \beta/(\alpha^2 + \beta^2)^{1/2}$ and $\sin\theta = N_x = \alpha/(\alpha^2 + \beta^2)^{1/2}$. Thus, the fields are decomposed as follows: for the TE in y ($\tilde{H}_y$, $\tilde{E}_v$, $\tilde{H}_u$) and for the TM in y ($\tilde{E}_y$, $\tilde{E}_u$, $\tilde{H}_v$).

In order to obtain the numerical values of the eigenvalues efficiently and with a high computational velocity and convergence a careful selection of the base functions for the expansion of the fields in the slots was done. These base functions satisfy the boundary conditions and have the following Fourier transforms:

$$\tilde{E}_{xj}(\alpha) = \sum_m a_{xjm}\frac{\pi W_j}{4}\cos[\alpha(x_j+0.5W_j)]\{J_o[0.5W_j(\alpha-m\pi)]+J_o[0.5W_j(\alpha+m\pi)]\} \qquad (3)$$

$$\tilde{E}_{zj}(\alpha) = \sum_m a_{zjm}\frac{\pi W_j}{4}\sin[\alpha(x_j+0.5W_j)]\{J_o[0.5W_j(\alpha-m\pi)]-J_o[0.5W_j(\alpha+m\pi)]\} \qquad (4)$$

where x_j is the distance from the side wall of the waveguide to the jth slot, W_j is the width of the jth slot and $J_o(\nu)$ is the Bessel function of the first kind and of zero order of the argument ν.

The current densities $\tilde{J}_u$ and $\tilde{J}_v$ [6] that generate the fields of the TM and TE waves, respectively, are then introduced. Using the theory of the equivalent transmission line, the current densities and electric fields are related as follows:

$$\tilde{J}_u(\alpha,t) = Y^e_{11}\ \tilde{E}_u(\alpha,t) \tag{5}$$

$$\tilde{J}_v(\alpha,t) = Y^h_{11}\ \tilde{E}_v(\alpha,t) \tag{6}$$

where,

$$Y^e_{11} = Y^e_{e1} + Y^e_{d1} \tag{7}$$

is the driving point admittance in interface 1 (see Fig. 1b) for the TM modes, with

$$Y^e_{e1} = Y_{TM(2)}\ \coth(\gamma_{(2)}f) \tag{8}$$

$$Y^e_{d1} = Y_{TM(1)}\left[Y_{TM(1)} + Y^e_{d2}\coth(\gamma_{(1)}g)\right]\left[Y^e_{d2} + Y_{TM(1)}\coth(\gamma_{(1)}g)\right]^{-1} \tag{9}$$

$$Y^e_{d2} = Y_{TM(3)}\left[Y_{TM(3)} + Y^e_{d3}\coth(\gamma_{(3)}p)\right]\left[Y^e_{d3} + Y_{TM(3)}\coth(\gamma_{(3)}p)\right]^{-1} \tag{10}$$

$$Y^e_{d3} = Y_{TM(2)}\ \coth(\gamma_{(2)}r) \tag{11}$$

$$Y_{TM(i)} = (\tilde{H}_v/\tilde{E}_u) = (j\omega\varepsilon_o\varepsilon_{r(i)}/\gamma_{(i)}) \tag{12}$$

$$\gamma_{(i)} = (\alpha^2 + \beta^2 - \varepsilon_{r(i)}k_o^2)^{1/2} \tag{13}$$

Here $Y_{TM(i)}$ is the admittance of the TM wave and $\gamma_{(i)}$ is the propagation constante in the ith dielectric layer. Similar expression is obtained for Y^h_{11} of the TE modes with the admittance for the TE wave obtained from

$$Y_{TE(i)} = \gamma_{(i)}/(j\omega\mu_o) \tag{14}$$

The current densities in the (x,z) coordinates are given by:

$$\begin{bmatrix}\tilde{J}_x(\alpha,t)\\ \tilde{J}_z(\alpha,t)\end{bmatrix} = \begin{bmatrix}Y^{11}_{xx} & Y^{11}_{xz}\\ Y^{11}_{zx} & Y^{11}_{zz}\end{bmatrix}\begin{bmatrix}\tilde{E}_x(\alpha,t)\\ \tilde{E}_z(\alpha,t)\end{bmatrix} \tag{15}$$

where

$$Y^{11}_{xx} = N_x^2 Y^e_{11} + N_z^2 Y^h_{11} \tag{16}$$

$$Y^{11}_{xz} = Y^{11}_{zx} = N_x N_z (Y^e_{11} - Y^h_{11}) \tag{17}$$

$$Y_{zz}^{11} = N_x^2 Y_{11}^h + N_z^2 Y_{11}^e \tag{18}$$

The method of moments is then used for the solution of (15) using as the weighting functions, the base functions [8]. A homogeneous eigenvalue matrix equation is derived, from which the propagation constant, β, for each mode is obtained.

3. NUMERICAL RESULTS AND CONCLUSIONS

Numerical results were obtained by means of a digital computer program from a PDP-10 unit. These results have shown a high convergence and confirm the efficiency of the spectral domain method. The data were obtained for fin-lines consisting of a WR-28 waveguide (2a = 7.112 mm and 2b = 3.556 mm) and substrates with relative dielectric permittivities of 2.2 and 3.0 (Kapton) with a thickness of 0.125 mm.

In Fig. 2 curves of the effective dielectric constant, $\varepsilon_{eff}=(\beta/k_o)^2$, as a function of the slot width, W_1, in a symmetrical position are shown. Also shown in these curves are the results obtained from the curves presented by Schmidt et al. [6]. A perfect agreement of the results is observed.

The influence of the substrate position along the E-plane of the waveguide is shown in Fig. 3, for a slot width of 0.2 mm. Results for W_1 = 0.1 mm and for S = 3.4935 mm with the substrate symmetrically located along the E-plane of the waveguide are also shown for the purpose of comparison with values obtained by Schmidt et al. [6]. A perfect agreement is again observed. Not shown in this figure are the values that were obtained for W_1 = 0.25 mm and W_1 = 0.50 mm that coincided with those obtained in reference [6].

A unilateral fin-line structure with two adjacent dielectric layers asymmetrically located in the E-plane of the waveguide is considered in Fig. 4. Curves of effective dielectric constant, ε_{eff}, as a function of the slot center coordinate, S_1, for two slots having widths, $W_1 = W_2$, and symmetrically located are shown in this figure.

It may be observed that, for $W_1 = W_2$ = 0.25 mm, the effect caused by the addition of the dielectric layer with relative dielectric permittivity ε_{r3} consists of an increase in the magnitude of ε_{eff}. For $\varepsilon_{r3} = \varepsilon_o$, the curves are the same as those presented in reference [6].

Finally, in Fig. 5, the effective dielectric constant of unilateral fin lines with two opposite dielectric layers with respect to the fin metallization and asymmetrically located along the E-plane of the waveguide is shown as a function of the frequency and for various values of the slot width, W_1.

A comparison of the values in Fig. 5 for frequencies 20, 30, 40 and 50 GHz with those from Fig. 2 show that the addition of the second dielectric layer ($\varepsilon_{r3} \neq \varepsilon_o$) causes an increase in the effective dielectric constant, ε_{eff}. In Fig. 5 it is also noted that ε_{eff} increases with increases in frequency and with decreases in W_1.

This work was partially supported by CNPq, CAPES, FAPESP and TELEBRÁS.

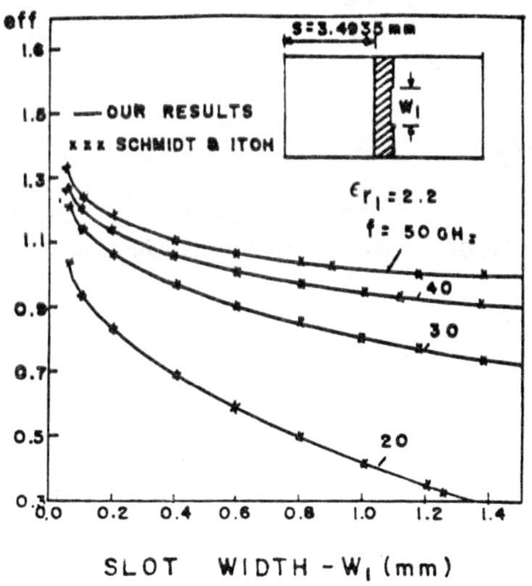

Figure 2 . Curves of the effective dielectric constant, $\varepsilon_{eff} = (\beta/k_o)^2$ as a function of the slot width, W_1, in a symmetrical position.

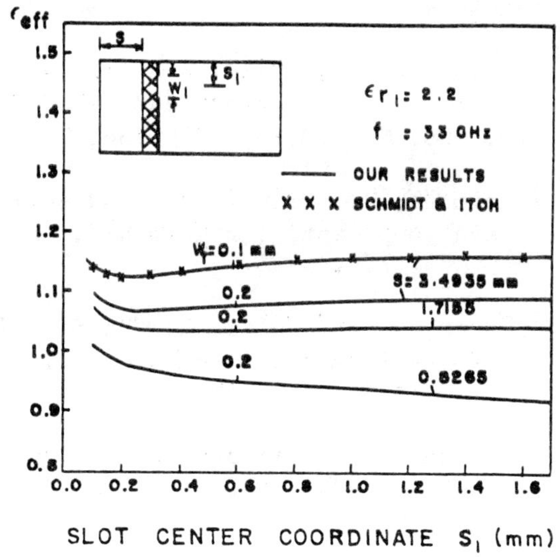

Figure 3 . Curves of the effective dielectric constant, ε_{eff}, as a function of the slot center coordinate, S_1, for unilateral fin-line symmetrically or asymmetrically located in the E-plane of the waveguide.

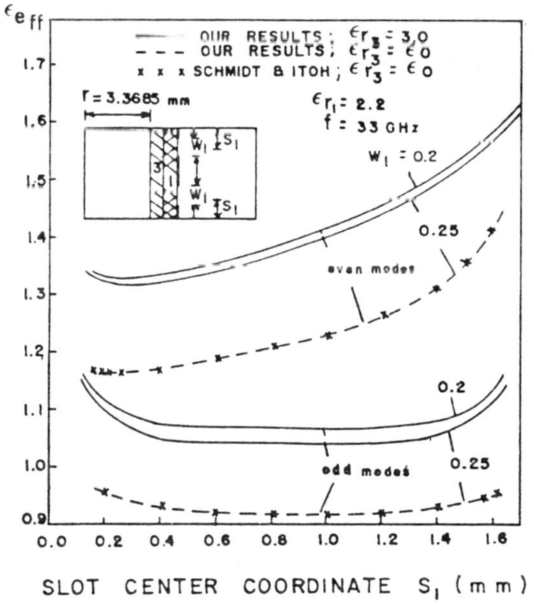

Figure 4 . Curves of the effective dielectric constant, ε_{eff}, as a function of the slot center coordinate, S_1, for a unilateral fin-line structure with two adjacent dielectric layers asymmetrically located and with two slots.

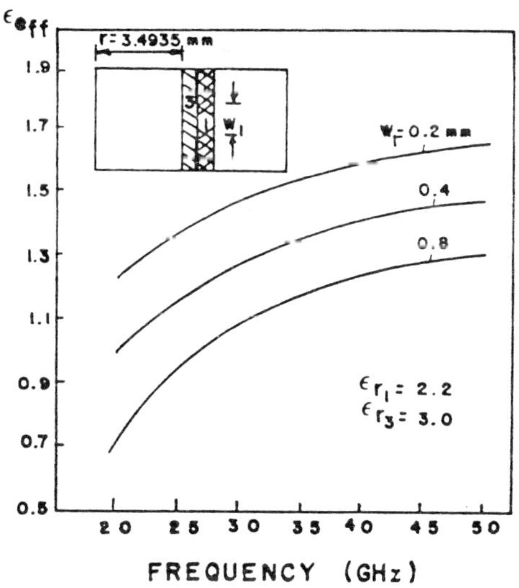

Figure 5 . Curves of the effective dielectric constant, ε_{eff}, as a function of frequency and for various values of the slot width, W_1, for a unilateral fin-line structure with two opposite dielectric layers, asymmetrically located along the E-plane of the waveguide.

REFERENCES

[1] Meier, P. J., "Millimeter integrated circuits suspended in the E-plane of rectangular waveguide", IEEE Trans. Microwave Theory Tech., Vol. MTT-26, No. 10, pp. 726-733, Oct. 1978.

[2] Hofmann, H., Meinel, H., and Adelseck, B., "Integration of millimeter wave components (in German)", Nachrichtentech. Z [Commun. J.] Vol. 31, pp. 752-757, 1978.

[3] Beyer, A. and Wolff, I., "A fin-line ferrite isolator and circulator for the R-band", 11th Europ. Microw. Conf., Amsterdam, Conf. Proc., pp. 321-326, Sep. 1981.

[4] Hofmann, H., "Dispersion of planar waveguides for millimeter-wave application", Arch. Elek. Übertragung [Electron. Commun. J.], Vol. 31, pp. 40-44, 1977.

[5] Knorr, J. B. and Shayda, P. M., "Millimeter-wave fin-line characteristics", IEEE Trans. Microwave Theory Tech., Vol. MTT-28, No. 7, pp. 737-743, July 1980.

[6] Schmidt, L. P., Itoh, T. and Hofmann, H., "Characteristics of unilateral fin-line structures with arbitrarily located slots", IEEE Trans. Microwave Theory Tech., Vol. MTT-29, No. 4, pp. 352-355, April 1981.

[7] Schmidt, L. P. and Itoh, T., "Characteristics of a generalized fin-line for millimeter wave integrated circuits", International J. of Infrared and Millimeter Waves, Vol. 2, No. 3, pp. 427-436, 1981.

[8] Harrington, R. F., "Field computation by moment method", The Macmillan Company, N.Y., 1968.

2.1.4

Millimeter-Wave Fin-Line Characteristics

JEFFREY B. KNORR, MEMBER, IEEE, AND PAUL M. SHAYDA, MEMBER, IEEE

Abstract—This paper presents an analysis of the fin line. The spectral-domain technique is used to determine both wavelength and characteristic impedance. Numerical results are compared with known data for ridged waveguide slab-loaded waveguide, and slotline. Design curves are presented for practical millimeter-wave fin line structures.

I. Introduction

IN THE PAST several years, fin line has been increasingly used as a medium for constructing millimeter wave circuits [1]–[4]. Fig. 1 shows a cross-sectional view of a fin line. The structure may be viewed as a slotline with a shield, a ridged waveguide with dielectric or a slab loaded waveguide with fins. A major attribute of the structure is its compatability with rectangular waveguide and the ease with which the fin pattern may be realized using printed circuit techniques.

In practice, when fin lines are constructed, the dielectric material is often allowed to pass through the broad wall of the shield [1]. An additional dielectric spacer may be used to provide complete dc isolation of one fin from the shield to allow biasing of solid-state devices mounted between the fins. In this case, the wall thickness is made equal to one quarter of a wavelength thereby causing an RF short circuit to appear between the fin and the inner wall of the shield. Thus, a practical fin line will have electrical characteristics similar to the idealized structure shown in Fig. 1.

Several recent papers [5]–[7] have described methods for the determination of fin line wavelength and an approximate method for calculating characteristic impedance is suggested in [1]. This paper presents an analysis of fin line using the spectral domain technique. The analysis covers both wavelength and characteristic impedance.

The paper first discusses the application of the spectral domain technique to fin line. A new matrix approach to the implementation of this method is described. Numerical results are then presented and compared with known data for ridged waveguide, slab loaded waveguide, and slotline. These comparisons establish the accuracy of the numerical results and illustrate the applicability of the method for the full range of structure parameters. It also provides perspective on the substructures which result in the various limits. Finally, several families of design curves are presented for a practical choice of fin-line parameters.

II. Fin-Line Analysis

The analysis of various structures using the spectral-domain technique has been discussed previously in several papers by one of the authors as well as other researchers [7]–[12]. Thus the application of the spectral-domain technique to the analysis of a fin line will be presented here in an abbreviated fashion.

The fin line supports a hybrid field. All transverse field components may be found from

$$E_z = k_c^2 \phi^e(x,y) e^{\Gamma z} \tag{1a}$$

$$H_z = k_c^2 \phi^h(x,y) e^{\Gamma z} \tag{1b}$$

where the scalar potential functions ϕ^e, ϕ^h satisfy the Helmholtz equation, and we assume lossless propagation so that $\gamma = \pm j\beta$. Further

$$k_{ci}^2 = k_i^2 - \beta^2 \tag{2}$$

with

$$k_i^2 = \omega^2 \mu_i \epsilon_i, \qquad i = 1,2,3 \tag{3}$$

for each of the three regions defined in Fig. 1. We will assume here that $\epsilon_1 = \epsilon_3 = \epsilon_0$ and $\epsilon_2 = \epsilon_0 \epsilon_r$.

The second-order partial differential equations for the unknown potentials ϕ^e, ϕ^h may be Fourier transformed with respect to x to obtain ordinary differential equations. When this is done and boundary conditions at the shield walls are applied, we obtain the solutions

$$\Phi_1^e(\alpha_n,y) = A^e(\alpha_n)\sinh\gamma_1(D+h_1-y) \tag{4a}$$

$$\Phi_2^e(\alpha_n,y) = B^e(\alpha_n)\sinh\gamma_2 y + C^e(\alpha_n)\cosh\gamma_2 y \tag{4b}$$

$$\Phi_3^e(\alpha_n,y) = D^e(\alpha_n)\sinh\gamma_3(h_2+y) \tag{4c}$$

$$\Phi_1^h(\alpha_n,y) = A^h(\alpha_n)\cosh\gamma_1(D+h_1-y) \tag{4d}$$

$$\Phi_2^h(\alpha_n,y) = B^h(\alpha_n)\sinh\gamma_2 y + C^h(\alpha_n)\cosh\gamma_2 y \tag{4e}$$

$$\Phi_3^h(\alpha_n,y) = D^h(\alpha_n)\cosh\gamma_3(h_2+y) \tag{4f}$$

where

$$\gamma_i^2 = \alpha_n^2 - k_{ci}^2 \tag{5}$$

and

$$\alpha_n = \begin{cases} n2\pi/b, & \phi^h \text{ even} \quad (6a) \\ (2n-1)\dfrac{\pi}{b}, & \phi^h \text{ odd} \quad (6b) \end{cases}$$

Manuscript received November 30, 1979; revised February 7, 1980. This work was supported in part by the Naval Ocean Systems Center, San Diego, CA, and in part by the Foundation Research Program, Naval Postgraduate School, Monterey, CA.

J. Knorr is with the Department of Electrical Engineering, Naval Postgraduate School, Monterey, CA 93940.

P. Shayda is with the 3D Radar Division, Surveillance System Subgroup, Naval Sea Systems Command, Washington DC.

J.B. Knorr and P.M. Shayda, *IEEE Transactions on Microwave Theory and Techniques*, Vol. MTT-28, No. 7, July 1980, pp. 737–743.

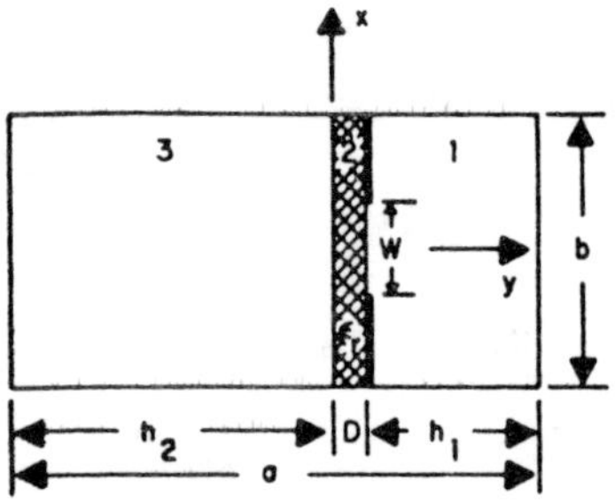

Fig. 1. Cross-sectional view of a fin line.

Through the application of boundary conditions at $y=0$ and $y=D$, the eight coefficients A^e through D^h may be related to each other, to the fin surface current, $j(x)$, and to the slot field $e(x)$. The resulting set of linear equations (really a set of transformed boundary equations) may be written in matrix form as follows:

$$[\mathcal{M}_{\mathcal{E}}]\begin{bmatrix} A^e \\ B^e \\ \vdots \\ D^h \end{bmatrix} = \begin{bmatrix} 0 \\ 0 \\ \vdots \\ \mathcal{E}_x \\ \mathcal{E}_z \end{bmatrix} \tag{7a}$$

$$[\mathcal{M}_{\mathcal{J}}]\begin{bmatrix} A^e \\ B^e \\ \vdots \\ D^h \end{bmatrix} = \begin{bmatrix} 0 \\ 0 \\ \vdots \\ \mathcal{J}_x \\ \mathcal{J}_z \end{bmatrix} \tag{7b}$$

where $\mathcal{E}_i(\alpha_n)$ is the transform of the slot field and $\mathcal{J}_i(\alpha_n)$ is the transform of the current. The matrices $[\mathcal{M}_{\mathcal{E}}]$ and $[\mathcal{M}_{\mathcal{J}}]$ differ in only the last two rows. Each is a square 8×8 matrix. Using (7a) and (7b), we may write

$$[\mathcal{M}_{\mathcal{J}}][\mathcal{M}_{\mathcal{E}}^{-1}]\begin{bmatrix} 0 \\ 0 \\ \vdots \\ \mathcal{E}_x \\ \mathcal{E}_z \end{bmatrix} = \begin{bmatrix} 0 \\ 0 \\ \vdots \\ \mathcal{J}_x \\ \mathcal{J}_z \end{bmatrix}. \tag{8}$$

From (8), using the four elements in the lower right-hand corner of the matrix $[\mathcal{M}_{\mathcal{J}}][\mathcal{M}_{\mathcal{E}}^{-1}]$, we obtain

$$[\mathcal{G}]\begin{bmatrix} \mathcal{E}_x \\ \mathcal{E}_z \end{bmatrix} = \begin{bmatrix} \mathcal{J}_x \\ \mathcal{J}_z \end{bmatrix} \tag{9}$$

where the 2×2 matrix $[\mathcal{G}]$ is the transform of the dyadic Green's function for this structure.

Previously, (9) has been obtained by extensive algebraic manipulation of the boundary equations. In the approach described here, the matrices (7) are defined directly from the boundary equations and (9) is arrived at by numerical computation. There is some sacrifice in numerical efficiency, but the formulation of the problem is more straightforward.

Equation (9) is exact, no approximations having been made so far. A solution to (9) is obtained using the Method of Moments as discussed elsewhere [7]–[12]. For this problem, we have chosen to approximate the field between the fins as

$$e_x(x)=\begin{cases} 1, & |x|<W/2 \\ 0, & \text{elsewhere} \end{cases} \tag{10a}$$

$$e_z=0. \tag{10b}$$

This choice has been shown to give accurate results for slotlines with $W/D<2$ [11], and for dielectric-loaded waveguide, $W/b=1$, it is exact. The results to be presented shortly indicate that (10) is a good choice for the fin line as well.

The dispersion problem is now reduced to the form

$$\sum_n g_{11}(\alpha_n,\beta)|\mathcal{E}_x(\alpha_n)|^2=0. \tag{11}$$

A numerical search for the value of β which satisfies (11) yields the propagation constant for the dominant fin-line mode.

The characteristic impedance of the fin line is not uniquely defined since the field is non-TEM. A useful definition, however, is

$$Z_0=\frac{V_0^2}{2P_{\text{avg}}} \tag{12}$$

where V_0 is the voltage between the fins and P_{avg} is the time-averaged power flow. V_0, the voltage across the interface, is determined by (10a) and

$$P_{\text{avg}}=\tfrac{1}{2}\operatorname{Re}\int_S\int(E_xH_y^*-E_yH_x^*)\,dx\,dy. \tag{13}$$

Parseval's theorem is applied to (13) to obtain

$$P_{\text{avg}}=\tfrac{1}{2}\operatorname{Re}\frac{1}{b}\sum_{n=-\infty}^{+\infty}\int_{-h_2}^{D+h_1}\left[\mathcal{E}_x(\alpha_n,y)\mathcal{H}_y^*(\alpha_n,y)-\mathcal{E}_y(\alpha_n,y)\mathcal{H}_x^*(\alpha_n,y)\right]dy. \tag{14}$$

The integration with respect to y can be carried out analytically in (14). This leaves an equation of the form

$$P_{\text{avg}}=\frac{1}{2b}\sum_{n=-\infty}^{+\infty}f(\alpha_n,\beta) \tag{15}$$

which is evaluated numerically in each of the three regions.

III. Numerical Results

A computer program was developed to solve the equations presented in the previous section. To check the accuracy of the numerical results, they were compared with data available in the literature as described below.

A. Ridged Waveguide

If $\epsilon_r=1$, the fin line becomes a ridged waveguide with zero-thickness ridges. The wavelength and impedance in this case are given by

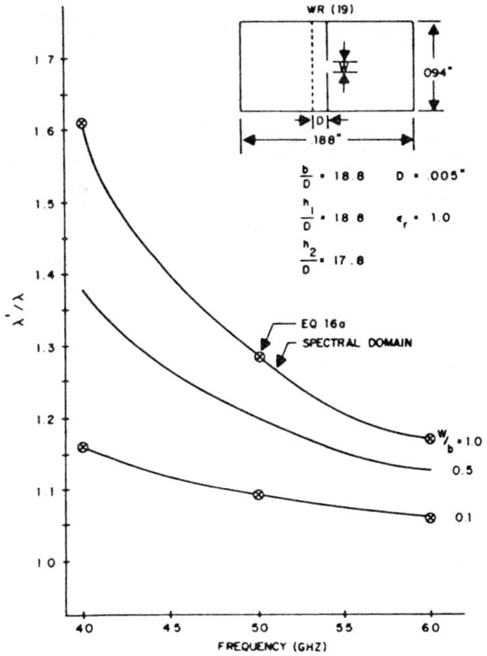

Fig. 2. Wavelength ratio λ'/λ versus frequency for a ridged waveguide. Ridges are centered and have zero thickness.

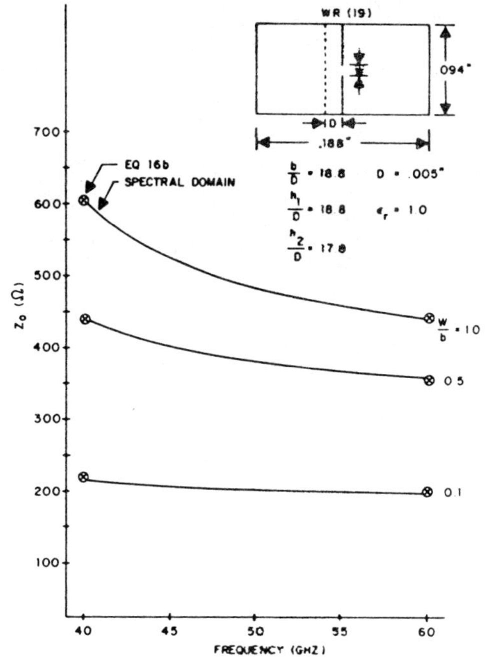

Fig. 3. Voltage impedance versus frequency for a ridged waveguide. Ridges are centered and have zero thickness.

$$\lambda'/\lambda = \frac{1}{\left[1-\left(\frac{\lambda}{\lambda_c}\right)^2\right]^{1/2}} \tag{16a}$$

$$Z_0 = \frac{Z_{0\infty}}{\left[1-\left(\frac{\lambda}{\lambda_c}\right)^2\right]^{1/2}}. \tag{16b}$$

The ridged waveguide has been treated by Hopfer [13] and Lagerlöf [14]. The cutoff wavelength λ_c was determined from [13] for several values of W/b. Equation (16a) was then used to calculate λ'/λ and these values were compared with results obtained using the spectral-domain method as shown in Fig. 2. It can be seen that the two results agree to within 0.5 percent. A similar comparison of impedance [14] as determined by both methods appears in Fig. 3. Again, the agreement is very good, with the difference being only 2 percent.

B. Slab-Loaded Waveguide

If $W/b=1$ and $\epsilon_r>1$, the fin line becomes a slab-loaded waveguide. The wavelength may easily be determined analytically by the transverse resonance procedure [15]. Fig. 4 shows the variation of λ'/λ with dielectric thickness D for several values of ϵ_r. The edge of the slab is placed at the midpoint of the broad wall of a WR(19) guide operating at 40 GHz. The agreement between the results obtained by the two methods is excellent.

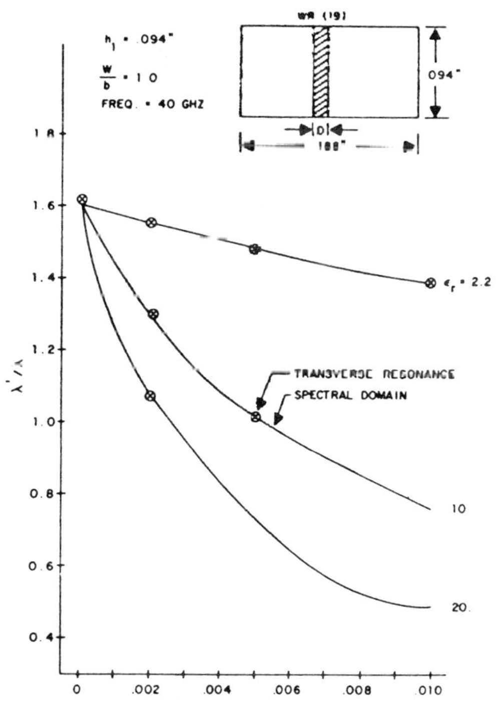

Fig. 4. Wavelength ratio λ'/λ versus dielectic thickness D for a slab-loaded WR(19) guide. Edge of slab is centered.

The slab-loaded waveguide has also been studied by Vartanian *et al.* [16]. They consider a guide with the slab centered and give an analytical expression for the (voltage) impedance Z_{pv} (using notation of [16]) at the center of the slab. The impedance at the edge of the slab (as it is

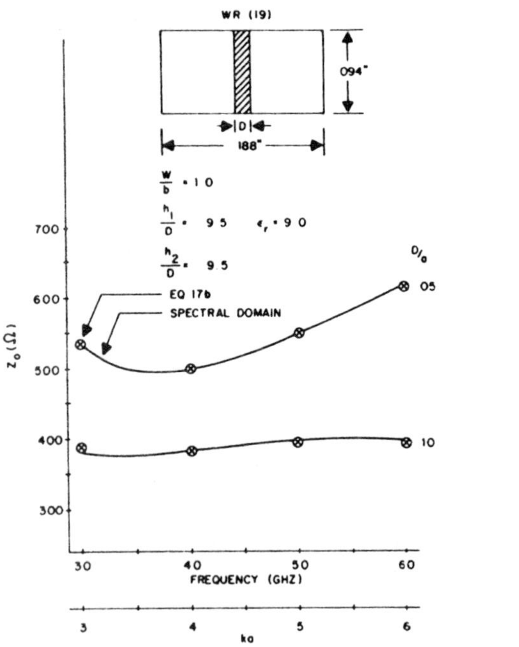

Fig. 5. Voltage impedance versus frequency for a slab-loaded WR(19) waveguide. Slab is centered.

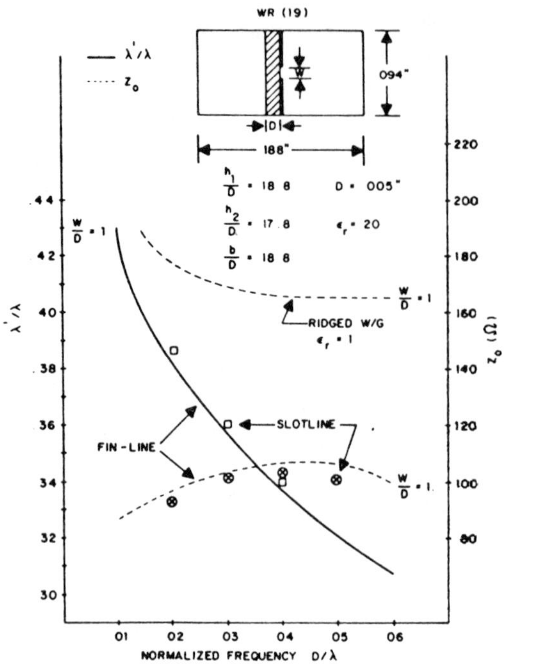

Fig. 6. Wavelength ratio λ'/λ and characteristic impedance Z_0 versus normalized frequency D/λ for a fin line and a slotline. The fin line has a WR(19) shield and fins are centered.

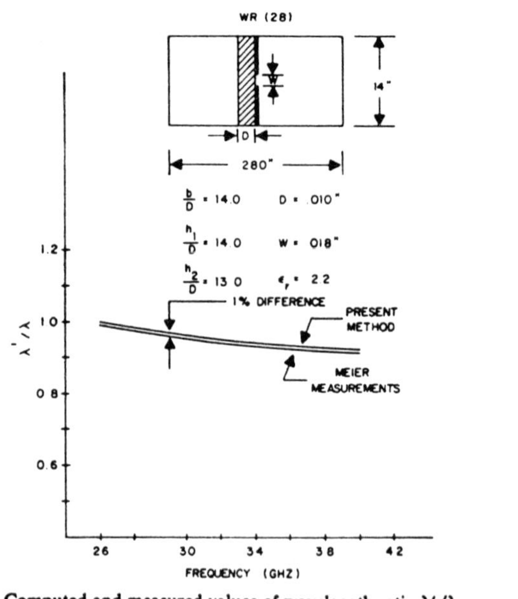

Fig. 7. Computed and measured values of wavelength ratio λ'/λ versus frequency for a fin line with WR(28) shield. Fins are centered.

defined here) may easily be obtained from Z_{pv} as

$$Z_0 = Z_{pv}\left(\frac{E_x^{\text{edge}}}{E_x^{\text{ctr}}}\right)^2 \tag{17a}$$

where E_x^{edge} is the field at the edge of the slab and E_x^{ctr} is the field at the center of the slab. Thus

$$Z_0 = Z_{pv}\cos^2\frac{qc}{2s} \tag{17b}$$

where the various quantities on the right-hand side of (17b) are defined in [16]. The characteristic impedance of a slab-loaded guide was computed using both the spectral-domain method and Vartanian's equations. The results are compared in Fig. 5 where the discrete points are calculated from (17a). It can be seen that the results agree very closely.

C. Slotline

If $W/D < 2$ and D/λ and ϵ_r are sufficiently large, the field is tightly bound to the slot. For this condition, the presence of the shield will have little effect if the walls are sufficiently far removed from the slot. In this case, the fin line will behave like a slotline. This behavior is illustrated in Fig. 6 where wavelength and characteristic impedance of a fin line with $W/D = 1$, $\epsilon_r = 20$ have been plotted. The discrete points are the values of the same parameters for a slotline with equal W/D as obtained from [17]. It can be seen that the results agree closely for the two structures. The agreement is within 1 percent for wavelength and 5 percent for impedance. The impedance of the ridged-waveguide substructure has also been plotted for reference.

D. Measurement

Fin-line wavelength measurements have been reported by Meier [1]. A computer run was made for a structure with the physical parameters reported in [1], and the results were compared with the measured values. This comparison appears in Fig. 7, where the values are seen to be in agreement to within about 1 percent.

E. Approximations

It has been suggested [1] that for low-dielectric-constant substrates the fin-line wavelength and impedance are approximated well by the relations

$$\lambda'/\lambda \doteq \frac{1}{\sqrt{k_e - \left(\frac{\lambda}{\lambda_c}\right)^2}} \tag{18a}$$

$$Z_0 \doteq \frac{Z_{0\infty}}{\sqrt{k_e - \left(\frac{\lambda}{\lambda_c}\right)^2}} \tag{18b}$$

where k_e is the effective dielectric constant and λ_c and $Z_{0\infty}$ are the cutoff wavelength and high-frequency limit impedance of the ridged-waveguide substructure. k_e is assumed invariant with frequency through the waveguide band and may be determined experimentally.

To check the accuracy of this approximation, k_e was determined for a fin line in WR(28) guide at $f=33$ GHz. This was accomplished using the numerical value of λ'/λ from Fig. 8 and the value of λ_c from [13] in (18a). The value of k_e thus determined was then assumed constant and (18) were used to calculate λ'/λ and Z_0 at $f=40$ GHz. These values were, in turn, compared with those in Fig. 8 for $W/b=0.1$, 0.5, and 1.0. Wavelength ratios agreed to within 0.2 percent while the difference in impedance values was 7–9 percent. The approximate expressions are thus useful for fin lines using low-dielectric-constant substrates. It is clear from Figs. 5 and 6, however, that there is no possibility that (18b) can be used to approximate the impedance if ϵ_r is high.

IV. Fin-Line Design Curves

In general, the publication of design curves for a structure like the fin line is problematic since there are several independently variable parameters which describe the structure. Practical considerations alleviate this difficulty to some extent, however. First, fin lines are generally enclosed with a shield that is compatible with the dimensions of the standard rectangular waveguides for the millimeter wavebands. Above 22 GHz (WR(34)), all these guides have aspect ratios $b/a=0.5$. Further, the fins are most often centered in the guide and printed using $D=$ 0.005-in substrate with $\epsilon_r=2.2$. It thus seems that it would be useful to provide design curves for structures with these parameters. This has been done for the 26.5–40-GHz, 40–60-GHz, and 60–90-GHz waveguide bands. The results appear in Figs. 8–10.

An inspection of Figs. 8–10 reveals that even with (centered) fin separations of a few mils, it is difficult to achieve low values of characteristic impedance. The lower limit for the parameters chosen here is in the range of 125–150 Ω. In some applications, it is desirable to have a lower impedance which raises the question as to how this might be achieved. One method is to increase the dielectric constant. This is clear from a comparison of Figs. 9 and 6. For a $W=0.005$-in slot, use $W/D=1$ in Fig. 6 and $W/b=0.053$ in Fig. 9 to find that the impedance is decreased from about 165 to 100 Ω when the dielectric constant is increased from $\epsilon_r=2.2$ to $\epsilon_r=20$. This approach has a disadvantage since the wavelength is decreased by a factor of 3 making already small circuit dimensions even smaller.

Another possibility for achieving a lower Z_0 might be to relocate the fins toward the sidewall of the guide. Fig. 11 illustrates the characteristics of a fin line with the fins located midway between the sidewall and the center of a WR(19) guide. Comparing Fig. 11 with Fig. 9, it can be seen that the relocation of the fins causes a small increase in the wavelength and pronounced changes in the impedance. The change in wavelength, for example, is 4 percent at 40 GHz and 2 percent at 60 GHz for $W/b=1$. The change is less for smaller values of W/b. With regard to characteristic impedance, it can be seen that the values are relatively unchanged for small values of W/b. For this situation, the line behaves like a slotline and there is little effect produced by the presence of the shield. For $W/b=1$, however, the impedance decreases significantly, from about 500 Ω for the centered configuration to about 275 Ω for the off-center configuration. In this case, the guide is also loaded, and since the low dielectric constant causes little change in the empty guide fields, the result agrees with expectations based on the voltage-impedance definition for an empty guide. The interesting feature of the results here is that the impedance for $W/b=1$ falls below that for $W/b=0.5$. Thus for the off-center configurations, the impedance first increases with W/b, then reaches a maximum, and decreases until $W/b=1$. Overall, it seems clear that relocation of the fins toward the side of the guide does not result in lower structure impedance when W/b is small. It should also be noted that the ridged waveguide data necessary for the evaluation of the approximate expressions given by (18) are not readily available in the literature.

Probably the best way to realize a lower structure impedance is to use a single-fin structure. If a single fin were separated from the bottom wall of the shield by a distance $W/2$, then for small values of W/b the impedance would be approximately one half that for a normal fin line with fin spacing W. For example, Fig. 8 shows that the impedance of a WR(28) fin line with $W/b=0.05$ ($W=0.007$ in) is approximately $Z_0=160$ Ω. Thus a single fin in the same shield would result in a structure impedance of approximately $Z_0=80$ Ω if located a distance $W=0.0035$ in from the bottom wall of the shield. This configuration has the additional advantage that the bottom wall of the shield may be used as a heat sink.

V. Conclusions

A transform method of obtaining the wavelength and characteristic impedance for the dominant mode of the fin line has been presented. It has been shown that a matrix formulation of the problem permits the elements of the

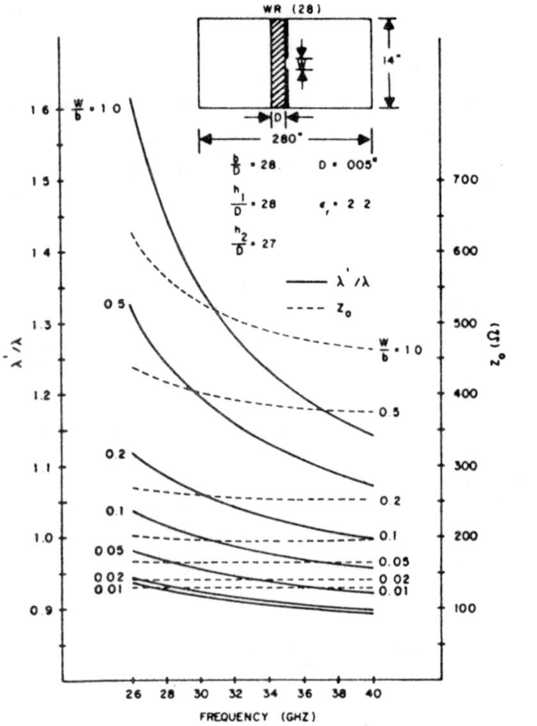

Fig. 8. Wavelength ratio λ′/λ and characteristic impedance Z_0 versus frequency for a fin line with WR(28) shield. Fins are centered.

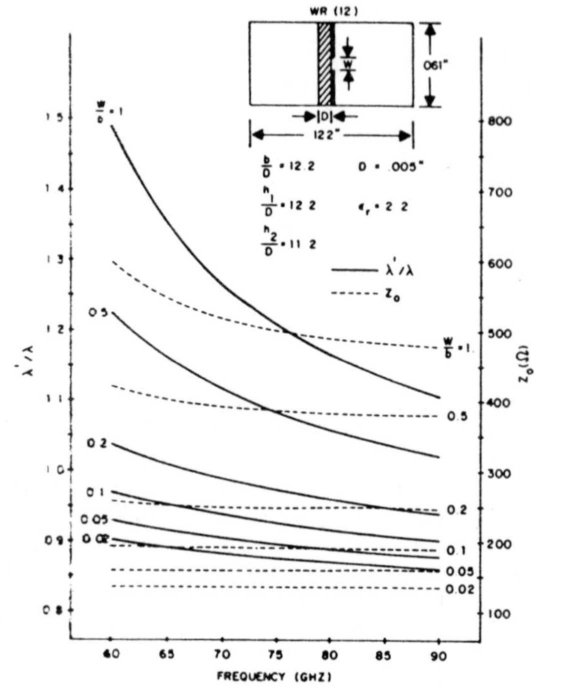

Fig. 10. Wavelength ratio λ′/λ and characteristic impedance Z_0 versus frequency for a fin line with WR(12) shield. Fins are centered.

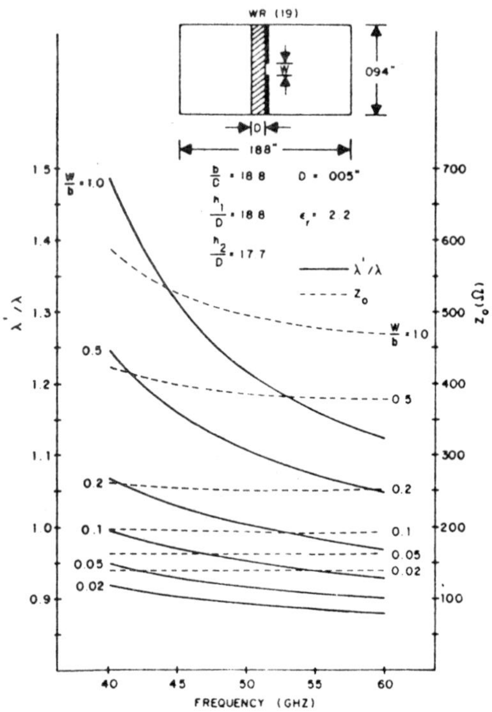

Fig. 9. Wavelength ratio λ′/λ and characteristic impedance Z_0 versus frequency for a fin line with WR(19) shield. Fins are centered.

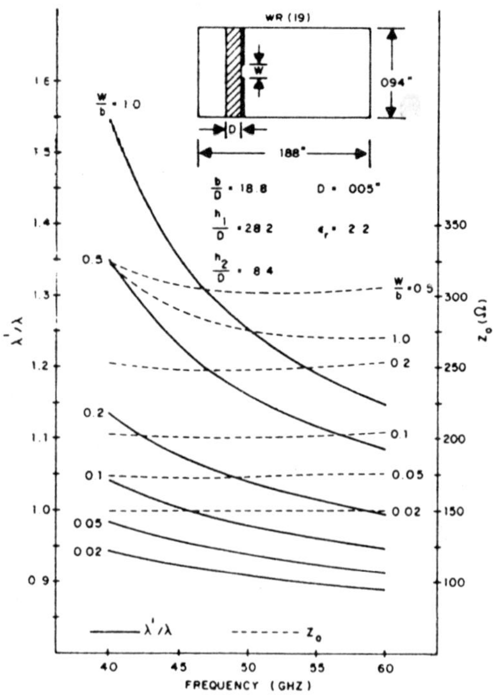

Fig. 11. Wavelength ratio λ′/λ and characteristic impedance Z_0 versus frequency for a fin line with WR(19) shield. Fins are located halfway between side wall and guide center.

dyadic Green's function to be calculated, and circumvents the extensive algebraic manipulation associated with formulations described previously. Numerical results obtained using this method have been presented and compared to other existing data. Good agreement was obtained in all cases thus establishing the accuracy and applicability of the method for the full range of structure parameters. Design curves have been included here for millimeter-wave fin lines of practical interest. Both center and off-center fin locations have been discussed, and the off-center location was shown to result in no significant change in impedance for small values of W/b. Lower impedance may be realized, however, by using a single-fin configuration.

It is clear from the results presented here that the fin line may exhibit the characteristics of a ridged waveguide, slotline, or dielectric slab-loaded waveguide, depending upon the values of the various fin-line parameters. All of these structures are fin-line substructures.

References

[1] P. J. Meier, "Integrated fin-line millimeter components," *IEEE Trans. Microwave Theory Tech.* vol. MTT-22, pp. 1209–1216, Dec. 1974.

[2] ——, "Millimeter integrated circuits suspended in the *E*-plane of rectangular waveguide," *IEEE Trans. Microwave Theory Tech.*, vol. MTT-26, pp. 726–733, Oct. 1978.

[3] ——, "Printed circuit balanced mixer for the 4 and 5 mm bands," in *IEEE MTT-S Symp. Dig.*, pp. 84–86, May 1979.

[4] W. Kpodzo *et al.*, "A quadriphase modulator in fin-line technique," in *IEEE MTT-S Symp. Dig.*, pp. 119–121, May 1979.

[5] A. M. K. Saad and K. Schunemann, "A simple method for analyzing fin-line structures," *IEEE Trans. Microwave Theory Tech.*, vol. MTT-26, pp. 1002–1007, Dec. 1978.

[6] W. Hoefer, "Fin-line parameters calculated with the TLM method," in *IEEE MTT-S Symp. Dig.*, pp. 341–343, May 1979.

[7] T. Itoh, "Spectral domain analysis of dominant and higher order modes in fin-lines," in *IEEE MTT-S Symp. Dig.*, pp. 344–345, May 1979.

[8] T. Itoh and R. Mittra, "Dispersion characteristics of slot lines," *Electron. Lett.*, vol. 7, pp. 364–365, July 1971.

[9] ——, "Spectral-domain approach for calculating the dispersion characteristics of microstrip lines," *IEEE Trans. Microwave Theory Tech.*, (Short Papers), vol. MTT-21, pp. 496–499, July 1973.

[10] ——, "A technique for computing dispersion characteristics of shielded microstrip lines," *IEEE Trans. Microwave Theory Tech.* (Short Papers), vol. MTT-22, pp. 896–898, Oct. 1974.

[11] J. B. Knorr and K -D. Kuchler, "Analysis of coupled slots and coplanar strips on dielectric substrate," *IEEE Trans. Microwave Theory Tech.*, vol. MTT-23, pp. 541–548, July 1975.

[12] J. B. Knorr and A. Tufekcioglu, "Spectral-domain calculation of microstrip characteristic impedance," *IEEE Trans. Microwave Theory Tech.*, vol. MTT-23, pp. 725–728, Sept. 1975.

[13] S. Hopfer, "The design of ridged waveguides," *IRE Trans. Microwave Theory Tech.*, vol. MTT-3, pp. 20–29, Oct. 1955.

[14] R. O. Lagerlöf, "Ridged waveguide for planar microwave circuits," *IEEE Trans. Microwave Theory Tech.*, vol. MTT-21, pp. 499–501, July 1973.

[15] R. Collin, *Field Theory of Guided Waves*. New York: McGraw-Hill, 1960.

[16] P. H. Vartanian *et al.*, "Propagation in dielectric slab loaded rectangular waveguide," *IRE Trans. Microwave Theory Tech.*, vol. MTT-6, pp. 215–222, Apr. 1958.

[17] E. A. Mariani *et al.*, "Slot line characteristics," *IEEE Trans. Microwave Theory Tech.* (1969 Symp. Issue), vol. MTT-17, pp. 1091–1096, Dec. 1969.

2.1.5

A Modified Mode-Matching Technique and Its Application to a Class of Quasi-Planar Transmission Lines

RUEDIGER VAHLDIECK, MEMBER IEEE, AND JENS BORNEMANN

Abstract —A rigorous and versatile hybrid-mode analysis is presented to determine the normalized propagation constants in a class of quasi-planar transmission-line structures.

The method is accurate and covers the finite metallization thickness, mounting grooves, and an arbitrary number of dielectric subregions.

Utilizing a modified mode-matching technique, one can derive discontinuity and transmission-line matrices for each homogeneous subregion. Successively multiplying matrix equations of all subregions leads to the characteristic matrix system. This procedure makes it possible to create a modularized computer program which can be conveniently extended to a wide spectrum of conceivable configurations simply by inserting the matrix equations of additional subregions in the multiplication process. To demonstrate the efficiency of the proposed method, dispersion characteristics of dominant and next higher order hybrid modes in earthed and insulated finlines, suspended microstrips, and coupled striplines with tuning septa, are given as examples.

I. Introduction

E-PLANE MILLIMETER-WAVE integrated circuits are of considerable interest in systems requiring single-mode broad-band operation. Earthed and insulated finlines, coupled strip- and coplanar lines are the most common structures which have been succesfully applied, for example, in the design of broad-band directional couplers, taper sections, and filters. Their propagation characteristics have been obtained by various methods. An early paper by Meier [1] describes the dominant mode in an earthed finline as a variation of the corresponding mode in a ridged waveguide, but test measurements are still necessary to determine the equivalent dielectric constant of the configuration. In order to avoid these time-consuming and expensive measurements, several theoretical methods have been proposed to evaluate the dispersion characteristics of the dominant and sometimes even higher order modes in an earthed finline or more complicated configurations like coupled strip or coplanar lines.

The two-dimensional transmission-line matrix (TLM) has been applied by Shih and Hoefer [2] to determine the dominant and second-order mode cutoff frequencies in a unilateral, bilateral, and so-called insulated finline. More complicated structures have been analyzed by using the numerically very efficient spectral-domain method (SDM) (Itoh and Schmidt [3]–[5]), but only results with zero metallization thickness and neglected mounting grooves have been published.

More recently, two different mathematical treatments have been presented which take into account the finite metallization thickness (Kitazawa and Mittra [13]) and, additionally, the mounting grooves (Vahldieck [6] and Bornemann [7]). In [13], a hybrid-mode formulation has been proposed in which Green's functions are derived using conventional circuit theory. Results are given for unilateral and bilateral finlines.

Vahldieck [6] and Bornemann [7] extended a modified mode-matching technique which utilizes a transverse resonance relation [8], [14] to reduce the number of required eigenvalue equations. This method, which includes both the finite metallization thickness and the mounting grooves, has been applied successfully to a generalized finline configuration with more than one dielectric subregion and different fin thicknesses [6]. In [7], structures containing both a strip and a slot were treated, and results were presented for coupled slotlines and suspended microstrips. This already indicates that the method is quite general and can be extended very easily to more complicated combinations of strip- and slotlines.

For the numerical computation of the propagation constants, it is an advantage of this method that the order of the characteristic matrix equation is only $2NH-1$ (where NH is the number of orthogonal eigenfunctions considered in each subregion) and remains constant even for an increasing number of transverse discontinuities. This is in contrast, for example, to the well-known mode-matching procedure used by Siegl [11] and Beyer [12] in which the size of the characteristic matrix equation usually increases with the number of subregions. Moreover, in the latter method, all boundary conditions are satisfied before the tangential E- and H-fields at each interface are matched. This leads to an inflexible procedure and makes it normally necessary to create a new computer program when investigating different configurations with additional subregions.

However, the modified mode-matching technique combined with a transverse resonance relation avoids this disadvantage. Initially, the boundary conditions at the waveguide sidewalls ($x=0$ and $x=a$, Fig. 1(a)–(c)) are neglected. The configuration can then be regarded as a

Manuscript received February 5, 1985; revised May 31, 1985.

R. Vahldieck is with the Department of Electrical Engineering, University of Ottawa, 770 King Edward Ave., Ottawa, Ontario K1N 6N5, Canada.

J. Bornemann is with the Microwave Department, University of Bremen, D-2800 Bremen 33, Kufsteiner Str., NW1, West Germany.

R. Vahldieck and J. Bornemann, *IEEE Transactions on Microwave Theory and Techniques*, Vol. MTT-33, No. 10, Oct. 1985, pp. 916–926.

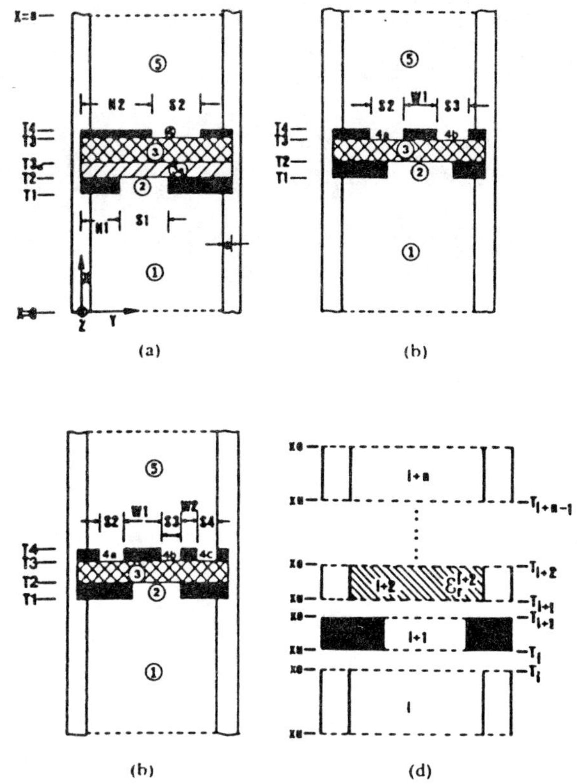

Fig. 1. Some general cross sections of quasi-planar transmission lines. (a) Earthed bilateral finline, (b) coupled slotline, (c) coupled suspended stripline, (d) subdivision of a cross section into an arbitrary number of homogeneous subregions.

parallel-plate line subdivided into homogeneous cross sections (Fig. 1(d)) in which partial wave amplitudes are defined, traveling in positive and negative x-directions with the still unknown propagation constant kx. After the elimination of the field coefficients in each subregion, a transverse transmission-line matrix results which relates the amplitudes at the lower $(x = xu)$ and the upper boundary $(x = xo$, cf., Fig. 1(d)) of a subregion. Satisfying a modified field-matching condition at the common interfaces (e.g., T_i, Fig. 1(d)) transforms the partial wave amplitudes of cross-section i into those of cross-section $i+1$. Successively multiplying each transformation matrix with the appropriate transmission-line matrix of the corresponding subregion, a relation only between the partial wave amplitudes at the lower $(x = 0)$ and upper $(x = a)$ boundary of the waveguide is obtained. Finally, the inhomogeneous waveguide cross section can be regarded as a line resonator in which the resonance condition is satisfied by inserting the up-to-now neglected boundary conditions at the metallic sidewalls $(x = 0$ and $x = a)$. This procedure reduces the size of the characteristic matrix equation to a quarter of the original size. Furthermore, it makes the method very flexible because an arbitrary number of subregions can be inserted easily in the matrix system simply by multiplying the additional transformation and transmission-line matrices with the previous.

The aim of this paper is to demonstrate the potential of this versatile and accurate treatment. As examples, earthed and insulated finlines will be analyzed, as well as coupled strip- and coupled slotlines. Finally, some detailed results on the effect of finite metallization thickness and mounting grooves on the dispersion characteristics of quasi-TEM, and next higher order hybrid modes will be presented.

II. Theory

Since the theoretical treatment has been described recently by Vahldieck [6] and Bornemann [7], only the principle steps will be explained. For further information, the reader is referred to [6] and [7].

It is well known that quasi-planar transmission lines support hybrid modes in which all six field components can occur. The transversely inhomogeneous configuration is divided into homogeneous subregions. In each of these subregions, the hybrid fields

$$\vec{E}^i = \nabla\times\nabla\times\vec{A}^i_e - j\omega\mu\nabla\times\vec{A}^i_m \qquad (1)$$

$$\vec{H}^i = \nabla\times\nabla\times\vec{A}^i_m + j\omega\epsilon^i\nabla\times\vec{A}^i_e \qquad (2)$$

can be expressed by a superposition of the axial z-components of two independent Hertzian vector potentials $\vec{A}_m$ and $\vec{A}_e$. The potential functions are a sum of suitable orthogonal eigenfunctions

$$\left.\begin{aligned} A^i_{mz} &= \sum_n^{\infty} f^i_{c(n)}(y)\cdot I^i_{m(n)}(x) \qquad (3)\\ A^i_{ez} &= \sum_n^{\infty} f^i_{s(n)}(y)\cdot U^i_{e(n)}(x) \qquad (4)\end{aligned}\right\}\cdot e^{-jk_z\cdot z}$$

which satisfy the boundary condition at the metallic surface and obeys the scalar Helmholtz equation. Initially, however, we neglect the boundary conditions at the waveguide sidewalls $(x = 0$ and $x = a$, cf., Fig. 1) and regard each subregion as a part of a parallel-plate line in which partial wave amplitudes I^i_m and U^i_e are defined and travel in positive and negative x-directions. This procedure results in a somewhat higher theoretical effort, but is necessary in order to reduce the size of the characteristic matrix equation. Finally, it makes it easy to implement any desired number of subregions in the formulation. The functions f_c and f_s in (3) and (4) (given in the Appendix) implicitly satisfy the y-dependent boundary conditions in their respective subregions. I^i_m and U^i_e, as well as their derivatives

$$U^i_m = \frac{dI^i_m}{dx} \quad \text{and} \quad I^i_e = \frac{dU^i_e}{dx}$$

combined as vectors U^i and I^i at the upper boundary $(x = xo)$ of each subregion, can be determined from the amplitudes at the lower boundary $(x = xu)$, see Fig. 1(d). Therefore, a generalized transmission-line matrix R^i can be found which relates U^i and I^i at the two coordinates as follows:

$$\begin{bmatrix} U^i \\ I^i \end{bmatrix}_{x=xo} = \underbrace{\begin{bmatrix} R^i_c & R^i_s \\ R^{i\prime}_s & R^i_c \end{bmatrix}}_{R^i} \cdot \begin{bmatrix} U^i \\ I^i \end{bmatrix}_{x=xo} \qquad (5)$$

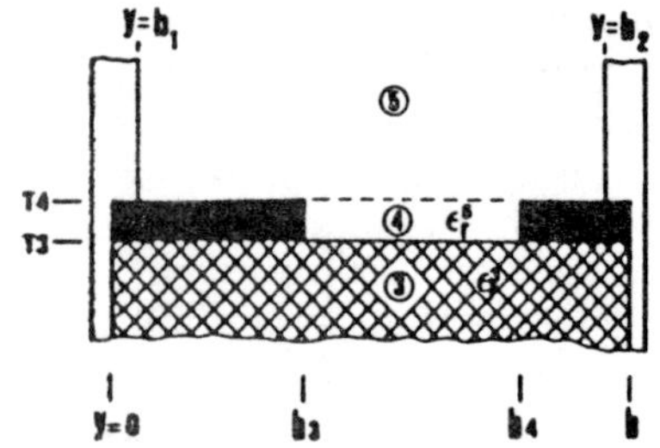

(a)

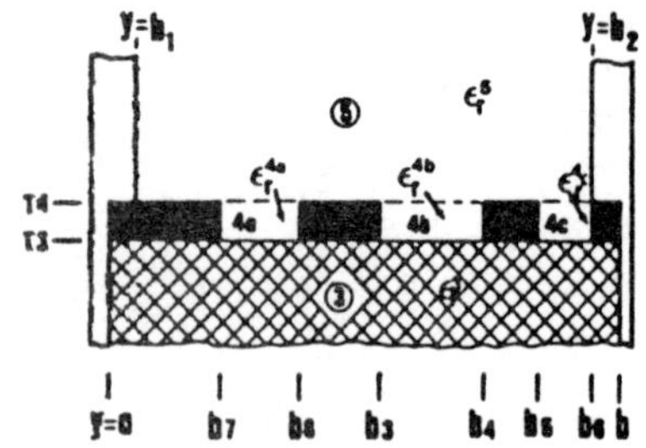

(b)

Fig. 2. Representative transition T4 and T3 for the configurations in Fig. 1. (a) Slot transition with only one slot. (b) Slot transition with three slots.

R^i_c, R^i_s, and R^r_c, are diagonal matrices containing sine and cosine functions. They are explained in the Appendix. Before a final relation between the amplitude vectors at the upper $(x = a)$ and the lower $(x = 0)$ sidewall of the waveguide can be given, it is necessary to modify the continuity equations at each transverse discontinuity, such that the amplitude vectors of subregion $i+1$, for example, are determined by those of region i.

The following mathematical derivation of this principle step is based on the representative transition $T4$ in Fig. 2. In the first case of Fig. 2(a), the continuity condition can be written as follows:

$$E^5_{y,z} = E^4_{y,z} \tag{6}$$

$$E^5_{y,z} = 0, \qquad b_1 \leq y < b_3, \quad b_4 < y \leq b_2 \tag{6a}$$

$$H^5_{y,z} = H^4_{y,z}, \qquad b_3 \leq y \leq b_4 \tag{7}$$

and in the second case of Fig. 2(b) in plane $T4$

$$E^5_{y,z} = \sum_\nu E^\nu_{y,z} \tag{8}$$

$$E^5_{y,z} = 0, \qquad b_1 \leq y < b_7, b_8 < y < b_3, b_4 < y < b_5 \tag{8a}$$

$$H^5_{y,z} = H^\nu_{y,z}, \qquad b_7 \leq y \leq b_8, b_3 \leq y \leq b_4, b_5 \leq y \leq b_2$$

$$(\nu \in 4a, 4b, 4c). \tag{9}$$

In order to determine the partial wave amplitudes of subregion $i = 5$ (Fig. 2(a)) from their corresponding amplitudes in the adjacent area $i = 4$, we first modify (6) and (7) such that the following expressions result:

$$E_y: \ \frac{\partial A^5_{mz}}{\partial x} = \frac{\partial A^4_{mz}}{\partial x} - K^4_\mu(\omega, kz)\frac{\partial A^4_{ez}}{\partial y} \tag{10}$$

$$E_z: \quad A^5_{ez} = K^4(\omega, kz)A^4_{ez} \tag{11}$$

$$H_z: \quad A^5_{mz} = K^4(\omega, kz)A^4_{mz} \tag{12}$$

$$H_y: \ \frac{\partial A^5_{ez}}{\partial x} = \frac{\epsilon^4_r}{\epsilon^5_r}\frac{\partial A^4_{ez}}{\partial x} + K^4_\epsilon(\omega, kz)\frac{\partial A^4_{mz}}{\partial y} \tag{13}$$

with

$$K^4(\omega, kz) = \frac{\epsilon^4_r - \beta^2}{\epsilon^5_r - \beta^2}, \qquad \beta = kz/ko$$

$$K^4_\mu(\omega, kz) = \frac{kz}{\omega\mu}(1 - K^4(\omega, kz))$$

$$K^4_\epsilon(\omega, kz) = \frac{kz}{\omega\epsilon^5}(1 - K^4(\omega, kz))$$

and similarly for (8) and (9)

$$E_y: \qquad \frac{\partial A^5_{mz}}{\partial x} = \sum_\nu\left[\frac{\partial A^\nu_{mz}}{\partial x} - K^\nu_\mu(\omega, kz)\frac{\partial A^\nu_{ez}}{\partial y}\right] \tag{14}$$

$$Ez: \qquad A^5_{ez} = \sum_\nu K^\nu(\omega, kz)A^\nu_{ez} \tag{15}$$

$$Hz: \ \bar{K}^\nu(\omega, kz)A^5_{mz} = A^\nu_{mz} \tag{16}$$

$$H_y: \ \frac{\epsilon^5_r}{\epsilon^\nu_r}\frac{\partial A^5_{ez}}{\partial x} + \bar{K}^\nu_\epsilon(\omega, kz)\frac{\partial A^5_{mz}}{\partial y} = \frac{\partial A^\nu_{ez}}{\partial x} \tag{17}$$

with

$$\bar{K}^\nu_\epsilon(\omega, kz) = \frac{kz}{\omega\epsilon^\nu}(1 - \bar{K}^\nu(\omega, kz))$$

$$\bar{K}^\nu(\omega, kz) = \frac{\epsilon^5_r - \beta^2}{\epsilon^\nu_r - \beta^2}$$

$$K^\nu_\mu(\omega, kz) = \frac{kz}{\omega\mu}(1 - K^\nu(\omega, kz)).$$

For the interface $T3$ (Fig. 2(a)), replace in (10)–(13) the index $i = 5$ with $i = 4$, and index $i = 4$ with $i = 3$. To obtain the interface expression for $T3$ in Fig. 2(b), replace in (14)–(17) $i = 5$ with $i = 3$. The discontinuities $T4$, $T3$ in Fig. 1(b) are then also included in (14)–(17). Moreover, the equation system (10)–(17) is very instructive. It shows that the coupling between TE and TM waves is automatically included, and that only in structures with homogeneously filled cross sections (e.g., ridged waveguide) or in inhomogeneously filled structures at cutoff $(kz = 0)$ are these wave types decoupled since $K\mu$ (ω, kz) and $K\epsilon$ (ω, kz) in (10) and (13) vanish as well as $K\mu$ (ω, kz) and $K\epsilon$ (ω, kz) in (14) and (17).

Multiplication of the above equation system with appropriate orthogonal eigenfunctions [6], [7] separates the partial wave amplitudes on the left-hand side of (10)–(13)

$$\begin{bmatrix} Um^5 \\ Ue^5 \end{bmatrix}_{x=T_4} = \underbrace{\begin{bmatrix} \frac{2}{b_2-b_1} Fc^4 & K_\mu^4(\omega, kz)\widetilde{Fc}^4 \\ 0 & \frac{2}{b_2-b_1} K^4(\omega, kz)\, Fs^4 \end{bmatrix}}_{V^E} \cdot \begin{bmatrix} Um^4 \\ Ue^4 \end{bmatrix}_{x=T_4} \tag{18}$$

$$\begin{bmatrix} Im^5 \\ Ie^5 \end{bmatrix}_{x=T_4} = \underbrace{\begin{bmatrix} \frac{b_4-b_3}{2} K^4(\omega, kz)(Fc^{4T})^{-1} & 0 \\ -K_\epsilon^4(\omega, kz)\left(\widetilde{Fs}^{4T}\right)^{-1} & \frac{b_4-b_3}{2}\frac{\epsilon_r^4}{\epsilon_r^5}(Fs^{4T})^{-1} \end{bmatrix}}_{V^H} \cdot \begin{bmatrix} Im^4 \\ Ie^4 \end{bmatrix}_{x=T_4} \tag{19}$$

and similary in (14)–(17)

$$\begin{bmatrix} Um^5 \\ Ue^5 \end{bmatrix}_{x=T_4} = \underbrace{\begin{bmatrix} \frac{2}{b_2-b_1} Fc^{4^a} & \frac{2}{b_2-b_1} Fc^{4^b} & \frac{2}{b_2-b_1} Fc^{4^c} & -K_\mu^{4^a}(\omega, kz)\widetilde{Fc}^{4^a} & -K_\mu^{4^b}(\omega, kz)\widetilde{Fc}^{4^b} & -K_\mu^{4^c}(\omega, kz)\widetilde{Fc}^{4^c} \\ 0 & 0 & 0 & \frac{2}{b_2-b_1} K^{4^a}(\omega, kz)\, Fs^{4^a} & \frac{2}{b_2-b_1} K^{4^b}(\omega, kz)\, Fs^{4^b} & \frac{2}{b_2-b_1} K^{4^c}(\omega, kz)\, Fs^{4^c} \end{bmatrix}}_{Vs^E} \cdot \begin{bmatrix} Um^{4^a} \\ Um^{4^b} \\ Um^{4^c} \\ Ue^{4^a} \\ Ue^{4^b} \\ Ue^{4^c} \end{bmatrix}_{x=T_4} \tag{20}$$

$$\underbrace{\begin{bmatrix} \bar{K}^{4^a}(\omega,kz)\dfrac{2}{b_8-b_7}Fc^{4^{a^T}} & 0 \\ \bar{K}^{4^b}(\omega,kz)\dfrac{2}{b_4-b_3}Fc^{4^{b^T}} & 0 \\ \bar{K}^{4^c}(\omega,kz)\dfrac{2}{b_6-b_5}Fc^{4^{c^T}} & 0 \\ \bar{K}_\epsilon^{4^a}(\omega,kz)\dfrac{2}{b_8-b_7}\widetilde{Fs}^{4^{a^T}} & \dfrac{\epsilon_{r^5}}{\epsilon_{r^{4a}}}\dfrac{2}{b_8-b_7}Fs^{4^{a^T}} \\ \bar{K}_\epsilon^{4^b}(\omega,kz)\dfrac{2}{b_4-b_3}\widetilde{Fs}^{4^{b^T}} & \dfrac{\epsilon_{r^5}}{\epsilon_{r^{4b}}}\dfrac{2}{b_4-b_3}Fs^{4^{b^T}} \\ \bar{K}_\epsilon^{4^c}(\omega,kz)\dfrac{2}{b_6-b_5}\widetilde{Fs}^{4^{c^T}} & \dfrac{\epsilon_{r^5}}{\epsilon_{r^{4c}}}\dfrac{2}{b_6-b_5}Fs^{4^{c^T}} \end{bmatrix}}_{Vs^H}$$

$$\cdot\underset{x=T_4}{\begin{bmatrix} Im^5 \\ Ie^5 \end{bmatrix}} = \underset{x=T_4}{\begin{bmatrix} Im^{4^a} \\ Im^{4^b} \\ Im^{4^c} \\ Ie^{4^a} \\ Ie^{4^b} \\ Ie^{4^c} \end{bmatrix}}. \qquad (21)$$

In the above equation system, F_c and F_s denote the coupling matrices. Combining vectors U_e and U_m into U as well as I_m and I_e into I finally yields the transformation matrix for the discontinuity $T4$ in Fig. 2(a)

$$\underset{x=T_4}{\begin{bmatrix} U^5 \\ I^5 \end{bmatrix}} = \underbrace{\begin{bmatrix} V^E & 0 \\ 0 & V^H \end{bmatrix}}_{V^4} \cdot \underset{x=T_4}{\begin{bmatrix} U^4 \\ I^4 \end{bmatrix}} \qquad (22)$$

or more generally

$$\underset{x=T_i}{\begin{bmatrix} U^{i+1} \\ I^{i+1} \end{bmatrix}} = \underbrace{\begin{bmatrix} V^E & 0 \\ 0 & V^H \end{bmatrix}}_{V^i} \cdot \underset{x=T_i}{\begin{bmatrix} U^i \\ I^i \end{bmatrix}}.$$

Isolating the left-hand side amplitude vector I^5 in (21) by multiplying with the inverse of matrix Vs^H yields a similar transformation matrix at the interface T^4 in Fig. 2(b).

$$\underset{x=T_4}{\begin{bmatrix} U^5 \\ I^5 \end{bmatrix}} = \underbrace{\begin{bmatrix} Vs^E & 0 \\ 0 & (Vs^H)^{-1} \end{bmatrix}}_{Vs^4} \cdot \underset{x=T_4}{\begin{bmatrix} U^{4^{a,b,c}} \\ I^{4^{a,b,c}} \end{bmatrix}}. \qquad (23)$$

Now, by successively multiplying the transmission-line matrix (5) of each subregion with the corresponding transformation matrix finally leads to the desired relation between amplitude vectors at the upper and lower waveguide sidewalls in the example of Fig. 1(c)

$$\underset{x=a}{\begin{bmatrix} U^5 \\ I^5 \end{bmatrix}} = \underbrace{\left[R^5\cdot Vs^4\cdot R^{4^{a,b,c}}\cdot Vs^3\cdot R^3\cdot V^2\cdot R^2\cdot V^1\cdot R^1\right]}_{G} \cdot \underset{x=0}{\begin{bmatrix} U^1 \\ I^1 \end{bmatrix}}. \qquad (24)$$

Inserting $V^4\cdot R^4\cdot V^3$ instead of $Vs^4\cdot R^{r}\cdot Vs^3$ in (24) yields the matrix equation for the structure in Fig. 1(a), and (24) then reads as follows:

$$\underset{x=a}{\begin{bmatrix} U^5 \\ I^5 \end{bmatrix}} = \underbrace{R^5\prod_{i=1}^{5} V^i\cdot R^i}_{G} \underset{x=0}{\begin{bmatrix} U^1 \\ I^1 \end{bmatrix}}. \qquad (25)$$

The partial wave amplitudes are directly proportional to the field components $E_{y,z} \sim U$ and $H_{z,y} \sim I$, respectively. Hence, it follows for the up-to-now neglected boundary conditions (the transverse resonance condition) at the waveguide sidewalls, that $U = 0$ (for $x = a$ and $x = 0$). Thus, the characteristic matrix equation is the upper-right quarter of the matrix product in both (25) and (24).

If NH is the number of summation terms in (3) and (4), G_{12} is of order $2NH - 1$ and remains constant even for an increasing number of discontinuities. The zeros of its determinant provide the desired propagation constants

$$\det(G_{12}(w,kz)) = 0. \qquad (27)$$

It has thus been demonstrated that this method can handle generalized quasi-planar transmission-line configurations in which strips and slots are located arbitrarily in the waveguide; more than one dielectric subregion and different strip thicknesses in the same structure can be considered, as well as the finite depth of the mounting grooves.

In many practical cases, however, symmetry conditions can be included in the formulation process which make this general treatment as efficient as specialized procedures. If we consider, for example, a symmetrical bilateral finline with a symmetry plane at $x = T3$ (Fig. 1(a)), (25) can be written as follows:

$$\underset{x=T_3}{\begin{bmatrix} U^3 \\ I^3 \end{bmatrix}} = R^3\prod_{i=4}^{3} V^i\cdot R^i \underset{x=0}{\begin{bmatrix} U^1 \\ I^1 \end{bmatrix}}. \qquad (28)$$

Thus, only three subregions ($i = 1,2,3$) have to be included in the entire multiplication procedure, which makes the numerical analysis more efficient. For a magnetic wall at $x = T3$, the new boundary condition can now be satisfied by setting $I^3 = 0$, and the corresponding resonant condition changes the characteristic matrix equation from the upper-right quarter in (25) to the lower right in (28)

$$0 = G_{22}\cdot I^1. \qquad (29)$$

For an electric wall at $x = T3$, however, the resonance condition is still the same as in the general case, and the characteristic matrix equation is again G_{12} in (28). Both submatriçes (G_{22} and G_{12}) can be solved independently and provide the propagation constants of all hybrid modes in symmetrical bilateral finlines. Other symmetry conditions may be implemented very easily so that solutions of a large spectrum of conceivable quasi-planar transmission-line configurations are available with only one general computer program.

For the numerical procedure, the infinite sum in (3) and (4) must be truncated after a certain number of terms. To

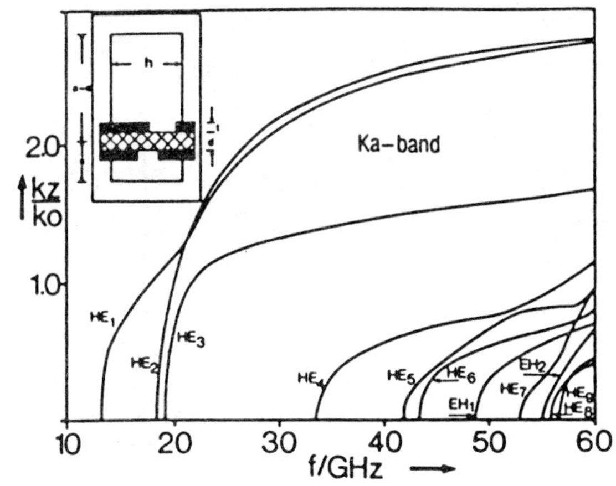

Fig. 3. Dispersion characteristic of an asymmetrical bilateral finline. $a = 7.112$ mm, $h = a/2$; $s = 2.2606$ mm, $H1 = 1.201$ mm, $s1 = 2.5$ mm, $H2 = 2.156$ mm, $s2 = 1.5$ mm, $e = 0.7$ mm; $d = 254.0$ μm, $\epsilon_r = 9.6$, $t = 70.0$ μm.

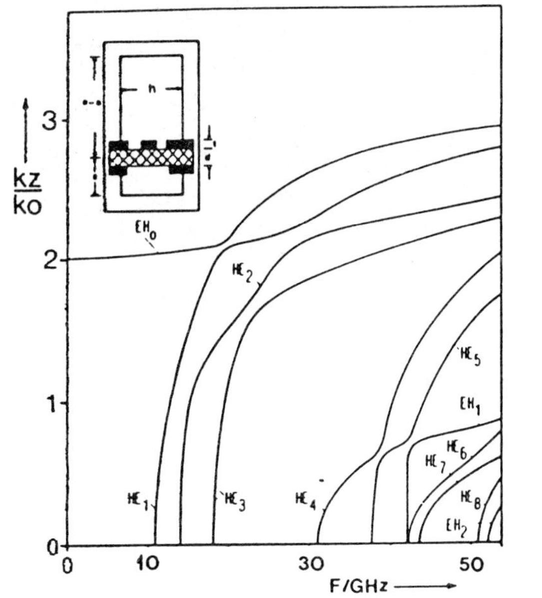

Fig. 4. Dispersion characteristic of a coupled slotline. $a = 7.112$ mm, $h = a/2$; $s = 1.778$ mm, $s1 = 2.2225$ mm, $H1 = 1.1557$ mm, $w1 = 0.592$ mm, $s2 = s3 = 0.592$ mm, $H2 = 1.3038$ mm; $d = 711.2$ μm, $\epsilon_r = 10.0$, $t = 71.0$ μm.

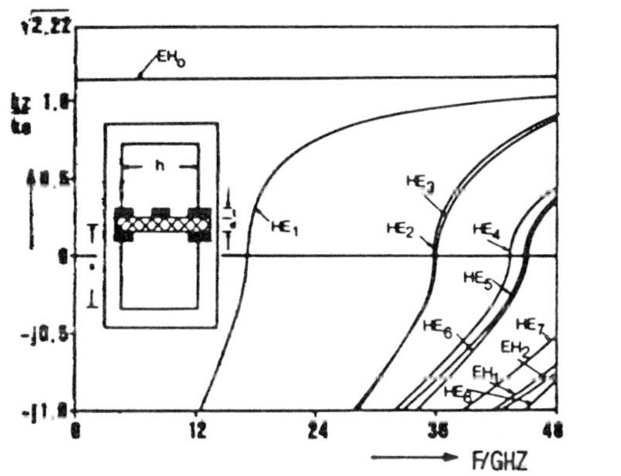

Fig. 5. Dispersion characteristic of a coupled slotline, centered in the waveguide. $a = 7.112$ mm, $h = a/2$; $s = 3.556$ mm, $s1 = 1.778$ mm, $w1 = 0.7112$ mm, $s2 = s3 = 0.7112$ mm; $d = 254.$ μm, $\epsilon_r = 2.22$, $t = 17.5$ μm.

achieve a 0.5-percent accuracy in the propagation constant and to overcome a relative convergence effect which usually occurs between 11–15 terms, a minimum of 17 and a maximum of 25 summation terms have been found to be sufficient for most configurations investigated. The amount of CPU time required to evaluate the propagation constant at one frequency sample is between 1–3 min on a VAX 11/750, but only a few seconds on a main-frame computer like a Siemens 7880. A higher accuracy is attainable by increasing the number of summation terms, but at the expense of increased CPU time.

The behavior of some dispersion curves presented below has been verified by increasing the truncation index NH to 30 to exclude numerical inaccuracies. Furthermore, all dispersion curves are labeled at cutoff. That implies that their notation beyond the cutoff frequencies remains the same even when some curves cross each other.

III. Results

First of all, the potential of the present method is demonstrated by calculating quasi-planar configurations arbitrarily located in the waveguide mount. Fig. 3 shows the dispersion behavior of a bilateral finline with a slot offset, asymmetric location, finite metallization thickness, and mounting grooves. All higher order hybrid modes are excited by an incident H_{10} wave, and it is a remarkable result that the HE_2 mode crosses the HE_1 mode resulting in a higher propagation constant than the intrinsic dominant mode for frequencies beyond 20 GHz. This is due to two causes. Firstly, both modes are strongly affected by the mounting grooves; secondly, the insert is stepped away from the center of the waveguide (towards the higher field concentration of the H_{20} mode of the empty waveguide).

In an asymmetrically suspended microstrip combined with a bilateral finline, this curious dispersion behavior is even more pronounced (Fig. 4). It is obvious that there exists an interaction between the hybrid eigenmodes resulting, for example, in nontypical dispersion curve of the EH_0 and HE_1 mode. However, this is no longer so when a still asymmetric transmission line is centered in the waveguide (Fig. 5).

Fig. 6 shows the dispersion of both quasi-TEM and first higher order hybrid modes in a shielded coupled stripline combined with a tuning septum. All strips and slots are arbitrarily located on the substrate. As shown in [3] and [5], the different phase velocities of both quasi-TEM modes can be equalized by tuning the slot on the opposite side of the strips (Fig. 7) which is a prior condition for the design of broad-band contradirectional couplers with high directivity. We have compared our results on the influence of slot width on the quasi-TEM-propagation constant with data given by Schmidt [5]. Agreement is very good, as

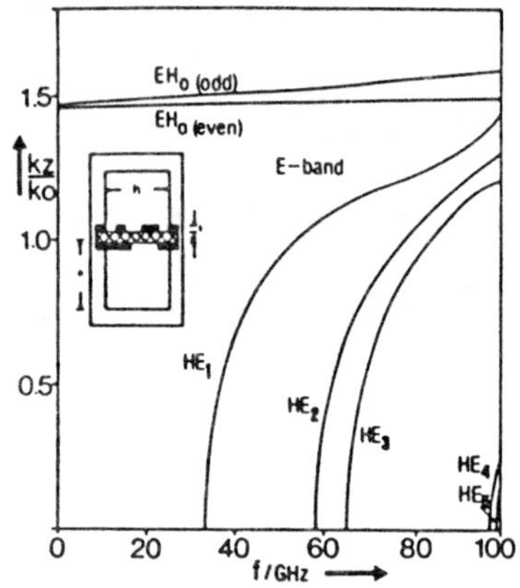

Fig. 6. Dispersion characteristic of a suspended coupled stripline centered in the waveguide but with arbitrarily located strips and slots. $a = 3.1$ mm, $h = a/2$; $s = a/2$, $s1 = 0.8$ mm, $s2 = 0.4$ mm, $s3 = 0.2$ mm, $s4 = 0.5$ mm, $w1 = 0.15$ mm, $w2 = 0.3$ mm, $H1 = 1.$ mm, $H2 = 0.5$ mm; $d = 110.$ μm, $\epsilon_r = 3.75$, $t = 10.$ μm.

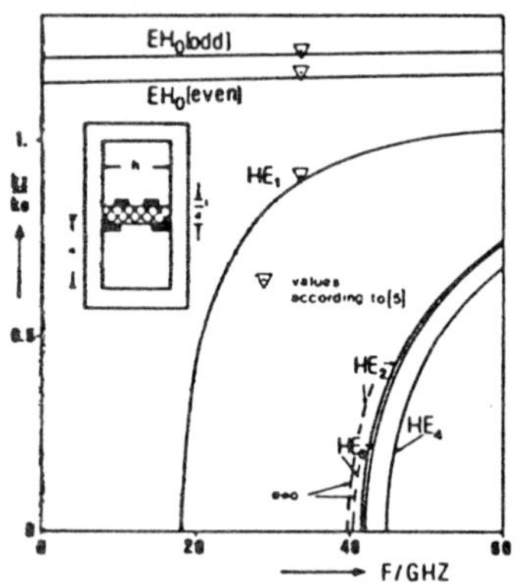

Fig. 7. Dispersion characteristic of a suspended coupled stripline compared with data given by [5]. $a = 7.112$ mm, $h = a/2$; $s = a/2$, $s1 = 0.2$ mm, $s2 = s4 = 1.378$ mm, $s3 = 0.4$ mm, $w1 = w2 = 0.2$ mm, $d = 254.$ μm, $\epsilon_r = 2.2$, $t = 5.$ μm; $e = 0$ (——), $e = 0.7$ mm (-----).

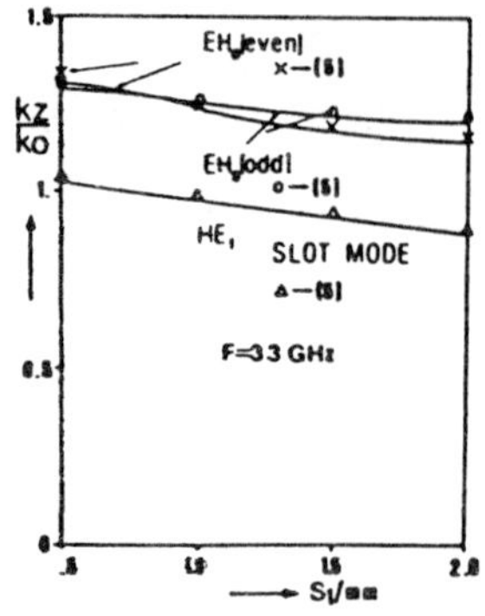

Fig. 8. Two coupled strips tuned by a slot on the opposite side of the dielectric. Dimensions identical to Fig. 7; $f = 33$ GHz; $e = 0$.

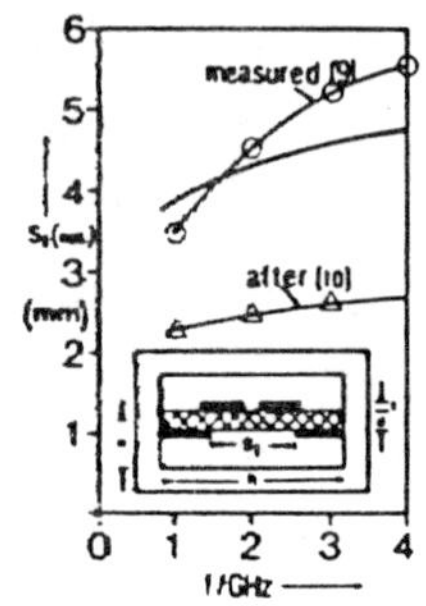

Fig. 9. Optimal slot width $S1$ (opt. in mm) of the tuning septum versus the frequency. $a = 7.27$ mm, $h = 25.$ mm; $s = a/2$, $w1 = w2 = 2.2$ mm, $s2 = s4 = 9.9$ mm, $s3 = 0.8$ mm; $d = 1.27$ mm, $\epsilon_r = 9.6$, $t = 17.5$ μm. measured [9], evaluated [10], —— this method.

shown in Fig. 8. Moreover, Fig. 7 shows the dispersion characteristics of the configuration investigated in Fig. 8. Differences in the propagation constant are less than 2 percent. The optimal slot width which equalizes both quasi-TEM modes was measured in [9] and is given in Fig. 9. While theoretical values published in [10] show only a poor agreement with these experimental results, our data agree relatively well.

Considering a symmetry plane with an electric or magnetic wall at discontinuity, T3 in Fig. 1(a) yields a so-called insulated finline [2], [6] (both fins are connected with the waveguide mount). For the same symmetry plane in Fig. 1(b), a single-side insulated finline is obtained in which only one fin is not connected with the waveguide mount, and Fig. 1(c) provides a dual-side insulated finline. The last two types of transmission lines play an important role when active components are used and one or both fins have to be insulated by a gasket to allow a dc voltage to be developed across the fins. The bias is introduced at the mounting grooves. To simplify the numerical procedure for determining the fundamental mode in these structures, the depth of the grooves is assumed to be a quarter wavelength at operating frequency and regarded as a choke section by which the RF continuity between the fins and the waveguide wall is achieved. Hence, the configuration has been investigated as an earthed finline. For practical application, however, these configurations support quasi-TEM modes, which can be seen in the following figures.

Fig. 10 presents the dispersion characteristics of a so-called insulated finline. A comparison with the configuration in Fig. 11, where only one fin is insulated, reveals the difference, which is essentially the occurence of the quasi-TEM mode (EH_0). In both cases, the HE_1 mode has almost the same cutoff frequency and is obviously not affected by the insulation of one fin. The insulation of both fins (Fig. 12) does not change the cutoff frequency of the HE_1 mode either, but a second quasi-TEM mode occurs as expected. Combining the single insulated with an earthed bilateral finline (Fig. 13) increases the propagation constant of EH_0 and decreases the cutoff frequencies of the higher order modes HE_2–HE_4. This tendency is also observed for the dual insulated finline in Fig. 14 and resembles somewhat the higher order mode behavior in an earthed bilateral finline [15]. Influences of the finite metallization thickness on the HE_1 and quasi-TEM modes are relatively small as can be seen from Fig. 15. A thicker metallization leads to decreasing quasi-TEM- and HE_1-mode propagation constants. This corresponds with the physical point of view that their electric fields are mainly concentrated in the air-filled region between the strips,

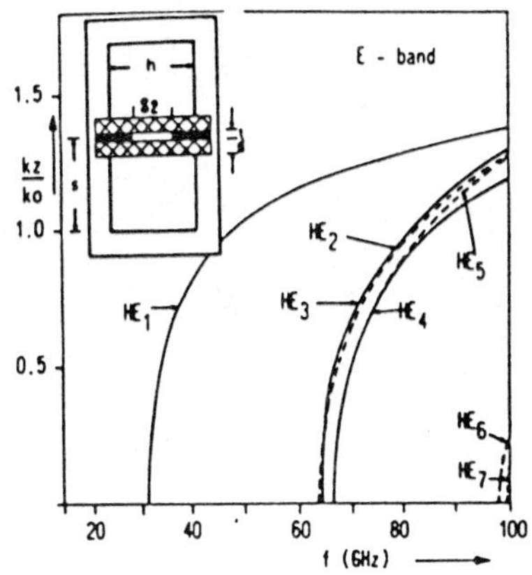

Fig. 10. Dispersion characteristic of a so-called insulated finline. $a = 3.1$ mm, $h = a/2$; $s = a/2$, $s2 = 0.6$ mm, $e = 0.5$ mm, $t = 5.$ μm, $d = 110.$ μm, $\epsilon_r = 3.75$. Symmetry plane at $s = a/2$: magnetic wall (——), electric wall (---------).

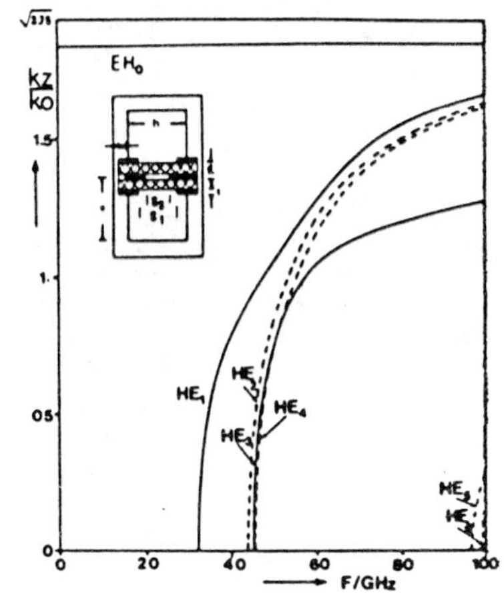

Fig. 13. Dispersion behavior of a single-side insulated finline combined with a bilateral finline. $s1 = 0.9$ mm, all other dimensions identical to Fig. 11.

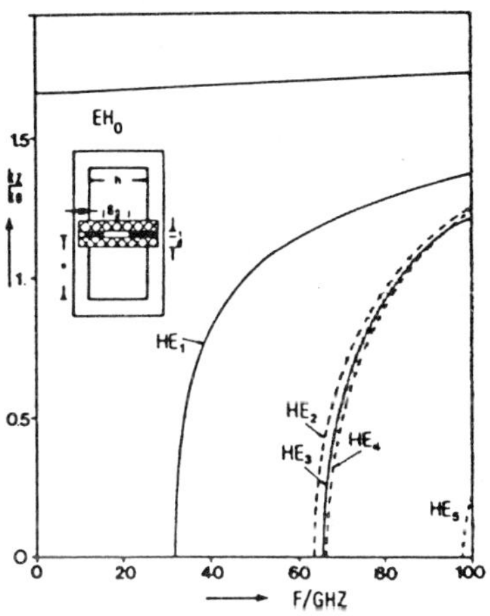

Fig. 11. Dispersion behavior of a single side insulated finline. $a = 3.1$ mm, $h = a/2$; $s = a/2$, $e = 0.5$ mm, $c = s2 = 0.3$ mm (cf. Fig. 1(b)), $s3 = 0.6$ mm; $d = 110.$ μm, $\epsilon_r = 3.75$, $t = 5.$ μm. Symmetry plane at $s = a/2$: magnetic wall (——), electric wall (------).

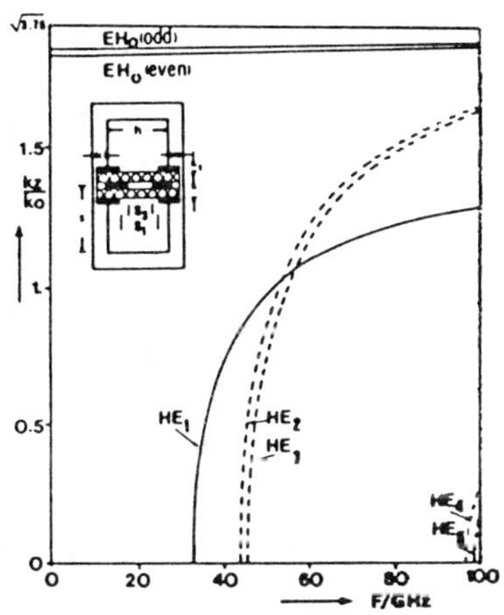

Fig. 14. Dispersion behavior of a dual-side insulated finline combined with a bilateral finline. Dimensions identical to Figs. 12 and 13.

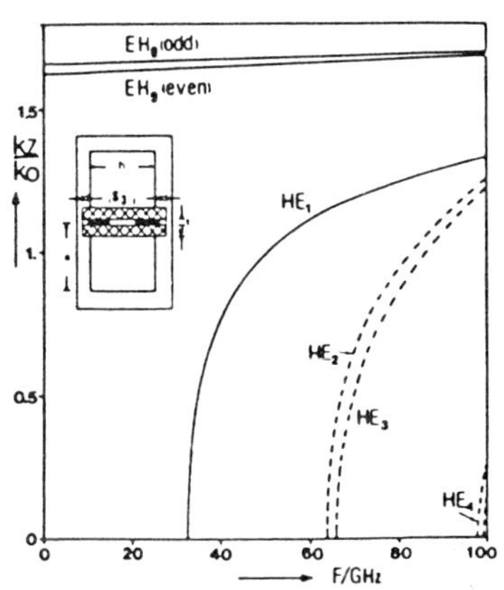

Fig. 12. Dispersion behavior of a dual-side insulated finline. $c = s2 = s4 = 0.3$ mm, all other dimensions identical to Fig. 11.

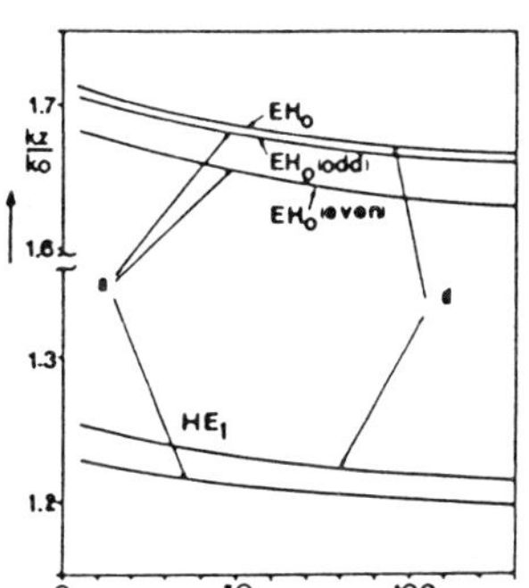

Fig. 15. Quasi-TEM and slot mode versus the metallization thickness, f-75 GHz. (a) Dual-side insulated finline (dimensions identical to Fig. 12. (b) Single-side insulated finline (dimensions identical to Fig. 11).

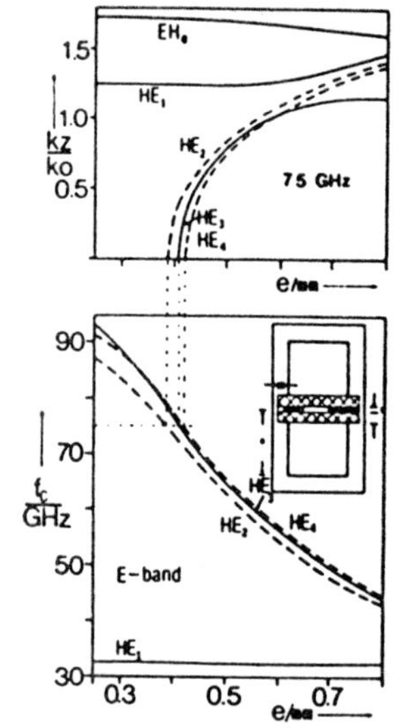

Fig. 16. Quasi-TEM and next higher order modes versus the groove depths e in a single-side insulated finline. f_c = cutoff frequency. Dimensions according to Fig. 11.

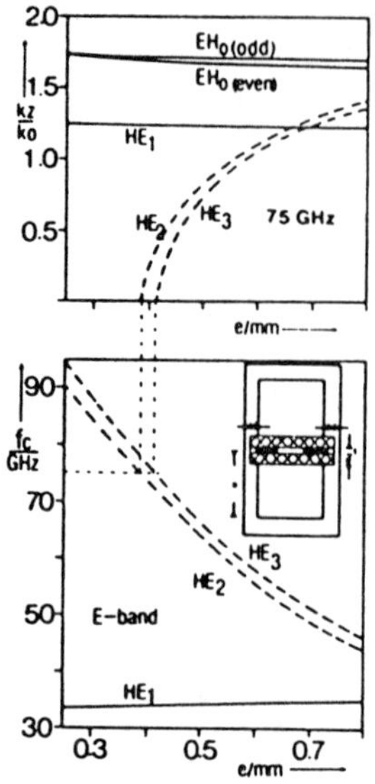

Fig. 17. Quasi-TEM and next higher order modes versus the groove depth e in a dual-side insulated finline. Dimensions according to Fig. 12.

resulting in a lower propagation constant when increasing the strip thickness.

Regarding the influence of the mounting grooves on the electrical characteristics of quasi-planar transmission lines, it was found out recently [6], [7] that higher order modes are strongly affected, whereas the HE_1 and the quasi-TEM modes are virtually insensitive to a wide range of changes in groove depth. In principle, this tendency is also observed for the single and dual insulated finline, and is shown in Figs. 16 and 17. For the single insulated configuration in Fig. 16, however, a mutual influence of the slot mode (HE_1) and the second higher order mode (HE_3) (which has the same field symmetry as the incident H_{10} mode) occurs, but was not observed for the dual-side insulated finline in Fig. 17. In that case (Fig. 17), both higher order modes (HE_2, HE_3) are obtained by considering an electric wall as the symmetry plane. It should be noted that, for operating frequencies beyond 75 GHz, the critical groove depth becomes shallower than given in Fig. 16, but can be increased a few percent by using a smaller substrate thickness.

IV. Conclusion

A versatile hybrid-mode analysis is presented taking into account the finite metallization thickness and mounting grooves in a class of various quasi-planar transmission lines. Using a modified mode-matching technique combined with a generalized transmission-line matrix exhibits the advantage of maintaining the size of the characteristic matrix equation constant even for an increasing number of discontinuities. Additionally, the method enables a user to create a modularized computer program which can easily be extended to a wide range of conceivable transmission-line configurations.

It is true that the method is not as fast as the spectral-domain technique; however, it includes the effect of finite metallization thickness and mounting grooves. Especially, the applications of quasi-planar transmission lines in the shorter millimeter-wave range have shown that the influence of the mounting grooves are more significant than the effect of finite metallization thickness. For example, a 0.7-mm-deep groove in a Ka-band waveguide mount causes a 2-GHz reduction of the cutoff frequency of the next higher order mode (Fig. 7), whereas the same groove depth reduces the next higher order mode in an E-band waveguide mount by at least 45 GHz (Fig. 16).

Appendix

Abbreviations for the partial wave amplitudes in (3) and (4) are

$$Im^i_{(n)}(x) = A^i_{(n)} e^{jkx^i_{(n)}x} + B^i_{(n)} e^{-jkx^i_{(n)}x}$$

$$Ue^i_{(n)}(x) = \frac{1}{jkx^i_{(n)}} \left\{ C^i_{(n)} e^{jkx^i_{(n)}x} - D^i_{(n)} e^{-jkx^i_{(n)}x} \right\}$$

$$Ie^i_{(n)}(x) = C^i_{(n)} e^{jkx^i_{(n)}x} + D^i_{(n)} e^{-jkx^i_{(n)}x}$$

$$Um^i_{(n)}(x) = jkx^i_{(n)} \left\{ A^i_{(n)} e^{jkx^i_{(n)}x} - B^i_{(n)} e^{-jkx^i_{(n)}x} \right\}.$$

The unknown coefficients $A^i_{(n)}$, $B^i_{(n)}$, $C^i_{(n)}$, and $D^i_{(n)}$ are eliminated by inserting the upper ($x = xo^i$) and the lower boundary ($x = xu^i$) values in the above equations. Thus, the generalized transmission-line matrix R^i for each subregion contains the diagonal matrices

$$Rc^i = \begin{bmatrix} \cos\left(kx^i_{(n)}xa^i\right) & 0 \\ 0 & \cos\left(kx^i_{(n)}xa^i\right) \end{bmatrix}$$

$$Rs^i = \begin{bmatrix} -kx_{(n)}\sin\left(kx^i_{(n)}xa^i\right) & 0 \\ 0 & \frac{1}{kx^i_{(n)}}\sin\left(kx^i_{(n)}xa^i\right) \end{bmatrix}$$

$$Rs^{i'} = \begin{bmatrix} \frac{1}{kx^i_{(n)}}\sin\left(kx^i_{(n)}xa^i\right) & 0 \\ 0 & -kx^i_{(n)}\sin\left(kx^i_{(n)}xa^i\right) \end{bmatrix} xa = xo - xu$$

and the propagation constant

$$kx^i_{(n)} = \sqrt{ko^2\epsilon^i_r - \left(ky^i_{(n)}\right)^2 - kz^2}$$

with $ko^2 = \omega^2\mu_0\epsilon_0$.

The functions $fc^i_{(n)}(y)$ and $fs^i_{(n)}(y)$ in (3) and (4) are determined by the boundary condition in the y-direction

$$fc^i_{(n)}(y) = \frac{\cos\widetilde{ky}^i_{(n)}}{\sqrt{1+\delta_{0n}}} \text{ and } fs^i_{(n)}(y) = \sin\widetilde{ky}^i_{(n)},$$

$$\delta_{0n} = \text{Kronecker delta}.$$

For example, $\widetilde{ky}^4_{(n)}$ in Fig. 2(a) means

$$\widetilde{Ky}^4_{(n)} = ky^4_{(n)}\cdot(y - b3), \qquad ky^4_{(n)} = \frac{n\cdot\pi}{b4-b3}.$$

The coupling integrals Fc^ν and Fs^ν in (20) can be written as follows:

$$Fc^\nu_{(nk)} = \int_{yu^\nu}^{yo^\nu} fc^5_{(n)}(y)fc^\nu_{(k)}(y)\,dy, \qquad \nu \in 4a, 4b, 4c$$

with $yu^\nu = (b7, b3, b5)$ and $yo^\nu = (b8, b4, b6)$.

For Fs^ν, replace fc with fs. The abbreviations for $\widetilde{Fc}^\nu$ and $\widetilde{Fs}^{\nu^T}$ in (20) and (21) are

$$\widetilde{Fc}^\nu = \frac{2}{b2-b1}\cdot Fc^\nu\cdot \text{Diag}\left(ky^\nu_{(n)}\right)$$

$$\widetilde{Fs}^{\nu^T} = Fs^{\nu^T}\cdot \text{Diag}\left(ky^\nu_{(n)}\right).$$

Fs^{ν^T} means the transposed of matrix Fs^ν.

Acknowledgment

The authors wish to thank Dr. W. J. R. Hoefer, University of Ottawa, and Dr. F. Arndt, University of Bremen, for helpful discussions and suggestions.

References

[1] P. J. Meier, "Integrated fin-line millimeter components," *IEEE Trans. Microwave Theory Tech.*, vol. MTT-22, pp. 1209–1216, Dec. 1974.

[2] Y.-C. Shih and W. J. R. Hoefer, "Dominant and second-order mode cutoff frequencies in fin-lines calculated with a two-dimensional TLM program," *IEEE Trans. Microwave Theory Tech.*, vol. MTT-28, pp. 1443–1448, Dec. 1980.

[3] T. Itoh and A. S. Hebert, "A generalized spectral domain analysis for coupled suspended microstriplines with tuning septums," *IEEE Trans. Microwave Theory Tech.*, vol. MTT-26, pp. 820–826, Oct. 1978.

[4] L. P. Schmidt and T. Itoh, "Characteristics of unilateral fin-line structures with arbitrarily located slots," *IEEE Trans. Microwave Theory Tech.*, vol. MTT-29, pp. 352–355, Apr. 1981.

[5] L. P. Schmidt, "A comprehensive analysis of quasiplanar waveguides for millimeter-wave application," in *Proc. 11th Eur. Microwave Conf.* (Amsterdam), 1981, pp. 315–320.

[6] R. Vahldieck, "Accurate hybrid-mode analysis of various finline configurations including multilayered dielectrics, finite metallization thickness, and substrate holding grooves," *IEEE Trans. Microwave Theory Tech.*, vol. MTT-32, pp. 1454–1460, Nov. 1984.

[7] J. Bornemann, "Rigorous field theory analysis of quasiplanar waveguides," *Proc. Inst. Elec. Eng.*, vol. 132, pt. H, pp. 1–6, Feb. 1985.

[8] F. Arndt and G. U. Paul, "The reflection definition of the characteristic impedance of microstrips," *IEEE Microwave Theory Tech.*, vol. MTT-27, Aug. 1979.

[9] J. P. Villotte, M. Aubourg, and Y. Garault, "Modified suspended striplines for microwave integrated circuits," *Electron. Lett.*, vol. 14, no. 18, pp. 602–603, Aug. 31, 1978.

[10] W. Schuhmacher, "Wellentypen und Feldverteilung auf unsymmetrischen planaren Leitungen mit inhomogenem Dielektrikum," *Arch. Elec. Übertragung.*, vol. 34, pp. 445–453, 1980.

[11] J. Siegl, "Phasenkonstante und Wellenwiderstand einer Schlitzleitung mit rechteckigem Schirm und endlicher Metallisierungsdicke," *Frequenz*, vol. 31, pp. 216–220, July 1977.

[12] A. Beyer, "Analysis of the characteristics of an earthed fin-line," *IEEE Trans. Microwave Theory Tech.*, vol. MTT-29, pp. 676–680, July 1981.

[13] T. Kitazawa and R. Mittra, "Analysis of finline with finite metallization thickness," *IEEE Trans. Microwave Theory Tech.*, vol. MTT-32, pp. 1484–1487, Nov. 1984.

[14] H. Fritzsche, "Die frequenzabhaengigen Uebertragungseigenschaften gekoppelter Streifenleitungen im geschichteten Dielektrikum," *Nachrichtentech. Z.*, vol. 26, no. 1, pp. 1–8, 1973.

[15] R. Vahldieck and W. J. R. Hoefer, "The influence of metallization thickness and mounting grooves on the characteristics of finlines,"in *IEEE MTT-S Int. Microwave Symp. Dig.*, pp. 182-184.

2.1.6

Formulation of the Singular Integral Equation Technique for Planar Transmission Lines

ABBAS SAYED OMAR AND KLAUS SCHÜNEMANN, MEMBER, IEEE

Abstract—The singular integral equation technique is used to determine the normal modes of propagation in general planar transmission lines. Taking finlines as an example, it is demonstrated how high-order modes can effectively and accurately be calculated. It is also shown that complex and backward-wave modes, which are known to exist in rectangular and circular waveguides with dielectric inserts, can also exist in finlines. Besides a discussion of their characteristic features, this paper describes the conditions under which complex and backward-wave modes are found in finlines.

I. Introduction

PLANAR TRANSMISSION LINES include a variety of structures showing one or more air–dielectric and/or dielectric–dielectric interfaces with metal strips printed at these interfaces. Determining the normal modes of propagation in these structures is of fundamental importance. Due to the completeness property of the set of normal modes [1], an arbitrary electromagnetic field can be expanded within this set so that the problem of determining the field, which can usually be formulated as a solution of integro-differential equations, is reduced to a solution of matrix equations.

Galerkin's method in the spectral domain has been successfully used in analyzing many of these structures (see, e.g., [2]–[4]). In fact, this method is superior over all other known methods if the dominant and the first few higher order modes are computed. A characteristic matrix of order 4 (which corresponds to just two basis functions in the expansion of each of the unknown field components) is quite sufficient to give accurate results. For the determination of still higher order modes, more basis functions are needed so that the order of the characteristic matrix is increased considerably. The singular integral equation technique (SIE), which has been used in [5]–[7] for solving many waveguide problems and in [8] for the analysis of microstrip lines, has the advantage of characterizing the problem by a relatively small-order matrix. Furthermore, it gives accurate results for the high-order modes as will be shown below. Another advantage is that the matrix elements are all given by analytical expressions so that neither infinite sums nor numerical integrations are involved.

Manuscript received March 6, 1985; revised June 26, 1985. This work was supported in part by the Deutsche Forschungsgemeinschaft.

The authors are with the Technische Universität Hamburg-Harburg, Arbeitsbereich Hochfrequenztechnik, Postfach 90 14 03, D-2100 Hamburg 90, West Germany.

Because the weight of the mode in a field expansion series becomes smaller as the mode order goes higher, the determination of the first, say N, modes is the actual need. Determining these N modes without loosing any intermediate ones is actually a problem for planar guiding structures, because it has been found that some pairs of modes become complex modes [9]–[15] within one or more ranges of certain structure parameters (e.g., the slot width in finlines) at a given frequency or, alternatively, within one or more frequency ranges at given structure parameters. This problem will also be investigated here taking the generalized unilateral finline as a case study.

Complex and backward-wave modes are known to exist in circular and rectangular waveguides with dielectric inserts [9]–[17]. It was reported in [9] that the existence of backward-wave modes in a shielded dielectric rod guarantees the existence of complex modes there. More theoretical as well as experimental investigations on complex modes in circular waveguides with dielectric inserts have been performed in [10]–[13]. Although the possibility of backward-wave modes in rectangular waveguides with dielectric inserts has been reported a long time ago [16], [17], complex modes in shielded rectangular dielectric image guides have just been reported [14], [15].

In fact, finlines can be regarded as ridged waveguides with dielectric inserts (the substrate), so that complex and backward-wave modes may also exist there. The nature of such modes in finlines should be similar to that of the corresponding ones in dielectric-loaded rectangular and circular waveguides.

II. Basic Formulation

The electromagnetic field in the generalized unilateral finline shown in Fig. 1 is a linear combination of LSM and LSE parts [1]. These two parts satisfy independently all interface conditions. They are only coupled in order to satisfy the edge conditions, as has been shown in [18]. The tangential electric field E_t and the surface current J_s at the interface $x=0$ can then be written as

$$E_t = E_t^e + E_t^h \quad J_s = J_s^e + J_s^h$$

where superscripts e and h refer to the LSM and LSE parts, respectively. The LSM part is, however, completely

A.S. Omar and K. Schunemann, *IEEE Transactions on Microwave Theory and Techniques*, Vol. MTT-33, No. 12, Dec. 1985, pp. 1313–1321.

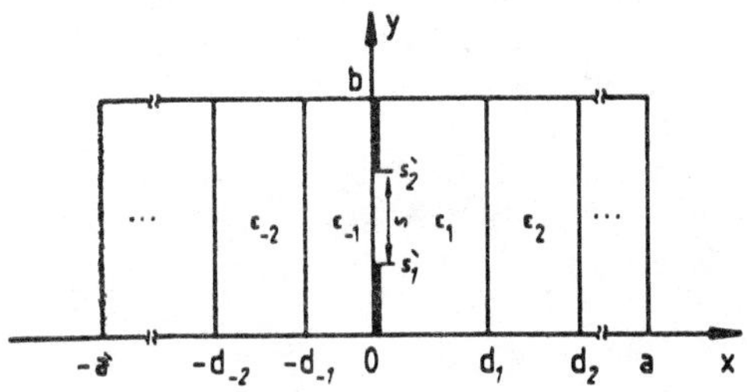

Fig. 1. Generalized unilateral finline.

characterized by the z-components (E_z^e, J_z^e) because

$$E_y^e = (-1/j\beta)\, dE_z^e/dy \quad J_y^e = (-1/j\beta)\, dJ_z^e/dy$$

whereas the LSE part is completely characterized by the y-components (E_y^h, J_y^h) because

$$E_z^h = (1/j\beta)\, dE_y^h/dy \quad J_z^h = (1/j\beta)\, dJ_y^h/dy.$$

With electric walls at $y=0$ and $y=b$, the four components E_z^e, J_z^e, E_y^h, and J_y^h can be written as

$$E_z^e = \sum_{n=1}^{\infty} A_n^e \sin(n\pi y/b)$$

$$J_z^e = j\omega\epsilon_0 \sum_{n=1}^{\infty} F_n^e A_n^e \sin(n\pi y/b)$$

$$E_y^h = \sum_{n=0}^{\infty} A_n^h \cos(n\pi y/b)$$

$$J_y^h = (1/j\omega\mu_0) \sum_{n=0}^{\infty} F_n^h A_n^h \cos(n\pi y/b) \qquad (1)$$

where F_n^e and F_n^h are the Fourier series expansion coefficients of the LSM and LSE Green's functions, respectively [19]. The SIE technique can now be formulated as follows.

1) Two cosine-series $f_1(y)$ and $f_2(y)$ are constructed in terms of the tangential electric field E_t so that they are unknown only in the slot ($s_1' \leqslant y \leqslant s_2'$). These cosine-series are then given by

$$f_1(y) = \sum_{n=0}^{\infty} A_n^{(1)} \cos(n\pi y/b)$$

$$f_2(y) = \sum_{n=0}^{\infty} A_n^{(2)} \cos(n\pi y/b) \qquad (2)$$

where the coefficients $A_n^{(1)}$ and $A_n^{(2)}$ are, in general, two independent linear combinations of A_n^e and A_n^h.

2) Two sine-series $f_3(y)$ and $f_4(y)$ are constructed in terms of the surface current J_s such that the asymptotic values (as $n \to \infty$) of their coefficients are $A_n^{(1)}$ and $A_n^{(2)}$, respectively. The expansion coefficients of the series $g_1(y)$ and $g_2(y)$ defined by

$$g_1(y) = \sum_{n=1}^{\infty} A_n^{(1)} \sin(n\pi y/b) - f_3(y)$$

$$g_2(y) = \sum_{n=1}^{\infty} A_n^{(2)} \sin(n\pi y/b) - f_4(y) \qquad (3)$$

are then vanishing asymptotically.

3) Applying the boundary conditions to be satisfied by $f_1(y)$ and $f_2(y)$, the coefficients $A_n^{(1)}$ and $A_n^{(2)}$ are determined in terms of integrals of $f_1(y)$ and $f_2(y)$, respectively, taken over the slot ($s_1' \leqslant y \leqslant s_2'$).

4) Substituting these integrals for $A_n^{(1)}$ and $A_n^{(2)}$ into (3) and applying the boundary conditions to be satisfied by $f_3(y)$ and $f_4(y)$, we get two integral equations relating $f_1(y)$ and $f_2(y)$ to $g_1(y)$ and $g_2(y)$, respectively. These integral equations are of the standard singular type. Their solutions are, e.g., given in [7].

5) Due to the asymptotic vanishing of the series coefficients of $g_1(y)$ and $g_2(y)$, these series can be truncated behind the Nth term, so that the functions $f_1(y)$ and $f_2(y)$ are then known in the slot in terms of $A_n^{(1)}$ and $A_n^{(2)}$, $n \leqslant N$.

6) The back substitution of $f_1(y)$ and $f_2(y)$ in the integrals determining $A_n^{(1)}$ and $A_n^{(2)}$, $n \leqslant N$ —step 3) —along with additional conditions (to be discussed later), results in a finite homogeneous system of linear equations, from which the propagation constants, as well as the field expansion coefficients of the different modes, are obtained.

Care should be taken, however, in constructing ($f_1(y)$, $f_2(y)$) and ($f_3(y)$, $f_4(y)$) which are related to $\boldsymbol{E}_t$ and $\boldsymbol{J}_s$, respectively. This is because E_y and J_z, as well as dE_z/dy and dJ_y/dy, are singular at the edges ($y=s_1'$, $y=s_2'$) [20]. The order of singularity is the same for all components, namely $|y-s'|^{-1/2}$. This is exactly the proper type of singularity, which can be described by the SIE technique. Hence, any y-differentiation of either E_y, J_z, dE_z/dy, or dJ_y/dy is not allowed in constructing $f_1(y)$, $f_2(y)$, $f_3(y)$, and $f_4(y)$. This leads to two alternatives. In the first case, $A_n^{(1)}$ and $A_n^{(2)}$ are linear combinations of A_n^e and A_n^h, so that the LSM and LSE parts of the field are coupled from the beginning. We call this formulation the coupling one. In the second case, $A_n^{(1)}$ and $A_n^{(2)}$ are proportional to A_n^e and A_n^h, respectively, so that the LSM and LSE parts of the field are decoupled. Coupling is performed as a final step in the form of additional conditions, as will be shown later. We call this formulation the decoupling one.

A. The Coupling Formulation

The functions $f_1(y)$ and $f_2(y)$ are constructed as

$$f_1(y) = dE_z/dy$$

$$f_2(y) = \left[(k_0^2 - \beta^2)E_y + j\beta(dE_z/dy)\right]/j\omega\mu_0 \qquad (4)$$

whereas $f_3(y)$ and $f_4(y)$ are constructed as

$$f_3(y) = \left[(K^h\beta^2 - K^e k_0^2)J_z + j\beta K^h(dJ_y/dy)\right]/j\omega\epsilon_0 K^e K^h$$

$$f_4(y) = \left[j\beta(K^e - K^h)J_z + K^h(dJ_y/dy)\right]/K^e K^h. \qquad (5)$$

Here K^e and K^h are the asymptotic limits of $(n\pi/b)\cdot F_n^e$ and $F_n^h/(n\pi/b)$, respectively.

The expansion coefficients $A_n^{(1)}$ and $A_n^{(2)}$ of $f_1(y)$ and $f_2(y)$, respectively, are then related to A_n^e and A_n^h by

$$\begin{bmatrix} A_n^{(1)} \\ A_n^{(2)} \end{bmatrix} = \begin{bmatrix} (n\pi/b) & -(n\pi/b)^2/j\beta \\ j\omega\epsilon_0(n\pi/b)/j\beta & \alpha_n^2/j\omega\mu_0 \end{bmatrix} \begin{bmatrix} A_n^e \\ A_n^h \end{bmatrix} \qquad (6)$$

where $\alpha_n^2 = k_0^2 - \beta^2 - (n\pi/b)^2$. The series expansions of $f_3(y)$ and $f_4(y)$ are readily proved to have the following forms:

$$f_3(y) = \sum_{n=1}^{\infty} \left[P_n^{(1)}A_n^{(1)} + P_n^{(2)}A_n^{(2)}\right] \sin(n\pi y/b)$$

$$f_4(y) = \sum_{n=1}^{\infty} \left[Q_n^{(1)}A_n^{(1)} + Q_n^{(2)}A_n^{(2)}\right] \sin(n\pi y/b) \quad (7)$$

where $P_n^{(1)}$, $P_n^{(2)}$, $Q_n^{(1)}$, and $Q_n^{(2)}$ are given in terms of F_n^e and F_n^h with their asymptotic limits being 1, 0, 0, and 1, respectively.

The series $g_1(y)$ and $g_2(y)$ defined in (3) are then given by

$$g_1(y) = \sum_{n=1}^{\infty} \left[\left(1 - P_n^{(1)}\right)A_n^{(1)} - P_n^{(2)}A_n^{(2)}\right] \sin(n\pi y/b)$$

$$g_2(y) = \sum_{n=1}^{\infty} \left[-Q_n^{(1)}A_n^{(1)} + \left(1 - Q_n^{(2)}\right)A_n^{(2)}\right] \sin(n\pi y/b). \quad (8)$$

Defining

$$\varphi = \pi y/b, \quad \varphi_1 = \pi s_1'/b, \quad \varphi_2 = \pi s_2'/b \quad (9)$$

we can write for the boundary conditions to be imposed on $f_1(\varphi), \cdots, f_4(\varphi)$

$$f_i(\varphi) = \begin{cases} 0 & (0 \leqslant \varphi \leqslant \varphi_1, \ \varphi_2 \leqslant \varphi \leqslant \pi) \\ h_i(\varphi) & (\varphi_1 \leqslant \varphi \leqslant \varphi_2) \end{cases}$$

$$f_{i+2}(\varphi) = 0 \qquad (\varphi_1 \leqslant \varphi \leqslant \varphi_2), \quad i = 1,2. \quad (10)$$

The coefficients $A_n^{(1)}$ and $A_n^{(2)}$ are then given by

$$A_0^{(i)} = [(i-1)/\pi] \int_{\varphi_1}^{\varphi_2} h_i(\varphi)\, d\varphi$$

$$A_n^{(i)} = [2/\pi] \int_{\varphi_1}^{\varphi_2} h_i(\varphi) \cos n\varphi\, d\varphi, \qquad i = 1,2. \quad (11)$$

Making use of the transformations

$$\cos\varphi = \cos[(\varphi_2 + \varphi_1)/2] \cos[(\varphi_2 - \varphi_1)/2] - \eta \sin[(\varphi_2 + \varphi_1)/2] \sin[(\varphi_2 - \varphi_1)/2]$$

$$G_i(\eta) = g_i(\varphi)/\sin\varphi$$

$$H_i(\eta) = h_i(\varphi)/\sin\varphi, \qquad i = 1,2 \quad (12)$$

we get the following standard singular integral equations:

$$G_i(\eta) = (1/\pi) \int_{-1}^{+1} H_i(\eta')/(\eta - \eta')\, d\eta',$$

$$-1 \leqslant \eta \leqslant +1, \ i = 1,2. \quad (13)$$

Their solutions are

$$H_i(\eta) = (1-\eta^2)^{-1/2} \left[A_0^{(i)}/\sin[(\varphi_2 + \varphi_1)/2] \cdot \sin[(\varphi_2 - \varphi_1)/2] + (1/\pi) \int_{-1}^{+1} (1-\eta'^2)^{1/2} \cdot G_i(\eta')/(\eta' - \eta)\, d\eta', \right. \qquad i = 1,2. \quad (14)$$

As may easily be seen, the vanishing of $f_1(y)$ and $f_2(y)$ on the metal fins and that of $f_3(y)$ and $f_4(y)$ in the slot, which are the considered boundary conditions, are only necessary conditions for the vanishing of E_t and J_s on their respective regions (i.e., E_t on the fins and J_s in the slot). This guarantees the vanishing of E_y and J_z on their respective regions but only the constancy of E_z and J_y on their respective regions. However, points $y = 0$ and $y = b$, at which $E_z = 0$, belong to the fins, so that the constant value of E_z on the fins is automatically zero. Hence, just one additional condition must be imposed, namely the vanishing of J_y at any point in the slot, so that the constant value of J_y over the slot is put to zero.

B. The Decoupling Formulation

Functions $f_1(y)$ and $f_2(y)$ are now constructed as

$$df_1/dy = \nabla_t \cdot E_t = -\Delta_t E_z^e/j\beta$$

$$f_2(y) = \hat{i} \cdot (\nabla_t \times E_t) = \Delta_t E_y^h/j\beta \quad (15)$$

whereas functions $f_3(y)$ and $f_4(y)$ are constructed according to

$$f_3(y) = -(\nabla_t \cdot J_s)/j\omega\epsilon_0 K^e = -\Delta_t J_z^e/\omega\epsilon_0\beta K^e$$

$$df_4/dy = j\omega\mu_0\{\hat{i} \cdot (\nabla_t \times J_s)\}/K^h = \omega\mu_0 \Delta_t J_y^h/\beta K^h. \quad (16)$$

Here ∇_t and Δ_t are the del and laplacian operators in the y–z-plane, respectively, and $\hat{i}$ is the unit vector in the x-directions. The boundary conditions to be imposed on $f_1(y)$, $f_2(y)$, $f_3(y)$, and $f_4(y)$ (which are $df_1/dy = 0 = f_2(y)$ on the fins and $f_3(y) = 0 = df_4/dy$ in the slot) are again necessary conditions for the vanishing of E_t and J_s on their respective regions. In this case, it is guaranteed that the individual components of E_t and J_s are harmonic functions on their respective regions [19], so that additional conditions must be imposed. These conditions can be shown to be the vanishing of only one component of E_t and J_s on the boundaries of their respective regions, which are the edges ($y = s_1'$, $y = s_2'$).

III. High-Order Modes Accuracy

The only approximation involved in the SIE technique is the truncation of the infinite series $g_1(y)$ and $g_2(y)$ behind the Nth term. However, it can be proved that the nth coefficients of $g_1(y)$ and $g_2(y)$ are almost zero if $(n\pi/b)^2 > (\hat{\epsilon}_r k_0^2 - \beta^2)$, where $\hat{\epsilon}_r$ is the largest dielectric constant involved. This limits the number of high-order modes, which can be calculated accurately, because the propagation constant of the highest order mode (which usually is evanescent) should satisfy

$$|\beta|^2 < [(N+1)\pi/b]^2 - \hat{\epsilon}_r k_0^2.$$

For normally used dimensions, dielectric constants, and operation frequencies, $N = 3$ (corresponding to a characteristic matrix of order 7) is sufficient to give the first 30 modes accurately.

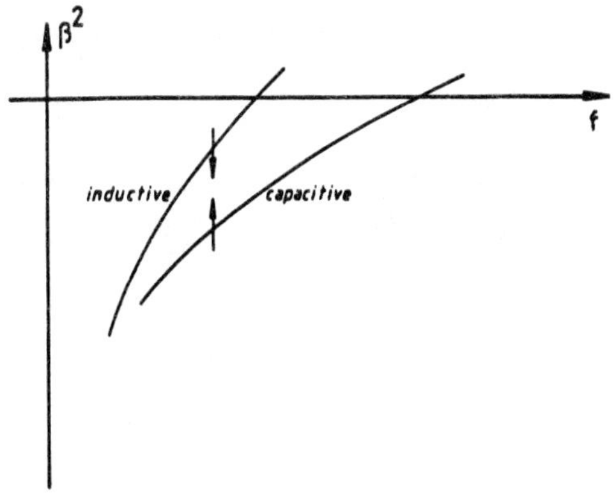

Fig. 2. Typical behavior of inductive and capacitive modes with increasing slot width.

IV. The Problem of Mode Disappearance

It has been shown in [18], that the modes of the generalized unilateral finline behave either inductively or capacitively if they are below cutoff. The interesting difference between the two types is the dependence of their propagation constants on the slot width. The propagation constant squared (β^2) (which is always a real number in the lossless case, whether the mode is below or above cutoff) of an inductive (a capacitive) mode decreases (increases) versus the slot width, as is shown in Fig. 2. Consider now the modes for a certain slot width being arranged according to their β^2. If one of any two successive modes is inductive with β_i^2 while the other is capacitive with a smaller β_c^2, and the slot width is increased, then β_i^2 and β_c^2 approach each other up to a certain slot width where $\beta_i^2=\beta_c^2$. For this slot width, we get two degenerate modes which are no longer orthogonal. They should be treated carefully because of the strong coupling between them. If the slot width is increased beyond this value both modes will disappear. This will be illustrated in the following.

The final step in nearly all mode analysis methods is the solution of a homogeneous system of linear equations which can be written as a matrix equation

$$[C]\cdot X=0.$$

The elements of the matrix $[C]$, usually called the characteristic matrix, depend on the unknown propagation constant β. The elements of the column vector X determine the electromagnetic-field configuration of the mode whose propagation constant is β. The elements of $[C]$ can always be normalized in such a way that they are all real functions of β^2 if the structure is lossless. The matrix equation is solved by looking for the values of β, making the matrix $[C]$ singular. Each of these values characterizes one of the structure modes.

For our case, we assume that the finline is lossless so that the determinant of the matrix $[C]$ is a real function $D(\beta^2)$. The different modes are then obtained by looking for the zeros of $D(\beta^2)$. The behavior of $D(\beta^2)$ near the aforementioned successive inductive and capacitive modes

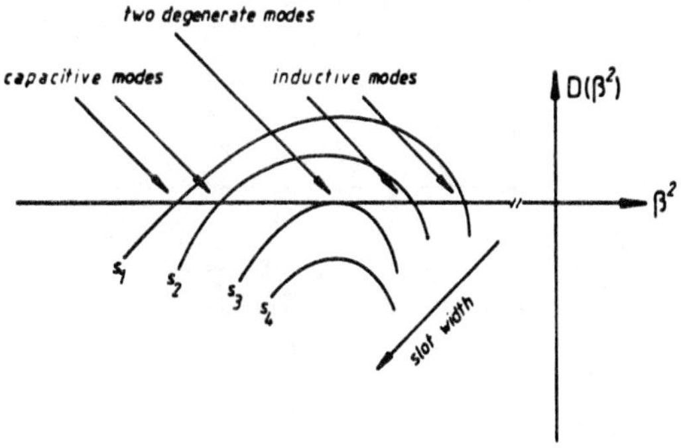

Fig. 3. Typical behavior of how two modes disappear.

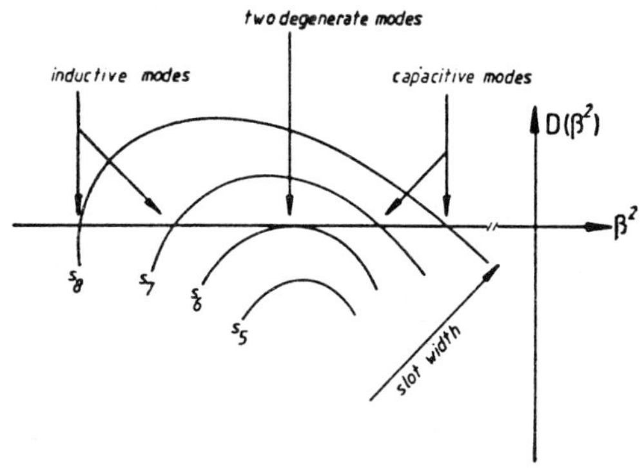

Fig. 4. Typical behavior of how two modes reappear.

for a certain range of the slot width typically is as shown in Fig. 3. As the slot width increases from s_1 to s_2, β^2 of the inductive mode decreases, while that of the capacitive mode increases. At s_3 the two modes become degenerate and at s_4 they disappear completely. This effect occurs within a certain range of slot widths. Afterwards, we get the situation shown in Fig. 4. At s_5 ($>s_4$), the two modes are still nonexisting; at s_6, two degenerate modes are obtained, and for $s>s_6$, two nondegenerate modes exist again: one is inductive (now to the left-hand side) and the other is capacitive (to the right-hand side). It is important to note that when the slot width increases from s_7 to s_8, β^2 of the inductive mode still decreases while that of the capacitive mode still increases.

The disappearance of some modes at a given frequency within a range of slot widths (namely for $s_3<s<s_6$) or, alternatively, at a given slot width within a range of frequencies is physically impossible. This can be explained by regarding the finline discontinuity shown in Fig. 5. Assuming that $\delta s \ll s_3$, this discontinuity is very weak, so that any incident mode in finline "a" can be nearly matched at z_0 to a single mode in finline "b" of nearly the same field configuration (whose propagation constant is very near that of the incident mode). Let us hypothetically assume that the incident field in finline "a" consists only of the two degenerate modes (shown in Fig. 3). If there were no modes in finline "b" with propagation constants

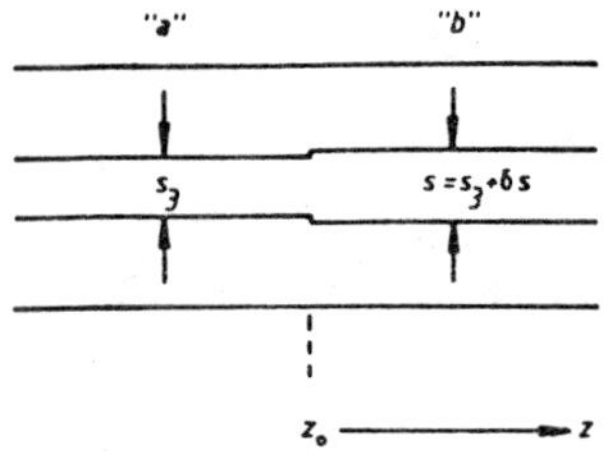

Fig. 5. A weak finline discontinuity.

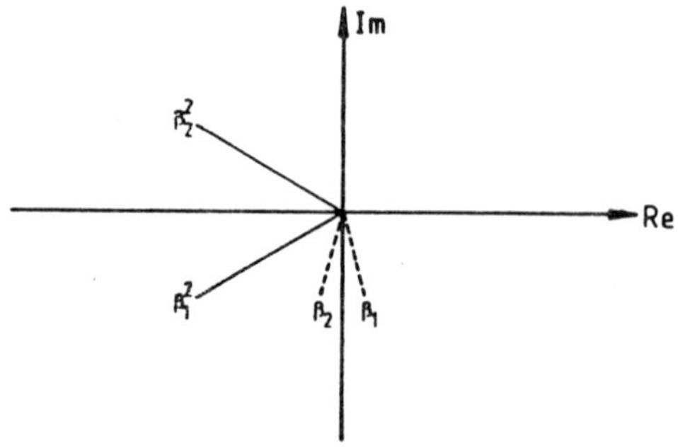

Fig. 6. The propagation constants of two complex modes.

which are near to that of the incident modes in finline "a", we should expect that there are also no modes in finline "b" with field configurations which are similar to those of the incident modes in finline "a". This means that the matching at z_0 requires a large number of excited modes at both sides of discontinuity, which contradicts with the weakness of the discontinuity.

V. Finline Complex Modes

Modes with complex propagation constants in lossless guiding structures have been known for a long time as "complex modes" [9]. They have been reported in [9] to occur as a continuation of backward-wave modes in the modal spectrum of shielded dielectric rods. They can also exist even if there is no region of the spectrum in which backward-wave modes can propagate [10]. The same has been reported in [14] and [15] with respect to shielded rectangular dielectric image guides. In both cases, backward-wave and complex modes can exist in a waveguide with a dielectric insert as modes of "hybrid-type." They cannot exist in a dielectric loaded circular waveguide without azimuthal dependence, because modes with no azimuthal dependence are either TE-type or TM-type modes. They can also not exist in the dielectric slab loaded rectangular waveguide because the modes of this structure are either of the LSE type or LSM type [1].

Finlines are, in fact, dielectric-loaded ridged waveguides, the modes of which are all of hybrid-type. Complex and backward-wave modes are consequently possible in finlines.

Now, we return to the problem of mode disappearance. The only possibility for the disappearance of the aforementioned two modes is that their β^2 are no longer real. Hence, zeros of $D(\beta^2)$ must be looked for in the complex plane rather than on the real axis. Although $D(\beta^2)$ is a complicated function and not at all a "polynomial with real coefficients," its zeros for the particular slot width range are found to be always a complex conjugate pair.

Now, let us investigate a pair of complex modes in some details. As shown in Fig. 6, let the corresponding complex conjugate zeros of $D(\beta^2)$ be β_1^2 and β_2^2. Their square roots, which are physically possible for a z-dependence $e^{-j\beta z}$ and a time dependence $e^{j\omega t}$, are also shown and can be written as

$$\beta_1 = \beta' - j\alpha' \quad \beta_2 = -\beta' - j\alpha'$$

where α' and β' are positive numbers. This means that one mode propagates in the $+z$-direction and is attenuated in the same direction. The other mode is also attenuated in this direction; it propagates, however, in the $-z$-direction. From the first view, it is easily stated that a mode with the complex propagation constant β_1 propagates in the same direction, in which it is attenuated; this means a continuous energy loss, although the structure has been assumed lossless. The other mode with propagation constant β_2 propagates opposite to the direction in which it is damped; this means a continuous energy gain, although the structure is passive. This point of view would be correct only if the two modes were not coupled. In fact, it has been found as a result of extensive numerical investigations that the two modes are so strongly coupled, that the electric field of each mode does not couple to its own magnetic field, but to the magnetic field of the other mode, i.e.,

$$\int_S (\boldsymbol{e}_1 \times \boldsymbol{h}_1^*) \cdot d\boldsymbol{S} = 0 = \int_S (\boldsymbol{e}_2 \times \boldsymbol{h}_2^*) \cdot d\boldsymbol{S} \tag{17}$$

$$\int_S (\boldsymbol{e}_1 \times \boldsymbol{h}_2^*) \cdot d\boldsymbol{S} = p \neq 0 \quad \int_S (\boldsymbol{e}_2 \times \boldsymbol{h}_1^*) \cdot d\boldsymbol{S} = -p^*. \tag{18}$$

Here $\boldsymbol{e}_1(\boldsymbol{h}_1)$ and $\boldsymbol{e}_2(\boldsymbol{h}_2)$ are the transverse electric- (magnetic-) field vectors of the modes with propagation constants β_1 and β_2, respectively, and S is the finline cross section. It has also been found from the same numerical investigations that, although the two modes are coupled in energy sense, they are still orthogonal in the sense of the following orthogonality relations:

$$\int_S (\boldsymbol{e}_1 \times \boldsymbol{h}_2) \cdot d\boldsymbol{S} = 0 = \int_S (\boldsymbol{e}_2 \times \boldsymbol{h}_1) \cdot d\boldsymbol{S} \tag{19}$$

$$\int_S (\boldsymbol{e}_1 \times \boldsymbol{h}_1) \cdot d\boldsymbol{S} = p \quad \int_S (\boldsymbol{e}_2 \times \boldsymbol{h}_2) \cdot d\boldsymbol{S} = -p^*. \tag{20}$$

A similar result has also been obtained for the shielded dielectric rod [13] and the shielded rectangular dielectric image guide [15]. Equations (17) and (18) mean that each mode cannot exist alone; both should always exist together, if they exist. As will be shown below, (18) plays an important role in the evanescent nature of the two coupled modes.

Now we investigate the problem from the energy point of view. Let us assume that the two questionable modes are excited (by, e.g., a certain discontinuity). Because each of these modes is not coupled to the other modes which may also be excited, it is sufficient to study the energy contained in these two modes only. Let E and H be the transverse electric- and magnetic-field vectors, respectively, of the two superposed modes; i.e., let

$$E = A_1 e^{-j\beta_1 z} e_1 + A_2 e^{-j\beta_2 z} e_2$$
$$H = A_1 e^{-j\beta_1 z} h_1 + A_2 e^{-j\beta_2 z} h_2.$$

Integrating the Poynting vector over the finline cross section, and making use of (17) and (18), one obtains

$$P = \int_S (E \times H^*) \cdot dS = jW \sin(2\beta' z + \vartheta) e^{-2\alpha' z}$$

where $A_1 A_2^* p = -(W/2) e^{-j\vartheta}$. The vanishing of the real part of P guarantees that the two superposed modes carry no active power, i.e., they behave as a whole evanescently. The energy stored in these superposed modes is shown in Fig. 7. It oscillates along the line once being inductive and once being capacitive in nature with exponential decay. This behavior is not, however, unexpected because one of these two modes was inductive and the other one was capacitive before their β^2 disappeared on the real axis.

The vanishing of the active power transmitted by a pair of complex modes has also been reported in [13] and [15] for the shielded dielectric rod and image guide, respectively, although using a different reasoning, namely, the total active powers transmitted inside and outside the dielectric region, are of equal magnitudes but of opposite directions.

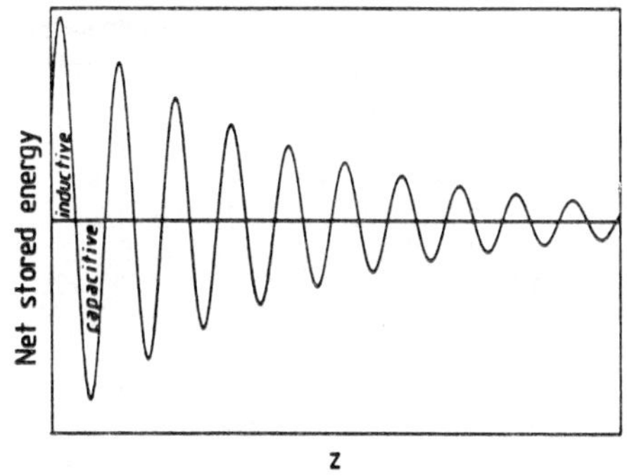

Fig. 7. Net stored energy of the superposed complex modes.

VI. Backward-Wave Modes in Finlines

Although pairs of complex modes can exist in finlines for any value of the substrate dielectric constant ϵ_r, backward-wave modes can only exist when the value of ϵ_r is relatively high. Small values of ϵ_r, which are usually used for normal finlines (typically, $\epsilon_r = 2.22$), lead to a very small (β'/α') ratio of any pair of complex modes which can exist. No backward-wave modes have been found for small values of ϵ_r. However, if both ϵ_r and the substrate thickness are considerably increased, the (β'/α') ratio of low-order complex modes increases considerably. In addition, if a pair of complex modes is near cutoff, it may be continued into a pair of forward- and backward-wave modes.

Finally, we would like to state that, although finlines have only been investigated, we believe that complex and backward-wave modes can also exist in any planar guiding structure with closed conducting boundaries.

VII. Numerical Results

To illustrate the fast convergence of truncating the characteristic matrix, the propagation constants of the first 30 modes of a unilateral finline have been calculated for three different slot widths using a matrix order of 5, 7, and 9 (i.e., truncating the infinite series $g_1(y)$ and $g_2(y)$ behind the 2nd, 3rd, and 4th term, respectively. The results in Tables I, II, and III show that, whatever the slot width, a matrix order of 7 is quite sufficient to give accurate results for the first 30 modes.

The four coupling integrals given in (17) and (18) have been numerically calculated by using a 100-term Fourier expansion for each field component. The results are tabulated in Table IV for the slot width ranging between 0.5 mm and 1.5 mm. For 0.5 mm $\leqslant s \leqslant$ 0.7 mm, we have two ordinary evanescent modes. The first (second) is inductive (capacitive) with β^2 decreasing (increasing) versus s. The values of the coupling integrals are in good agreement to the exact values given, e.g, in [1] for general inhomogeneously filled waveguides. For 0.8 mm $\leqslant s \leqslant$ 1.3 mm, we have a pair of complex modes. The coupling integrals are again in good agreement with the values given by (17) and (18). The smallness of the (β'/α') ratio is due to the smallness of the substrate dielectric constant ($\epsilon_r = 2.22$) and the substrate thickness. Higher (β'/α') values can be obtained by increasing these parameters as will be shown later. Finally, for 1.4 mm $\leqslant s \leqslant$ 1.5 mm, we have two ordinary evanescent modes again. The first (second) is now capacitive (inductive) with β^2 increasing (decreasing) versus s. The quantitative values of the coupling integrals are again in good agreement to the exact values given in [1].

The frequency dependencies of β' and α' of the complex modes existing in the same finline with a slot width of 1.0 mm are plotted in Fig. 8. For $F < 15$ GHz, we have two ordinary evanescent modes. Their dispersion curves are not distinugishable in the figure. The values of their α' at, e.g., $F = 10$ GHz are (3.1608 mm^{-1}) and (3.1673 mm^{-1}). For 15 GHz $\leqslant F \leqslant$ 117 GHz, we have a pair of complex modes. An interesting observation is the wideness of the frequency band over which the complex modes exist (more than 100 GHz). This wideness is, however, related to the smallness of the (β'/α') ratio (note that the upper plot of Fig. 8 is for 100 times β', whereas the lower one is for α') because this band decreases considerably when the (β'/α') ratio increases as will be shown below. For $F \geqslant 118$ GHz, the two complex modes are continued into two ordinary evanescent modes again.

Complex modes in a finline with a high value of the substrate dielectric constant ($\epsilon_r = 20$) and a relatively thick

TABLE I
THE PROPAGATION CONSTANTS OF THE FIRST 30 MODES IN A UNILATERAL FINLINE OF 0.2-mm SLOT WIDTH

mode no. / matrix order	1	2	3	4	5	6	7	8	9	10
5 x 5	+0.6837	-j 0.6067	-j 0.7448	-j 1.5955	-j 1.6489	-j 1.6801	-j 1.7472	-j 1.9938	-j 2.4747	-j 2.5509
7 x 7	+0.6824	-j 0.6067	-j 0.7448	-j 1.5955	-j 1.6489	-j 1.6801	-j 1.7470	-j 1.9926	-j 2.4747	-j 2.5495
9 x 9	+0.6820	-j 0.6067	-j 0.7448	-j 1.5955	-j 1.6489	-j 1.6800	-j 1.7470	-j 1.9922	-j 2.4747	-j 2.5491

mode no. / matrix order	11	12	13	14	15	16	17	18	19	20
5 x 5	-j 2.7083	-j 3.0745	-j 3.1136	-j 3.2065	-j 3.2384	-j 3.3653	-j 3.4763	-j 3.4976	-j 3.6058	-j 3.6683
7 x 7	-j 2.7073	-j 3.0738	-j 3.1136	-j 3.2065	-j 3.2375	-j 3.3649	-j 3.4763	-j 3.4971	-j 3.6057	-j 3.6656
9 x 9	-j 2.7070	-j 3.0735	-j 3.1136	-j 3.2065	-j 3.2372	-j 3.3648	-j 3.4763	-j 3.4969	-j 3.6057	-j 3.6648

mode no. / matrix order	21	22	23	24	25	26	27	28	29	30
5 x 5	-j 3.8140	-j 3.8787	-j 3.9308	-j 3.9683	-j 4.0196	-j 4.0854	-j 4.2335	-j 4.3530	-j 4.3621	-j 4.4326
7 x 7	-j 3.8130	-j 3.8787	-j 3.9308	-j 3.9663	-j 4.0196	-j 4.0786	-j 4.2318	-j 4.3483	-j 4.3635	-j 4.4326
9 x 9	-j 3.8127	-j 3.8787	-j 3.9308	-j 3.9658	-j 4.0196	-j 4.0770	-j 4.2314	-j 4.3475	-j 4.3637	-j 4.4326

Parameters: $a=2$ $b=3.556$ mm, substrate thickness = 0.254 mm, substrate dielectric constant = 2.22, frequency = 30 GHz.

TABLE II
THE PROPAGATION CONSTANTS OF THE FIRST 30 MODES IN A UNILATERAL FINLINE OF 1.0-mm SLOT WIDTH

mode no. / matrix order	1	2	3	4	5	6	7	8	9	10
5 x 5	+0.5976	-j 0.6154	-j 0.8564	-j 1.6031	-j 1.6489	-j 1.6847	-j 1.7697	-j 1.8877	-j 2.0873	-j 2.4642
7 x 7	+0.5976	-j 0.6154	-j 0.8564	-j 1.6031	-j 1.6489	-j 1.6847	-j 1.7697	-j 1.8877	-j 2.0873	-j 2.4640
9 x 9	+0.5975	-j 0.6154	-j 0.8563	-j 1.6031	-j 1.6489	-j 1.6847	-j 1.7696	-j 1.8877	-j 2.0872	-j 2.4640

mode no. / matrix order	11	12	13	14	15	16	17	18	19	20
5 x 5	-j 2.4754	-j 2.6055	-j 2.8094	-j 3.2019	-j 3.2757	-j 3.4238	-j 3.4764	-j 3.5024	-j 3.5950	-j 3.6260
7 x 7	-j 2.4754	-j 2.6054	-j 2.8094	-j 3.2014	-j 3.2757	-j 3.4230	-j 3.4764	-j 3.5021	-j 3.5948	-j 3.6234
9 x 9	-j 2.4754	-j 2.6054	-j 2.8089	-j 3.2012	-j 3.2756	-j 3.4221	-j 3.4764	-j 3.5021	-j 3.5947	-j 3.6225

mode no. / matrix order	21	22	23	24	25	26	27	28	29	30
5 x 5	-j 3.6920	-j 3.8520	-j 3.8605	-j 3.9270	-j 3.9783	-j 4.0130	-j 4.1337	-j 4.2822	-j 4.4232	-j 4.4794
7 x 7	-j 3.6916	-j 3.8518	-j 3.8563	-j 3.9265	-j 3.9750	-j 4.0120	-j 4.1326	-j 4.2815	-j 4.4212	-j 4.4729
9 x 9	-j 3.6916	-j 3.8516	-j 3.8540	-j 3.9263	-j 3.9746	-j 4.0115	-j 4.1326	-j 4.2806	-j 4.3498	-j 4.3543

Parameters: As in Table I.

TABLE III

THE PROPAGATION CONSTANTS OF THE FIRST 30 MODES IN A UNILATERAL FINLINE OF 3.0-mm SLOT WIDTH

mode no. \ matrix order	1	2	3	4	5	6	7	8	9	10
5 x 5	+0.4931	-j 0.6207	-j 1.1356	-j 1.6489	-j 1.6501	-j 1.7080	-j 1.7391	-j 1.8721	-j 1.8777	-j 2.0731
7 x 7	+0.4930	-j 0.6207	-j 1.1356	-j 1.6489	-j 1.6501	-j 1.7080	-j 1.7382	-j 1.8721	-j 1.8776	-j 2.0723
9 x 9	+0.4930	-j 0.6207	-j 1.1355	-j 1.6489	-j 1.6501	-j 1.7080	-j 1.7379	-j 1.8721	-j 1.8776	-j 2.0719

mode no. \ matrix order	11	12	13	14	15	16	17	18	19	20
5 x 5	-j 2.4167	-j 2.4415	-j 2.5705	-j 2.9917	-j 3.1180	-j 3.1648	-j 3.4764	-j 3.4816	-j 3.5054	-j 3.5621
7 x 7	-j 2.4166	-j 2.4415	-j 2.5703	-j 2.9903	-j 3.1178	-j 3.1647	-j 3.4764	-j 3.4815	-j 3.5054	-j 3.5584
9 x 9	-j 2.4166	-j 2.4414	-j 2.5702	-j 2.9897	-j 3.1177	-j 3.1647	-j 3.4764	-j 3.4814	-j 3.5054	-j 3.5573

mode no. \ matrix order	21	22	23	24	25	26	27	28	29	30
5 x 5	-j 3.5867	-j 3.5914	-j 3.7128	-j 3.7903	-j 3.8597	-j 3.9001	-j 3.9082	-j 3.9162	-j 3.9609	-j 4.0982
7 x 7	-j 3.5868	-j 3.5913	-j 3.7121	-j 3.7793	-j 3.8555	-j 3.9002	-j 3.9080	-j 3.9160	-j 3.9607	-j 4.0964
9 x 9	-j 3.5869	-j 3.5913	-j 3.7118	-j 3.7762	-j 3.8534	-j 3.9002	-j 3.9079	-j 3.9159	-j 3.9607	-j 4.0953

Parameters: As in Table I.

TABLE IV

THE COUPLING INTEGRALS ($I_{ij} = \int_S (e_i \times h_j^*) \cdot dS$) BETWEEN TWO MODES OF A UNILATERAL FINLINE

slot width (m.m.)	0.5	0.7	0.8	1.0
β_1	(0.00000)-j(3.08835)	(0.00000)-j(3.09792)	+(0.00368)-j(3.10463)	+(0.00855)-j(3.10558)
β_2	(0.00000)-j(3.11208)	(0.00000)-j(3.10892)	-(0.00368)-j(3.10463)	-(0.00855)-j(3.10558)
I_{11}	(0.00000)+j(1.00000)	(0.00000)+j(1.00000)	+(0.00009)+j(0.00012)	-(0.00023)+j(0.00017)
I_{12}	(0.00000)+j(0.00038)	(0.00000)+j(0.00024)	+(0.90786)+j(0.41927)	+(0.28358)+j(0.95895)
I_{21}	(0.00000)+j(0.00011)	(0.00000)-j(0.00007)	-(0.90786)+j(0.41927)	-(0.28358)+j(0.95895)
I_{22}	(0.00000)-j(1.00000)	(0.00000)-j(1.00000)	-(0.00009)+j(0.00012)	+(0.00023)+j(0.00017)

slot width (m.m.)	1.1	1.3	1.4	1.5
β_1	+(0.00912)-j(3.10489)	+(0.00192)-j(3.09950)	(0.00000)-j(3.08082)	(0.00000)-j(3.06148)
β_2	-(0.00912)-j(3.10489)	-(0.00192)-j(3.09950)	(0.00000)-j(3.10667)	(0.00000)-j(3.10869)
I_{11}	-(0.00048)+j(0.00011)	-(0.00102)-j(0.00369)	(0.00000)-j(1.00000)	(0.00000)-j(1.00000)
I_{12}	-(0.08344)+j(0.99651)	-(0.98287)+j(0.18427)	(0.00000)-j(0.00251)	(0.00000)-j(0.00259)
I_{21}	+(0.08344)+j(0.99651)	+(0.98287)+j(0.18427)	(0.00000)-j(0.00003)	(0.00000)+j(0.00010)
I_{22}	+(0.00048)+j(0.00011)	+(0.00102)-j(0.00369)	(0.00000)+j(1.00000)	(0.00000)+j(1.00000)

Parameters: As in Table I.

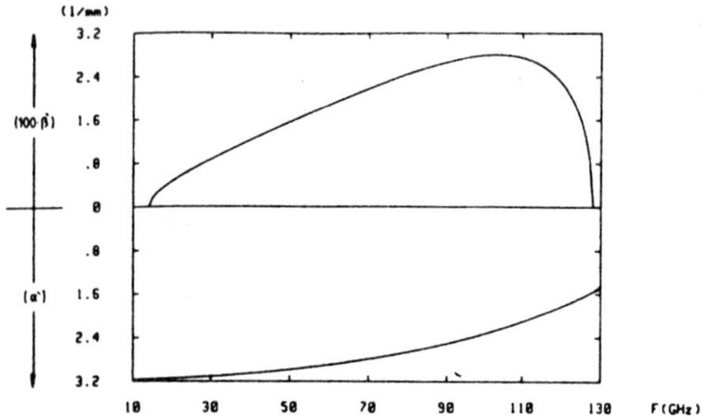

Fig. 8. Dispersion characteristics of a pair of complex modes in a unilateral finline. Parameters: $a = 2$ $b = 3.556$ mm, substrate thickness $= 0.254$ mm, substrate dielectric constant $= 2.22$, slot width $= 1.0$ mm.

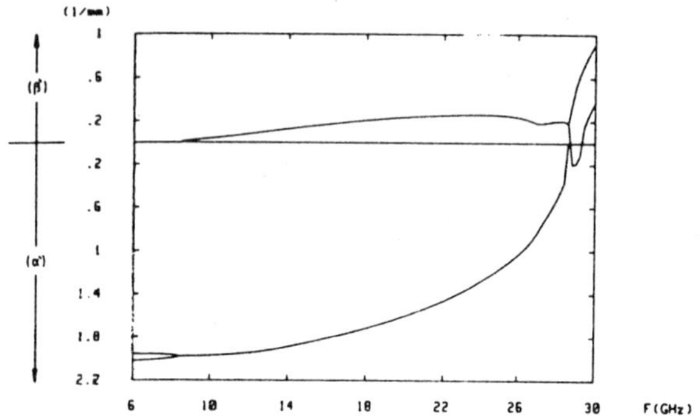

Fig. 9. Dispersion characteristics of a pair of complex modes in a unilateral finline. Parameters: $a = 2$ $b = 3.556$ mm, substrate thickness $= 1.067$ mm, substrate dielectric constant $= 20$, slot width $= 0.4$ mm.

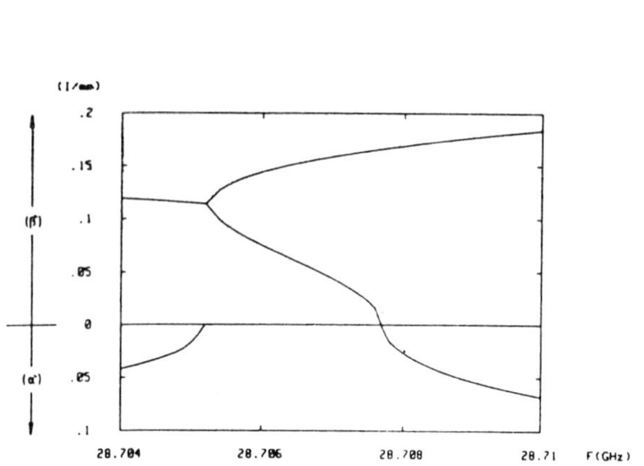

Fig. 10. Dispersion characteristics of a pair of forward- and backward-wave modes in a unilateral finline. Parameters: As in Fig. 9.

substrate (0.3 times the housing height) are finally investigated. The dispersion curves of β' and α' are plotted in Fig. 9. The (β'/α') ratio is much larger than in the previous case, whereas the frequency band over which the complex modes exist is much smaller now (about 20 GHz). The complex modes in this case are continued into a pair of forward- and backward-wave modes. The backward-wave mode exists over an extremely narrow frequency band. This is shown in more detail in Fig. 10.

ACKNOWLEDGMENT

The authors thank C. Bühst for preparing the manuscript, and E. Jensen for helpful discussions.

REFERENCES

[1] R. E. Collin, *Field Theory of Guided Waves*. New York: McGraw-Hill, 1960.

[2] J. B. Davies and D.M.-Syahkal,"Spectral domain solution of arbitrary coplanar transmission line with multilayer substrate," *IEEE Trans. Microwave Theory Tech.*, vol. MTT-25, pp. 143–146, 1977.

[3] L.-P. Schmidt and T. Itoh, "Spectral domain analysis of dominant and higher order modes in fin-lines," *IEEE Trans. Microwave Theory Tech.*, vol. MTT-28, pp. 981–985, 1980.

[4] L.-P. Schmidt, T. Itoh, and H. Hofmann, "Characteristic of unilateral fin-line structures with arbitrary located slots," *IEEE Trans. Microwave Theory Tech.*, vol. MTT-29, pp. 352–355, 1981.

[5] L. Lewin, *Advanced Theory of Waveguides*. London: Iliffe, 1951.

[6] L. Lewin, "The use of singular integral equations in the solution of waveguide problems," in *Advances of Microwaves*, vol. 1, Leo Young, Ed. New York: Academic Press, 1966.

[7] L. Lewin, *Theory of Waveguides*. London: Newnes Butterworths, 1975.

[8] R. Mittra and T. Itoh, "A new technique for the analysis of the dispersion characteristics of microstrip lines," *IEEE Trans. Microwave Theory Tech.*, vol. MTT-19, pp. 47–56, 1971.

[9] P. J. B. Clarricoats and B. C. Taylor, "Evanescent and propagating modes of dielectric-loaded circular waveguide," *Proc. Inst. Elec. Eng.*, vol. 111, pp. 1951–1956, 1964.

[10] P. J. B. Clarricoats and K. R. Slinn, "Complex modes of propagation in dielectric loaded circular waveguides," *Electron Lett.*, vol. 1, pp. 145–146, 1965.

[11] S. B. Rayevskiy, "Some properties of complex waves in a double-layer circular, shielded waveguides," *Radio Eng. Electron Phys.*, vol. 21, pp. 36–39, 1976.

[12] V. A. Kalmyk, S. B. Rayevskiy, and V. P. Ygryumov, "An experimental verification of existence of complex waves in a two-layer, circular, shielded, waveguide," *Radio Eng. Electron Phys.*, vol. 23, pp. 16–19, 1978.

[13] H. Katzier and F. J. K. Lange, "Grundlegende Eigenschaften komplexer Wellen am Beispiel der geschirmten kreiszylindrischen dielektrischen Leitung," *Arch. Elek. Übertragung.*, vol. AEÜ-37, pp. 1–5, 1983.

[14] U. Crombach. "Complex waves on shielded lossless rectangular dielectric image guide," *Electron Lett.*, vol. 19, pp. 557–558, 1983.

[15] J. Strube and F. Arndt, "Rigorous hybrid-mode analysis of the transition from rectangular waveguide to shielded dielectric image guide," *IEEE Trans. Microwave Theory Tech.*, vol. MTT-33, pp. 391–401, 1985.

[16] A. Lavik and H.-G. Unger, "Der Rechteckhohlleiter mit rechteckigem Stoffeinsatz," *Arch. Elek. Übertragung.*, vol. AEÜ-18, pp. 25–34, 1964.

[17] A. Lavik and H.-G. Unger, "Rückwärtswellen in homogenen Wellenleitern," *Arch. Elek. Übertragung.*, vol. AEÜ-18, pp. 35–42, 1964.

[18] A. S. Omar and K. Schünemann, "Space-domain decoupling of LSE and LSM fields in generalized planar guiding structures," *IEEE Trans. Microwave Theory Tech.*, vol. MTT-32, pp. 1626–1632, 1984.

[19] A. S. Omar and K. Shcünemann, "Generalized representation of Green's functions in multi-layer planar structures," in *Proc. 14th EuMC* (Liege), 1984, pp. 436–441.

[20] R. Mittra and S. W. Lee, *Analytical Techniques in the Theory of Guided Waves*. New York: Macmillan, 1971.

2.1.7

The Transmission-Line Matrix Method—Theory and Applications

WOLFGANG J. R. HOEFER, SENIOR MEMBER, IEEE

(Invited Paper)

Abstract—This paper presents an overview of the transmission-line matrix (TLM) method of analysis, describing its historical background from Huygens's principle to modern computer formulations. The basic algorithm for simulating wave propagation in two- and three-dimensional transmission-line networks is derived. The introduction of boundaries, dielectric and magnetic materials, losses, and anisotropy are discussed in detail. Furthermore, the various sources of error and the limitations of the method are given, and methods for error correction or reduction, as well as improvements of numerical efficiency, are discussed. Finally, some typical applications to microwave problems are presented.

I. Introduction

Before the advent of digital computers, complicated electromagnetic problems which defied analytical treatment could only be solved by simulation techniques. In particular, the similarity between the behavior of electromagnetic fields, and of voltages and currents in electrical networks, was used extensively during the first half of the twentieth century to solve high-frequency field problems [2]–[4].

When modern computers became available, powerful numerical techniques emerged to predict directly the behavior of the field quantities. The great majority of these methods yield harmonic solutions of Maxwell's equations in the space or spectral domain. A notable exception is the transmission-line matrix (TLM) method of analysis which represents a true computer simulation of wave propagation in the time domain.

In this paper, the theoretical foundations of the TLM method are reviewed, its basic algorithm for simulating the propagation of waves in unbounded and bounded space is derived, and it is shown how the eigenfrequencies and field configurations of resonant structures can be determined with the Fourier transform. Sources and types of errors are discussed, and possible pitfalls are pointed out. Then, various methods of error correction are presented, and the most significant improvements to the conventional TLM approach are described. A referenced list of typical applications of the method is included as well. In the conclusion, the advantages and disadvantages of the method are summarized, and it is indicated under what circumstances it is appropriate to select the TLM method rather than other numerical techniques for solving a particular problem.

Manuscript received February 22, 1985; revised May 31, 1985.
The author is with the Department of Electrical Engineering, University of Ottawa, Ottawa, Ontario, Canada K1N 6N5.

II. Historical Background

Two distinct models describing the phenomenon of light were developed in the seventeenth century: the corpuscular model by Isaak Newton and the wave model by Christian Huygens. At the time of their conception, these models were considered incompatible. However, modern quantum physics has demonstrated that light in particular, and electromagnetic radiation in general, possess both granular (photons) and wave properties. These aspects are complementary, and one or the other usually dominates, depending on the phenomenon under study.

At microwave frequencies, the granular nature of electromagnetic radiation is not very evident, manifesting itself only in certain interactions with matter, while the wave aspect predominates in all situations involving propagation and scattering. This suggests that the model proposed by Huygens, and later refined by Fresnel, could form the basis for a general method of treating microwave propagation and scattering problems.

Indeed, Johns and Beurle [5] described in 1971 a novel numerical technique for solving two-dimensional scattering problems, which was based on Huygens's model of wave propagation. Inspired by earlier network simulation techniques [2]–[4], this method employed a Cartesian mesh of open two-wire transmission lines to simulate two-dimensional propagation of delta function impulses. Subsequent papers by Johns and Akhtarzad [6]–[16] extended the method to three dimensions and included the effect of dielectric loading and losses. Building upon the groundwork laid by these original authors, other researchers [17]–[34] added various features and improvements such as variable mesh size, simplified nodes, error correction techniques, and extension to anisotropic media.

The following section describes briefly the discretized version of Huygens's wave model which is suitable for implementation on a digital computer and forms the algorithm of the TLM method. A detailed description of this model can be found in a very interesting paper by P. B. Johns [9].

III. Huygen's Principle and Its Discretization

According to Huygens [1], a wavefront consists of a number of secondary radiators which give rise to spherical wavelets. The envelope of these wavelets forms a new

W.J.R. Hoefer, *IEEE Transactions on Microwave Theory and Techniques*, Vol. MTT-33, No. 10, Oct. 1985, pp. 882–893.

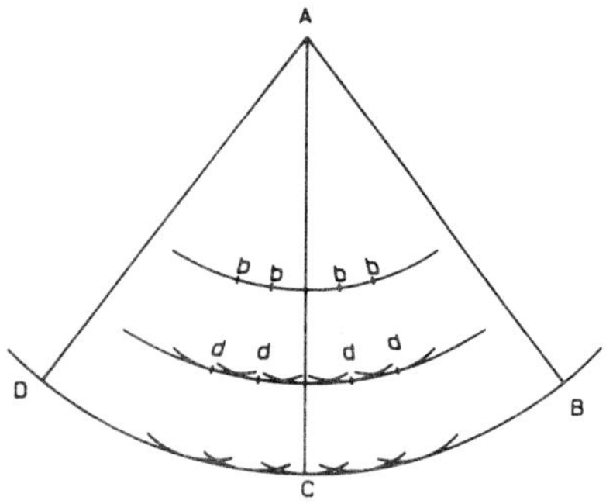

Fig. 1. Huygens's principle and formation of a wavefront by secondary wavelets.

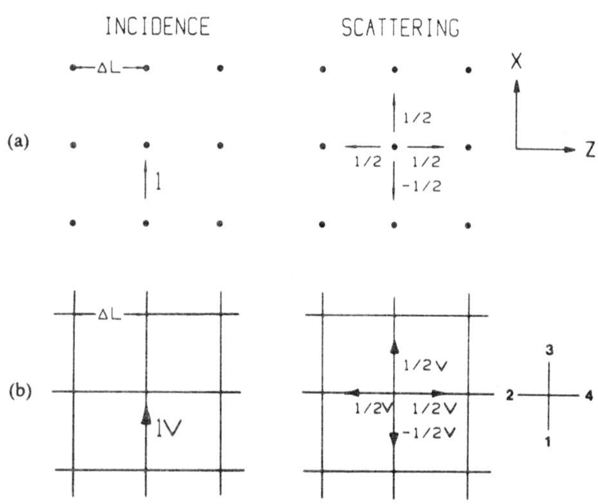

Fig. 2. The discretized Huygens's wave model (a) in two-dimensional space and (b) in an equivalent Cartesian mesh of transmission lines (after Johns [9]).

wavefront which, in turn, gives rise to a new generation of spherical wavelets, and so on (Fig. 1). In spite of certain difficulties in the mathematical formulation of this mechanism, its application nevertheless leads to an accurate description of wave propagation and scattering, as will be shown below.

In order to implement Huygens's model on a digital computer, one must formulate it in discretized form. To this end, both space and time are represented in terms of finite, elementary units Δl and Δt, which are related by the velocity of light such that

$$\Delta t = \Delta l / c. \tag{1}$$

Accordingly, two-dimensional space is modeled by a Cartesian matrix of points or nodes, separated by the mesh parameter Δl (see Fig. 2(a)). The unit time Δt is then the time required for an electromagnetic pulse to travel from one node to the next.

Assume that a delta function impulse is incident upon one of the nodes from the negative x-direction. The energy in the pulse is unity. In accordance with Huygen's principle, this energy is scattered isotropically in all four directions, each radiated pulse carrying one fourth of the incident energy. The corresponding field quantities must then be 1/2 in magnitude. Furthermore, the reflection coefficient "seen" by the incident pulse must be negative in order to satisfy the requirement of field continuity at the node.

This model has a network analog in the form of a mesh of orthogonal transmission lines, or transmission-line matrix (Fig. 2(b)), forming a Cartesian array of shunt nodes which have the same scattering properties as the nodes in Fig. 2(a). It can be shown that there is a direct equivalence between the voltages and currents on the line mesh and the electric and magnetic fields of Maxwell's equations [5].

Consider the incidence of a unit Dirac voltage-impulse on a node in the TLM mesh of Fig. 2(b). Since all four branches have the same characteristic impedance, the reflection coefficient "seen" by the incident impulse is indeed $-1/2$, resulting in a reflected impulse of -0.5 V and three transmitted impulses of $+0.5$ V.

The more general case of four impulses being incident on the four branches of a node can be obtained by superposition from the previous case. Hence, if at time $t = k\,\Delta t$, voltage impulses ${}_kV_1^i$, ${}_kV_2^i$, ${}_kV_3^i$, and ${}_kV_4^i$ are incident on lines 1–4, respectively, on any junction node, then the total voltage impulse reflected along line n at time $(k+1)\,\Delta t$ will be

$${}_{k+1}V_n^r = \frac{1}{2}\left[\sum_{m=1}^{4} {}_kV_m^i\right] - {}_kV_n^i. \tag{2}$$

This situation is conveniently described by a scattering matrix equation [7] relating the reflected voltages at time $(k+1)\,\Delta t$ to the incident voltages at the previous time step $k\,\Delta t$

$${}_{k+1}\begin{pmatrix} V_1 \\ V_2 \\ V_3 \\ V_4 \end{pmatrix}^r = 1/2 \begin{pmatrix} -1 & 1 & 1 & 1 \\ 1 & -1 & 1 & 1 \\ 1 & 1 & -1 & 1 \\ 1 & 1 & 1 & -1 \end{pmatrix} \times {}_k\begin{pmatrix} V_1 \\ V_2 \\ V_3 \\ V_4 \end{pmatrix}^i . \tag{3}$$

Furthermore, any impulse emerging from a node at position (z, x) in the mesh (reflected impulse) becomes automatically an incident impulse on the neighboring node. Hence

$$\begin{aligned} {}_{k+1}V_1^i(z,x) &= {}_{k+1}V_3^r(z, x-1) \\ {}_{k+1}V_2^i(z,x) &= {}_{k+1}V_4^r(z-1, x) \\ {}_{k+1}V_3^i(z,x) &= {}_{k+1}V_1^r(z, x+1) \\ {}_{k+1}V_4^i(z,x) &= {}_{k+1}V_2^r(z=1, x). \end{aligned} \tag{4}$$

Consequently, if the magnitudes, positions, and directions of all impulses are known at time $k\,\Delta t$, the corresponding values at time $(k+1)\,\Delta t$ can be obtained by operating (3) and (4) on each node in the network. The impulse response of the network is then found by initially fixing the magni-

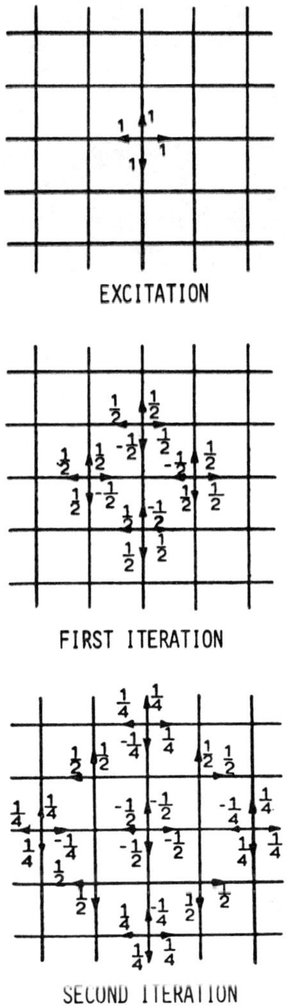

Fig. 3. Three consecutive scatterings in a two-dimensional TLM network excited by a Dirac impulse.

tudes, directions, and positions of all impulses at $t = 0$ and then calculating the state of the network at successive time intervals.

The scattering process described above forms the basic algorithm of the TLM method. Three consecutive scatterings are shown in Fig. 3, visualizing the spreading of the injected energy across the two-dimensional network.

This sequence of events closely resembles the disturbance of a pond due to a falling drop of water. However, there is one obvious difference, namely the discrete nature of the TLM mesh which causes dispersion of the velocity of the wavefront. In other words, the velocity of a signal component in the mesh depends on its direction of propagation as well as on its frequency.

In order to appreciate the importance of this dispersion, note that the process in Fig. 3 depicts a short episode of the response of the TLM network to a single impulse which contains all frequencies. Thus, harmonic solutions to a problem are obtained from the impulse response via the

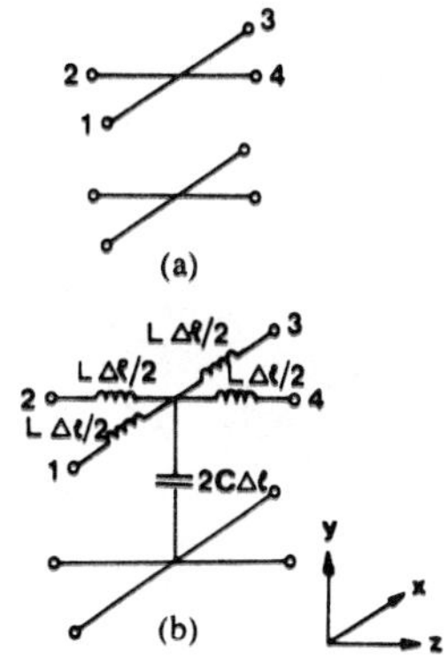

Fig. 4. The building block of the two-dimensional TLM network. (a) Shunt node. (b) Equivalent lumped-element model.

Fourier transform. Accurate solutions will be obtained only at frequencies for which the dispersion effect can be neglected. This aspect will be discussed in Section IV.

The TLM mesh can be extended to three dimensions, leading to a rather complex network containing series as well as shunt nodes. Each of the six field components is simulated by a voltage or a current in that mesh. Three-dimensional TLM networks will be discussed in Section V.

IV. The Two-Dimensional TLM Method

A. *Wave Properties of the TLM Network*

The basic building block of a two-dimensional TLM network is a shunt node with four sections of transmission lines of length $\Delta l/2$ (see Fig. 4(a)). Such a configuration can be approximated by the lumped-element model shown in Fig. 4(b). Comparing the relations between voltages and currents in the equivalent circuit with the relations between the H_z-, H_x-, and E_y-components of a TE_{m0} wave in a rectangular waveguide, the following equivalences can be established [5]:

$$\begin{aligned} E_y &\equiv V_y \qquad -H_z \equiv (I_{x3} - I_{x1}) \\ -H_x &\equiv (I_{z2} - I_{z4}) \quad \mu \equiv L \quad \epsilon \equiv 2C. \end{aligned} \tag{5}$$

For elementary transmission lines in the TLM network, and for $\mu_r = \epsilon_r = 1$, the inductance and capacitance per unit length are related by

$$1/\sqrt{LC} = 1/\sqrt{\epsilon_0\mu_0} = c \tag{6}$$

where $c = 3\times10^8$ m/s.

Hence, if voltage and current waves on each transmission-line component travel at the speed of light, the complete network of intersecting transmission lines represents a medium of relative permittivity twice that of free space. The means that as long as the equivalent circuit in Fig. 4 is valid, the propagation velocity in the TLM mesh is $1/\sqrt{2}$ the velocity of light.

Note that the dual nature of electric and magnetic fields also allows us to simulate, for example, the longitudinal

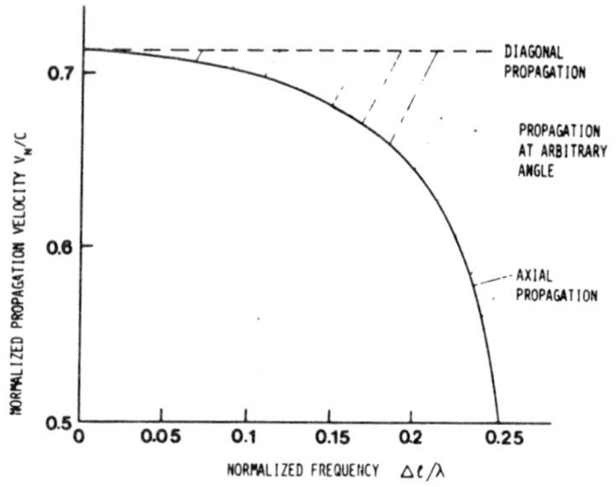

Fig. 5. Dispersion of the velocity of waves in a two-dimensional TLM network (after Johns and Beurle [5]).

magnetic field of TE modes by the network voltage, while the network currents simulate the transverse electric-field components. Whatever the relationship between field and network variables, the wave properties of the mesh, which will be discussed next, remain the same. Considering the mesh as a periodic structure, Johns and Beurle [5] calculate the following dispersion relation for propagation along the main mesh axes:

$$\sin(\beta_n \Delta l/2) = \sqrt{2}\sin[\omega \Delta l/(2c)] \tag{7}$$

where β_n is the propagation constant in the network. The resulting ratio of velocities on the matrix and in free space, $v_n/c = \omega/(\beta_n c)$, is shown in Fig. 5. It appears that a first cutoff occurs for $\Delta l/\lambda = 1/4$ (λ is the free-space wavelength). However, no cutoff occurs in the diagonal direction, where the velocity is frequency-independent, while in intermediate directions, the velocity ratio lies somewhere between the two curves shown in Fig. 5.

In conclusion, the TLM network simulates an isotropic propagating medium only as long as all frequencies are well below the network cutoff frequency, in which case the network propagation velocity may be considered constant and equal to $c/\sqrt{2}$.

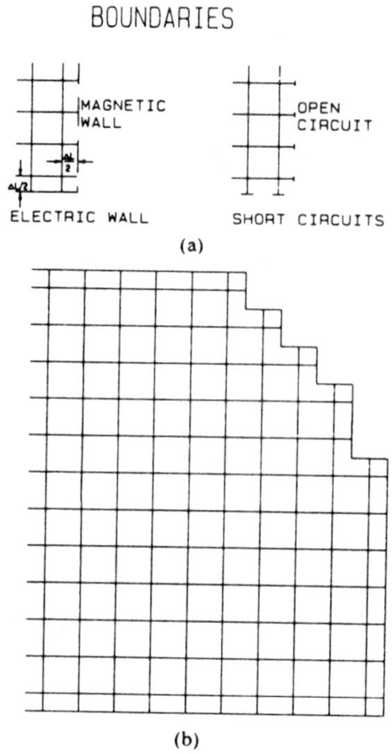

Fig. 6. Representation of boundaries in the TLM mesh. (a) Electric and magnetic walls. (b) Curved wall represented by a piecewise straight boundary.

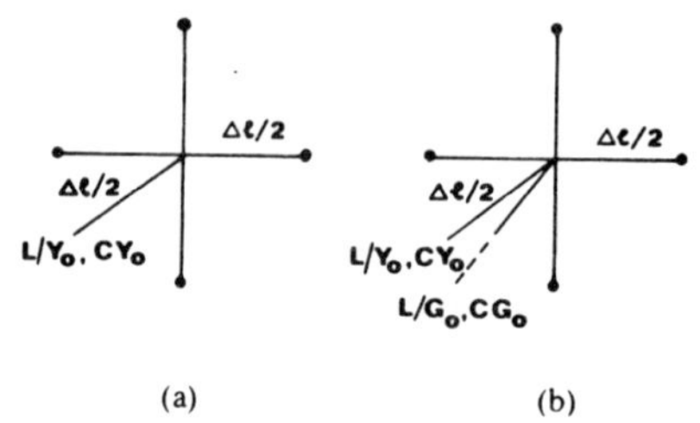

Fig. 7. Simulation of permittivity and losses. (a) Permittivity stub. (b) Permittivity stub and loss stub.

B. Representation of Lossless and Lossy Boundaries

Electric and magnetic walls are represented by short and open circuits, respectively, at the appropriate positions in the TLM mesh. To ensure synchronism, they must be placed halfway between two nodes. In practice, this is achieved by making the mesh parameter Δl an integer fraction of the structure dimensions. Curved walls are represented by piecewise straight boundaries as shown in Fig. 6.

In the computation, the reflection of an impulse at a magnetic or electric wall is achieved by returning it, after one unit time step Δt, with equal or opposite sign to its boundary node of origin.

Lossy boundaries can be represented in the same way as lossless boundaries, with the difference that the reflection coefficient in each boundary branch is now

$$\rho = (R-1)/(R+1) \tag{8}$$

instead of unity. R is the normalized surface resistance of the boundary.

For a good but imperfect conductor of conductivity σ, the reflection coefficient ρ is approximately

$$\rho \simeq -1 + 2[\epsilon_0\omega/(2\sigma)]^{1/2}. \tag{9}$$

Note that since ρ depends on the frequency ω, the loss calculations are accurate only for that frequency which has been selected in determining ρ.

C. Representation of Dielectric and Magnetic Materials

The presence of dielectric or magnetic material (for example, in partial dielectric or magnetic loading of a waveguide) can be taken into account by loading inside nodes with reactive stubs of appropriate characteristic impedance and a length equal to half the mesh spacing [7], as shown in Fig. 7(a). For example, if the network voltage

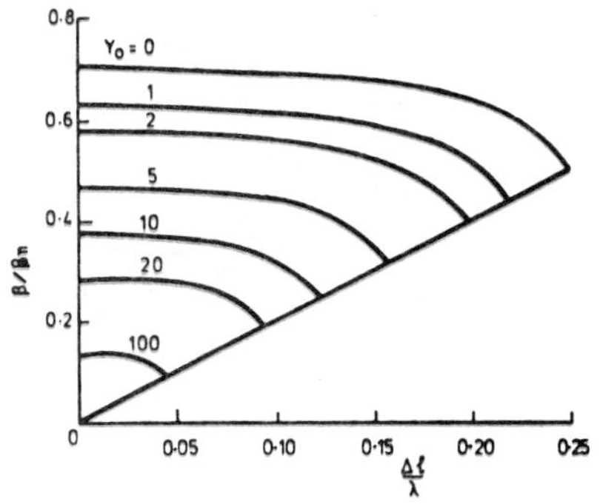

Fig. 8. Dispersion of the velocity of waves in a two-dimensional stub-loaded TLM network with the characteristic admittance of the reactive stubs as a parameter (after Johns [7]).

simulates an electric field, an open-circuited shunt stub of length $\Delta l/2$ will produce the effect of additional capacity at the nodes. This reduces the phase velocity in the structure and, at the same time, satisfies the boundary conditions at the air–dielectric interface [7]. At low frequencies, the velocity of waves in the stub-loaded TLM mesh is given by

$$v_n^2 = 0.5c^2/(1 + Y_0/4) \tag{10}$$

where c is the free-space velocity, and Y_0 is the characteristic admittance of the stubs, normalized to the admittance of the main network lines. Note that the velocity in the network is now made variable by altering the single constant Y_0. The relationship between ϵ_r of the simulated space and Y_0 is

$$\epsilon_r = 2(1 + Y_0/4). \tag{11}$$

The velocity characteristic along the main axes of the stub-loaded network is shown in Fig. 8 for various values of Y_0. Again, for relatively low frequencies, the mesh velocity is practically the same in all directions.

In cases where the voltage on the TLM mesh represents a magnetic field, the open shunt stubs describe a permeability. The velocity of waves in a magnetically loaded medium will be simulated correctly by such a mesh. However, the interface conditions are not satisfied, and a correction must be introduced in the form of local reflection and transmission coefficients at the interface between the media, as described in [7].

D. Description of Dielectric Losses

Losses in a dielectric can be accounted for in two different ways. One can either consider the TLM mesh to consist of lossy transmission lines, or one can load the nodes of a lossless mesh with so-called loss-stubs (Fig. 7(b)).

In the first case, the magnitude of each pulse is reduced by an appropriate amount while traveling from one node to the next, and the ensuing change in velocity is accounted for by increasing the time required to reach the next node [8]. This method is particularly suited for homogeneous structures.

In the second case, each node is resistively loaded with a matched transmission line of appropriate characteristic admittance G_0, extracting energy from each node at every iteration [10]. This technique is more suitable for inhomogeneous structures since it describes the interface conditions as well as the loss mechanism.

The normalized admittance of the loss-stub is related to the conductivity σ of the lossy medium by

$$G_0 = \sigma \Delta l (\mu_0/\epsilon_0)^{1/2}. \tag{12}$$

E. Computation of the Frequency Response of a Structure

The previous sections have described how the wave properties of two-dimensional unbounded and bounded space can be simulated by a two-dimensional mesh of transmission lines, and how the impulse response of such a mesh can be computed by iteration of (3) and (4). Any node (or several nodes) may be selected as input and/or output points. The output function is an infinite series of discrete impulses of varying magnitude, representing the response of the system to an impulsive excitation (see Fig. 12). The output corresponding to any other input may be obtained by convolving it with this impulse response.

Of particular interest is the response to a sinusoidal excitation which is obtained by taking the Fourier transform of the impulse response. Since the latter is a series of delta functions, the Fourier integral becomes a summation, and the real and imaginary parts of the output spectrum are

$$\mathrm{Re}\left[F(\Delta l/\lambda)\right] = \sum_{k=1}^{N} {}_kI \cos(2\pi k \Delta l/\lambda) \tag{13}$$

$$\mathrm{Im}\left[F(\Delta l/\lambda)\right] = \sum_{k=1}^{N} {}_kI \sin(2\pi k \Delta l/\lambda) \tag{14}$$

where $F(\Delta l/\lambda)$ is the frequency response, ${}_kI$ is the value of the output response at time $t = k\Delta l/c$, and N is the total number of time intervals for which the calculation has been made, henceforth called the "number of iterations."

In the case of a closed structure, this frequency response represents its mode spectrum. A typical example is Fig. 9(a), which shows the cutoff frequencies of the modes in a WR-90 waveguide.

Note that, as in a real measurement, the position of input and output points as well as the nature of the field component under study will affect the magnitudes of the spectral lines. For example, if input and output nodes are situated close to a minimum of a particular mode field, the corresponding eigenfrequency will not appear in the frequency response. This feature can be used either to suppress or enhance certain modes.

F. Computation of Fields and Impedances

Since the network voltages and currents are directly proportional to field quantities in the simulated structure, the TLM method also yields the field distribution. In order to obtain the configuration of a particular mode, its eigen-

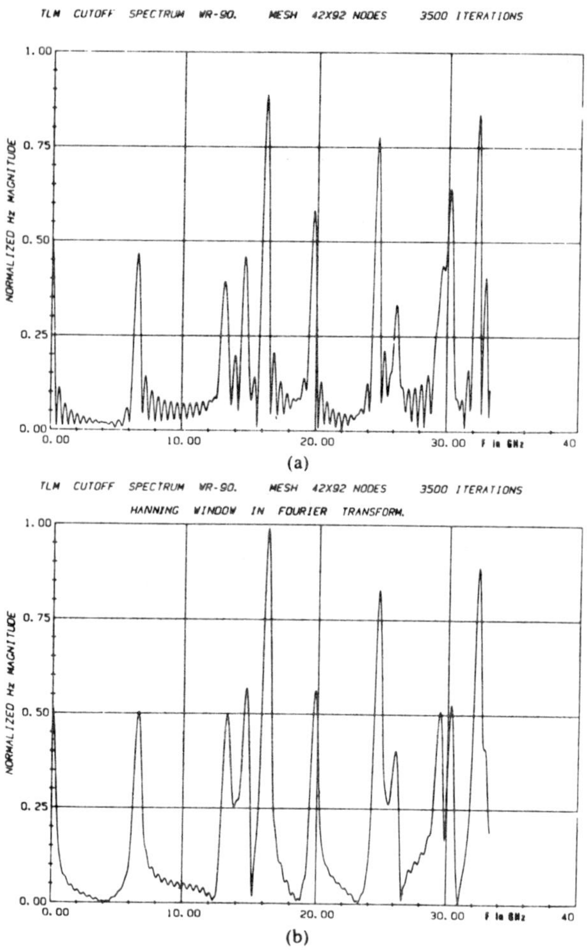

Fig. 9. Typical output from a two-dimensional TLM program. (a) Cutoff spectrum of a WR-90 waveguide. (b) The same spectrum after convolution of the output impulse function with a Hanning window (TLM) mesh: 42×92 nodes, 3500 iterations).

frequency must be computed first. Then the Fourier transform of the network variable representing the desired field component is computed at each node during a second run. In this process, (13) and (14) are computed for each node, with $\Delta l/\lambda$ corresponding to the eigenfrequency of the mode. The field between nodes can be obtained by interpolation techniques.

Impedances can, in turn, be obtained from the field quantities. Local field impedances can be found directly as the ratio of voltages and currents at a node, while impedances defined on the basis of particular field integrals (such as the voltage–power impedance in a waveguide) are computed by stepwise integration of the discrete field values. This procedure is identical to that used in finite-element and finite-difference methods of analysis.

V. The Three-Dimensional TLM Method

The two dimensional method described above can be extended to three dimensions at the expense of increased complexity [10] to [15]. In order to simultaneously describe all six field components in three-dimensional space, the basic shunt node must be replaced by a hybrid TLM cell consisting of three shunt and three series nodes as shown in Fig. 10. The side of the cell is $\Delta l/2$. The voltages at the three series nodes represent the three electric-field components, while the currents at the series nodes represent the magnetic-field components.

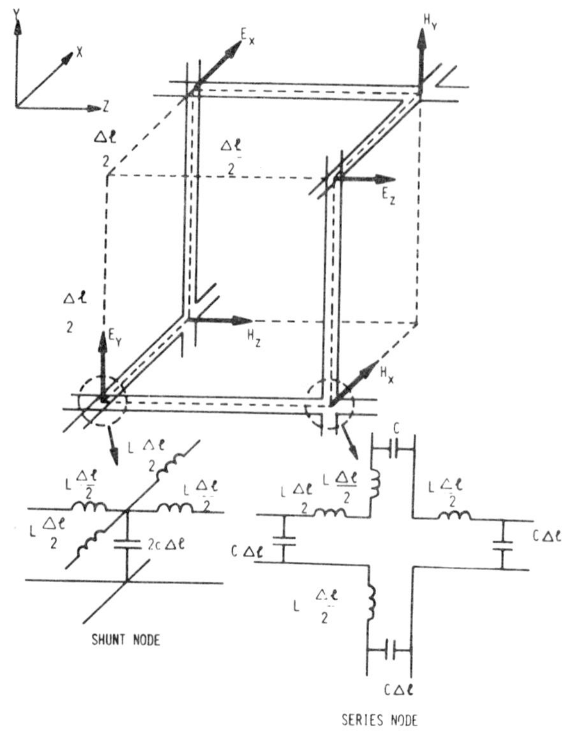

Fig. 10. The three-dimensional TLM cell featuring three series and three shunt nodes.

The wave properties of the three-dimensional mesh are similar to that of its two-dimensional counterpart with the difference that the low-frequency velocity is now $c/2$ instead of $c/\sqrt{2}$ [15].

Boundaries are simulated by short-circuiting shunt nodes (electric wall) or open-circuiting shunt nodes (magnetic wall) situated on a boundary. Wall losses are included by introducing imperfect reflection coefficients.

Magnetic and dielectric materials may be introduced by adding short-circuited $\Delta l/2$ series stubs at the series nodes and open-circuited $\Delta l/2$ shunt stubs at the shunt nodes, respectively. Furthermore, losses are taken into account by resistively loading the shunt nodes in the network (see Fig. 11). Even anisotropic materials may be simulated by introducing at each of the three series or shunt nodes of a cell a stub with a different characteristic admittance [17]. Finally, losses as well as permittivities and permeabilities can be varied in space and in time by controlling the admittances of the dissipative and reactive stubs. The relationships between material parameters and stub admittances are the same as in the two-dimensional case.

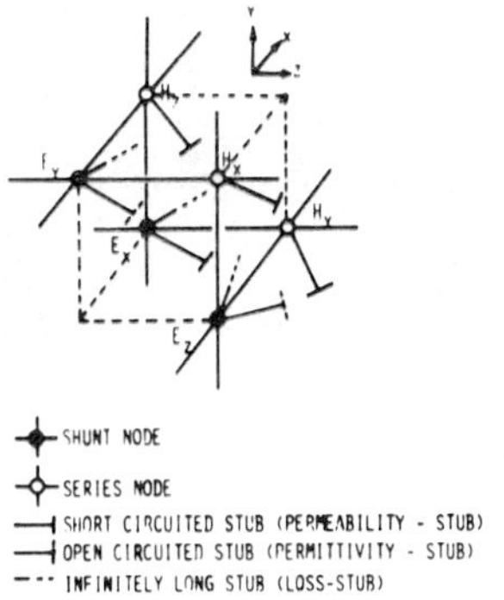

Fig. 11. Simulation of permittivity, permeability, and losses in a three-dimensional TLM network (after Akhtarzad [13]).

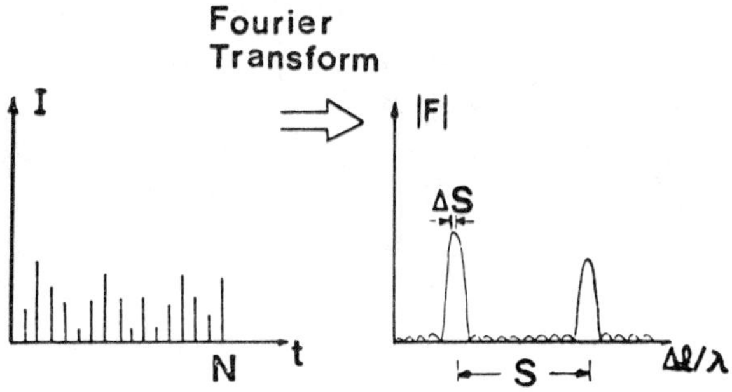

(a) (b)

Fig. 12. (a) Truncated output impulse response and (b) resulting truncation error in the frequency domain.

The impulse response of a three-dimensional network is found in the same way as in the two-dimensional case, and everything that has been said about the computation of eigenfrequencies, fields, and impedances, applies here as well.

VI. Errors and Their Correction

Like all other numerical techniques, the TLM method is subject to various sources of error and must be applied with caution in order to yield reliable and accurate results. The main sources of error are due to the following circumstances:

a) The impulse response must be truncated in time.
b) The propagation velocity in the TLM mesh depends on the direction of propagation and on the frequency.
c) The spatial resolution is limited by the finite mesh size.
d) Boundaries and dielectric interfaces cannot be aligned in the 3-D TLM model.

The resulting errors will be discussed below, and ways of eliminating or, at least, significantly reducing these errors will be described.

A. *Truncation Error*

The need to truncate the output impulse function leads to the so-called truncation error: Due to the finite duration of the impulse response, its Fourier transform is not a line spectrum but rather a superposition of $\sin x/x$ functions (Gibbs's phenomenon) which may interfere with each another such that their maxima are slightly shifted. The resulting error in the eigenfrequency, or truncation error, is given by

$$E_T \leqq \Delta S/(\Delta l/\lambda_c) = 3\lambda_c/(SN^2\pi^2\Delta l) \qquad (15)$$

where N is the number of iterations and S is the distance in the frequency domain between two neighboring spectral peaks (see Fig. 12).

This expression shows that the truncation error decreases with increasing separation S and increasing number of iterations N. It is thus desirable to suppress all unwanted modes close to the desired mode by choosing appropriate input and output points in the TLM network. Another technique, proposed by Saguet and Pic [20], is to use a Hanning window in the Fourier transform, resulting in a considerable attenuation of the sidelobes.

In this process, the output impulse response is first convolved with the Hanning profile

$$f_k(k) = 0.5(1+\cos \pi k/N), \qquad k = 1,2,3,\cdots, N \qquad (16)$$

where k is the iteration variable or counter. The filtered impulse response is then Fourier transformed. The resulting improvement can be appreciated by comparing Figs. 9(a) and 9(b).

Finally, the number of iterations may be made very large, but this leads to increased CPU time. It is recommended that the number of iterations be chosen such that the truncation error given by (16) is reduced to a fraction of a percent and can be neglected.

B. *Velocity Error*

If the wavelength in the TLM network is large compared with the network parameter Δl, it can be assumed that the fields propagate with the same velocity in all directions. However, when the wavelength decreases, the velocity depends on the direction of propagation (see Fig. 5). At first glance, the resulting velocity error can be reduced only by choosing a very dense mesh, unless propagation occurs essentially in an axial direction (e.g., rectangular waveguide), in which case the error can be corrected directly using the dispersion relation (7). Fortunately, the velocity error responds to the same remedial measures as the coarseness error (which will be described next), and it therefore does not need to be corrected separately.

C. *Coarseness Error*

The coarseness error occurs when the TLM mesh is too coarse to resolve highly nonuniform fields as can be found at corners and wedges. This error is particularly cumbersome when analyzing planar structures which contain such regions. A possible but impractical measure would be to choose a very fine mesh. However, this would lead to large memory requirements, particularly for three-dimensional

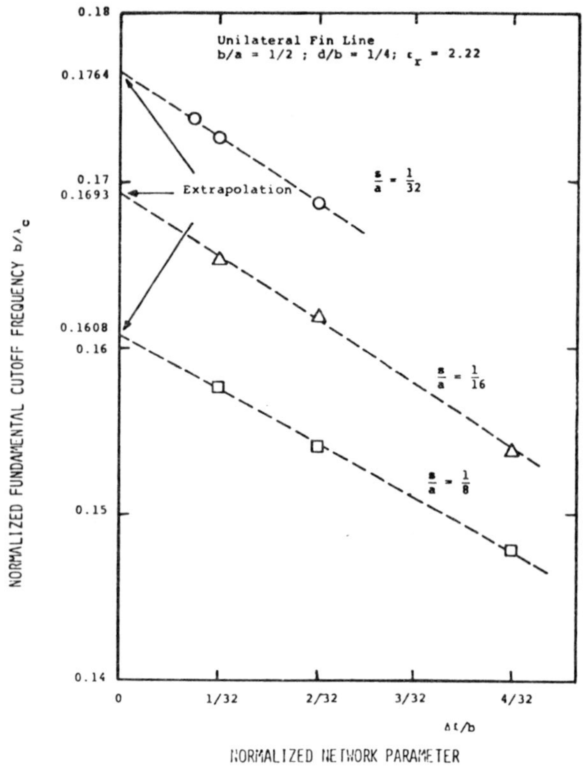

Fig. 13. Elimination of coarseness error by linear extrapolation of results obtained with TLM meshes of different parameter $\Delta l/b$ (after Shih and Hoefer [25]).

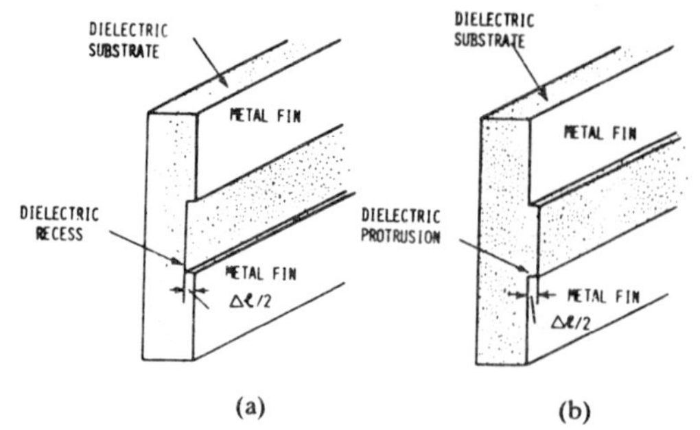

Fig. 14. Misalignment of conducting boundaries and dielectric interfaces in the three-dimensional TLM simulation of planar structures (after Shih [26]).

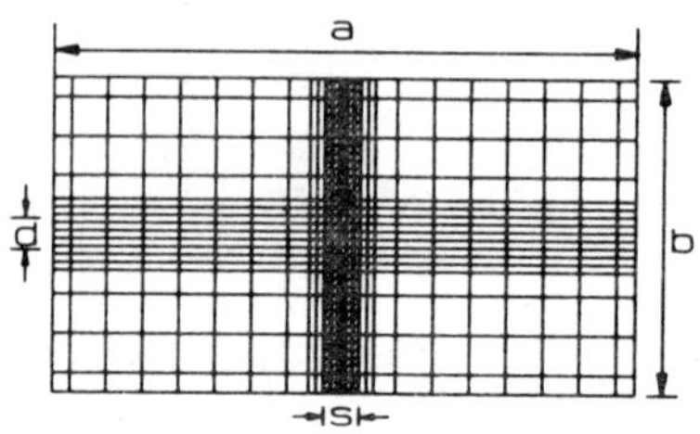

Fig. 15. Two-dimensional TLM network with variable mesh size for the computation of cutoff frequencies of finlines (after Saguet and Pic [28]).

problems. A better response is to introduce a network of variable mesh size to provide higher resolution in the nonuniform field region [27]–[29]. This approach is described in the next section; however, it requires more complicated programming. Yet another approach, proposed by Shih and Hoefer [25], is to compute the structure response several times using coarse meshes of different mesh parameter Δl, and then to extrapolate the obtained results to $\Delta l = 0$ as shown in Fig. 13.

Both measures effectively reduce the error by one order of magnitude and simultaneously correct the velocity error.

D. *Misalignment of Dielectric Interfaces and Boundaries in Three-Dimensional Inhomogeneous Structures*

Due to the particular way in which boundaries are simulated in a three-dimensional TLM network, dielectric interfaces appear halfway between nodes, while electric and magnetic boundaries appear across such nodes. This can be a problem when simulating planar structures such as microstrip or finline. In the TLM model, the dielectric either protrudes or is undercut by $\Delta l/2$, as shown in Fig. 14. Unless the resulting error is acceptable, one must make two computations, one with recessed and one with protruding dielectric, and take the average of the results. The problem does not occur in a variation of the three-dimensional TLM method involving an alternative node configuration proposed by Saguet and Pic [31], and described in the next section.

VII. Variations of the TLM Method

A number of modifications of the conventional TLM method have been proposed over the last few years with the aim of reducing errors, memory requirements, and CPU time. Some of them have already been mentioned, such as the introduction of a Hanning window [20], and extrapolation from coarse mesh calculations [21], [25]. Some effort has also been directed towards improving the efficiency of programing techniques [24].

In the following, three other interesting and significant innovations will be discussed briefly.

A. *TLM Networks with Nonuniform Mesh*

In order to ensure synchronism, the conventional TLM network uses a uniform mesh parameter throughout. This can lead to considerable numerical expenditure if the structure contains sharp corners or fins producing highly nonuniform fields and thus demands a high density mesh. Saguet and Pic [28] and Al-Mukhtar and Sitch [29] have independently proposed ways to implement irregularly graded TLM meshes which, as in the finite-element method, allow the network to adapt its density to the local nonuniformity of the fields. Fig. 15 shows such a network as proposed by Saguet and Pic [28] for the computation of cutoff frequencies in a finline. Note, however, that the size of the mesh cells is not arbitrary as in the case of finite elements; the length of each side is an odd integer multiple P of the smallest cell length in the network. To keep the

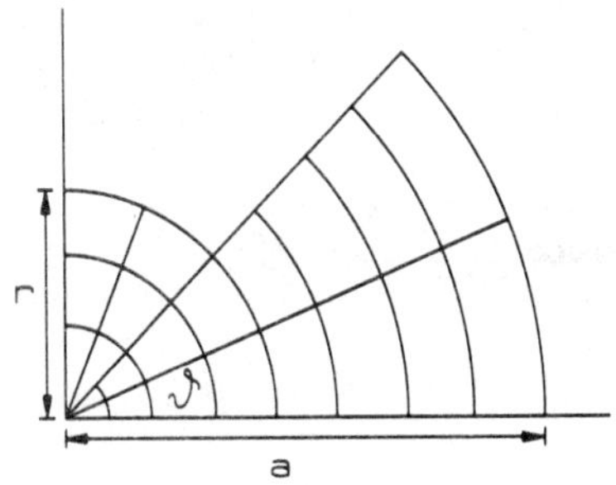

Fig. 16. A radial TLM mesh for the treatment of circular ridged waveguides (after Al-Mukhtar and Sitch [29]).

velocity of traveling impulses the same in all branches, the inductivity per unit length of the longer mesh lines is increased by a factor P while their capacity per unit length is reduced by $1/P$. This, in turn, increases their characteristic impedance by a factor P, and the scattering matrix of nodes connecting cells of different size must be modified accordingly.

To preserve synchronism, impulses traveling on longer branches are kept in store for P iterations before being reinjected at the next node.

For the configuration shown in Fig. 15, Saguet and Pic found that computing time was reduced between 3.5 and 5 times over a uniform mesh, depending on the relative size of the larger cells.

A different approach has been proposed by Al-Mukhtar and Sitch [27], [29]. They describe two possible ways to modify the characteristics of mesh elements in order to ensure synchronism, one involving the insertion of series stubs between nodes and loading of nodes by shunt stubs, the other involving modification of inductivity and capacity per unit length in such a way that propagation velocity in a branch becomes proportional to its length. The work by Al-Mukhtar and Sitch also covers the representation of radial meshes (see Fig. 16) as well as three-dimensional inhomogeneous structures. They report an economy of 45 percent in computer expenditure for a two-dimensional ridged waveguide problem, and a 40-percent reduction in storage and an 80-percent reduction in run time for a three-dimensional finline problem thanks to mesh grading.

B. A Punctual Node for Three-Dimensional TLM Networks

Conventional three-dimensional TLM networks require three shunt and three series nodes for the representation of one single cell (see Fig. 10). Saguet and Pic [31] have proposed an alternative method of interconnection. Representation of the short transmission-line sections by two rather than three lumped elements (see Fig. 17) makes it possible to realize both shunt and series connections in one point, resulting in a punctual node with 12 branches. This node is equivalent to a cell, such as that in Fig. 10, in which the inner connections have been eliminated. Losses and dielectric or magnetic loading can be simulated with stubs in the same way as discussed earlier.

This new node representation reduces, according to Saguet and Pic [31], the computation time by about 30

Fig. 17. Alternative lumped-element network for the two-dimensional shunt node. (a) Classical representation. (b) Representation proposed by Saguet and Pic [31].

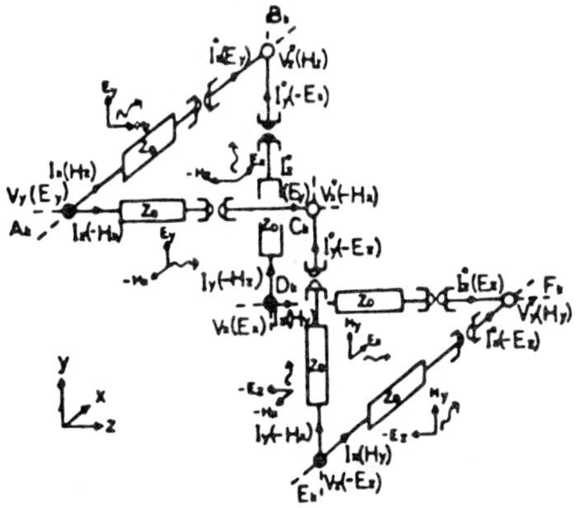

(a)

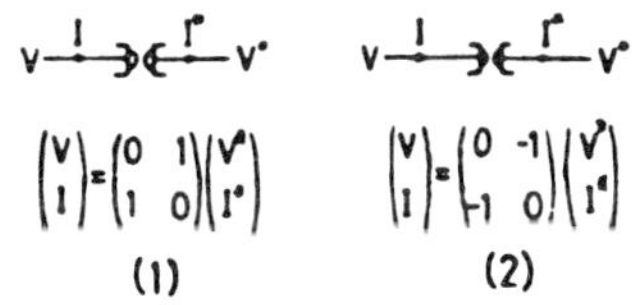

(b)

Fig. 18. Alternative network for three-dimensional TLM analysis proposed by Yoshida *et al.* [33]. (a) Equivalent circuit of an alternative three-dimensional TLM cell. (b) Definition of gyrators in (a): 1) positive gyrator, 2) negative gyrator.

percent. By employing both the punctual node and variable mesh size in a three-dimensional program, Saguet [32] has computed the resonant frequencies of a finline cavity 35 times faster than with a program based on the traditional TLM method.

C. Alternative Network Simulating Maxwell's Equations

Yoshida, Fukai, and Fukuoka [19], [23], [30], [33] have described a network similar to the TLM mesh, differing only in the way the basic cell element has been modeled. Instead of series and shunt nodes, this network contains so-called electric and magnetic nodes which are both "shunt-type nodes": while at the electric node, the voltage variable represents an electric field, it symbolizes a magnetic field at the magnetic node. The resulting ambivalence in the nature of the network voltage and current must be removed by inserting gyrators between the two types of nodes, as shown in Fig. 18. The wave properties of this network are identical with that of the conventional TLM

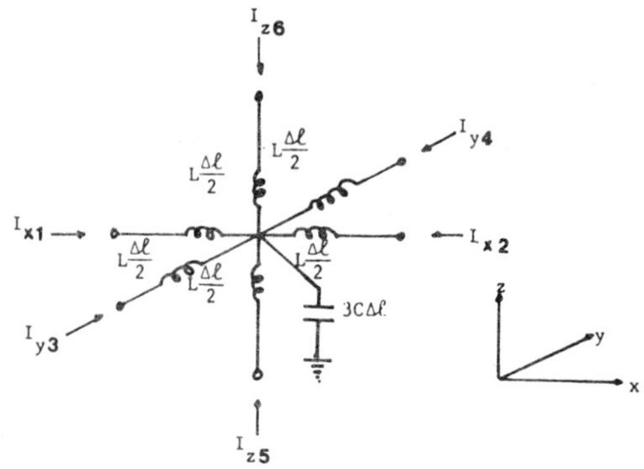

Fig. 19. Basic node of a three-dimensional scalar TLM network (after Choi and Hoefer [34]).

mesh. Errors and limitations are the same, and so are the possibilities of introducing losses and isotropic as well as anisotropic dielectric and magnetic materials.

D. *The Scalar TLM Method*

In those cases where electromagnetic fields can be decomposed into TE and TM modes (or LSE and LSM modes), it is only necessary to solve the scalar wave equation. Choi and Hoefer [34] have described a scalar TLM network to simulate a single field component or a Hertzian potential in three-dimensional space. The scalar TLM mesh can be thought of as a two-dimensional network to which additional transmission lines are connected orthogonally at each node as shown in Fig. 19. Such a structure could be realized in the form of a three-dimensional grid of coaxial lines.

The voltage impulses traveling across such a network represent the scalar variable to be simulated. Boundary reflection coefficients depend on both the nature of the boundary and that of the quantity to be simulated. For example, impulses will be subject to a reflection coefficient of -1 at a lossless electric wall if they represent either a tangential electric- or a normal magnetic-field component. A normal electric or a tangential magnetic field will be reflected with a coefficient of $+1$ in the same circumstances.

The slow-wave velocity in the three-dimensional scalar mesh is $c/\sqrt{3}$ as opposed to $c/2$ in the conventional TLM network. Dielectric or magnetic material as well as losses may be simulated using reactive and dissipative stubs.

The scalar method requires only 1/4 of the memory space and is seven times faster than the conventional method for a commensurate problem. However, its application is severely restricted, as it can be applied to scalar wave problems only.

VIII. Applications of the TLM Method

In the previous sections, the flexibility, versatility, and generality of the transmission-line matrix method has been demonstrated. In the following, an overview of potential applications of the method is given, and references describing specific applications are indicated. This list is not exhaustive, and many more applications can be found, not only in electromagnetism, but also in other fields dealing with wave phenomena, such as optics and acoustics.

For completeness, it should be mentioned that the TLM procedure can also be used to model and solve linear and nonlinear lumped networks [35]–[37] and diffusion problems [38]. Readers with a special interest in these applications should consult these references for more details.

Wave problems can be simulated in unbounded and bounded space, either in the time domain or—via Fourier analysis—in the frequency domain. Arbitrary homogeneous or inhomogeneous structures with anisotropic, space- and time-dependent electrical properties, including losses, can be simulated in two and three dimensions.

Below are some typical application examples.

A. *Two-Dimensional Scattering Problems in Rectangular Waveguides (Field Distribution of Propagating and Evanescent Modes, Wave Impedance, Scattering Parameters)*

- Open-circuited rectangular waveguide (TE_{n0}) [5].
- Bifurcation in rectangular waveguide (TE_{n0}) [5].
- Scattering by arbitrarily-shaped two-dimensional discontinuities in rectangular waveguide (TE_{n0}) including losses.

B. *Two-Dimensional Eigenvalue Problems (Eigenfrequencies, Mode Fields)*

- Cutoff frequencies and mode fields in homogeneous waveguides of arbitrary cross section, such as ridged waveguides [6], [8], [13], [26], [27].
- Cutoff frequencies and mode fields in inhomogeneous wavegudies of arbitrary cross section, such as dielectric loaded waveguides, finlines, image lines [7], [13], [16], [18], [21], [22], [25], [26], [28].

C. *Three-Dimensional Eigenvalue and Hybrid Field Problems (Dispersion Characteristics, Wave Impedances, Losses, Eigenfrequencies, Mode Fields, Q-Factors)*

- Characteristics of dielectric loaded cavities [10], [13], [14], [15], [18], [26], [27], [29], [32], [34].
- Dispersion characteristics and scattering in inhomogeneous planar transmission-line structures, including anisotropic substrate [11], [12], [13], [14], [17], [26], [27], [29], [30], [32].
- Transient analysis of transmission-line structures [19], [23], [33].

General-purpose two-dimensional and three-dimensional TLM programs can be found in Akhtarzad's Ph.D. thesis [13]. They can be adapted to most of the applications described above. If the various improvements and modifications described in Section VII are implemented in these programs, versatile and powerful numerical tools for the solution of complicated field problems are indeed obtained.

IX. Discussion and Conclusion

This paper has described the physical principle, the formulation, and the implementation of the transmission-line matrix method of analysis. Numerous features and applications of the method have been discussed, in particular the principal sources of error and their correction, the inclusion of losses, inhomogeneous and anisotropic properties of materials, and the capability to analyze transient as well as steady-state wave phenomena.

A general-purpose two-dimensional TLM program can be written in about 80 lines of FORTRAN, while a three-dimensional program is about 110 lines long. [13].

The method is limited only by the amount of memory storage required, which depends on the complexity of the structure and the nonuniformity of fields set up in it. In general, the smallest feature in the structure should at least contain three nodes for good resolution. The total storage requirement for a given computation can be determined by considering that each two-dimensional node requires five real number storage places, and an additional number equal to the number of iterations is needed to store the output impulse function. A basic three-dimensional node requires twelve number locations; if it is completely equipped with permittivity, permeability, and loss stubs, the required number of stores goes up to 26. Again, one real number must be stored per output function and per iteration. The number of iterations required varies between several hundred and several thousand, depending on the size and complexity of the TLM mesh.

As far as computational expenditure is concerned, the TLM method compares favorably with finite-element and finite-difference methods. Its accuracy is even slightly better by virtue of the Fourier transform, which ensures that the field function between nodes is automatically circular rather than linear as in the two other methods.

The main advantage of the TLM method, however, is the ease with which even the most complicated structures can be analyzed. The great flexibility and versatility of the method reside in the fact that the TLM network incorporates the properties of the electromagnetic fields and their interaction with the boundaries and materials. Hence, the electromagnetic problem need not to be reformulated for every new structure; its parameters are simply entered into a general-purpose program in the form of codes for boundaries, losses, permeability and permittivity, and excitation of the fields. Furthermore, by solving the problem through simulation of wave propagation in the time domain, the solution of large numbers of simultaneous equations is avoided. There are no problems with convergence, stability, or spurious solutions.

Another advantage of the TLM method resides in the large amount of information generated in one single computation. Not only is the impulse response of a structure obtained, yielding, in turn, its response to any excitation, but also the characteristics of the dominant and higher order modes are accessible in the frequency domain through the Fourier transform.

In order to increase the numerical efficiency and reduce the various errors associated with the method, more programing effort must be invested. Such an effort may be worthwhile when faced with the problem of scattering by a three-dimensional discontinuity in an inhomogeneous transmission medium, or when studying the overall electromagnetic properties of a monolithic circuit.

Finally, the TLM method may be adapted to problems in other areas such as thermodynamics, optics, and acoustics. Not only is it a very powerful and versatile numerical tool, but because of its affinity with the mechanism of wave propagation, it can provide new insights into the physical nature and the behavior of electromagnetic waves.

References

[1] C. Huygens, "Traité de la Lumière" (Leiden, 1690).

[2] J. R. Whinnery and S. Ramo, "A new approach to the solution of high-frequency field problems," *Proc. IRE*, vol. 32, pp. 284–288, May 1944.

[3] G. Kron, "Equivalent circuit of the field equations of Maxwell—I," *Proc. IRE*, vol. 32, pp. 289–299, May 1944.

[4] J. R. Whinnery, C. Concordia, W. Ridgway, and G. Kron, "Network analyzer studies of electromagnetic cavity resonators," *Proc. IRE*, vol. 32, pp. 360–367, June 1944.

[5] P. B. Johns and R. L. Beurle, "Numerical solution of 2-dimensional scattering problems using a transmission-line matrix," *Proc. Inst. Elec. Eng.*, vol. 118, no. 9, pp. 1203–1208, Sept. 1971.

[6] P. B. Johns, "Application of the transmission-line matrix method to homogeneous waveguides of arbitrary cross-section," *Proc. Inst. Elec. Eng.*, vol. 119, no. 8, pp. 1086–1091, Aug. 1972.

[7] P. B. Johns, "The solution of inhomogeneous waveguide problems using a transmission-line matrix," *IEEE Trans. Microwave Theory Tech.*, vol. MTT-22, pp. 209–215, Mar. 1974.

[8] S. Akhtarzad and P. B. Johns, "Numerical solution of lossy waveguides: T.L.M. computer program," *Electron. Lett.*, vol. 10, no. 15, pp. 309–311, July 25, 1974.

[9] P. B. Johns, "A new mathematical model to describe the physics of propagation," *Radio Electron. Eng.*, vol. 44, no. 12, pp. 657–666, Dec. 1974.

[10] S. Akhtarzad and P. B. Johns, "Solution of 6-component electromagnetic fields in three space dimensions and time by the T.L.M. method," *Electron. Lett.*, vol. 10, no. 25/26, pp. 535–537, Dec. 12, 1974.

[11] S. Akhtarzad and P. B. Johns, "T.L.M. analysis of the dispersion characteristics of microstrip lines on magnetic substrates using 3-dimensional resonators," *Electron. Lett.*, vol. 11, no. 6, pp. 130–131, Mar. 20, 1975.

[12] S. Akhtarzad and P. B. Johns, "Dispersion characteristic of a microstrip line with a step discontinuity," *Electron. Lett.*, vol. 11, no. 14, pp. 310–311, July 10, 1975.

[13] S. Akhtarzad, "Analysis of lossy microwave structures and microstrip resonators by the TLM method," Ph.D dissertation, Univ. of Nottingham, England, July 1975.

[14] S. Akhtarzad and P.B. Johns, "Three-dimensional transmission-line matrix computer analysis of microstrip resonators," *IEEE Trans. Microwave Theory Tech.*, vol. MTT-23, pp. 990–997, Dec. 1975.

[15] S. Akhtarzad and P. B. Johns, "Solution of Maxwell's equations in three space dimensions and time by the T.L.M. method of analysis," *Proc. Inst. Elec. Eng.*, vol. 122, no. 12, pp. 1344–1348, Dec. 1975.

[16] S. Akhtarzad and P. B. Johns, "Generalized elements for T.L.M. method of numerical analysis," *Proc. Inst. Elec. Eng.*, vol. 122, no. 12, pp. 1349–1352, Dec. 1975.

[17] G. E. Mariki, "Analysis of microstrip lines on inhomogeneous anisotropic substrates by the TLM numerical technique," Ph.D. thesis, Univ. of California, Los Angeles, June 1978.

[18] W. J. R. Hoefer and A. Ros, "Fin line parameters calculated with the TLM method," in *IEEE MTT Int. Microwave Symp. Dig.* (Orlando, FL), Apr. 28–May 2, 1979.

[19] N. Yoshida, I. Fukai, and J. Fukuoka, "Transient analysis of two-dimensional Maxwell's equations by Bergeron's method," *Trans. IECE Japan*, vol. J62B, pp. 511–518, June 1979.

[20] P. Saguet and E. Pic, "An improvement for the TLM method," *Electron. Lett.*, vol. 16, no. 7, pp. 247–248, Mar. 27, 1980.

[21] Y.-C. Shih, W. J. R. Hoefer, and A. Ros, "Cutoff frequencies in fin lines calculated with a two-dimensional TLM-program," in *IEEE MTT Int. Microwave Symp. Dig.* (Washington, DC), June 1980, pp. 261–263.

[22] W. J. R. Hoefer and Y.-C. Shih, "Field configuration of fundamental and higher order modes in fin lines obtained with the TLM method," presented at URSI and Int. IEEE-AP Symp., Quebec, Canada, June 2–6, 1980.

[23] N. Yoshida, I. Fukai, and J. Fukuoka, "Transient analysis of three-dimensional electromagnetic fields by nodal equations," *Trans. IECE Japan*, vol. J63B, pp. 876–883, Sept. 1980.

[24] A. Ros, Y.-C. Shih, and W. J. R. Hoefer, "Application of an accelerated TLM method to microwave systems," in *10th Eur. Microwave Conf. Dig.* (Warszawa, Poland), Sept. 8–11, 1980, pp. 382–388.

[25] Y.-C. Shih and W. J. R. Hoefer, "Dominant and second-order mode cutoff frequencies in fin lines calculated with a two-dimensional TLM program," *IEEE Trans. Microwave Theory Tech.*, vol. MTT-28, pp. 1443–1448, Dec. 1980.

[26] Y.-C. Shih, "The analysis of fin lines using transmission line matrix and transverse resonance methods," M.A.Sc. thesis, Univ. of Ottawa, Canada, 1980.

[27] D. Al-Mukhtar, "A transmission line matrix with irregularly graded space," Ph.D. thesis, Univ. of Sheffield, England, Aug. 1980.

[28] P. Saguet and E. Pic, "Le maillage rectangulaire et le changement de maille dans la méthode TLM en deux dimensions," *Electron. Lett.*, vol. 17, no. 7, pp. 277–278, Apr. 2, 1981.

[29] D.A. Al-Mukhtar and J. E. Sitch, "Transmission-line matrix method with irregularly graded space," *Proc. Inst. Elec. Eng.*, vol. 128, pt. H, no. 6, pp. 299–305, Dec. 1981.

[30] N. Yoshida, I. Fukai, and J. Fukuoka, "Application of Bergeron's method to anisotropic media," *Trans. IECE Japan*, vol J64B, pp. 1242–1249, Nov. 1981.

[31] P. Saguet and E. Pic, "Utilisation d'un nouveau type de noeud dans la méthode TLM en 3 dimensions," *Electron. Lett.*, vol. 18, no. 11, pp. 478–480, May 1982.

[32] P. Saguet, "Le maillage parallelepipédique et le changement de maille dans la méthode TLM en trois dimensions," *Electron. Lett.*, vol. 20, no. 5, pp. 222–224, Mar. 15, 1984.

[33] N. Yoshida and I. Fukai, "Transient analysis of a stripline having a corner in three-dimensional space," *IEEE Trans. Microwave Theory Tech.*, vol. MTT-32, pp. 491–498, May 1984.

[34] D. H. Choi and W. J. R. Hoefer, "The simulation of three-dimensional wave propagation by a scalar TLM model," in *IEEE MTT Int. Microwave Symp. Dig.* (San Francisco), May 1984.

[35] P. B. Johns, "Numerical modelling by the TLM method," in *Large Engineering Systems*, A. Wexler, Ed. Oxford: Pergamon Press, 1977.

[36] J. W. Bandler, P. B. Johns, and M. R. M. Rizk, "Transmission-line modeling and sensitivity evaluation for lumped network simulation and design in the time domain," *J. Franklin Inst.*, vol. 304, no. 1, pp. 15–23, 1977.

[37] P. B. Johns and M. O'Brien, "Use of the transmission-line modelling (T.L.M.) method to solve non-linear lumped networks," *Radio Electron. Eng.*, vol. 50, no. 1/2, pp. 59–70, Jan./Feb. 1980.

[38] P. B. Johns, "A simple explicit and unconditionally stable numerical routine for the solution of the diffusion equation," *Int. J. Num. Meth. Eng.*, vol. 11, pp. 1307–1328, 1977.

2.1.8

Analysis of Finline with Finite Metallization Thickness

TOSHIHIDE KITAZAWA AND RAJ MITTRA, FELLOW, IEEE

Abstract —In this paper, we present a method for analyzing finline structures with finite metallization thickness. The method is based on a hybrid mode formulation but it by-passes the lengthy process of formulating the determinantal equation for the unknown propagation constant. Some numerical results are presented to show the effect of the metallization thickness for unilateral and bilateral finlines.

Manuscript received October 6, 1983; revised June 11, 1984. This work was supported in part by DAAG29-82-K-0084 and JSEP N000-1479C-0424.

The authors are with the Electrical Engineering Department, University of Illinois, Urbana, IL. T. Kitazawa is on leave from the Kitami Institute of Technology, Kitami, Japan.

I. Introduction

Finline structures have received considerable attention because of their usefulness as millimeter-wave integrated-circuit components. Recently, two efficient numerical methods for analyzing the propagation characteristics of finline structures were presented. The first of these employs the spectral-domain technique [1], [2], whereas the second utilizes network analytical methods for electromagnetic fields [3]. Both of these methods are based on the hybrid mode formulation, but they neglect the effect of the metallization thickness, which increases with higher operating frequencies and narrower gaps in the metallization. An eigenvalue equation for a unilateral finline with a finite metallization thickness has been previously derived [4] using the hybrid mode formulation, but only approximate results based on the TE-approximation have been presented in the above paper. In this paper, we discuss an efficient hybrid mode formulation for the finite metallization problem and derive the solution to the problem without resorting to the TE-approximation. Although the method is an extension of the treatment in [3], [5], and [6], it derives Green's functions using the conventional circuit theory rather than by directly solving the differential equations with boundary conditions.

II. The Network Formulation of the Problem

The unilateral finline shown in Fig. 1 is used to illustrate the formulation procedure, but the method itself is quite general.

As a first step, we express the transverse (to z) fields in each region by the following spectral representation:

$$\left.\begin{matrix}\bar{E}_t^{(i)}(x,y,z)\\ \bar{H}_t^{(i)}(x,y,z)\end{matrix}\right\} = \sum_{l=1}^{2}\sum_{n=0}^{\infty}\left\{\begin{matrix}V_{ln}^{(i)}(z)\bar{f}_{ln}^{(i)}(x)\\ I_{ln}^{(i)}(z)\bar{g}_{ln}^{(i)}(x)\end{matrix}\right\} e^{-j\beta_0 y},\quad i=1,2,3,4 \tag{1}$$

where the vector mode functions $\bar{f}_{ln}^{(i)}$ and $\bar{g}_{ln}^{(i)}$ in each region are given as

A) region (1), (3), and (4)

$$\begin{aligned}
\bar{f}_{1n}^{(i)} &= \frac{-1}{K_A}\sqrt{\frac{\eta_n}{2A}}\{\bar{x}_0\alpha_A\cos(\alpha_A x) - \bar{y}_0 j\beta_0\sin(\alpha_A x)\}\\
\bar{f}_{2n}^{(i)} &= \frac{1}{K_A}\sqrt{\frac{\eta_n}{2A}}\{\bar{x}_0 j\beta_0\cos(\alpha_A x) - \bar{y}_0\alpha_A\sin(\alpha_A x)\}\\
\bar{g}_{ln}^{(i)} &= \bar{z}_0\times\bar{f}_{ln}^{(i)}\qquad (l=1,2)\\
\alpha_A &= \frac{n\pi}{A},\quad K_A = \sqrt{\alpha_A^2+\beta_0^2}\\
\eta_n &= \begin{cases}1\ (n=0)\\ 2\ (n\neq 0)\end{cases}.
\end{aligned} \tag{2}$$

B) region (2)

$$\begin{aligned}
\bar{f}_{1n}^{(2)} &= \frac{-1}{K_W}\sqrt{\frac{\eta_n}{2W}}\{\bar{x}_0\alpha_W\cos(\alpha_W x) - \bar{y}_0 j\beta_0\sin(\alpha_W x)\}\\
\bar{f}_{2n}^{(2)} &= \frac{1}{K_W}\sqrt{\frac{\eta_n}{2W}}\{\bar{x}_0 j\beta_0\cos(\alpha_W x) - \bar{y}_0\alpha_W\sin(\alpha_W x)\}\\
\bar{g}_{ln}^{(2)} &= \bar{z}_0\times\bar{f}_{ln}^{(2)}\qquad (l=1,2),\ \alpha_W = \frac{n\pi}{W},\ K_W = \sqrt{\alpha_W^2+\beta_0^2}
\end{aligned} \tag{3}$$

where β_0 is the propagation constant, and $\bar{x}_0$, $\bar{y}_0$, and $\bar{z}_0$ are the x-, y-, and z-directed unit vectors, respectively. It should be noted that the vector mode functions $\bar{f}_{ln}^{(i)}, \bar{g}_{ln}^{(i)}$ satisfy the

T. Kitazawa and R. Mittra, *IEEE Transactions on Microwave Theory and Techniques*, Vol. MTT-32, No. 11, Nov. 1984, pp. 1484–1487.

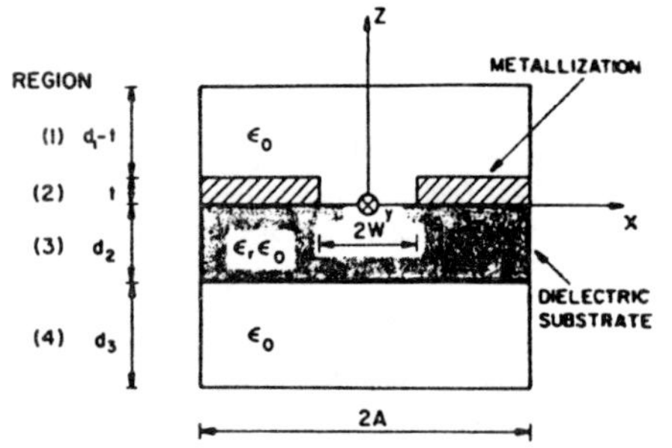

Fig. 1. Unilateral finline.

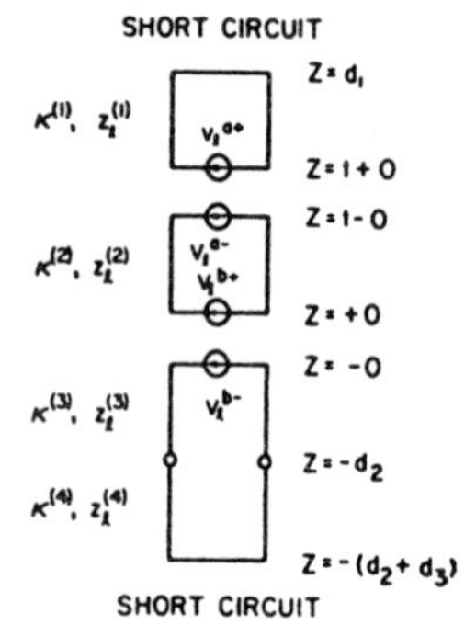

Fig. 2. Equivalent circuits for transverse section of finline.

boundary conditions at $x = \pm W$ and $\pm A$ and the orthonormal properties.

Substituting (1) into Maxwell's field equations, we obtain the differential equations for the modal voltages and currents

$$-\frac{d}{dz}V_{ln}^{(i)} = j\kappa^{(i)} z_l^{(i)} I_{ln}^{(i)}$$

$$-\frac{d}{dz}I_{ln}^{(i)} = j\kappa^{(i)} y_l^{(i)} V_{ln}^{(i)} \tag{4}$$

where

$$\kappa^{(i)} = \sqrt{\omega^2\epsilon^{(i)}\mu_0 - K^{(i)2}}$$

$$\epsilon^{(i)} = \begin{cases} \epsilon_r\epsilon_0 & \text{(region (3))} \\ \epsilon_0 & \text{(otherwise)} \end{cases} \qquad K^{(i)} = \begin{cases} K_W & \text{(region (2))} \\ K_A & \text{(otherwise)} \end{cases}$$

$$z_1^{(i)} = \frac{\kappa^{(i)}}{\omega\epsilon^{(i)}}, \qquad z_2^{(i)} = \frac{\omega\mu_0}{\kappa^{(i)}}$$

$$y_l^{(i)} = \frac{1}{z_l^{(i)}} \qquad (l = 1,2). \tag{5}$$

Equivalent circuits in the z-direction can be derived by considering the differential equations (4) together with the associated boundary conditions (Fig. 2). The equivalent voltage sources $\mathscr{V}_l^{a\pm}, \mathscr{V}_l^{b\pm}$ shown in this figure are given by

$$\left.\begin{matrix} \mathscr{V}_l^{a+} \\ \mathscr{V}_l^{a-} \end{matrix}\right\} = \int_{-W}^{W} \begin{Bmatrix} \bar{f}_{ln}^{(1)*}(x') \\ \bar{f}_{ln}^{(2)*}(x') \end{Bmatrix} \cdot \bar{e}_a(x')\,dx' \tag{6a}$$

$$\left.\begin{matrix} \mathscr{V}_l^{b+} \\ \mathscr{V}_l^{b-} \end{matrix}\right\} = \int_{-W}^{W} \begin{Bmatrix} \bar{f}_{ln}^{(2)*}(x') \\ \bar{f}_{ln}^{(3)*}(x') \end{Bmatrix} \cdot \bar{e}_b(x')\,dx' \tag{6b}$$

and where $\bar{e}_a$ and $\bar{e}_b$ are the transverse electric fields at $z = t$ and $z = 0$, respectively

$$\bar{e}_{\substack{a\\b}} = \bar{x}_0 e_{\substack{ax\\bx}} + \bar{y}_0 e_{\substack{ay\\by}}. \tag{7}$$

The modal voltages $V_{ln}^{(i)}$ and currents $I_{ln}^{(i)}$ in each region can be obtained by using conventional circuit theory. Electromagnetic fields in each region can then be derived by substituting $V_{ln}^{(i)}$ and $I_{ln}^{(i)}$ into (1). Finally, the application of the continuities of magnetic fields results in the following set of equations for the unknown electric fields $\bar{e}_a$ and $\bar{e}_b$ at $z = t$ and $z = 0$, and for the unknown propagation constant β_0:

$$\sum_{l=1}^{2}\sum_{n=0}^{\infty}\int_{-W}^{W} Y_l^{(1)}(t+0|t)\,\bar{g}_{ln}^{(1)}(x)\,\bar{f}_{ln}^{(1)*}(x')\cdot\bar{e}_a(x')\,dx'$$

$$= \sum_{l=1}^{2}\sum_{n=0}^{\infty}\int_{-W}^{W}\left\{Y_l^{(2)}(t-0|t)\,\bar{g}_{ln}^{(2)}(x)\,\bar{f}_{ln}^{(2)*}(x')\cdot\bar{e}_a(x') + Y_l^{(2)}(t-0|0)\,\bar{g}_{ln}^{(2)}(x)\,\bar{f}_{ln}^{(2)*}(x')\cdot\bar{e}_b(x')\right\}dx' \tag{8a}$$

$$\sum_{l=1}^{2}\sum_{n=0}^{\infty}\int_{-W}^{W}\left\{Y_l^{(2)}(+0|t)\,\bar{g}_{ln}^{(2)}(x)\,\bar{f}_{ln}^{(2)*}(x')\cdot\bar{e}_a(x') + Y_l^{(2)}(+0|0)\,\bar{g}_{ln}^{(2)}(x)\,\bar{f}_{ln}^{(2)*}(x')\cdot\bar{e}_b(x')\right\}dx'$$

$$= \sum_{l=1}^{2}\sum_{n=0}^{\infty}\int_{-W}^{W} Y_l^{(3)}(-0|0)\,\bar{g}_{ln}^{(3)}(x)\,\bar{f}_{ln}^{(3)*}(x')\cdot\bar{e}_b(x')\,dx' \tag{8b}$$

where $Y_l^{(i)}(z|z')$ are the Green's functions which relate the modal currents $I_l^{(i)}(z)$ to the voltage source in each of the three regions. The set of equations (8) is rigorous. Numerical solution of the equations is discussed in the next section.

The equivalent circuit concept, which is used to derive the Green's functions, is similar to that introduced in [2], but is extended here to treat the geometry with a finite metallization thickness. We remark here that the method itself is quite general and can be applied equally well to other finline structures. For example, Fig. 3(b) shows the equivalent circuits for the bilateral finline whose geometry is given in Fig. 3(a).

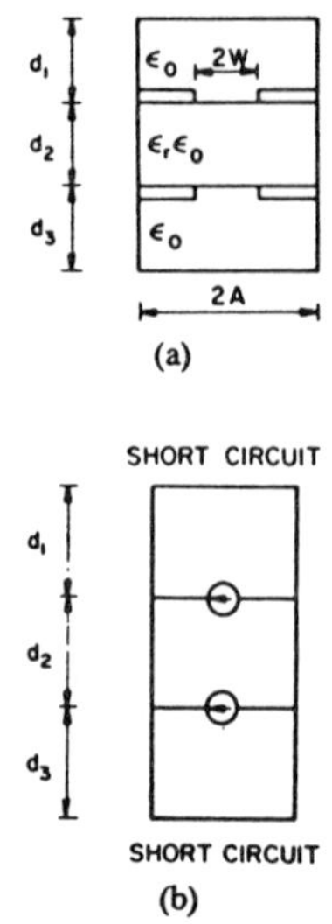

Fig. 3. Bilateral finline (metallization thickness is neglected in this figure).

III. Numerical Computations

The numerical procedure for solving (8) is analogous to that used in [3], [5], and [6]; therefore, only a summary of the steps will be given below.

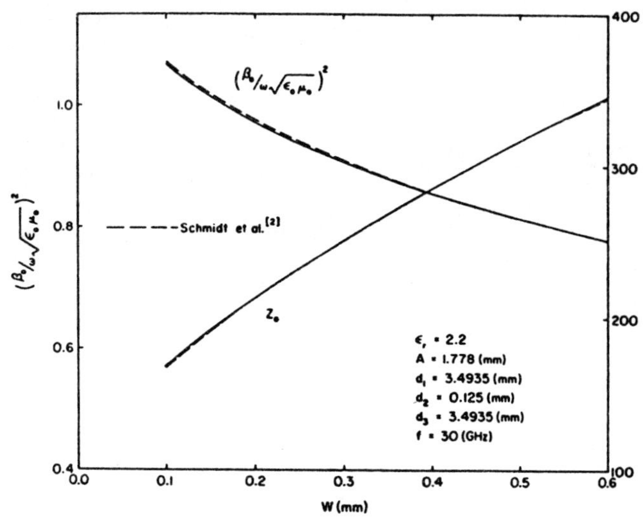

Fig. 4. The comparison of this method with other method (the unilateral finline with zero metallization thickness).

The first step is to expand the unknown electric fields $\bar{e}_a$ and $\bar{e}_b$ at $z = t$ and 0, in terms of an appropriate set of basis

$$\left.\begin{matrix} e_{xa}(x) \\ e_{xb}(x) \end{matrix}\right\} = \sum_{k=1}^{2N_x} \left\{\begin{matrix} a_{xk} \\ b_{xk} \end{matrix}\right\} f_{xk}(x)$$

$$\left.\begin{matrix} e_{ya}(x) \\ e_{yb}(x) \end{matrix}\right\} = j\sum_{k=1}^{2N_y} \left\{\begin{matrix} a_{yk} \\ b_{yk} \end{matrix}\right\} f_{yk}(x) \tag{9}$$

where $a_{\substack{xk\\y}}$ and $b_{\substack{xk\\y}}$ are the unknown coefficients. The second step is to apply the Galerkin's procedure to (8), which results in the determinantal equation for the propagation constant β_0. Finally, the determinantal equation is solved for β_0 in the finline structure. Accurate solutions can be obtained with only a small number of basis functions, if these functions incorporate the edge effect. The following basis functions were used for the numerical results presented in this paper:

$$f_{x1}(x) = 1,$$

$$f_{xk}(x) = \frac{T_{k-2}\left(\frac{x}{W}\right)}{\sqrt{1-\left(\frac{x}{W}\right)^2}}, \qquad k = 2, 3, \cdots$$

$$f_{yk}(x) = U_k\left(\frac{x}{W}\right), \qquad k = 1, 2, 3, \cdots \tag{10}$$

where $T_k(y)$ and $U_k(y)$ are the Chebyshev's polynomials of the first and second kind, respectively.

Preliminary computations show that $N_x = N_y = 2$ in (9) is sufficient for deriving accurate results for finite metallization thickness, just as in the case of zero metallization thickness [3].

The numerical results obtained by this method are plotted in Fig. 4 together with the published data [4] for the unilateral finline with zero metallization thickness ($t = 0$) and close agreement between the two sets of results is seen for a wide range of slot widths. The effective dielectric constant and the characteristic impedance Z_0 are defined as

$$\left(\beta_0/\omega\sqrt{\epsilon_0\mu_0}\right)^2$$

$$Z_0 = \frac{V_0^2}{2P_{av}} \tag{11}$$

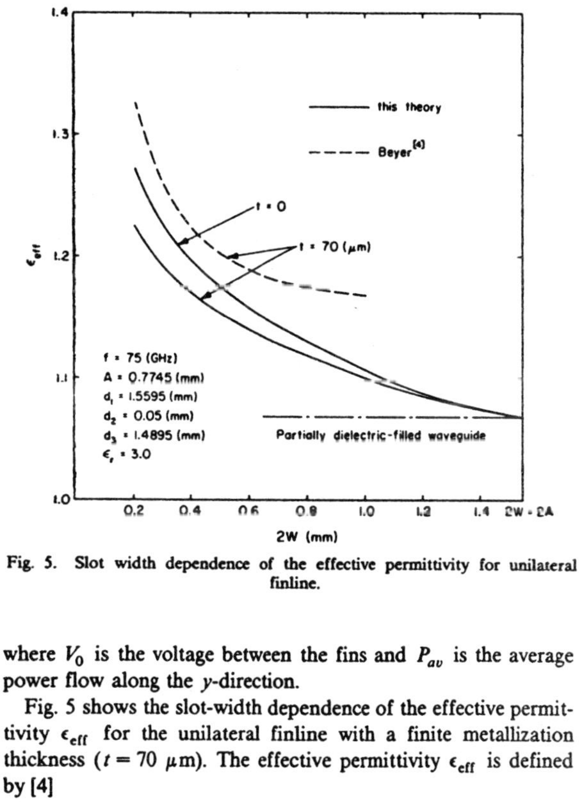

Fig. 5. Slot width dependence of the effective permittivity for unilateral finline.

where V_0 is the voltage between the fins and P_{av} is the average power flow along the y-direction.

Fig. 5 shows the slot-width dependence of the effective permittivity ϵ_{eff} for the unilateral finline with a finite metallization thickness ($t = 70\ \mu$m). The effective permittivity ϵ_{eff} is defined by [4]

$$\lambda_g = \lambda_0\left\{\epsilon_{eff} - \left(\frac{\lambda_0}{\lambda_c}\right)^2\right\}^{-1/2} \tag{12}$$

where λ_g is the wavelength in finline, λ_0 is the free-space wavelength, and λ_c is the cutoff wavelength of a ridged waveguide of identical dimensions. The values for the case with zero metallization thickness ($t = 0$) are presented for comparison. It is seen that the effect of the metallization thickness becomes smaller as the slot width becomes larger. We also note that the results for the limiting case of $W = A$ converge to those for the partially

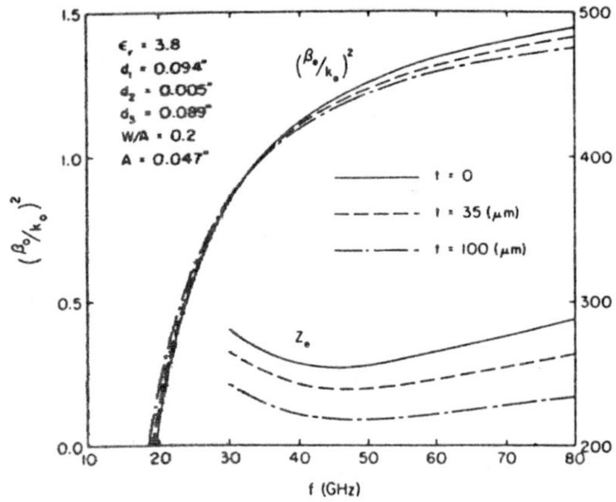

Fig. 6. Propagation characteristics of unilateral finline.

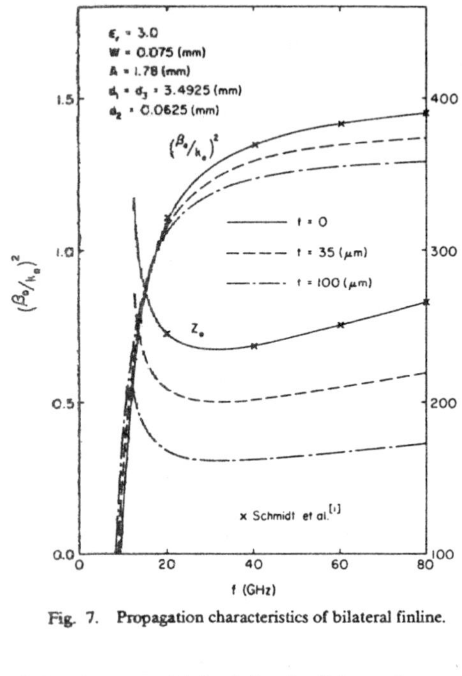

Fig. 7. Propagation characteristics of bilateral finline.

filled dielectric waveguide both for the finite and zero thickness of the metallization. However, when our results are compared with the published data by Beyer [4], some discrepancies are noted. The values obtained by Beyer are larger than those for the case of infinitely thin metallization ($t = 0$), and they do not appear to converge to a correct limiting value as W approaches A, as they should.

Fig. 6 shows the frequency dependence of the effect of the metallization thickness on the effective dielectric constant, and on the characteristic impedance in a unilateral finline. The finite thickness reduces the propagation constant in the higher frequency range as it does in the open slot line [5], because in these frequency ranges the fields are concentrated near the gap in the finline and it acts similar to an open slot line. In contrast, a finline behaves as a ridged waveguide near the cutoff frequency, and consequently, the thicker its diaphragm, the lower its cutoff frequency [7].

Fig. 7 shows the effect of the metallization thickness of a bilateral finline. The results for the limiting case of $t = 0$, i.e., zero metallization thickness, are compared with those published by Schmidt and Itoh [1], and the agreement is quite good.

IV. Conclusions

In this paper, the hybrid mode formulation was used to analyze finline structures with finite metallization thicknesses. This formulation used in conjunction with the equivalent circuit analysis is considerably simpler than the conventional Green's function approach. This method itself is quite general and can be applied to different types of finline structures by a simple modification of equivalent circuits which can be obtained easily.

Numerical results are presented to show the effects of finite metallization thickness on the propagation characteristics of unilateral and bilateral finline structures.

References

[1] L. P. Schmidt and T. Itoh, "Spectral domain analysis of dominant and higher order modes in fin-lines," *IEEE Trans. Microwave Theory Tech.*, vol. MTT-28, pp. 981–985, Sept. 1980.

[2] L. P. Schmidt, T. Itoh, and H. Hofman, "Characteristics of unilateral fin-line structures with arbitrary located slots," *IEEE Trans. Microwave Theory Tech.*, vol. MTT-29, pp. 352–355, Apr. 1981.

[3] Y. Hayashi, E. Farr, S. Wilson, and R. Mittra, "Analysis of dominant and higher-order modes in unilateral finlines," *Arch. Elek. Übertragung.*, vol. 37, no. 3/4, pp. 117–122, Mar.–Apr. 1983.

[4] A. Beyer, "Analysis of characteristics of an earthed finline," *IEEE Trans. Microwave Theory Tech.*, vol. MTT-29, pp. 676–680, July 1981.

[5] T. Kitazawa, Y. Hayashi, and M. Suzuki, "Analysis of the dispersion characteristic of slot line with thick metal coating," *IEEE Trans. Microwave Theory Tech.*, vol. MTT-28, pp. 387–392, Apr. 1980.

[6] T. Kitazawa, Y. Hayashi, and M. Suzuki, "A coplanar waveguide with thick metal coating," *IEEE Trans. Microwave Theory Tech.*, vol. MTT-24, pp. 604–608, Sept. 1976.

[7] J. R. Pyle, "The cutoff wavelength of the TE_{10} mode in ridged rectangular waveguide of any aspect ratio," *IEEE Trans. Microwave Theory Tech.*, vol. MTT-14, pp. 175–183, Apr. 1966.

2.1.9

EFFICIENT ANALYSIS OF FINLINE WITH FINITE METALLIZATION THICKNESS

Jerzy K. Piotrowski

Warsaw University of Technology,
Institute of Electron Technology,
Koszykowa 75, 00-662 Warsaw, Poland

ABSTRACT

The transverse resonance method is used for deriving dispersion characteristics of fundamental and higher order modes in finline with arbitrary slot widths and locations with regard to the finite thickness of metallization. In addition to the propagation constant, the characteristic impedance is calculated on the basis of the power-voltage definition.

INTRODUCTION

Finline structures have found frequent applications in millimeter-wave components. The problem of finline propagation characteristics has been treated by many exact and approximate methods. The most general, rigorous and systematic one is the mode matching technique, which has been used by Vahldieck (1) to show the influence of metallization thickness and mounting grooves on these characteristics. The analysis of finlines with finite mettalization thickness utilizing the network analytical method for electromagnetic fields has been presented in (2). The spectral domain technique for an idealized finline structures (zero metallization thickness and absence of mounting grooves) has been described, e.g., in (3)-(5). The idealized structure of this transmission line has also been analyzed in terms of the singular integral equation technique (6).

All the four rigorous methods are based on the hybrid mode formulation and they are complicated, computer time consuming and cumbersome. On the other hand, the need for the design of finline circuits in a lucid and tractable manner has caused the development of closed-form design equations like, e.g., in (7),(8). However, the design equations are appropriated for the idealized finline structure and accurate enough only for a definite range of its parameters. Similar disadvantades can be found in the analysis reported in (9) and based on the transverse resonance method (TRM) which was previously used for slotline by Cohn (10).

Until recently there have been no efficient analysis of finline with finite metallization thickness, and less complicated than the above mentioned rigorous mathods. The purpose of this paper is to present such a formulation of TRM, which should partly fill in this gap. The proposed method used for the lossless unilateral finline shown in Fig.1 will be presented here as an example illustrating the problem. Slot and dielectric layers are arbitrarily located in waveguide housing.

BASIS OF SOLUTION

The dispersion characteristics of modes in finline can be obtained by solving the boundary value problem of the half - wavelength finline resonator. Hence, the length of resonator in the x direction (Fig.1), equal to π/β_x (β_x-phase constant for fundamental or higher order mode in finline), is an unknown function of the resonant frequency f. The analysis is limited to such modes in finline which are excited (in the x direction) by an incident TE_{10}-wave of the empty waveguide.

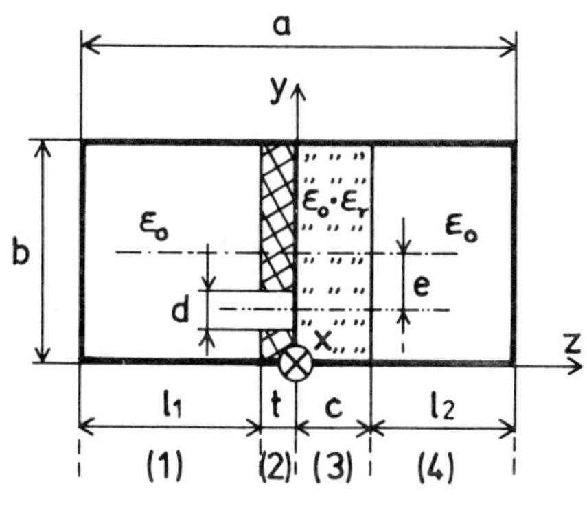

Fig.1: Unilateral finline.

J.K. Piotrowski, *IEEE MTT-S International Microwave Symposium Digest*, 1986, pp. 213–216.

The electromagnetic fields in the four regions shown in Fig.1 are superpositions of waveguide modes "propagated" in the z direction, which have π variation in the x direction when the half-wavelength resonator is considered. Therefore, the full set of modes satisfying the boundary conditions is as follows: TE_{1n} (to z axis) for n an integer $n \geq 0$ and TM_{1n} for $n \geq 1$. The coupling of TE- and TM-modes is automatically involved by assumption that only the y component of the E-field exists in the slot at z=0 (T_0 plane). Additionally, the analysis is simplified by assuming that only TE_{10}-mode is present in the region (2). The amplitude of each mode in each region must be such, that when the full set of modes is superimposed, the boundary and continuity conditions in the planes: T_2 and T_0 will be met, and all fields components in plane T_3 will be matched. That is an equivalent condition that transverse resonance will occur in the circuit presented in Fig.2. The boundary value problem has been expressed by the following real equation:

$$B_t = B_+ + B_- = 0 \tag{1}$$

where B_t is the total susceptance in the plane T_0, and B_+, B_- are susceptances in this plane looking in the +z and -z directions, respectively.

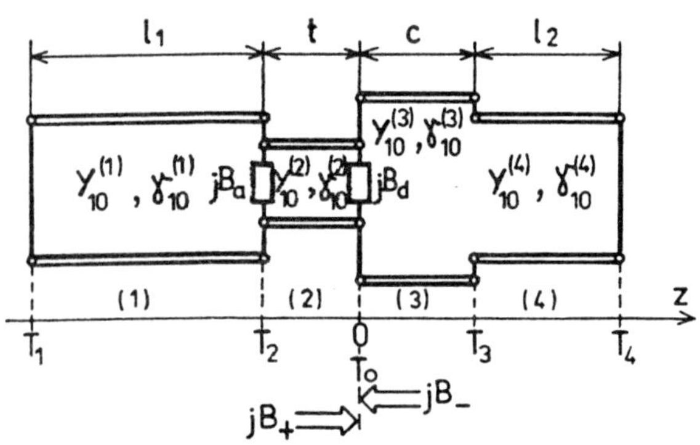

Fig.2: Equivalent circuit for transverse section of half-wavelength unilateral finline resonator.

The B_t is a function of f and β_x for the given set of parameters: ε_r, a, b, c, d, e, t, l_1 and l_2 which are indicated in Fig.1. The roots of equation (1), at a certain frequency, are phase constants of the fundamental and higher order modes in finline. It is self-evident that the above equation has no solution when $\beta_x = \beta_0$ (β_0-free space phase constant). However, since propagation characteristics in finline are monotonic, their values, at the discontinuity points of equation (1), can be calculated with ordered accuracy.

FORMULAS FOR ELEMENTS OF THE RESONATOR EQUIVALENT CIRCUIT

Components B_+ and B_- of the resonance equation are easily derived by using the conventional circuit theory, and the elements of the circuit in Fig.2 are analytically determined below. This circuit consists of the segments of TE_{10}-mode air- or dielectric-filled waveguide, which are described by propagation constants $\gamma_{10}^{(i)}$ i=1,2,3,4 and characteristic admittances $Y_{10}^{(i)}$ based on the power-voltage definition. These parameters are defined as follows:

$$\left(\gamma_{10}^{(i)}\right)^2 = \begin{cases} \varepsilon_r \cdot \gamma_0^2 - \gamma_x^2 & \text{region (3)} \\ \gamma_0^2 - \gamma_x^2 & \text{otherwise} \end{cases}$$

$$Y_{10}^{(i)} = \begin{cases} \dfrac{1}{2\eta_0} \cdot \dfrac{b}{d} \cdot \dfrac{\gamma_y \cdot \gamma_{10}^{(i)}}{\gamma_x \cdot \gamma_0} & \text{region (2)} \\ \dfrac{1}{2\eta_0} \cdot \dfrac{\gamma_y \cdot \gamma_{10}^{(i)}}{\gamma_x \cdot \gamma_0} & \text{otherwise} \end{cases}$$

where $\gamma_0 = j\cdot\beta_0$, $\gamma_x = j\cdot\beta_x$, $\eta_0 = \sqrt{\mu_0/\varepsilon_0}$, $\gamma_y = j\beta_y = j\frac{\pi}{b}$.

The susceptance B_a represents all higher order modes (TE_{1n}, TM_{1n} for $n \geq 1$) in the T_2 plane from region (1). These modes in regions (3) and (4) are equalized in the plane T_0 by the susceptance B_d. Both equivalent susceptances have been derived by using a properly modified variational technique, which has been applied previously by Collin (11) for the capacitive aperture in waveguide.

According to the earlier assumptions, there is only the y component of the E-field in the plane T_2. The electromagnetic field in region (1) is TE to x axis and can be constructed as the superposition of TEx_{1n} modes. Fig.3 shows the network representation of these hybrid modes for $n \geq 1$.

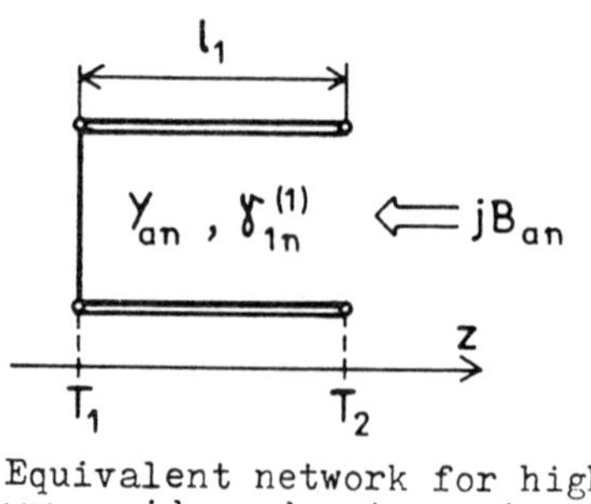

Fig.3: Equivalent network for higher order waveguide modes in region (1).

Their characteristic admittances based on the power-voltage definition are as follows:

$$Y_{an} = Y_{10}^{(1)} \cdot \gamma_{10}^{(1)} / \gamma_{1n}^{(1)}$$

and the mode propagation constants are expressed as

$$\left(\gamma_{1n}^{(1)}\right)^2 = \gamma_o^2 - \gamma_x^2 - \left(n \cdot \gamma_y\right)^2 .$$

The E-field in the slot at plane T_2 is constant along the y axis, therefore susceptance B_a is described by the equation:

$$B_a = B_{aG} + 2 \cdot \sum_{n=1} \left[B_{an} - \frac{1}{n} \cdot Y_{10}^{(1)} \cdot \frac{\gamma_{10}^{(1)}}{\gamma_y} \right] \cdot P_{n0}^2 \qquad (2)$$

with $B_{aG} = Y_{10}^{(1)} \cdot \frac{\gamma_{10}^{(1)}}{\gamma_y} \cdot (-\ln \alpha_2)$

where $\alpha_2 = \sin\left(\frac{\pi}{2} \cdot \frac{d}{b}\right) \cdot \sin\left[\frac{\pi}{2} \cdot \left(1 - 2 \cdot \frac{e}{b}\right)\right]$,

and B_{an} is the input susceptance of the TEx_{1n}-mode as in Fig.3. Coefficients P_{n0} can be obtained from the tables of Chebyshev polynomials.

The formula for B_d is as follows:

$$B_d = B_{dG} + 2 \cdot \sum_{n=1} \left[B_{dn} - \frac{1}{n} \cdot Y_{10}^{(3)} \cdot \frac{\gamma_{10}^{(3)}}{\gamma_y} \right] \cdot P_{n0}^2 \qquad (3)$$

with $B_{dG} = Y_{10}^{(3)} \cdot \frac{\gamma_{10}^{(3)}}{\gamma_y} \cdot (-\ln \alpha_2)$.

Susceptance B_{dn} involves the coupling between TE_{1n}(to z) and TM_{1n}(to z) waves in regions (3) and (4) and is determined as follows:

$$B_{dn} = \left[\left(n \cdot \gamma_y\right)^2 \cdot B_{dn}^{TM} + \gamma_x^2 \cdot B_{dn}^{TE} \right] \Big/ \left[\left(n \cdot \gamma_y\right)^2 + \gamma_x^2 \right]$$

where B_{dn}^{TM} and B_{dn}^{TE} are input susceptances at T_0 as in Fig.4 for TM_{1n} and TE_{1n} modes, respectively. Propagation constants and characteristic admittances of the transmission lines in Fig.4 are equal:

$$\left(\gamma_{1n}^{(3)}\right)^2 = \varepsilon_r \cdot \gamma_o^2 - \gamma_x^2 - \left(n \cdot \gamma_y\right)^2 \ ; \ \gamma_{1n}^{(4)} = \gamma_{1n}^{(1)} \ ;$$

$$\left(Y_{1n}^{(3)}\right)^{TM} = Y_{10}^{(3)} \cdot \varepsilon_r \cdot \gamma_o^2 \Big/ \left(\gamma_{10}^{(3)} \cdot \gamma_{1n}^{(3)}\right) ;$$

$$\left(Y_{1n}^{(4)}\right)^{TM} = Y_{10}^{(4)} \cdot \gamma_o^2 \Big/ \left(\gamma_{10}^{(4)} \cdot \gamma_{1n}^{(4)}\right) ;$$

$$\left(Y_{1n}^{(i)}\right)^{TE} = Y_{10}^{(i)} \cdot \gamma_{1n}^{(i)} \Big/ \gamma_{10}^{(i)} \qquad \text{with } i=3,4 .$$

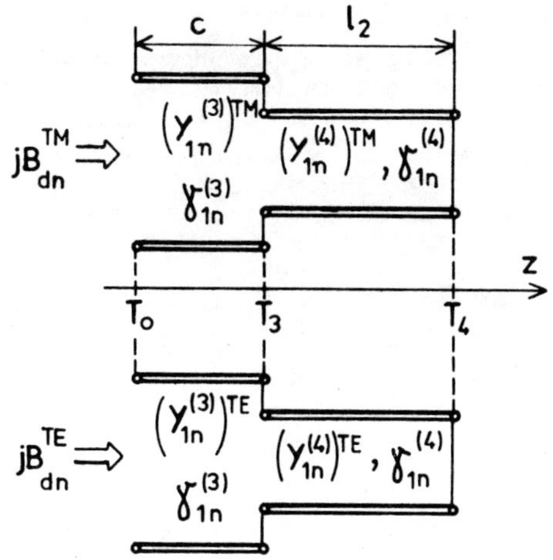

Fig.4: Network representation for higher order modes in regions (3) and (4).

The above formulas allow us to construct the resonance condition (1).

The characteristic impedance of finline on the basis of power-voltage definition can be calculated by using the procedure given in (10).

NUMERICAL RESULTS

The bisection method has been used to solve equation (1). The first twelve terms of

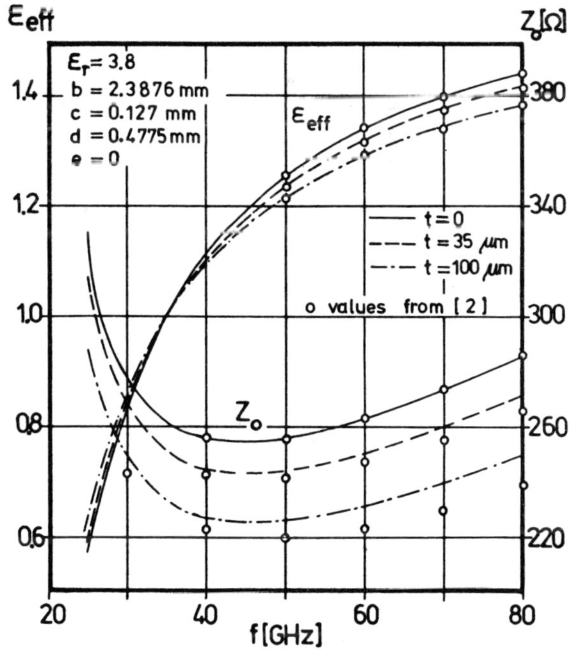

Fig.5: Characteristics of unilateral finline (fundamental mode).

the series in equations (2) and (3) have been taken into account.

Fig.5 shows frequency dependence of the effect of the metallization thickness on the effective dielectric constant ε_{eff}, and on the characteristic impedance Z_0 in a unilateral finline. The published data (2) are also plotted in Fig.5 and close agreement between the two sets of results is seen for effective dielectric constant. The results for Z_0 (for t=0) presented in Fig.5 show good agreement. Comparing the characteristic impedance for the case of finite metallization thickness to results published in (2) yields deviations of less than 3 per cent for t=35µm and less than 6 per cent for t=100µm.

The comparison of the numerical results obtained by this method (TRM) and by the singular integral equation technique (SIET) (12), for propagation constants of fundamental and higher order modes in unilateral finline with zero metallization thickness, is presented in Table 1. The agreement is very good.

Table 1 - The propagation constants (β_x) of the first three modes in a unilateral finline. parameters: ε_r=3.75 , a=1.65mm, b=0.825mm, c=0.11mm, d=0.3mm, l_1=0.77mm, e=t=0 .

f [GHz]	mode no.	β_x [rad/mm]	
		SIET	TRM
60	1	0.17908	0.17894
120	1	2.85901	2.85913
180	1	4.79879	4.79910
	2	0.63598	0.63617
230	1	6.45114	6.45071
	2	3.12472	3.12480
	3	2.05299	2.05570
250	1	7.13004	7.12945
	2	3.76650	3.76664
	3	2.97324	2.97567

ACKNOWLEDGMENT

The author is extremely grateful to Mr. Abbas S. Omar who is with Technische Universität Hamburg-Harburg, for helpful discussions and also for providing computer data on finlines, obtained via singular integral equation technique.

REFERENCES

(1) R.Vahldieck,"Accurate hybrid-mode analysis of various finline configurations including multilayered dielectrics, finite metallization thickness, and substrate holding grooves",IEEE Trans. Microwave Theory Tech.,vol.MTT-32, pp. 1454-1460, Nov. 1984.

(2) T.Kitazawa and R.Mittra,"Analysis of finline with finite metallization thickness", ibid.,vol.MTT-32, pp.1484-1487, Nov. 1984.

(3) L.P.Schmidt and T.Itoh,"Spectral domain anlysis of dominant and higher order modes in finlines", ibid.,vol.MTT-28, pp.981-985, Sept. 1980.

(4) L.P.Schmidt, T.Itoh and H.Hofman,"Characteristics of unilateral finline structures with arbitrary located slots", ibid.,vol.Mtt-29,pp.352-355, Apr. 1981.

(5) J.B.Knorr and P.M.Shayda,"Millimeter--wave finline characteristics", ibid., vol.MTT-28, pp.737-743, Juli 1980.

(6) A.S.Omar and K.Schünemann,"Formulation of the singular integral equation technique for general planar transmission lines", in Proc. IEEE MTT-S Symp.(St. Louis), 1985, pp.135-138.

(7) J.K.Piotrowski,"Accurate and simple formulas for dispersion in finlines", in Proc. IEEE MTT-S Symp.(San Francisco), 1984, pp.333-336.

(8) P.Pramanick and P.Bhartia,"Accurate analysis equations and synthesis technique for unilateral finlines", IEEE Trans. Microwave Theory Tech.,vol.MTT-33, pp.24-30, Jan. 1985.

(9) R.N.Simons and M.Tech,"Analysis of millimetre-wave integrated finline", IEE Proc.,vol.130, Pt.H, pp.166-169, March 1983.

(10) S.B.Cohn,"Slot line on a dielectric substrate", IEEE Trans. Microwave Theory Tech.,vol.MTT-17, pp.768-778, Oct. 1969.

(11) R.E.Collin,"Field theory of guided waves", New York, McGraw-Hill, 1960.

(12) A.S.Omar, private communication.

2.1.10

Systematic Characterization of the Spectrum of Unilateral Finline

C. A. OLLEY AND T. E. ROZZI, SENIOR MEMBER, IEEE

Abstract—The fundamental mode of finline is well documented. A sufficiently complete knowledge of the higher mode spectrum, however, is necessary for the treatment of discontinuities. This paper is concerned with a rigorous approach to the problem of the spectrum of unilateral finline on the basis of a variational solution in the space domain and transverse equivalent network considerations. We call this method transverse resonance diffraction. Having satisfied the correct edge condition *a priori*, the analysis produces highly accurate dispersion characteristics calculated at each "spot frequency" for the fundamental and higher order modes with matrices of very low order. By exploiting the quasi-analytical character of the model, it is possible to derive a systematic and complete characterization of the spectrum in terms of "mode families." Moreover, simple-to-use wide-band approximations to the dispersion characteristics are also obtained which are suitable for evaluation by a desktop calculator.

I. Introduction

WITH THE INCREASED DEMAND for millimeter systems, *E*-plane structures such as finline have become established for use in millimeter-wave integrated circuits. Originally proposed by Meier [1], finline avoids unnecessary miniaturization at the millimeter frequencies while offering the potential for low-cost batch production and a degree of compatibility with active devices. Finline may be regarded as a form of planar ridged waveguide; this was recognized by Saad and Begemann [2], who produced approximate solutions for the fundamental mode using a simple equivalent circuit. Although the first fully rigorous solution was given earlier by Hofmann [3], this was time consuming and suffered from relative convergence problems. Saad and Schünemann [4] presented an analysis in terms of an equivalent set of rectangular waveguides to obtain a first-order design theory for simple finline filter sections, while Shih and Hoefer [5] applied the general TLM method with a view to obtaining the equivalent fin susceptance for a transverse resonance solution of the cutoff.

However, since 1979, the spectral-domain technique developed by Itoh and subsequently applied to finline [6] has been the accepted analysis. Subsequent papers [7], [8] have used this analysis to present results on the general properties of finline, including an assessment of loss, and have [9] extended the method to more general unilateral structures.

Manuscript received January 21, 1986; revised June 9, 1986. This work was supported in part by SERC under Grant 86352 and by Marconi Research.

The authors are with the School of Electrical Engineering, University of Bath, Claverton Down, Bath, England.

IEEE Log Number 8610284.

Work has since been directed at the effect of finite fin thickness and substrate holding grooves using more general analyses, e.g., Beyer [10] and Wahldieck [11]. But, with certain exceptions [5], [6], most work has been directed at the fundamental mode. In addition, considerable computational effort is required even with the spectral-domain analysis, and this has led other authors to adopt approximate techniques, such as curve fitting (Pramanick and Bhartia [12]), and modified transverse resonance (Saad and Schünemann [13]). This motivates work towards rigorous approaches that yield simple and easy-to-use equivalent circuits.

Actual finline components, however, involve discontinuities. The solution of discontinuity problems, in turn, involves field matching with higher order modes. Therefore, in order for work to proceed on the analysis of finline discontinuities, such as in [15]–[16], a deeper understanding of the higher order modes is required. The identification of inductive and capacitive modes by Omar and Schünemann [17] is useful. Although LSE and LSM decoupling has been previously employed in the spectral-domain work, its significance was not fully appreciated.

We present an electromagnetic treatment which is rigorous but results in easy-to-use equivalent networks for the general solution and for cutoffs of various types of higher order mode. It also provides an accurate but simplified dispersion relation. The method operates in the space domain making use of transverse circuits and taking into explicit account the fins and their associated edge conditions. Restricting ourselves to the idealized unilateral finline structure as given by Fig. 1, the thickness of the fin metallization is ignored, as are any means of supporting the substrate. The transverse Green's admittances as seen at the fin aperture are obtained in the space domain; use is then made of the mapping, originally introduced by Schwinger [18], for the analysis of capacitive irises in waveguide, in order to introduce basis functions which implicitly satisfy the correct edge conditions. This judicious choice of basis yields rapid convergence with an exceptionally low order of required expansion.

With a view to characterizing the spectrum and obtaining simple but accurate broad-band equivalent circuits, we recognize two fundamental properties of hybrid modes in closed waveguide.

1) At cutoff, TE and TM contributions are decoupled, so that the Green's admittance becomes block diagonal and the resulting dispersion equation depends only on the

C.A. Olley and T.E. Rozzi, *IEEE Transactions on Microwave Theory and Techniques*, Vol. MTT-34, No. 11, Nov. 1986, pp. 1147–1156.

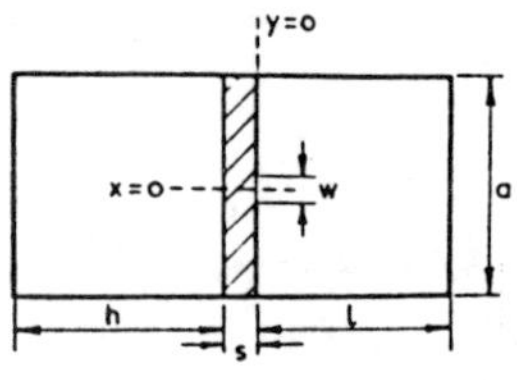

Fig. 1. Idealized unilateral finline.

cutoff wavenumber $k_0 = k_c$ rather than on k *and* the propagation constant β. This leads to a systematic and quasi-analytical determination of the cutoff wavenumbers k_c. A natural grouping of modes in "mode families" ensues, each family being characterized by a distinct quasi-static field in the gap.

2) Once the cutoff wavenumber of a mode is found, the propagation constant near cutoff must be expressible as an odd Laurent series of $\sqrt{k_0^2 - k_c^2}$ [19]. This realization leads naturally to an approximation for the dispersion equation, the simplest form of which is of the type

$$\beta = \eta\sqrt{k_0^2 - k_c^2}$$

with a real $\eta > 0$ virtually independent of frequency. This approximation we prefer to use over the "effective dielectric constant," which is still considerably frequency dependent.

II. Rigorous Analysis

The fields in finline will be described by y-directed Hertzian vector potentials coupled together in some way. Thus

$$E = -j\omega\mu\nabla \times \Pi_h + k^2\Pi_e + \nabla\nabla\cdot\Pi_e \tag{1}$$

$$H = k^2\Pi_h + \nabla\nabla\cdot\Pi_h + j\omega\epsilon\nabla \times \Pi_e \tag{2}$$

where

$$\Pi_h = \psi_h(x, y)e^{-j\beta z} \tag{3}$$

$$\Pi_e = \psi_e(x, y)e^{-j\beta z} \tag{4}$$

and propagation has been assumed in the z-direction with phase coefficient β.

For the analysis, the y-dependence of functions $\psi(x, y)$ is expressed in terms of the TE, TM (to y) eigenfunctions of slab-loaded waveguide $\chi_{hn}(y)$, $\chi_{en}(y)$. These are normalized so that $\chi_n(0) = 1$

$$\psi_h(x, y) = \sum_{n=0}^{\infty} U_{hn}\phi_{hn}(x)\chi_{hn}(y) \tag{5a}$$

$$\psi_e(x, y) = \sum_{n=1}^{\infty} U_{en}\phi_{en}(x)\chi_{en}(y) \tag{5b}$$

where

$$\phi_{hn}(x) = \sqrt{\frac{2}{a}}\cos\frac{n\pi}{a}x, \qquad n = 2,4,6,\cdots,$$

$$-\frac{a}{2} \leqslant x \leqslant \frac{a}{2} \tag{6a}$$

$$= \sqrt{\frac{1}{a}}, \qquad n = 0$$

$$\phi_{en}(x) = \sqrt{\frac{2}{a}}\sin\frac{n\pi}{a}x. \tag{6b}$$

Moreover, at $y = 0$, it is possible to express four independent x and z field components in terms of the set $\{\phi_{hn}(x)\}$ only, by means of the choice below:

$$E_x(x,0) = \sum_{n=0}^{\infty} E_{xn}\phi_{hn}(x) \tag{7}$$

$$\frac{a}{\pi}\frac{\partial}{\partial x}E_z(x,0) = \sum_{n=1}^{\infty} E_{zn}\phi_{hn}(x) \tag{8}$$

$$\frac{\pi}{a}\int H_x(x,0)\,dx = \sum_{n=1}^{\infty} H_{xn}\phi_{hn}(x) \tag{9}$$

$$H_z(x,0) = \sum_{n=1}^{\infty} H_{zn}\phi_{hn}(x). \tag{10}$$

An additional advantage of using modified versions of E_z and H_x, namely, to produce proper convergence of the Green's functions, is discussed further in Appendix I.

The as yet unknown coefficients may be obtained from the modal amplitudes of transversely propagating pure TE and TM modes in the regions left and right of $y = 0$ using a now classical axis rotation due to Itoh [20]. The hybrid field is thus expressed in terms of voltages and currents upon TE and TM transmission lines. Eliminating voltages and currents leads to a direct relationship between the E and H fields. For instance, to the right of $y = 0$

$$\begin{bmatrix} -H_{zn} \\ H_{xn} \end{bmatrix} = NT_n^{-1}\begin{bmatrix} h_{nR} & 0 \\ 0 & g_{nR} \end{bmatrix} T_n N \begin{bmatrix} E_{xn} \\ E_{zn} \end{bmatrix} \tag{11}$$

where

$$T_n = \begin{bmatrix} \cos\tau_n & -j\sin\tau_n \\ -j\sin\tau_n & \cos\tau_n \end{bmatrix} N \begin{bmatrix} 1 & 0 \\ 0 & \frac{1}{n} \end{bmatrix}$$

$$\sin\tau_n = \frac{\beta}{\sqrt{\left[\frac{n\pi}{a}\right]^2 + \beta^2}}.$$

Here, h_{nR}, g_{nR} are the TM and TE admittances, respectively, seen by the nth mode looking right of $y = 0$, given by

$$h_{nR} = \frac{j\omega\epsilon}{k_{ny}}\coth k_{ny}l$$

$$g_{nR} = \frac{k_{ny}}{j\omega\mu}\coth k_{ny}l$$

where

$$k_{ny} = \sqrt{\left[\frac{n\pi}{a}\right]^2 + \beta^2 - k_0^2}.$$

From the above, by superposition, a relationship between the total fields left and right of the fin aperture may be formed. Thus, at $y = 0$ for the right-hand side of the fin, we have in operator form,

$$\begin{bmatrix} -H_z \\ \frac{a}{\pi}\int H_x\,dx \end{bmatrix} = \begin{bmatrix} \tilde{Y}_{11}^R & \tilde{Y}_{12}^R \\ \tilde{Y}_{21}^R & \tilde{Y}_{22}^R \end{bmatrix} \cdot \begin{bmatrix} E_x \\ \frac{\pi}{a}\frac{\partial}{\partial x}E_z \end{bmatrix}. \quad (12)$$

It is noted that no distinction between fields right and left of the aperture need be made inasmuch as the fields are continuous there. On the left-hand side, we have

$$\begin{bmatrix} H_z \\ -\frac{a}{\pi}\int H_x\,dx \end{bmatrix} = \begin{bmatrix} \tilde{Y}_{11}^L & \tilde{Y}_{12}^L \\ \tilde{Y}_{21}^L & \tilde{Y}_{22}^L \end{bmatrix} \cdot \begin{bmatrix} E_x \\ \frac{\pi}{a}\frac{\partial}{\partial x}E_z \end{bmatrix}. \quad (13)$$

The sign difference on the left-hand sides of (12) and (13) is noted. This is a consequence of the opposite orientation of the Poynting vector at the two sides of the aperture.

The operator notation is defined as follows. For example,

$$-H_z|_{E_z=0} = \tilde{Y}_{11}^R \cdot E_x = \int_{-w/2}^{w/2} Y_{11}^R(x,x')E_x(x',0)\,dx'$$

where $Y_{11}^R(x,x')$, etc., are defined in Appendix I. It is noted here that $\tilde{Y}_{11}$ denotes the integral operator whose kernel is given by $Y_{11}(x,x')$. The matrix representation of Y_{11} in terms of a discrete basis will be indicated below by Y_{11}, and elementwise by $(Y_{11})mn$.

We now apply the continuity of fields at the fin aperture. By adding (12) and (13), one derives the following system of integral equations for E_x and $(\partial/\partial x)E_z$ involving the total cross-sectional admittance operators $\tilde{Y}_{11} = \tilde{Y}_{11}^R + \tilde{Y}_{11}^L$, etc:

$$\begin{bmatrix} \tilde{Y}_{11} & \tilde{Y}_{12} \\ \tilde{Y}_{21} & \tilde{Y}_{22} \end{bmatrix} \cdot \begin{bmatrix} E_x \\ \frac{a}{\pi}\frac{\partial}{\partial x}E_z \end{bmatrix} = 0 \quad (14)$$

where the corresponding Green's admittance functions are of the form

$$Y_{11}(x,x') = \sum_{n=0}^{\infty} Y_{11n}\phi_{hn}(x)\phi_{hn}(x')$$

$$Y_{12}(x,x') = \sum_{n=1}^{\infty} Y_{12n}\phi_{hn}(x)\phi_{hn}(x')$$

$$Y_{21}(x,x') = -Y_{12}(x,x')$$

$$Y_{22}(x,x') = \sum_{n=1}^{\infty} Y_{22n}\phi_{hn}(x)\phi_{hn}(x')$$

with Y_{11n}, etc., given in Appendix I.

In order to solve (14) by the Ritz–Galerkin method in real space, we expand the fields onto the orthonormal basis $\{f_m\}$ which satisfies the $r^{-1/2}$ singularity at the fin edges, as follows:

$$\begin{bmatrix} E_x(x) \\ \frac{a}{\pi}\frac{\partial}{\partial x}E_z(x) \end{bmatrix} \frac{dx}{d\theta} = \sum_{m=0}^{N} \begin{bmatrix} X_m \\ Z_m \end{bmatrix} f_m(\theta). \quad (15)$$

A suitable choice for the functions f_m is that originally introduced by Schwinger for the analysis of capacitive irises in waveguide [18] (see Appendix II).

The Schwinger functions result from the conformal mapping of the static fields between a parallel-plane waveguide and a waveguide with fin. Neglecting the slight perturbing effect of the dielectric, the quasi-static behavior of the fields in the fin gap is described exactly in terms of the above functions. The functions thus lend themselves as an ideal basis for use in a variational solution for finline. The perturbations due to the presence of the dielectric substrate and the proximity of the end walls, however, do not allow an exact solution directly from the mapping.

In the fin gap, we can also expand each ϕ_{hn} in terms of f_m, i.e.,

$$\phi_{hn} = \sum^{N} P_{mn} f_m. \quad (16)$$

The coefficients P_{mn} are defined in Appendix II. By application of (15) and (16) in (14), the latter is therefore transformed into a matrix equation

$$\begin{bmatrix} Y_{11} & Y_{12} \\ Y_{21} & Y_{22} \end{bmatrix} \begin{bmatrix} X \\ Z \end{bmatrix} = 0 \quad (17)$$

where

$$(Y_{11})_{mk} = \sum_{n=0}^{\infty} y_{11n}P_{mn}P_{kn}$$

$$(Y_{12})_{mk} = \sum_{n=1}^{\infty} y_{12n}P_{mn}P_{kn}$$

$$(Y_{21})_{mk} = -(Y_{12})_{mk}$$

$$(Y_{22})_{mk} = \sum_{n=1}^{\infty} y_{22n}P_{mn}P_{kn}.$$

III. Transverse Equivalent Circuit in the Fundamental Finline Mode

The term $n=0$ in Y_{11} is the contribution of the fundamental "box" mode transverse to y (fundamental transverse mode). This is not yet the fundamental finline mode. By making this contribution explicit, however, it is possible to develop a convenient transverse equivalent circuit in the fundamental finline mode itself.

H_{z0} is the transverse magnetic field of the fundamental transverse mode (as $H_{x0}=0$). Due to the presence of the iris at $y=0$, the above is discontinuous by an amount $H_{z0}^L - H_{z0}^R = \Delta H_{z0}$. Let us extract the term $n=0$ from the summation in Y_{11} and write in place of the integral equation (17) an equivalent integral equation of the form

$$\Delta H_{z0}\begin{bmatrix} P_0 \\ 0 \end{bmatrix} + Y \cdot \begin{bmatrix} X \\ Z \end{bmatrix} = 0 \quad (18)$$

where P denotes the column vector P_m and

$$Y = \begin{bmatrix} \sum_{n=1}^{\infty} y_{11n}P_nP_n^T & Y_{12} \\ Y_{21} & Y_{22} \end{bmatrix}.$$

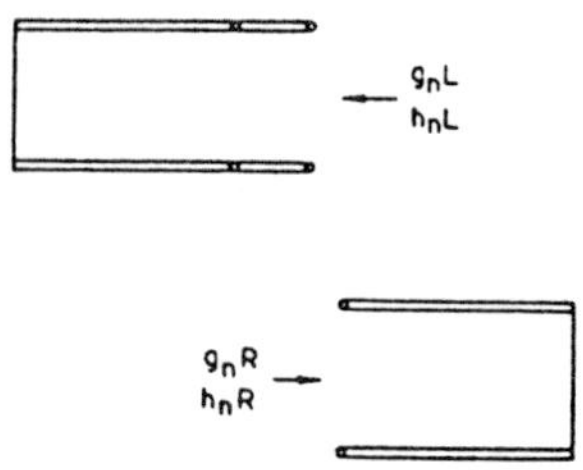

Fig. 2. Transverse admittances as seen by the fin.

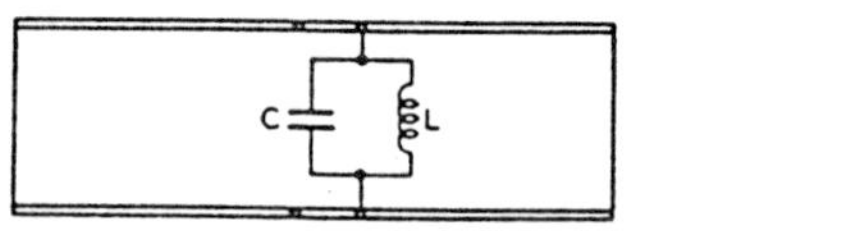

Fig. 3. Simplified equivalent circuit for the fundamental transverse mode.

Solving for X gives

$$X = -\Delta H_{z0}\left[Y_{11} - Y_{12}Y_{22}^{-1}Y_{21}\right]^{-1} \cdot P_0. \tag{19}$$

The amplitude E_{x0} of the transverse field of the fundamental transverse mode (its voltage at $y=0$ in the equivalent circuit of Fig. 3) is given by

$$P_0^T \cdot X = E_{x0}.$$

This is related to the discontinuity in the transverse magnetic field (the current) by the total admittance seen by the mode at $y=0$, i.e.,

$$\Delta H_{z0} = (g_{0L} + g_{0R})E_{x0}$$

where g_{0L} and g_{0R} are the admittances seen by the fundamental transverse mode on either side of the fin (see Fig. 2). The fin itself is seen by the fundamental transverse mode as a transverse discontinuity whose admittance is given by

$$g = \left(P_0^T \cdot \left[Y_{11} - Y_{12}Y_{22}^{-1}Y_{21}\right]^{-1} \cdot P_0\right)^{-1}. \tag{20}$$

The condition of transverse resonance at $y=0$ gives therefore

$$g_{0L}(\beta, k_0) + g_{0R}(\beta, k_0) + g(\beta, k_0) = 0.$$

A solution for β at a given frequency k_0 is thus obtained at the resonance of the network given in Fig. 3. When solving for the fundamental mode, the higher order transverse modes are cutoff and therefore essentially localized about the fin. Thus, they can be represented as a fin admittance in terms of lumped components.

The order of matrix employed in the expression for the fin admittance g is determined by the size of expansion. But in view of the nature of the expansion set, the matrix order need not be increased beyond 2×2. In fact, the zeroth order is sufficient for all but very large fin gaps; furthermore, the term $Y_{12} \cdot (Y_{22})^{-1} \cdot Y_{21}$ arising from E_x to E_z coupling may be ignored for small gaps and thin irises at least. Hence, by virtue of the Schwinger mapping,

$$g = y_{11}^s \ln\left|\csc^2\left[\frac{\pi w}{2a}\right]\right| + \sum_{n=1}^{nd}\left[y_{11n} - \frac{y_{11}^s}{n}\right]\frac{P_{n0}^2}{P_{00}^2}$$

TABLE I
COMPARISON OF ZEROTH-, FIRST-, SECOND-, AND THIRD-ORDER SOLUTIONS

Frequency/Ghz	Beta/ko			
	(0)	(1)	(2)	(3)
8.0	0.850961	0.849011	0.849001	0.849001
8.2	0.861708	0.859725	0.859716	0.859716
8.4	0.871585	0.869568	0.869558	0.869560
8.6	0.880684	0.878635	0.878627	0.878627
8.8	0.889091	0.887007	0.887000	0.887000
9.0	0.896874	0.894764	0.894754	0.894753
9.2	0.904111	0.901963	0.901936	0.901937
9.4	0.910834	0.908655	0.908678	0.908625
9.6	0.917112	0.914893	0.914884	0.914877
9.8	0.922993	0.920718	0.920709	0.920709
10.0	0.928497	0.926166	0.926158	0.926158
10.2	0.933679	0.931271	0.931263	0.931263
10.4	0.938332	0.936062	0.936055	0.936055
10.6	0.942864	0.940565	0.940559	0.940558
10.8	0.947132	0.944805	0.944798	0.944798
11.0	0.951156	0.948801	0.948794	0.948794
11.2	0.954954	0.952571	0.952564	0.952563
11.4	0.958546	0.956133	0.956126	0.956126
11.6	0.961945	0.959502	0.959495	0.959494
11.8	0.965167	0.962692	0.962684	0.962683
12.0	0.968224	0.965711	0.965703	0.965702

where y_{11}^s/n is the limit of y_{11n} for large n, i.e., as $n \to \infty$, $y_{11n} \to y_{11}^s/n$. Its value is obtained by taking the above limit in Appendix I, which yields

$$y_{11}^s = \frac{2\beta^2 - (1+\epsilon_r)k_0^2}{\sqrt{\epsilon_r k_0^2 - \beta^2}\left[\dfrac{\pi}{a}\right]}.$$

It can be shown that the correction series converges rapidly and can be truncated after a few terms ($n > nd$).

For very small fin gaps, the correction series is negligible and in terms of the "effective frequency" variable

$$u = \sqrt{\epsilon_r k_0^2 - \beta^2}\left[\frac{a}{\pi}\right]$$

we can write

$$g = \frac{1}{juL} + juC$$

where

$$\frac{1}{L} = (\epsilon_r - 1)\left[\frac{a}{\pi}\right]^2 k_0^2 \ln\left|\csc^2\left[\frac{\pi w}{2a}\right]\right|$$

$$C = 2\ln\left|\csc^2\left[\frac{\pi w}{2a}\right]\right|$$

(see Fig. 3). Table I illustrates the convergence of the propagation constant of the fundamental finline mode for increasing order of the expansion for a moderately large fin gap ($w/a = 0.2$). The first column ($N=0$) corresponds to the above analytical formula including the correction series with $nd = 12$.

IV. CUTOFF PROPERTIES

It is apparent from (11) that at cutoff ($k = k_c$, $\beta = 0$) the off-diagonal blocks of the 2×2 Green's function vanish completely, so that the TE and TM components of the hybrid mode are decoupled. Also, at that point, only the cutoff wavenumber k_c enters the dispersion equation. The cutoff is, therefore, more easily determined as well as analytically approximated than the general case. Once the cutoff is found, it is also possible to identify specific

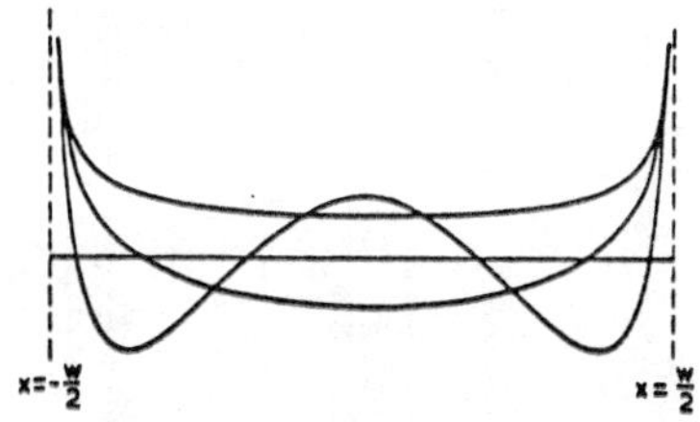

Fig. 4. Basis functions for LSE modes.

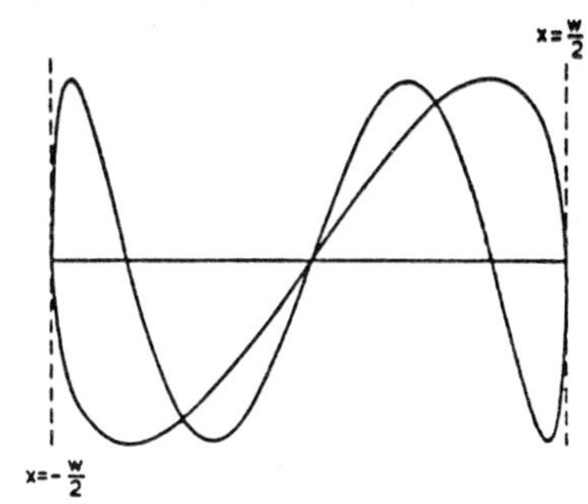

Fig. 5. Basis functions for LSM modes.

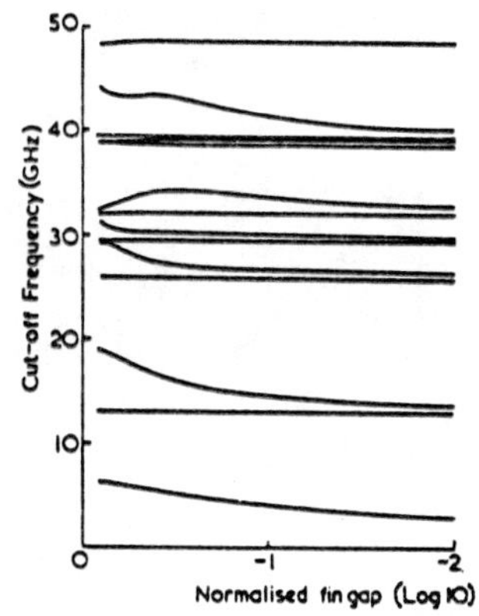

Fig. 6. Cutoff frequency versus fin gap for LSE(0) modes. Guide dimensions: $a = 10.16$ mm, $l = h + s = 11.43$ mm, $s = 0.254$ mm, $\epsilon_r = 2.20$.

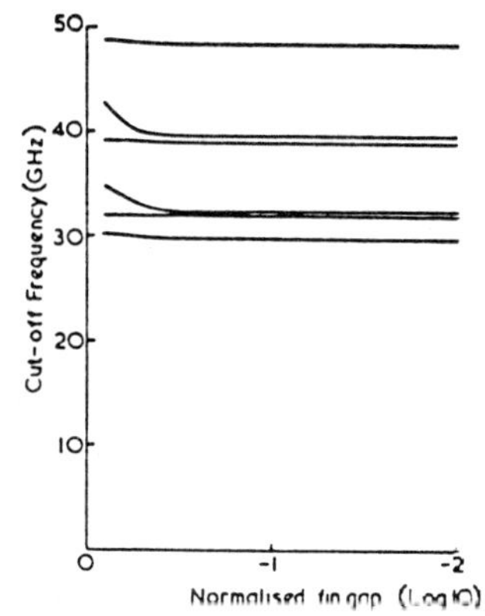

Fig. 7. Cutoff frequency versus fin gap of LSE(1) modes, dimensions as in Fig. 6.

families of finline modes, and their individual $\beta - k$ characteristics not too far from cutoff can be approximated by means of a simple square root behavior. For these reasons, we will now consider the problem of determining cutoff in some detail.

As noted, at cutoff, the coupling matrices Y_{12} and Y_{21} vanish, and (17) reduces to

$$Y_{11}X = 0 \tag{21a}$$

$$Y_{22}Z = 0. \tag{21b}$$

Modes which are TE to y, i.e., LSE at cutoff, are obtained by setting $E_z = 0$ (i.e., $Z = 0$).

Now orthogonal solutions of the quasi-static problem in the fin gap are given by

$$f_m(\theta) = \sqrt{\frac{2}{\pi}} \cos m\theta, \qquad m > 0$$

$$f_0(\theta) = \frac{1}{\sqrt{\pi}}.$$

The relationship between x and θ is established in (A3). The f_m's are in fact solutions to the quasi-static problem of the infinite parallel-plate waveguide with fin. In the x domain, these yield fin-gap fields, as shown in Figs. 4 and 5. Under quasi-static conditions, no coupling takes place between these solutions. Each value of m distinguishes therefore a whole family of modes (LSE(m) or LSM(m)) with the same quasi-static field in the gap at cutoff. Each member of the family, however, is a distinct solution of the dispersion equation arising from assuming that field distribution. Mathematically, this is expressed by the matrix Y_{11} becoming diagonal. Therefore, the transverse resonance condition (21a) reduces to each diagonal element being individually zero, i.e.,

$$(Y_{11})_{mm} = 0$$

or

$$\sum_{n=m}^{\infty} y_{11n}(k_{cm}) P_{mn}^2 = 0 \tag{22}$$

(since $P_{mn} = 0$ for $n < m$), which gives a very good approximation to the cutoff of the LSE(m) mode family.

For a given m, each solution of (22) (an infinite discrete set) corresponds to a member of the LSE(m) family. Similarly, the LSM(m) mode cutoffs are obtained from

$$\sum_{n=m}^{\infty} y_{22n}(k_{cm}) P_{mn}^2 = 0. \tag{23}$$

Note, in particular, that $m = 0$ variations for the LSM case are not acceptable since, from the definition of field quantities, this would represent a dc field in the fin gap.

From the solutions to (22) and (23), it is found that the fundamental finline mode is given by the first member of the LSE(0) family, corresponding to $m = 0$ field variation in the slot. This mode is therefore pure LSE at the cutoff point, although in general it is a hybrid HE mode.

Figs. 6 and 7 show the variation of cutoff frequencies versus the fin gap for the LSE(0) and LSE(1) mode fami-

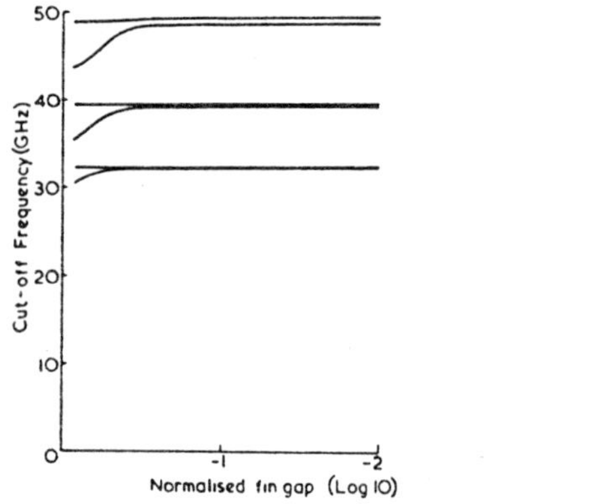

Fig. 8. Cutoff frequency versus fin gap of LSM(1) modes, dimensions as in Fig. 6.

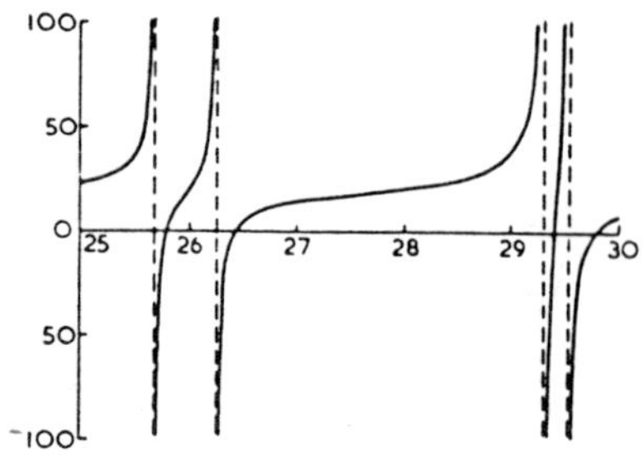

Fig. 9. Typical occurrence of roots to the cutoff equation, dimensions as in Fig. 6.

TABLE II
POSITION OF POLES OCCURRING IN SOLUTION FOR CUTOFF

LSE(0)	LSE(1)	LSM(1)
12.8375	-	-
13.1234	-	-
25.6719	-	-
26.2467	-	-
29.3099	29.3099	-
29.5275	29.5275	-
31.8687	31.8687	32.1964
32.3125	32.3125	32.3125
38.5000	-	-
38.8595	38.8595	39.1233
39.3701	-	-
39.5065	39.5065	39.5065
48.3023	48.3023	48.5125
49.2126	49.2126	49.2126

lies occurring below 50 GHz in an X-band finline. Notice that modes which display maximum E_x at $y = 0$ experience a pronounced reduction in cutoff as the fins intrude into the guide. This is due to their increased interaction. Those with a node at $y = 0$, however, starting with the first higher order finline mode, remain essentially unaffected by the fins. This is particularly so in the case of the highly symmetrical guide chosen in the analysis, where the central node of E_x occurs very near the plane of the fins.

Fig. 8 gives cutoffs for the first family of inductive modes, the LSM(1). As reported by Omar and Schünemann [17], the inductive modes experience an increase in cutoff as the fin gap is reduced. Although these changes should more appropriately be regarded as shifts away from modes of dielectrically loaded waveguide as the fin modes become established.

From Figs. 6–8, we therefore conclude that finline is LSE dominated (HE dominant in general). The first LSM-type mode typically occurs as the tenth higher order mode; however, the exact position in the mode spectrum depends upon the fin gap, since LSE and LSM cutoffs can cross over, leading to degeneracy when they coincide.

V. Asymptotic Behavior of Cutoff

We shall now examine (22) and (23) in some detail to discover any asymptotic behavior in the cutoff values which might be of use in any further work.

Now the elements y_{11n}, y_{22n} appearing in (22) and (23) correspond to field variations with x of order n. As the frequency is increased, a point is reached at which the corresponding parallel-plate mode becomes propagating and the modal admittances seen at either side of the fin begin to introduce tangent-type variations into the overall susceptance functions ((22) and (23)). The behavior of these functions therefore becomes progressively more complicated with cutoff frequency. However, the cutoffs given by zeros are straddled by poles of the admittance at either side of the fin. Thus, locating these poles leads to systematic solution for cutoffs by computer. Furthermore, poles and zeros increasingly cluster together, so that when searching for very high cutoffs (> 100 GHz in the present case) it is sufficient to locate just the poles. Even for low cutoff values, the same effect of clustering is observed for very small fin gaps, as a consequence of a very large number of transverse modes being coupled to the fin-gap field.

Fig. 9 illustrates the occurrence of pole–zero pairs in the LSE(0) solution. Since the pole positions can be easily determined, the asymptotic cutoff frequencies can be obtained in a straightforward manner. Accurate solutions to (22) and (23) are only needed to provide a small number of corrections appropriate to the particular fin gap. Table II gives a list of poles relevant for the LSE(0), LSE(1), and LSM(1) solutions, confirming the above observations.

VI. Simplified Dispersion Equations

To complete the characterization of the finline modes, we return to the problem of describing their dispersion relationships by a simple expression of the form

$$\beta = \eta\sqrt{k_0^2 - k_c^2} \tag{24}$$

where k_c is the cutoff wavenumber previously obtained and η is now to be determined.

It has been shown that in a lossless, reciprocal, uniform waveguide, the propagation constant as a function of frequency can be expanded near cutoff as an odd Laurent series of $\sqrt{k_0^2 - k_c^2}$ [19]. In its simplest form, this expansion is just (24) where η is a constant.

As we now assume that η is a constant quantity, we therefore find the limit $k_0^2 \to 0$ of interest. At this point, the problem becomes quasi-static and since all the finline modes are cutoff in this limit, η can be determined from

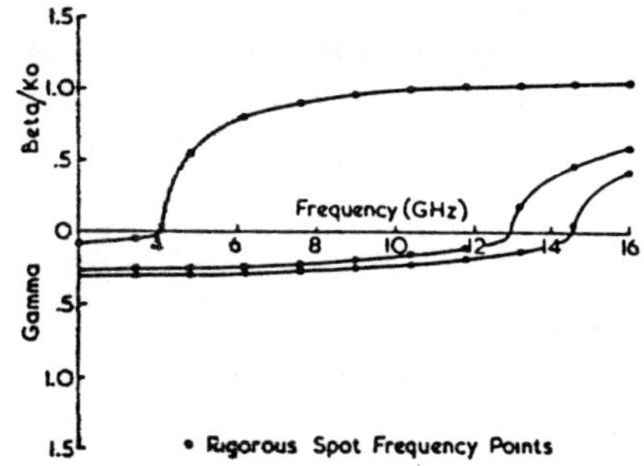

Fig. 10. Approximate dispersion curves with exact points for the first three modes, dimensions as in Fig. 6.

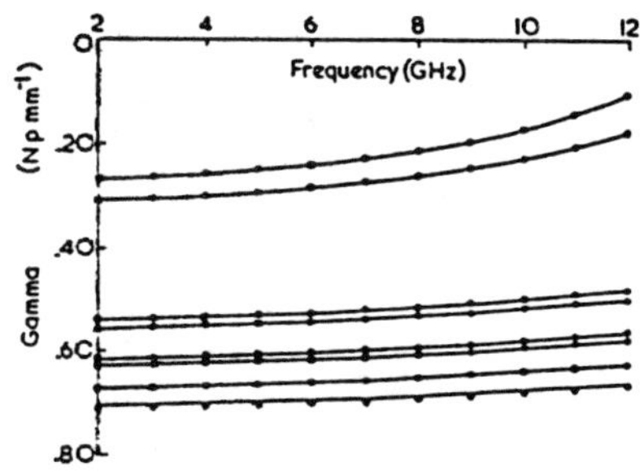

Fig. 11. Approximate dispersion of the following first eight higher order modes (with exact points), dimensions as in Fig. 6.

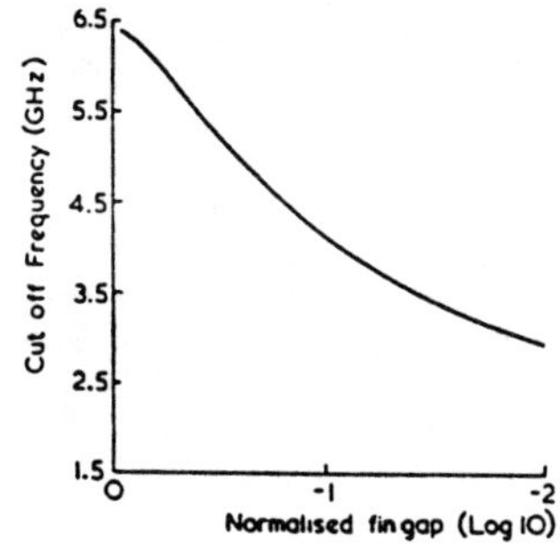

Fig. 12. Cutoff frequency for the fundamental mode in X-band finline, dimensions as in Fig. 6.

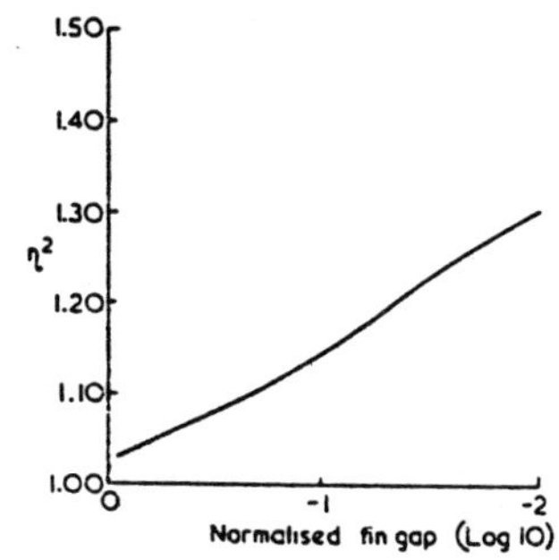

Fig. 13. η for the fundamental mode in X-band finline, dimensions as in Fig. 6.

the decay coefficient γ_0 and the cutoff wavenumber k_c

$$\eta = \frac{\gamma_0}{k_c}.$$

Before solving for γ_0, closer examination of the Green's admittance yields some useful simplifications, which greatly ease correlating the γ_0 with its respective k_c. As $k_0^2 \to 0$ we observe that all TM admittances vanish; furthermore, the air and substrate regions become indistinguishable to the TE modes. Thus, the effect of the dielectric is not felt and the problem becomes that of the static case in a finned waveguide. If, moreover, the fins are centrally placed in the finline enclosure, then it is noted that the coupling matrices Y_{12} and Y_{21} vanish, so that the solution for γ_0 of the LSE(m) and LSM(m) mode families decouples and can proceed as before.

Thus, γ_0 for the LSE(m) modes is given by

$$\sum_{n=m}^{\infty} y_{11n}(\gamma_0) P_{mn}^2 = 0$$

and for the LSM(m) modes by

$$\sum_{n=m}^{\infty} y_{22n}(\gamma_0) P_{mn}^2 = 0.$$

The assumption of a centrally placed fin leads to discrepancies with the higher order modes. But since they tend to propagate much more in the air region of the enclosure, η is sufficiently close to unity not to warrant further evaluation. However, an exception arises in the case of the third mode, which, like the fundamental, has strong interaction with the fins, and assuming $\eta = 1.0$ may not be sufficiently accurate. The second mode, as mentioned earlier, vanishes at $y = 0$ and is little affected by the fin; hence, η is very close to unity in any case.

Dispersion curves calculated for the first three modes give excellent agreement with the "spot frequency" results obtained with the rigorous analysis of Section II, since for these the symmetry assumption has little effect. Results for the dispersion of modes of even higher order assuming $\eta = 1.0$ are similarly very good, as shown in Figs. 10 and 11.

Finally, Figs. 12 and 13 give k_c and η for the fundamental mode for any given fin gap on a particular substrate. Such results can be readily incorporated into a curve fit for design purposes.

VII. Experimental and Theoretical Comparison

In order to assess validity, measurements of guided wavelength were performed using X-band finline with an adjustable fin gap. The method of observing the frequency response of a resonant length of line was employed, and in order to maintain a high Q factor, the short-circuited line was excited by small loop probes. Table III gives results for the fundamental mode, along with a comparison of two sets of theoretical predictions, one obtained with the method of Section II, the other by the established spectral-domain analysis. In both, a two-term expansion in the fin gap was used. However, the additional calculations required in the spectral-domain technique imply that the present method has a clear computational advantage, par-

TABLE III
COMPARISON OF EXPERIMENTAL RESULTS AND SPECTRAL-DOMAIN METHOD WITH PRESENT THEORY FOR THE FUNDAMENTAL MODE

Frequency/GHz	Guided wavelength /mm		
		Theoretical	
	Measured	Present method	Spectral domain
4.920	300.00	299.82	330.26
5.020	200.00	207.59	216.91
5.200	150.00	148.75	151.41
5.390	120.00	119.51	121.18
5.610	100.00	100.29	101.26
5.857	85.70	86.44	87.05
6.141	75.00	75.54	75.94
6.447	66.66	67.11	67.38
6.772	60.00	60.38	60.57
7.113	54.54	54.90	55.03
7.472	50.00	50.29	50.38
7.843	46.15	46.40	46.47
8.223	42.85	43.08	43.13
8.612	40.00	40.22	40.25
9.010	37.50	37.70	37.70
9.414	35.29	35.49	35.49
9.825	33.33	33.52	33.52
10.241	31.58	31.75	31.75
10.663	30.00	30.16	30.16
11.091	28.57	28.72	28.70
11.516	27.27	27.42	27.41
11.946	26.08	26.24	26.22
12.388	25.00	25.14	25.14

TABLE IV
EXPERIMENTAL MEASUREMENT OF THE FIRST HIGHER ORDER MODE COMPARED WITH SIMPLIFIED DISPERSION THEORY

Frequency/GHz	Guided wavelength /mm	
	Measured	Calculated
14.5034	46.15	46.22
14.7190	42.85	[illegible]
14.9634	40.00	[illegible]

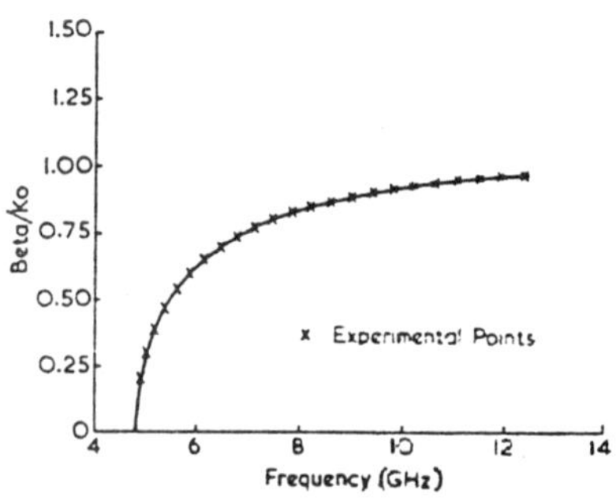

Fig. 14. Theoretical dispersion curve with experimental points, dimensions as in Fig. 6.

ticularly so when the accurate approximation of Section V is employed. Fig. 14 illustrates the excellent agreement between experimental and theoretical values. Finally, by operating the resonant line in the 14-GHz region, it was possible to discern resonances due to the first higher order mode. These compare very successfully with the theoretical predictions using $\eta = 1.0$ and $f_c = 12.97$ GHz in Table IV.

VIII. CONCLUSIONS

We have introduced a general and rigorous space-domain solution of unilateral finline based on transverse resonance diffraction. The affinity of the problem to that of a capacitive iris in waveguide is highlighted, and by satisfying *a priori* the appropriate edge condition, convergence is achieved with very low-order matrices. The quasi-analytic character of the treatment allows an accurate wide-band scalar dispersion equation for the fundamental mode as well as accurate approximate dispersion equations for the higher order modes to be derived. Finally, consideration of the cutoff properties of these modes results in a natural division of the spectrum in mode families and their characterization.

APPENDIX I

The admittances appearing in the sums under (14) are given by

$$y_{11n} = \cos^2 \tau_n (h_{nR} + h_{nL}) + \sin^2 \tau_n (g_{nR} + g_{nL})$$

$$y_{12n} = \frac{\cos \tau_n \sin \tau_n}{n} (h_{nL} - h_{nR} - g_{nL} + g_{nR})$$

$$y_{22n} = \frac{\cos^2 \tau_n}{n^2} (g_{nR} + g_{nL}) + \frac{\sin^2 \tau_n}{n^2} (h_{nR} + h_{nL}). \quad \text{(A1)}$$

It is now noted that the choice of (8) and (9) has the additional advantage of producing by integration by parts the factors n^{-1} in y_{12n} and n^{-2} in y_{22n} that ensure proper convergence. The exclusion of the $n = 0$ term from all but $y_{11}(x, x')$ arises naturally from the aforementioned definition of field quantities (1)–(10).

APPENDIX II

The Schwinger functions are derived from the conformal mapping between a parallel-plate waveguide and a system containing an iris. Such a mapping is of the general form

$$\cosh W = \alpha_1 + \alpha_2 \cosh T.$$

For the symmetrical case in question, the above mapping restricted to the x axis is

$$\cos \frac{2\pi x}{a} = \alpha_1 + \alpha_2 \cos \theta \quad \text{(A2)}$$

such that

$$x = 0 \rightarrow \theta = 0$$

$$x = \frac{w}{2} \rightarrow \theta = \pi.$$

Thus

$$\alpha_1 = \cos^2 \frac{\pi w}{2a} \quad \text{(A3a)}$$

$$\alpha_2 = \sin^2 \frac{\pi w}{2a}. \quad \text{(A3b)}$$

We then find that the required singularity at $x = w/2$ is found in the term $\partial\theta/\partial x$ linking functions of θ to functions of x. Furthermore, the mapping now allows us to analytically determine certain infinite sums.

In the θ domain, we chose the basis set $\{f_m\}$, where

$$f_m(\theta) = \sqrt{\frac{2}{\pi}} \cos m\theta, \qquad m > 0$$

$$= \frac{1}{\sqrt{\pi}}, \qquad m = 0. \quad \text{(A4)}$$

This basis may be linked to the set (ϕ_{hn}) in the x domain via the coefficients P_{mn}, such that

$$\phi_{hn}(x) = \sum_{m=0}^{n} P_{mn} f_m(\theta). \tag{A5}$$

It is clear from (6a), (A4), and (A2) that the first few coefficients are obtained as

$$P_{00} = \sqrt{\frac{\pi}{a}}\, P_{01} = \sqrt{\frac{2\pi}{a}} \cos^2 \frac{\pi w}{2a}$$

$$P_{11} = \sqrt{\frac{\pi}{a}} \sin^2 \frac{\pi w}{2a}.$$

By considering Chebyshev polynomials, we can derive a recursive formula for generating further coefficients. Since

$$\cos \frac{2n\pi}{a} x = T_n\left[\cos \frac{2\pi}{a} x\right] = T_n(\alpha_1 + \alpha_2 \cos\theta) \tag{A6}$$

using the recursive formula

$$T_n(Z) = 2ZT_{n-1}(Z) - T_{n-2}(Z) \tag{A7}$$

with the trigonometric identity

$$\cos\theta \cos m\theta = \frac{1}{2}[\cos(m-1)\theta + \cos(m+1)\theta]$$

equation (A7) becomes

$$\sum_{m=0}^{n} P'_{mn} \cos p\theta$$

$$= 2\alpha_1 \sum_{m=0}^{n-1} P'_{m,n-1} \cos m\theta - \sum_{m=0}^{n-2} P'_{m,n-1} \cos m\theta$$

$$+ \alpha_2 \sum_{m=-1}^{n-2} P'_{m+1,n-1} \cos m\theta$$

$$+ \alpha_2 \sum_{m=1}^{n} P'_{m-1,n-1} \cos m\theta \tag{A8}$$

where P'_{mn} are the coefficients linking the unnormalized sets

$$\left\{\cos \frac{2n\pi}{a} x\right\} \text{ and } \{\cos m\theta\}.$$

Hence

$$P'_{mn} = 2\alpha_1 P'_{m,n-1} - P'_{m,n-2} + \alpha_2 P'_{m+1,n-1}$$

$$+ \alpha_2 \delta_{1,m} P'_{0,n-1} + \alpha_2 P'_{m-1,n-1} \tag{A9}$$

$$\delta_{1,m} = \begin{cases} 0, & m \neq 1 \\ 1, & m = 1 \end{cases}$$

since, from (A2), (A3a), and (A3b), we have

$$P'_{00} = 1 \qquad P'_{01} + \cos^2 \frac{\pi w}{2a} \qquad P'_{11} = \sin^2 \frac{\pi w}{2a}$$

and the coefficients P'_{mn} can be generated as required. The coefficients P_{mn} are obtained directly from P'_{mn} by applying the normalization

$$P_{mn} = \sqrt{\frac{\delta_n}{\delta_m}} \sqrt{\frac{\pi}{a}}\, P'_{mn} \tag{A10}$$

where

$$\delta_n = 2, \qquad n = 0$$

$$\delta_n = 1, \qquad n > 0.$$

Acknowledgment

The authors wish to thank their colleagues C. Ryecroft, P. N. Sargeant, and S. F. Clarke of the Marconi Research Laboratories, Chelmsford, Essex, for many helpful discussions and their encouragement.

References

[1] P. J. Meier, "Integrated finline millimeter components," *IEEE Trans. Microwave Theory Tech.*, vol. MTT-22, pp. 1209–1216, Dec. 1974.

[2] A. M. K. Saad and G. Begemann, "Electrical performance of finlines of various configurations," *IEEE Trans. Microwaves, Opt., Acoust.*, vol. 1, pp. 81–88, Jan. 1977.

[3] H. Hofmann, "Finline dispersion," *Electron. Lett.*, vol. 12, no. 17, pp. 428–429, Aug. 1976.

[4] A. M. K. Saad and K. Schünemann, "A simple method for analyzing fin-line structures," *IEEE Trans. Microwave Theory Tech.*, vol. MTT-26, pp. 1002–1011, Dec. 1978.

[5] H.-C. Shih and W. Hoefer, "Dominant and second-order mode cutoff frequencies in fin lines calculated with a two-dimensional TLM program," *IEEE Trans. Microwave Theory Tech.*, vol. MTT-28, pp. 1443–1448, Dec. 1980.

[6] L.-P. Schmidt and T. Itoh, "Spectral domain analysis of dominant and higher order modes in fin-lines," *IEEE Trans. Microwave Theory Tech.*, vol. MTT-28, pp. 981–985, Sept. 1980.

[7] J. B. Knorr and P. M. Shayda, "Millimeter-wave fin-line characteristics," *IEEE Trans. Microwave Theory Tech.*, vol. MTT-28, pp. 737–743, July 1980.

[8] D. M-Syahkal and J. B. Davies, "An accurate, unified solution to various fin-line structures, of phase constant, characteristic impedance, and attenuation," *IEEE Trans. Microwave Theory Tech.*, vol. MTT-30, pp. 1854–1861, Nov. 1982.

[9] L. P. Schmidt, T. Itoh, and H. Hofmann, "Characteristics of unilateral fin-line structures with arbitrarily located slots," *IEEE Trans. Microwave Theory Tech.*, vol. MTT-29, pp. 352–355, Apr. 1981.

[10] A. Beyer, "Analysis of the characteristics of an earthed fin line," *IEEE Trans. Microwave Theory Tech.*, vol. MTT-29, pp. 676–680, July 1981.

[11] R. Vahldiek, "Accurate hybrid-mode analysis of various finline configurations including multilayered dielectrics, finite metallization thickness, and substrate holding grooves," *IEEE Trans. Microwave Theory Tech.*, vol. MTT-32, pp. 1454–1460, Nov. 1984.

[12] P. Pramanick and P. Bhartia, "Accurate analysis equations and synthesis technique for unilateral finlines," *IEEE Trans. Microwave Theory Tech.*, vol. MTT-33, pp. 24–29, Jan. 1985.

[13] A. M. K. Saad and K. Schünemann, "Closed-form approximations for fin-line eigenmodes," *Proc. Inst. Elec. Eng.*, pt. H. (MOA), vol. 129, pp. 253–261, Oct. 1982.

[14] R. Sorrentino and T. Itoh, "Transverse resonance analysis of finline discontinuities," *IEEE Trans. Microwave Theory Tech.*, vol. MTT-12, pp. 1633–1638, Dec. 1984.

[15] K. J. Webb and R. Mittra, "Solution of the finline step-discontinuity problem using the generalized variational technique," *IEEE Trans. Microwave Theory Tech.*, vol. MTT-33, pp. 1004–1100, Sept. 1985.

[16] A. S. Omar and K. Schünemann, "Transmission matrix representation of finline discontinuities," *IEEE Trans. Microwave Theory Tech.*, vol. MTT-33, pp. 765–770, Sept. 1985.

[17] ——, "Space domain decoupling of LSE and LSM fields in generalized planar guiding structures," *IEEE Trans. Microwave Theory Tech.*, vol. MTT-32, pp. 1626–1632, Dec. 1984.

[18] R. E. Collin, *Field Theory of Guided Waves*. New York: McGraw-Hill, 1960, ch. 8.

[19] J. D. Rhodes, "General constraints on propagation characteristics of electromagnetic waves in uniform inhomogeneous waveguides," *Proc. Inst. Elec. Eng.*, vol. 118, no. 7, pp. 849–856, 1971.

[20] T. Itoh, "Spectral domain immittance approach for dispersion characteristics of generalised printed transmission lines," *IEEE Trans. Microwave Theory Tech.*, vol. MTT-28, pp. 733–736, July 1980.

2.2.1

EQUIVALENT RELATIVE PERMITTIVITY AND UNLOADED Q FACTOR OF INTEGRATED FINLINE

Indexing terms: High-frequency transmission lines, Integrated circuits, Permittivity, Q factor, Microwave measurements

To experimentally demonstrate the capabilities of integrated finline, a cavity test fixture was constructed and tested. The results were analysed to determine the manner in which the guide wavelength and unloaded Q factor varied. The measured unloaded Q factor of integrated finline was found to be two to three times better than that of microstrip. Owing to the low loss of integrated finline, its freedom from tight tolerances and compatibility with hybrid devices, this new transmission line is well suited to millimetre integrated circuits.

Although standard microstrip techniques can be applied to millimetre components,[1-3] several problems arise. These problems include critical tolerances, fragile substrates, thin conductor strips that are not completely compatible with hybrid devices, and difficulty in obtaining a simple transition to conventional waveguide. A new integrated-circuit transmission line, called integrated finline,[4] can overcome or alleviate these problems. This paper will summarise the performance characteristics of integrated finline, as determined by cavity tests.

In integrated finline, metal fins are printed on a dielectric substrate, which bridges the broad walls of a rectangular waveguide. This adaptation of ridge-loaded waveguide permits circuit elements to be fabricated by low-cost printing, and is compatible with thin-film hybrid techniques. The degree of miniaturisation can be limited, which is an advantage at millimetre wavelengths. The fins increase the separation between the first two modes of propagation, thereby providing a wider useful bandwidth than conventional waveguides. Owing to the similarity between integrated finline and conventional ridged waveguide, a wealth of design information[5,6] is available. For thin moderate-ε substrates, the dielectric will have a minor effect,[7] and the single-mode bandwidth and attenuation of integrated finline may be estimated from existing data.[6,8] Such estimates lead to the conclusion that integrated finline can provide bandwidths in excess of an octave, with less attenuation than microstrip.

To experimentally demonstrate the capabilities of integrated finline, the cavity-test fixture shown in Fig. 1 was constructed. The cavity was designed to operate at X band frequencies, since wideband sweepers are readily available in the 8–12 GHz range. The substrate thickness, however, was chosen to be sufficiently large (1·6 mm) to permit direct scaling to higher frequencies. The cavity consists of split housing, which accommodates various substrate samples, three of which are shown at the right foreground. The halves of the housing are held together by four nylon screws, which run parallel to the broad walls of the cavity. Metal irises provide loose coupling (30 dB or better) between the input and output ports of the cavity, thereby permitting a direct determination of the unloaded Q factor from the observed bandwidth at resonance. With the aid of beam-lead contacts, r.f. continuity is maintained between each iris and its mating finline sample.

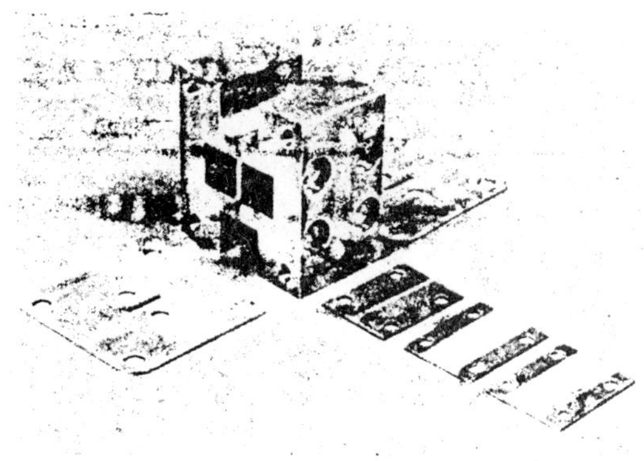

Fig. 1 *Cavity-test fixture*

Two different finline configurations, as shown in Fig. 2A, were evaluated. In the configuration of Fig. 2A(i), the fins are printed on opposite sides of a single dielectric substrate. Since the fins are directly earthed to the metal waveguide walls, this configuration is applicable only to passive devices.

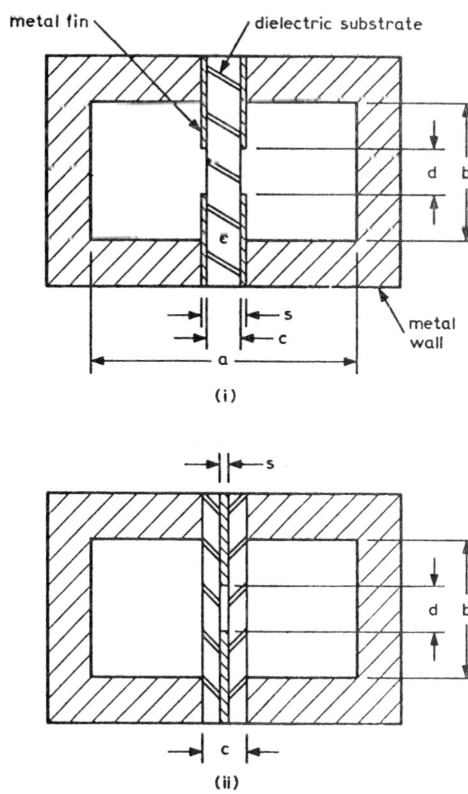

Fig. 2A *Finline configurations*
(i) Earthed fins
(ii) Insulated fins

In the configuration of Fig. 2A(ii), however, the fins are printed on the mating surfaces of two dielectric substrates. Since the fins are insulated from the waveguide for direct current, bias may be introduced for active components. R.F. continuity between the fins and the waveguide wall is obtained by choosing the thickness of the broad walls to be a quarter wavelength in the dielectric medium.[4]

Polytetrafluorethylene-glass-fibre* substrates, with normalised gaps in the range $d/b = 0{\cdot}1$–$1{\cdot}0$, were evaluated in the cavity whose aspect ratio (b/a) was 0·45. The centre frequency

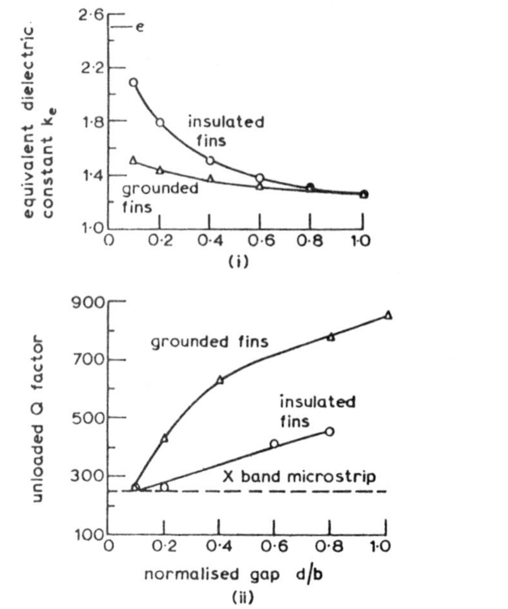

Fig. 2B *Equivalent K and unloaded Q factors of integrated finline*

(i) Equivalent relative permittivity

$$\lambda_g = \frac{\lambda_0}{\sqrt{\{k_e - (\lambda_0/\lambda_c)^2\}}}$$

$c/a = 0{\cdot}07;\ b/a = 0{\cdot}45$

(ii) Measured Q factor

and 3 dB bandwidth were recorded for each sample, and the results were analysed to determine the manner in which the guide wavelength and unloaded Q factor varied. It was determined that the guide wavelength could be approximately calculated from the simple equation given in Fig. 2B(i) in terms of the free-space wavelength λ_0, the equivalent relative permittivity k_e, and the cutoff wavelength λ_c. The values of k_e plotted in Fig. 2B(i) were calculated from the simple equation, by substituting the λ_0 observed at the full-wave resonance and the corresponding λ_c for an airfilled ridge-loaded waveguide of the same dimensions.[6] The general validity of the equation was confirmed by predicting λ_0 at the next highest $(3\lambda_g/2)$ resonance to an accuracy of 15%. Although a more rigorous analysis will be required to calculate the exact variation of λ_g with respect to λ_0, the equation given in Fig. 2B(i) is sufficiently accurate for many engineering applications. It is interesting to note that the value of k_e varies little with respect to d/b when the fins are earthed directly to the waveguide. When d/b approaches unity, the value of k_e determined from the cavity agrees with the published value[8] for both the earthed and insulated cases. As the gap between the insulated fins becomes smaller, however, the value of k_e approaches the relative permittivity of the substrate material.

The unloaded Q factor of integrated finline was determined from the halfpower bandwidth of the cavity resonances. Fig. 2B(ii) shows how the measured Q factor (including the effect of losses within the irises) varies with respect to the gap between the fins. Note that, for values of d/b above 0·1, the Q factor of integrated finline is superior to microstrip in the same X band range.[9] For moderate loading $(d/b \geq 0{\cdot}6)$, the Q factor of integrated finline exceeds 400 for insulated fins and is better than 700 for the earthed-fin configuration. Owing to the low loss of integrated finline, its freedom from tight tolerances and compatibility with hybrid devices, this new transmission line is well suited to millimetre integrated circuits.

The direction and encouragement provided by C. C. Gauder, F. T. Hutson and C. A. Hines of the US Air Force Avionics Laboratory, and J. G. Smith and Jesse J. Taub of AIL is gratefully acknowledged.

The work described was sponsored by the US Air Force Avionics Laboratory, on contract F33615–C–1837.

PAUL J. MEIER

26th February 1973

Applied Electronics Division
AIL, Cutler-Hammer
Melville, NY 11746, USA

References

1 MAO, S., JONES, S., and VENDELIN, G. D.: 'Millimeter-wave integrated circuits', *IEEE Trans.*, 1968, **MTT-16**, pp. 451–461
2 DOBRAMYSL, D.: 'Integrated mixers for 18 and 26 GHz'. IEEE G-MTT symposium digest, 1971, pp. 18–19
3 OXLEY, T. H. *et al.*: 'Hybrid microwave integrated circuits for millimeter wavelengths'. IEEE G-MTT symposium digest, 1972, pp. 224–226
4 MEIER, P. J.: 'Two new integrated-circuit media with special advantages at millimeter wavelengths'. IEEE G-MTT symposium digest, 1972, pp. 221–223
5 COHN, S. B.: 'Properties of ridged waveguide', *Proc. Inst. Radio Eng.*, 1947, **35**, pp. 783–788
6 HOPFER, S.: 'The design of ridged waveguides', *IRE Trans.*, 1955, **MTT-3**, pp. 20–29
7 GARDIOL, F. E.: 'Higher-order modes in dielectrically loaded rectangular waveguides', *IEEE Trans.*, 1968, **MTT-16**, pp. 919–924
8 VARTANIAN, P. H., AYRES, W. P., and HELGESSON, A. L.: 'Propagation in dielectric slab loaded rectangular waveguide', *IRE Trans.*, 1958, **MTT-6**, pp. 215–222
9 HINTE, A. F., KOPCSAY, G. V., and TAUB, J. J.: 'Choosing a transmission line', *Microwaves*, 1971, pp. 46–50

* Teflon-Fiberglas (3M type FL-GT)

2.2.2

Accurate Analysis Equations and Synthesis Technique for Unilateral Finlines

PROTAP PRAMANICK AND PRAKASH BHARTIA, SENIOR MEMBER, IEEE

Abstract —Accurate analysis equations and synthesis techniques are presented for unilateral finlines, valid over a wide range of structural parameters and substrate dielectric constants ($1 \leqslant \epsilon_r \leqslant 3.75$). These expressions are usable for computing the cutoff wavelength to within ± 0.6 percent, the guided wavelength to within ± 2 percent, and the characteristic impedance (based on the power–voltage definition) to within ± 2 percent, of the spectral-domain method, over the normalized frequency range $0.25 \leqslant b/\lambda \leqslant 0.6$.

I. Introduction

FINLINE IS AN ideal transmission line for millimeter-wave circuits because it avoids miniaturization and offers the potential for low-cost production through batch processing techniques [1], [2]. It is also easily compatible with semiconductor devices. It has wide bandwidth for single-mode operation, moderate attenuation, and low dispersion in the frequency range of interest. These properties have made it more popular than microstrip about 30 GHz.

Dispersion in finline has been accurately analyzed by Hofmann [3], Knorr and Shayda [4], Schmidt and Itoh [5], Beyer and Wolff [6], Sharma, Costache, and Hoefer [7], Shih and Hoefer [8], and Saad and Schunemann [9]. These analyses use the eigenmode analysis in space or the spectral-domain, finite-element method, or a two-dimensional transmission-line matrix. The network analytical method of electromagnetic fields, which is similar to the spectral-domain technique, was extended to the more general case of higher order modes by Hayashi, Farr, and Mittra [10]. Although the above-mentioned methods are highly accurate, they require considerable analytical effort and lead to complicated computer programming.

Besides the rigorous analyses above, the propagation constant in finlines has been approximated by various methods. Many authors have treated finlines as ridged waveguides [11], [12]. But the resulting expressions are of poor accuracy for the guided wavelength and the characteristic impedance. For an adequately accurate expression for the effective dielectric constant of finlines, one has to depend on experimental data [1] from expensive and time-consuming sample measurements. Therefore, in spite of all the advantages of a novel transmission line, the basic problem faced by the designers is the cumbersome design procedure.

Consequently, there remains a strong need for accurate closed-form expressions for the equivalent dielectric constant and characteristic impedance for finlines. Recently, Sharma and Hoefer [13] have presented purely empirical expressions for the cutoff wavelength of unilateral and bilateral finlines, which were developed by curve fitting to numerical results obtained by the spectral-domain technique [7]. Because of their purely empirical nature, these expressions are valid for a small range of finline geometries. For example, the equations are valid for $1/16 \leqslant d/b \leqslant 1/4$, $b/a = 0.5$, and $\epsilon_r = 2.22$ and 3.00 only (see Fig. 1(a)). Moreover, different equations are required for dielectric substrates of different permittivity values.

Manuscript received May 3, 1984; revised July 29, 1984. This work was supported in part by the Natural Sciences and Engineering Research Council of Canada under Grant A-0001.

P. Pramanick is with the Department of Electrical Engineering, University of Ottawa, Ottawa, Canada, K1N 6N5.

P. Bhartia is with the Electronics Division, Defense Research Establishment of Ottawa, Ottawa, Canada, K1A 0Z4.

P. Pramanick and P. Bhartia, *IEEE Transactions on Microwave Theory and Techniques*, Vol. MTT-33, No. 1, Jan. 1985, pp. 24–30.

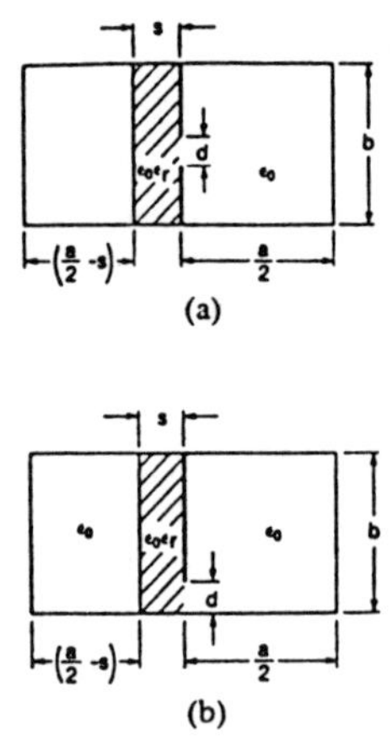

Fig. 1. (a) Unilateral finline. (b) Complementary unilateral finline.

Hence, the equations are of limited use and inappropriate for other dielectric substrates.

In the present work, accurate closed-form expressions are developed for the theoretical prediction of equivalent dielectric constants at cutoff and the cutoff wavelength of unilateral finlines. The expressions are accurate to within ± 0.6 percent. They have been derived using a stationary formula and curve fitting to numerical results of the spectral-domain method. The expressions are valid over the range (see Fig. 1(a))

$$0 < b/a \leqslant 1 \qquad 1/32 \leqslant d/b \leqslant 1.00$$
$$1/64 \leqslant s/a \leqslant 1/4 \qquad 1 \leqslant \epsilon_r \leqslant 3.75.$$

In addition, unlike Sharma and Hoefer's [13] expressions, the present expressions are less arbitrary and the constants appearing in them are independent of substrate dielectric constants for the most commonly used substrate materials.

II. Expressions for Guided Wavelength and Frequency-Dependent Characteristic Impedance

The guided wavelength in finline is defined as

$$\lambda_g = \lambda/\sqrt{\epsilon_e(f)} \tag{1}$$

where λ is the free-space wavelength and $\epsilon_e(f)$, the frequency-dependent effective dielectric constant of the finline, is given by [1]

$$\epsilon_e(f) = k_e - (\lambda/\lambda_{ca})^2 \tag{2}$$

where k_e is the equivalent dielectric constant at frequency f corresponding to wavelength λ, and λ_{ca} is the cutoff wavelength of a finned waveguide of identical dimensions and completely filled with air. It has been shown in [14] that for moderate $\epsilon_r (\epsilon_r \leqslant 2.5)$ and thin substrates one can make a first-order approximation by equating k_e to its value k_c at cutoff frequency. Otherwise, k_e must be considered frequency-dependent and has the general form [13]

$$k_e = k_c f(d/b, s/a, b/\lambda, \epsilon_r) \tag{3}$$

where

$$k_c = (\lambda_{cf}/\lambda_{ca})^2 \tag{4}$$

and λ_{cf} is the cutoff wavelength of the finline.

An empirical expression for the function $f(d/b, s/a, b/\lambda, \epsilon_r)$ in (3) is given in [13]. Having defined the effective dielectric constant $\epsilon_e(f)$, the characteristic impedance of the finline is defined as

$$Z_0 = Z_{0\infty}/\sqrt{\epsilon_e(f)} \tag{5}$$

where $Z_{0\infty}$, according to the ridged guide model [1], is the characteristic impedance at infinite frequency of the equivalent finned waveguide of identical dimensions.

III. Derivation of the Form of the Expression

The unilateral finline shown in Fig. 1(a) can be thought of as a combination of a finned waveguide and a waveguide loaded with a centered dielectric slab in the E-plane. The cutoff wavelength λ_{ca} of the finned waveguide can be determined accurately using the following equation [15]:

$$\frac{b}{\lambda_{ca}} = \frac{b}{2a}\left[1 + \frac{4}{\pi}\left(\frac{b}{a}\right)\left(1 + 0.2\sqrt{\frac{b}{a}}\right)\ln \operatorname{cosec}\left(\frac{\pi}{2}\frac{d}{b}\right)\right]^{-1/2} \tag{6}$$

The cutoff wavelength λ_{cd} of the dielectric loaded waveguide can be obtained, for small s/a and ϵ_r, using the stationary formula [16], obtained from the variational technique, as

$$\frac{b}{\lambda_{cd}} = \frac{b}{2a}\left[1 + 0.5\left(\frac{2s}{a} + \frac{1}{\pi}\sin\left(\frac{2\pi s}{a}\right)\right)(\epsilon_r - 1)\right]^{-1/2}. \tag{7}$$

The derivation of (7) using the variational formula [16] assumes a sinusoidal field distribution in the transverse cross section of the guide. The field distribution has the form

$$\overline{E_y} = \bar{y}\sin\left(\frac{\pi x}{a}\right) \tag{8}$$

where $\bar{y}$ is the unit vector in the y-direction.

For a general field distribution in the transverse cross section of the waveguide, (7) can be written as

$$\frac{b}{\lambda_{cd}} = \frac{b}{2a}\left[1 + F\left(\frac{s}{a}\right)(\epsilon_r - 1)\right]^{-1/2} \tag{9}$$

where the function $F(s/a)$ depends upon the nature of the field distribution. Equation (9) has a stationary form. Therefore, it remains unchanged for small variations in ϵ_r, s/a, and frequency. For example, the inaccuracy of (7) is less than ± 0.8 percent for $s/a \leqslant 1/4$, which may be the thickest substrate used for finlines.

Now, consider the combination of the finned waveguide with the dielectric loaded waveguide. Due to the presence of the fins, the fields tend to concentrate in the vicinity of the fins, and the field distribution in the transverse cross section of the guide no longer remains a function of s/a only but also becomes a function of d/b.

Therefore, (9) assumes the form

$$\frac{b}{\lambda_{cd}}=\frac{b}{2a}\left[1+F\left(\frac{s}{a},\frac{d}{b}\right)(\epsilon_r-1)\right]^{-1/2}. \tag{10}$$

Once the function $F(s/a, d/b)$ is known, the cutoff wavelength λ_{cf} of the unilateral finline of Fig. 1(a) can be obtained using (2), (6), and (10) as

$$\frac{b}{\lambda_{cf}}=\frac{b}{2a}\left[\left\{1+\frac{4}{\pi}\left(\frac{b}{a}\right)\left(1+0.2\sqrt{\frac{b}{a}}\right)\ln\operatorname{cosec}\left(\frac{\pi}{2}\frac{d}{b}\right)\right\}\cdot\left\{1+F\left(\frac{s}{a},\frac{d}{b}\right)(\epsilon_r-1)\right\}\right]^{-1/2}. \tag{11}$$

IV. Derivation of the Function $F(s/a, d/b)$

Equation (11), when solved for $F(s/a, d/b)$, gives

$$F\left(\frac{s}{a},\frac{d}{b}\right)=\frac{\left[\left\{\left(\frac{\lambda_{cf}^2}{4a^2}\right)\Big/\left(1+\frac{4}{\pi}\left(\frac{b}{a}\right)\left(1+0.2\sqrt{\frac{b}{a}}\right)\ln\operatorname{cosec}\left(\frac{\pi}{2}\frac{d}{b}\right)\right)\right\}-1\right]}{(\epsilon_r-1)}. \tag{12}$$

Expressing

$$F\left(\frac{s}{a},\frac{d}{b}\right)=\left\{a_1\left(\frac{s}{a}\right)\ln\operatorname{cosec}\left(\frac{\pi}{2}\frac{d}{b}\right)+b_1\left(\frac{s}{a}\right)\right\}\frac{s}{a} \tag{13}$$

and using the accurately computed values of λ_{cf} from the spectral-domain technique [7] and (12), the function $F(s/a, d/b)$ can be computed for several combinations of s/a and d/b. The computed results are plotted in Fig. 2 ($(F(s/a, d/b)(a/s)$ versus $\ln\operatorname{cosec}((\pi/2)(d/b))$ with s/a as a parameter). From the plots, we find that the curves are almost linear and the slopes and intercepts are functions of the s/a values. Slopes are larger for smaller s/a.

Using least-squares curve fitting gives the following set of equations for four different values of s/a for $\epsilon_r=2.22$:

$$\left.\begin{aligned} a_1\left(\frac{s}{a}\right)&=0.1006616\\ b_1\left(\frac{s}{a}\right)&=1.6926529\end{aligned}\right\}\quad \text{for } \frac{s}{a}=\frac{1}{4} \tag{14}$$

$$\left.\begin{aligned} a_1\left(\frac{s}{a}\right)&=0.5339579\\ b_1\left(\frac{s}{a}\right)&=2.1643506\end{aligned}\right\}\quad \text{for } \frac{s}{a}=\frac{1}{8} \tag{15}$$

$$\left.\begin{aligned} a_1\left(\frac{s}{a}\right)&=1.353632\\ b_1\left(\frac{s}{a}\right)&=2.4213244\end{aligned}\right\}\quad \text{for } \frac{s}{a}=\frac{1}{16} \tag{16}$$

$$\left.\begin{aligned} a_1\left(\frac{s}{a}\right)&=2.5611088\\ b_1\left(\frac{s}{a}\right)&=2.3070609\end{aligned}\right\}\quad \text{for } \frac{s}{a}=\frac{1}{32}. \tag{17}$$

$a_1(s/a)$ and $b_1(s/a)$ are plotted as functions of $\ln(s/a)$ in Fig. 3. The two graphs can be represented by the equations

$$a_1\left(\frac{s}{a}\right)=0.4020974\left(\ln\left(\frac{a}{s}\right)\right)^2-0.7684487\ln\left(\frac{a}{s}\right)+0.3932021 \tag{18}$$

$$b_1\left(\frac{s}{a}\right)=2.42\sin\left(0.556\ln\left(\frac{a}{s}\right)\right). \tag{19}$$

From (10), the equivalent dielectric constant at cutoff is given by

$$k_c=1+\frac{s}{a}\left[a_1\left(\frac{s}{a}\right)\ln\operatorname{cosec}\left(\frac{\pi}{2}\frac{d}{b}\right)+b_1\left(\frac{s}{a}\right)\right](\epsilon_r-1) \tag{20}$$

which shows that k_c depends linearly on ϵ_r.

From (20), the dielectric filling factor can be written as

$$q_f=\frac{k_c-1}{\epsilon_r-1}=\frac{s}{a}\left[a_1\left(\frac{s}{a}\right)\ln\operatorname{cosec}\left(\frac{\pi}{2}\frac{d}{b}\right)+b_1\left(\frac{s}{a}\right)\right]. \tag{21}$$

Once k_c has been obtained, the frequency-dependent effective dielectric constant can be obtained using (2) and (3).

V. Characteristic Impedance

The definition of characteristic impedance for finlines is not unique. It depends upon the application. According to Meinel and Rembold [17], the characteristic impedance of a finline should be defined in terms of voltage and current in the finline, where voltage is defined as the line integral over the electric field between the fins taken along the shortest path on the substrate surface, and current is the total longitudinal current in the structure. This definition is useful for switching applications and particularly suitable for beam-lead devices. On the other hand, Meier [1] defines the characteristic impedance on a power–voltage basis and uses a ridged waveguide model for the finline. For his model, the term $Z_{0\infty}$ in (5) is frequency-independent. But, in practice, $Z_{0\infty}$ rises slowly with frequency as the crowding of the electric field causes the fin gap voltage to rise. This has been observed by computation of Z_0 using the network analytical technique [10]. Keeping the above facts in mind, the following expression for Z_0 has been derived by curve fitting to the spectral-domain results:

$$Z_0=\frac{240\pi^2(px+q)(b/a)}{(0.385x+1.762)^2\sqrt{\epsilon_e(f)}} \tag{22}$$

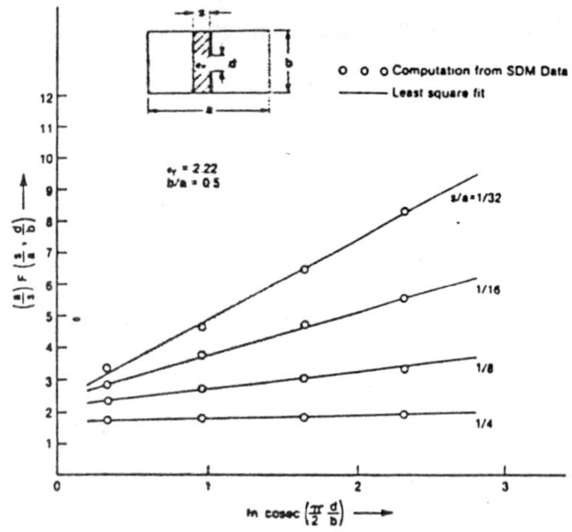

Fig. 2. Least-squares curve fitting to SDM data.

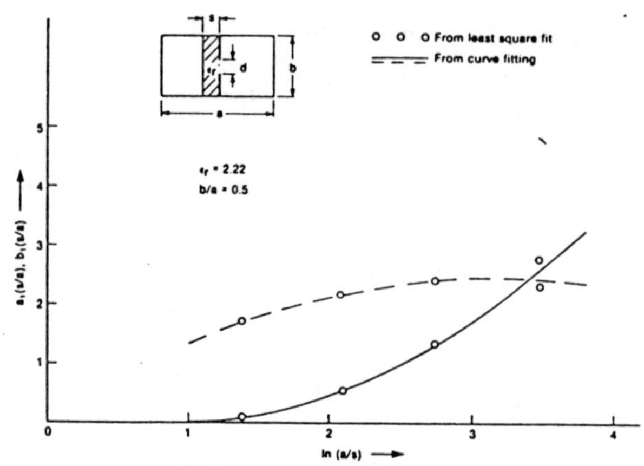

Fig. 3. Curves representing $a_1(s/a)$ and $b_1(s/a)$ a function of $\ln(a/s)$(—— $a_1(s/a)$ - - - - $b_1(s/a)$).

where

$$p = -0.763\left(\frac{b}{\lambda}\right)^2 + 0.58\left(\frac{b}{\lambda}\right) + 0.0775\left[\ln\left(\frac{a}{s}\right)\right]^2 - 0.668\left[\ln\left(\frac{a}{s}\right)\right] + 1.262 \quad (23)$$

$$q = 0.372\left(\frac{b}{\lambda}\right) + 0.914, \qquad \text{for } d/b > 0.3 \quad (24)$$

and

$$p = 0.17\left(\frac{b}{\lambda}\right) + 0.0098 \quad (25)$$

$$q = 0.138\left(\frac{b}{\lambda}\right) + 0.873, \qquad \text{for } d/b \leqslant 0.3 \quad (26)$$

and

$$x = \ln \operatorname{cosec}\left(\frac{\pi}{2}\frac{d}{b}\right). \quad (27)$$

Equation (22) is accurate within ± 2 percent for $s/a \leqslant 1/20$ and within ± 3 percent for $s/a > 1/20$, while Meier's ridged waveguide model [1] has inaccuracies of the order of ± 9 percent [4].

VI. Synthesis

To fabricate the appropriate finline structure, one requires the normalized fin-gap d/b for a given set of s/a, b/a, and ϵ_r. This can be obtained from (22) in the following way.

For low dielectric constant and thin substrate, the frequency-dependent effective dielectric constant can be written as

$$\epsilon_e(f) = k_c - \left(\frac{\lambda}{\lambda_{ca}}\right)^2. \quad (28)$$

Using (20) and noting that

$$\frac{\frac{s}{a}a_1\left(\frac{s}{a}\right)(\epsilon_r - 1)x - \left(\frac{\lambda}{\lambda_{ca}}\right)^2}{1 + \frac{s}{a}b_1\left(\frac{s}{a}\right)(\epsilon_r - 1)} \ll 1$$

for $0 < b/a \leqslant 1$, $1/32 \leqslant d/b \leqslant 1$, $s/a \leqslant 1/20$, $1 \leqslant \epsilon_r \leqslant 3.75$, and $0.1 \leqslant b/\lambda \leqslant 0.6$, one can write

$$\sqrt{\epsilon_e(f)} = \frac{Gx^2 + Hx + I}{Fx + E} \quad (29)$$

where

$$E = 8\left[1 + \frac{s}{a}b_1\left(\frac{s}{a}\right)(\epsilon_r - 1)\right]^{1/2} \quad (30)$$

$$F = \left(\frac{4}{\pi}\right)\left(\frac{b}{a}\right)\left(1 + 0.2\sqrt{\frac{b}{a}}\right)E \quad (31)$$

$$G = 0.5\left(\frac{s}{a}\right)a_1\left(\frac{s}{a}\right)(\epsilon_r - 1)F \Big/ \left[1 + \frac{s}{a}b_1\left(\frac{s}{a}\right)(\epsilon_r - 1)\right]^{1/2} \quad (32)$$

$$H = E(F/8 + G/F) \quad (33)$$

$$I = E^2/8 - \left(\frac{b}{a}\right)^2(\lambda/b)^2. \quad (34)$$

Combining (22) and (29) gives

$$Z_0 = \frac{240\pi^2(px + q)(Fx + E)(b/a)}{(0.385x + 1.7621)^2(Gx^2 + Hx + I)} \quad (35)$$

for a given set of Z_0, b/a, s/a, ϵ_r, and b/λ, (35) reduces to a quartic equation in x of the form

$$x^4 + C_3x^3 + C_2x^2 + C_1x + C_0 = 0 \quad (36)$$

where

$$C_3 = H/G + 9.156 \quad (37)$$

$$C_2 = I/G + 9.156(H/G) + 20.95 - 6.748\frac{pF}{G\bar{Z}} \quad (38)$$

$$C_1 = 9.156(I/G) + 20.95(H/G) - 6.748(pE + qF)/G\bar{Z} \quad (39)$$

$$C_0 = 20.95(I/G) - 6.748(qE/G\bar{Z}) \quad (40)$$

$$\bar{Z} = Z_0/\{(240\pi^2)(b/a)\}. \quad (41)$$

Equation (36) can be solved easily using a suitable iterative

TABLE I

Normalized substrate Thickness s/a	Normalized Gap Width d/b	Cutoff Frequency b/λ_{cf} of the dominant mode $\epsilon_r = 3.00$ SDT*	SH**	PM***
1/4	1/2	0.16269	-	0.16363
	1/4	0.13908	0.13840	0.13961
	1/8	0.12146	0.12233	0.12169
	1/16	0.10874	0.10814	0.10883
1/8	1/2	0.17706	-	0.17731
	1/4	0.14756	0.14673	0.14853
	1/8	0.12684	0.12800	0.12761
	1/16	0.11244	0.11167	0.11208
1/16	1/2	0.19114	-	0.19147
	1/4	0.15799	0.15640	0.15853
	1/8	0.13410	0.13502	0.13448
	1/16	0.11755	0.11657	0.11712
1/32	1/2	0.20275	-	0.20396
	1/4	0.16881	0.16766	0.16844
	1/8	0.14285	0.14364	0.14253
	1/16	0.12409	0.12306	0.12386

*Spectral domain technique [7]
**Empirical equation by Sharma and Hoefer [13]
***Present method

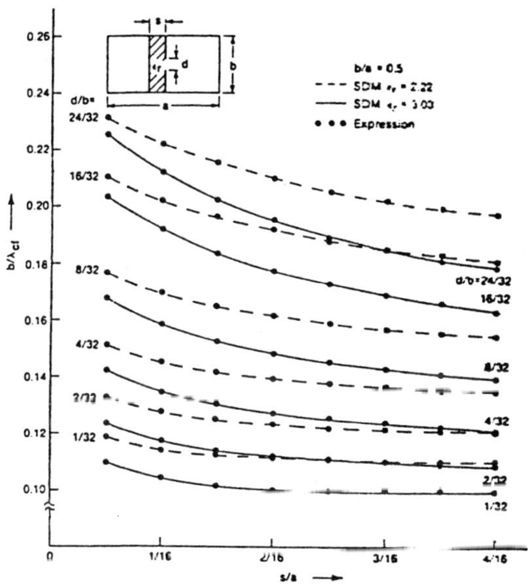

Fig. 4. Normalized cutoff frequency (b/λ_{cf}) as a function of normalized fin-gap $(d/b)\epsilon_r = 2.22$ and 3.00. $b/a = 0.5$.

technique to determine x, and hence d/b, since

$$d/b = \left(\frac{2}{\pi}\right)\sin^{-1}[\exp(-x)]. \tag{42}$$

VII. Computed Results

The results computed using (11), (18), (19), and (20) are shown in Table I and compared with those obtained using the spectral-domain technique [7] and the empirical expressions of Sharma and Hoefer [13]. The results show excellent agreement. Since the expressions were derived using the spectral-domain technique data for $\epsilon_r = 2.22$, the excellent agreement of the expression when $\epsilon_r = 3.00$ confirms the stationary nature of (20). Hence, (20) is a general equation for low ϵ_r values. The expressions have been found valid for $\epsilon_r = 3.75$ within ± 1 percent.

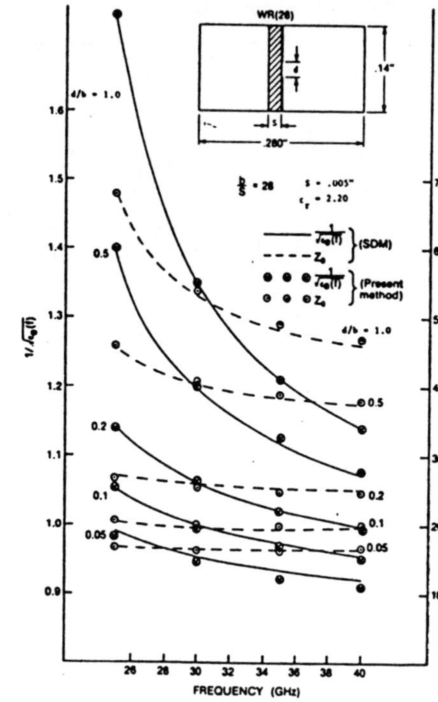

Fig. 5. Normalized phase velocity $1/\sqrt{\epsilon_e(f)}$ versus frequency and characteristic impedance Z_0 versus frequency in WR(28) waveguide. $s/a = 0.01786$.

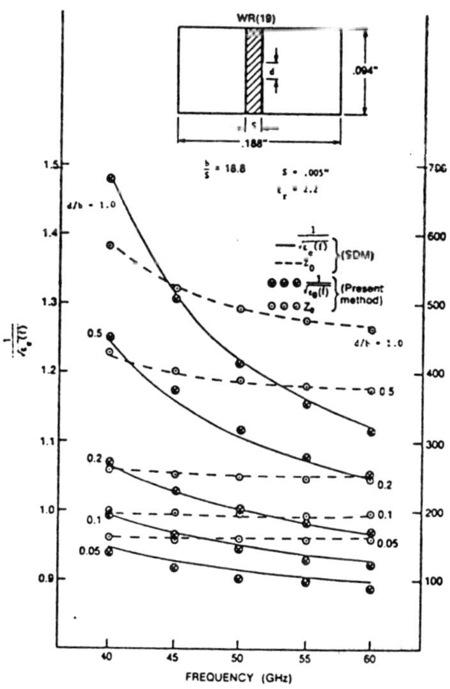

Fig. 6. Normalized phase velocity $1/\sqrt{\epsilon_e(f)}$ versus frequency and characteristic impedance versus frequency in WR(19) waveguide. $s/a = 0.0265957$.

TABLE II

d/b	Frequency (GHz)	Z_0 (D) (Ohm)	Z_0 (S) (Ohm)	$\frac{1}{\sqrt{\epsilon_e(f)}}$ (D)	$\frac{1}{\sqrt{\epsilon_e(f)}}$ (S)
	40	258.754	256.689	1.0773	1.0743
	45	252.733	248.420	1.0337	1.0292
0.2	50	249.338	244.012	1.00495	0.99968
	55	247.180	241.716	0.98462	0.9788
	60	245.556	240.695	0.96949	0.96321
	65	244.078	240.499	0.95777	0.95105
	40	429.924	433.563	1.2768	1.2986
	45	403.359	405.294	1.1797	1.1864
0.5	50	389.174	391.762	1.1222	1.1239
	55	381.068	381.429	1.0843	1.0840
	60	376.327	375.360	1.0578	1.0565
	65	373.585	371.724	1.03822	1.0363

D ≙ Desired
S ≙ Synthesized

Fig. 4 compares the normalized dominant-mode cutoff wavelength (b/λ_{cf}) in unilateral finlines, computed by the present method, with those obtained by the spectral-domain technique [7] for $\epsilon_r = 2.22$ and 3.00. Figs. 5 and 6 show the variations of normalized guided wavelength and characteristic impedance with frequency and compare with the results of Knorr and Shayda [4]. The agreement is within ± 2 percent. This inspires confidence in the above expressions.

The results of synthesis are shown in Table II for 127-μm RT Duroid substrate in WR(19) waveguide. Z_0 values were calculated, over the useful frequency band, using (22). The computed values of Z_0 were then used in (36)–(41) to compute x. The computed values of x were then used to obtain the synthesized Z_0 and $\epsilon_e(f)$ values from the analysis equations. It is found that the synthesized Z_0 are within ± 3 percent and guided wavelength is within ± 2 percent of design specifications. Therefore, the synthesis technique may be used to obtain an initial design, after which the more accurate analysis equation could be employed to correct the design dimensions, if required.

VIII. Application to Complementary Unilateral Finline

The closed-form expressions derived above for analysis of the unilateral finline can be easily applied to the complementary unilateral finline, shown in Fig. 1(b), with the following interpretations: b is twice the height of the housing of the complementary finline, and d is twice the gap between the tip of the fin and the bottom wall of the housing. With this, (20) remains unchanged for the complementary unilateral finline.

The interpretation remains the same for the characteristic impedance. But, the characteristic impedance of the complementary unilateral finline is half that of the unilateral finline.

IX. Conclusions

In the preceding sections, accurate closed-form expressions are developed for the cutoff wavelength and the equivalent dielectric constant at cutoff of unilateral finlines. The derivation starts with the assumption of a stationary form for the equivalent dielectric constant at cutoff. The stationary function is subsequently determined by least-squares curve fitting to accurate numerical data, obtained by the spectral-domain technique. The expressions are accurate within ± 0.6 percent over the complete practical range of interest. They can be used for quick and easy evaluation of the dispersion characteristic by hand or calculator within ± 2 percent, for $\epsilon_r \leq 2.50$ and $b/\lambda \leq 0.6$. For higher ϵ_r, a frequency correction of the equivalent dielectric constant is required.

An expression is developed for the characteristic impedance, based on the power–voltage definition, by curve fitting to results obtained by the spectral-domain technique. The expression is accurate to within ± 2 percent for all practical purposes.

The first-order synthesis technique gives the phase velocity within ± 2 percent and the characteristic impedance within ± 3 percent of originally specified values. They can be further corrected using the more accurate analysis formulas. The present analysis and synthesis equations will be useful in computer-aided design and optimization of unilateral finline circuits [18].

Acknowledgment

The authors are grateful to Dr. H. Hofmann of AEG Telefunken, West Germany, and Dr. P. M. Shayda of the U.S. Department of Navy, Washington, DC, for providing data and helpful discussions.

References

[1] P. J. Meier, "Integrated fin-line millimeter components," *IEEE Trans. Microwave Theory Tech.*, vol. MTT-22, pp. 1209–1216, Dec. 1974.

[2] P. J. Meier, "Microwave transmission line," U.S. Patent 3 825 863, July 1974.

[3] H. Hofmann, "Dispersion of planar waveguides for millimeter-wave applications," *Arch. Elec. Übertragung.*, vol. 31, pp. 40–44, Jan. 1977.

[4] J. B. Knorr and P. M. Shayda, "Millimeter wave fin-line characteristics," *IEEE Trans. Microwave Theory Tech.*, vol. MTT-28, pp. 737–743, July 1980.

[5] L.-P. Schmidt and T. Itoh, "Spectral domain analysis of dominant and higher order modes in fin-lines," *IEEE Trans. Microwave Theory Tech.*, vol. MTT-28, pp. 981–985, Sept. 1980.

[6] A. Beyer and I. Wolff, "A solution of the earthed fin-line with finite metalization thickness," in *1980 IEEE MTT-S Int. Microwave Symp. Dig.*, (Washington, DC), pp. 258–260.

[7] A. K. Sharma, G. I. Costache, and W. J. R. Hoefer, "Cutoff in finlines evaluated with the spectral domain technique and with the finite element method," in *1981 IEEE AP-S Int. Antenna Propagation Symp. Dig.*, (Los Angeles, CA), pp. 308–311.

[8] Y. C. Shih and W. J. R. Hoefer, "Dominant and second order mode cutoff frequencies in fin-lines calculated with two-dimensional TLM program," *IEEE Trans. Microwave Theory Tech.*, vol. MTT-28, pp. 1443–1448, Dec. 1980.

[9] A. K. Saad and K. Schunemann, "Efficient eigenmode analysis for planar transmission lines," *IEEE Trans. Microwave Theory Tech.*, vol. MTT-30, pp. 2125–2132, Dec. 1982.

[10] Y. Hayashi, E. Farr, and R. Mittra, "Analysis of dominant and higher order modes in unilateral fin-lines," Interim Tech. report, U.S. Army Research Office, Contract No. DAAG29-82-k-0084, Electromagnetics Laboratory Department of Electrical Engineering, Engineering Experiment Station, University of Illinois at Urbana-Champaign, August 1982.

[11] A. M. K. Saad and G. Begemann, "Electrical performance of fin-lines of various configurations," *Microwaves. Opt. Acoust.*, vol. 1, pp. 81–88, Jan. 1977.

[12] W. J. R. Hoefer, "Fin-line design made easy," in *1978 IEEE MTT-S Int. Microwave Symp. Dig.*, (Ottawa, Canada), p. 471.

[13] A. K. Sharma and W. J. R. Hoefer, "Empirical expressions for finline design," *IEEE Trans. Microwave Theory Tech.*, vol. MTT-31, pp. 350–356, Apr. 1983.

[14] A. M. K. Saad and K. Schunemann, "A simple method for analysing fin-line structures," *IEEE Trans. Microwave Theory Tech.*, vol. MTT-26, Dec. 1978.

[15] M. N. Burton and W. J. R. Hoefer, "Closed-form expressions for the parameters of finned and ridged waveguides," *IEEE Trans. Microwave Theory Tech.*, vol. MTT-30, pp. 2190–2194, Dec. 1982.

[16] R. F. Harrington, *Time Harmonic Electromagnetic Fields.* New York: McGraw-Hill, 1961, p. 345.

[17] H. Meinel and B. Rembold, "New millimeter wave fin-line attenuators and switches," in *1979 IEEE MTT-S Int. Microwave Symp. Dig.*, (Orlando, FL), pp. 249–252.

[18] P. Pramanick and P. Bhartia, "Analysis and synthesis of tapered fin-lines," presented at IEEE MTT-S, Int. Symp., San Francisco, CA, 1984.

2.2.3

Simple Formulae for Dispersion in Bilateral Fin-Lines

by Protap Pramanick* and Prakash Bhartia**

Accurate analysis equations are presented for dispersion in bilateral fin-lines, valid over a wide range of structural parameters and the substrate dielectric constant ($1 \leqq \varepsilon_r \leqq 3.80$). The expressions are accurate to within $\pm$ 0.8% for the cutoff frequency, $\pm$ 1% for the guided wavelength and $\pm$ 2% for the characteristic impedance (based on voltage current definition) over the frequency band $0.1 \leqq b/\lambda \leqq 0.65$, which is suitable for most practical applications.

Einfache Dispersionsbeziehungen für bilaterale Flossenleitungen

Für die Dispersion von bilateralen Flossenleitungen werden genaue Beziehungen angegeben, die für weite Bereiche der Strukturparameter und für Dielektrizitätszahlen des Substrates von 1 bis 3,8 gelten. Für die Grenzfrequenz sind sie auf $\pm$ 0,8% genau, für die Leitungswellenlänge auf $\pm$ 1% und für den Wellenwiderstand aus Spannung und Strom auf $\pm$ 2%, und zwar für ein Verhältnis von Hohlleiterhöhe zu Wellenlänge im Bereich von 0,1 bis 0,65, so wie es praktisch meist vorkommt.

1. Introduction

Fin-line structures [1] are quite popular due to their favorable properties in millimeter-wave *E*-plane circuit applications. These include low dispersion, relatively wide single-mode range of frequency, easy compatibility with semi-conductor devices, moderate loss compared to other planar transmission lines and avoidance of miniaturization.

Among the various types of fin-lines available todate, unilateral and bilateral fin-lines are the most suitable for circuit applications [1], [2].

Dispersion in fin-line has been well analysed by Hofmann [3], Knorr and Shayda [4], Schmidt and Itoh [5], Bayer and Wolff [6], Sharma, Costache and Hoefer [7], Shih and Hoefer [8], Saad and Schünemann [9], Sorrentino and Itoh [10] and Hayashi, Farr and Mittra [11]. These analyses use eigenmode analysis in space or spectral domain, finite element method or transmission line matrix (TLM) analysis. All these methods are mathematically rigorous but require considerable analytical effort and complicated computer programming.

For ease of design, the phase velocity in fin-line has been approximated by various methods. They have been treated as ridged waveguides by Hoefer [12], and Saad and Begemann [13]. But such analysis does not result in good accuracy. Recently, Piotrowski [14] has presented simple formulae for unilateral and bilateral fin-lines, using transverse resonance technique. But the formulae are not in closed form and require solution of two transcendental equations. Moreover the accuracies of the formulae are restricted to $\pm$ 3%. For an adequately accurate expression for the equivalent dielectric constant of the fin-line one has to depend on experimental data [1] of expensive and time-consuming sample measurement.

As a result there is an urgent need for accurate closed-form expressions for the phase velocity and the characteristic impedance of fin-lines. Sharma and Hoefer [15] have presented purely empirical expressions for the cutoff wavelength of unilateral and bilateral fin-lines. But those formulae are valid for small slots and the accuracy is on the order of five per cent. Moreover, different formulae are needed for different values of the substrate dielectric constant.

In this paper, accurate closed-form expressions are presented for bilateral fin-lines, for prediction of the phase velocity within $\pm$ 1% of spectral domain results. The expressions are valid over the range (see Fig. 1)

$$0 \leqq b/a \leqq 1, \quad 1/64 \leqq s/a \leqq 1/8,$$
$$1/32 \leqq d/b \leqq 1, \quad 1 \leqq \varepsilon_r \leqq 3.80.$$

The equations have been derived by the least squares curve fitting to accurate results of spectral domain analysis [7].

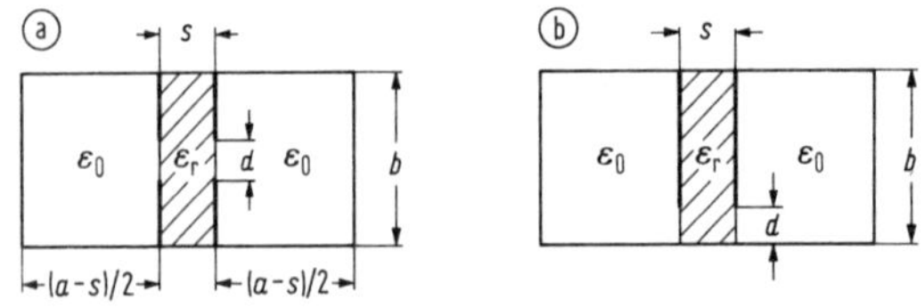

Fig. 1. Bilateral fin-lines.

2. The Ridged Waveguide Model for Guide Wavelength

The guide wavelength in the ridged waveguide model of fin-line [1] is defined as

$$\lambda_g = \lambda / \sqrt{\varepsilon_e(f)} \tag{1}$$

---

* Dr. P. Pramanick, Avionics Division, Canadian Marconi Co., Kanata, Ontario, Canada K2K 2B2.

** Dr. P. Bhartia, Electronics Division, Defence Research Establishment of Ottawa, Ottawa, Ontario, Canada, K1A 0Z4.

P. Pramanick and P. Bhartia, *Archiv fur Elektronik und Ubertragungstechnik*, Vol. 39, No. 6, Nov.-Dec. 1985, pp. 383–386.

where λ is the free space wavelength and $\varepsilon_e(f)$ the frequency dependent effective dielectric constant, is defined according to Meier [1] as

$$\varepsilon_e(f) = k_e(f) - (\lambda/\lambda_{ca})^2 \qquad (2)$$

where $k_e(f)$ is the frequency dependent equivalent dielectric constant and λ_{ca} is the cutoff wavelength of an air filled finned waveguide of identical dimensions.

3. Cutoff in Air Filled Fin-Line

The cutoff wavelength in air filled fin-line is obtained from, with the help of the least squares curve fitting to spectral domain results,

$$\frac{b}{\lambda_{ca}} = \frac{b}{2a}\left\{1 + N\left[\left(1 + \frac{s}{a}\,p\left(\frac{s}{a}\right)\right)x + \frac{s}{a}\,q\left(\frac{s}{a}\right)\right]\right\}^{-1/2} \qquad (3)$$

where

$$p(s/a) = -3.185[\ln(a/s)]^3 + 28.86[\ln(a/s)]^2 - 65.51\ln(a/s) + 53.40$$

$$q(s/a) = 0$$

for $3/4 \leqq d/b \leqq 1$,

$$p(s/a) = 0.3[\ln(a/s)]^3 - 3.155[\ln(a/s)]^2 + 13.46\ln(a/s) - 13.34$$

$$q(s/a) = -0.33[\ln(a/s)]^3 + 2.95[\ln(a/s)]^2 - 7.8\ln(a/s) + 7.1$$

for $1/6 \leqq d/b \leqq 3/4$,

$$p(s/a) = -1.156[\ln(a/s)]^3 + 11.4[\ln(a/s)]^2 - 23.98\ln(a/s) + 17.63$$

$$q(s/a) = 1.834[\ln(a/s)]^3 - 18.18[\ln(a/s)]^2 + 44.86[\ln(a/s)] - 33.56$$

for $1/32 \leqq d/b \leqq 1/6$ and $1/64 \leqq s/a \leqq 1/8$,

$$N = (4/\pi)(b/a)\left(1 + 0.2\sqrt{b/a}\right), \qquad (4)$$

$$x = \ln\operatorname{cosec}\left(\frac{\pi}{2}\,\frac{d}{b}\right). \qquad (5)$$

4. Equivalent Dielectric Constant

The frequency dependent equivalent dielectric constant of the fin-line is given by [16]

$$k_e(f) = k_c + \frac{k_1 - k_c}{\frac{b}{\lambda_1} - \frac{b}{\lambda_{cf}}}\left(\frac{b}{\lambda} - \frac{b}{\lambda_{cf}}\right) \qquad (6)$$

where k_c, the equivalent dielectric constant at cutoff, is given by

$$k_c = 1 + \frac{s}{a}\left[a_1\left(\frac{s}{a}\right)x + b_1\left(\frac{s}{a}\right)\right](\varepsilon_r - 1) \qquad (7)$$

where

$$a_1(s/a) = 1.666\ln(a/s) - 2.125$$

$$b_1(s/a) = 0.1114\ln(a/s) + 1.8$$

for $1/2 \leqq d/b \leqq 1$,

$$a_1(s/a) = 0.392[\ln(a/s)]^2 - 0.8\ln(a/s) + 0.366$$

$$b_1(s/a) = 2.554\sin[0.567\ln(a/s)]$$

for $1/32 \leqq d/b \leqq 1/2$

and

$$\lambda_{cf} = \lambda_{ca}\sqrt{k_c}.$$

k_1 is the equivalent dielectric constant at the wavelength λ_1 at which the air-filled portion of the fin-line goes to cutoff. λ_1 is given by

$$\frac{\lambda_1}{\lambda_{ca}} = \pi\left[\frac{\bar{x}(3 - 2\bar{x})(\varepsilon_r - 1)}{12}\right]^{1/2} \qquad (8)$$

and

$$k_1 = 1 + \frac{\pi^2}{12}\bar{x}(3 - 2\bar{x})(\varepsilon_r - 1) \qquad (9)$$

where

$$\bar{x} = \left(\frac{2q_f}{\pi^2} - s_1^3\right)^{1/3} + s_1 + \frac{q_f}{3}, \qquad (10)$$

$$s_1 = [(0.0666 + 0.04669\,q_f^2 + 0.015\,q_f^4 - 0.00137\,q_f^6)^{1/2} + q_f/\pi^2]^{1/3} \qquad (11)$$

where q_f, the dielectric filling factor at cutoff, is given by

$$q_f = \frac{k_c - 1}{\varepsilon_r - 1}. \qquad (12)$$

Eq. (2), when used together with eq. (6), is accurate to within ± 1%.

5. Characteristic Impedance

As in any other dispersive transmission line the definition of characteristic impedance for fin-lines is not unique. It depends upon the application. According to Meinel and Rembold [17] the characteristic impedance Z_0 of a fin-line should be defined in terms of voltage and current in the fin-line, where voltage is defined as the line integral over the shortest path between the fins on the substrate and current is the total longitudinal current in the structure. This definition is particularly suitable for switching applications and beam lead-devices.

Sharma and Hoefer [15] have shown that the following expression for Z_0 gives results within ± 2% of Hofmann's [3] results:

$$Z_0 = Z_{0\infty}/\sqrt{\varepsilon_e(f)} \qquad (13)$$

where

$$Z_{0\infty} = \frac{120\,\pi^2\left(\frac{b}{\lambda_{ca}}\right)}{\frac{b}{d}\sin\left(\frac{\pi s}{\lambda_{ca}}\right) + \left[\frac{B_0}{Y_0} + \tan\frac{\pi(a-s)}{2\lambda_{ca}}\right]\cos\left(\frac{\pi s}{\lambda_{ca}}\right)}, \qquad (14)$$

$$\frac{B_0}{Y_0} \approx \left(\frac{2b}{\lambda_{ca}}\right)\ln\operatorname{cosec}\left(\frac{\pi}{2}\,\frac{d}{b}\right). \qquad (15)$$

6. Application to Single Ridge Fin-Line

The foregoing analysis equations are also applicable to the single ridged bilateral fin-line shown in Fig. 1b. With the interpretation that b is twice the height of the fin-line housing and d is twice the gap between the tip of the fin and the broad wall of the housing. In addition, the characteristic impedance must be divided by two to obtain the characteristic impedance of the single ridge structure.

7. Computed Results

The computed results are shown in Tables 1 and 2 and compared with those obtained with the spectral domain technique. Figs. 2, 3 and 4 display the results graphically. The results agree within ±1%, which inspires confidence in the above expressions.

8. Conclusions

In the foregoing sections, accurate closed-form equations are presented for the cutoff wavelength and equivalent dielectric constant of bilateral fin-line, with symmetrically located fins. The equations are correct to within ± 0.8% of the results obtained from rigorous spectral domain analysis. An equation

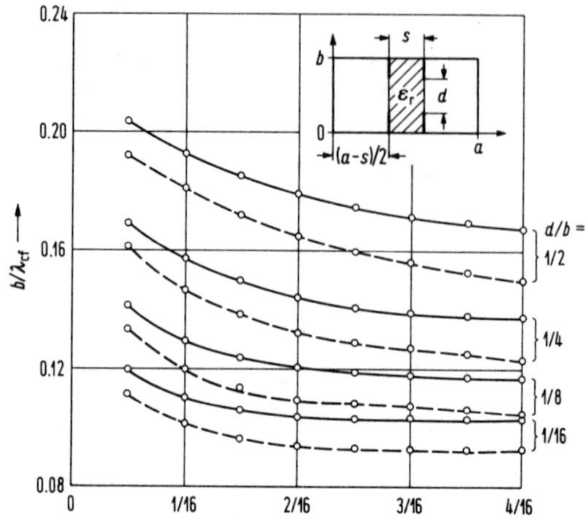

Fig. 2. Normalized cutoff frequencies in bilateral fin-lines; $b/a = 0.5$, $\varepsilon_r = 2.22$ (——) and 3.00 (– – –): spectral domain method; ○ expressions.

Table 1.

Normalized separation between the fins	Normalized fin gap	Cutoff frequency b/λ_{ca} of the dominant mode $\varepsilon_r = 1.00$		
s/a	d/b	SDM*	SH**	PM***
1/8	1/2	0.21029	–	0.20914
	1/4	0.17318	0.19276	0.17325
	1/8	0.14597	0.17097	0.14533
	1/16	0.12747	0.15165	0.12699
1/16	1/2	0.21251	–	0.21136
	1/4	0.17762	0.19276	0.17610
	1/8	0.15021	0.17097	0.14967
	1/16	0.13023	0.15165	0.12941
1/32	1/2	0.21601	–	0.21484
	1/4	0.18223	0.19276	0.18090
	1/8	0.15586	0.17097	0.15314
	1/16	0.13573	0.15165	0.13497

* Spectral domain method [5],
** Sharma and Hoefer [15],
*** Present method.

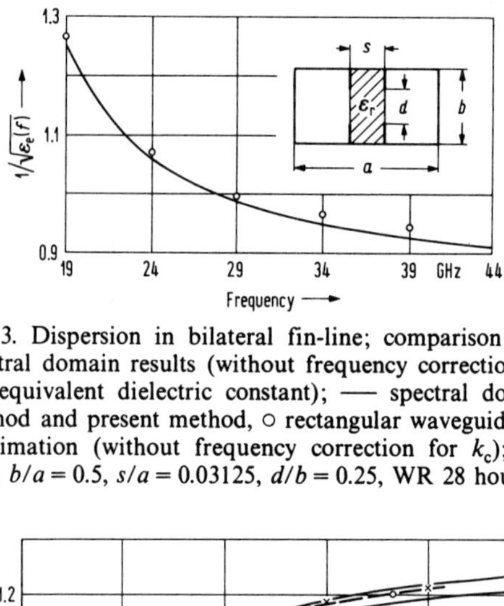

Fig. 3. Dispersion in bilateral fin-line; comparison with spectral domain results (without frequency correction for the equivalent dielectric constant); —— spectral domain method and present method, ○ rectangular waveguide approximation (without frequency correction for k_c); $\varepsilon_r = 3.00$, $b/a = 0.5$, $s/a = 0.03125$, $d/b = 0.25$, WR 28 housing.

Table 2. Propagation characteristics of bilateral fin-line; $b/a = 0.5$, $s/a = 0.03125$, $d/b = 0.25$, $\varepsilon_r = 3.00$, $b/\lambda_{ca} = 0.18224$, $b/\lambda_{cf} = 0.16159$, $k_c = 1.2719161$ (see Fig. 1b for definition of dimensions).

Frequency	λ_g/λ	λ_g/λ	λ_g/λ
GHz	SDM*	PM**	RWA***
19	1.2493	1.2586	1.2729
24	1.0584	1.0607	1.0774
29	0.9858	0.98467	1.0046
34	0.9479	0.9444	0.9679
39	0.9246	0.91926	0.9464
44	0.9085	0.9017	0.9326

* Spectral domain method [5],
** Present method,
*** Rectangular waveguide approximation (without frequency correction of k_c).

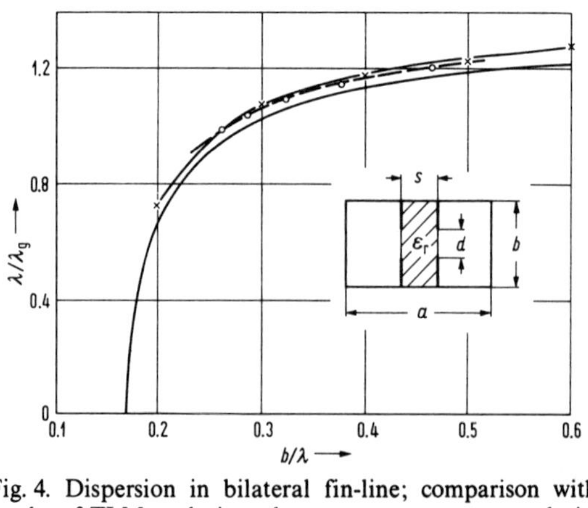

Fig. 4. Dispersion in bilateral fin-line; comparison with results of TLM analysis and transverse resonance analysis; – –○– – TLM method, —— transverse resonance analysis, —×— eq. (2); $\varepsilon_r = 2.22$, $b/a = 0.5$, $d/b = 0.5$, $s/a = 0.25$.

is presented which estimates the linear dependence of the equivalent dielectric constant with frequency. The ridged guide model, together with frequency dependence of equivalent dielectric constant taken into consideration, gives the phase velocity within ± 1% of the spectral domain results and the characteristic impedance within ± 2% of Hofmann's [3] results. The present expressions are given in such a form that they can be implemented easily on a programmable calculator or a desktop computer. In terms of accuracy, the described closed-form expressions will be vital to computer aided design techniques and helpful in the proliferation of basic know-how concerning bilateral fin-line techniques.

Acknowledgements

This work has been supported by a Natural Sciences and Engineering Research Council of Canada, Grant A-0001. Authors are thankful to Dr. Holger Hofmann of AEG Telefunken, West Germany, for providing helpful discussions.

(Received October 11th, 1984.)

References

[1] Meier, P. J., Integrated fin-line millimeter components. Transact. IEEE MTT-**22** [1974], 1209–1216.

[2] Meier, P. J., Microwave transmission line. US Patent 3 825-863, July 1974.

[3] Hofmann, H., Dispersion of planar waveguides for millimeter-wave application. AEÜ **31** [1977], 40–44.

[4] Knorr, J. B. and Shayda, P. M., Millimeter-wave fin-line characteristics. Transact. IEEE MTT-**28** [1980], 737–743.

[5] Schmidt, L. P. and Itoh, T., Spectral domain analysis of dominant and higher order modes in fin-lines. Transact. IEEE MTT-**28** [1980], 981–985.

[6] Bayer, A. and Wolff, I., A solution of the earthed fin-line with finite metallization thickness. IEEE MTT-S Internat. Microwave Symposium Digest, Washington, D.C., 1980, pp. 258–260.

[7] Sharma, A. K., Costache, G. I., and Hoefer, W. J. R., Cutoff in fin-lines evaluated with the spectral domain technique and with finite element method. IEEE AP-S Internat. Antenna and Propagation Symposium Digest, Los Angeles, CA, 1981, pp. 308–311.

[8] Shih, Y. C. and Hoefer, W. J. R., Dominant and second order mode cutoff frequencies in fin-lines calculated with two-dimensional TLM program. Transact. IEEE MTT-**28** [1980], 1443–1448.

[9] Saad, A. K. and Schünemann, K., Efficient eigenmode analysis for planar transmission lines. Transact. IEEE MTT-**31** [1983], 2125–2132.

[10] Sorrentino, R. and Itoh, T., Transverse resonance analysis of fin-line discontinuities. IEEE MTT-S Internat. Microwave Symposium Digest, San Francisco, CA, 1984, pp. 414–416.

[11] Hayashi, Y., Farr, E., and Mittra, R., Analysis of dominant and higher order modes in unilateral fin-lines. Interim Tech. Rep. U.S. Army Research Office, Contract No. DAAG 29-83-K-0084, Electromagnetics Laboratory, Dept. of Electrical Engineering, Engineering Experiment Station, University of Illinois, Urbana, IL, Aug. 1983.

[12] Hoefer, W. J. R., Fin-line design made easy. IEEE MTT-S Internat. Microwave Symposium Digest, Ottawa, Canada, 1978, p. 471.

[13] Saad, A. M. K. and Begemann, G., Electrical performance of fin-lines of various configurations. Microwave Opt. and Acoust. **1** [1977], 81–88.

[14] Piotrowski, J. K., Accurate and simple formulas for dispersion in fin-lines. IEEE MTT-S Internat. Microwave Symposium Digest, San Francisco, CA, 1984, pp. 333–335.

[15] Sharma, A. K. and Hoefer, W. J. R., Empirical expressions for fin-line design. Transact. IEEE MTT-**31** [1983], 350–356.

[16] Pramanick, P. and Bhartia, P., Analysis and synthesis of tapered fin-lines. IEEE MTT-S Internat. Microwave Symposium Digest, San Francisco, CA, 1984, pp. 336–338.

[17] Meinel, H. and Rembold, B., New millimeter-wave fin-line attenuators and switches. IEEE MTT-S Internat. Microwave Symposium Digest, Orlando, FL, 1979, pp. 249–252.

2.2.4

Accurate Analysis and Synthesis Equations for Insulated Fin-Lines

by Protap Pramanick* and Prakash Bhartia*

Accurate analysis equations and synthesis technique are presented for insulated fin-lines valid over a wide range of structural parameters and the substrate dielectric constant ($1 \leqq \varepsilon_r \leqq 3.75$) within $\pm 0.6\%$ of the TLM method. The expressions are usable for computing the effective dielectric constant and characteristic impedance within $\pm 2\%$ over the normalized frequency range $0.1 \leqq b/\lambda \leqq 0.6$, which is suitable for most practical applications. The first order synthesis technique presented yields results within $\pm 2\%$ of the originally specified design values.

Genaue Gleichungen zur Analyse und Synthese von isolierten Flossenleitungen

Für isolierte Flossenleitungen werden genaue Analysegleichungen und Syntheseverfahren angegeben, die für weite Bereiche der Strukturparameter und Substrat-Dielektrizitätszahlen zwischen 1 und 3,75 gelten und innerhalb $\pm 0{,}6\%$ der Ergebnisse mit der Methode der TEM-Leitungs-Matrix (TLM) liegen. Die Ausdrücke sind nützlich, um die effektive Dielektrizitätszahl und die charakteristische Impedanz mit einer Genauigkeit von $\pm 2\%$ im normierten Frequenzbereich $0{,}1 \leqq b/\lambda \leqq 0{,}6$ zu berechnen, der für die meisten praktischen Anwendungen vorliegt. Das angegebene Syntheseverfahren erster Ordnung liefert Ergebnisse, deren Genauigkeit innerhalb von $\pm 2\%$ der ursprünglich vorgegebenen Entwurfsziele liegt.

1. Introduction

Fin-line, structures [1] are quite popular currently due to their favourable properties in millimeter-wave circuit applications. These include low dispersion, relatively wide single-mode range of frequency, easy compatibility with semi-conductor devices, moderate loss compared to other planar transmission lines and avoidance of miniaturization.

Among the various types of fin-lines available to date, unilateral and bilateral fin-lines are the most suitable for passive circuits because of their grounded fin configuration [1], [2]. However, in one case of the insulated fin-line, shown in Fig. 1, the fins are insulated at dc. Therefore bias may be applied for active circuit components.

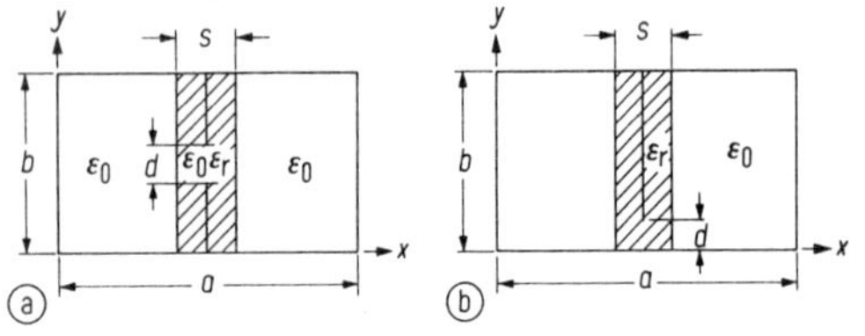

Fig. 1. (a) Insulated fin-line, (b) complementary insulated fin-line.

Dispersion in fin-line has been well analyzed by Hofmann [3], Knorr and Shayda [4], Schmidt and Itoh [5], Beyer and Wolff [6], Sharma, Costache and Hoefer [7], Shih and Hoefer [8] and Saad and Schünemann [9]. These analyses use eigenmode analysis in space or spectral domain, finite element method or two dimensional transmission line matrix (TLM) analysis [8]. Hayashi, Farr and Mittra [10] further extended Knorr and Shayda's method to the more general case of higher order modes. All these methods are highly accurate but require considerable analytical effort and lead to complicated computer programming.

For ease of design, the propagation constant for fin-lines has been approximated by various methods. Fin-lines have been treated as ridged waveguides by Hoefer [11], and Saad and Begemann [12]. But such analysis does not result in good accuracy. For an adequately accurate expression for the equivalent dielectric constant of fin-line, one has to depend on experimental data [1] of expensive and time consuming sample measurement. Therefore, despite all the advantages of fin-lines, the basic problem faced by fin-line circuit designers is the cumbersome design procedure.

As a result there is an urgent need for accurate closed-form expressions for the equivalent dielectric constant and the characteristic impedance of fin-lines. Though Sharma and Hoefer [13] have presented purely empirical expressions for the cut-off wavelength of unilateral and bilateral fin-lines by curve fitting to the data obtained by the spectral domain technique, the expressions have a very small range of validity as far as the fin-line geometry and substrate dielectric constant are concerned. Using a stationary formula approach and curve fitting to spectral domain technique data, Pramanick and Bhartia [14] have presented accurate closed form expressions for the unilateral fin-line. The equations are valid over a wide range of fin-line parameters and substrate permittivity values.

* Dr. P. Pramanick, Prof. Dr. P. Bhartia, Department of Electrical Engineering, University of Ottawa, Ottawa, Ontario, Canada K1N 6N5.

P. Pramanick and P. Bhartia, *Archiv fur Elektronik und Ubertragungstechnik*, Vol. 39, No. 1, Jan. 1985, pp. 31–36.

In the present work, accurate closed-form expressions are developed for the theoretical prediction of the equivalent dielectric constant and the cutoff frequency of insulated fin-lines. The expressions are accurate to within 0.6%. They have been derived assuming a stationary formula for the equivalent dielectric constant of the fin-line and then determining the formula exactly by curve fitting to numerical data from two dimensional transmission line matrix analysis [8]. The expressions are valid over the range (see Fig. 1a).

$$0 < b/a \leqq 1,$$
$$1/32 \leqq s/a \leqq 1/4,$$
$$1/32 \leqq d/b \leqq 1$$
and
$$1 \leqq \varepsilon_r \leqq 3.75.$$

Currently a synthesis technique for the design of insulated fin-lines is not available and computer-aided techniques with optimization must be resorted to. However, using the analysis equations it is possible to derive a synthesis procedure for the insulated fin-line. The procedure uses the initial design specifications for the parameters of interest to the designer, i.e., equivalent-dielectric constant cutoff wavelength and characteristic impedance, and leads to an equation for the dimensions of the fin-line.

2. Expressions for Guide Wavelength and Frequency-Dependent Characteristic Impedance

The guide wavelength in fin-line is defined as

$$\lambda_g = \lambda / \sqrt{\varepsilon_e(f)} \tag{1}$$

where λ is the free-space wavelength and $\varepsilon_e(f)$, the frequency dependent effective dielectric constant of the fin-line is given by [1]

$$\varepsilon_e(f) = k_e - (\lambda/\lambda_{ca})^2 \tag{2}$$

where k_e is the equivalent dielectric constant, and λ_{ca} is the cutoff wavelength of a finned waveguide of identical dimensions but completely filled with air. In Meier's [1] expression k_e is a constant. But this is usually the case only when ε_r is less than 2.50 and the substrate is thin. Otherwise k_e, considered frequency dependent, can be obtained from the dispersion model of Pramanick and Bhartia [15] as

$$k_e = k_c\, G(d/b, s/a, b/\lambda, \varepsilon_r) \tag{3}$$

where

$$k_c = (\lambda_{cf}/\lambda_{ca})^2. \tag{4}$$

In eq. (4) λ_{cf} is the cutoff wavelength of the fin-line. $G = 1$ for $\varepsilon_r \leqq 2.5$ and $b/\lambda \leqq 0.6$ within 2%.

The frequency dependent characteristic impedance for a fin-line is defined as

$$Z_0 = Z_{0\infty} / \sqrt{\varepsilon_e(f)} \tag{5}$$

where $Z_{0\infty}$ is the characteristic impedance of the equivalent air-filled finned waveguide at infinite frequency.

3. Derivation of the Form of the Expression for Equivalent Dielectric Constant

The insulated fin-line shown in Fig. 1a can be thought of as a combination of an air-filled finned waveguide and a dielectric loaded waveguide. The cutoff wavelength λ_{ca} of the finned waveguide can be determined accurately using the equation [16]

$$\frac{b}{\lambda_{ca}} = \frac{b}{2a}\left[1 + \frac{4}{\pi}\left(\frac{b}{a}\right)\left(1 + 0.2\sqrt{\frac{b}{a}}\right) \cdot \ln \operatorname{cosec}\left(\frac{\pi d}{2b}\right)\right]^{-1/2}. \tag{6}$$

The cutoff wavelength λ_{cd} of the dielectric loaded waveguide can be obtained, for small s/a and ε_r, using the stationary formula [17] obtained by the variational technique,

$$\frac{b}{\lambda_{cd}} = \frac{b}{2a}\left\{1 + \left[\frac{s}{a} + \frac{1}{\pi}\sin\left(\frac{\pi s}{a}\right)(\varepsilon_r - 1)\right]\right\}^{-1/2}. \tag{7}$$

The derivation of eq. (7), using a variational technique [17], assumes a sinusoidal field distribution in the transverse cross section of the guide. The assumed field distribution has the form

$$\overline{E_y} = \hat{y}\sin(\pi x/a) \tag{8}$$

where $\hat{y}$ is the unit vector in the y-direction.

For a general field distribution in the transverse cross section of the waveguide, eq. (7) can be written as

$$\frac{b}{\lambda_{cd}} = \frac{b}{2a}\left[1 + F\left(\frac{s}{a}\right)(\varepsilon_r - 1)\right]^{-1/2} \tag{9}$$

where the function $F(s/a)$ depends upon the nature of the field distribution. Eq. (9) has a stationary form. Therefore, it remains unchanged for small variations in ε_r, s/a and frequency. For example, the inaccuracy of eq. (7) is less than $\pm 0.8\%$ for $s/a \leqq 1/4$, which is the thickest substrate used for insulated fin-lines.

Now consider the combination of the finned waveguide with the dielectric loaded waveguide. Due to presence of the fins, the fields try to concentrate in their vicinity, and the field distribution in the transverse cross section of the guide no longer remains a function of s/a only, but also becomes a function of d/b. Therefore eq. (9) assumes the form

$$\frac{b}{\lambda_{cd}} = \frac{b}{2a}\left[1 + F\left(\frac{s}{a}, \frac{d}{b}\right)(\varepsilon_r - 1)\right]^{-1/2}. \tag{10}$$

Once the function $F(s/a, d/b)$ is known, the cutoff wavelength of the insulated fin-line of Fig. 1 can be obtained using eqs. (6) and (10) and is given by

$$\frac{b}{\lambda_{cf}} = \frac{b}{2a}\left\{\left[1 + \frac{4}{\pi}\left(\frac{b}{a}\right)\left(1 + 0.2\sqrt{\frac{b}{a}}\right) \cdot \ln \operatorname{cosec}\left(\frac{\pi}{2}\frac{d}{b}\right)\right]\left[1 + F\left(\frac{s}{a}, \frac{d}{b}\right)(\varepsilon_r - 1)\right]\right\}^{-1/2}. \tag{11}$$

4. Derivation of the Function $F(s/a, d/b)$

Eq. (11) when solved for $F(s/a, d/b)$ gives

$$F\left(\frac{s}{a}, \frac{d}{b}\right) = \qquad (12)$$

$$= \frac{\left\{\left[\left(\frac{\lambda_{cf}}{2a}\right)^2 \Big/ \left[1 + \frac{4}{\pi}\left(\frac{b}{a}\right)\left(1 + 0.2\sqrt{\frac{b}{a}}\right) \ln \operatorname{cosec}\left(\frac{\pi}{2}\frac{d}{b}\right)\right]\right] - 1\right\}}{\varepsilon_r - 1}.$$

Expressing (13)

$$F\left(\frac{s}{a}, \frac{d}{b}\right) = \left[a_1\left(\frac{s}{a}\right) \ln \operatorname{cosec}\left(\frac{\pi}{2}\frac{d}{b}\right) + b_1\left(\frac{s}{a}\right)\right] \frac{s}{a}$$

where $a_1(s/a)$ and $b_1(s/a)$ are functions of s/a, and using the accurately computed values of λ_{cf} from the two dimensional TLM analysis [8] and eq. (12), the function $F(s/a, d/b)$ can be computed for several combinations of s/a and d/b. Using the computed results, the equivalent dielectric constant k_c at cutoff can be written as (14)

$$k_c = 1 + \frac{s}{a}\left[a_1\left(\frac{s}{a}\right) \ln \operatorname{cosec}\left(\frac{\pi}{2}\frac{d}{b}\right) + b_1\left(\frac{s}{a}\right)\right](\varepsilon_r - 1)$$

where

$$a_1(a/s) = -0.2311338\,[\ln(a/s)]^3 + 1.324553\,[\ln(a/s)]^2 - 0.8372744\,[\ln(a/s)] + 0.112267, \qquad (15)$$

$$b_1(a/s) = 0.0209695\,[\ln(a/s)]^3 - 0.170823\,[\ln(a/s)]^2 + 0.60167\,[\ln(a/s)] + 1.4772971 \qquad (16)$$

for $1/4 \leq d/b \leq 1$, and

$$a_1(a/s) = 0.479574\,[\ln(a/s)]^3 - 2.2755107\,[\ln(a/s)]^2 + 4.4691399\,[\ln(a/s)] - 2.61026, \qquad (17)$$

$$b_1(a/s) = -0.6700156\,[\ln(a/s)]^3 + 3.318579\,[\ln(a/s)]^2 - 4.54300\,[\ln(a/s)] + 4.0156906 \qquad (18)$$

for $1/32 \leq d/b \leq 1/4$.

From eq. (14) the dielectric filling factor can be written as

$$q_f = \frac{k_c - 1}{\varepsilon_r - 1} = \qquad (19)$$

$$= \frac{s}{a}\left[a_1\left(\frac{s}{a}\right) \ln \operatorname{cosec}\left(\frac{\pi}{2}\frac{d}{b}\right) + b_1\left(\frac{s}{a}\right)\right].$$

Eq. (14) shows a linear dependence of k_c on ε_r.

5. Characteristic Impedance

Like for any other dispersive transmission line the definition of characteristic impedance for fin-lines is not unique. It depends upon the application. According to Meinel and Rembold [18] the characteristic impedance of a fin-line should be defined in terms of voltage and current in the fin-line, where voltage is defined as the line integral over the substrate surface and current is the total longitudinal surface current in the structure. This definition is particularly useful in switching applications. But Meier [1] defines the characteristic impedance of fin-lines on a power-voltage basis.

The characteristic impedance based on the voltage-current basis of an equivalent air-filled ridged waveguide, at infinite frequency, is given by [16]

$$Z_{0\infty} = \qquad (20)$$

$$= \frac{120\pi^2\left(\frac{b}{\lambda_{ca}}\right)}{\frac{b}{d}\sin\left(\frac{\pi s}{\lambda_{ca}}\right) + \left[\frac{B_0}{Y_0} + \tan\left(\frac{\pi(a-s)}{2\lambda_{ca}}\right)\right]\cos\left(\frac{\pi s}{\lambda_{ca}}\right)}$$

where s is the thickness of the ridge and

$$\frac{B_0}{Y_0} \approx \left(\frac{2b}{\lambda_{ca}}\right) \ln \operatorname{cosec}\left(\frac{\pi}{2}\frac{d}{b}\right). \qquad (21)$$

The characteristic impedance of the finned waveguide is obtained by setting s to zero in eq. (20), which gives

$$Z_{0\infty} = \frac{120\pi^2\left(\frac{b}{\lambda_{ca}}\right)}{\frac{B_0}{Y_0} + \tan\left(\frac{\pi a}{2\lambda_{ca}}\right)}. \qquad (22)$$

Therefore from eq. (5), the frequency dependent characteristic impedance of the insulated fin-line is given by

$$Z_0 = \frac{120\pi^2\left(\frac{b}{\lambda_{ca}}\right)}{\left[\frac{B_0}{Y_0} + \tan\left(\frac{\pi a}{2\lambda_{ca}}\right)\right]\left[k_e - \left(\frac{\lambda}{\lambda_{ca}}\right)^2\right]^{1/2}}. \qquad (23)$$

6. Synthesis

To fabricate the appropriate fin-line structure one requires a, b, s, d and ε_r. Normally since conventional waveguide is used for the fine-line housing, a and b are known. Similarly ε_r is known and only s and d remain to be determined. Using these and the k_e, λ_{cf} and Z_0 values specified, the transcendental equation (23) can be solved exactly to determine d for various specified values of s. Thus the fine-line is solved completely.

However, eq. (23) is cumbersome and a simpler synthesis equation can be deemed without sacrificing accuracy. For low dielectric constant substrate the frequency dependent effective dielectric constant can be written as

$$\varepsilon_e(f) = k_c - (\lambda/\lambda_{ca})^2. \qquad (24)$$

Using eq. (14) and noting that

$$\frac{\frac{s}{a}a_1\left(\frac{s}{a}\right)(\varepsilon_r - 1)\ln \operatorname{cosec}\left(\frac{\pi}{2}\frac{d}{b}\right) - \left(\frac{\lambda}{\lambda_{ca}}\right)^2}{1 + \frac{s}{a}b_1\left(\frac{s}{a}\right)(\varepsilon_r - 1)} \ll 1$$

for $0<b/a\leqq 1$, $1/32\leqq d/b\leqq 3/4$, $1/32\leqq s/a\leqq 1/4$, $1\leqq \varepsilon_r\leqq 3.75$ and $0.1\leqq b/\lambda\leqq 0.6$ one can write

$$\sqrt{\varepsilon_e(f)}=\left(\frac{F\,x+E}{G\,x^2+H\,x+I}\right)^{-1} \tag{25}$$

where

$$E=8\left[1+\frac{s}{a}b_1\left(\frac{s}{a}\right)(\varepsilon_r-1)\right]^{1/2}, \tag{26}$$

$$F=NE \tag{27}$$

$$\text{with } N=\frac{4}{\pi}\left(\frac{b}{a}\right)\left(1+0.2\sqrt{\frac{b}{a}}\right), \tag{28}$$

$$G=0.5F\left(\frac{s}{a}\right)a_1\left(\frac{s}{a}\right)(\varepsilon_r-1)\Big/\left[1+\frac{s}{a}b_1\left(\frac{s}{a}\right)(\varepsilon_r-1)\right]^{1/2},$$

$$H=E(F/8+G/F), \tag{29}$$

$$I=\frac{E^2}{8}-\left(\frac{b}{a}\right)^2\left(\frac{\lambda}{b}\right)^2, \tag{30}$$

$$x=\ln\operatorname{cosec}\left(\frac{\pi}{2}\frac{d}{b}\right). \tag{31}$$

For $d/b\leqq 0.75$.

$$\pi a/\lambda_{ca}\leqq 0.76372 \tag{32}$$

and hence making the approximation

$$\tan\left(\frac{\pi a}{\lambda_{ca}}\right)\approx\frac{1}{2\dfrac{\lambda_{ca}}{\pi a}-\dfrac{\pi a}{6\lambda_{ca}}} \tag{33}$$

in eq. (23) the characteristic impedance can be written as

$$Z_0=\frac{(S\,x+T)(F\,x+E)}{(U\,x^2+Vx+1)(G\,x^2+H\,x+I)} \tag{34}$$

where

$$S=240\pi\left(\frac{b}{a}\right)N,\quad T=120\pi^2\left(\frac{2}{\pi}-\frac{\pi}{24}\right)\left(\frac{b}{a}\right), \tag{35), (36}$$

$$U=\left(\frac{4}{\pi}\right)\left(\frac{b}{a}\right)N,\quad V=2\left(\frac{b}{a}\right)\left(\frac{2}{\pi}-\frac{\pi}{24}\right)+N. \tag{37), (38}$$

For a given set of parameters Z_0, b/a, s/a, ε_r and b/λ, eq. (33) reduces to a quartic equation of the form

$$x^4+a_3\,x^3+a_2\,x^2+a_1\,x+a_0=0 \tag{39}$$

where

$$a_3=H/G+V/U,$$

$$a_2=[Z_0(UI+VH+G)-SF]/(Z_0\,UG),$$

$$a_1=[Z_0(VI+H)-(SE+TF)]/(Z_0\,UG),$$

$$a_0=(Z_0\,I-TE)/(Z_0\,UG).$$

This can be easily solved to determine x and hence d/b, since

$$d/b=(2/\pi)\sin^{-1}[\exp(-x)]. \tag{40}$$

Table 1. Normalized cut-off frequencies b/λ_{cf} for the dominant mode in insulated fin-line; $b/a=0.5$, $\varepsilon_r=2.22$ (see Fig. 1a for definition of dimensions).

	d/b	b/λ_{cf}(TLM.M)	b/λ_{cf}(FEM)	b/λ_{cf}(PM)
$s/a=1/8$	1/8	0.1285	0.1299	0.1285
	1/4	0.1512	0.1529	0.15116
	1/2	0.1853	0.1868	0.1864
	3/4	0.2090	0.2098	0.2088
$s/a=1/16$	1/8	0.1373	0.1394	0.1373
	1/4	0.1630	0.1647	0.1629
	1/2	0.1989	0.2003	0.1996
	3/4	0.2229	0.2235	0.2227
	1	0.2325	0.2325	0.2308

TLM.M ≙ Transmission line matrix method, FEM ≙ Finite element method, PM ≙ Present method.

Table 2. $b/a=0.5$, $\varepsilon_r=3.00$.

	d/b	b/λ_{cf} (TLM.M)	b/λ_{cf}(PM)
$s/a=1/4$	1/8	0.1056	0.1057
	1/4	0.1237	0.1241
	1/2	0.1511	0.1518
$s/a=1/16$	1/8	0.1249	0.1244
	1/4	0.1501	0.1502
	1/2	0.1884	0.1879

Table 3. $b/a=0.5$, $\varepsilon_r=9.00$.

	d/b	b/λ_{cf} (TLM.M)	b/λ_{cf}(PM)
$s/a=1/4$	1/8	0.06267	0.06288
	1/4	0.07416	0.07491
	1/2	0.09220	0.09120
$s/a=1/16$	1/8	–	0.08074
	1/4	0.10354	0.10212
	1/2	0.13503	0.13715

7. Computed Results

The results computed using eqs. (14) and (15) through (18) are shown in Tables 1, 2 and 3. The results show excellent agreement except when ε_r is too high, i.e., $\varepsilon_r=9.00$. Since the derivation of the equations used the two dimensional TLM analysis data for $\varepsilon_r=2.22$, their excellent agreement in case of $\varepsilon_r=3.00$ basically proves the stationary nature of the derived expressions. Fig. 2 compares the normalized dominant mode cut-off wavelength (b/λ_{cf}), computed by the present method with those obtained by the two-dimensional TLM analysis [8] for $\varepsilon_r=2.22$ and $b/a=0.5$. For low ε_r we can neglect the frequency dispersion in k_e and make a first order approximation for the effective dielectric constant as shown in Fig. 3. This is accurate to within $\pm 2\%$ for $b/\lambda\leqq 0.6$. For $\varepsilon_r>2.22$ the frequency dependence of k_e should be evaluated using dispersion model of Pramanick and Bhartia [15].

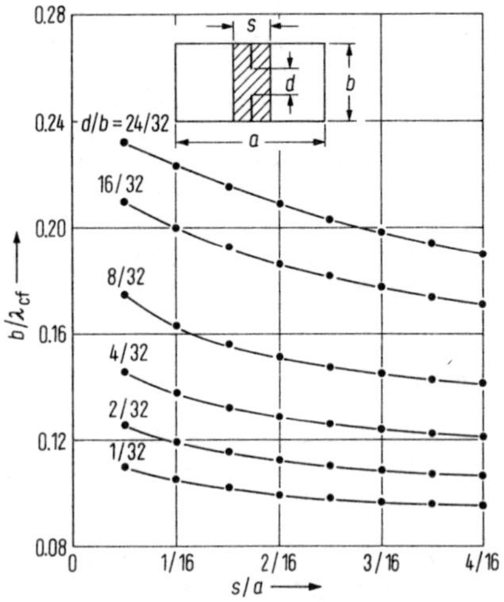

Fig. 2. Comparison of b/λ_{cf} values obtained by the present method and 2-D TLM analysis; —— TLM, ● expression; $\varepsilon_r = 2.22$, $b/a = 0.5$.

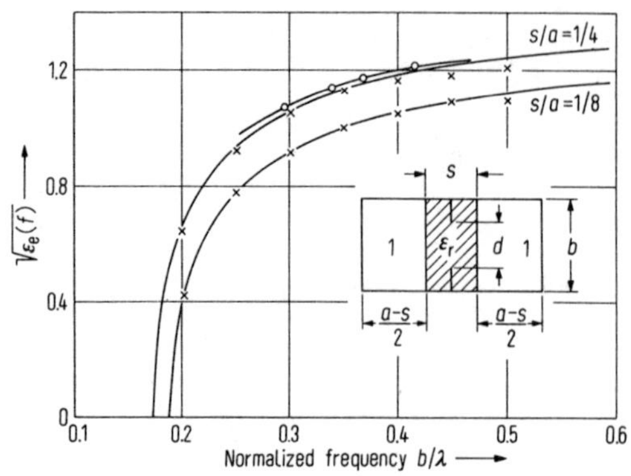

Fig. 3. Dispersion characteristics; insulated fin-line; $b/a = 0.5$, $d/b = 0.5$, $\varepsilon_r = 2.22$;
—○— three-dimensional TLM [8],
—— transverse resonance method [11],
× eq. (2).

The variation of characteristic impedance is shown in Fig. 4 for $b/\lambda = 0.3556$.

The results computed using the approximate synthesis equations are shown in Fig. 4. Table 4 depicts the computed results. The fin-line structural parameters were synthesized for specified k_c, λ_{cf} and Z_0 values. Excellent agreement can be seen over the total useful range of fin-line dimensions. The slight disagreement at the larger d/b values can be attributed to the approximation in (32) becoming less valid for larger d/b values. However, these values can be corrected by using the analysis equations.

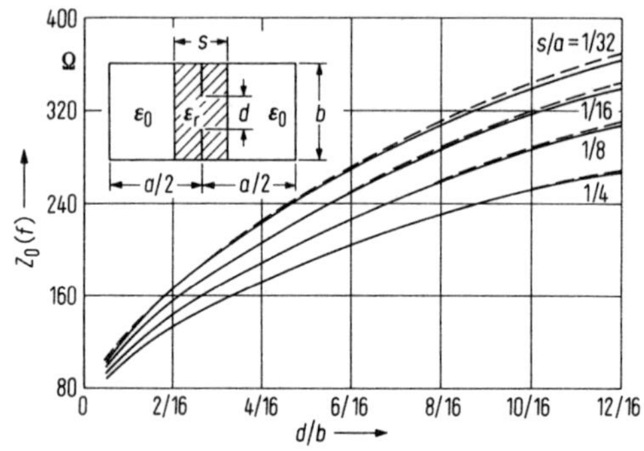

Fig. 4. Z_0 versus d/b; $\varepsilon_r = 2.22$, $b/a = 0.5$, $b/\lambda = 0.3556$;
—— analysis, - - - synthesis.

8. Application to Complementary Insulated Fin-Lines

The forgoing analysis and synthesis equations are also applicable to the complementary insulated fin-line shown in Fig. 1 b with the interpretation that b is twice the height of the fin-line housing and d is twice the gap between the tip of the fin and the broad wall of the housing. In addition the characteristic impedance must be divided by two to obtain the value for the complementary structure.

Table 4. Desired and synthesized values of b/λ_{ca}, Z_0, $\varepsilon_e(f)$ for the dominant mode in insulated fin-line; $b/a = 0.5$, $\varepsilon_r = 2.22$ (see Fig. 1 a for the definition of dimensions).

s/a	d/b	b/λ_{ca} (D)	b/λ_{ca} (S)	Z_0 (D)	Z_0 (S)	$\varepsilon_e(f)$ (D)	$\varepsilon_e(f)$ (S)
1/16	1/32	0.1400	0.1405	98.957	99.967	0.1049	0.1056
	1/16	0.1525	0.1527	121.614	122.094	0.1188	0.1191
	1/8	0.1690	0.1690	154.695	154.635	0.1373	0.1373
	1/4	0.1919	0.1917	205.493	205.281	0.1630	0.1628
	1/2	0.2234	0.2240	285.663	287.203	0.1997	0.2005
	3/4	0.2431	0.2455	339.637	344.563	0.2227	0.2258
1/4	1/32	0.1400	0.1401	89.169	89.414	0.0953	0.0955
	1/16	0.1525	0.1525	107.619	107.670	0.1064	0.1065
	1/8	0.1690	0.1689	133.800	133.673	0.1211	0.1210
	1/4	0.1919	0.1917	172.310	172.118	0.1411	0.1410
	1/2	0.2234	0.2234	230.578	230.522	0.1708	0.1708
	3/4	0.2431	0.2436	266.685	267.382	0.1897	0.1902

D ≙ Desired, S ≙ Synthesized, Z_0 is in ohms.

9. Conclusions

In the foregoing sections, accurate closed-form expressions are developed for the cut-off wavelength and equivalent dielectric constant of insulated fin-lines. The derivation starts with the assumption of a stationary form for the expression for the equivalent dielectric constant at cutoff. The stationary function is subsequently determined by least squares curve fitting to accurate numerical data, obtained by the two-dimensional TLM analysis. The expressions are accurate to within $\pm$ 0.6% over the complete practical range of interest. They can be used for quick and easy evaluation dispersion characteristic by hand or calculator within $\pm$ 2%, for $\varepsilon_r \leqq 2.22$ and $b/\lambda \leqq 0.6$. For higher ε_r a frequency correction the equivalent dielectric constant is required.

Using the analysis equations, a first-order synthesis technique is developed which synthesizes the equivalent dielectric constant within $\pm$ 0.4% and the characteristic impedance within $\pm$ 2% of the originally specified values. The present analysis and synthesis equations will be useful in computer aided design and optimization of insulated fin-line circuits.

Acknowledgements

This research has been supported by the Natural Sciences and Engineering Research Council of Canada Grant A-0001. Authors are also thankful to Prof. W. J. R. Hoefer and his research group for providing the necessary data.

(Received February 8th, 1984.)

References

[1] Meier, P. J., Integrated fin-line millimeter components. Transact. IEEE MTT-**22** [1974], 1209–1216.

[2] Meier, P. J., Microwave transmission line. U.S. Patent 3825 863, July 1974.

[3] Hofmann, H., Dispersion of planar waveguides for millimeter-wave application. AEÜ **31** [1977], 40–44.

[4] Knorr, J. B. and Shayda, P. M., Millimeter-wave fin-line characteristics. Transact. IEEE MTT-**28** [1980], 737–743.

[5] Schmidt, L. P. and Itoh, T., Spectral domain analysis of dominant and higher order modes in fin-lines. Transact. IEEE MTT-**28** [1980], 981–985.

[6] Beyer, A. and Wolff, I., A solution of the earthed fin-line with finite metallization thickness. IEEE MTT-S Internat. Microwave Symp. Digest, Washington, D.C. 1980, pp. 258–260.

[7] Sharma, A. K., Costache, G. I., and Hoefer, W. J. R., Cutoff in fin-lines evaluated with the spectral domain technique and with the finite element method. IEEE AP-S Internat. Antenna and Propagation Symp. Digest, Los Angeles, CA., 1981, pp. 308–311.

[8] Shih, Y. C. and Hoefer, W. J. R., Dominant and second order mode cutoff frequencies in fin-lines calculated with two-dimensional TLM program. Transact. IEEE MTT-**28** [1980], 1443–1448.

[9] Saad, A. K. and Schünemann, K., Efficient eigen-mode analysis for planar transmission lines. Transact. IEEE MTT-**30** [1982], 2125–2132.

[10] Hayashi, Y., Farr, E., and Mittra, R., Analysis of dominant and higher order modes in unilateral fin-lines. Interim. Tech. Rep. U.S. Army Research Office, Contract No. DAAG29-82-K-0084, Electromagnetics Lab., Dept. of Elect. Engrg., University of Illinois, Urbana, Ill., Aug. 1982.

[11] Hoefer, W. J. R., Fin-line design made easy. IEEE MTT-S. Internat. Microwave Symp. Digest. Ottawa, Canada, 1978, p. 471.

[12] Saad, A. M. K. and Begemann, G., Electrical performance of fin-lines of various configurations. Microwaves Opt. Acoust. **1** [1977], 81–88.

[13] Sharma, A. K. and Hoefer, W. J. R., Empirical expressions for fin-line design. IEEE MTT-**31** [1983], 350–356.

[14] Pramanick, P. and Bhartia, P., Accurate analysis equations and synthesis technique for unilateral fin-lines. Transact. IEEE MTT-**33** [1985], No. 1.

[15] Pramanick, P. and Bhartia, P., Fin-line dispersion model. Accepted for presentation at IEEE MTT-S Internat. Symp., June 1984.

[16] Burton, M. N. and Hoefer, W. J. R., Closed-form expression for the parameters of finned and ridged waveguides. Transact. IEEE MTT-**30** [1982], 2190–2194.

[17] Harrington, R. F., Time harmonic electromagnetic fields. McGraw-Hill Book Co., New York 1961, p. 345.

[18] Meinel, H. and Rembold, B., New millimeter-wave fin-line attenuators and switches. IEEE MTT-S Internat. Symp. Digest, Orlando, FL, 1979.

[19] Shih, Y. C., Analysis of fin-line using transmission line matrix and transverse resonance method. Ph.D. Thesis, Dept. of Elect. Engrg., University of Ottawa, Canada, 1980.

2.2.5

FREQUENCY DEPENDENCE OF EQUIVALENT DIELECTRIC CONSTANT IN FINLINES

Protap Pramanick

Canadian Marconi Company
Avionics Division
Legget Drive
Kanata, Ontario, Canada

and
Prakash Bhartia
Electronics Division
Defense Research Establishment Ottawa
Ottawa, Ontario, Canada

Received September 8, 1985

ABSTRACT

This paper develops a simple but accurate LSE model for accurate prediction of phase velocity in fin-lines. The model is general and equally applicable to unilateral, bilateral and insulated fin-lines. Closed-form expressions, derived from the model, give the phase velocity within to ±0.8 percent of sophisticated numerical techniques.

P. Pramanick and P. Bhartia, *International Journal of Infrared and Millimeter Waves*, Vol. 7, No. 2, Feb. 1986, pp. 251–268.

I. INTRODUCTION

Within the past several years, a number of publications have appeared dealing with the rigorous analysis of fin-line eigenmodes. An exhaustive account of such publications is available in a recently published review article by Solbach [1] on the present status of E-plane circuits. Common to all investigations is that the method of computation is cumbersome and that the computer time involved with the numerical evaluation is extremely large. Therefore it is concluded from this that there is a need for an accurate design theory with lucid and tractable formulae. A first indication of how such a method could be established appears in [2],[3]. In these attempts the rigorous HE and EH modes, supported by fin-lines, have been approximated by their dominating TE and TM counterparts respectively. This basically establishes an approximate rectangular waveguide equivalence for fin-line. References [2] and [3] show that such an equivalence is valid for the analysis of fin-line discontinuities within 3% and in some instances the error may shoot up to ±5%. Moreover, such a model is valid for low ε_r and thin substrates ($s/a \ll 1$ and $\varepsilon_r < 2.22$)(See Fig. 1). The invalidity of the model is due to the fact that the uniform rectangular waveguide model makes no provision for the frequency dependence of the equivalent dielectric constant k_c at the fin-line cutoff frequency.

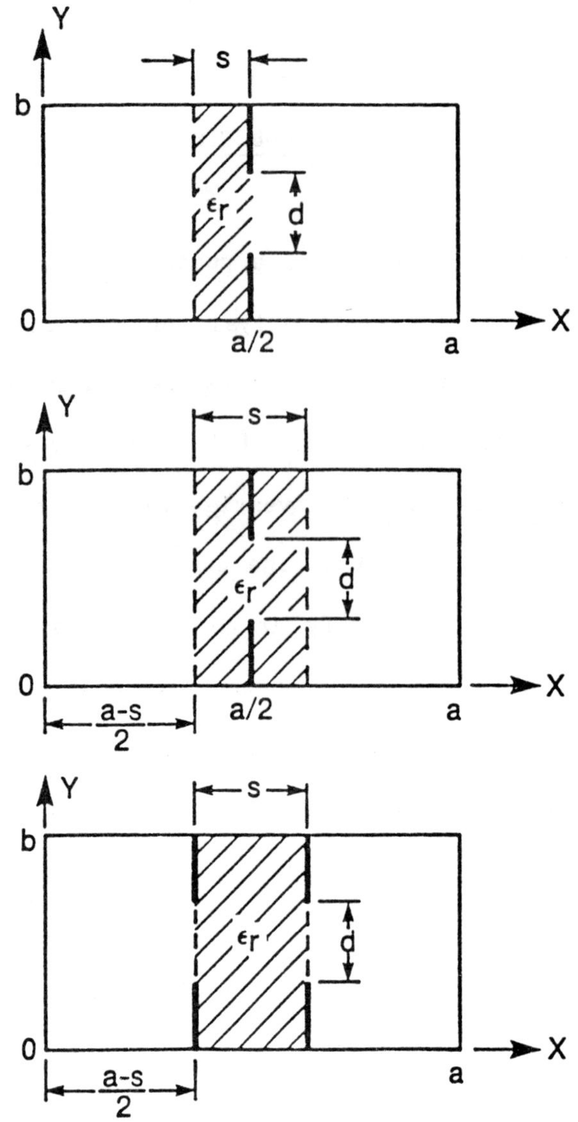

Fig. 1. Various types of finlines.

However, such frequency dependence becomes considerable for thicker and higher ε_r substrates. In the present work the fin-line is represented by an LSE model of a rectangular waveguide, longitudinally and centrally loaded with a dielectric slab of dielectric constant ε_r as shown in Fig. 2. Since such a structure can support both the longitudinal components of electric and magnetic fields, it is assumed to give a closer approximation to fin-line than what a uniform rectangular waveguide equivalence [2] gives. Using this LSE model, closed-form equations have been derived which predict the phase velocity of a fin-line to within ±0.8% of spectral domain results [4] over the range

$$1 \leqslant \varepsilon_r \leqslant 3.80, \; 0 \leqslant b/a \leqslant 1$$

$$\frac{1}{64} \leqslant s/a \leqslant 1/4, \; 1/32 \leqslant d/b \leqslant 1$$

II. THE ANALYTICAL MODEL

The conventional fin-line structures, e.g. unilateral, bilateral and insulated, are shown in Fig. 1. For the dominant HE mode the fields are concentrated around the edges of the fins and the dielectric slab supporting them. Near the fins, the magnetic field is predominantly normal to and the electric field predominantly tangential to the air dielectric interfaces. This is characteristic of the LSE mode

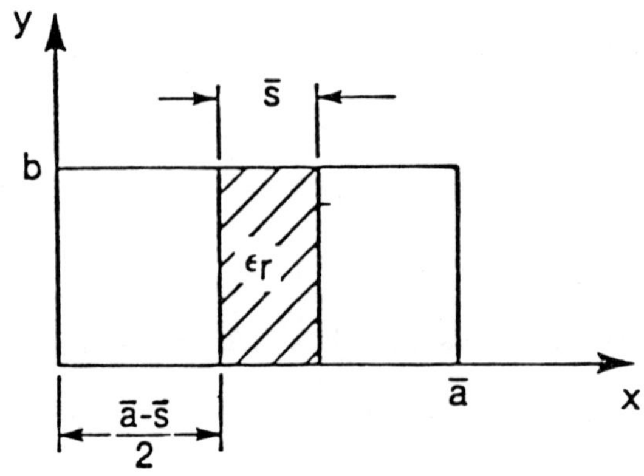

Fig. 2. LSE-mode dispersion model.

[5],[6]. The structures shown in Figures 1a, 1b and 1c are intractable to direct analysis on this basis, but its boundaries can be distorted to result in the model, shown in Fig. 2, that can be analyzed. The model consists of a rectangular wave guide of height b and width $\bar{a}$, centrally loaded with a dielectric slab of thickness $\bar{s}$ and dielectric ε_r.

The model is an intuitive aid in setting up the simplest mathematics that leads to a useful dispersion relation; it need not be physically realizable.

The analysis proceeds by forcing the model to have the same cutoff frequencies as the fin-line without and with the dielectric. Accurate closed-form expressions for these frequencies can be found from [7][8]. Next, a transverse resonance analysis of the model relates the propagation constant (or effective dielectric constant) to frequency. A closed-form approximation of this function is then found and compared with the data obtained by measurement and sophisticated numerical techniques.

Cutoff Frequency Relationships

The cutoff wavelengths λ_{ca} of the airfilled fin-line and λ_{cf} of the fin-line can be obtained from [7][8]. Forcing the LSE model in Fig. 2 to have the same λ_{ca} and λ_{cf} gives the transcendental equation

$$\cot\left(\frac{\pi}{2}\,\bar{x}\sqrt{\frac{\varepsilon_r}{k_c}}\right) = \sqrt{\varepsilon_r}\,\tan\left(\frac{\pi}{2}\,(1 - \bar{x})/\sqrt{k_c}\right) \tag{1}$$

where

$$\begin{aligned} \bar{x} &= \bar{s}/\bar{a} \\ \bar{a} &= \lambda_{ca}/2 \end{aligned} \tag{2}$$

and

$$k_c = (\lambda_{cf}/\lambda_{ca})^2 \tag{4}$$

is the equivalent dielectric constant of the fin-line at cutoff.

Transverse Resonance Solution

Once $\bar{x}$ is known, the characteristic equation for the fundamental LSE mode of the model can be written as

$$\begin{aligned} &\frac{1}{\left(\varepsilon_r - \overline{\varepsilon_e(f)}\right)^{\frac{1}{2}}}\cot\left\{\frac{\pi}{2}\,\bar{x}\,\frac{\lambda_{ca}}{\lambda}\left(\varepsilon_r - \overline{\varepsilon_e(f)}\right)^{\frac{1}{2}}\right\} \\ &= \frac{1}{\left(1 - \overline{\varepsilon_e(f)}\right)^{\frac{1}{2}}}\tan\left\{\frac{\pi}{2}\,(1 - \bar{x})\,\frac{\lambda_{ca}}{\lambda}\left(1 - \overline{\varepsilon_e(f)}\right)^{\frac{1}{2}}\right\} \end{aligned} \tag{5}$$

where

$\overline{\varepsilon_e(f)} = \beta^2/k_0^2$, the effective dielectric constant of LSE model

$k_0 = 2\pi/\lambda$, free space wave number,

and λ is the ree space wavelength.

At a particular wavelength λ, equation (5) can be solved numerically for $\varepsilon_e(f)$. Results obtained by such a solution for s/a=0.0265957 and ε_r=3.80, 2.20 in WR-19 housing are shown in Fig. 3 and compared with those obtained by the network analytical method [9] and uniform rectangular waveguide approximation [2]. It is observed from the plots that the model agrees accurately with the numerical technique up to a little above the cutoff of the airfilled portion of the fin-line. It is found that the following equations agree closely with numerical results [9]

$$\frac{\lambda}{\lambda_g} = \frac{\lambda}{\lambda_{gm}} \text{ ; for } \frac{b}{\lambda} \leq 0.4 \tag{6}$$

$$= \frac{1}{\sqrt{2}}\left[\left(\frac{\lambda}{\lambda_{gm}}\right)^2 + \left(\frac{\lambda}{\lambda_{10}}\right)^2\right]^{\frac{1}{2}} \text{ ; for } b/\lambda > 0.4 \tag{7}$$

where λ_{gm} is the guided wavelength in the model, λ_{10} is that of the TE_{10} mode of the model homogeneously filled with an equivalent dielectric of constant k_c [2]. The accuracy of the above expressions is within ±0.8% when compared with numerical techniques [9].

Closed-form Equations

Although the above solution procedure reduces the computational effort to a great extent when compared with that required by sophisticated numerical techniques, it still requires the solutions of transcendental equations (1) and (5) and fails to give any closed-form solution.

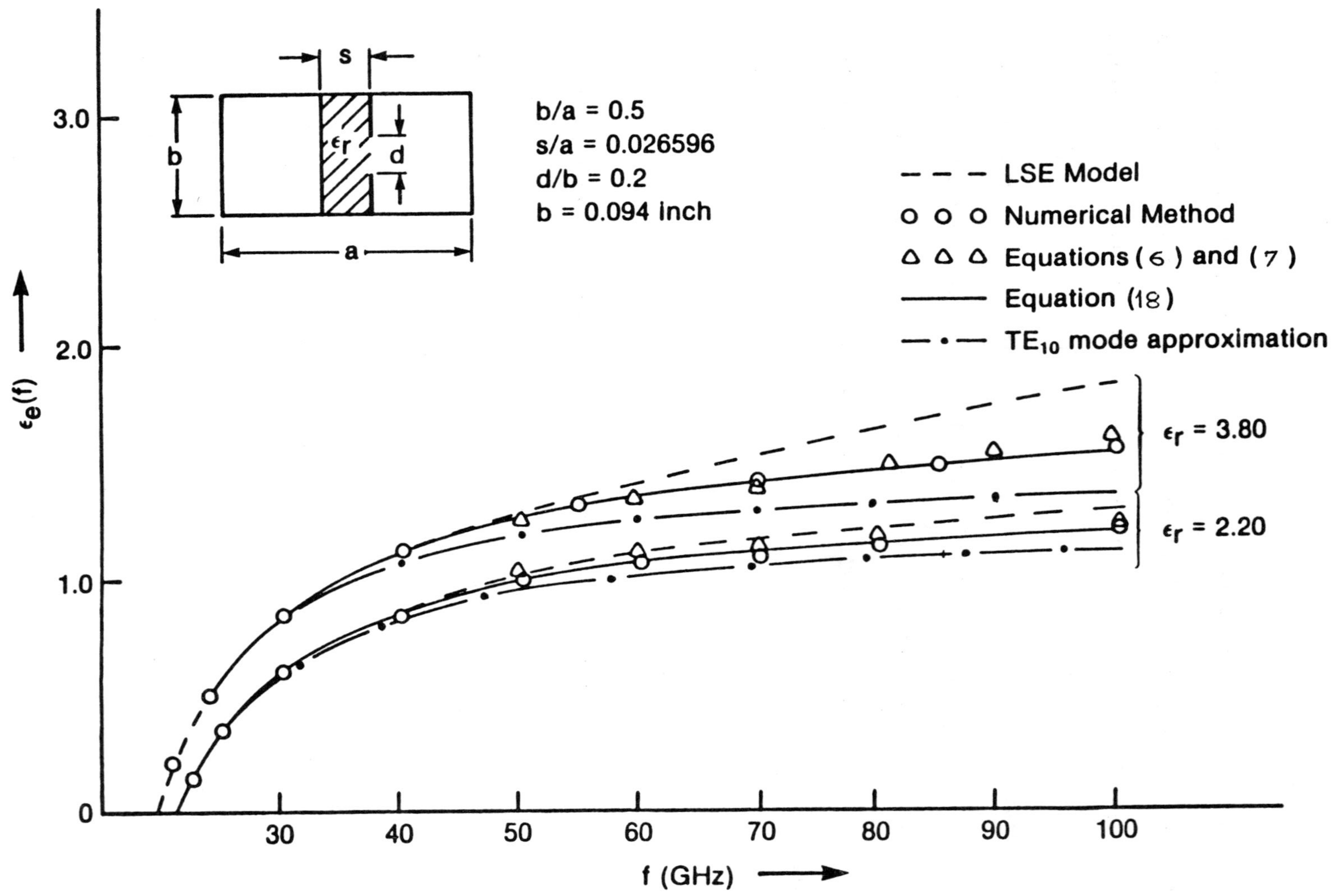

Fig. 3. Comparison with numerical results for unilateral finline.

A closed-form equation can be obtained from the LSE model if the following facts are kept in mind:

(1) the equivalent dielectric constant $k_e(f)$ of the fin-line is a linear function of frequency [10],

(2) the present model approximates $k_e(f)$ accurately up to $\varepsilon_e(f) = 1$.

Equation for $\bar{x}$

The variational expression for $k_e(f) = k_c$ at the cutoff frequency f_c obtained from the model [11] is given by

$$k_c = 1 + (\varepsilon_r - 1)\left(\bar{x} + \frac{1}{\pi}\sin(\pi\bar{x})\right) \tag{8}$$

using

$$\operatorname{cosec}(\Theta) = \frac{1}{\Theta} + \frac{\Theta}{6}, \text{ for } \Theta \leqslant 0.707$$

in (8) gives

$$\bar{x}^3 - q_f\bar{x}^2 + (12/\pi^2)\bar{x} - (6q_f/\pi^2) = 0 \tag{9}$$

or

$$\bar{x} = (2q_f/\pi^2 - s_1^3)^{1/3} + s_1 + q_f/3 \tag{10}$$

where

$$s_1 = \{(0.0666 + 0.0466q_f^2 + 0.015q_f^4 - 0.00137q_f^6)^{\frac{1}{2}} + q_f/\pi^2\}^{1/3} \tag{11}$$

and q_f, the dielectric filling factor at cutoff, is given by

$$q_f = \frac{k_c - 1}{\varepsilon_r - 1} \tag{12}$$

Equation for Phase Velocity

For $\overline{\varepsilon_e(f)} = 1$, equation (5) is reduced to

$$\cot\left\{\frac{\pi}{2}\,\bar{x}\,\frac{\lambda_{ca}}{\lambda_1}(\varepsilon_r - 1)^{\frac{1}{2}}\right\} = \frac{\pi}{2}(\varepsilon_r - 1)^{\frac{1}{2}}(1 - \bar{x})\,\frac{\lambda_{ca}}{\lambda_1} \tag{13}$$

For all practical purposes, it can be shown that

$$\frac{\pi}{2}\,\bar{x}\,\frac{\lambda_{ca}}{\lambda_1}(\varepsilon_r - 1)^{\frac{1}{2}} \leqslant 1.$$

Therefore making the approximation

$$\cot\Theta = \frac{1}{\Theta} - \frac{\Theta}{3} \text{ in (13) gives}$$

$$\frac{\lambda_1}{\lambda_{ca}} = \pi\left\{\bar{x}\,\frac{(3-2\bar{x})(\varepsilon_r-1)}{12}\right\}^{\frac{1}{2}} \tag{14}$$

Let the equivalent dielectric constant at $\lambda = \lambda_1$ be $k_e(1)$. Therefore

$$k_e(1) - \left(\frac{\lambda_1}{\lambda_{ca}}\right)^2 = 1 \tag{15}$$

and (14) and (15) give

$$k_e(1) = 1 + \frac{\pi^2}{12}\,\bar{x}(3-2\bar{x})(\varepsilon_r-1) \tag{16}$$

A linear interpolation through the points $(b/\lambda_{cf}, k_c)$ and $(b/\lambda_1, k_e(1))$ gives

$$k_e(f) = k_c + \frac{k_e(1)-k_c}{\dfrac{b}{\lambda_1} - \dfrac{b}{\lambda_{cf}}}\left(\frac{b}{\lambda} - \frac{b}{\lambda_{cf}}\right) \tag{17}$$

Once $k_e(f)$ is known from equation (17) the guided wavelength can be written as

$$\frac{\lambda_g}{\lambda} = \left[k_e(f) - \left(\frac{\lambda}{\lambda_{ca}}\right)^2\right]^{-\frac{1}{2}} \tag{18}$$

III. COMPARISON WITH NUMERICAL AND EXPERIMENTAL RESULTS

The theory developed in this paper will first be compared with the theoretical prediction of Hayashi, Farr and Mittra [9] using the network analytical method, for a unilateral fin-line having s/a = 0.0266 d/b = 0.2, ε_r = 2.22 and ε_r = 3.8 in WR-19 housing. The computed results are shown in Fig. 3. It is observed that the LSE model agrees accurately with the numerical results up to $b/\lambda \approx 0.4$ the LSE model deviates considerably from the numerical results. In this case the results obtained by using equations (6) and (7) are found to be in excellent agreement (within ±0.8%) with numerical results [9]. The solid line obtained by using the closed-form equation (18) is found to give even better agreement. Fig. 4 compares the present equations with Hofmann's [12] numerical and Meier's experimental results for s/a=0.036, d/b=0.13, b/a=0.5 and ε_r=2.22. The present equations are also compared with the results of 3-dimensional transmission line matrix analysis [13] and the approximate transverse resonance method [13]. Once again the agreement is found to be excellent.

Fig. 5 compares the present equations in case of a bilateral fin-line, with s/a=0.25, d/b=0.5, b/a=0.5, and ε_r=2.22, with the results of TLM [13] and transverse resonance [13] analyses. A more detailed eval-

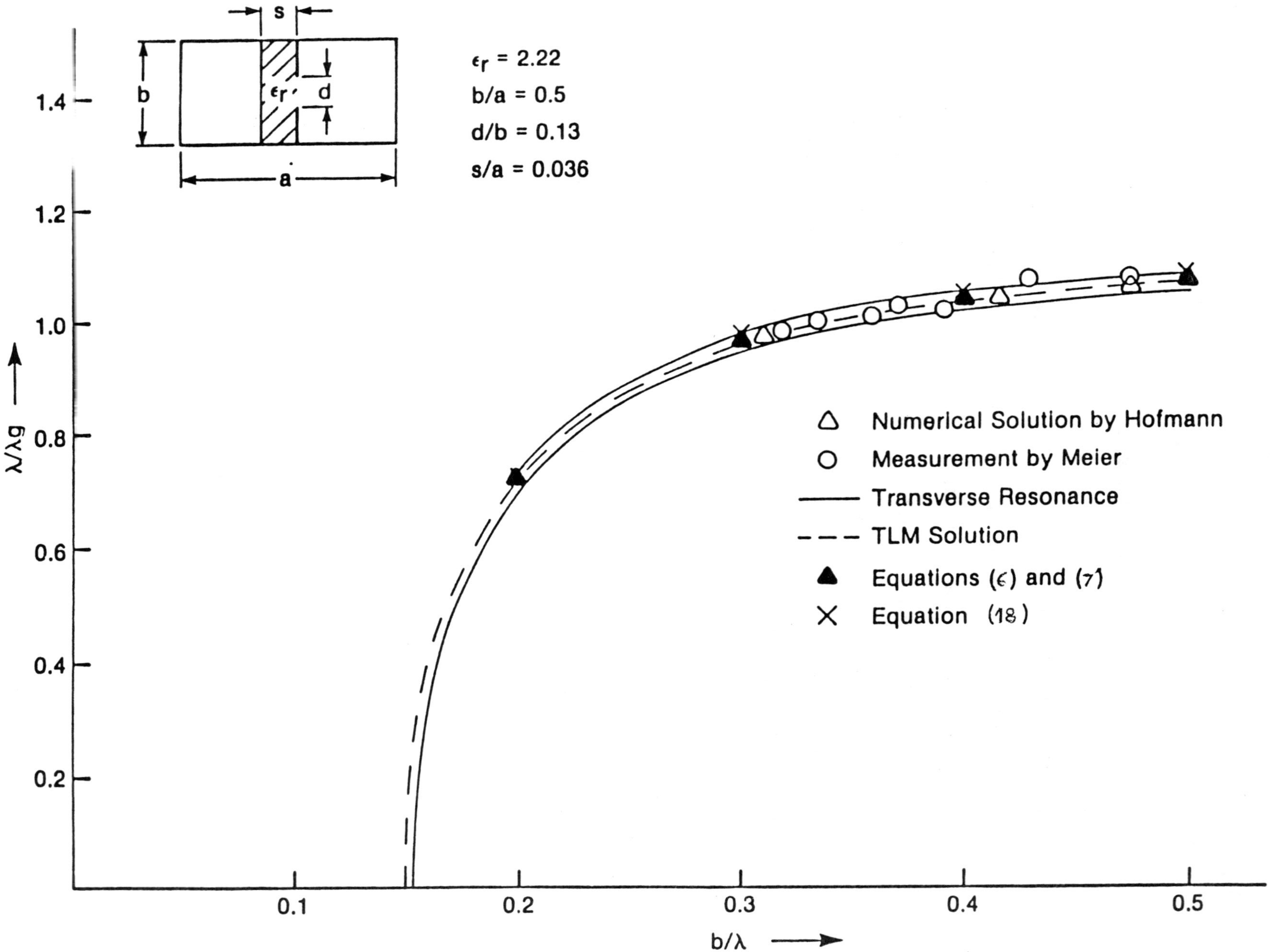

Fig. 4. Comparison with numerical and experimental results for unilateral finline.

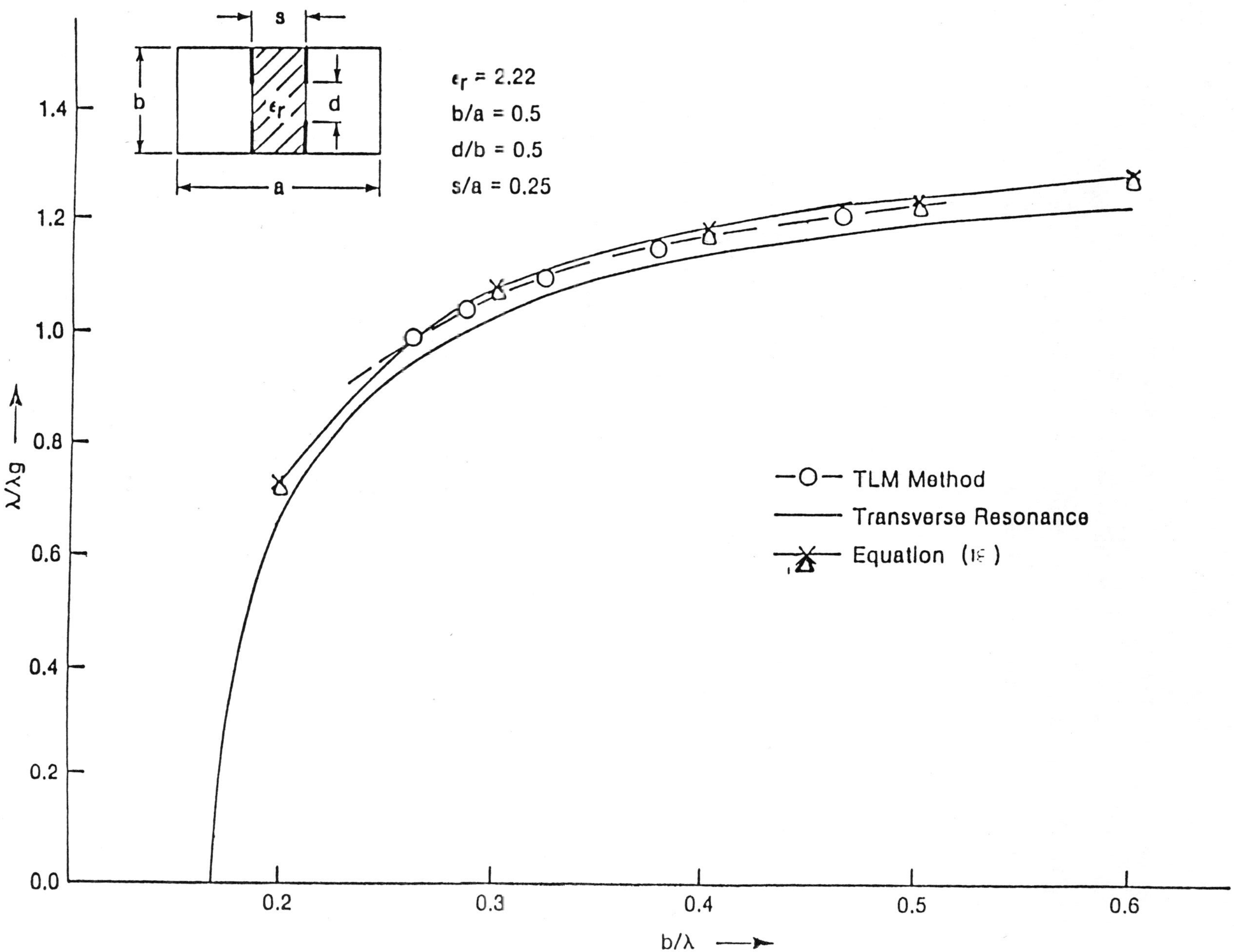

Fig. 5. Comparison with numerical results for bilateral finline.

uation of this agreement can be made by studying Table I which compares results from the proposed design equations with those of the spectral domain technique [4]. The proposed equations have also been verified for insulated fin-line but the results are not presented here due to lack of space.

IV DISCUSSION

The LSE model developed in this paper from the cutoff frequencies in fin-lines is found to give a better and more realistic approximation of HE mode than the quasi TE mode approximation of Saad and Schunemann [3]. The final closed-form expressions derived are lucid and easily tractable. The accuracy of the equations are sufficient for most practical fin-line applications.

V CONCLUSION

Novel frequency-dependent analytical expressions have been derived, for the frequency dependent equivalent dielectric constant of unilateral, bilateral and insulated fin-lines, using an LSE model determined from cutoff conditions. The expressions are given in such a form that they can be implemented easily on a programmable calculator or a desktop computer. In terms of accuracy, the described closed-form equations together with those in [7] will be vital to CAD techniques and helpful in the proliferation of basic know-how concerning fin-line techniques.

ACKNOWLEDGEMENT

This work has been supported by a Natural Sciences and Engineering Research Council of Canada, Grant A-0001.

[1] K. Solbach, "The status of printed millimeter-wave E-plane circuits", Trans. IEEE, Vol. MTT-31, No. 2, pp. 107-121, 1983.

[2] A.M.K. Saad and K. Schunemann, "A simple method for analysing fin-line structures", Trans. IEEE, Vol. MTT-26, No. 12, pp. 1002-1007, 1978.

[3] A.M.K. Saad, and K. Schunemann, "A rectangular waveguide equivalent for bilateral and unilateral fin-lines", Arch. fur. Elektron. Und Ubertra., Vol. 35, pp. 287-292, 1981.

[4] L.P. Schmidt and T. Itoh, "Spectral domain analysis of dominant and higher order modes in fin-lines", Trans. IEEE, Vol. MTT-28, pp. 981-985, Sept. 1980.

[5] R.E. Collin, Field Theory of Guided Waves, New York: McGraw-Hill, 1960, p. 224.

[6] C. Montgomery, R. Dicke and E. Purcell, Principles of Microwave Circuits, (M.I.T. Radiation Laboratory Series), Vol. 8, New York: McGraw-Hill, 1948.

[7] P. Pramanick and P. Bhartia, "Accurate analysis and synthesis equations for fin-lines", Symposium Digest, 8th Annual IEEE International Conference on Infrared and Millimeter Waves, Miami (FL), 1983.

[8] P. Pramanick and P. Bhartia, "Accurate analysis equations and synthesis technique for unilateral fin-lines", IEEE Trans. on MTT, Vol. 33, No. 1, pp. 24-30, Jan. 1985.

[9] Y. Hayashi, E. Farr and R. Mittra, "Analysis of dominant and higher order modes in unilateral fin-lines", Interim Technical Report, U.S. Army Research Office, Contract No. DAAG29-82-K-0084, Electromagnetics Laboratory, Department of Electrical Engineering, Engineering Experiment Station, University of Illinois at Urbana, Champaign, IL, August 1982.

[10] A.M.K. Saad and K. Schunemann, "Efficient eigenmode analysis for planar transmission lines", Trans. IEEE, Vol. MTT-30, No. 12, pp. 2125-2132, 1982.

[11] R.F. Harrington, Time Harmonic Electromagnetic Fields, McGraw-Hill Book Company, New York: p. 345, 1961.

[12] H. Hofmann, "Calculation of quasiplanar lines for mm wave application", Int. IEEE MTT Symposium, San Diego, California, pp. 381-383, 1977.

[13] Y.C. Shih, "Analysis of fin-lines using transmission line matrix and transverse resonance methods", M.A.Sc. Dissertation Dept. of Electrical Engineering, University of Ottawa, Canada, 1980.

[14] P.J. Meier, "New developments with integrated fin-line and related printed millimeter circuits", IEEE MTT-S, Intl. Microwave Symposium Digest, pp. 143-145, Palo Alto, CA, 1975.

2.2.6

Sensitivity Analysis of the Fin-Line Housing

A fin-line housing can be fabricated at low cost, because there are no tight tolerance limits for the dimensions of the housing. This is proved by calculating the variations of the cutoff wave number, the effective dielectric constant, and the wave impedance with respect to the dimensions of the waveguide housing.

Empfindlichkeitsanalyse des Gehäuses der Flossenleitung

Das Gehäuse der Flossenleitung kann billig hergestellt werden, weil an seine Abmessungen keine hohen Toleranzanforderungen gestellt werden müssen. Dies wird dadurch bewiesen, daß die Änderungen der Eigenwerte, effektiven Dielektrizitätszahl und des Wellenwiderstandes mit den Gehäuseabmessungen berechnet werden.

1. Introduction

Integrated fin-line is an attractive solution for the realization of millimeterwave integrated circuits, when microstrip shows inherent shortcomings as, e.g., dispersion, radiation loss, spurious coupling, higher mode propagation, and critical tolerances. Moreover, fin-line has potential applications even at centimeter-wavelengths, if circuits with medium Q-factor are to be realized. Examples are narrow-band filters and stable (or low-noise) oscillators. Unloaded Q-factors of up to 1000 are obtainable with unilateral fin-lines at X-band. It is then interesting to know, to what extent the fin-line characteristics are influenced by tolerances in the fabrication of the waveguide housing. The fin-line could favourably compete with microstrip even in the microwave region, if this influence were small. We will therefore present a sensitivity analysis of the fin-line housing.

2. Analysis

The characteristics of the fundamental mode in both bilateral and unilateral fin-line have been investigated in [1]. Closed-form solutions were derived for the cutoff frequency, the propagation constant, the equivalent dielectric constant, and the wave impedance which agree within 3% with "exact" (numerical) results published elsewhere. We will utilize here these approximate formulas in order to calculate the partial derivatives of the interesting quantities with respect to the dimensions of the housing.

The *bilateral fin-line* will be considered first. Its cross-section is shown in Fig. 1a (the nomenclature is the same as in [1]). It is convenient to define an equivalent dielectric constant k_e by [1], [2]

$$k_e = k_{c\,air}^2 / k_{c\,d}^2 \qquad (1)$$

where $k_{c\,d}$ is the cutoff wave number with and $k_{c\,air}$ without dielectric. The frequency dependence of both the guided wavelength λ_g and the wave impedance Z can then be written as

$$\lambda_g = \frac{\lambda_0}{\sqrt{k_e - (\lambda_0/\lambda_{c\,air})^2}}, \quad Z = \frac{Z_\infty}{\sqrt{k_e - (\lambda_0/\lambda_{c\,air})^2}}. \qquad (2)$$

λ_0 means free-space wavelength, $\lambda_{c\,air} = 2\pi/k_{c\,air}$, and Z_∞ impedance at infinite frequency. It has been observed experimentally [2] and theoretically [3] that k_e is nearly independent of frequency.

Using the transverse resonance method, closed-form expressions have been derived in [1] for k_e (eq. (16)), $k_{c\,air}$ (eq. (17)), and Z_∞ based on a power-voltage definition (eq. (25)). (As concerns the latter, the amplitude of $\cos^2$ should read $2C_d/\varepsilon_0$ in the first line of eq. (25).) We will investigate the sensitivity of these formulas with respect to variations in waveguide width a, height b, and position d_s of the fin-line slots above the bottom of the waveguide.

Normalizing the variations in the cutoff wave number $k_{c\,air}$ against those of a rectangular waveguide $k_{c\,wg}$, we get

$$\frac{\partial k_{c\,air}^2/\partial a}{\partial k_{c\,wg}^2/\partial a} = \frac{H}{2(1-c/a)}\left[1 + H\left(1-\frac{c}{a}\right)^2\right] \approx$$
$$\approx \frac{H}{2}(1+H) \qquad (3)$$

for $c \ll a$. H means the ratio of the respective cutoff wave numbers squared:

$$H = k_{c\,air}^2 / k_{c\,wg}^2 . \qquad (4)$$

The variations with respect to the waveguide height are

$$\frac{\partial k_{c\,air}^2/\partial b}{\partial k_{c\,wg}^2/\partial a} = \frac{a}{b}\,\frac{H}{2}\left[1 - H\left(1-\frac{c}{a}\right)^2\right] - \frac{\pi^2 c l}{4da} H^2 \approx$$
$$\approx \frac{a}{b}\,\frac{H}{2}(1-H) - \frac{\pi^2 c}{8d} H^2 . \qquad (5)$$

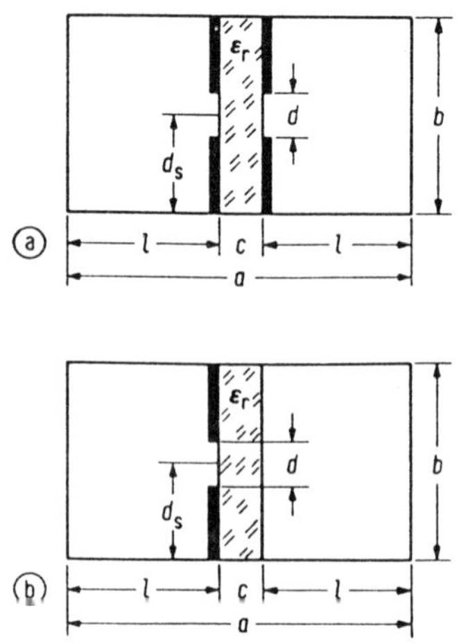

Fig. 1. Cross section of the bilateral (a) and unilateral (b) fin-line.

H. El Hennawy and K. Schunemann, *Archiv fur Elektronik und Ubertragungstechnik*, Vol. 35, No. 5, Sept.-Oct. 1981, pp. 220–222.

Similarly, we get for the variations of the equivalent dielectric constant

$$\frac{\partial k_e}{\partial a} = H(1 - k_e)\frac{a-c}{a^2} \approx \frac{H}{a}(1 - k_e),$$

$$\frac{\partial k_e}{\partial b} = \frac{H}{b}\left[\left(1 - \frac{c}{a}\right)^2 (k_e - 1) + \frac{\pi^2 l b c}{2 d a^2}(k_e - \varepsilon_r)\right] \approx$$

$$\approx \frac{H}{b}\left[k_e - 1 + \frac{\pi^2}{4}\frac{cb}{ad}(k_e - \varepsilon_r)\right] \qquad (6)$$

and of the wave impedance

$$\frac{\partial Z_\infty}{\partial a} = -\frac{Z_\infty^2}{240\,\pi}\frac{\partial x}{\partial l}, \quad \frac{\partial Z_\infty}{\partial b} = -\frac{Z_\infty^2}{120\,\pi}\frac{\partial x}{\partial b} \qquad (7)$$

with abbreviations

$$x = 120\,\pi/Z_\infty, \quad \varphi = k_{c\,d}\,l, \quad h = c\sqrt{\varepsilon_r}/2,$$

$$\frac{\partial x}{\partial l} = \frac{2}{b}\left[1 - \frac{\varphi}{\sin^3\varphi}\left(1 - \frac{\sin 2\varphi}{2\varphi}\right) + K_1\frac{\partial k_{c\,d}}{\partial a}\right],$$

$$\frac{\partial x}{\partial b} = -\frac{x}{b} + \frac{K_1}{b}\frac{\partial k_{c\,d}}{\partial b}, \qquad (8)$$

$$K_1 = \frac{l^2}{\sin^2\varphi}\left(\frac{\sin 2\varphi}{2\varphi^2} - \frac{1}{\varphi} - 2\cot\varphi\right) +$$

$$+ 2h^2 k_{c\,d}\left(\frac{2hb}{d} - bx\right).$$

The variations with respect to the position of the slots are

$$\frac{\partial k^2_{c\,air}}{\partial d_s} - k^4_{c\,air}\,l\,\frac{\partial D}{\partial d_s},$$

$$\frac{\partial k_e}{\partial d_s} = \frac{1}{k^2_{c\,d}}\left(\frac{\partial k^2_{c\,air}}{\partial d_s} - k_e\frac{\partial k^2_{c\,d}}{\partial d_s}\right) = \qquad (9)$$

$$= k^2_{c\,air}(1 - k_e)\,l\,\frac{\partial D}{\partial d_s}$$

with $D = bC_d/\varepsilon_0$. C_d represents the fringing field capacitance at the fins, which must (according to the calculations in [1]) be taken from [4]. It is shown in [4] that $D \sim \sin(\pi d_s/b)$. Hence $\partial D/\partial d_s = 0$ for a symmetrically located slot ($d_s = b/2$). The variations of both $k^2_{c\,air}$ and k_e with respect to d_s disappear in this case.

The analysis has been repeated for the ***unilateral fin-line***, whose cross-section is given in Fig. 1b. The cutoff wave number and the equivalent dielectric constant are given by eqs. (21) and (20), respectively, of [1]. The corresponding variations with respect to a, b and d_s are

$$\frac{\partial k^2_{c\,air}/\partial a}{\partial k^2_{c\,wg}/\partial a} = \frac{H}{2\left(1+\frac{c}{a}\right)}\left[1 + H\frac{\left(1-\frac{c}{a}\right)^2\left(1+\frac{4c}{a}\right)}{1+\frac{c}{a}}\right] \approx$$

$$\approx \frac{H}{2}(1 + H),$$

$$\frac{\partial k^2_{c\,air}/\partial b}{\partial k^2_{c\,wg}/\partial b} = \frac{a}{b}\frac{H}{2}\cdot$$

$$\cdot\left\{1 - \frac{H}{2}\left[\frac{\pi^2 l c}{a^2} + \frac{4}{a^2}\left(c^2 + 2l^2 + 2lc + \frac{l^2 c}{l+c}\right)\right]\right\} \approx$$

$$\approx \frac{a}{b}\frac{H}{2}(1 - H) - \frac{\pi^2 c}{8b}H^2,$$

$$\frac{\partial k_e}{\partial a} = \frac{H}{a+c}\left[\frac{\pi^2}{2}\frac{c^2}{a^2}(\varepsilon_r - k_e) + 1 - k_e\right] \approx$$

$$\approx \frac{H}{a}(1 - k_e),$$

$$\frac{\partial k_e}{\partial b} = \frac{H}{b}\left[\frac{2}{a^2}(k_e - 1)\left(c^2 + 2l^2 + 2lc + \frac{l^2 c}{l+c}\right) + \right.$$

$$\left. + \frac{\pi^2 l c}{2a^2}(k_e - \varepsilon_r)\right] \simeq \qquad (10)$$

$$\simeq \frac{H}{b}\left[k_c - 1 + \frac{\pi^2 c}{4a}(k_e - \varepsilon_r)\right].$$

The variations with respect to d_s are the same as in eqs. (9). They disappear again for a symmetrically

Table 1. Numerical results of the tolerance analysis for X-band with a = 22.85 mm, b = 10.15 mm, c = 0.25 mm, ε_r = 2.22.

	bilateral fin-line		unilateral fin-line	
$d/b =$	0.05	0.2	0.05	0.2
$\frac{\partial k^2_{c\,air}/\partial a}{\partial k^2_{c\,wg}/\partial a} =$	0.20	0.39	0.24	0.39
$\frac{\partial k_{c\,air}/\partial a}{\partial k_{c\,wg}/\partial a} =$	0.36	0.54	0.40	0.54
$\frac{\partial k^2_{c\,air}/\partial b}{\partial k^2_{c\,wg}/\partial a} =$	0.18	0.24	0.25	0.26
$\frac{\partial k_{c\,air}/\partial b}{\partial k_{c\,wg}/\partial a} =$	0.32	0.34	0.41	0.37
$\partial k_e/\partial a =$	−0.0027 mm^{-1}	−0.0018 mm^{-1}	−0.0061 mm^{-1}	−0.0066 mm^{-1}
$\partial k_e/\partial b =$	−0.00089 mm^{-1}	−0.00044 mm^{-1}	−0.0122 mm^{-1}	−0.01246 mm^{-1}
$\partial x/\partial l =$	0.178	0.161	0.080	0.087
$\partial x/\partial b =$	−0.403	−0.196	−0.277	−0.167

located slot. The variations of the wave impedance are as given in eqs. (7) with

$$\frac{\partial x}{\partial l} = \frac{2}{b \sin^2 \varphi}\left(1 - \varphi \cot \varphi - K_2 \frac{\partial k_{cd}}{\partial a}\right),$$

$$\frac{\partial x}{\partial b} = -\frac{x}{b} - \frac{K_2}{b \sin^2 \varphi} \frac{\partial k_{cd}}{\partial b},$$

$$K_2 = \frac{1}{k_{cd}} (\cos 2\varphi + 2\varphi \cot \varphi + 4 k_{cd}^2 h^2) - \\ - \frac{\sin 2\varphi}{2\varphi}\left(\frac{l}{k_{cd}} + 2 l^2 \cot \varphi + 4 k_{cd} h^2 l\right). \qquad (11)$$

3. Results

An evaluation of the above equations yields: The sensitivity of the waveguide characteristics of a fin-line with respect to variations in the dimensions of the waveguide housing is much less than for a rectangular waveguide provided that the fin-line slot corresponds to an impedance of about 100 Ω. It is typically reduced to about 20% due to the tight coupling of the wave to the slot. It is hence not necessary that the fin-line housing is machined carefully. The fin-line might therefore be preferable to the microstrip even at X-band frequencies. These statements are proved by some numerical results presented in Table 1.

Acknowledgement

The authors are indebted to the Deutsche Forschungsgemeinschaft for financial support.

(Received January 28th, 1981.)

Dipl.-Ing. Hadia El Hennawy
Dr.-Ing. Klaus Schünemann
Institut für Hochfrequenztechnik
Technische Universität
Postfach 3329, D-3300 Braunschweig

References

[1] Saad, A. M. K. and Begemann, G., Electrical performance of fin-lines of various configurations. IEE J. Microwaves, Optics, and Acoustics 1 [1977], 81–88.

[2] Meier, P. J., Integrated fin-line millimeter components. Transact. IEEE MTT-22 [1974], 1209–1216.

[3] Saad, A. M. K. and Schünemann, K., A simple method for analyzing fin-line structures. Transact. IEEE MTT-26 [1978], 1002–1007.

[4] Whinnery, J. R. and Jamieson, H. W., Equivalent circuits for discontinuities in transmission lines. Proc. IRE 32 [1944], 98–114.

Chapter 3
Synthesis Techniques

To fabricate the appropriate fin-line structure, we require the normalized fin gap for a given set of the housing aspect ratio, substrate dielectric constant, substrate thickness, and the operating frequency. In general, a fin line is synthesized for a specified characteristic impedance. However, because a fin line is not a TEM line, its characteristic impedance is not uniquely defined. Usually, the choice of definition depends on the application. If we wish to match one fin gap to another fin gap (e.g., matching one fin gap to a full-height, dielectric, slab-loaded waveguide), we should use the wave impedance of the fin line [1]. We apply the value of the characteristic impedance only when we need to match an active or a passive device to the fin line. Many researchers have used the power-current definition of the characteristic impedance. Some are in favor of the voltage-current definition, while others prefer the power-voltage definition. Thus, we still do not know which definition is the most appropriate.

Sections 2.2.2 and 2.2.4 presented closed-form synthesis equations for unilateral and insulated fin lines. However, no such synthesis equations are available for bilateral fin lines.

In this chapter, the first paper presents closed-form synthesis equations based on an apparently uncontroversial definition of the fin-line characteristic impedance.

3.1 THEORETICAL CONSIDERATION

3.1.1 "A New Model for the Apparent Characteristic Impedance of Finned Waveguide and Finlines" by Pramanick and Bhartia

This reprint paper presents a new model for the apparent characteristic impedance of finned waveguides and fin lines, in terms of capacitance per unit length. We obtain the fin capacitance from the results of conformal mapping, and the housing capacitance from the cutoff and wavelength. This model strongly supports the power-voltage definition of the characteristic impedance. Incidentally, the synthesis equations for unilateral fin lines have a few typographical errors, which were corrected in subsequent printings.

3.2 FABRICATIONAL CONSIDERATIONS

3.2.1 "Fabricational Considerations in Fin-Lines" by Pramanick and Bhartia

This previously unpublished technical note briefly discusses a few points on the mechanical aspects of the fin-line structure. We present the advantages and disadvantages of two types of housings, as well as the advantages and disadvantages of using soft and hard substrates.

BIBLIOGRAPHY

1. K. Solbach, "The Status of Printed Millimeter-Wave E-Plane Circuits," *IEEE Trans. Microwave Theory and Techniques*, Vol. MTT-31, No. 2, Feb. 1983, pp. 91–107.

3.1.1

A New Model for the Apparent Characteristic Impedance of Finned Waveguide and Finlines

PROTAP PRAMANICK, MEMBER, IEEE, AND PRAKASH BHARTIA, SENIOR MEMBER, IEEE

***Abstract* —This paper presents a new model for the apparent characteristic impedance of finned waveguide and finlines from the standpoint of per-unit-length capacitance of the fins, obtained by conformal mapping, and of the waveguide, obtained from cutoff wavelengths, without going into the controversy of definition. The model strongly supports the power–voltage definition. Closed-form synthesis equations have also been derived which are within 1.5 percent of the analysis equations. The model will be useful in computer-aided design of millimeter-wave finned waveguide and finlines.**

I. Introduction

THE CONCEPT OF IMPEDANCE is indispensable in microwave circuit design since the distribution of microwave energy among the elements, through the interconnections, is decided by the impedance. Therefore, the exact prediction of circuit performance depends upon how accurately the impedance of its elements is known.

Microwave circuits are made of distributed networks using pieces of transmission lines which are characterized in terms of their phase constants and characteristic impedances.

As discussed by Getsinger [1] for microstrip line, we can also define the "apparent" characteristic impedance of finned waveguide or finline. This describes how finned waveguide or finline exchanges power with a TEM line or a two-terminal device, just as the characteristic impedance is the parameter which determines how a TEM line exchanges power with another transmission line or with a two-terminal device.

The definition of characteristic impedance is unique in the case of a purely TEM line. The controversy regarding the definition of characteristic impedance arises when the transmission line supports a non-TEM mode. The problem of the apparent characteristic impedance of quasi-TEM line (like microstrip) has been solved by experimental modeling [1] and by theoretical analysis [2]. For purely non-TEM planar transmission lines such as finned waveguide and finline, the definition of the apparent characteristic impedance still remains an open question. Till now there have been no adequately conclusive experimental results or theoretical arguments in favor of any of the three possible definitions, i.e., voltage/current, power/voltage and power/current. Moreover, neither the voltage nor the current can be uniquely defined in a finned waveguide or in a finline. However, while the experimental findings of Meinel and Rembold [3] tend towards the voltage/current definition, those of Willing and Spcilman [4] are in favor of the power/voltage definition. In the above definition, voltage is defined as the line integral of the electric field over the shortest path between the fins, and the current is the total longitudinal current flowing in the line. Still, the experimental findings show a wide divergence from the numerical computations [5], [6]. Approximate ridge guide definitions of Meier [7] differ by 8 to 9 percent from numerically computed values [5], [6]. On the other hand, the slow rise in the characteristic impedance with increasing frequency at the upper edge of the usable band, as predicted by numerical computation, has not been experimentally confirmed. Keeping all this in mind, the present work derives the characteristic impedance from the per-unit-length capacitance standpoint.

Manuscript received April 1, 1986; revised June 24, 1986.

P. Pramanick is with the Satellite Communication Department, COM DEV Ltd., Cambridge, Ontario, Canada N1R 7H6.

P. Bhartia is with the Department of National Defence, Ottawa, Canada.

IEEE Log Number 8610834.

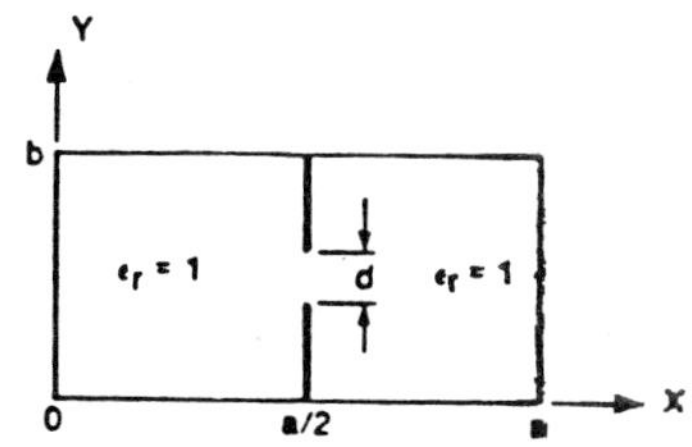

Fig. 1. Finned waveguide.

II. Theory

A. Analysis of Finned Waveguide

The finned waveguide in Fig. 1 can be considered to be a combination of a finned parallel plate of infinite width, as shown in Fig. 2(a), and a rectangular waveguide, as shown in Fig. 2(b).

The per-unit-length capacitance between the fins of the parallel-plate waveguide is uniquely defined and is obtained accurately from conformal mapping as [8]

$$C_f = \left(\frac{2}{\pi}\right) \ln \csc\left(\frac{\pi}{2}\frac{d}{b}\right). \tag{1}$$

P. Pramanick and P. Bhartia, *IEEE Transactions on Microwave Theory and Techniques*, Vol. MTT-34, No. 12, Dec. 1986, pp. 1437–1441.

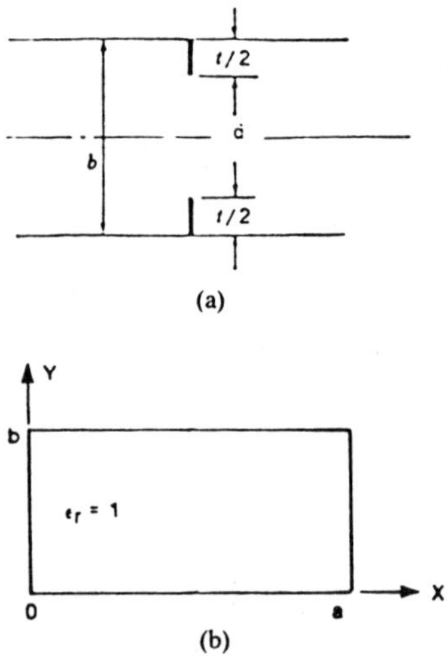

Fig. 2. (a) Finned parallel plate. (b) Rectangular waveguide.

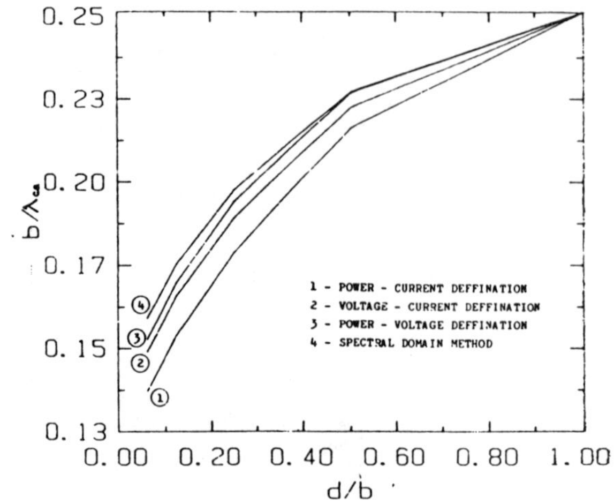

Fig. 3. Comparison of cutoff wavelengths.

Now, if the capacitance between the midpoints of the broadwalls of the waveguide is C_w, then it follows from the definition of characteristic impedance [9], that the characteristic impedance of the finned waveguide at frequency f will be given by

$$Z_0(f) = \frac{Z_{0\infty}}{\sqrt{1-(\lambda/\lambda_{ca})^2}} \tag{2}$$

where λ is the wavelength corresponding to frequency f, and λ_{ca} is the cutoff wavelength of the finned waveguide. $Z_{0\infty}$, the characteristic impedance at infinite frequency, is given by [9]

$$Z_{0\infty} = \frac{\eta_0}{C_w + C_f} \text{ ohms} \tag{3}$$

where $\eta_0 = 120\pi\ \Omega$ (free-space impedance).

Since the characteristic impedance of the rectangular waveguide $Z_{0\infty}$ cannot be defined uniquely, C_w can have three different values depending upon the definition. For example, the power–voltage definition gives, (TE_{10} mode)

$$Z^{PV}_{0\infty w} = 2\eta_0(b/a) \text{ ohms}. \tag{4}$$

The voltage–current definition gives

$$Z^{VI}_{0\infty w} = \left(\frac{\pi}{2}\right)\eta_0(b/a) \text{ ohms} \tag{5}$$

while the power–current definition gives

$$Z^{PI}_{0\infty w} = \left(\frac{\pi^2}{8}\right)\eta_0(b/a) \text{ ohms}. \tag{6}$$

Corresponding to each definition of $Z_{0\infty w}$, a C_w can be defined. The question now arises which one should be used in (3) to define $Z_{0\infty}$. Although the characteristic impedance could not be defined uniquely, the propagation constant of the finned waveguide can be defined uniquely. Or, in other words, the cutoff wavelength can be defined uniquely. Using the perturbation theory, the cutoff wavelength of the finned waveguide can be written as

$$\lambda_{ca} = 2a\left[1 + \frac{\Delta C}{C_w}\right]^{1/2} \tag{7}$$

where

$$\Delta C = C_f \tag{8}$$

and

$$C_w = C_{wPV} = \frac{1}{2}\frac{a}{b} \quad \text{(from (4))} \tag{9}$$

$$= C_{wVI} = \frac{2}{\pi}\frac{a}{b} \quad \text{(from (5))} \tag{10}$$

$$= C_{wPI} = \frac{8}{\pi^2}\frac{a}{b} \quad \text{(from (6))}. \tag{11}$$

Let us compute λ_{ca} using all three of the above definitions of C_w and compare the results with accurately computed values of λ_{ca} from the spectral-domain method [10]. Fig. 3 shows that the best results are obtained by using the power–voltage definition or $C_w = C_{wPV}$. The difference between the perturbational formula and the spectral-domain results increases for smaller d/b values because the field distribution, in the waveguide cross section, tends to shift from the sinusoidal form as d/b decreases. Consequently, (9) needs to be modified to

$$C_{wPV} = \left(\frac{1}{G}\right)\left(\frac{1}{2}\right)\left(\frac{a}{b}\right). \tag{12}$$

The factor G has been determined empirically to be

$$G = 1 + (0.129\sqrt{x} + 0.00915)\sqrt{\frac{b}{a}} \tag{13}$$

where

$$x = \ln\csc\left(\frac{\pi}{2}\frac{d}{b}\right). \tag{14}$$

Equation (7) together with (12), (13), and (14) gives λ_{ca} within 0.4 percent of spectral-domain results.

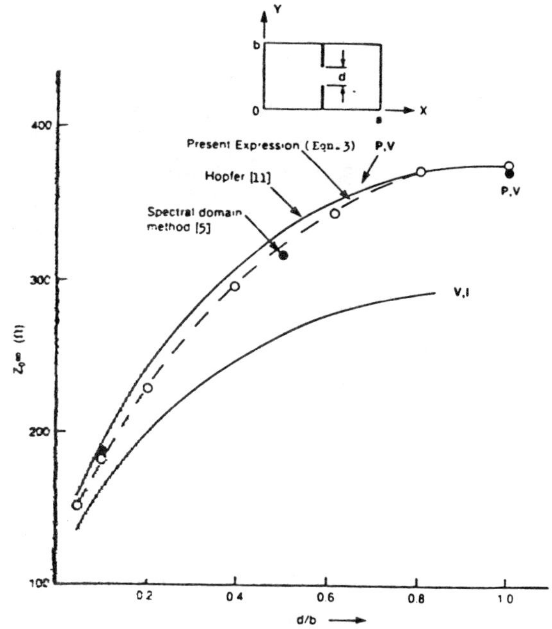

Fig. 4. Comparison of present model with ridged guide definition and spectral-domain results.

Therefore, using (1), (2), (3) and (12), (13), (14), the characteristic impedance of the finned waveguide in Fig. 1 is given by

$$Z_0(f)=\frac{2\eta_0(b/a)G}{\left[1+\frac{4}{\pi}\left(\frac{b}{a}\right)Gx\right]\sqrt{1-(\lambda/\lambda_{ca})^2}}. \qquad (15)$$

Fig. 4 compares results obtained by using (15) with those obtained from Hopfer's [11] ridged waveguide model and the spectral-domain method. A close agreement among the results is observed. The results agree exactly for $d/b=1$.

B. Synthesis of Finned Waveguide

The factor G appearing in (15) can be approximated within 1 percent in the numerator as

$$G=1+(r_1x+r_2)\sqrt{b/a} \qquad (16a)$$

where

$$(r_1,r_2)=(0.16814,0.00894),\quad \text{for } d/b\geqslant 1/3 \qquad (16b)$$

$$=(0.051496,0.08739),\quad \text{for } d/b\leqslant 1/3 \qquad (16c)$$

and in the denominator as

$$G=1+0.2\sqrt{b/a}. \qquad (16d)$$

Then, solving (15) for x gives

$$x=\frac{1}{2}\left\{-p+(p^2-4q)^{1/2}\right\} \qquad (17)$$

$$p=\left[N\left\{2-\frac{1}{4}(b/a)^2(\lambda/b)^2\right\}Z_0^2(f)-2\alpha_1\alpha_2\right]/\left[N^2Z_0^2(f)-\alpha_1^2\right] \qquad (18a)$$

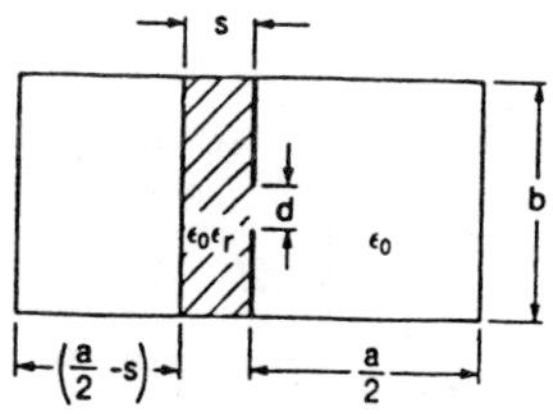

Fig. 5. Unilateral finline.

and

$$q=\left[Z_0^2(f)\left\{1-\frac{1}{4}(b/a)^2(\lambda/b)^2\right\}-\alpha_2^2\right]/\left[N^2Z_0^2(f)-\alpha_1^2\right] \qquad (18b)$$

where

$$N=\frac{4}{\pi}\left(\frac{b}{a}\right)\left(1+0.2\sqrt{b/a}\right) \qquad (19)$$

$$\alpha_1=2\eta_0r_1(b/a)^{3/2} \qquad (20a)$$

$$\alpha_2=2\eta_0(b/a)\left(1+r_2\sqrt{b/a}\right). \qquad (20b)$$

The above synthesis equation (17) gives results within 1.5 percent of the analysis equation. Therefore, an accurate synthesis is possible by improving the results of the synthesis equation using the analysis equation.

III. Application to Unilateral Finline

A. Analysis of Unilateral Finline

The effective dielectric constant in unilateral finline shown in Fig. 5 can be written as

$$\epsilon_e(f)=k_e-\left(\frac{\lambda}{\lambda_{ca}}\right)^2 \qquad (21)$$

where k_e, the equivalent dielectric constant of the finline, can be obtained from [12] and [13], and λ_{ca} is given by (7) together with (1) and (12) through (14). Therefore, the frequency-dependent characteristic impedance of unilateral finline is given by

$$Z_0(f)=\frac{2\eta_0(b/a)G}{\left[1+\frac{4}{\pi}(b/a)Gx\right]\sqrt{k_e-(\lambda/\lambda_{ca})^2}}. \qquad (22)$$

Equation (22), however, does not necessarily fully agree with numerically computed $Z_0(f)$ [5] based on the power–voltage definition, as shown in Figs. 6 and 7. It has been found that the numerically computed $Z_0(f)_{PV}$ can be accurately modeled as (see Figs. 6 and 7)

$$Z_0(f)_{PV}=Z_0(f)\left\{\frac{1-1/k_e}{1-1/k_c}\right\}^{1/2}\left(\frac{k_e}{k_c}\right) \qquad (23)$$

where $Z_0(f)$ is given by (22), and k_c is the value of k_e at cutoff.

Equation (23) shows a rise in $Z_0(f)_{PV}$ with frequency. However, the rise is slow and significant only outside the useful frequency band.

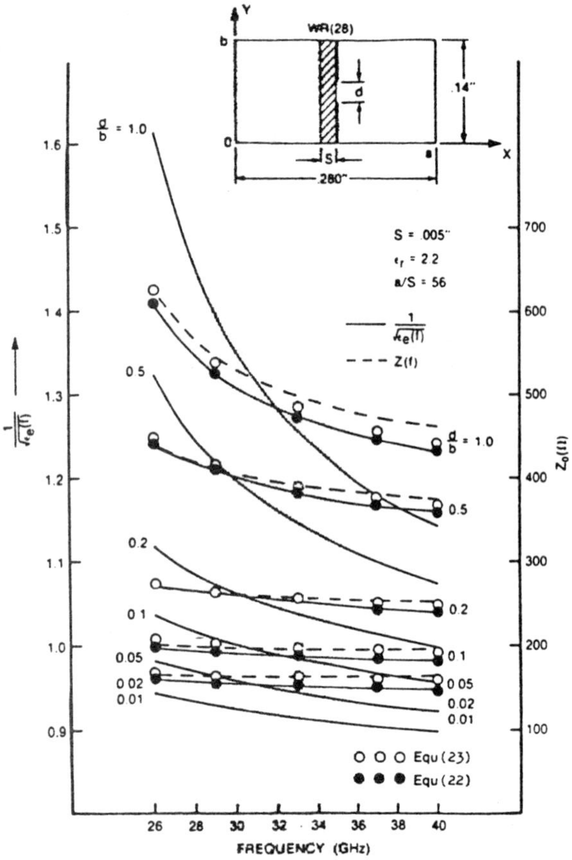

Fig. 6. Comparison with spectral-domain results.

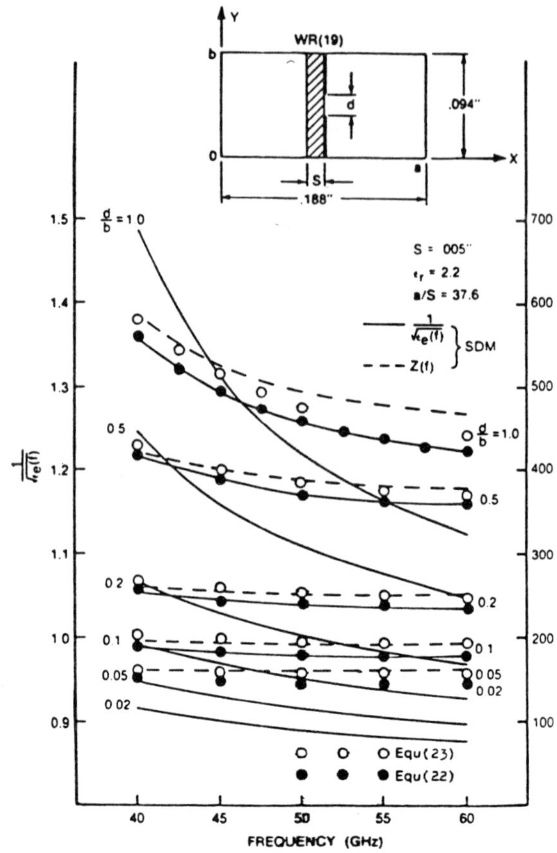

Fig. 7. Comparison with spectral-domain results.

B. Synthesis of Unilateral Finline

For most practical finline applications $\epsilon_r < 2.5$ and $s/a \ll 1$ (see Fig. 5). Therefore, one can make a first-order approximation by considering the equivalent dielectric constant to be frequency-independent and given by [12] its value k_c at the cutoff frequency as

$$k_e = k_c = 1 + \frac{s}{a}(a_1 x + b_1)(\epsilon_r - 1) \tag{24}$$

where

$$a_1 = [0.6177\ln(a/s) - 0.5457]^2$$

$$b_1 = 2.42\sin[0.556\ln(a/s)].$$

Substituting (24) into (22) gives the following cubic equation for x:

$$x^3 + px^2 + qx + r = 0 \tag{25}$$

where

$$p = [AN^2 + 2BN - \alpha_1^2]/BN^2$$

$$q = \left[B + AN - \frac{1}{4}(b/a)^2(\lambda_b/b)^2 - 2\alpha_1\alpha_2\right]/BN^2$$

$$r = \left[A - \frac{1}{4}(b/a)^2(\lambda_b/b)^2 - \alpha_2^2\right]/BN^2$$

$$A = 1 + b_1(s/a)(\epsilon_r - 1)$$

$$B = a_1(s/a)(\epsilon_r - 1)$$

and α_1 and α_2 are given by (20a) and (20b), respectively.

The values of x obtained from (25), when substituted into (22), yield results within 2 percent of the design value. The accuracy can be improved to within 1 percent by using the analysis equation iteratively with the x obtained from (25) as the starting value. Such an iterative technique can also be used for $\epsilon_r \nless 2.5$ and $s/a \not\ll 1$. One can also use (23) in such an iterative synthesis technique if one has more confidence in the power–voltage definition of the characteristic impedance. A general computer program has been developed to implement the above technique and is being used as a CAD tool [14]. The present modeling can also be used for the insulated finline using closed-form equations [15].

A similar model has been developed for bilateral finline and will be reported soon.

IV. Application to Single-Ridge Finline

The model derived herein for the analysis and synthesis of finned waveguide and finline can also be applied to a single-ridge finline with the following interpretation: b is twice the height of the housing of the single-ridge finline, and d is twice the gap between the tip of the fin and the broadwall of the housing. The interpretation remains the same for the characteristic impedance; however, the characteristic impedance of the single-ridge finline is half that of the finline.

V. Conclusions

This paper defines the characteristic impedance of finned waveguides and finlines from the per-unit-length capacitance standpoint, without going into the controversy of definition used in numerical analysis of the structures. The total per-unit-length capacitance of the fins is obtained from the results of conformal mapping. The new definition strongly supports the power–voltage definition. Simple synthesis equations have been derived under practical assumptions. The model throws new light on the controversial definition of the apparent characteristic impedance of finline and will be vital to the CAD of millimeter-wave circuits.

Acknowledgment

The authors are indebted to A. Swarup and Dr. R. S. Tomar for helpful discussions and Ms. Pauline Fowler of COM DEV Ltd. for a critical review of the manuscript.

References

[1] W. J. Getsinger, "Measurement and modelling of the apparent characteristic impedance of microstrip," *IEEE Trans. Microwave Theory Tech.*, vol. MTT-31, pp. 624–632, Aug. 1983.

[2] R. H. Jansen and M. Kirsching, "Arguments and an accurate model for the power–current formulation of microstrip characteristic impedance," *Arch. Elek. Übertragung.*, vol. 37, pp. 108–112, 1983.

[3] H. Meinel and B. Rembold, "New millimeter-wave finline attenuators and switches," in *IEEE MTT-S Int. Symp. Dig.*, 1979, pp. 249–252.

[4] H. A. Willing and B. E. Spielman, "Experimental assessment of bilateral finline impedance for device matching," in *IEEE MTT-S Int. Symp. Dig.*, 1981, pp. 105–107.

[5] J. B. Knorr and P. M. Shayda "Millimeter-wave finline characteristics," *IEEE Trans. Microwave Theory Tech.*, vol. MTT-28, pp. 737–473, July 1980.

[6] L. P. Schmidt and T. Itoh, "Spectral domain analysis of dominant and higher order modes in finlines," *IEEE Trans. Microwave Theory Tech.*, vol. MTT-28, pp. 981–985, Sept. 1980.

[7] P. J. Meier, "Integrated finline millimeter components," *IEEE Trans. Microwave Theory Tech.*, vol. MTT-22, pp. 1209–1216, Dec. 1974.

[8] R. E. Collin, *Field Theory of Guided Waves*. New York: McGraw-Hill, 1960.

[9] ——, *Foundations for Microwave Engineering*. New York: McGraw-Hill, 1966, p. 85.

[10] A. K. Sharma, G. I. Costache, and W. J. R. Hoefer, "Cutoff in finlines evaluated with the spectral domain technique and with the finite element method," in *IEEE AP-S Int. Symp. Dig.*, 1981, pp. 308–311.

[11] S. Hopfer, "Design of ridged waveguide," *IRE Trans. Microwave Theory Tech.*, vol. 3, pp. 20–29, Oct. 1955.

[12] P. Pramanick and P. Bhartia, "Accurate analysis equations and synthesis technique for unilateral finline," *IEEE Trans. Microwave Theory Tech.*, vol. MTT-33, pp. 24–80, Jan. 1985.

[13] ——, "Frequency dependence of equivalent dielectric constant in finlines," *Int. J. Infrared and Millimeter Waves*, vol. 7, no. 2, pp. 251–264, Feb. 1982.

[14] ——, "A unified approach to the exact design of optimum Chebyshev non-TEM transmission line tapers," to be published.

[15] ——, "Accurate analysis and synthesis equations for insulated finlines," *Arch. Elek. Übertragung.*, vol. 39, no. 1, pp. 31–36, Jan. 1985.

3.2.1

Fabricational Considerations in Fin Lines

Protap Pramanick and Prakash Bhartia

ABSTRACT

This technical note discusses the fabricational aspects of fin lines in different housings and on different substrates. The advantages and the disadvantages of hard and soft substrates are discussed.

FIN-LINE HOUSING

There are two different types of structures used for fin-line housing. The type used by Meier [1] is shown in Fig. 1. The housing will have identical inner dimensions of millimeter-wave waveguides, i.e., WR(28), WR(19), WR(10) *et cetera.* The substrate includes a number of mounting holes and two stepped edges. The stepped edges protrude into the abutting waveguides and serve as quarter-wavelength transformers.

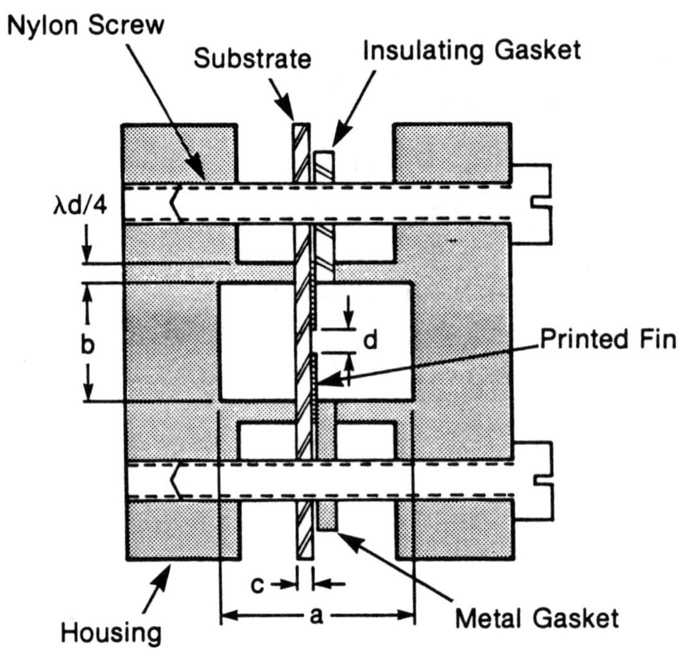

Fig. 1

P. Pramanick and P. Bhartia, Technical note, University of Ottawa, 1987, 2 pp.

The quarter-wavelength should be in the medium of dielectric constant ϵ_r. The transition between the fin line and dielectric loaded waveguide is obtained by tapering the fin gap up to the full-height waveguide. The main disadvantage of this structure is the unwanted power flow through the clamping regions. This is generally prevented by milling a choke-slot in the waveguide block, or using a serration in the metalization pattern on the substrate.

The structure shown in Fig. 2 is more compact than the one described above. The structure was used by Adelseck *et al.* [2]. It differs at the clamping region, where the circuit board is clamped in the housing. The structure has a number of disadvantages. Its attenuation rate is high due to transverse current flow at the clamps. This may well lead to spurious resonances in the case of a bad electrical contact. This imposes a tight tolerance, on the order of half a mil, for the slit width d in order that there may be adequate contacting pressure at the meeting faces of the split blocks. Besides, unlike Meier's structure, it is difficult to make dc connections for biasing the active devices, to the metalization pattern on the substrate. The connections are fed through the clamping region, or a strip of the circuit board is allowed to contact the waveguide block to make such connections. The arrangement is shown in Fig. 3. Despite its many disadvantages, this structure is favored due to its compactness in size and ease of handling.

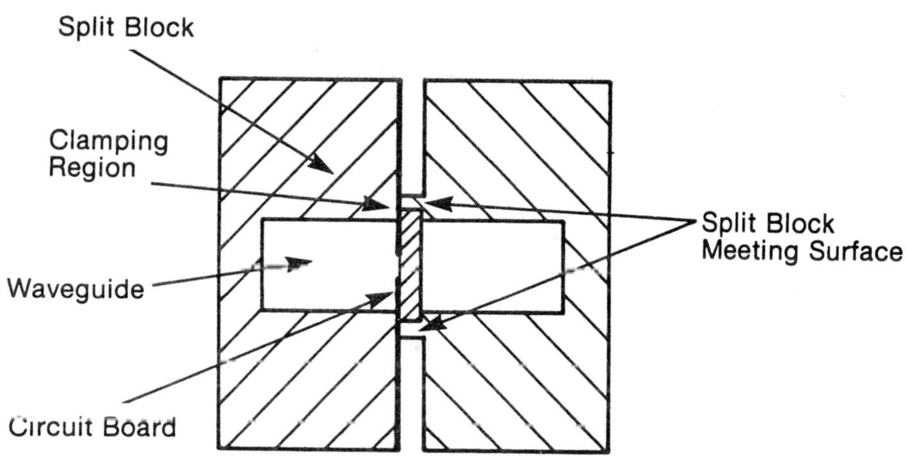

Fig. 2

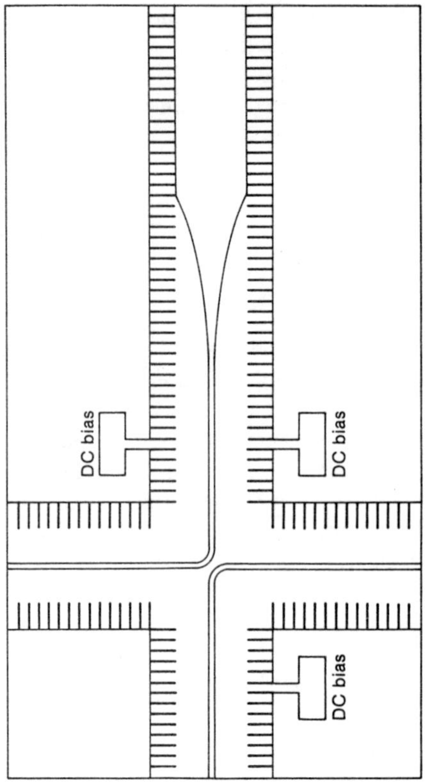

Fig. 3

SUBSTRATES

The dielectric substrate in a fin line supports the metal fins so that the photo-etching technique may be used to render the fin-gap pattern for a desired circuit performance. The substrate in principle should be as thin as possible and should have the lowest possible dielectric constant. Substrates which are most commonly used are RT Duroid 5880™ (ϵ_r = 2.22) and fused quartz (ϵ_r = 3.78). Usually the chosen substrate thicknesses are 254 μm for Ka band operating and 127 μm for E-band operations.

RT Duroid™ satisfies the basic requirement of low dielectric constant. However, its low mechanical strength makes it unsuitable for many applications, such as where bonding the fins with delicate leads of beam-lead devices is required. Such bonding often breaks under production handling, temperature cycling, and high *g* forces. On the other hand, hard fused quartz substrate does not have any such shortcomings. A compromise between the two substrates may be effected by using small chips of fused quartz for mounting the semiconductor devices, which may be integrated into the soft substrate circuit as an upside down overlay. However, such a structure may not be as good as a quartz substrate, which withstands acceleration levels up to 3000g's for 2 ms without any problem. Soft substrates can withstand up to 2000g's for 0.5 ms without damage.

REFERENCES

1. P.J. Meier, "Integrated Finline Millimeter Components," *IEEE Trans. Microwave Theory and Techniques*, Vol. MTT-22, No. 10, Dec. 1974, pp. 1209–1216.
2. B. Adelseck et al., "A Survey of Planar Integrated Millimeter-Wave Components," *The Radio and Electronic Engineer*, Vol. 52, No. 1, 1982, pp. 46–50.

Chapter 4
Discontinuities

No transmission line is useful in circuit theory if it has no discontinuities, and the fin line is no exception. In E-plane integrated circuits, discontinuities are introduced so that the appropriate circuit functions may be performed. Typical discontinuities include steps in the fin gap, small capacitive strips, and inductive notches. A capacitive strip is used for mounting semiconductor devices. A capacitive strip and an inductive notch, when appropriately cascaded, give rise to a low-pass filter. The strip inductor, an important discontinuity, has been successfully used in matching bandpass filters up to D-band.

Much has been published on the various methods for numerically analyzing fin-line discontinuities. Such methods include the variational technique, spectral-domain analysis, mode-matching technique, generalized scattering-parameter technique and the transverse resonance analysis.

In this chapter, we will present papers on the numerical and experimental modeling of fin-line discontinuities, and some on useful analytical models.

4.1 NUMERICAL ANALYSIS

4.1.1 "Impedance Transformation in Fin Lines" by El Hennawy and Schunemann

This paper, based on modal analysis, is the first successful attempt to characterize fin-line discontinuities theoretically. The authors analyze virtually all types of fin-line discontinuities, which we usually encounter while designing matching networks, filters, and semiconductor mounts. The paper is adequately thorough, and supplies much useful design data.

4.1.2 "A Variational Solution of the Fin-Line Discontinuity Problem" by Webb and Mittra

This paper concentrates on step discontinuities only, and obtains the numerical solution using the generalized variational method.

4.1.3 "Transverse Resonance Analysis of Finline Discontinuities" by Sorrentino and Itoh

The authors use a slightly different method to analyze uniform fin lines as well as discontinuities in fin lines. They first expand the fin-line fields in terms of LSE and LSM modes of a rectangular waveguide, and then match the fields at various junctions. This procedure substantially reduces computer time. The authors claim that the accuracy of the uniform fin-line propagation constant is good. However, they offer no quantitative comparison with other existing methods.

4.1.4 "Spectral-Domain Analysis of Scattering from E-Plane Circuit Elements" by Zhang and Itoh

This paper shows how the spectral-domain technique can be used efficiently and accurately for analyzing the scattering phenomenon in E-plane obstacles. We use the method to analyze scattering from nontouching E-plane septa in unilateral fin lines. This method can also be applied to the cases of bilateral and insulated fin lines. The paper presents a wide range of practical, useful design data. However, the data are valid only for zero fin thickness.

4.2 EXPERIMENTAL CHARACTERIZATION

4.2.1 "A Combined Experimental and Theoretical Characterisation of Discontinuities in Unilateral Finlines" by Beyer, Kother, and Wolff

This paper presents an optimization procedure based on the generalized scattering-matrix formulation of fin-line step discontinuities. The scattering-matrix representation is deduced from a combined experimental and numerical procedure. The measurement setup is controlled by a desk-top computer. This method takes into consideration second-order effects like loss and finite fin thickness. This paper also presents detailed measured data on a Ku-band fin line.

4.2.2 "An Improved Model for Short- and Open-Circuited Series Stubs in Fin Lines" by Burton and Hoefer

This paper models the useful discontinuity series stubs (open- and short-circuited) by an equivalent circuit, which takes into account the field disturbances at the branching points and at the far end of the stub. We obtain the expressions for different circuit parameters by fitting curves to experimental results. The expressions are valid over a single waveguide band and for single-ridge, unilateral fin lines with a polystyrene-type substrate and small fin gaps. The expressions are independent of housing dimensions, and are particularly useful in low-pass filter design.

BIBLIOGRAPHY

1. K.J. Webb and R. Mittra, "Solution of Finline Step Discontinuity Problem Using the Generalized Variational Technique," *IEEE Trans. Microwave Theory and Techniques*, Vol. MTT-33, No. 10, Oct. 1985, pp. 1004–1010.
2. Y.Y. Chen and S.F. Li, "Analysis of Finline Discontinuities," *IEEE MTT-S Symposium Digest*, 1986, pp. 229–232.
3. K. Uhde, "Discontinuities in Finlines on Semiconductor Substrate," *IEEE Trans. Microwave Theory and Techniques*, Vol. MTT-34, No. 12, Dec. 1986, pp. 1490–1498.
4. A.S. Omar and K. Schunemann, "Transmission Matrix Representation of Finline Discontinuities," *IEEE Trans. Microwave Theory and Techniques*, Vol. MTT-33, No. 9, Sept. 1985, pp. 765–770.
5. A.S. Omar and K. Schunemann, "The Effect of Complex Modes at Finline Discontinuities," *IEEE Trans. Microwave Theory and Techniques*, Vol. MTT-34, No. 12, Dec. 1986, pp. 1508–1509.
6. R. Safavi-Naini and R.H. Macphie, "On Solving Waveguide Junction Scattering Problems by the Conservation of Complex Power Technique," *IEEE Trans. Microwave Theory and Techniques*, Vol. MTT-29, No. 4, April 1981, pp. 337–343.
7. R.R. Monsour and R.H. Macphie, "An Improved Transmission Matrix Formulation of Cascaded Discontinuities and Its Application to E-Plane Circuits," *IEEE Trans. Microwave Theory and Techniques*, Vol. MTT-34, No. 12, Dec. 1986, pp. 1490–1498.

4.1.1

Impedance transformation in fin lines

Hadia El Hennawy, Ph.D., Mem. I.E.E.E., and Prof. Klaus Schünemann, Dr.-Ing., Mem. I.E.E.E., M.V.D.E.

Indexing terms: *Semiconductor devices and materials, Microwave circuits and networks, Microwave components, Numerical analysis*

Abstract: Various discontinuities in the slot width of fin lines are accurately analysed. These include a step in the slot width, an inductive notch, a small capacitive strip and a longitudinal stripe for mounting a semiconductor diode. Equivalent circuits are derived and related to the geometrical parameters. The results can directly be applied to a design of impedance transforming and filter networks.

1 Introduction

Various passive and active components have been realised in integrated fin line throughout a broad frequency spectrum ranging from 10 to 140 GHz. These include oscillators [1, 2], mixers [3–5], digital modulators [6], switches and attenuators [7], detectors [8], couplers [6, 9], filters [10, 11] and tapers [12]. Nearly all of these components have been developed by trial and error, except for the filters and tapers reported in References 11 and 12. Theoretical investigations published so far deal almost exclusively with the calculation of the propagation constants and wave impedances of the fundamental and higher-order modes: this topic has been solved by approximations (leading to closed-form solutions [13]) and by rigorous calculations [14, 15]. There is, however, a lack of fundamental theoretical research concerning an analytical description of almost all passive circuit elements used for either mounting semiconductor devices or impedance transformation, or for both. It is hoped that the present work will be a first step in filling this gap.

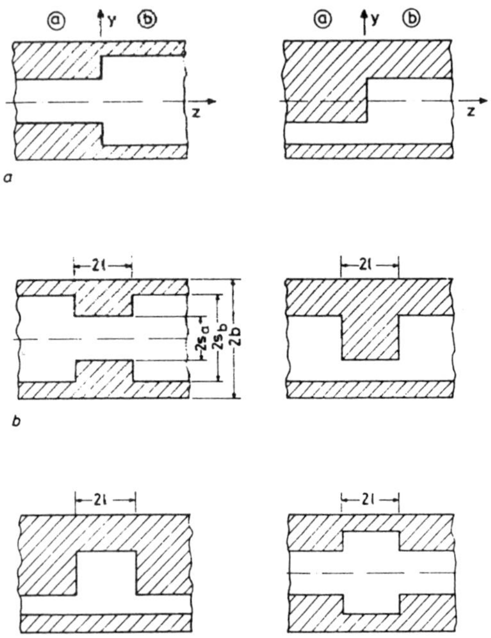

Fig. 1 *Slot patterns of fin-line discontinuities*

a step, *b* strip, *c* notch

Paper 2234H, received 8th December 1981

The authors are with the Institut fuer Hochfrequenztechnik, Technische Universitaet, Postfach 3329, D-3300 Braunschweig, West Germany

The structures to be investigated are shown in Fig. 1. The basic building block is a step discontinuity in the slot width (Fig. 1*a*) which may or may not be symmetric with respect to a plane $y =$ constant. This discontinuity acts as an impedance transformer shunted by a susceptance. Two such steps can be combined to form either a strip (Fig. 1*b*) or a notch (Fig. 1*c*). Potential applications depend on the electrical length of the strip or of the notch. A short strip can be used for realising a capacitance, whereas a short notch can be modelled by an inductance [6]. A long, unsymmetric strip (which will be called a longitudinal stripe) is often used for mounting a semiconductor device [1]. This mount resembles a post mount in rectangular waveguides. A narrow slit in one fin, which can be utilised as a DC separation in a complex structure, can be regarded as a special case of the notch: it forms a very short, unsymmetric notch with a slot width equal to the height of the waveguide housing. Finally, strips and notches can be cascaded for the design of a lowpass filter.

2 Modal analysis

The step discontinuity in the slot width of Fig. 1*a* will be treated by a modal analysis [16]. In Reference 16, two uniform cylindrical waveguides *a* and *b* with different cross-sections and distributions of enclosed electrical properties are considered which are joined end to end forming a junction. The boundary conditions are satisfied by a suitable infinite series of modes appropriate to each side of the junction. We will assume, now, that the modes in both waveguides and the scattering coefficients of successive discontinuities are known. The properties of the junction can then be calculated, if one finds out how power is distributed between the various scattered modes.

The solution of this basic problem will first be described with the help of a modal analysis. The co-ordinate system is chosen so that $z = 0$ is the plane of the junction, and x and y are transverse co-ordinates. The transverse fields of each mode are written as

$$e_i(x,y,z) = a_i\, e_i(x,y)\, e^{\pm \gamma_i z}$$
$$h_i(x,y,z) = a_i\, h_i(x,y)\, e^{\pm \gamma_i z} \quad (1)$$

e_i and h_i are transverse vector functions of the electric and magnetic fields, respectively. Expanding the transverse fields in terms of these modes just to the left of the junction (waveguide *a*), yields

$$E_a = (1+\rho)\, a_1\, e_{a1} + \sum_{i=2}^{\infty} a_i e_{ai}$$

$$H_a = (1-\rho)\, a_1\, h_{a1} - \sum_{i=2}^{\infty} a_i h_{ai} \quad (2)$$

Similarly, we can write, just to the right of the junction

(waveguide b),

$$E_b = \sum_{j=1}^{\infty} b_j(e_{bj} + \sum_{k=1}^{\infty} s_{jk}\, e_{bk})$$

$$H_b = \sum_{j=1}^{\infty} b_j(h_{bj} - \sum_{k=1}^{\infty} s_{jk} h_{bk}) \tag{3}$$

The subscripts a and b denote the individual waveguides, ρ means reflection coefficient of the incident mode, which is characterised by $i = 1$ and amplitude a_1. s_{jk} are the scattering coefficients of the next discontinuity, which is located at $z > 0$. The magnetic and electric fields have been expressed independently, because a unique wave impedance cannot be defined for a fin line.

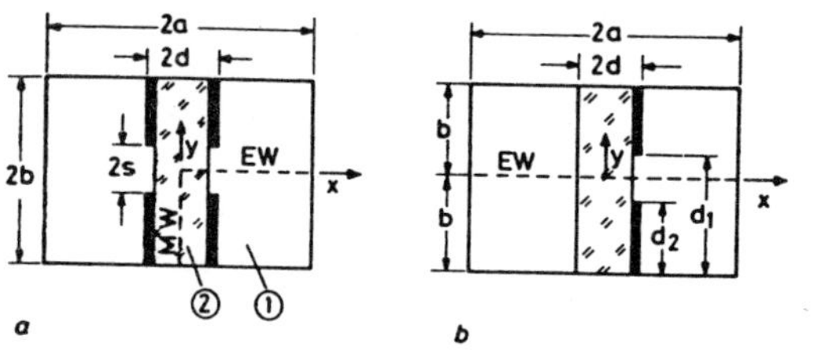

Fig. 2 *Cross-section of bilateral and unilateral fin lines*

EW = electric wall, MW = magnetic wall
***a* Bilateral fin line, *b* unilateral fin line**

The following calculations will be illustrated for a bilateral fin line with a symmetrically located slot as an example. Its cross-section is shown in Fig. 2*a*. It is obvious that only modes which are symmetrical with respect to $x = 0$ will be excited. One may hence assume a magnetic wall (MW) at $x = 0$ and an electric wall (EW) at $y = 0$. The transverse vector functions e_i and h_i can then be composed of the transverse field components in subregions 1 and 2 of Fig. 2*a* at $z = 0$, according to:

$$E_{ti}^{(1)} = \sum_{n=1}^{\infty} A_n^{(1)} \cos\alpha_n(x-a) \sin\alpha_n y \cdot u_x$$

$$+ \sum_{n=0}^{\infty} S_n^{(1)} \sin\alpha_n(x-a) \cos\alpha_n y \cdot u_y$$

$$E_{ti}^{(2)} = \sum_{n=1}^{\infty} A_n^{(2)} \sin\tilde{\alpha}_n x \sin\alpha_n y \cdot u_x$$

$$+ \sum_{n=0}^{\infty} S_n^{(2)} \cos\tilde{\alpha}_n x \cos\alpha_n y \cdot u_y \tag{4}$$

and

$$H_{ti}^{(1)} = \sum_{n=0}^{\infty} B_n^{(1)} \sin\alpha_n(x-a) \cos\alpha_n y \cdot u_x$$

$$+ \sum_{n=1}^{\infty} M_n^{(1)} \cos\alpha_n(x-a) \sin\alpha_n y \cdot u_y$$

$$H_{ti}^{(2)} = \sum_{n=0}^{\infty} B_n^{(2)} \cos\tilde{\alpha}_n x \cos\alpha_n y \cdot u_x$$

$$+ \sum_{n=1}^{\infty} M_n^{(2)} \sin\tilde{\alpha}_n x \sin\alpha_n y \cdot u_y \tag{5}$$

u_x and u_y are unit vectors in the x- and y-directions, respectively. Owing to the dielectric substrate, the fields are neither TE nor TM but hybrid. Further details concerning the fin line eigenmodes are given in Appendix 7.2.

The boundary conditions for the junction of Fig. 1*a* (continuity of the transverse fields through all apertures and zero tangential electric field at conducting obstacles), which can be formulated with eqns. 2 and 3, are now satisfied in the following way: The crossproduct of the electric-field component of eqn. 2 with h_{am} is taken and integrated over the cross-section of waveguide a. For the unknown aperture field on the left-hand side, E_b is inserted from eqn. 3. The crossproduct of the magnetic-field component of eqn. 2 with e_{bn} is also taken and integrated over the cross-section of waveguide b. The unknown aperture field H_b is inserted from eqn. 3. Both operations yield:

$$(1+\rho)a_1 \int_{(a)} e_{a1} \times h_{am} \cdot u_z\, dxdy$$

$$+ \sum_{i=2}^{\infty} a_i \int_{(a)} e_{ai} \times h_{am} \cdot u_z\, dxdy = \sum_{j=1}^{\infty} \left[b_j \left(\left(\int_{(b)} e_{bj} \times h_{am} \right.\right.\right.$$

$$\left.\left.\left. \cdot\, u_z\, dxdy + \sum_{k=1}^{\infty} s_{jk} \int_{(b)} e_{bk} \times h_{am} \cdot u_z\, dxdy \right)\right)\right] \tag{6a}$$

$$(1-\rho)a_1 \int_{(b)} e_{bn} \times h_{a1} \cdot u_z\, dxdy - \sum_{i=2}^{\infty} a_i \int_{(b)} e_{bn} \times h_{ai}$$

$$\cdot\, u_z\, dxdy = \sum_{j=1}^{\infty} \left[b_j \left(\left(\int_{(b)} e_{bn} \times h_{bj} \cdot u_z\, dxdy \right.\right.\right.$$

$$\left.\left.\left. - \sum_{h=1}^{\infty} s_{jk} \int_{(b)} e_{bn} \times h_{bk} \cdot u_z\, dxdy \right)\right.\right] \tag{6b}$$

u_z is the unit vector in the z-direction. Both equations assure the continuity of the transverse fields. Because E_a exists over the aperture only and vanishes elsewhere, the integrals on the right-hand side of eqn. 6*a* must only be taken over waveguide b. The solutions of the integrals in eqns. 6 are given in Appendix 7.1.

The next task is to calculate the scattering coefficients s_{jk}. In a waveguide system with more than one discontinuity, interactions occur between dominant (fundamental) and higher-order modes. In our case of an abrupt transition between waveguides a and b, it is necessary to know the scattering properties of the termination of waveguide b, before eqns. 6 can be solved. It is sufficient to consider simple terminations such as a matched load, a short-circuit, or an open circuit, which cause an independent reflection of each mode incident upon it regardless of the amplitude and phase of any other mode. In the special case of a single discontinuity, waveguide b can be thought to be terminated with a matched load; hence $s_{jk} = 0$. In the other case shown in Fig. 1 (two abrupt transitions in cascade), we can take advantage of the symmetry of the structure with respect to the plane $z = l$. The equivalent circuit of the symmetric strip or notch is usually found from a symmetrical and an antisymmetrical excitation of both ports, which produces either a short-circuit or an open circuit in the plane of symmetry. Hence $s_{jk} = 0$ for $j \neq k$ and

$$s_{jj} = (1 - y_{bj})/(1 + y_{bj}) = \pm \exp(-2j\beta_{bj} l) \tag{7}$$

y_{bj} is the normalised input admittance of the jth mode in waveguide b measured at $z = 0$, β_{bj} is the propagation constant of this mode. The solution to the general case (more than two discontinuities or two unsymmetrical transitions) can be composed as sketched in Reference 16.

We are now in a position to solve the system of eqns. 6. It is convenient to normalise the wave amplitudes on a_1. Setting

$$a'_i = a_i/a_1\,,\; b'_j = b_j/a_1\,,\; I'_{ambj} = I_{ambj}/(F_{am}F_{bj}) \qquad (8)$$

with the Is and Fs taken from Appendix 7.1 we obtain, for the case that the exciting wave a_1 impinges on waveguide b from waveguide a,

$$\rho I'_{a1bn} + \sum_{i=2}^{L} a'_i I'_{aibn} - \sum_{j=1}^{L} b'_j (1 + s_{jj}) I'_{bjbn} = -I'_{a1bn}$$

$$\rho I'_{ama1} + \sum_{i=2}^{L} a'_i I'_{amai} + \sum_{j=1}^{L} b'_j (1 - s_{jj}) I'_{ambj} = -I'_{ama1} \qquad (9)$$

The slot width of waveguide b is assumed to be larger than that of waveguide a: $s_b > s_a$. This junction (which corresponds to a boundary enlargement problem in the terminology of Reference 16), in this case, will be called a step-up junction. In the complementary case called a step-down junction (excited from waveguide b) $s_b > s_a$ (boundary reduction problem), we obtain

$$\rho I'_{b1bm} + \sum_{i=2}^{L} a'_i I'_{bibm} - \sum_{j=1}^{L} b'_j (1 + s_{jj}) I'_{ajbm} = -I'_{b1bm}$$

$$\rho I'_{anb1} + \sum_{i=2}^{L} a'_i I'_{anbi} + \sum_{j=1}^{L} b'_j (1 - s_{jj}) I'_{anaj} = I'_{anb1} \qquad (10)$$

The infinite series are truncated at $i = j = L$, thus taking into account L modes in each waveguide. Then eqns. 9 or 10 generate $2L$ linear equations for $1 + (L - 1) + L$ unknowns.

The discontinuities are characterised either by a transformer (Fig. 3*a*) or by a π-network (Fig. 3*b*), for a notch, and by a T-network (Fig. 3*c*), for a strip. The former is calculated from the reflection coefficient ρ, the latter from B_{sc} (X_{sc}), the short-circuit admittance (impedance), and B_{oc} (X_{oc}), the open-circuit input admittance (impedance). Strictly speaking, the equivalent circuit for the step discontinuity (Fig. 3*a*) must be completed by two series reactances. The numerical calculations have, however, shown that both are extremely small and can be neglected for any parameter combination.

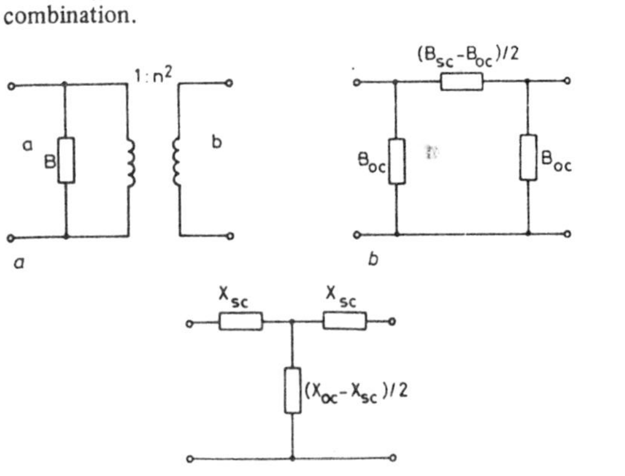

Fig. 3 *Equivalent circuits of a step, a notch and a strip discontinuity*
a Step discontinuity, *b* notch discontinuity, *c* strip discontinuity

3 Results

The impedance transforming potential of a single step, a notch and a strip will be illustrated in the following (some preliminary results have already been given in Reference 22): The characteristics of a single step in both unilateral and bilateral fin lines are shown in Figs. 4 and 5. The shunt susceptance is of course capacitive; it increases against the slot-width ratio s_b/s_a. Furthermore, it is larger for a bilateral than for a unilateral fin line, while the turns ratio behaves in an opposite way. The normalised series reactances (which have been neglected in the simplified equivalent circuit of Fig. 3*a*) are shown against frequency in Fig. 5*b*. They are indeed very small, except near the cutoff frequency and far above cutoff. The normalised shunt susceptance also shows a pole at cutoff. This can be attributed to the rapid increase of the wave impedance. All three parasitic elements increase almost linearly with frequency well above cutoff.

The characteristics of various notches have been displayed in Figs. 6–8. The series element of the π-equivalent circuit

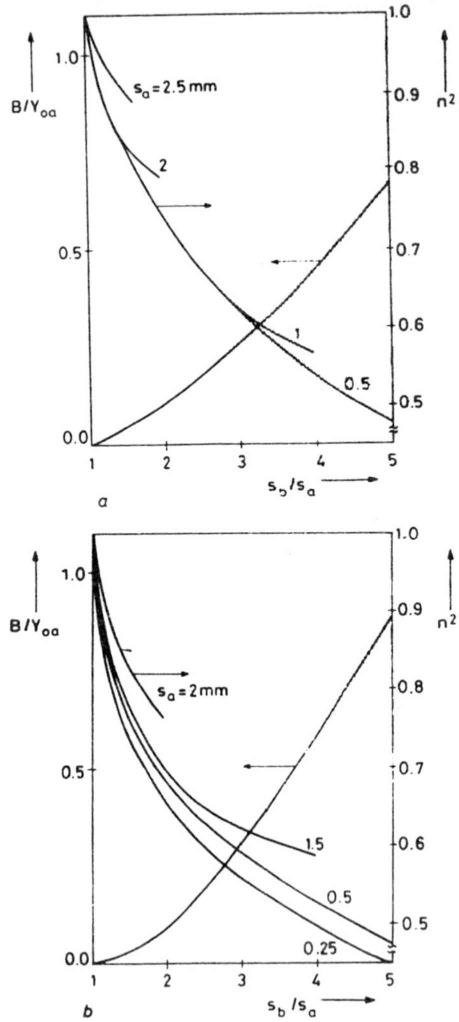

Fig. 4 *Elements of equivalent circuit of step discontinuity (Fig. 3a) against slot width ratio s_b/s_a*

a unilateral fin line, *b* bilateral fin line
Common parameters: symmetrically located slots of width $2s$, f = 15 GHz, ϵ_r = 2.22 (RT-Duroid), $2a$ = 15.8 mm, $2b$ = 7.9 mm, $2d$ = 0.254 mm, Y_{oa} = wave admittance of waveguide a

turns out to be always inductive, as long as the notch length is less than half a wavelength, while the shunt element is capacitive. This behaviour can be explained if higher-order mode coupling between the two step discontinuities of the notch is neglected. Then the equivalent circuit can be thought to be composed of the equivalent circuit of the two steps with a combined transmission line of length l. Hence the series element $B_{sc} - B_{oc}$ is proportional to $n^2/\sin(2\beta l)$, with β as the propagation constant of the fundamental mode. This relation easily explains the shape of the curves in Figs. 6–8.

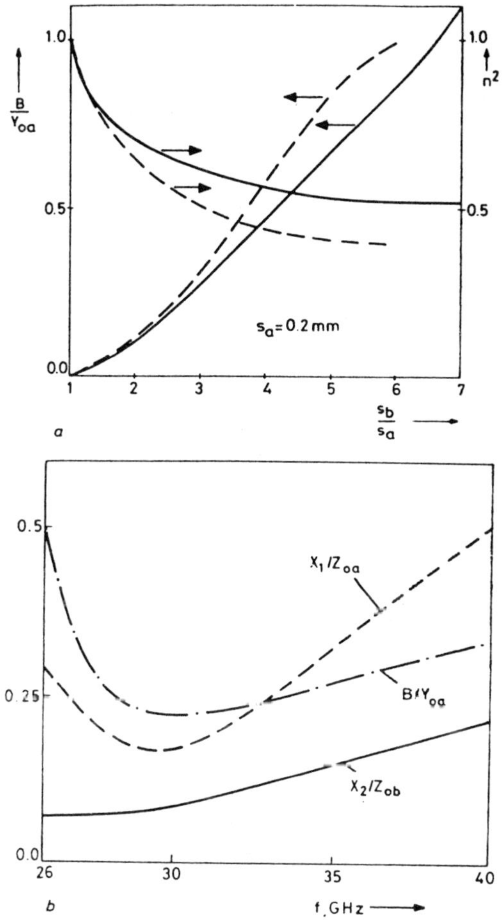

Fig. 5 *Elements of equivalent circuit of step discontinuity (Fig. 3a)*

a Unilateral (solid lines) and bilateral fin line (dashed lines) at 30 GHz
b Shunt susceptance B and series reactances X_1 (port *a*) and X_2 (port *b*), for $s_b/s_a = 1.778$ and $2s_a = 1$ mm
Common parameters: symmetrically located slots of width $2s$, $\epsilon_r = 2.22$, $2a = 7.112$ mm, $2b = 3.556$ mm, $2d = 0.254$ mm, $Y_0(X_0)$ wave admittance (impedance)

The inductive susceptance increases with increasing ratio of the slot widths, because the turns ratio increases similarly. The susceptance is very large for both $l \to 0$ and $l \to \lambda/4$, where λ is the wavelength of the fundamental mode. In between these two extremes, we observe a flat response against l. The shunt element B_{oc} is proportional to $n^2\{B + Y_{ob}\tan(\beta l)\}$. Hence the susceptance increases against l up to a maximum (pole) at $l = \lambda/4$. Increasing the ratio of the slot widths means increasing B (see Fig. 4), and thus increasing B_{oc}. The overlapping of the curves in Fig. 7 can be understood, if we take into account the fact that the wavelength λ decreases when the slot width s_b, within the notch, decreases. Then, $l = \lambda/4$ must be achieved for a decreasing ratio l/λ_0 also.

Higher-order mode coupling between the step discontinuities can be neglected if $\gamma_i l > 2$, where γ_i is the real propagation constant of the ith mode. The numerical calculations yielded as a rule of thumb that this inequality is always fulfilled for $l/\lambda_0 > 0.05$. The error introduced by the neglect of higher-order mode coupling is then less than 10%.

The characteristics of an unsymmetrically located notch whose slot width equals the waveguide height are depicted in Fig. 8. A very short notch of this type ($l \to 0$) is of practical interest, because it can be used for DC separation in a circuit containing semiconductor devices; then the shunt susceptance should be small and the series susceptance large. The curves in Fig. 8 show that extremely short notches must be realised in order not to influence the high frequency performance.

The characteristics of various strips have been displayed in Figs. 9–11. The series element is always inductive while the shunt element is capacitive for $l/\lambda_0 < 1/4$. As for the notch, impedances within a large range can be realised. The shape of the curves can again be explained with the help of a simplified equivalent circuit neglecting higher-order

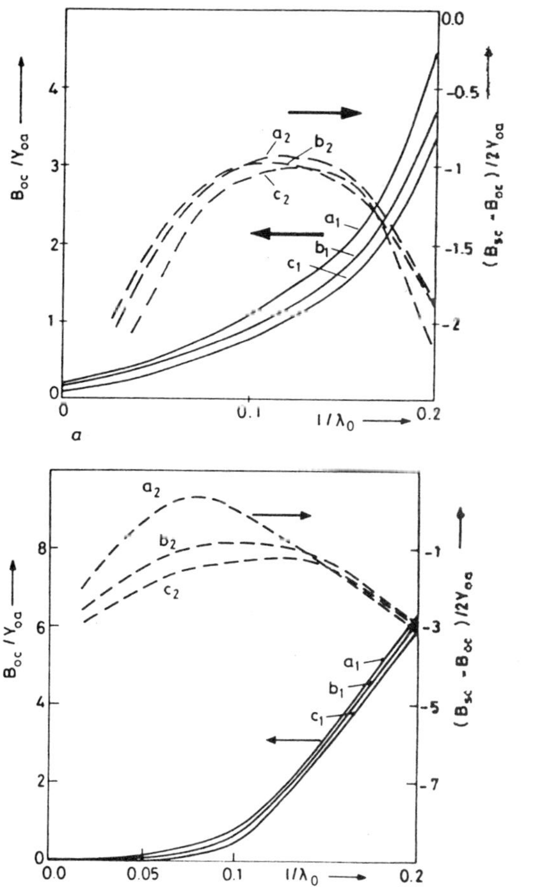

Fig. 6 *Elements of equivalent circuit of symmetrically located notch (Fig. 3b) against normalised notch length*

a Unilateral fin line, *b* bilateral fin line
Common parameters as in Fig. 4. Furthermore, $2s_a = 1.5$ mm and $2s_b = 3$ mm (curves a_1, a_2), 2.5 mm (curves b_1, b_2) and 2 mm (curves c_1, c_2),
λ_0 = free-space wavelength

mode coupling between the two step discontinuities of the strip. Then the series element X_{sc} is proportional to

$$X_{sc} = 1/[(Y_{oa}/n^2)\cot(\beta l) - B]$$

whereas the shunt element $(X_{oc} - X_{sc})/2$ is proportional to

$$(X_{oc} - X_{sc})/2 = -Y_{oa}/[(Y_{oa}^2/n^2 - n^2B^2)\sin(2\beta l) + 2BY_{oa}\cos(2\beta l)]$$

The series element represents the length of the strip. It increases against l, while its variation with the slot-width ratio is small. The shunt element, on the other hand, varies strongly with the slot-width ratio, whereas its response against the strip length is flat.

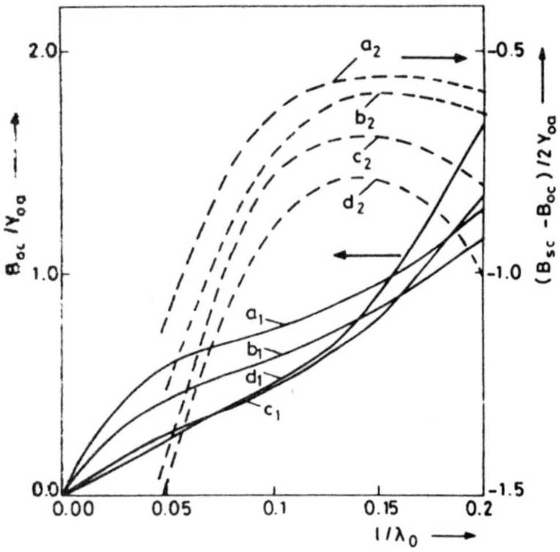

Fig. 7 *Elements of equivalent circuit of symmetrically located notch (Fig. 3b) against normalised notch length for a bilateral fin line at 30 GHz*

Common parameters as in Fig. 5. Furthermore, $2s_a = 0.5$ mm and $2s_b = 1.786$ mm (curves a_1, a_2), 1.5 mm (curves b_1, b_2), 1 mm (curves c_1, c_2) and 0.665 mm (curves d_1, d_2), λ_0 = free-space wavelength

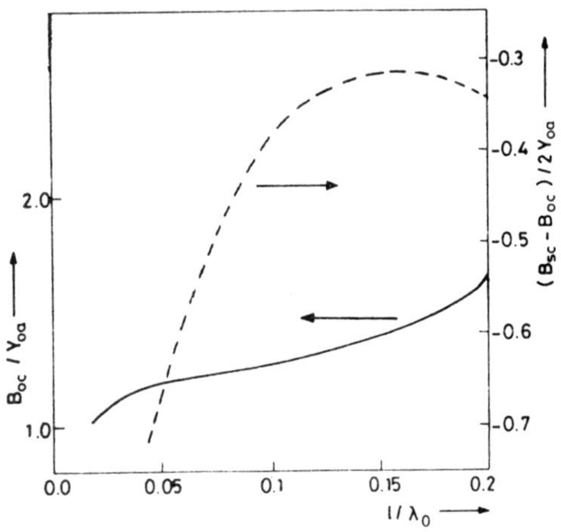

Fig. 8 *Elements of equivalent circuit of unsymmetrically located notch (Fig. 3b) against normalised notch length for a unilateral fin line at 30 GHz*

$s_a = 0.5$ mm, $s_b = 3.56$ mm, WR-28 waveguide, λ_0 = free-space wavelength

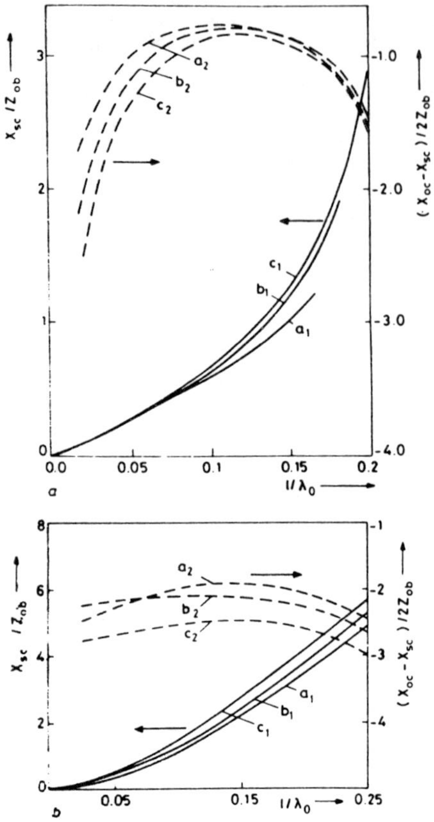

Fig. 9 *Elements of equivalent circuit of symmetrically located strip (Fig. 3c) against normalised strip length*

a Unilateral fin line, *b* bilateral fin line

Common parameters as in Fig. 4. The curves have been labelled as in Fig. 6.

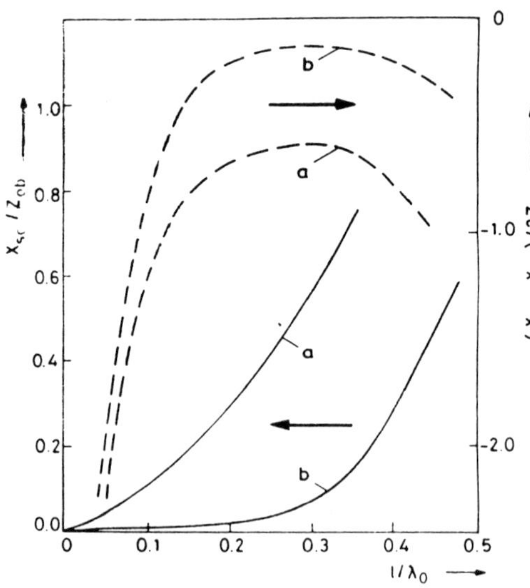

Fig. 10 *Elements of equivalent circuit of symmetrically located strip (Fig. 3c) against normalised strip length for a bilateral fin line at 30 GHz*

Common parameters as in Fig. 5. Furthermore, $s_b = 1.778$ mm with $s_a = 1$ mm (curves a) and $s_a = 0.5$ mm (curves b)

Relatively short strips are used for narrowing a fin-line slot to solder a beam-lead diode directly on the metallic fins. The length of these diodes amounts to some 0.1 mm in the lower millimetre-wave region. The characteristics of such a strip are shown in Fig. 11. The series element cannot be neglected except for extremely short strips. In practice, such a diode mount will also perform an impedance transformation, so that it must be taken into account in the circuit design.

Finally, Fig. 12 shows that the computed results agree well with measurements, which have been performed in K_u-band in order to use a network analyser.

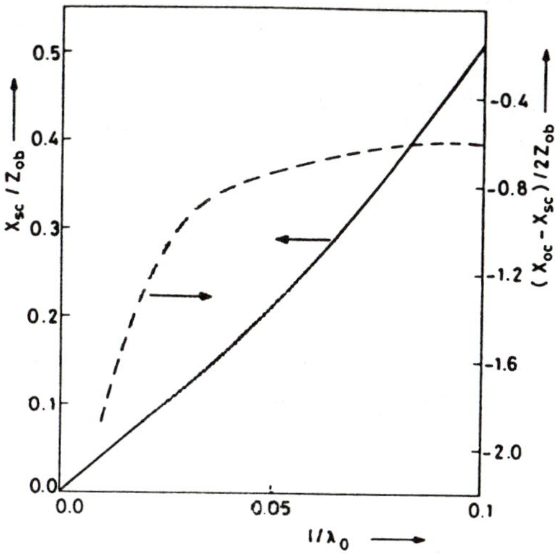

Fig. 11 *Elements of equivalent circuit of unsymmetrically located strip (Fig. 3c) against normalised strip length for a unilateral fin line at 30 GHz*

$s_a = 0.2$ mm, $s_b = 0.6$ mm, WR-28 waveguide, λ_0 = free-space wavelength

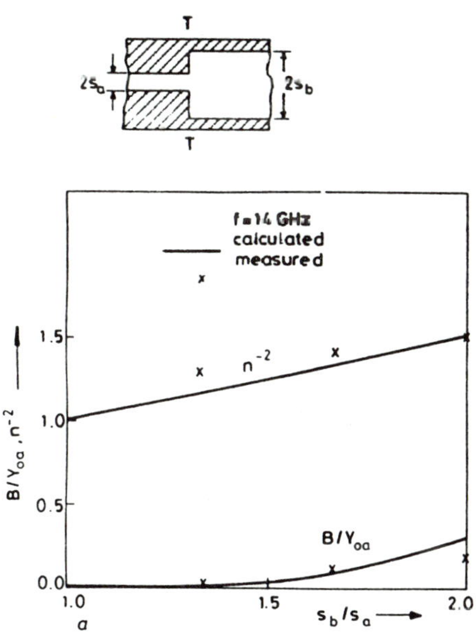

Fig. 12 *Comparison between theory and experiment*

a Step discontinuity, $2s_a = 3.0$ mm
b Notch with $B_{in} = \text{Im}(Y_{in})$, $2s_a = 3.0$ mm, $2s_b = 6.0$ mm
c Strip with $B_{in} = \text{Im}(Y_{in})$, $2s_a = 3.0$ mm, $2s_b = 6.0$ mm
Common parameters as in Fig. 4
(WR-62 waveguide), unilateral fin line

4 Conclusions

A modal analysis has been applied to various discontinuities in the slot width of unilateral and bilateral fin lines. Equivalent circuits have been calculated for a single step discontinuity and for composite structures like notches, slits and strips. These structures offer a large flexibility to the circuit engineer for impedance transformation and semiconductor mounts. A computer-aided circuit design should now be possible, based on the numerous illustrative examples given in this paper.

5 Acknowledgment

The authors thank the Deutche Forschungsgemeinschaft for financial support.

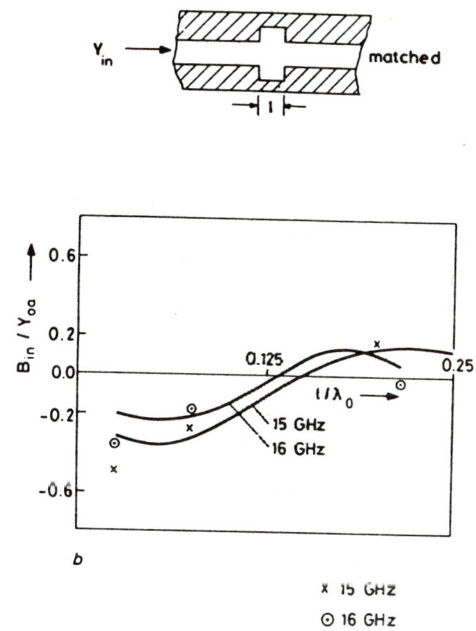

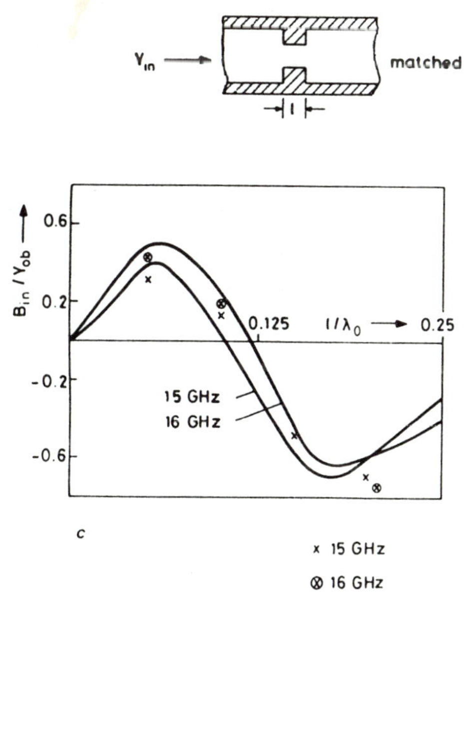

6 References

1 KNÖCHEL, R.: 'Design and performance of microwave oscillators in integrated fin-line technique', *IEE J. Microwaves, Opt. & Acoust.*, 1979, **3**, (3), pp. 115–120

2 COHEN, L.D.: 'Advances in printed millimeter-wave oscillator circuits'. IEEE MTT Symposium digest, Washington DC, USA, 1980, pp. 264–266

3 BEGEMANN, G.: 'An X-band balanced fin-line mixer', *IEEE Trans.*, 1978, **MTT-26**, pp. 1007–1011

4 TROOST, A.: 'Fortschritte auf dem Gebiet der Millimeterwellen im Bereich 35 bis 90 GHz', *Nach. Elektron.*, 1976, **12**, pp. 295–297

5 KNÖCHEL, R., and SCHLEGEL, A.: 'Octave-band double-balanced integrated fin-line mixers at mm-wavelengths'. Proceedings of the 10th European Microwave Conference, Warsaw, 1980, pp. 722–726

6 KPODZO, E., SCHÜNEMANN, K., and BEGEMANN, G.: A quadriphase fin-line modulator', *IEEE Trans.*, 1980, **MTT-28**, pp. 747–752

7 MEINEL, H., and REMBOLD, B.: 'New mm-wave fin-line attenuators and switches'. IEEE MTT Symposium digest, Orlando, USA, 1979, pp. 249–252

8 MEINEL, H., and SCHMIDT, L.-P.: 'High sensitivity mm-wave detectors using fin-line technology'. Fifth international conference on Infrared and Millimeter Waves, Würzburg, Deutschland, 1980

9 COHEN, L.D., and MEIER, P.J.: 'Advances in E-plane printed millimeter-wave circuits', IEEE MTT Symposium digest, Ottawa, Canada, 1978, pp. 27–29

10 MEIER, P.J.: 'Integrated fin-line millimeter components', *IEEE Trans.*, 1974, **MTT-22**, pp. 1209–1216

11 SAAD, A.M.K., and SCHÜNEMANN, K.: 'Design and performance of fin-line bandpass filters'. Proceedings of the 9th European Microwave Conference, Brighton, England, 1979, pp. 397–401

12 BEYER, A., and WOLFF, I.: 'Calculation of the transmission properties of inhomogeneous fin-lines'. Proceedings of the 10th European Microwave Conference, Warsaw, 1980, pp. 322–326

13 SAAD, A.M.K., and BEGEMANN, G.: 'Electrical performance of finlines of various configurations', *IEE J. Microwaves, Opt. & Acoust.*, 1977, **1**, (2), pp. 81–88

14 SCHMIDT, L.P., ITOH, T., and HOFMANN, H.: 'Characteristics of unilateral finline structures with arbitrarily located slots'. IEEE MTT Symposium digest, Washington DC, USA, 1980, pp. 255–257

15 BEYER, A., and WOLFF, I.: 'A solution of the earthed fin-line with finite metallization thickness'. IEEE MTT Symposium digest, Washington DC, USA, 1980, pp. 258–260

16 WEXLER, A.: 'Solution of waveguide discontinuities by modal analysis', *IEEE Trans.*, 1967, **MTT-15**, pp. 508–517

17 HOFMANN, H.: 'Dispersion of planar waveguides for millimetrewave applications', *Arch. Elektron. & Ueberrragungstech.*, 1977, **31**, pp. 40–44

18 MITTRA, R., ITOH, T., and LI, T.: 'Analytical and numerical studies of the relative convergence phenomenon arising in the solution of an integral equation by the moment method', *IEEE Trans.*, 1972, **MTT-20**, pp. 96-104

19 EL-SHERBINY, A., and SAID, M.: 'Spectral domain calculation of the fin-line dispersion characteristics', 3rd international congress for Statistics and Computer Science, Cairo, Egypt, 1978

20 CHANG, C., and ITOH, T.: 'Spectral domain analysis of dominant and higher-order modes in fin-lines', IEEE MTT Symposium digest, Orlando, USA, 1979, pp. 344–346

21 SCHMIDT, L.P., and ITOH, T.: 'Spectral domain analysis of dominant and higher-order modes in fin-lines', *IEEE Trans.*, 1980, MTT-28, pp. 981–985

22 EL HENNAWY, H., and SCHÜNEMANN, K.: 'Analysis of fin-line discontinuities'. Proceedings of the 9th European Microwave Conference, Brighton, 1979, pp. 448–452

7 Appendix

7.1 Some integrals

For solving the various integrals appearing in eqns. 6, the transverse vector functions have to be inserted from eqns. 4 and 5. This yields

$$I_{aibj} = \int_{(a,b)} \boldsymbol{e}_{ai} \times \boldsymbol{h}_{bj} \cdot \boldsymbol{u}_z \, dxdy$$

$$= \frac{b}{2} \sum_{n=0}^{\infty} \Bigg\{ [\delta_n A_{na}^{(1)} M_{nb}^{(1)} - \eta_1 \epsilon_n B_{nb}^{(1)} S_{na}^{(1)}] \frac{\sin(\alpha_{nai} - \alpha_{nbj})(a-d)}{2(\alpha_{nai} - \alpha_{nbj})}$$

$$+ [\delta_n A_{na}^{(1)} M_{nb}^{(1)} + \eta_1 \epsilon_n B_{nb}^{(1)} S_{na}^{(1)}] \frac{\sin(\alpha_{nai} + \alpha_{nbj})(a-d)}{2(\alpha_{nai} + \alpha_{nbj})} + [\eta_2 \delta_n A_{na}^{(2)} M_{nb}^{(2)} - \epsilon_n B_{nb}^{(2)} S_{na}^{(2)}] \frac{\sin(\tilde{\alpha}_{nai} - \tilde{\alpha}_{nbj})d}{2(\tilde{\alpha}_{nai} - \tilde{\alpha}_{nbj})}$$

$$- [\eta_2 \delta_n A_{na}^{(2)} M_{nb}^{(2)} + \epsilon_n B_{nb}^{(2)} S_{na}^{(2)}] \frac{\sin(\tilde{\alpha}_{nai} + \tilde{\alpha}_{nbj})d}{2(\tilde{\alpha}_{nai} + \tilde{\alpha}_{nbj})} \Bigg\} \qquad (11)$$

with

$$\eta_1 = \begin{cases} 1 \text{ for } \alpha_{nbj} \text{ real} \\ -1 \text{ for } \alpha_{nbj} \text{ imaginary} \end{cases} \qquad \delta_n = \begin{cases} 0 \text{ for } n = 0 \\ 1 \text{ for } n \neq 0 \end{cases}$$

$$\eta_2 = \begin{cases} 1 \text{ for } \tilde{\alpha}_{nbj} \text{ real} \\ -1 \text{ for } \tilde{\alpha}_{nbj} \text{ imaginary} \end{cases} \qquad \epsilon_n = \begin{cases} 1 \text{ for } n \neq 0 \\ 2 \text{ for } n = 0 \end{cases} \qquad (12)$$

$$F_{ai}^2 = \int_{(a,b)} \boldsymbol{e}_{ai} \times \boldsymbol{e}_{ai}^* \cdot \boldsymbol{u}_z \, dxdy$$

$$= \frac{b}{2} \sum_{n=0}^{\infty} \Bigg\{ [\delta_n |A_n^{(1)}|^2 + \eta_1 \epsilon_n |S_n^{(1)}|^2] \frac{a-d}{2}$$

$$+ [\delta_n |A_n^{(1)}|^2 - \eta_1 \epsilon_n |S_n^{(1)}|^2] \frac{\sin 2\alpha_n(a-d)}{4\alpha_n}$$

$$+ [\delta_n \eta_2 |A_n^{(2)}|^2 + \epsilon_n |S_n^{(2)}|^2] \frac{d}{2}$$

$$- [\eta_2 \delta_n |A_n^{(2)}|^2 - \epsilon_n |S_n^{(2)}|^2] \frac{\sin 2\tilde{\alpha}_n d}{4\tilde{\alpha}_n} \Bigg\} \qquad (13)$$

$$I_{aiai} = \int_{(a,b)} \boldsymbol{e}_{ai} \times \boldsymbol{h}_{ai} \cdot \boldsymbol{u}_z \, dxdy$$

$$= \frac{b}{2} \sum_{n=0}^{\infty} \Bigg\{ [\delta_n A_{na}^{(1)} M_{na}^{(1)} - \eta_1 \epsilon_n S_{na}^{(1)} B_{na}^{(1)}] \frac{a-d}{2}$$

$$+ [\delta_n A_{na}^{(1)} M_{na}^{(1)} + \eta_1 \epsilon_n S_{na}^{(1)} B_{na}^{(1)}] \frac{\sin 2\alpha_n(a-d)}{4\alpha_n}$$

$$+ [\eta_2 \delta_n A_{na}^{(2)} M_{na}^{(2)} - \epsilon_n B_{na}^{(2)} S_{na}^{(2)}] \frac{d}{2}$$

$$- [\eta_2 \delta_n A_{na}^{(2)} M_{na}^{(2)} + \epsilon_n B_{na}^{(2)} S_{na}^{(2)}] \frac{\sin 2\tilde{\alpha}_n d}{4\tilde{\alpha}_n} \Bigg\} \qquad (14)$$

$$I_{aiaj} = \int_{(a,\,b)} e_{ai} \times h_{aj} \cdot u_z \, dxdy$$

$$= \frac{b}{2} \sum_{n=0}^{\infty} \left\{ [\delta_n A^{(1)}_{nai} M^{(1)}_{naj} - \epsilon_n B^{(1)}_{naj} S^{(1)}_{nai}] \frac{\sin(\alpha_{ni} - \alpha_{nj})(a-d)}{2(\alpha_{ni} - \alpha_{nj})} \right.$$

$$+ [\delta_n A^{(1)}_{nai} M^{(1)}_{naj} + \epsilon_n B^{(1)}_{naj} S^{(1)}_{nai}] \frac{\sin(\alpha_{ni} + \alpha_{nj})(a-d)}{2(\alpha_{ni} + \alpha_{nj})} + [\delta_n A^{(2)}_{nai} M^{(2)}_{naj} - \epsilon_n B^{(2)}_{naj} S^{(1)}_{nai}] \frac{\sin(\tilde{\alpha}_{ni} - \tilde{\alpha}_{nj}) d}{2(\tilde{\alpha}_{ni} - \tilde{\alpha}_{nj})}$$

$$- [\delta_n A^{(2)}_{nai} M^{(2)}_{naj} + \epsilon_n B^{(2)}_{naj} S^{(2)}_{nai}] \frac{\sin(\tilde{\alpha}_{ni} + \tilde{\alpha}_{nj}) d}{2(\tilde{\alpha}_{ni} + \tilde{\alpha}_{nj})} \qquad (15)$$

Some of the indices in eqns. 11 to 15 have been omitted, as they were not necessary for a unique characterisation. The integral F^2_{ai} of eqn. 13 is later used for normalisation. The infinite series in the above formulas should be truncated at $N = Mb/s_b$ for $s_b > s_a$ and at $N = Mb/s_a$ for $s_b < s_a$, in order to make accurate use of the edge condition [18]. M is the number of slot field components used in eqn. 25. (We have chosen $M = 4$.)

7.2 Fin-line eigenmodes

The eigenmodes have been calculated by modifying the analysis given in Reference 17 in such a way that the edge conditions [18] can be taken into account without considerably increasing the computation time. This has been achieved by using the slot field distribution suggested in Reference 19 for the first time. (More recently, the method has also been described in Reference 20.) The procedure will be briefly described in the following:

The fin-line fields are derived from the E_x- and H_x-components which must satisfy the Helmholtz equation

$$\left[\frac{\partial^2}{\partial x^2} + \frac{\partial^2}{\partial y^2} + (k_0^2 \epsilon_r - \beta^2)\right] \begin{pmatrix} E_x \\ H_x \end{pmatrix} = 0 \qquad (16)$$

The tangential field components are calculated from

$$\begin{pmatrix} E_t \\ H_t \end{pmatrix} = \frac{1}{k_0^2 \epsilon_r - k_x^2} \begin{pmatrix} \partial/\partial x & -k_0 u_x x \\ -k_0 \epsilon_r u_x x & \partial/\partial x \end{pmatrix} \nabla_t \begin{pmatrix} E_x \\ H_x \end{pmatrix} \qquad (17)$$

with $\nabla_t = \partial/\partial y(u_y) - j\beta u_z$, β = the propagation constant, and $k_0^2 = \omega^2 \mu_0 \epsilon_0$.

E_x and H_x are given in eqns. 4 and 5 for the odd modes of a bilateral fin line with symmetrically located slot. The various constants read

$$\alpha_n = \sqrt{k_0^2 - \beta^2 - \alpha_n^2} \qquad \tilde{\alpha}_n = \sqrt{\epsilon_r k_0^2 - \beta^2 - \alpha_n^2}$$

$$\alpha_n = n\pi/b \qquad (18)$$

Applying eqn. 17 to the x-components of the electric- and magnetic-field strength yields relations between the Fourier coefficients of the x- and y-components:

$$S_n^{(1)} = (-A_n^{(1)} \alpha_n \alpha_n - j k_0 \beta B_n^{(1)})/(\beta^2 + \alpha_n^2)$$

$$S_n^{(2)} = (A_n^{(2)} \alpha_n \tilde{\alpha}_n - j k_0 \beta B_n^{(2)})/(\beta^2 + \alpha_n^2)$$

$$M_n^{(1)} = (-B_n^{(1)} \alpha_n \alpha_n - j k_0 \beta A_n^{(1)})/(\beta^2 + \alpha_n^2)$$

$$M_n^{(2)} = (B_n^{(2)} \alpha_n \tilde{\alpha}_n - j k_0 \epsilon_r \beta A_n^{(2)})/(\beta^2 + \alpha_n^2) \qquad (19)$$

It follows from the continuity of the tangential field components that:

$$A_n^{(2)} = \frac{-\alpha_n}{\tilde{\alpha}_n} \frac{\sin \alpha_n(d-a)}{\cos \tilde{\alpha}_n d} A_n^{(1)} \qquad B_n^{(2)} = \frac{\sin \alpha_n(d-a)}{\cos \tilde{\alpha}_n d} B_n^{(1)} \qquad (20)$$

The remaining two coefficients $A_n^{(1)}$ and $B_n^{(1)}$ can be expressed in terms of the slot field components $E_y(y)$ and $E_z(y)$. This is done, in the spectral domain, by Fourier transforming all field components and applying the interface conditions. One obtains the field coefficients:

$$A_n^{(1)} = -2(\alpha_n L_{2n} + j\beta L_{1n})/[\alpha_n b \sin \alpha_n (d-a)]$$

$$B_n^{(1)} = 2(\alpha_n L_{1n} + j\beta L_{2n})/[k_0 b \sin \alpha_n (d-a)] \qquad (21)$$

and the coupled algebraic equations

$$\sum_{n=0}^{\infty} \Lambda_{11} L_{2n} \cos\alpha_n y + \sum_{n=1}^{\infty} \Lambda_{12} L_{1n} \cos\alpha_n y = \frac{b}{2} I_y(y)$$

$$\sum_{n=1}^{\infty} \Lambda_{21} L_{2n} \sin\alpha_n y + \sum_{n=1}^{\infty} \Lambda_{22} L_{1n} \sin\alpha_n y = \frac{b}{2} I_z(y) \qquad (22)$$

L_{1n} and L_{2n} are Fourier transforms of the slot field components

$$L_{1n} = \int_0^s E_z \sin\alpha_n y \, dy \qquad L_{2n} = \int_0^s E_y \cos\alpha_n y \, dy \qquad (23)$$

and I_y and I_z are transforms of the current on the fins. These two unknowns are of no importance for the solution of eqns. 22, because they can easily be eliminated.

The various constants in eqns. 22 read:

$$\Lambda_{11} = -\frac{[\alpha_n^2 k_0 H_{1n}/(\alpha_n \tilde{\alpha}_n) + \beta^2/k_0 H_{2n}]}{(\alpha_n^2 + \beta^2) \sin \alpha_n (d-a)}$$

$$\Lambda_{12} = -\Lambda_{21} = \frac{-j\alpha_n \beta k_0 H_{1n}/(\alpha_n \tilde{\alpha}_n) + j\alpha_n \beta H_{2n}/k_0}{(\alpha_n^2 + \beta^2) \sin \alpha_n (d-a)}$$

$$\Lambda_{22} = -\frac{[\alpha_n^2 H_{2n}/k_0 + \beta^2 k_0 H_{1n}/(\alpha_n \tilde{\alpha}_n)]}{(\alpha_n^2 + \beta^2) \sin \alpha_n (d-a)} \qquad (24)$$

The abbreviations H_{1n} and H_{2n} will be given below.

The coupled equations (eqns. 22) for the transformed-slot-field components are solved by applying Ritz-Galerkin's method. To this end, the fields are expanded in terms of known basis functions which have been chosen according to Reference 19:

$$E_y(y) = \sum_{m=0}^{M} c_m y^{2m}(d^2 - y^2)^{-1/2} \text{ for } |y| \leq s$$

$$E_z(y) = \sum_{m=1}^{M} d_m y^{(2m-1)}(d^2 - y^2)^{+1/2} \text{ for } |y| \leq s \quad (25)$$

Substituting these values into eqns. 23 yields

$$L_{1n}^{(m)} = \pi/2\, s^{(2m+1)}(-1)^m \left[\left(\frac{\partial}{\partial \alpha_n s}\right)^{2m-1} J_0(\alpha_n s) + \left(\frac{\partial}{\partial \alpha_n s}\right)^{2m+1} J_0(\alpha_n s)\right]$$

$$L_{2n}^{(m)} = \pi/2\, s^{2m}(-1)^m \left(\frac{\partial}{\partial \alpha_n s}\right)^{2m} J_0(\alpha_n s) \quad (26)$$

with J_0 Bessel function of order zero.

The Fourier transforms of the assumed-slot-field components are substituted into eqns. 22. Eliminating the fin currents by taking the inner products of these equations with L_{1n} and L_{2n}, one obtains a linear system of homogeneous equations for the unknown expansion coefficients c_m and d_m:

$$\sum_{k=0}^{M} c_k \sum_{n=0}^{N} \Lambda_{11} L_{2n}^{(m)} L_{2n}^{(k)} + \sum_{k=1}^{M} d_k \sum_{n=1}^{N} \Lambda_{12} L_{2n}^{(m)} L_{1n}^{(k)} = 0$$

$$\sum_{k=0}^{M} c_k \sum_{n=1}^{N} \Lambda_{21} L_{1n}^{(m)} L_{2n}^{(k)} + \sum_{k=1}^{M} d_k \sum_{n=1}^{N} \Lambda_{22} L_{1n}^{(m)} L_{1n}^{(k)} = 0$$

$$m = 0, 1, 2 \ldots M \quad (27)$$

M is the total number of basis-function components and N is the number of fringing-field components. They are both related to each other via the edge conditions. $M = 4$ has been chosen here, which gives accurate results up to the 30th mode. N has been specified at the end of Appendix 7.1.

The dispersion relation for the propagation constant of a particular mode is obtained from eqns. 27 by equating the determinant of the coefficient matrix to zero.

The basis functions have to be modified for the case when the fin-line slot is not symmetrically located in the cross-section (i.e. it is not around $y = 0$). Then, we set:

$$E_y(y) = \sum_{m=0}^{M} c_m y^m (s^2 - y'^2)^{-1/2}$$

$$E_z(y) = \sum_{m=0}^{M} d_m y^m (s^2 - y'^2)^{+1/2} \quad (28)$$

Here $y' = y - (d_1 + d_2)/2$ and d_1 and d_2 are defined in Fig. 2. The Fourier transforms are given by:

$$L_{1n}^{(m)} = \frac{\pi}{2} s^2 \left(\frac{\partial}{\partial \alpha_n}\right)^m \left[\sin\left(\alpha_n \frac{d_1 + d_2}{2} + \frac{m\pi}{2}\right) [J_0(\alpha_n s) + J_2(\alpha_n s)]\right]$$

$$L_{2n}^{(m)} = \pi \left(\frac{\partial}{\partial \alpha_n}\right)^m \left[\cos\left(\alpha_n \frac{d_1 + d_2}{2} + \frac{m\pi}{2}\right) J_0(\alpha_n s)\right] \quad (29)$$

Up to now, we have only treated a bilateral fin line with both a symmetrically and an unsymmetrically located slot. Both cases differ only in the choice of basis functions for the slot fields. These functions are given by eqns. 25 for a symmetrically located slot and by eqns. 28 for an unsymmetrically located slot. The results must still be completed by defining the symbols H_{1n} and H_{2n} in eqns. 24. They read, for a bilateral fin line,

$$H_{1n} = [\tilde{\alpha}_n \cos \alpha_n (d-a) \cos \tilde{\alpha}_n d - \epsilon_r \alpha_n \times \sin \tilde{\alpha}_n d \sin \alpha_n (d-a)] / \cos \tilde{\alpha}_n d$$

$$H_{2n} = [\alpha_n \cos \tilde{\alpha}_n d \cos \alpha_n (d-a) + \tilde{\alpha}_n \sin \tilde{\alpha}_n d \sin \alpha_n (d-a)] / \cos \tilde{\alpha}_n d \quad (30)$$

for the odd modes, and

$$H_{1n} = [\tilde{\alpha}_n \cos \alpha_n (d-a) \sin \tilde{\alpha}_n d - \epsilon_r \alpha_n \times \cos \tilde{\alpha}_n d \sin \alpha_n (d-a)] / \sin \tilde{\alpha}_n d$$

$$H_{2n} = [\alpha_n \cos \alpha_n (d-a) \sin \tilde{\alpha}_n d - \tilde{\alpha}_n \cos \tilde{\alpha}_n d \sin \alpha_n (d-a)] / \sin \tilde{\alpha}_n d \quad (31)$$

for the even modes.

For a unilateral fin line, one must also distinguish between a symmetrically and an unsymmetrically located slot. All results given above, for the bilateral fin line, remain valid except for H_{1n} and H_{2n}. They are given for a unilateral fin line, as

$$H_{1n} = \frac{-2\epsilon_r \alpha_n \tilde{\alpha}_n + [\tilde{\alpha}_n^2 \cot \alpha_n (d-a) - \epsilon_r^2 \alpha_n^2 \tan \alpha_n (d-a)] \tan 2\tilde{\alpha}_n d}{\epsilon_r \alpha_n \tan \alpha_n (a-d) + \tilde{\alpha}_n \tan 2\tilde{\alpha}_n d}$$

$$H_{2n} = \frac{-2\alpha_n \tilde{\alpha}_n - [\alpha_n^2 \cot \alpha_n (a-d) - \tilde{\alpha}_n^2 \tan \alpha_n (a-d)] \tan 2\tilde{\alpha}_n d}{\alpha_n \tan 2\tilde{\alpha}_n d - \tilde{\alpha}_n \tan \alpha_n (d-a)} \quad (32)$$

For an antipodal fin line [17], the basis functions are given by eqns. 25 and the symbols H_{1n} and H_{2n} are given by

$$H_{1n} = \left[\tilde{\alpha}_n \cot \alpha_n (d-a) + \epsilon_r \alpha_n \frac{\delta_0 \sin \tilde{\alpha}_n d + \delta_e \cos \tilde{\alpha}_n d}{\delta_0 \cos \tilde{\alpha}_n d - \delta_e \sin \tilde{\alpha}_n d}\right] \sin \alpha_n (d-a)$$

$$H_{2n} = \left[\alpha_n \cot \alpha_n (d-a) + \tilde{\alpha}_n \frac{\delta_0 \sin \tilde{\alpha}_n d - \delta_e \cos \tilde{\alpha}_n d}{\delta_0 \cos \tilde{\alpha}_n d - \delta_e \sin \tilde{\alpha}_n d}\right] \sin \alpha_n (d-a) \quad (33)$$

with $\delta_0 = 1$, $\delta_e = 0$, for $n = 0, 2, 4 \ldots$, and $\delta_0 = 0$, $\delta_e = 1$, for $n = 1, 3, 5 \ldots$.

A detailed description of the eigenmode analysis for a bilateral fin line has recently been published in Reference 21.

4.1.2

A VARIATIONAL SOLUTION OF THE FIN-LINE DISCONTINUITY PROBLEM

Kevin J. Webb [+] and *Raj Mittra*

ABSTRACT

The fin-line step discontinuity is an essential component in millimeter-wave integrated circuits. The discontinuity problem is formulated using an unknown magnetic current in the transverse junction plane. A numerical solution is obtained by a method termed the Generalized Variational Method, which is an extension of the traditional variational method. Illustrative examples of scattering parameter calculations are given.

INTRODUCTION

The general fin-line discontinuity problem is fundamental to many millimeter-wave integrated circuit components, such as filters and matching structures. The traditional mode-matching approach has been applied to the discontinuity problem with some success [1]-[3]. Recently several iterative numerical solution procedures, such as the Conjugate Gradient Method [4], have appeared in the literature.

A novel approach for solving this problem using unknowns in the transverse junction plane is presented here. The method is called the Generalized Variational Technique (GVT). The Generalized Variational method is an extension of the classical variational method, which represents the unknown in terms of one function [5]. Techniques presented here allow an improved solution by representing the unknown in terms of a number of characteristic functions, which are generated by means of prescribed routines. The method of formulation and numerical solution together with some example computed results are given.

FORMULATION OF THE DISCONTINUITY PROBLEM

A discontinuity consisting of a step change in the slot width of unilateral fin-line is shown in Fig.1. Consider the junction of two waveguides, with waveguide A for $z<0$ and waveguide B for $z>0$. The transverse field for $z<0$ and $z>0$ can be expressed in terms of the mode functions in waveguides A and B. Consider that I modes are used in this representation, with one propagating mode. A representation for the junction can employ equivalent magnetic currents ($\overline{K}$) over the transverse junction plane backed by perfect electric conductors, which radiate into both waveguides [6]. For the discontinuity of Fig. 2, the equivalent magnetic currents exist over the complete cross-section of the fin-line shielding enclosure. An integral operator equation can be derived using the Green's function for each waveguide [7]. This integral equation is of the form $Lf = g$, where L is an integral operator, f is the unknown (corresponding to $\overline{K}$), and g is the known incident field.

+ Electrical Engineering Department, University of Maryland, College Park, Maryland 20742, USA.

* Department of Electrical and Computer Engineering, University of Illinois, Urbana, Illinois 61801, USA.

K.J. Webb and R. Mittra, *Proceedings of 15th European Microwave Conference*, 1985, pp. 311–316.

The modal solutions can be determined by firstly employing a spectral domain formulation in the plane of the fin, and solving for the electric field. The electric field is expressed in terms of a set of basis functions in the context of the Galerkin method [8]. The resulting equation is first solved for the phase constants. The relative weights of the basis functions, and hence the slot electric field, are determined next. The knowledge of the modal slot electric field can then be used to derive the modal fields over the entire waveguide cross-section by using a transverse resonance technique [8]. The number of modes derived from this procedure is usually limited to a small number owing to the numerical difficulty of finding a large number of orthogonal modes.

Following the solution for $\overline{K}$, the mode coefficients can be found by utilizing the orthogonal property of the modes [9]. Normalization of the mode coefficients for unit power results in the scattering parameters for the fin-line discontinuity.

SOLUTION USING THE GENERALIZED VARIATIONAL TECHNIQUE

The approach for solving the discontinuity problem presented here considers the fields at the plane of the junction, $z = 0$, as the unknowns. Consider now the integral operator equation $Lf = g$, where L is in general not positive definite. The unknown quantity f may be expressed as a superposition of characteristic functions f_n, which are recursively generated by using a procedure given as follows. The unknown f can be expressed as

$$f = \sum_{n=0}^{N} c_n f_n . \tag{1}$$

Consider an inner product defined by

$$<\overline{F}, \overline{G}> = \iint_s \overline{F}^* \cdot \overline{G} \, ds. \tag{2}$$

The generation routine for f_n may be summarized as

$$f_o = \frac{u_o}{||u_o||^2}$$

$$\eta_n = <f_n, Lf_n>$$

$$c_n = \frac{<f_n, g>}{\eta_n}$$

$$u_{n+1} = u_n - \frac{1}{\eta_n} Lf_n$$

$$f_{n+1} = \frac{u_{n+1}}{||u_{n+1}||^2} + f_n. \tag{3}$$

The Generalized Variational Technique may also be applied to the equation $L^A Lf = L^A g$, where L^A is the adjoint integral operator. It should be pointed out that $L^A L$ is positive definite, hence the convergence of this procedure is guaranteed to be monotonic. A generation routine analogous to Eq (3) can be realized for this positive definite operator.

These routines can be used to solve for the unknown magnetic current in the junction plane of the fin-line discontinuity. A good choice for the auxiliary function u_o is $\hat{z}\times\bar{e}_{a1}$, where $\bar{e}_{a1}$ is the incident transverse E-field mode function, with a possible alternative being $\hat{z}\times(\bar{e}_{a1}+\bar{e}_{b1})$.

NUMERICAL RESULTS

The results given here were generated using three waveguide modes for construction of the Green's function. These modes were generated in turn by using two basis functions for E_x and E_z in the plane of the fin. Two basis functions were used to represent each of the slot fields. These functions consisted of trigonometric functions modified by the edge condition. Let us now turn to the characteristic function solution. Referring to Eq. (3), the initial auxiliary function u_0 was chosen as $\hat{z}\times\bar{e}_{a1}$. This results in the first characteristic function f_0. Subsequent characteristic functions f_n, and coefficients c_n, can then be generated according to Eq. (3). The fields over the waveguide cross-section were represented discretely at (m,n) points. That is, the magnetic currents, characteristic functions and waveguide modes were represented by a matrix of points over the shielded enclosure cross-section. The mode coefficients and scattering parameters can then be found.

There is a shortage of data in the literature for comparative purposes. Consider as an example fin-line in a WR-28 shield, for which mode matching data is given in the paper by Helard et. al. [3]. The fin-line cross-section shown in Fig. 1 gives the dimensional parameters. Figure 3 shows the computed scattering parameters for a step discontinuity in fin-line with a WR-28 shield. The equation $Lf = g$ is used here with one and two characteristic functions. As the operator L is not positive definite, a monotonically convergent solution as a function of the number of characteristic functions is not guaranteed. It is important to verify that for the single propagating mode the numerical results satisfy

$$|S_{11}|^2+|S_{21}|^2=1. \tag{4}$$

In drawing a comparison with the mode-matching data, it should be pointed out that Eq. (4) is not satisfied well in [3]. Quite good agreement for S_{11} using one characteristic function was achieved. There is a significant change in the results for one and two characteristic functions. This change can be attributed to the small number of modes available for constructing the Green's function, and the fact that the solution does not converge monotonically with an increasing number of characteristic functions.

A comparison between the equation $Lf = g$ with one and two characteristic functions, and the equation $L^A Lf = L^A g$ with two characteristic functions, is given in Fig. 4. Use of the positive definite operator improves the behavior of the solution, but a larger number of waveguide modes are required to improve the accuracy significantly.

It should be pointed out that the maximum number of characteristic functions, and hence the accuracy of the procedure, is limited by the number of waveguide modes that have been used to construct the Green's function. The accuracy of the solution cannot be increased arbitrarily due to the lack of availability of a large number of satisfactory higher order modes. An approximate convergence criterion for the maximum number of characteristic functions p, with q modes in the Green's function, is $p \leq q-1$.

CONCLUSION

The fin-line step discontinuity problem has been formulated with an unknown magnetic current over the transverse junction plane. A solution for this magnetic current was obtained using an approach termed the Generalized Variational Technique. This resulted in the scattering parameters for the fin-line discontinuities. The computed scattering parameters for example discontinuities compared favorably with existing data. The approaches outlined can be applied to other metallized substrate waveguides. Further work remains to be done in order to improve the accuracy of the solution, and it is important to obtain experimental verification. If a larger number of good solutions for the fin-line modes are used to construct the Green's function, more characteristic functions can be used, resulting in a more accurate solution. It is therefore necessary to improve the accuracy of the higher order mode solutions.

REFERENCES

[1] H. El Hennawy and K. Schunemann, "Hybrid fin-line matching structures," *IEEE Trans. Microwave Theory Tech.*, vol. MTT-30, pp. 2132-2138, Dec. 1982.

[2] H. El Hennawy and K. Schunemann, "Impedance transformation in fin-lines," *IEE Proc. Pt. H, Microwaves Opt. Acoust.*, vol. 129, pp. 342-350, Dec. 1982.

[3] M. Helard, J. Citerne, O. Picon, and V. Fouad Hanna, "fin-line," *Electron. Lett.* vol. 19, pp. 537-539, 7th. Jul. 1983.

[4] P. M. van den Berg, "Iterative computational techniques in scattering based upon the integrated square error criterion," *IEEE Trans. Antennas Propagat.*, vol. AP-32, no. 10, pp. 1063-1071, Oct. 1984.

[5] R. F. Harrington, *Time Harmonic Electromagnetic Fields.* New York, NY: McGraw-Hill, 1961, Chap. 7.

[6] H. Auda and R. Harrington, "A moment solution for waveguide junction problems," *IEEE Trans. Microwave Theory Tech.*, vol. MTT-31, pp. 515-519, Jul. 1983.

[7] R. Collin and F. Zucker, Eds., *Antenna Theory.* New York, NY: McGraw-Hill, 1969, Pt. 1, Chap 14.

[8] L. Schmidt, T. Itoh, and H. Hofmann, "Characteristics of unilateral fin-line structures with arbitrarily located slots," *IEEE Trans. Microwave Theory Tech.*, vol. MTT-29, no. 4, pp. 352-355, Apr. 1981.

[9] R. E. Collin, *Field Theory of Guided Waves,* New York, NY: McGraw-Hill, 1960, Chap. 5.

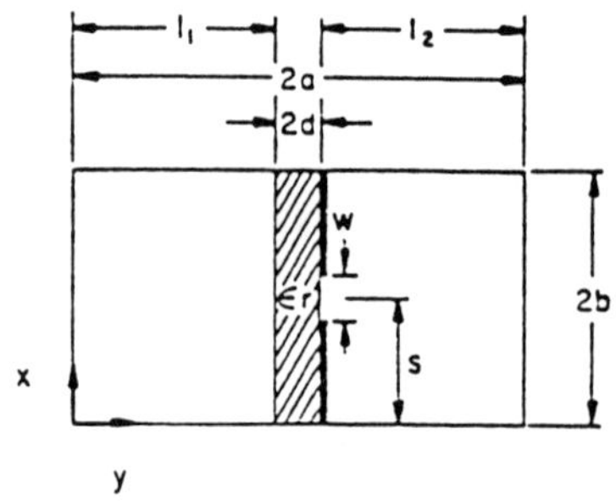

Figure 1. Unilateral fin-line structure.

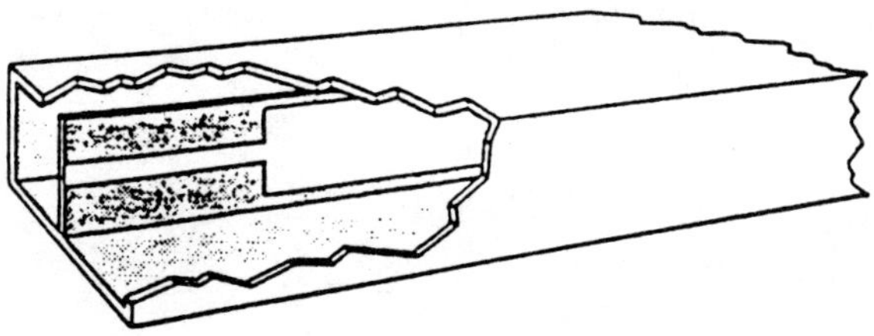

Figure 2. Single step discontinuity in fin-line.

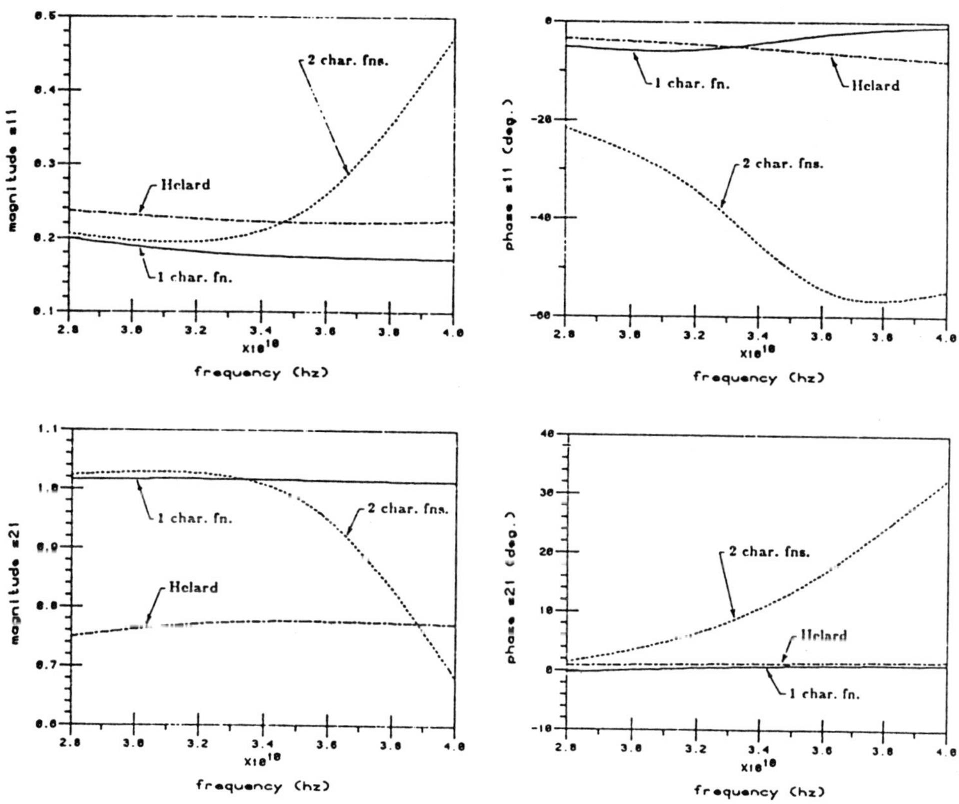

Figure 3. Scattering parameters for a discontinuity in fin-line with a WR-28 shield using $Lf = g$. $l_1 = l_2 = 3.429$ mm, $2d = 0.254$ mm, $\epsilon_r = 2.22$, $2b = 2.556$ mm, $w_1 = 1$ mm, $w_2 = 2$ mm, $m = 17$, $n = 33$.

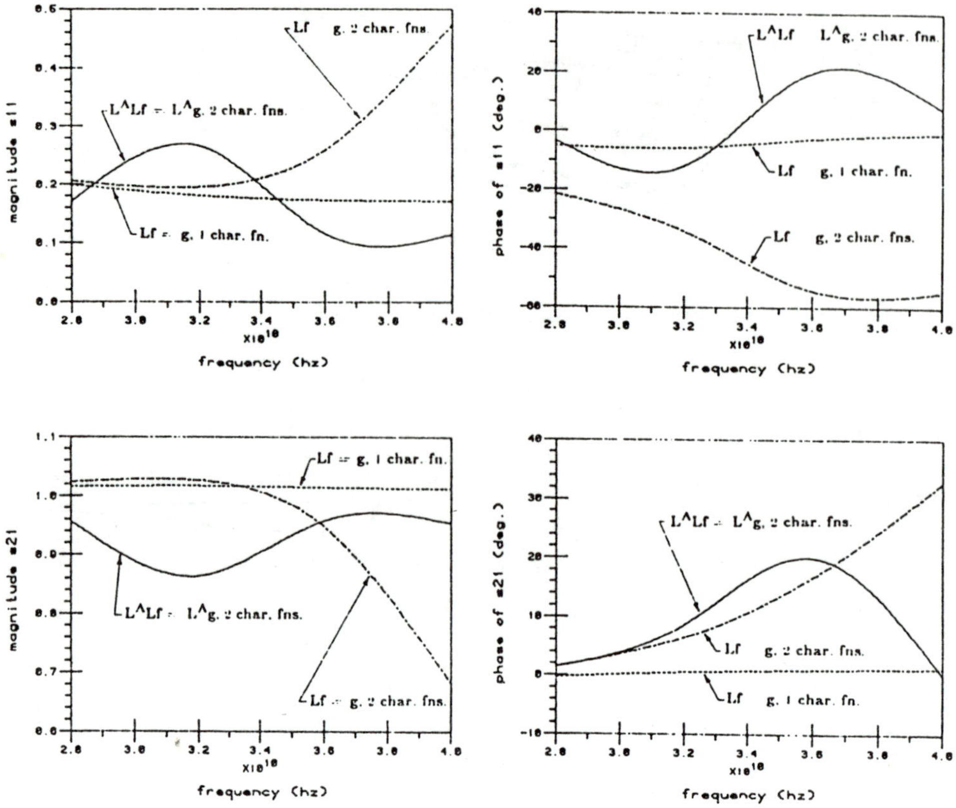

Figure 4. Comparison of scattering parameters for a discontinuity in fin-line with a WR-28 shield using $Lf = g$ and $L^A Lf = L^A g$. $l_1 = l_2 = 3.429$ mm, $2d = 0.254$ mm, $\epsilon_r = 2.22$, $2b = 2.556$ mm, $w_1 = .1$ mm, $w_2 = 2$ mm, $m = 17$, $n = 33$.

4.1.3

Transverse Resonance Analysis of Finline Discontinuities

ROBERTO SORRENTINO, MEMBER, IEEE, AND TATSUO ITOH, FELLOW, IEEE

Abstract—**A method of analysis is proposed for characterizing finline discontinuities. Two conducting or magnetic planes are inserted at some distances away from the discontinuity so as to obtain a closed resonant structure. A transverse resonance technique is then used to compute the resonant frequencies and, from these, the equivalent circuit parameters of the discontinuity. In the particular case when the discontinuity is removed, the method can be used to characterize uniform finlines.**

I. Introduction

FINLINES are now recognized as a suitable technology of millimeter-wave integrated circuits. While much theoretical work has been done concerning the analysis and characterization of uniform finline structures [1], [2], a relatively small number of analyses of finline discontinuities have been developed [3], [4].

This paper presents a new method of analysis and characterization of both uniform finlines and finline discontinuities. The method consists of computing the resonant frequencies of a resonator obtained by inserting two conducting or magnetic planes apart from the discontinuity; using a transverse resonance technique, the electromagnetic (EM) fields are expanded in terms of longitudinal section magnetic (LSM) and electric (LSE) modes of the rectangular waveguide. With respect to other approaches based on the field expansion in terms of finline modes [3], [4], the present one has the advantage of a substantial reduction of computer time. In this paper, this new method is applied to the simple step discontinuity as well as to the cascade of step discontinuities.

II. Characterization of the Discontinuity

The characterization of a finline discontinuity is obtained with the resonant frequencies of resonators which are obtained by introducing two shorting planes at some distances away from the discontinuity. The resultant structure is shown in Fig. 1 along with the dimensions and the coordinate system.

As long as the frequency is such that only dominant modes can propagate in the two finline sections and the higher order modes excited at the discontinuity have

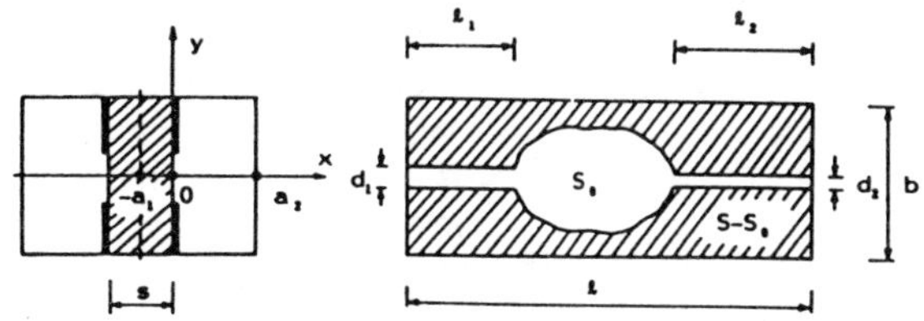

Fig. 1. Transverse and longitudinal cross sections of a finline discontinuity in a shorted cavity.

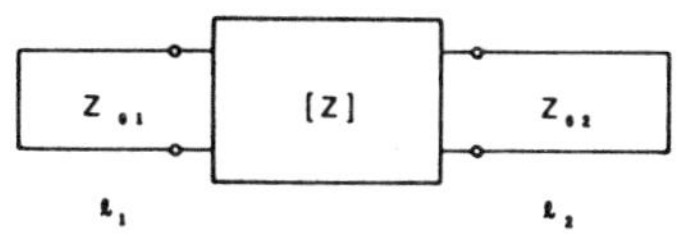

Fig. 2. Equivalent circuit of Fig. 1.

negligible amplitudes at the shorting planes, the discontinuity can be modeled as an equivalent two-port network, as shown in Fig. 2.

The resonance condition in terms of the impedance parameters of the discontinuity is

$$(Z_{11}+Z_1)(Z_{22}+Z_2)-Z_{12}^2=0 \tag{1}$$

where

$$Z_i=jZ_{oi}\tan(\beta_i l_i), \qquad i=1,2.$$

Z_{oi} is the characteristic impedance of the ith finline, and β_i is the corresponding phase constant. Alternatively, (1) can also be formulated in terms of the scattering parameters of the discontinuity.

If the same resonant frequency ω_r rad/s is obtained for three different pairs of l_1, l_2, (1) allows the evaluation of the three impedance parameters of the discontinuity at ω_r.

In the absence of the discontinuity, $\beta_1=\beta_2=\beta$, (1) then reduces to

$$\beta(\omega_r)=n\pi/l$$

with $l=l_1+l_2$. Thus, the length l corresponding to the resonant frequency ω_r yields the phase constant of a uniform finline at ω_r.

With simple modifications, the above procedure can be applied to other types of finline discontinuity problems, such as end effects in open- or short-circuited finline sections. In such cases, the equivalent circuit will consist of a line section terminated at one end with an unknown reactance. Its value can be computed by way of the resonant frequencies of a resonator obtained by short-circuit-

Manuscript received April, 26 1984; revised July 31, 1984. This work was supported in part by the U.S. Army Research Office under Contract DAAG 29-81-K-0053, and in part by the Joint Services Electronics Program.

R. Sorrentino is with the Department of Electronics, University of Rome La Sapienza, Via Eudossiana 18, 00184 Rome, Italy.

T. Itoh is with the Department of Electrical Engineering, University of Texas at Austin, Austin, TX 78712.

R. Sorrentino and T. Itoh, *IEEE Transactions on Microwave Theory and Techniques*, Vol. MTT-32, No. 12, Dec. 1984, pp. 1633–1638.

ing the waveguide at some distance away from the line termination.

III. Computation of the Resonant Frequencies

The method for computing the resonant frequencies of the structure will be illustrated in the case of bilateral finline, shown in Fig. 1. The metallic fins are assumed to be infinitely thin, although the method can easily be modified to account the finite thickness of metallization.

Because of symmetry, a longitudinal magnetic plane can be inserted at the symmetric plane $x = -a_1$, so that only the region $x \geqslant -a_1$ has to be analyzed. The extension to nonsymmetrical structures, such as unilateral finlines, is straightforward and will not be considered here.

The EM field in the dielectric region (region 1: $-a_1 \leqslant x \leqslant 0$) and in the air region (region 2: $0 \leqslant x \leqslant a_2$) can be expanded in terms of TE and TM modes of a rectangular waveguide with inner dimensions l and b. We obtain the following expressions for the transverse E- and H-field components in the two regions:

Dielectric Region: $-a_1 \leqslant x \leqslant 0$

$$\begin{aligned} \boldsymbol{E}_{t1} &= \sum_{mn} A'_{mn} \cos k'_{mn}(x+a_1)\hat{\boldsymbol{x}} \times \nabla_t \psi_{mn} \\ &\quad + \frac{1}{j\omega\epsilon_0\epsilon_r} \sum_{mn} B'_{mn} k'_{mn} \cos k'_{mn}(x+a_1) \nabla_t \varphi_{mn} \\ \boldsymbol{H}_{t1} &= \frac{-1}{j\omega\mu_0} \sum_{mn} A'_{mn} k'_{mn} \sin k'_{mn}(x+a_1) \nabla_t \psi_{mn} \\ &\quad + \sum_{mn} B'_{mn} \sin k'_{mn}(x+a_1) \nabla_t \varphi_{mn} \times \hat{\boldsymbol{x}}. \end{aligned} \tag{2}$$

Air Region: $0 \leqslant x \leqslant a_2$

$$\begin{aligned} \boldsymbol{E}_{t2} &= \sum_{mn} A_{mn} \sin k_{mn}(x-a_2)\hat{\boldsymbol{x}} \times \nabla_t \psi_{mn} \\ &\quad - \frac{1}{j\omega\epsilon_0} \sum_{mn} B_{mn} k_{mn} \sin k_{mn}(x-a_2) \nabla_t \varphi_{mn} \\ \boldsymbol{H}_{t2} &= \frac{1}{j\omega\mu_0} \sum_{mn} A_{mn} k_{mn} \cos k_{mn}(x-a_2) \nabla_t \psi_{mn} \\ &\quad + \sum_{mn} B_{mn} \cos k_{mn}(x-a_2) \nabla_t \varphi_{mn} \times \hat{\boldsymbol{x}} \end{aligned} \tag{3}$$

where

$$\begin{aligned} \psi_{mn} &= P_{mn} \cos\frac{m\pi z}{l} \cos\frac{n\pi y}{b} \\ \varphi_{mn} &= P_{mn} \sin\frac{m\pi z}{l} \sin\frac{n\pi y}{b} \\ P_{mn} &= \sqrt{\frac{\delta_m \delta_n}{lb}} \frac{1}{\gamma_{mn}} \qquad \delta_i = \begin{cases} 1, & i=0 \\ 2, & i \neq 0 \end{cases} \\ \gamma_{mn}^2 &= \left(\frac{m\pi}{l}\right)^2 + \left(\frac{n\pi}{b}\right)^2 \\ k_{mn}^2 &= k_0^2 - \gamma_{mn}^2 \qquad k'^2_{mn} = k_0^2\epsilon_r - \gamma_{mn}^2 \\ k_0^2 &= \omega^2\mu_0\epsilon_0 \end{aligned} \tag{4}$$

where ψ_{mn} and φ_{mn} are the TE and TM scalar potentials, and m and n are integers with starting values of 0 or 1, depending on whether the TE or TM mode is being considered. P_{mn} are determined from the normalization conditions for ψ_{mn}, φ_{mn}

$$\int_S |\nabla_t \psi_{mn}|^2 \, dS = 1$$

$$\int_S |\nabla_t \varphi_{mn}|^2 \, dS = 1.$$

Equations (2)–(4) already satisfy the boundary conditions at $x = -a_1$ and $x = a_2$. The boundary conditions at $x = 0$ are

$$\boldsymbol{E}_{t1} = \boldsymbol{E}_{t2} = \begin{cases} \boldsymbol{E}_{t0}, & \text{on } S_0 \\ 0, & \text{on } S - S_0 \end{cases} \tag{5}$$

$$\boldsymbol{H}_{t1} = \boldsymbol{H}_{t2} = \boldsymbol{H}_{t0}, \qquad \text{on } S_0 \tag{6}$$

where $\boldsymbol{E}_{to}$ and $\boldsymbol{H}_{to}$ are unknown functions of z, y. These functions are expanded in terms of a set of orthonormal vector functions $\boldsymbol{e}_\nu$, or $\boldsymbol{h}_\mu$ defined over aperture region S_o (see Appendix)

$$\boldsymbol{E}_{to} = \sum_\nu V_\nu \boldsymbol{e}_\nu \tag{7}$$

$$\boldsymbol{H}_{to} = \sum_\mu I_\mu \boldsymbol{h}_\mu. \tag{8}$$

Inserting (2), (3), (7), and (8) into (5) and (6), and making use of the orthogonal properties of $\psi_{mn}, \varphi_{mn}, \boldsymbol{e}_\nu$, and $\boldsymbol{h}_\mu$, we obtain a homogeneous system of equations in terms of unknown coefficients V_ν

$$\sum_\nu V_\nu \Bigg[\xi_{mn\nu}\zeta_{mn\mu}\left(k'_{mn}\tan k'_{mn}a_1 - k_{mn}\operatorname{cotan} k_{mn}a_2\right) + \chi_{mn\nu}\theta_{mn\mu}k_0^2\left(\epsilon_r \frac{\tan k'_{mn}a_1}{k'_{mn}} - \frac{\operatorname{cotan} k_{mn}a_2}{k_{mn}}\right)\Bigg] = 0, \qquad \mu = 1, 2 \cdots \tag{9}$$

where

$$\begin{aligned} \xi_{mn\nu} &= \int_{S_0} \hat{\boldsymbol{x}} \times \nabla_t \psi_{mn} \cdot \boldsymbol{e}_\nu \, dS & \chi_{mn\nu} &= \int_{S_0} \nabla_t \varphi_{mn} \cdot \boldsymbol{e}_\nu \, dS \\ \zeta_{mn\mu} &= \int_{S_0} \nabla_t \psi_{mn} \cdot \boldsymbol{h}_\mu \, dS & \theta_{mn\mu} &= \int_{S_0} \nabla_t \varphi_{mn} \times \hat{\boldsymbol{x}} \cdot \boldsymbol{h}_\mu \, dS. \end{aligned} \tag{10}$$

The condition for nontrival solutions determines the characteristic equation of the given structure. This equation may be regarded as a real function of ω, l_1, and l_2 equated to zero

$$f(\omega, l_1, l_2) = 0. \tag{11}$$

For a given value of $\omega = \omega_r$, (11) can be solved to evaluate three different pairs of l_1 and l_2 yielding the same resonant frequency ω_r. These values of l_1 and l_2 can be used for computing the discontinuity parameters discussed in the previous section.

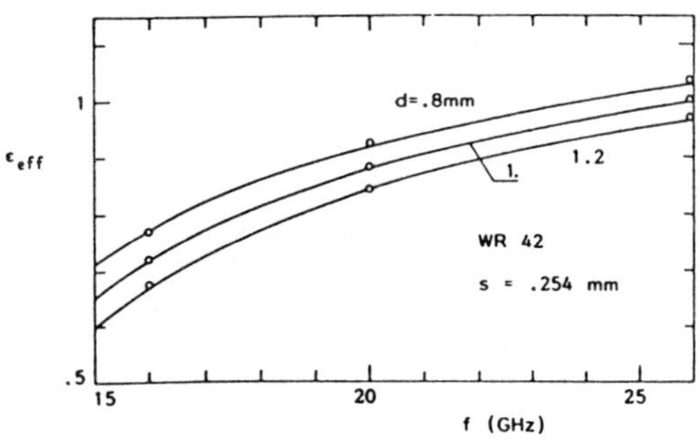

Fig. 3. Effective permittivity of a uniform finline. ∘ Spectral-domain method.

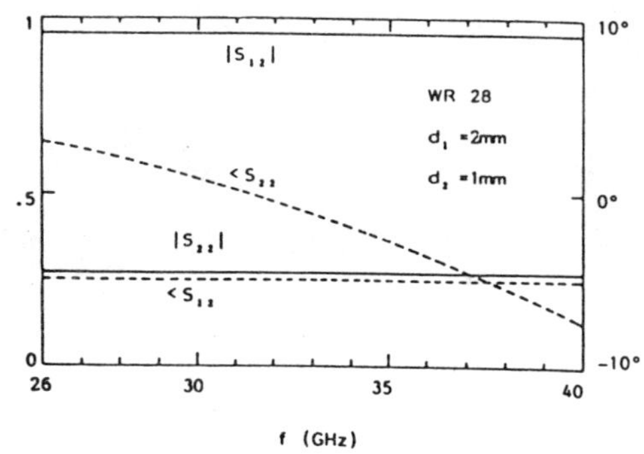

Fig. 5. Scattering parameters of the step discontinuity of Fig. 4.

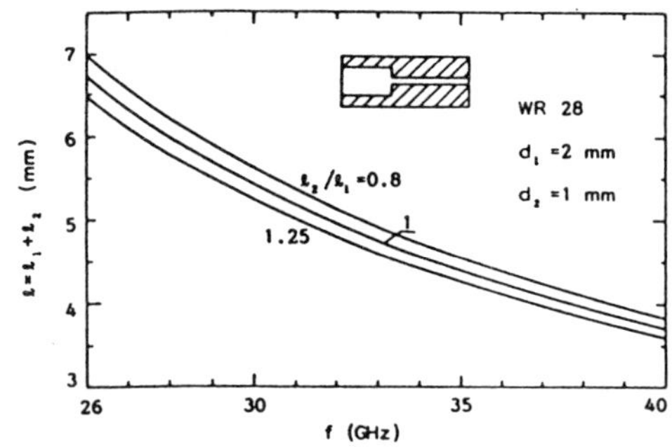

Fig. 4. Resonant frequency of a step discontinuity in a shorted cavity.

IV. Computed Results

According to the above-described technique, the EM fields in the air region, in the dielectric region, and in the aperture region of the resonator are expressed in terms of the series expansions (2), (3), (7), and (8). In the numerical computations, only a finite number of terms of each series can be retained. In order to obtain a proper convergent behavior of the solution, the number of terms in adjacent regions was chosen in such a way that the highest spatial frequencies of the EM field were about the same in the two regions [5], [6]. The method was first tested for computing the propagation characteristics of a uniform finline. In the absence of the discontinuity, the vector basis functions $\boldsymbol{e}_\nu$ and $\boldsymbol{h}_\mu$ on the aperature region (see Appendix) simply reduce to the transverse components of the normal modes of a rectangular waveguide with inner dimensions l and b.

The computed frequency behavior of the effective permittivity

$$\epsilon_{\text{eff}} = (\beta/\beta_0)^2$$

for different gap widths is shown in Fig. 3. Increasing the number of basis functions from 1 to 10, only small differences (less than 1 percent) have been obtained. The time required for computing one resonant frequency using four basis functions was typically 0.3 s on a Univac 1100 computer. The comparison with the results obtained using the spectral-domain approach is quite satisfactory.

In the presence of a step discontinuity, the vector basis functions $\boldsymbol{e}_\nu$ and $\boldsymbol{h}_\mu$ required to represent the EM field over the aperture, have a more complicated spatial distribution, and were evaluated as shown in Appendix. This required some additional computer time.

Fig. 4 shows the resonant frequency of the finline resonator containing a step discontinuity as a function of the total length $l = l_1 + l_2$, with the ratio l_2/l_1 as a parameter. Utilizing these data, the scattering parameters of the discontinuity have been computed using the procedure outlined in Section II, and are shown in Fig. 5. The computed scattering parameters of a unilateral finline discontinuity are compared in Fig. 6 with those computed by Schmidt [5] using the mode-matching procedure.

Although the procedure described above applies to a more complicated discontinuity structure, a certain simplification can be introduced if the discontinuity is longitudinally symmetric, such as the cascaded step discontinuities shown in Fig. 7. For instance, because of the symmetry, the analysis of the structure in Fig. 7(a) is reduced to the two equivalent structures containing a single step terminated by either a magnetic wall or an electric wall, as shown in Fig 8. The equivalent circuits of the original and the two reduced structures are also shown there.

With obvious modifications of expressions (4) for ψ_{mn} and φ_{mn}, and of the basis functions $\boldsymbol{e}_\nu$ and $\boldsymbol{h}_\mu$ (see Appendix), the field analysis procedure described in Section III can be applied to the case of magnetic walls to obtain Z_{11}, Z_{22}, and Z_{21} by way of the resonant frequencies.

Fig. 9 shows the computed results at 26 GHz for the capacitive strips. The normalized reactance parameters of the equivalent T-network are shown as a function of the fin gap d_2 and the distance h. As expected, the capacitance associated with the shunt branch X_{12} increases with both h and the ratio d_1/d_2. On the contrary, the series branches have an inductive reactance whose value is much less sensitive to variations with respect to d_1/d_2. It can be shown that increasing h or d_1/d_2 results in an increase in the magnitudes of the reflection coefficient s_{11}. The phase of s_{11} varies almost linearly with h.

The dual case of inductive notches is shown in Fig. 10, where the normalized admittance parameters of the equivalent π-network are shown as functions of h and d_2/d_1. In this case, the inductance associated with the series branch

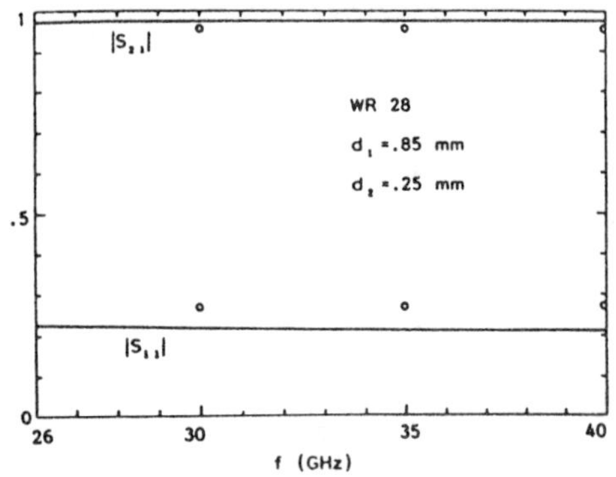

(a)

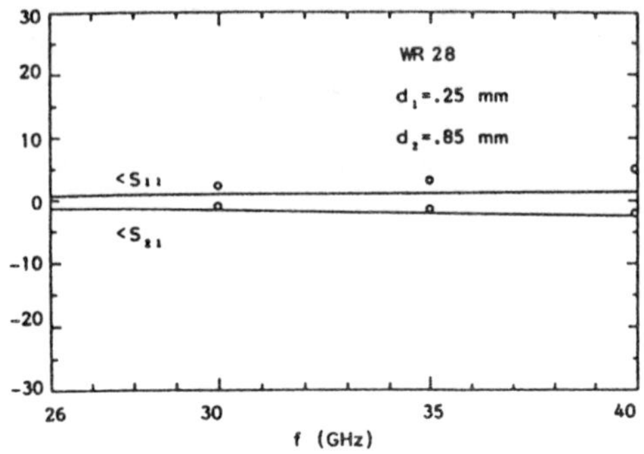

(b)

Fig. 6. Scattering parameters of a unilateral finline step discontinuity. ∘ Schmidt [7].

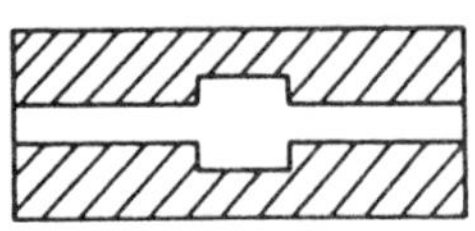

(a)

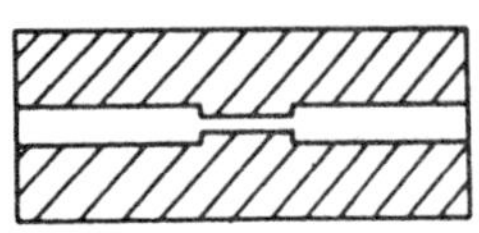

(b)

Fig. 7. Cascaded step discontinuities: (a) inductive notch and (b) capacitive strip.

increases with h and d_2/d_1, while the capacitance of the shunt branches increases only slightly as a function of these parameters.

V. Conclusions

A new method of analysis has been proposed for the characterization of uniform finlines and finline discontinuities. The method is based on the computation of the resonant frequencies of a resonator obtained by short- (or open-) circuiting a finline section containing the discontinuity. The analysis procedure consists of a field expansion in terms of LSM and LSE modes of the rectangular

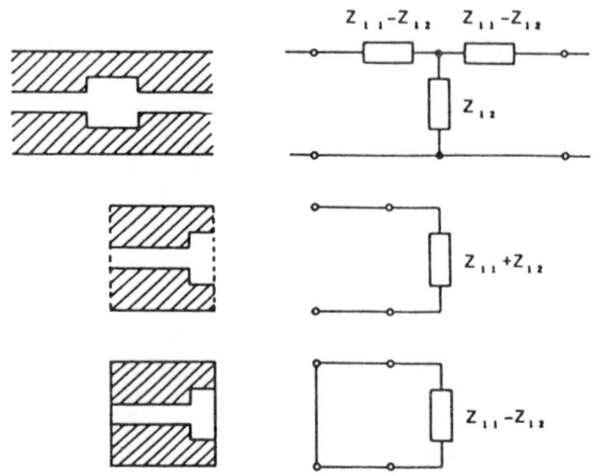

Fig. 8. Evaluation of the Z-parameters of a longitudinally symmetric discontinuity.

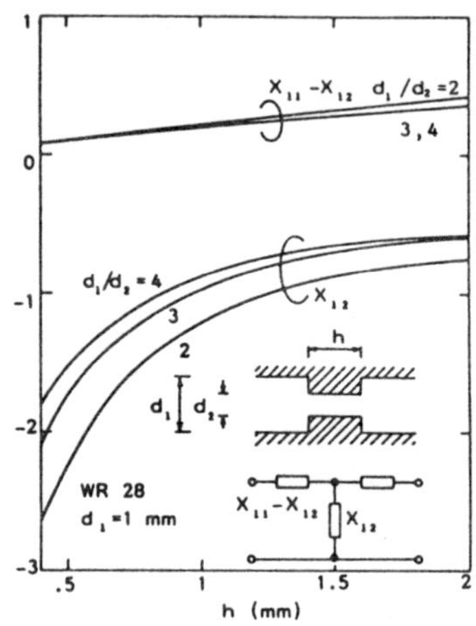

Fig. 9. Normalized reactance parameters of capacitive strips.

waveguide. These expressions are matched with the field distribution in the plane of the fins. With respect to other approaches based on the field expansion in terms of finline modes, this procedure reduces computer time. The results are in good agreement with the numerical values obtained with other techniques.

Appendix

The two sets of orthonormalized vector functions $\boldsymbol{e}_\nu$, $\boldsymbol{h}_\mu$ used in (7) and (8) for expanding the EM field at $x=0$ in the aperture region are derived in this Appendix in the case of a step discontinuity between two finline sections of different slot widths. Because of symmetry considerations, a longitudinal electric plane can be placed at $y=0$ (see Fig. 1), so reducing the longitudinal section to that of Fig. 11.

The aperture region $S_0 \equiv (S_1 \cup S_2)$ may be viewed as the cross section of a waveguide having a stepped cross section. We can therefore expand the EM-field components $\boldsymbol{E}_{t0}$, $\boldsymbol{H}_{t0}$ lying in the yz plane in terms of the TE and TM scalar

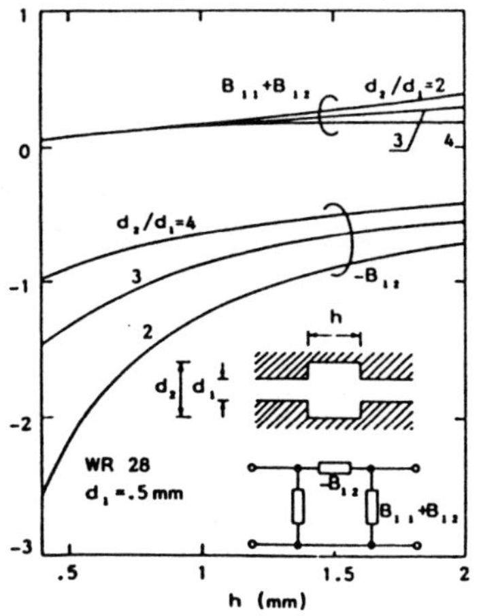

Fig. 10. Normalized admittance parameters of inductive notches.

potentials

$$E_{t0} = \sum_n V_n \hat{x} \times \nabla_t \psi_n + \sum_\nu V_\nu \nabla_t \varphi_\nu \quad \text{(A1)}$$

$$H_{t0} = \sum_n I_n \nabla_t \psi_n + \sum_\nu I_\nu \nabla_t \varphi_\nu \times \hat{x}. \quad \text{(A2)}$$

ψ_n and φ_ν represent the transverse potentials for TE and TM modes, respectively, satisfying the eigenvalue equations

$$\nabla_t^2 \psi_n + k_{cn}^2 \psi_n = 0 \quad \text{(A3)}$$

$$\nabla_t^2 \varphi_\nu + k_{c\nu}^2 \varphi_\nu = 0 \quad \text{(A4)}$$

in S_0 together with proper boundary conditions.

For the sake of brevity, only the solution of (A3) will be illustrated. Moreover, in order to simplify the notation, the index n will be dropped.

In order to solve (A3), the function ψ can be expressed as follows:

$$\psi = \begin{cases} \psi_1 = \sum_r A_r \psi_r^{(1)}, & \text{in } S_1 \\ \psi_2 = \sum_s B_s \psi_s^{(2)}, & \text{in } S_2 \end{cases} \quad \text{(A5)}$$

where

$$\psi_r^{(1)} = \cos k_{1r}(z + l_1) \cos \frac{r\pi y}{d_1/2} \quad \text{(A6)}$$

$$\psi_s^{(2)} = \cos k_{2s}(z - l_2) \cos \frac{s\pi y}{d_2/2} \quad \text{(A7)}$$

$$k_{ir}^2 = k_c^2 - \left(\frac{r\pi}{d_i/2}\right)^2, \quad i = 1,2. \quad \text{(A8)}$$

Expressions (A5)–(A8) are such that (A3) is satisfied together with the boundary conditions at $z = -l_1, l_2$ and $y = 0, d_1/2, d_2/2$. The boundary conditions at $z = 0$

$$\psi_1 = \psi_2, \quad 0 \leqslant y \leqslant d_2/2 \quad \text{(A9)}$$

$$\frac{\partial \psi_1}{\partial z} = \begin{cases} \dfrac{\partial \psi_2}{\partial z}, & 0 \leqslant y \leqslant d_2/2 \\ 0, & d_2/2 \leqslant y \leqslant d_1/2 \end{cases} \quad \text{(A10)}$$

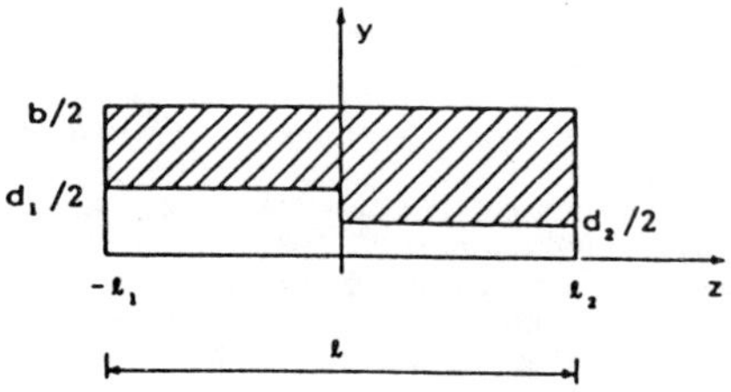

Fig. 11. Reduced geometry of the step discontinuity.

through the orthogonal properties of cosine functions, lead to a homogeneous system of equations in the expansion coefficients A_r, B_s

$$\sum_r A_r f_{rs} \cos k_{1r} l_1 - \frac{d_2}{2\delta_s} B_s \cos k_{2s} l_z = 0, \quad s = 0,1,2 \cdots \quad \text{(A11)}$$

$$\frac{d_1}{2\delta_r} A_r k_{1r} \sin k_{1r} l_1 + \sum_s B_s f_{rs} k_{2s} \sin k_{2s} l_2 = 0, \quad r = 0,1,2 \cdots \quad \text{(A12)}$$

where

$$\delta_r = \begin{cases} 1, & r = 0 \\ 2, & r \neq 0 \end{cases}$$

$$f_{rs} = \int_0^{d_2/2} \cos \frac{r\pi y}{d_1/2} \cos \frac{s\pi y}{d_2/2} \, dy.$$

The condition for nontrivial solutions of (A11)–(A12) constitutes the characteristic equation from which the eigenvalues k_c^2 can be computed. For each k_c^2, the expansion coefficients A_r, B_s are determined using (A11) (A12) and imposing the normalization condition

$$\int_{S_0} |\nabla_t \psi|^2 \, dS = 1.$$

Finally, it can be easily demonstrated that the ψ_n's so obtained satisfy the orthogonality condition

$$\int_{S_0} \nabla_t \psi_n \cdot \nabla_t \psi_m \, dS = 0, \quad n \neq m$$

even if, for numerical reasons, the series in (A5) will be truncated to a finite number of terms.

A similar procedure can be applied to the evaluation of the φ_ν's. The right-hand side of (A1) and (A2) finally provide the required expansions in terms of orthonormal vector functions.

If the resonator is terminated at $z = -l_1, l_2$ by magnetic walls, (A6) and (A7) are modified corresponding, in order to satisfy the open-circuit boundary conditions. Moreover, the eigenfunction φ_0, corresponding to the eigenvalue $k_c^2 = 0$, must also be included in expansions (A1) and (A2). This eigenfunction corresponds to the TEM mode of the stepped waveguide with mixed conducting and magnetic boundaries.

REFERENCES

[1] H. Hofmann, "Dispersion of planar waveguides for millimeter-wave application," *Arch. Elek. Übertragung.*, vol. 31, pp. 40–44, Jan. 1977.

[2] L.-P. Schmidt and T. Itoh, "Spectral domain analysis of dominant and higher order modes in fin-lines," *IEEE Trans. Microwave Theory Tech.* vol. MTT-28, pp. 981–985, Sept. 1980.

[3] A. Beyer, "Calculation of discontinuities in grounded finlines taking into account the metallization thickness and the influence of the mount-slits," in *Proc. of the 12th European Microwave Conf.* (Helsinki, Finland), 1982, pp. 681–686.

[4] H. El Hennaway and K. Schunemann, "Analysis of fin-line discontinuities," in *Proc. of the 9th European Microwave Conf.* (Brighton, England), 1979, pp. 448–452.

[5] S. W. Lee, W. R. Jones, and J. J. Campbell, "Convergence of numerical solutions of iris-type discontinuity problems," *IEEE Trans. Microwave Theory Tech.*, vol. MTT-19, pp. 528–536, June 1971.

[6] Y. C. Shih and K. G. Gray, "Convergence of numerical solutions of step-type waveguide discontinuity problems by modal analysis," in *IEEE MTT-S Int. Symp. Dig.* (Boston, MA), 1983, pp. 233–235.

[7] L.-P. Schmidt, private communication.

4.1.4

Spectral-Domain Analysis of Scattering from *E*-Plane Circuit Elements

QIU ZHANG AND TATSUO ITOH, FELLOW, IEEE

Abstract **—The spectral-domain method is used for analysis of the scattering characteristics of *E*-plane circuit components such as nontouching *E*-plane fins. The method deals with inhomogeneous algebraic equations instead of integral equations. It provides a number of attractive features. Numerical results have been compared with those reported in the literature for special cases. Several data items useful for *E*-plane configurations are included.**

I. Introduction

RECENTLY, finlines [1], [2] and other E-plane structures [3]–[5] have found wide applications in millimeter-wave integrated circuits. A number of passive, active, and nonreciprocal components have been developed with the E-plane technique. One of the key elements for passive E-plane components is the E-plane strip. A comprehensive design process of E-plane bandpass filters has been reported [6]. The analysis of the E-plane fin connecting the top and bottom walls is relatively straightforward [6], because the problem is a two-dimensional one. On the other hand, no extensive and accurate characterizations of nontouched E-plane fins seem to exist. A method based on a variational technique has been introduced for a special case where there is no dielectric substrate inserted in a waveguide [7]. The method in [7] is useful for a narrow strip, because only one current component along the E-plane direction is used and the assumed current distribution is constant in the axial (Z) direction.

This paper introduces a new analytical technique to characterize the scattering phenomena of a number of planar E-plane obstacles. For instance, it can handle a wide nontouching E-plane fin on a dielectric substrate. Unlike the method based on the variational technique, scattering coefficients of the dominant, as well as higher order, modes can be derived. The incident mode can be either dominant or a higher order.

The method in this paper is an extension of the spectral-domain method commonly used for characterizations of eigenmodes in a printed transmission line. It is extended to the excitation problem and hence provides a set of algebraic equations corresponding to coupled integral equations that would be derived in the space domain. Compared to the integral equation method, the new technique has a number of advantages. For instance, algebraic equations are easier to handle numerically. Also, the spectral-domain Green's functions have simple closed-form expressions. Compared to the variational method [7], the present method is not only more versatile but is also attractive from a computational point of view. In the new method, it is necessary to calculate the eigenvalue of only the particular scattered mode of interest. The variational method requires evaluations of all eigenvalues. Furthermore, the method in [7] assumes that only the TE modes are scattered. By the nature of the formulation, the present method contains both TE and TM modes in its formulation.

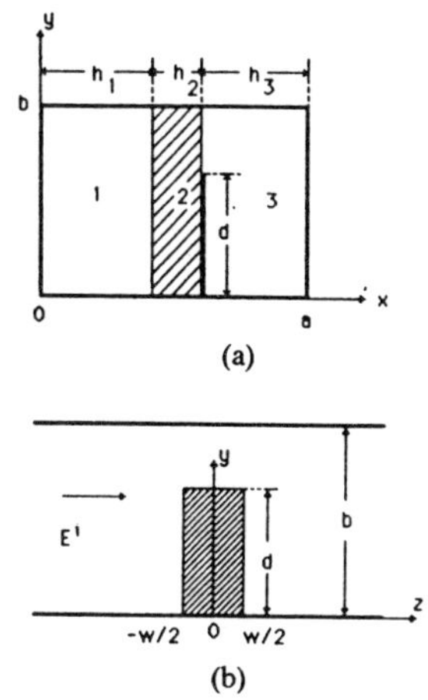

Fig. 1. E-plane strip in a rectangular waveguide. (a) End view. (b) Side view.

The calculated results for a special case ($\epsilon_1 = \epsilon_2 = \epsilon_3 = 1$) are compared with experimental and computed data in [7] to check the accuracy. Several curves of normalized input admittance and equivalent circuit element values are presented for a number of different parameters of the structure.

II. Formulation

In this paper, only the unilateral E-plane fin is treated. An application to a bilateral configuration is straightforward.

With reference to Fig. 1, the strip is assumed to be perfectly conducting and infinitesimally thin. For a given incident field E^i, the field scattered by the strip can be expressed by modal expansion. For instance, the E_y com-

Manuscript received June 12, 1986; revised September 6, 1986. This work was supported in part by the U.S. Army Research Office under Contract DAAG-29-84-K-0076.

The authors are with the Department of Electrical and Computer Engineering, University of Texas, Austin, TX 78712.

IEEE Log Number 8611630.

Q. Zhang and T. Itoh, *IEEE Transactions on Microwave Theory and Techniques*, Vol. MTT-35, No. 2, Feb. 1987, pp. 138–150.

ponent is

$$E_y^S(x,y,z) = E_{y\,\mathrm{LSE}}^S + E_{y\,\mathrm{LSM}}^S$$

$$= \begin{cases} \sum_{m=1}^{\infty}\sum_{n=0}^{\infty} C_{mn}^{-}\phi_m(x)\cos(\alpha_n y)e^{j\beta_{mn}(z+W/2)}, & z<-W/2 \\ \sum_{m=1}^{\infty}\sum_{n=0}^{\infty} \phi_m(x)\cos(\alpha_n y)\left(E_{mn}^{-}e^{j\beta_{mn}z}+E_{mn}^{+}e^{-j\beta_{mn}z}\right), & |z|<W/2 \\ \sum_{m=1}^{\infty}\sum_{n=0}^{\infty} C_{mn}^{+}\phi_m(x)\cos(\alpha_n y)e^{-j\beta_{mn}(z-W/2)}, & z>W/2 \end{cases}$$

$$+\begin{cases} \sum_{m=1}^{\infty}\sum_{n=1}^{\infty} D_{mn}^{-}\psi_m(x)\sin(\alpha_n y)e^{j\beta'_{mn}(z+W/2)}, & z<-W/2 \\ \sum_{m=1}^{\infty}\sum_{n=1}^{\infty} \psi_m(x)\sin(\alpha_n y)\left(F_{mn}^{-}e^{j\beta'_{mn}z}+F_{mn}^{+}e^{-j\beta'_{mn}z}\right), & |z|<W/2 \\ \sum_{m=1}^{\infty}\sum_{n=1}^{\infty} D_{mn}^{+}\psi_m(x)\sin(\alpha_n y)e^{-j\beta'_{mn}(z-w/2)}, & z>W/2 \end{cases} \tag{1}$$

where

$$\phi_m(x) = \begin{cases} C_m\sinh(\gamma_{1m}x), & 0<x<h_1 \\ A_m\sinh[\gamma_{2m}(x-h_1)] + B_m\cosh[\gamma_{2m}(x-h_1)], & h_1<x<h_1+h_2 \\ \sinh[\gamma_{3m}(a-x)], & h_1+h_2<x<a \end{cases} \tag{2}$$

$$\psi_m(x) = \begin{cases} C'_m\cosh(\gamma'_{1m}x), & 0<x<h_1 \\ A'_m\sinh[\gamma'_{2m}(x-h_1)] + B'_m\cosh[\gamma'_{2m}(x-h_1)], & h_1<x<h_1+h_2 \\ \cosh[\gamma'_{3m}(a-x)], & h_1+h_2<x<a \end{cases} \tag{3}$$

$$\alpha_n = \frac{n\pi}{b} \tag{4}$$

$$\alpha_n^2 + \beta_{mn}^2 = k_i^2 + \gamma_{im}^2 \tag{5a}$$

$$\alpha_n^2 + \beta_{mn}'^2 = k_i^2 + \gamma_{im}'^2 \tag{5b}$$

$$k_i = \omega\sqrt{\epsilon_0\mu_0\epsilon_i}, \qquad i=1,2,3. \tag{6}$$

Here, ϵ_i is the relative dielectric constant; γ_{im} and γ'_{im} are the mth eigenmodes of the LSE and LSM modes in the partially filled waveguide in region i, and they can be obtained by solving the eigenvalue equations [8]. Similar equations can be written for E_z. On the other hand, the scattered fields E_y^S and E_z^S can be expressed as follows, if the induced current components $J_y(y,z)$ and $J_z(y,z)$ are provided:

$$E_y^S(x,y,z) = \int_0^b dy' \int_{-\infty}^{\infty} dz' \left[Gyy(x,y-y',z-z')J_y(y',z') + G_{yz}(x,y-y',z-z')J_z(y',z')\right] \tag{7a}$$

$$E_z^S(x,y,z) = \int_0^b dy' \int_{-\infty}^{\infty} dz' \left[G_{zy}(x,y-y',z-z')J_y(y',z') + G_{zz}(x,y-y',z-z')J_z(y',z')\right]. \tag{7b}$$

One way to find J_y and J_z is by applying the integral equations which require the total tangential electric-field components to be zero on the strip

$$E_y^i(x,y,z) + E_y^S(x,y,z) = 0 \tag{8a}$$

$$(x = h_1+h_2,\ y, z \text{ on strip}).$$

$$E_z^i(x,y,z) + E_z^S(x,y,z) = 0 \tag{8b}$$

The integral equations are

$$E_y^i(h_1+h_2,y,z) + \int_0^b dy' \int_{-W/2}^{W/2} dz' \big[G_{yy}(h_1+h_2,y-y',z-z')J_y(y',z') + G_{yz}(h_1+h_2,y-y',z-z')J_z(y',z')\big] = 0 \qquad \left(0<y<b, |z|<\frac{W}{2}\right) \tag{9a}$$

$$E_z^i(h_1+h_2,y,z) + \int_0^b dy' \int_{-W/2}^{W/2} dz' \big[G_{zy}(h_1+h_2,y-y',z-z')J_y(y',z) + G_{zz}(h_1+h_2,y-y',z-z')J_z(y',z')\big] = 0 \qquad \left(0<y<b, |z|<\frac{W}{2}\right). \tag{9b}$$

If those equations are solved, J_y and J_z can be obtained. These J_y and J_z are then substituted into (7), so that E_y^S and E_z^S are available everywhere. If the scattered field coefficient of a particular mode is needed, the E_y^S and E_z^S can be used in (1). Each coefficient may be found from orthogonality of the expansion functions.

Although the bove formulation is correct, we do not follow such an approach. Instead, we adopt a corresponding procedure in the Fourier-transformed domain. There are two reasons for using this new technique.

1) In the Fourier-transformed (spectral) domain, we deal with coupled algebraic equations instead of the coupled integral equations (9a) and (9b).

2) Derivations of the Green's functions in the space domain are very complicated. In the spectral domain, the Fourier-transformed Green's functions are given in closed form.

Let us introduce the Fourier transform as

$$\tilde{F}(\alpha_n,\beta) = \int_{-b}^{b} e^{j\alpha_n y}\, dy \int_{-\infty}^{\infty} f(y,z) e^{j\beta z}\, dz$$
$$\alpha_n = \frac{n\pi}{b}. \tag{10}$$

The Fourier transform of (1) at $x = h_1 + h_2$ is given by

$$\tilde{E}_y^S(\alpha_n,\beta) = \tilde{E}_{y\mathrm{LSE}}^S + \tilde{E}_{y\mathrm{LSM}}^S = \sum_{m=1}^{\infty} b\delta_n \sinh(\gamma_{3m}h_3)\Bigg\{ C_{mn}^- \left[\frac{e^{jW\beta/2}}{j(\beta+\beta_{mn})} - \pi\delta(\beta+\beta_{mn})e^{j\beta_{mn}W/2}\right]$$
$$+\left[E_{mn}^- \frac{\sin\frac{(\beta+\beta_{mn})W}{2}}{(\beta+\beta_{mn})} + E_{mn}^+ \frac{\sin\frac{(\beta-\beta_{mn})W}{2}}{(\beta-\beta_{mn})}\right] + C_{mn}^+\left[\pi\delta(\beta-\beta_{mn})e^{j\beta_{mn}W/2} - \frac{e^{j\beta W/2}}{j(\beta-\beta_{mn})}\right]\Bigg\}$$
$$+\sum_{m=1}^{\infty} k_n jb\cosh(\gamma'_{3m}h_3)\Bigg\{ D_{mn}^-\left[\frac{e^{-j\beta W/2}}{j(\beta+\beta'_{mn})} - \pi\delta(\beta+\beta'_{mn})e^{j\beta'_{mn}W/2}\right]$$
$$+\left[F_{mn}^- \frac{\sin\frac{(\beta+\beta'_{mn})W}{2}}{\beta+\beta'_{mn}} + F_{mn}^+ \frac{\sin\frac{(\beta-\beta'_{mn})W}{2}}{\beta-\beta'_{mn}}\right]$$
$$+ D_{mn}^+\left[\pi\delta(\beta-\beta'_{mn})e^{j\beta'_{mn}W/2} - \frac{e^{j\beta W/2}}{j(\beta-\beta'_{mn})}\right]\Bigg\} \qquad n=0,1,2,\cdots \tag{11}$$

where

$$\delta_n = \begin{cases} 2 & n=0 \\ 1 & n\neq 0 \end{cases} \tag{12a}$$

$$k_n = \begin{cases} 1 & n\neq 0 \\ 0 & n=0. \end{cases} \tag{12b}$$

The Fourier transform of (9) at $x = h_1 + h_2$ is given by

$$\tilde{E}_y^i(\alpha_n,\beta) + \tilde{E}_y^S(\alpha_n,\beta) = \tilde{E}_y^t(\alpha_n,\beta) \tag{13a}$$

$$\tilde{E}_z^i(\alpha_n,\beta) + \tilde{E}_z^S(\alpha_n,\beta) = \tilde{E}_z^t(\alpha_n,\beta) \tag{13b}$$

where

$$\tilde{E}_y^S(\alpha_n,\beta) = \tilde{G}_{yy}(\alpha_n,\beta)\tilde{J}_y(\alpha_n,\beta) + \tilde{G}_{yz}(\alpha_n,\beta)\tilde{J}_z(\alpha_n,\beta) \tag{14a}$$

$$\tilde{E}_z^S(\alpha_n,\beta) = \tilde{G}_{zy}(\alpha_n,\beta)\tilde{J}_y(\alpha_n,\beta) + \tilde{G}_{zz}(\alpha_n,\beta)\tilde{J}_z(\alpha_n,\beta) \tag{14b}$$

are the scattered electric fields in the spectral domain. $\tilde{G}_{yy}$, $\tilde{G}_{yz}$, and $\tilde{G}_{zz}$ can be obtained by the immittance approach [9] (see Appendix). Notice that the right sides of (13) are not zero. This is because the application of the Fourier transform requires the use of fields not only on the strip but also the one outside. Hence, (13) contains four unknowns, $\tilde{J}_y$, $\tilde{J}_z$, $\tilde{E}_y$, and $\tilde{E}_z$. In the process of solution by Galerkin's method, $\tilde{E}_y$ and $\tilde{E}_z$ are eliminated. To this end, let us expand $\tilde{J}_y$ and $\tilde{J}_z$ as follows:

$$\tilde{J}_y(\alpha_n,\beta) = \sum_{i=1}^{I} a_i \tilde{J}_{yi}(\alpha_n,\beta) \tag{15a}$$

$$\tilde{J}_z(\alpha_n,\beta) = \sum_{j=1}^{J} b_j \tilde{J}_{zj}(\alpha_n,\beta). \tag{15b}$$

We substitute (15) into (13) and take the inner product of the resultant equations with basis functions. Use of the Parseval's relation eliminates $\tilde{E}_y^t$ and $\tilde{E}_z^t$. The inner products of $\tilde{E}_y^t$ and $\tilde{J}_{yi}$ and of $\tilde{E}_z^t$ and $\tilde{J}_{zj}$ are zero, because $\tilde{J}_{yi}$ and $\tilde{J}_{zj}$ are Fourier transforms of the functions nonzero on the strip while $\tilde{E}_y^t$ and $\tilde{E}_z^t$ are transforms of the functions nonzero outside the strip.

The results are

$$\sum_{i=1}^{I} K_{pi}^{yy} a_i + \sum_{j=1}^{J} K_{pj}^{yz} b_j = -S_{yp}, \qquad p=1,2,\cdots I \tag{16a}$$

$$\sum_{i=1}^{I} K_{qi}^{zy} a_i + \sum_{j=1}^{J} K_{qj}^{zz} b_j = -S_{zq}, \qquad q=1,2,\cdots J \tag{16b}$$

where

$$K_{pi}^{yy} = \sum_{n=-\infty}^{\infty} \int_{-\infty}^{\infty} \tilde{J}_{yp}^{*}(\alpha_n,\beta)\tilde{G}_{yy}(\alpha_n,\beta)\tilde{J}_{yi}(\alpha_n,\beta)\,d\beta \tag{17a}$$

$$K_{pj}^{yz} = K_{qi}^{zy} = \sum_{n=-\infty}^{\infty} \int_{-\infty}^{\infty} \tilde{J}_{yp}^{*}(\alpha_n,\beta)\tilde{G}_{yz}(\alpha_n,\beta)\tilde{J}_{zj}(\alpha_n,\beta)\,d\beta \tag{17b}$$

$$K_{qj}^{zz} = \sum_{n=-\infty}^{\infty} \int_{-\infty}^{\infty} \tilde{J}_{zq}^{*}(\alpha_n,\beta)\tilde{G}_{zz}(\alpha_n,\beta)\tilde{J}_{zj}(\alpha_n,\beta)\,d\beta \tag{17c}$$

$$S_{py} = \sum_{n=-\infty}^{\infty} \int_{-\infty}^{\infty} \tilde{J}_{yp}^{*}(\alpha_n,\beta)\tilde{E}_y^{i}(\alpha_n,\beta)\,d\beta \tag{18a}$$

$$S_{qz} = \sum_{n=-\infty}^{\infty} \int_{-\infty}^{\infty} \tilde{J}_{zq}^{*}(\alpha_n,\beta)\tilde{E}_z^{i}(\alpha_n,\beta)\,d\beta \tag{18b}$$

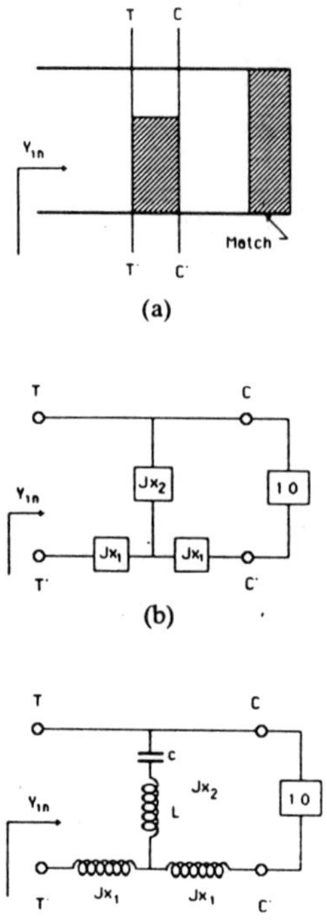

Fig. 2. (a) An E-plane strip circuit with a matched termination. (b) Two-port equivalent circuit. (c) Equivalent T-network for a narrow E-plane strip.

The scattering coefficients $C_{mn}^{\pm}$ and $D_{mn}^{\pm}$ can now be obtained. Let us express the left-hand side of (11) with (14a). Since $\tilde{G}_{yy}$ and $\tilde{G}_{yz}$ are given in closed forms and $\tilde{J}_y$ and $\tilde{J}_z$ are now known, the left-hand side of (11) is completely known. Furthermore, the left-hand side contains poles at $\beta = \pm\beta_{mn}$ and $\beta = \pm\beta'_{mn}$, since they are zeros of the denominators of $\tilde{G}_{yy}$ and $\tilde{G}_{yz}$. These values provide the eigenvalues of the LSE and LSM modes, γ_{im} and γ'_{im}. The right-hand side of (11) contains LSE poles at $-\beta_{mn}$ in the C_{mn}^{-} term and at $+\beta_{mn}$ in the C_{mn}^{+} term. It contains LSM poles at $-\beta'_{mn}$ in the D_{mn}^{-} term and at $+\beta'_{mn}$ in the D_{mn}^{+} term. Therefore, $C_{mn}^{\pm}$ and $D_{mn}^{\pm}$ can be obtained by residue calculus.

$$C_{mn}^{\pm} = \pm \lim_{\beta \to \pm\beta_{mn}} \left[(\beta \mp \beta_{mn}) \frac{\tilde{G}_{yy}(\alpha_n,\beta)\tilde{J}_y(\alpha_n,\beta) + \tilde{G}_{yz}(\alpha_n,\beta)\tilde{J}_z(\alpha_n,\beta)}{jb\delta_n \sinh(\gamma_{3m}h_3)} e^{\mp jW\beta/2} \right]$$

$$\left(n=0,1,2,\cdots, m=1,2,3\cdots \delta_n = \begin{cases} 2 & n=0 \\ 1 & n\neq 0 \end{cases} \right) \tag{19a}$$

$$D_{mn}^{\pm} = \mp \lim_{\beta \to \pm\beta'_{mn}} \left[(\beta \mp \beta'_{mn}) \frac{\tilde{G}_{yy}(\alpha_n,\beta)\tilde{J}_y(\alpha_n,\beta) + \tilde{G}_{yz}(\alpha_n,\beta)\tilde{J}_z(\alpha_n,\beta)}{b\cosh(\gamma'_{3m}h_3)} e^{\mp jW\beta/2} \right]$$

$$(n=1,2,3\cdots, m=1,2,3\cdots). \tag{19b}$$

where $*$ indicates the complex conjugate.

For a given incident field, we solve (16) and find a_i and b_j. Hence, $\tilde{J}_y$ and $\tilde{J}_z$ are now given.

This process corresponds to the use of the orthogonality relationship to find the modal coefficients in the space domain.

TABLE I
CONFIRMATION OF POWER CONSERVATION

d (mm)	R		T		$\|R\|^2 + \|T\|^2$
0.5	-.02008	-.02598 j	.79080	-.61119 j	1.00000
1.0	-.14712	-.13549 j	.66376	-.72071 j	1.00000
1.5	-.70508	-.10766 j	.10580	-.69288 j	1.00000
2.0	-.78485	.61825 j	.02602	.03304 j	1.00000
2.5	-.53859	.77455 j	.27229	.18934 j	1.00000
3.0	-.45482	.79016 j	.35606	.20495 j	1.00000
3.5	-.38694	.79227 j	.42394	.20705 j	1.00000

$h_1 = h_3 = 3.43$ mm; $h_2 = 0.254$ mm; $\epsilon_1 = \epsilon_3 = 1$; $b = 3.56$ mm; $w = 1.0$ m; $f = 35$ GHz; $\epsilon_2 = 2.2$.

It should be noted that in the case of E-plane fins connecting the top and bottom wall, the above equations are simplified. Since there is no field variations in the y-direction and only TE_{n0} modes are scattered for a TE_{10} excitation, we have only the one equation (13a). All the Fourier-transformed quantities are functions of β only.

III. NUMERICAL RESULTS

To compare the present method with the experimental and computed data in the previous publication [7], the special case $\epsilon_1 = \epsilon_2 = \epsilon_3 = 1$ is considered first. We assume that a dominant-mode incident electric field $E_{y10} = -j\beta_{10}\phi_1(x)e^{-\gamma\beta_{10}z}$, where $\phi_1(x)$ is defined in (2), comes from the left of the waveguide. The E-plane fin shown in Fig. 2(a) may be represented by an equivalent T-network, as shown in Fig. 2(b). When the waveguide is terminated with a matched load, the normalized input admittance may be represented by

$$Y_{in} = G_{in} + jB_{in} = \frac{1-R}{1+R} \tag{20}$$

where R is the reflection coefficient for the dominant mode, and can be determined in the present method by means of (19). In this study, the current basis functions are chosen as

$$J_{yi}(y,z) = \frac{\cos\left[(k-1)\left(\frac{\pi z}{W}+\frac{\pi}{2}\right)\right]}{\sqrt{1-\left(\frac{2z}{W}\right)^2}} \cdot \frac{\sin\left[(2l-1)\left(\frac{\pi y}{2d}+\frac{\pi}{2}\right)\right]}{\sqrt{1-\left(\frac{y}{d}\right)^2}} \tag{21a}$$

$$J_{zj}(y,z) = \frac{\sin\left[k\left(\frac{\pi z}{W}+\frac{\pi}{2}\right)\right]}{\sqrt{1-\left(\frac{2z}{W}\right)^2}} \frac{\cos\left[(2l-1)\left(\frac{\pi y}{2d}+\frac{\pi}{2}\right)\right]}{\sqrt{1-\left(\frac{y}{d}\right)^2}} \tag{21b}$$

where k and l are integers, and i and j are given by a combination of k and l. The equivalent circuit for a nontouching fin can be expressed as in Fig. 2(c) if the strip is not too wide. We can determine the normalized reactances X_1 and X_2 after Y_{in} in (20) is obtained. The expressions of X_1 and X_2 are given by

$$X_2 = \pm\left[\left(\frac{B_{in}}{B_{in}^2+G_{in}^2-G_{in}}\right)^2 + \frac{G_{in}-1}{B_{in}^2+G_{in}^2-G_{in}}\right]^{1/2} \tag{22a}$$

$$X_1 = \frac{B_{in}}{G_{in}-\left(B_{in}^2+G_{in}^2\right)} \mp X_2. \tag{22b}$$

The sign in (22a) and (22b) can be determined by the Foster reactance theorem in which $dx/dw > 0$ has to be satisfied for a lossless element. We can also determine the values of C and L in the equivalent circuit Fig. 2(c), once X_2 is found, under the assumption that the variations of C and L with frequency are small. The capacitance and inductance can be obtained by solving the coupled equations

$$\omega L - \frac{1}{C\omega} = X_2 \tag{23a}$$

$$L + \frac{1}{C\omega^2} = \frac{dX_2}{d\omega}. \tag{23b}$$

The numerical result has been checked by the power conservation law in which the equation $|R|^2 + |T|^2 = 1$ has to be satisfied. R and T are the reflection and transmission coefficients for the dominant mode, and can be determined from C_{10}^- and C_{10}^+ which are given by (19). The calculated results for different parameters are given in Table I. It shows that $|R|^2 + |T|^2$ is essentially 1. The convergence test with different numbers of basis functions has been performed. Although the accuracy can be increased with a larger number of basis functions, computation efforts also increase.

A comparison of the normalized susceptance versus frequency between the numerical results obtained by the present approach and those given in [7] is shown in Fig. 3(a) and (b). It is found that the numerical results agree

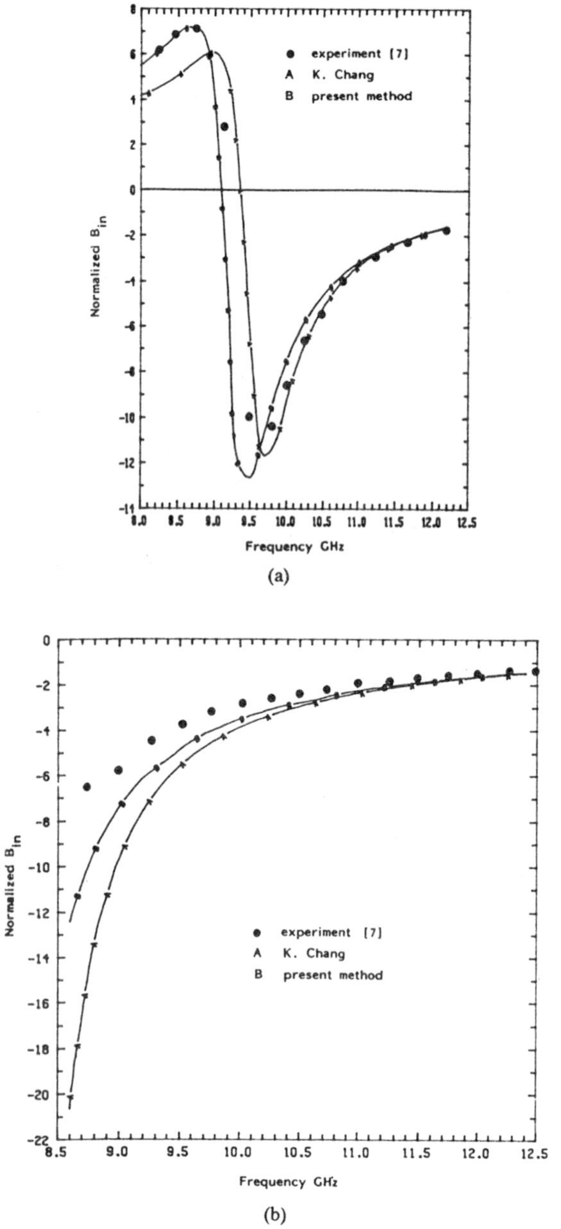

Fig. 3. Normalized susceptance of an E-plane strip in an X-band waveguide terminated with a matched load for $\epsilon_1 = \epsilon_2 = \epsilon_3 = 1$; $a = 22.86$ mm and $b = 10.16$ mm. (a) $h_1 + h_2 = 12.57$ mm, $d = 7.37$ mm, $w = 3.38$ mm. (b) $h_1 + h_2 = 11.43$ mm, $d = 9.19$ mm, $w = 1.7$ mm.

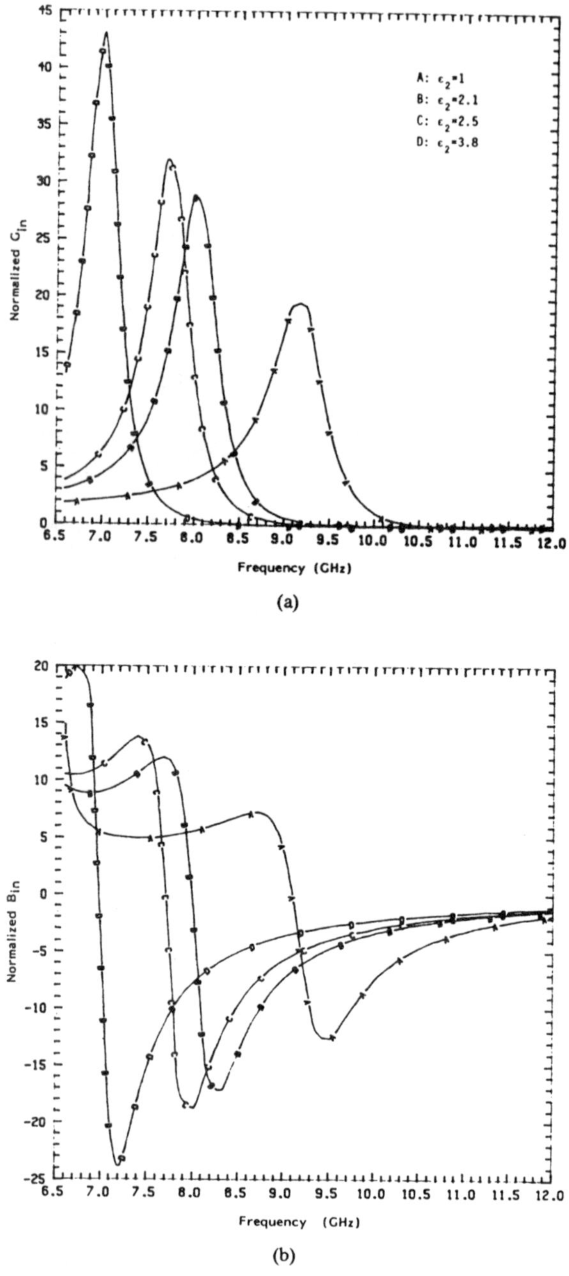

Fig. 4. Normalized admittance of an E-plane strip in an X-band waveguide terminated with a matched load versus frequency for different values of dielectric constant of the substrate. $\epsilon_1 = \epsilon_3 = 1$, $h_1 = 11.57$ mm, $h_2 = 1$ mm, $d = 7.37$ mm, and $w = 3.38$ mm. (a) Normalized conductance versus frequency. (b) Normalized susceptance versus frequency.

well with the experimental data and Chang's data. It is believed that the present method is more accurate and can be improved systematically with the use of more basis functions. Fig. 4(a) and (b) show the variations in normalized admittance versus frequency with different values of dielectric constant of the substrate in region 2. As the dielectric constant of the substrate increases, the resonant frequency at which B_{in} becomes zero decreases. This phenomenon happens because as the dielectric constant of the substrate increases, the wavelengths corresponding to each mode become shorter. The resonant frequency and the characteristics of input admittance of the considered structure can be controlled by the dielectric constant of the substrate.

Figs. 5 and 6 show the variations of normalized admittance with the height and width of the strip at different frequencies for an E-plane fin inserted in a Ka-band rectangular waveguide. Figs. 7 and 8 show the values of X_1 and X_2 versus the height d and the width w of the strip in an X-band rectangular waveguide. We note that for a

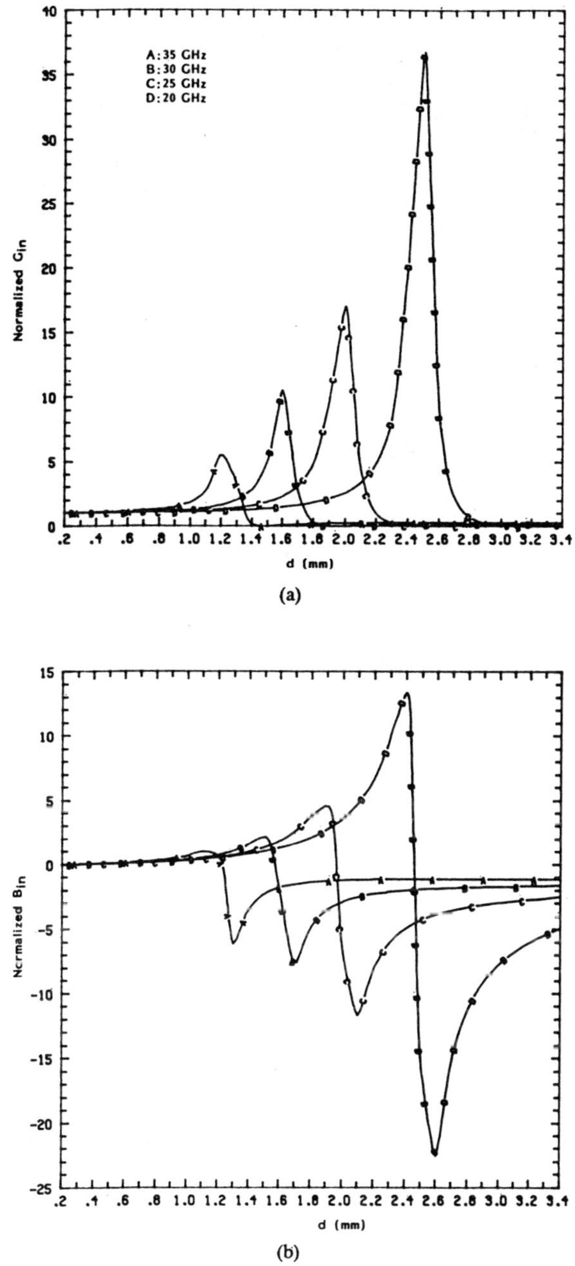

Fig. 5. Normalized admittance of an E-plane strip in a Ka-band waveguide terminated with a matched load versus strip height for different values of frequency. $\epsilon_1 = \epsilon_3 = 1$, $\epsilon_2 = 2.2$, $h_1 = 3.43$ mm, $h_2 = 0.254$ mm, $a = 7.11$ mm, $b = 3.56$ mm, $w = 1$ mm. (a) Normalized conductance versus strip height. (b) Normalized susceptance versus stri height.

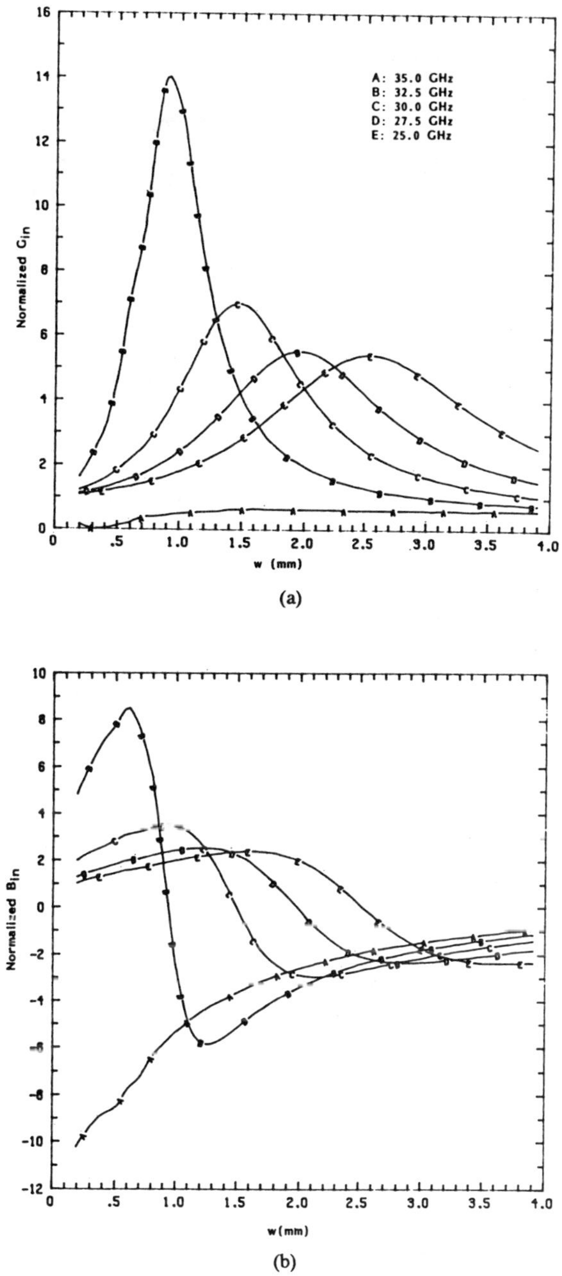

Fig. 6. Normalized admittance of an E-plane strip in a Ka-band waveguide terminated with a matched load versus strip width for different values of frequency. $\epsilon_1 = \epsilon_3 = 1$, $\epsilon_2 = 2.2$, $h_1 = 3.43$ mm, $h_2 = 0.254$ mm, and $d = 1.78$ mm. (a) Normalized conductance versus strip width. (b) Normalized susceptance versus strip width.

given frequency, X_1 is not sensitive to d whereas X_2 increases with d. When w increases, X_1 increases. On the other hand, X_2 decreases with w for a higher frequency and increases in a certain region of w for lower frequency (curve A in Fig. 8(b)). Figs. 9 and 10 are the data corresponding to those in Figs. 7 and 8 except that the frequencies are in the Ka-band. Figs. 11 and 12 show X_1 and X_2 versus frequency with different values of height and width for the Ka-band waveguide. In Fig. 12, it is seen that for a narrow strip, X_1 varies slowly as the frequency is increased, while X_2 varies faster for a narrower strip than for a wider one. Fig. 13 shows the variation of normalized capacitance C and normalized inductance L with d. We note that there are two regions. In one of them (approximately corresponding to $d > w$, in this calculation $w = 1.0$ mm) C increases and L decreases as the frequency in-

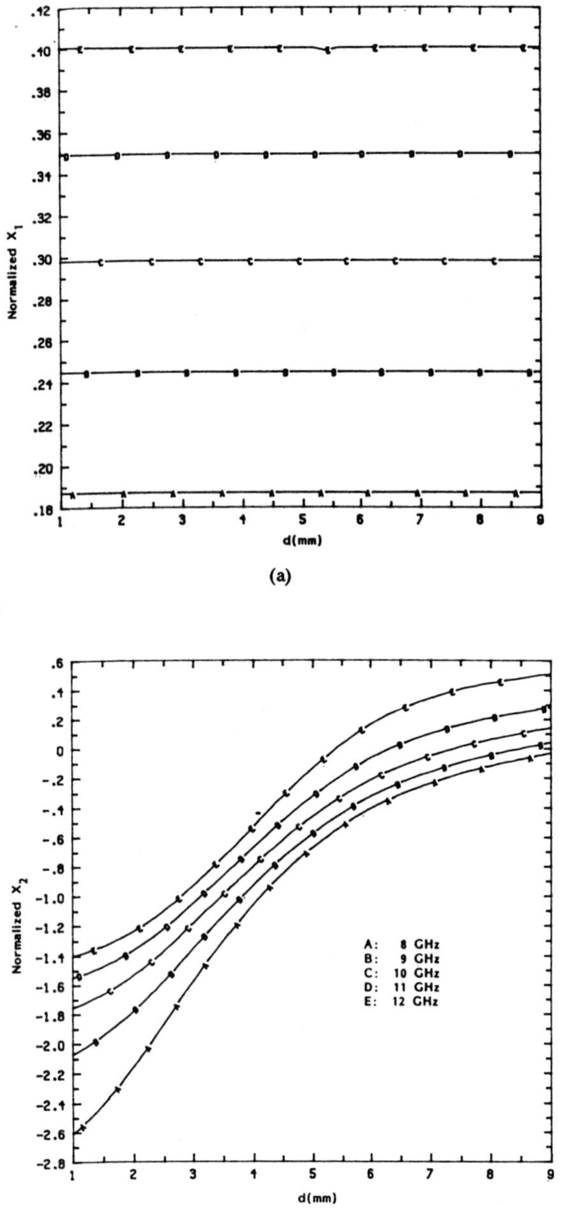

Fig. 7. Normalized reactance of equivalent circuit of an E-plane strip in an X-band waveguide versus strip height for different values of frequency. $\epsilon_1 = \epsilon_3 = 1$, $\epsilon_2 = 2.1$, $h_1 = 11.57$ mm, $h_2 = 1$ mm, and $w = 3.38$ mm. (a) Normalized X_1 versus strip height. (b) Normalized X_2 versus strip height.

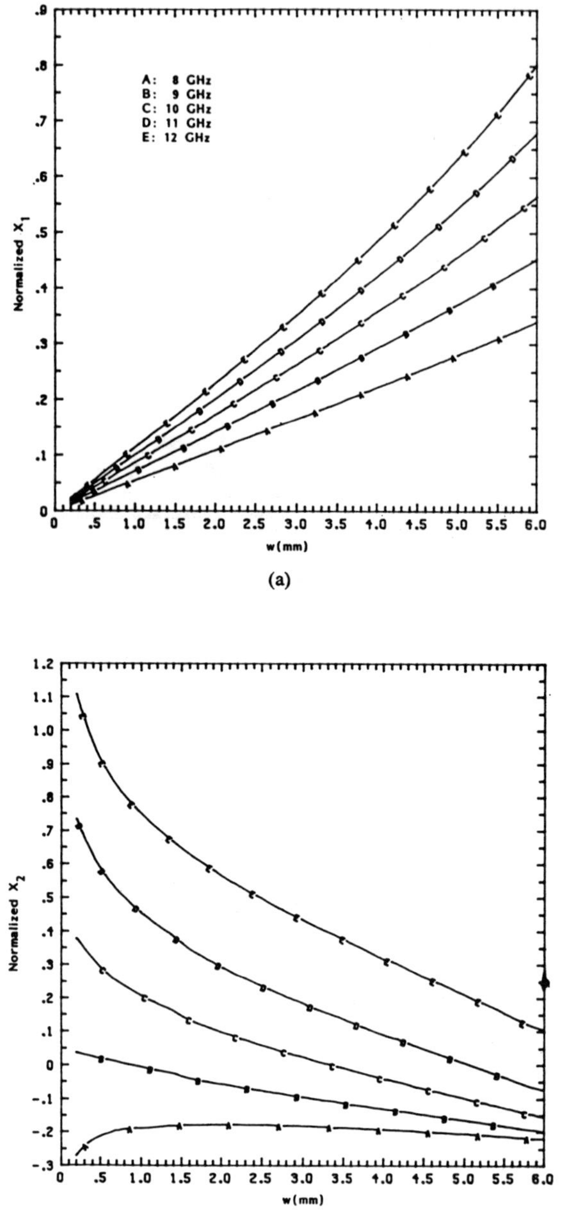

Fig. 8. Normalized reactance of equivalent circuit of an E-plane strip in an X-band waveguide versus strip width for different values of frequency. $\epsilon_1 = \epsilon_3 = 1$, $\epsilon_2 = 2.1$, $h_1 = 11.57$ mm, $h_2 = 1$ mm, and $d = 7.37$ mm. (a) Normalized X_1 versus strip width. (b) Normalized X_2 versus strip width.

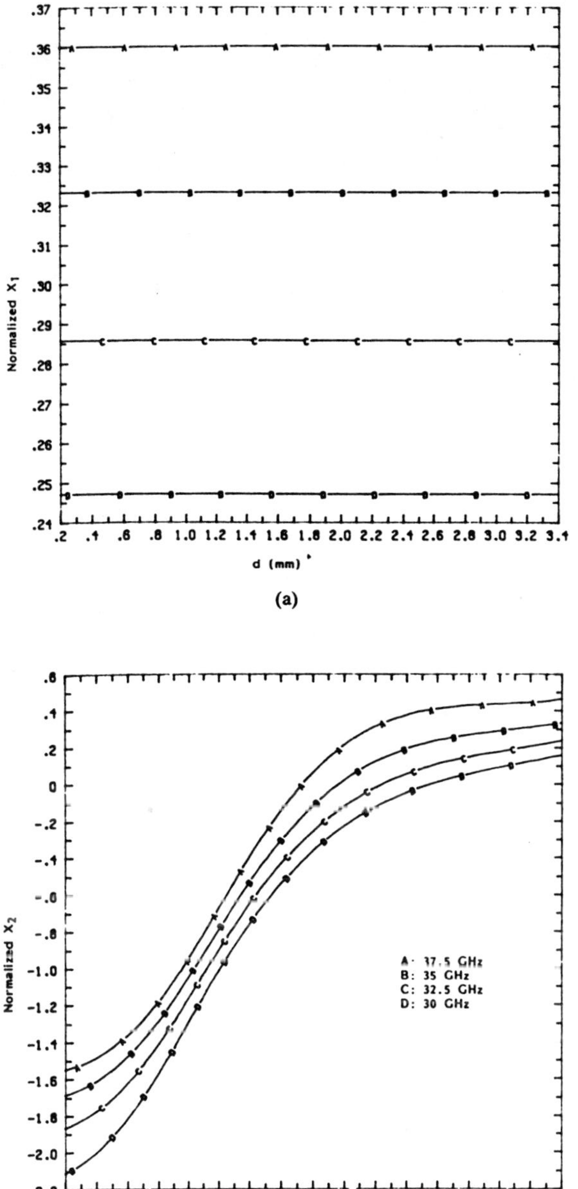

Fig. 9. Normalized reactance of equivalent circuit of an E-plane strip in a Ka-band waveguide versus strip height for different values of frequency. $\epsilon_1 = \epsilon_3 = 1$, $\epsilon_2 = 2.2$, $h_1 = 3.43$ mm, $h_2 = 0.254$ m, and $w = 1$ mm. (a) Normalized X_1 versus strip height. (b) Normalized X_2 versus strip height.

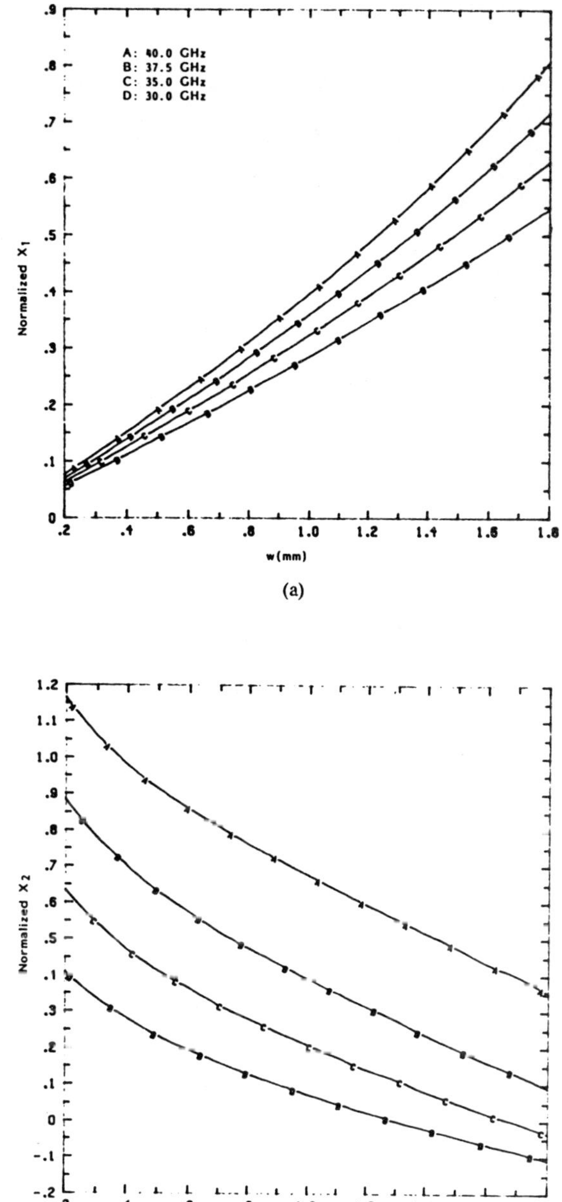

Fig. 10. Normalized reactance of equivalent circuit of an E-plane strip in a Ka-band waveguide versus strip width for different values of frequency. $\epsilon_1 = \epsilon_3 = 1$, $\epsilon_2 = 2.2$, $h_1 = 3.43$ mm, $h_2 = 0.254$ mm, and $d = 2.45$. (a) Normalized X_1 versus strip width. (b) Normalized X_2 versus strip width.

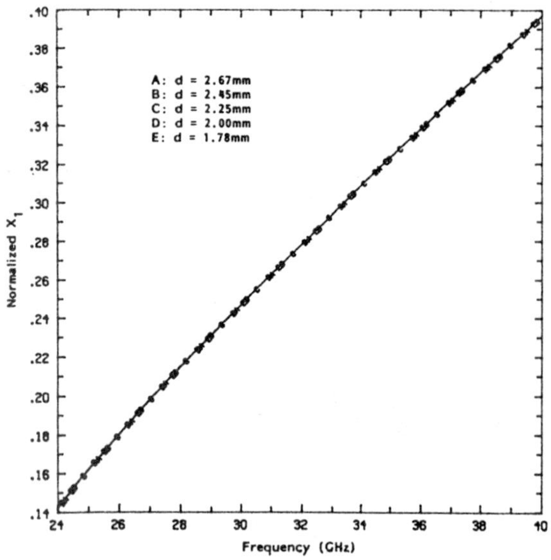

(a)

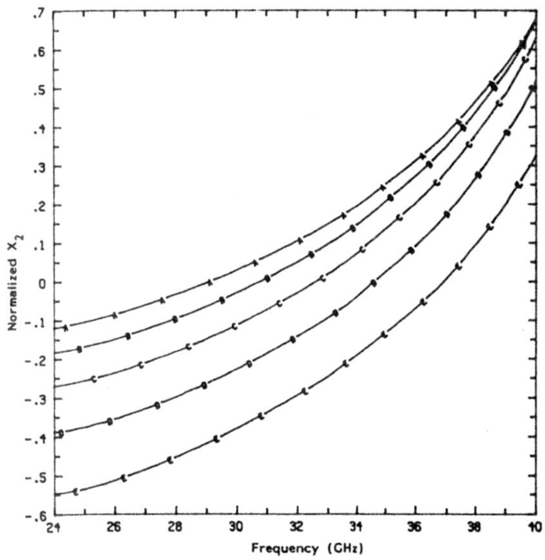

(b)

Fig. 11. Normalized reactance of equivalent circuit of an *E*-plane strip in a *Ka*-band waveguide versus frequency for different strip heights. $\epsilon_1 = \epsilon_3 = 1$, $\phi e_2 = 2.2$, $h_1 = 3.43$ mm, $h_2 = 0.254$ mm, and $w = 1$ mm. (a) Normalized X_1 versus frequency. (b) Normalized X_2 versus frequency.

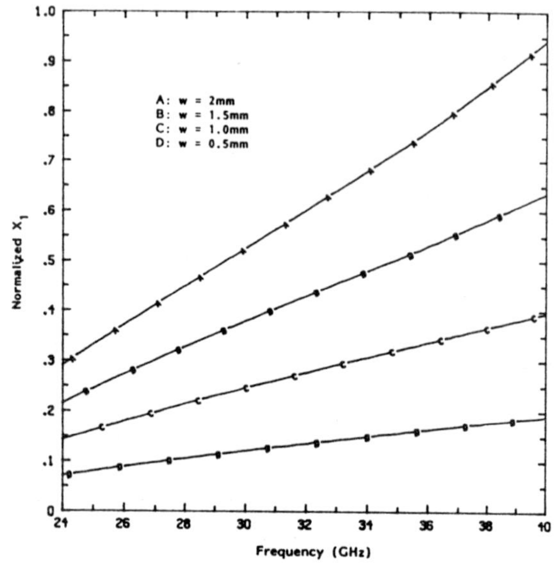

(a)

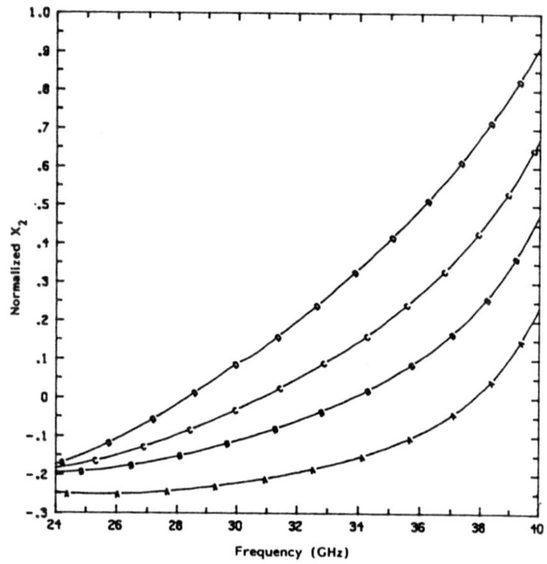

(b)

Fig. 12. Normalized reactances of equivalent circuit of an *E*-plane strip in a *Ka*-band waveguide versus frequency for different strip widths. $\epsilon_1 = \epsilon_3 = 1$, $\epsilon_2 = 2.2$, $h_1 = 3.43$ mm, $h_2 = 0.254$ mm, and $d = 2.45$ mm. (a) Normalized X_1 versus frequency. (b) Normalized X_2 versus frequency.

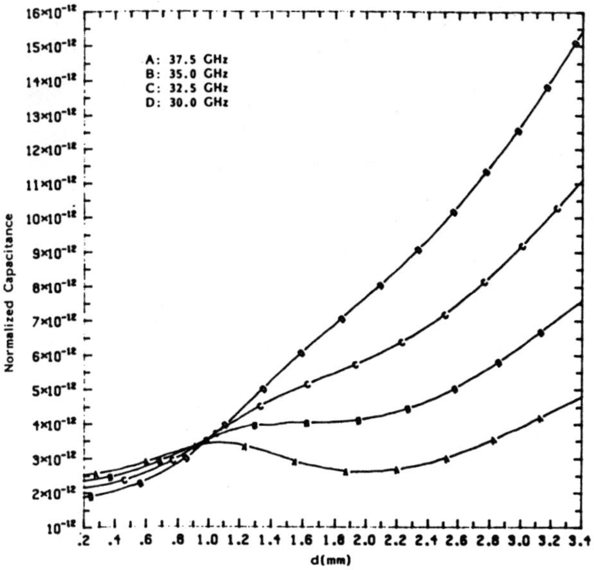

(a)

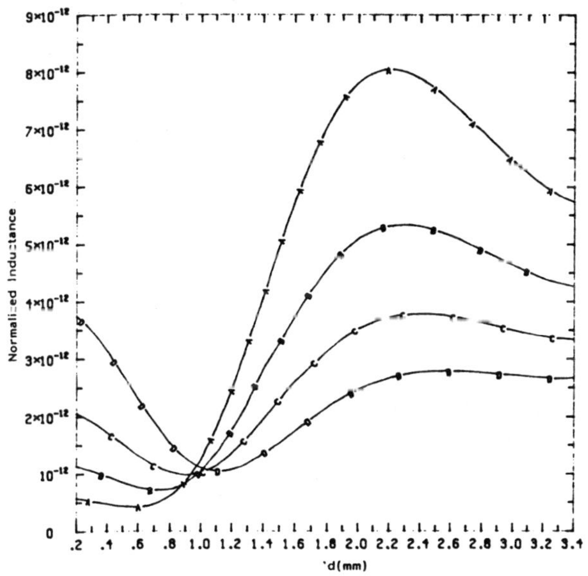

(b)

Fig. 13. Normalized equivalent circuit element values C and L of a narrow E-plane strip as a function of strip height. $\epsilon_1 = \epsilon_3 = 1$, $\epsilon_2 = 2.2$, $h_1 = 3.43$ mm, $h_2 = 0.254$ mm, and $w = 1$ mm. (a) Normalized C versus strip height. (b) Normalized L versus strip height.

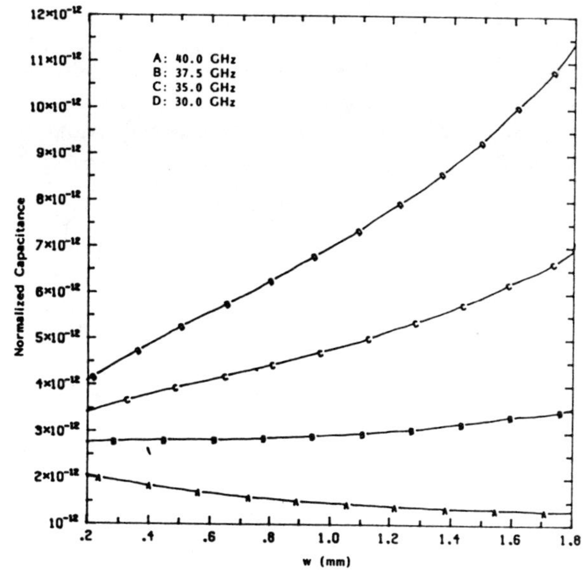

(a)

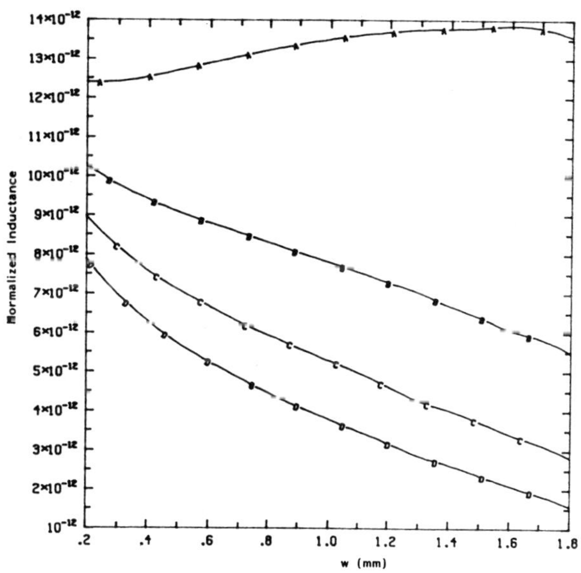

(b)

Fig. 14. Normalized equivalent circuit element values C and L of a narrow E-plane strip as a function of strip width. $\epsilon_1 = \epsilon_3 = 1$, $\epsilon_2 = 2.2$, $h_1 = 3.43$ mm, $h_2 = 0.254$ mm, and $d = 2.45$ mm. (a) Normalized C versus strip width. (b) Normalized L versus strip width.

creases. In the other region, C decreases and L increases as frequency becomes higher. It is conjectured that this phenomenon is related to the field distributions and to the use of the equivalent circuit chosen here. Fig. 14 shows the variation of C and L with w for different values of frequency.

IV. Conclusions

A new analytical technique has been developed here which can be used for characterizations of the scattering phenomena of planar E-plane obstacles. The numerical results for a special case agree well with experimental and published data. The curves of normalized input admittance and equivalent circuit element values are presented for a number of different parameters of this structure. This technique is believed to be useful in the design of microwave filters and other planar circuit components.

Appendix
Closed-Form Expression of Green's Functions in the Spectral Domain

According to [9], the TM-to-X (LSM) and TE-to-X (LSE) equivalent transmission lines for the E-plane strip described in Fig. 1 can be drawn in Fig. 15. Here,

$$\gamma_i = \sqrt{\alpha_n^2 + \beta^2 - \epsilon_i k^2} \tag{A1}$$

$$Z_{\mathrm{TM}i} = \frac{\gamma_i}{j\omega\epsilon_0\epsilon_i} \tag{A2a}$$

$$Z_{\mathrm{TE}i} = -\frac{j\omega\mu}{\gamma_i} \qquad i=1,2,3, \quad n=0,1,2,3\cdots. \tag{A2b}$$

The driving point input impedance Z^e for the TM equivalent circuit is given by

$$Z^e = \frac{1}{y_2^e + y_{2\gamma}^e} \tag{A3a}$$

where

$$y_{2\gamma}^e = y_{\mathrm{TM}3} \cdot \coth \gamma_3 h_3 \tag{A3b}$$

$$y_2^e = y_{\mathrm{TM}2} \frac{y_{\mathrm{TM}2} + y_1^e \coth \gamma_2 h_2}{y_1^e + y_{\mathrm{TM}2} \coth \gamma_2 h_2} \tag{A3c}$$

$$y_1^e = y_{\mathrm{TM}1} \coth \gamma_1 h_1. \tag{A3d}$$

Notice that y_1^e and y_2^e are input admittances looking left at $x = h_1$ and $x = h_1 + h_2$, respectively, while $y_{2\gamma}^e$ is the one looking right at $x = h_1 + h_2$. Similar equations can be written for the TE equivalent circuit

$$Z^h = \frac{1}{y_2^h + y_{2\gamma}^h} \tag{A4a}$$

$$y_{2\gamma}^h = y_{\mathrm{TE}3} \coth \gamma_3 h_3 \tag{A4b}$$

$$y_1^h = y_{\mathrm{TE}1} \coth \gamma_1 h_1 \tag{A4c}$$

$$y_2^h = y_{\mathrm{TE}2} \frac{y_{\mathrm{TE}2} + y_1^h \coth \gamma_2 h_2}{y_1^h + y_{\mathrm{TE}2} \coth \gamma_2 h_2}. \tag{A4d}$$

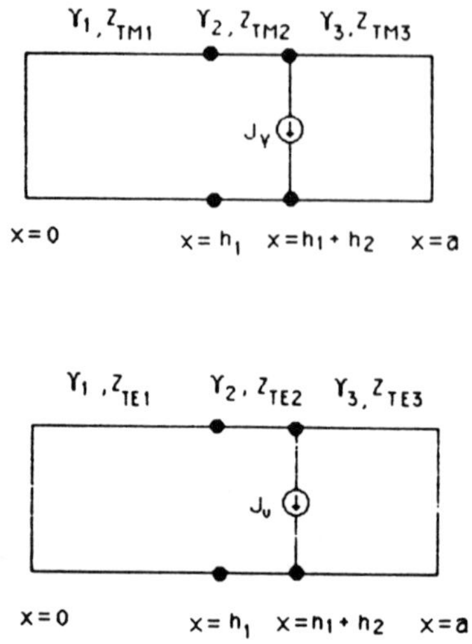

Fig. 15. The TM and TE equivalent transmission lines for the E-plant strip.

Finally, the Green's functions in the spectral domain can be represented by Z^h and Z^e as follows [8]:

$$\tilde{G}_{yy} = Z^e N_z^2 + Z^h N_y^2 \tag{A5a}$$

$$\tilde{G}_{yz} = \tilde{G}_{zy} = (Z^e - Z^h) N_z N_y \tag{A5b}$$

$$\tilde{G}_{zz} = Z^h N_z^2 + Z^e N_y^2 \tag{A5c}$$

where

$$N_z = \frac{\beta}{\sqrt{\alpha_n^2 + \beta^2}} \tag{A6a}$$

$$N_y = \frac{\alpha_n}{\sqrt{\alpha_n^2 + \beta^2}}. \tag{A6b}$$

Note that the denominators of Z^e and Z^h are the eigenvalue equations of the LSE and LSM modes.

References

[1] P. J. Meier, "Two new integrated circuit media with special advantages at millimeter wavelengths," in *IEEE-GMTT Int. Microwave Symp. Dig.*, pp. 221–223, 1972.
[2] P. J. Meier, "Integrated fin-line millimeter components," *IEEE Trans. Microwave Theory Tech.*, vol. MTT-22, pp. 1209–1216, Dec. 1974.
[3] Y. Konishi, K. Uenakada, and N. Hoshino, "The design of planar circuit mounted in waveguide and the application to low-noise 12-GHz converter," in *IEEE S-MTT Int. Microwave Symp. Dig.*, pp. 168–170, 1974.
[4] Y. Konishi and K. Uenakada, "The design of a bandpass filter with inductive strip-planar circuit mounted in waveguide," *IEEE Trans. Microwave Theory Tech.*, vol. MTT-22, pp. 869–873, Oct. 1974.
[5] Y. Konishi, "Planar circuit mounted in waveguide used as a downconverter," *IEEE Trans. Microwave Theory and Tech.*, vol. MTT-26, pp. 716–719, Oct. 1978.
[6] Y. C. Shih, T. Itoh, and L. Bui, "Computer-aided design of millimeter-wave E-plane filter," *IEEE Trans. Microwave Theory Tech.*, vol. MTT-31, pp. 135–141, Feb. 1983.
[7] K. Chang and P. J. Khan, "Equivalent circuit of a narrow axial strip in waveguide," *IEEE Trans. Microwave Theory Tech.*, vol. MTT-24, pp. 611–615, Sept. 1976.
[8] R. Collin, *Field Theory of Guided Waves.* New York: McGraw-Hill, 1960, pp. 224–232.

[9] T. Itoh, "Spectral domain immittance approach for dispersion characteristics of generalized printed transmission lines," *IEEE Trans. Microwave Theory Tech.*, vol. MTT-28, pp. 733–736, July 1980.

4.2.1

A COMBINED EXPERIMENTAL AND THEORETICAL CHARACTERISATION OF DISCONTINUITIES IN UNILATERAL FINLINES

Adalbert Beyer, Dietmar Köther and Ingo Wolff

Department of Electrical Engineering, Duisburg University,
Bismarckstraße 81, D-4100 Duisburg 1, FRG

Abstract

An optimization procedure for characterizing finline discontinuities has been developed using the generalized scattering matrix formulation for steps in finline technique. The general waveguide representation of a two-taper finline section with a step is investigated first. The scattering matrix representation of the step is deduced from a combined measurement-numerical optimization procedure. Experimental results will verify the feasibility of the method.

Introduction

Many techniques are available to calculate the effects of a discontinuity in an unilateral finline /3,4,5,11/. Among these, the scattering matrix solution is perhaps the most commonly used. One of the scattering matrix methods, for analysing discontinuities, involves the determination of equivalent circuits of the discontinuities using the transmission properties of the considered finlines.

Up to now some authors have performed calculations on equivalent circuits of discontinuities in finlines such as: steps, small capacitive strips and inductive notches /3,6,7/, but their theoretical results have been derived for an idealized cross sectional model of the finlines /8/, i.e. the second order effects such as e.g. the effects of the finite metallization thickness of the fins and the effects of the slits in the finline mounts have been neglected; also the losses caused by the no-ideal dielectric substrate and the finite conductibility of the metallization have not been taken into account.

Theoretical Background

In this paper, a concept for an efficient calculation method of the scattering matrix of finline discontinuities will be presented, taking into account the above-mentioned second order effects.

A simple step between two finitely long, uniform finlines (1) and (2) which are terminated by smooth tapers (T1) and (T2) on the left- and on the right-hand sight is considered (Fig. 1).

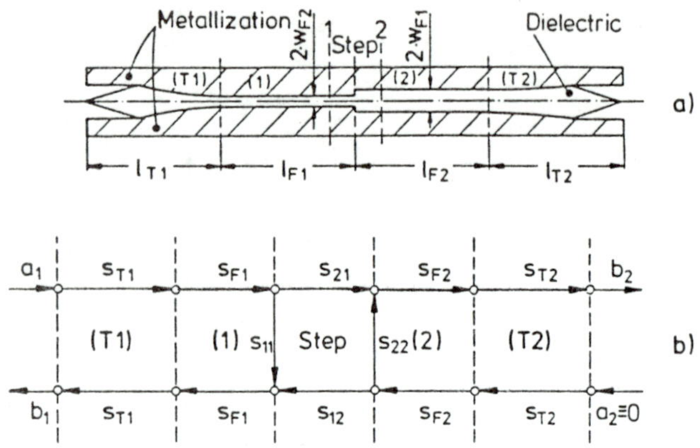

Fig. 1: Longitudinal section of a finline structure with a step.
a) The considered finline structure.
b) The appertaining signal flow diagram.

For the step shown in Fig. 1 the reciprocity theorem is valid:

$$|S_{12}| = |S_{21}| \quad , (1)$$

$$\varphi_{12} = \varphi_{21} \quad . (2)$$

From the symmetry it follows that:

$$|S_{11}| = |S_{22}| \quad . (3)$$

For the finlines yield:

$$S_{F1} = \exp(-a_{F1} l_{F1})\exp(-j\beta_{F1} l_{F1}) \quad , (4)$$

$$S_{F2} = \exp(-a_{F2} l_{F2})\exp(-j\beta_{F2} l_{F2}) \quad . (5)$$

From the signal flow diagram shown in Fig. 1b) the transmission loss of the complete finline structure shown in Fig. 1a) follows as:

$$a_1 = S_{T1} S_{F1} S_{12} S_{F2} S_{T2} \quad (6)$$

and for the reflection loss:

$$r_1 = S_{T1}^2 S_{F1}^2 S_{11} \quad (7)$$

with the transmission losses S_{T1} and S_{T2} of the two tapers (T1) and (T2).

Further, let measurements of the reflection- as well as of the transmission-coefficients have been made for the step feeded at port (1) and port (2) /1/. Then the customary definition of the scattering parameters which describe the step, leads to the equation for the absolute magnitudes of these quantities; e.g. for S_{11} yields:

$$|S_{11}| = \frac{|r_1|_M}{|S_{T1}|^2 |S_{F1}|^2} \qquad (8)$$

with r_{1M} the measured value of the magnitude of the reflection loss r_1 from (7). Provided that the transmission properties of the applied impedance transformers S_{T1}, S_{T2} between the rectangular waveguide and the homogeneous finline sections and the transmission properties S_{F1}, S_{F2} of the homogeneous finline sections are known, the magnitudes of the other elements of the scattering matrix of the step can similarily be calculated from the measurements.

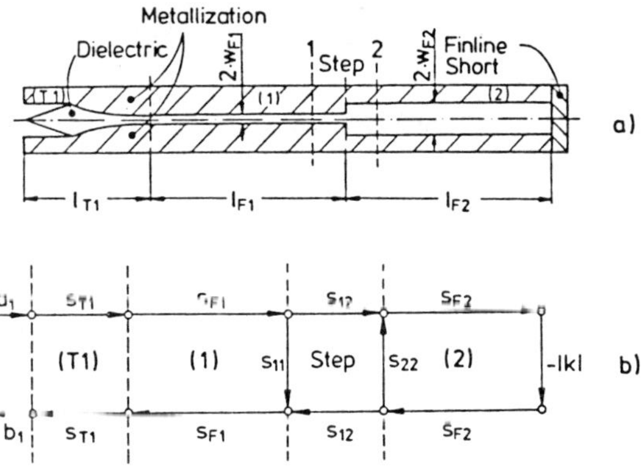

Fig. 2: Longitudinal section of a finline structure with a step and a short circuit.
a) The considered finline structure.
b) The appertaining signal flow diagram.

Now, the finline structure shown in Fig. 2a) will be inspected. The reflection loss of the arrangement follows from the appertaining signal flow diagram (Fig. 2b)):

$$r_2' = S_{T1}^2 S_{F1}^2 \left(S_{11} - \frac{|k_2| S_{12}^2 S_{F2}^2}{1 + |k_2| S_{22} S_{F2}^2}\right) \qquad (9)$$

whereby $|k_2| \leq 1$ describes the properties of the finline short.

The phase angles φ_{11}, φ_{12} and φ_{22} of the considered step scattering parameters will be calculated from the measurements of $|r_2|$ by the aid of an optimization criterion. Hereby the following equations are valid:

$$E_1 = |r_2'|_C^2 - |r_2'|_M^2 \quad , (10)$$

$$E_2 = |r_2''|_C^2 \quad |r_2''|_M^2 \qquad (11)$$

with E_1 and E_2 two error functions which must fulfill the condition:

$$E = E_1^2 + E_2^2 = \text{min. value} \quad , (12)$$

whereby $|r_2'|$ is defined by eq. (9) and $|r_2''|$ is the reflection loss of the structure shown in Fig. 2 inspected by the right-hand side. The indeces C and M stand for the calculated and the measured values of $|r_2|$, respectively.

Numerical and Experimental Results

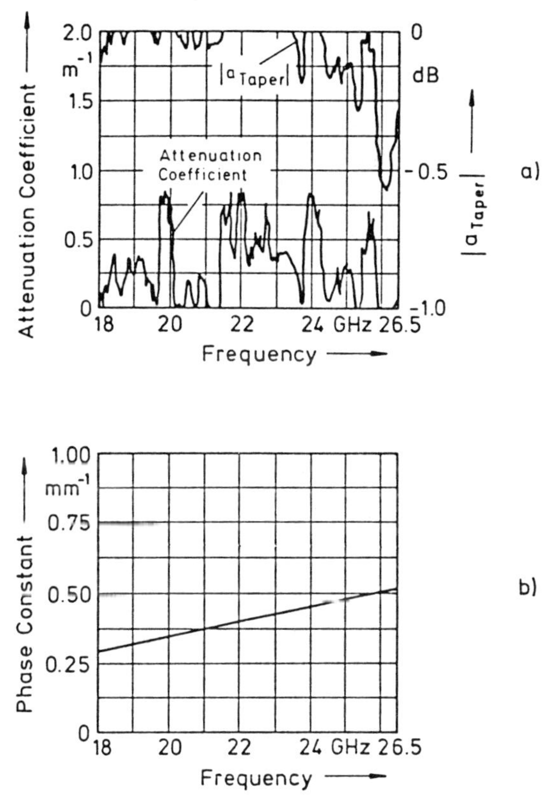

Fig. 3: Transmission properties of a homogeneous finline section with the length l_{F1}=240mm and the slotwidth $2W_F$=100µm and a single taper section.
a) The attenuation coefficient of the homogeneous finline section and the damping of a single taper.
b) The phase constant of the homogeneous finline section.

To calculate the elements of the scattering matrix the properties of the homogeneous finline and of the tapers are needed. Fig. 3a) shows the measured attenuation coefficient for a homogeneous finline with the length l_{F1}=240mm and a slot width of $2W_F$=100μm which is determined in dependence on the frequency. In Fig. 3b) the calculated values of the phase constant are shown for the same finline. The calculations took into account the finite metallization thickness and the influence of the slits in the finline mounts /2/. Fig. 3a) shows the calculated attenuation for a taper with a length of l_T=30mm which is used between the rectangular waveguide and the above described finline structure. Using the results in Fig. 3 and the corresponding results of another finline slot width of $2W_F$=200μm and the parameters of the used tapers the magnitudes of the scattering parameters of the finline step can be calculated as a function of the frequency. In Fig. 4 these parameters are shown in dependence on the frequency.

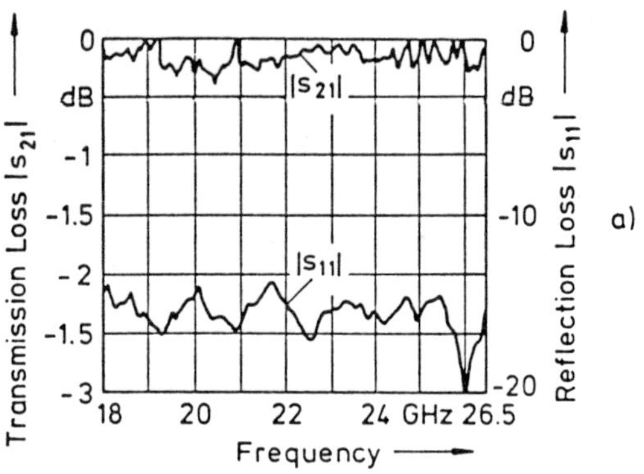

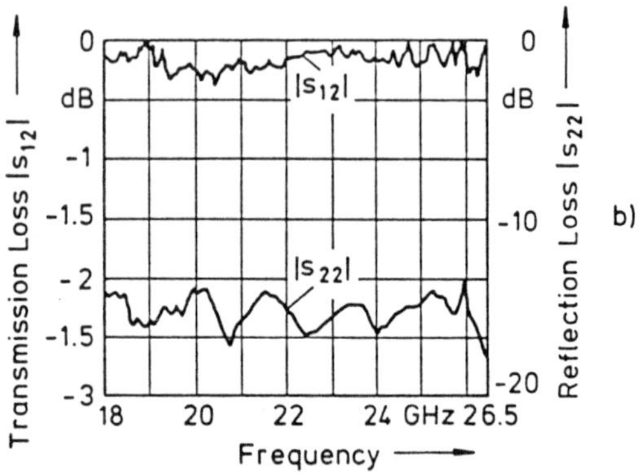

Fig. 4a)-b): The magnitudes of the scattering parameters of a finline step with the slot widths $2W_{F1}$=100μm and $2W_{F2}$=200μm.

Fig. 5 shows the measured properties of the used finline short, which are used for the optimization process to determine the phase angles. It can be realized that the finline short has a strong frequency dependence.

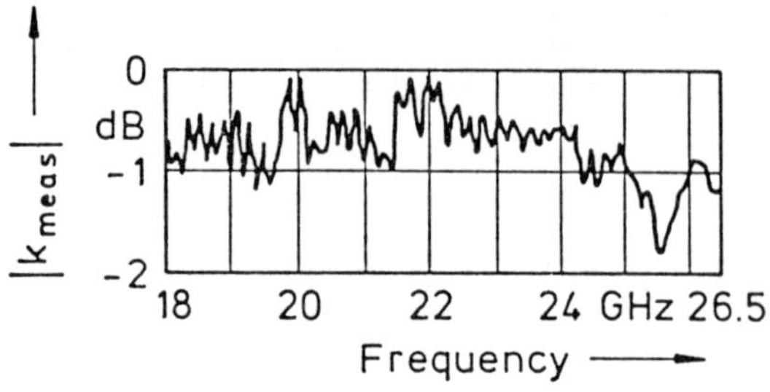

Fig. 5: The reflection loss of the used finline short (see Fig. 2).

The optimized values of the phase angles of the finline step due to the quantities in Fig. 4 are shown in Fig. 6.

In a similar way discontinuities such as capacitive strips or inductive notches can be investigated and filter structures can be developed. In the continuation of these activities it is scheduled to present and discuss some discontinuities and filter structures.

The Measurement Set-Up

For the design of high-quality millimetre-wave filters in finline technique the exact knowledge of the equivalent circuit elements at the operating frequency is needed /9,10/. Therefore, a measurement technique was developed for determining the transmission properties of the considered discontinuities. The measurement set-up is an automatic system which is steered by a small desk-top computer. The computer controls the measuring frequency, the input power and it stores the measured values. Additionally, the complete optimizing procedure for the evaluation of the measuring results is programmed on the same computer. It should be mentioned that the considered finline discontinuities are lossy two ports. Consequently, all the phase angles of the scattering parameters of the investigated discontinuity must be optimized separately.

Conclusions

A design method is described which allows a more realistic description of finline discontinuities. In contrast to other methods for equivalent circuits of the finline discontinuities, the present technique takes into account the influence on the transmission properties of both the finite thickness of the fins and the finite longitudinal slit in the waveguide mount as well as the losses of the finline discontinuities.

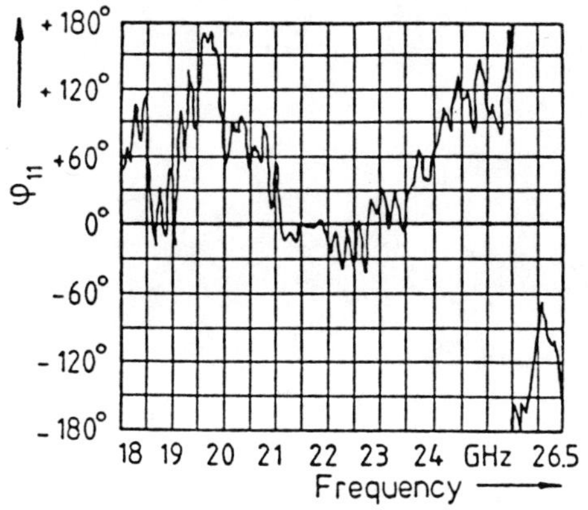

a)

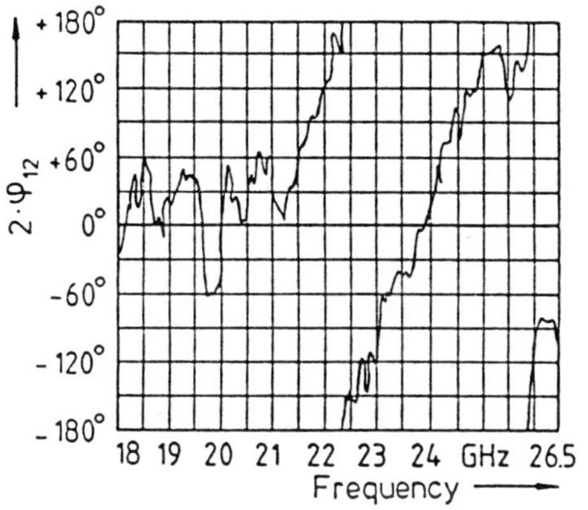

b)

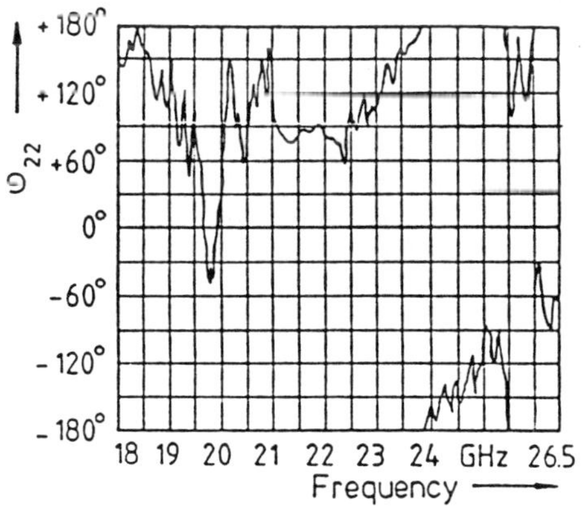

c)

Fig. 6: The phase angles of a finline step with the slot widths of $2W_{F1}$=100μm and $2W_{F2}$=200μm.
a) The phase angle φ_{11}.
b) The phase angle $2\varphi_{12}$.
c) The phase angle φ_{22}.

References

1) Auda, H. and Harrington, R.: A moment solution for wave guide junctions. IEEE Transactions on Microwave Theory and Techniques, Vol. MTT-31, No. 7, July 1983, pp. 515-519.

2) Beyer, A. and Wolff, I.: A solution of the earthed fin line with finite metallization thickness. 1980 IEEE MTT-S International Symposium Digest, Washington, D.C., pp. 258-260.

3) El Hennawy, H. and Schünemann, K.: Hybrid finline matching structures. IEEE Transactions on Microwave Theory and Techniques, Vol. MTT-30, December 1982, pp. 2132-2138.

4) El Hennawy, H. and Schünemann, K.: Impedance transformation in fin-lines. IEE Proc. Part H, No. 6, Microwaves, Optics and Antennas, Vol.129, December 1982, pp. 342-350.

5) Helard, M., Citerne, J., Picon, O. and Fouad Hanna, V.: Exact calculations of scattering parameters of a step slot width discontinuity in an unilateral fin-line. Electronics Letters, 7th July 1983, Vol. 19, No. 14, pp. 537-539.

6) Konishi, Y. and Uenakada, K.: The design of a bandpass filter with inductive strip-planar circuits mounted in waveguide. IEEE Transactions on Microwave Theory and Techniques, Vol. MTT-22, October 1974, pp. 869-873.

7) Meier, P.J.: Integrated fin-line millimeter components. IEEE Transaction on Microwave Theory and Techniques, Vol. MTT-22, December 1974, pp. 1209-1216.

8) Schmidt, L.-P.: A comprehensive analysis of quasiplanar waveguides for millimeter-wave applications. Proc. of the 11th European Microwave Conference, Amsterdam (NL), 1981, pp. 315-320.

9) Tajima, Y. and Sawayama, Y.: Design and analysis of a waveguide-sandwich microwave filter. IEEE Transaction on Microwave Theory and Techniques, Vol. MTT-22, September 1974, pp. 839-841.

10) Taub, J.J.: Design of minimum loss band-pass filters. Microwave Journal, Vol. 6, 1963, pp. 67-76.

11) Webb, K. J.: Mittra, R.: A variational solution of the fin-line discontinuity problem. Proc. of the 15th European Microwave Conference, Paris (F), 1985, pp. 311-316.

4.2.2

AN IMPROVED MODEL FOR SHORT- AND OPEN-CIRCUITED SERIES STUBS IN FIN LINES

Miles BURTON and Wolfgang J.R. HOEFER

Department of Electrical Engineering
University of Ottawa
Ottawa, Ontario, Canada K1N 6N5

ABSTRACT

This paper presents empirical closed-form expressions for the scattering parameters of series stubs in single-ridge, unilateral fin line with narrow gap width. These expressions show the effect of fin gap width, stub length, and frequency and are suitable for computer-aided design.

INTRODUCTION

The series stub has been a popular circuit element in the design of oscillators and narrow-band filters in fin line [1], [2]. However, the available information is neither accurate nor versatile enough for most practical design problems. We therefore saw the need for an improved model which incorporated the effects of the frequency, substrate thickness, fin line gap width and stub length.

Hence, this paper significantly increases the available design information for short- and open-circuited series stubs in fin line.

The stubs are modeled by an equivalent circuit which takes into account the field disturbances at the branching points and at the far end. Expressions for the circuit elements have been derived from measurements. They are valid over one waveguide band and are suitable for unilateral fin lines on low permittivity substrates (ε_r=2.22) with narrow gap widths (d/b << 1).

THE STUB MODEL

A fin line stub can be described either by its S-parameters or by an equivalent circuit with lumped elements. The latter has two advantages. Firstly, it directly depicts the circuit behaviour of the discontinuity. Secondly, it separately describes the various field disturbances at the branching points and at the far end of the stub.

These effects were incorporated into the equivalent transmission line circuits shown in Fig.1(a) and (b). Then, measurements were made with an automatic network analyzer to determine the dependence of the equivalent circuit parameters on frequency, stub length, and fin line gap width.

Each measured stub was centered between two identical waveguide to fin line tapers which were etched on 0.76 mm thick Rt/duroid (ε_r=2.22), and mounted in WR-90 waveguide. (See Fig.2). All stubs had a standard width of 1 mm.

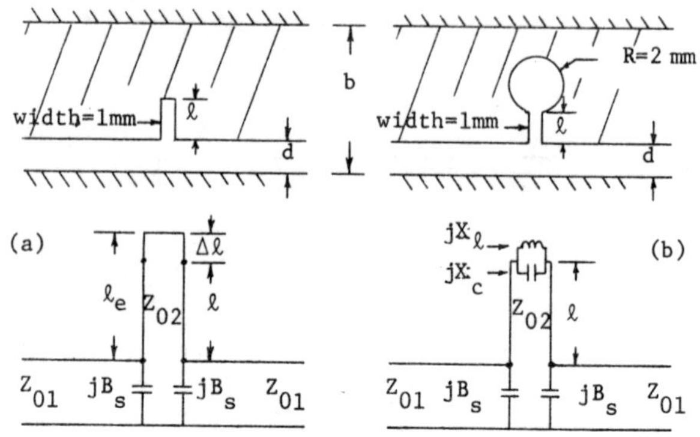

Fig. 1 - Equivalent transmission line circuits for the short-(a) and open-(b) circuited series stubs in unilateral single-ridge fin line.

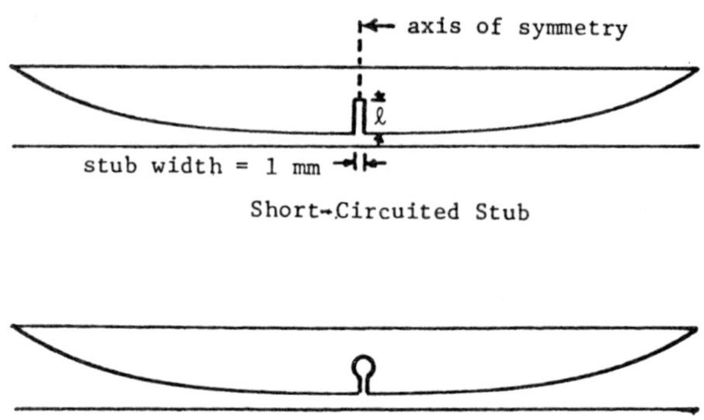

Fig. 2 - Metalization patterns used to measure the scattering parameters of the stub discontinuities in unilateral fin line (WR-90) (ε_r=2.22, substrate thickness s=0.762 mm.)

DE-EMBEDDING THE STUBS

To de-embed the stubs, the scattering parameters of the tapers were found by a method used to characterize connections to a transmission medium in which no standard loads are available [3], [4]. In this method transmission and reflection measurements are made on two similar two-ports. Each is composed of two identical connectors (waveguide to fin line tapers in our case) separated by a length of transmission medium of known electrical length.

The intervening transmission medium of one two-port is approximately one quarter wavelength shorter than that of the other two-port. The method yields scattering information for the taper in the form of S_{11}, S_{22} and the product $S_{21} \cdot S_{12}$.

In [3, 4], a quasi-closed form expression for each of S_{11}, S_{22}, and $S_{21} \cdot S_{12}$ are derived. Higher order terms in S_{11} and S_{22} also appear in these expressions but their effects are ignored by assuming $|S_{11}|, |S_{22}| \simeq 0$. This assumption restricts the application of the method to cases involving nearly reflectionless connectors.

We extend the method to cases involving reflective connections by using an iterative computation. Here, the assumption above is the starting point in an algorithm that merely repeatedly computes the same expressions mentioned above. Each iteration after the first invokes the higher order terms by using the previous results for S_{11} and S_{22}. The iterations are stopped when a test for convergence is satisfied.

The stubs were de-embedded by using the explicit formulas of Kruppe and Sodomsky [5] originally intended to correct for the errors in imperfect network analyser test-sets. The test-set error terms in these formulae were replaced by the scattering parameters of the two identical fin line to waveguide transitions.

DERIVING THE EMPIRICAL EXPRESSIONS

The scattering parameters of an ideal series stub in a transmission-line (i.e., one with no losses and zero width and hence no branching point and end effects) are

$$S_{11s} = \frac{z_{in}}{2 + z_{in}} \tag{1}$$

$$S_{21s} = \frac{2}{2 + z_{in}} \tag{2}$$

where z_{in} is the impedance seen looking into the stub normalized to the characteristic impedance of the main transmission line.

$$z_{in} = z \left(\frac{z_t + \tanh \gamma \ell_e}{1 + z_t \tanh \gamma \ell_e}\right) \tag{3}$$

where z is the characteristic impedance of the stub divided by the characteristic impedance of the main transmission line.

$\gamma = \alpha + j\beta$ is the propagation constant in the stub.

ℓ_e = length of the stub in the equivalent circuit.

z_t is the terminating impedance of the stub divided by the characteristic impedance of the stub.

When a shunt admittance jB_s is added as shown in Figs. 1(a) and 1(b) the overall scattering parameters can be found from Mason's non-touching loop rule [6]

$$S_{11} = S_{11b} + \frac{S_{21b}^2 (S_{11s} - \Delta S_s S_{11b})}{1 - 2S_{11s}S_{11b} + \Delta S_s S_{11b}^2} \tag{4}$$

$$S_{21} = \frac{S_{21b}^2 S_{21s}}{1 - 2S_{11s}S_{11b} + \Delta S_s S_{11b}^2} \tag{5}$$

where $S_{11b} = \frac{-jb_s}{2+jb_s}$ is the reflection coefficient of a shunt admittance $jb_s = jB_s Z_{01}$ (normalized to the characteristic admittance of the main transmission line), and $S_{21b} = 1 + S_{11b}$ is the transmission coefficient of this shunt admittance, and $\Delta S_s = S_{11s}^2 - S_{21s}^2$. The following formulae were obtained by optimizing the fit of equations (4) and (5) to the experimental data for series stubs. For convenience of scaling, all variables were normalized to the substrate thickness s.

$$b_s = 7.56 \, (s/\lambda) \exp\,(-0.3117 \, d/s) \tag{6}$$

$$z = 2.0 \, (1.524 \, d/s)^{-0.594} \tag{7}$$

$$\alpha = (0.08 \, s/\lambda) \text{ Nepers/mm} \tag{8}$$

Short-circuited stubs are terminated by an excess length of line $\Delta\ell$

$$\Delta\ell/s = 0.1082 \, (d/s)^2 \, [1 - \exp(-0.762 \, \ell/s)] \tag{9}$$

Open circuited stubs are terminated by a parallel combination of an inductive and a capacitive reactance, x_ℓ and x_c (See Fig. 1(b)). These reactances are normalized to the characteristic impedance of the stub and given by (10) and (11).

$$x_\ell = (s/\lambda) \, (84.12 - 2.474 \, \ell/s) \tag{10}$$

$$1/x_c = (s/\lambda)(30.93 - 7.422 \, \ell/s)(0.026 - 0.17 \, d/s) \tag{11}$$

(10) and (11) can be used to find the normalized terminating impedance

$$z_t = \frac{x_\ell x_c}{x_\ell + x_c} \tag{12}$$

Expressions (6) to (11) are valid in the following ranges:

$$1.6 < \ell/s < 5.3 \tag{13}$$

$$0.6 < d/s < 2.7 \tag{14}$$

$$32 < \lambda/s < 49 \tag{15}$$

When the error is computed as the magnitude of the difference between the fitted and measured scattering parameters, the standard deviation for short circuit fit is 0.03 and that for the open circuit fit is 0.045. The reference planes of the scattering parameters are coincident with the stub edges.

The restriction to constant stub width allows simple expressions and should have little effect on the flexibility of practical designs. Also, expressions (6) to (11) do not depend on the waveguide dimensions because small gap ratios ($d/b \ll 1$) and narrow substrates ($s/a \ll 1$) are assumed.

The expressions operate on normalized dimensions to allow convenient scaling to frequencies higher than x-band. Losses are expected to increase linearly with frequency as indicated by (8).

The wavelength in a stub can be computed from the following formula for wavelength in single ridge unilateral fin line on substrates with $\varepsilon_r = 2.22$:

$$\lambda_g/s = a_1 + a_2(s/\lambda) + a_3(s/\lambda)^2 + a_4(s/\lambda)^3 \tag{16}$$

where $a_1 = 150.4 + 24.29\ (d/s)$ (17)

$a_2 = -9092 - 2098\ (d/s)$ (18)

$a_3 = 236500 + 64900\ (d/s)$ (19)

$a_4 = -2269000 - 690800\ (d/s)$ (20)

This formula is accurate to ±1 percent over the same ranges as (14) and (15) and for

$$s/a < 0.1$$

The wavelength in the stub is the same as the wavelength in a single ridge unilateral fin line when the gap is equal to half the stub width.

CONCLUSION

S-parameter measurements were made to derive expressions for the parameters of open- and short-circuited series stubs in single ridge unilateral fin line. The expressions include losses and show the effect of fin line gap, stub length, and frequency and are valid for a complete waveguide bandwidth, for a specified range of low fin line gap ratios, and for a specified range of narrow substrate thicknesses.

REFERENCES

[1] Hofmann, H.: "MM-Wave Gunn Oscillator with Distributed Feedback Fin-Line Circuit", Proc. 1980 IEEE MTT-S International Microwave Conference, pp. 59-61.

[2] Meier, P.J.: "Wideband Subharmonically Pumped W-Band Mixer in Single-Ridge Fin-Line", 1982 IEEE MTT-S Digest, pp. 201-203.

[3] Souza, J.R. and Talboys, E.C.: "S-Parameter Characterization of Coaxial to Microstrip Transition", IEEE Proc., Vol. 129, Pt. H., No. 1, Feb. 1982, pp. 37-40.

[4] Gupta, K.C., Garg, R., and Chadha, R.: "Computer-Aided Design of Microwave Circuits", Artech House, 1981, pp. 319-321.

[5] Kruppe, W. and Sodomsky, K.F.: "An Explicit Solution for the Scattering Parameters of a Linear Two-Port Measured with an Imperfect Test-Set", IEEE Trans. MTT, Jan. 1971, pp. 122-123.

[6] Mason, S.J.: "Feedback Theory-Further Properties of Signal Flow Graphs", Proc. IRE, 1956, 44, pp. 920-926.

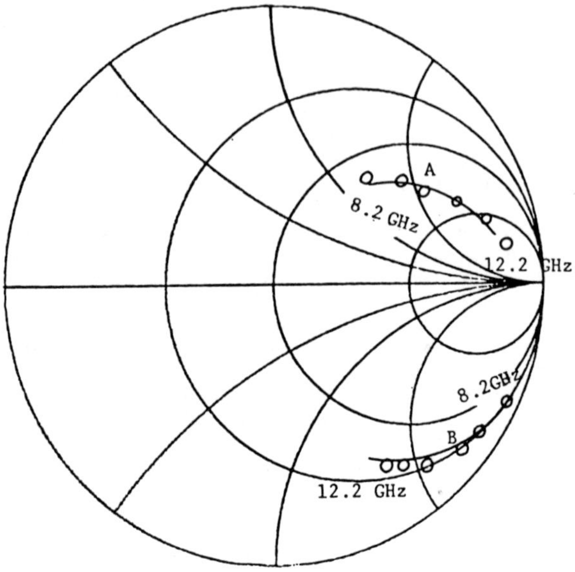

Fig. 3 Comparison of measured data (circles) to fitted curves (continuous lines) for (A) a short circuited stub and (B) an open circuited stub. Each stub is 4 mm long and in a single ridge unilateral fin line with fin line gap of 0.85 mm.

Chapter 5
Measurements

In the case of transmission lines, particularly those that do not support TEM modes, we need to define the lines' characteristics and performance through measurements, and then use these measurements to verify the theoretical concepts. In fact, Paul Meier, inventor of the fin line, used only measurements to show the potential of E-plane circuits for millimeter-wave technology. Measurements characterized most of the useful discontinuities, and now verify many theoretical analyses. Until recently, the fin line has been used to characterize material properties and even to realize six-port network analyzers.

This chapter is devoted to several important papers on discontinuity measurements, material measurements, and fin-line six-ports.

5.1 DISCONTINUITY MEASUREMENTS

5.1.1 "Integrated Fin-Line Millimeter Components" by Meier

This was the second paper on fin lines to be published in a regular journal. The paper reviews the experimentally obtained results on the characteristics of the fin line. The results show the superiority of fin lines to microstrips at millimeter wavelengths. Meier shows the various kinds of active and passive E-plane circuits that were designed from experimentally measured data of fin-line propagation constants, the *Q*-factor, and the equivalent network of inductive septa. Unlike any other subsequent paper, this paper supports the power-current definition of the fin-line characteristic impedance.

5.1.2 "Equivalent Reactance of a Shorting Septum in a Fin-Line" by Knorr

The paper analyzes unilateral fin-line shorting septa using the spectral-domain method. It confirms the analysis experimentally and discusses various sources of uncertainty in the experiment. It also provides practical, useful design data over a wide range of structural parameters on polystyrene-type substrates.

5.1.3 "Scattering Coefficients of an Inductive Strip in a Finline: Theory and Experiment" by Knorr and Deal

Inductive strips are useful in designing fin-line bandpass filters. This paper analyzes such septa in unilateral fin lines using the spectral-domain method, and experimentally verifies the theoretical data. The magnitudes of the scattering coefficients were measured using an HP8409B network analyzer, whereas the phase angle of the reflection coefficient was measured using a slotted line to obtain the best results. However, the experimental accuracy was limited to 5%.

5.1.4 "Theoretical and Experimental Investigation of Finline Discontinuities" by Helard, Citerne, Picon, and Hanna

This paper shows that the spectral-domain approach, combined with direct modal analysis, is a very promising technique for determining the scattering matrix of E-plane discontinuities. Sufficient accuracy is obtained by using accurate values of the first six eigenmodes. We obtain satisfactory convergence on the scattering parameters of a step discontinuity by using relatively simple aperture field descriptions.

Unlike the previous two articles, this article gives no details on either the measurement system or the method used. Instead, the authors analyze and measure three types of step discontinuities. Figures 9 and 10 lend validity to the analysis method.

5.2 MEASUREMENT SYSTEMS

5.2.1 "A Computer-Controlled Integrated Microwave Measurement System in Finline Technique for Automatic Material Parameter Measurements" by Beyer, Gronau, and Wolff

This unique paper presents a fully automatic, microcomputer-controlled, E-plane measurement system that characterizes complex material properties of dielectrics and ferrites. The system works at X-band. The typical frequency accuracy is 0.5 MHz, while the Q-factor is obtained within 3% of manually obtained values. The system is much less expensive than its conventional waveguide counterpart.

5.3 E-PLANE SIX-PORT

5.3.1 "Integrated Ka-Band Fin-Line Six-Port" by Malkomes and Walsdorf

This paper presents the concept and the successful realization of an integrated, E-plane, six-port reflectometer at Ka-band, using a cheap, fin-line planar circuit and beam-lead diodes. The high precision measurement capability of the reflectometer equals that of its conventional waveguide counterpart.

BIBLIOGRAPHY

1. H.A. Willing and B.E. Spielman, "Experimental Assessment of Bilateral Finline Impedance for Device Matching," *IEEE MTT-S Symposium Digest*, 1981, pp. 105–107.
2. M. Malkomes and R. Walsdorf, "Integrated Finline Six-Ports for Millimeter-Wave Network Analysis," *IEEE MTT-S Symposium Digest*, 1986, pp. 669–672.
3. M. Malkomes, R. Walsdorf, F. Reiffert, D. Seindberg, and P. Pogalzki, "A Portable Six-Port Network Analyser for X- and Ka-Band," *Proc. European 16th Microwave Conference*, 1986, pp. 773–778.

5.1.1

Integrated Fin-Line Millimeter Components

PAUL J. MEIER, SENIOR MEMBER, IEEE

***Abstract*—This paper reviews the characteristics of integrated fin-line, a low-loss transmission line which is compatible with batch-processing techniques and superior to microstrip in several respects at millimeter wavelengths. Relative to microstrip, fin-line can provide less stringent tolerances, greater freedom from radiation and higher mode propagation, better compatibility with hybrid devices, and simpler interfaces with waveguide instrumentation. Examples of solid-state and passive components are presented which illustrate the potential of integrated fin-line at millimeter wavelengths. The examples include a p-i-n attenuator which has demonstrated the capability of constructing low-loss semiconductor mounts in fin-line. A four-pole bandpass filter, which performs in close agreement with theory, is also discussed.**

INTRODUCTION

INCREASED activity in the spectrum above 30 GHz has recently stirred interest in the development of millimeter integrated circuits. Much of the enthusiasm associated with integrated circuits can be traced to the clear advantages that such circuits provide below 3 GHz, namely, reduced size, weight, and cost combined with improved electrical performance, production uniformity, and reliability. However, those who have worked with integrated circuits at centimeter wavelengths (3–30 GHz) have encountered some fundamental problems which have limited the utility of such circuits. These problems include the critical tolerances and questionable production uniformity that can occur when miniaturization is carried too far. Although enhanced performance is possible in centimeter integrated circuits through the reduction of parasitics and the elimination of superfluous interfaces, poorer overall performance is also possible. The fundamental microstrip problems, which generally increase in severity as the operating frequency is raised, include radiation loss, spurious coupling, dispersion, and higher mode propagation. Attempts to integrate a large number of components in a single housing have generally demonstrated the need for "mode barriers" and "box-resonance absorbers." Radiation and related problems can be controlled by choosing progressively thinner substrates as the operating frequency is raised, but this only serves to degrade the Q factor, compound tolerance problems, and restrict the range over which the characteristic impedance can be varied.

Although standard microstrip techniques can be applied to millimeter components [1], [2], the problems listed

Manuscript received May 8, 1974; revised August 30, 1974.
The author is with AIL, a Division of Cutler-Hammer, Melville, N. Y. 11746.

P.J. Meier, *IEEE Transactions on Microwave Theory and Techniques*, Vol. MTT-22, No. 12, Dec. 1974, pp. 1209–1216.

previously can be expected to become more severe. As the operating frequency is raised and the microstrip dimensions are decreased, a limit will be reached where the strip width is no longer compatible with chip and beam-lead devices. In addition, millimeter integrated circuits must be tailored to requirements that are generally different from those which apply at lower frequencies. For example, the ability to construct a simple transition to waveguide becomes important at millimeter wavelengths where coaxial instrumentation is not practical. Moreover, the miniaturization that proved to be an asset at centimeter wavelengths can become a liability in millimeter applications. Designers of millimeter equipment have historically selected quasi-optical approaches to *increase* the physical size of components and thereby ease tolerance problems and improve performance [3]. Thus the ideal transmission line for millimeter integrated circuits is one which avoids miniaturization and yet offers the potential for low-cost production through batch-processing techniques. Integrated fin-line [4]–[6] is such a transmission line.

After introducing the symbols to be utilized throughout this paper, the body of the text will review the characteristics of integrated fin-line and demonstrate how this transmission line can be profitably applied to solid-state and passive millimeter components. Conclusions will then be presented regarding the potential application of integrated fin-line to future millimeter systems.

NOMENCLATURE

a	Major inner dimension of fin-line housing.
b	Minor inner dimension of fin-line housing.
c	Thickness of dielectric substrate.
d	Gap between fins.
s	Equivalent ridge thickness.
B	Susceptance magnitude.
k_e	Equivalent dielectric constant.
Q	Unloaded quality factor.
Y_0	Characteristic admittance of fin-line.
Z_0	Characteristic impedance of fin-line $= 1/Y_0$.
R_0	Resistance of load matched to Z_0.
Z_∞	Characteristic impedance at infinite frequency.
λ_g	Guide wavelength.
λ_0	Free-space wavelength.
λ_c	Cutoff wavelength of equivalent ridged waveguide.
C_j, R_j	Capacitance and resistance of p-i-n junction.
L_s, R_s	Series parasitic inductance and resistance of p-i-n diode.
C_p	Capacitance of diode package.
L_{s1}	Series inductance external to diode package.
W	Width of inductive strip.

CHARACTERISTICS OF INTEGRATED FIN-LINE

In an integrated fin-line structure [4], metal fins are printed on a dielectric substrate which bridges the broad walls of a rectangular waveguide. This adaptation of ridge-loaded waveguide permits circuit elements to be photoetched by low-cost batch techniques. The dimensions of practical fin-line circuits remain compatible with chip and beam-lead devices throughout the millimeter spectrum, thereby offering great potential for the construction of active and passive hybrid integrated circuits. Examples of such circuits and a simple low-reflection transition to rectangular waveguide will be presented.

In addition to serving as the bonding areas for hybrid devices, the printed fins increase the separation between the first two modes of propagation thereby providing a wider useful bandwidth than conventional waveguide. Owing to the similarity between integrated fin-line and conventional ridged waveguide, considerable design information [7], [8] is available. For thin substrates of moderate permittivity, the dielectric will have a minor effect [9] and single-mode bandwidth and attenuation of integrated fin-line may be estimated from existing data [8]–[10]. Such estimates lead to the conclusion that integrated fin-line can provide bandwidths in excess of an octave with less attenuation than microstrip. Larger single-mode bandwidths are feasible at the expense of attenuation.

The low-loss properties of integrated fin-line have been experimentally demonstrated through cavity tests at centimeter wavelengths [5] and transmission tests at millimeter wavelengths [11]. Preliminary tests were conducted at X band where wide-band sweepers were more readily available. The substrate thickness, however, was chosen to be sufficiently large (0.062 in) to permit direct scaling to higher frequencies.

Two different fin-line configurations, as shown in Fig. 1, were evaluated. In the configuration of Fig. 1(a), the fins are printed on opposite sides of a single dielectric substrate. Since the fins are directly grounded to the metal waveguide walls, this configuration is applicable only to passive devices. In the configuration of Fig. 1(b), however, the fins are printed on the mating surfaces of two dielectric substrates. Since the fins are insulated from the waveguide at dc, bias may be introduced for active components. RF continuity between the fins and the waveguide wall is obtained by choosing the thickness of the broad walls to be a quarter wavelength in the dielectric medium and by selecting $c \ll a$. The resultant choke prevents radiation or TEM propagation across a bandwidth in excess of an octave.

Teflon-fiber glass (3M type FL-GT) substrates with normalized gaps ranging from $d/b = 0.1$ to 1.0 were evaluated in the cavity whose aspect ratio (b/a) was 0.45. The center frequency and 3-dB bandwidth were recorded for each sample and the results were analyzed to determine the manner in which the guide wavelength and unloaded Q varied.

Fig. 2 shows how the measured Q (including the effect of losses within the irises) varies with respect to the gap between the fins. For moderate loading ($d/b \geq 0.6$), the measured Q of integrated fin-line exceeds 400 for insulated fins, and is better than 700 for the grounded-fin configura-

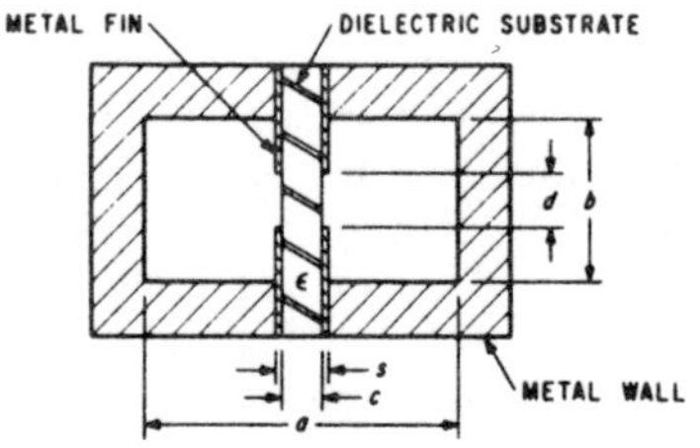

(a)

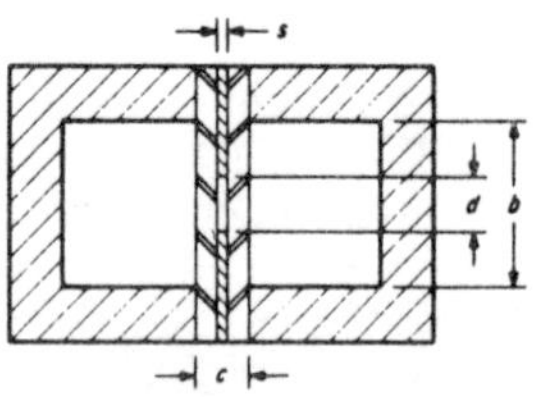

(b)

Fig. 1. Fin-line cavity configurations. (a) Grounded fins. (b) Insulated fins.

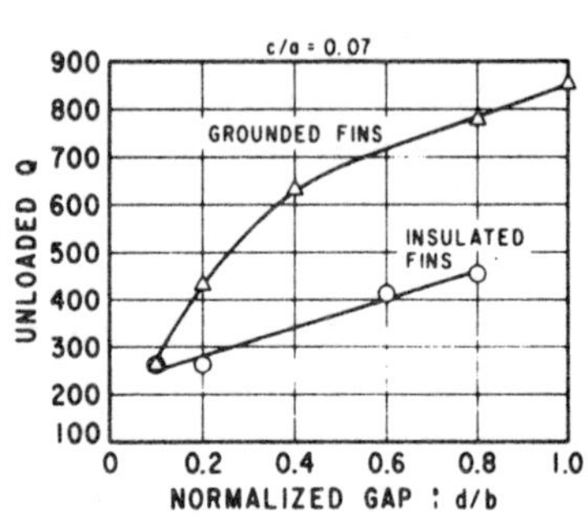

Fig. 2. Measured Q of integrated fin-line.

tion. These values compare favorably with those reported for practical microstrip lines at centimeter wavelengths [12].

To supplement the earlier measurements of unloaded Q performed in the 3-cm band, the insertion loss of integrated fin-line has been measured for a comparable fin-line structure in the 9-mm band [11]. A copper-clad laminate (Duroid 5880) with a thickness of 0.020 in was selected to maintain the ratio of c/a at 0.07. By accurately measuring the insertion loss of two fin-line transmission fixtures of different lengths, the loss was found to be 0.1 dB/wavelength for grounded fins with $d/b = 0.1$. This corresponds to an unloaded Q of 273, which is in agreement with the X-band cavity measurement of 260. Since the latter figure is somewhat pessimistic owing to loss in the coupling irises, it may be concluded that the loss in integrated fin-line holds fairly constant throughout the frequency range of 10–40 GHz. Although the loss can be expected to increase gradually as the frequency is raised above 40 GHz, the general utility of fin-line should be preserved throughout the millimeter spectrum.

Another important property of integrated fin-line is the guide-wavelength variation which can be calculated from the simple equation

$$\lambda_g = \frac{\lambda_0}{[k_e - (\lambda_0/\lambda_c)^2]^{1/2}} \tag{1}$$

where k_e is the equivalent dielectric constant, λ_0 is the free-space wavelength, and λ_c is the cutoff wavelength for an air-filled ridge-loaded waveguide of the same dimensions [8]. The value of k_e may be experimentally determined for a given fin-line configuration from a single cavity test [5]. It has been shown that the value of k_e varies little with respect to d/b when the fins are grounded directly to the waveguide. When d/b approaches unity, the value of k_e determined from the cavity agrees with the published value [10] for both the grounded and insulated cases. As the gap between the insulated fins becomes smaller, however, the value of k_e approaches the dielectric constant of the substrate material. Representative values of k_e appear in Table I.

To confirm the validity of (1) and demonstrate the capability of integrated fin-line at millimeter wavelengths, a transmission fixture was developed for operation in the frequency range of 26.5–40.0 GHz. Fig. 3 shows the cross section of this structure which is suitable for mounting semiconductor devices. In this structure, the upper fin is insulated from the housing at dc by a dielectric gasket, but is grounded at RF by choosing the thickness of the broad walls to be a quarter wavelength in the dielectric medium. The lower fin is grounded directly by a metal gasket to provide a dc return in solid-state applications. Nylon screws hold the halves of the housing together and align the substrate and its associated gaskets. Nylon is preferable to metal as the former has little effect on the RF choke formed within the broad walls.

The guide wavelength in the millimeter fin-line fixture has been measured across the band of 26.5–40.0 GHz by a sliding short-circuit technique. The fin-line was connected to a conventional slotted line through a low-reflection transition and the fin-line was internally terminated in a sliding short circuit of special design. The short consisted of a forked structure which straddled the dielectric substrate and provided a high standing wave ratio (SWR) by means of noncontacting chokes. The measurement was performed by recording the distance through which the short was moved in order to repeat nulls observed at a fixed location in the slotted line. Fig. 4 is a plot of the measured guide wavelength as a function of frequency for a typical structure ($d/b = 0.13$) suitable for mounting semiconductor devices. The value of k_e was determined by substituting into (1) the measured λ_g at 31 GHz ($\lambda_0 =$ 0.3807 in) and the published value of λ_c [8] for an air-filled ridged waveguide of the same dimensions ($s/a \approx 0$, $a = 0.280$ in). Based upon this single-frequency measurement, the value of k_e was determined to be 1.31. Also shown in Fig. 4 is the curve calculated by applying (1) throughout the band for $k_e = 1.31$. The measured values of λ_g agree with the calculation ± 2 percent across the 40-percent instrumentation band.

To predict the behavior of semiconductor devices in fin-line, it is helpful to know the absolute value of the

TABLE I

EQUIVALENT DIELECTRIC CONSTANT

Dielectric Constant of Substrate Material	Fin-Line Configuration	s/a	d/b	Equivalent Dielectric Constant k_e
2.5	Figure 1A	0.07	0.1	1.50
2.5	Figure 1A	0.07	1.0	1.25
2.5	Figure 1B	0.07	0.1	2.10
2.5	Figure 1B	0.07	1.0	1.25
2.2	Figure 3	0.036	0.08	1.33
2.2	Figure 3	0.036	0.13	1.31

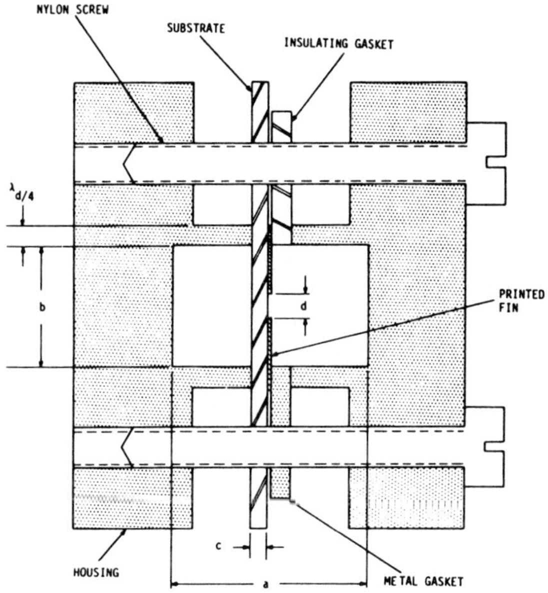

Fig. 3. Integrated fin-line semiconductor mount.

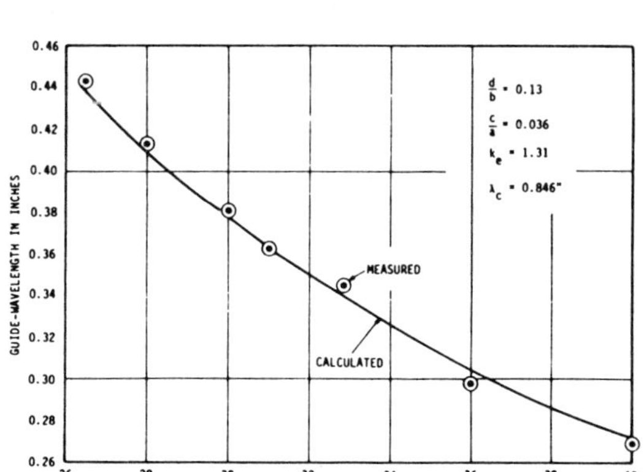

Fig. 4. Wavelength in insulated fin-line.

characteristic impedance ($Z_0 = 1/Y_0$). Values of the characteristic admittance of ridge-loaded waveguide, based upon a power-voltage definition, have been published for a wide range of parameters [8]. It has been found that these data are directly applicable to fin-line structures containing thin substrates ($c/a < 0.1$) of moderate permittivity. The characteristic impedance of fin-line may be approximated by

$$Z_0 \approx \frac{Z_\infty}{[k_e - (\lambda_0/\lambda_c)^2]^{1/2}} \tag{2}$$

where Z_∞ is the characteristic impedance of a ridge-loaded waveguide of the same dimensions at infinite frequency (that is, the reciprocal of the published admittance values). Appropriate factors may be applied to account for other aspect ratios (b/a), configurations (single versus double ridge), and impedance definitions [13].

Based upon a knowledge of the basic characteristics discussed previously, one may design a wide range of solid-state and passive components in integrated fin-line. The following sections will present examples of such components which serve to illustrate the advantages of fin-line construction techniques.

APPLICATION TO SOLID-STATE DEVICES

A millimeter demonstration of the compatibility of integrated fin-line with semiconductors has been performed with the aid of the previously discussed transmission fixture (Fig. 3). Additional features of this fixture appear in Fig. 5, wherein the fully assembled fixture is shown adjacent to a duplicate set of components. The housing mates directly with two standard WR28 waveguides and has identical inner dimensions (0.140 × 0.280 in). The substrate is cut from a 0.010-in sheet of laminate (Duroid 5880) and includes six mounting holes and two stepped edges. The latter protrude into the abutting WR28 waveguides and serve as quarter-wave transformers. After establishing a low-reflection transition between the WR28 waveguide and a slotted waveguide loaded by a dielectric slab ($d/b = 1.0$), the substrate metallization is tapered until the desired gap between the fins is obtained. The illustrated transitions at the ends of the substrate are each three wavelengths long near the middle of the instrumentation band (33 GHz). The measured VSWR of each transition is 1.2 or better across the 26.5–40-GHz band. It is believed that a more compact transition is feasible with quarter-wave transformers substituted for the fin-line taper. The substrate metallization also includes an RF blocking network connected to the upper fin.

To demonstrate the compatibility of integrated fin-line with semiconductor devices, two beam-lead p-i-n diodes (Alpha D5840B) were mounted across the fins near the center of the previously discussed substrate ($d/b = 0.13$). The diodes were spaced a quarter wavelength apart at a frequency near the upper end of the instrumentation band. The measured insertion loss of the p-i-n fin-line attenuator is plotted in Fig. 6 as a function of bias with frequency as a parameter. At the lower end of the band, where parasitics play a minor role, the reversed-bias insertion loss of the attenuator is only 0.3 dB, thereby demonstrating the capability of constructing low-loss semiconductor mounts in fin-line. As the bias is varied in the forward direction,

Fig. 5. Fin-line test fixture.

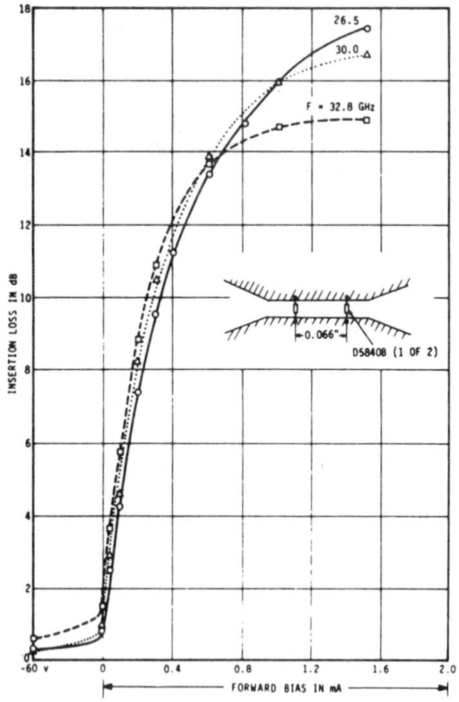

Fig. 6. Insertion loss of fin-line attenuator.

the attenuation varies smoothly over a 14-dB range throughout a 20-percent band.

Although the preliminary results of Fig. 6 are gratifying, it may be observed that the minimum loss and maximum attenuation deteriorate as the frequency is raised. To obtain a better understanding of the frequency limitations of this design, a program was launched to obtain an equivalent circuit for the attenuator, including the parasitic elements associated with the diode and its mounting structure. The SWR, as a function of bias and frequency, was measured for a single diode mounted across a matched fin-line. A computer-aided technique was then applied to find the parasitic elements which best reconciled the measured data with the general equivalent circuit shown in Fig. 7. Based upon the manufacturer's published data, a value of 1 Ω was assigned to R_s and R_j was represented as

$$R_j = 14.1 I^{-0.83} \tag{3}$$

where I is the diode current in milliamperes ($0.02 \leq I \leq 2$). All other parasitics were assigned plausible initial values and then optimized by an iterative process. Based upon a power-voltage definition of Z_0, the fit between the measured and calculated values of SWR proved to be only fair after exhaustive iterations. Moreover, the optimum value of C_j was found to be well below the specified value of 0.03 pF. Consequently, other definitions of Z_0 were examined; the best overall fit was obtained by adopting the power-current definition and thereby reducing Z_∞ by a factor of $\pi^2/16$ [13].

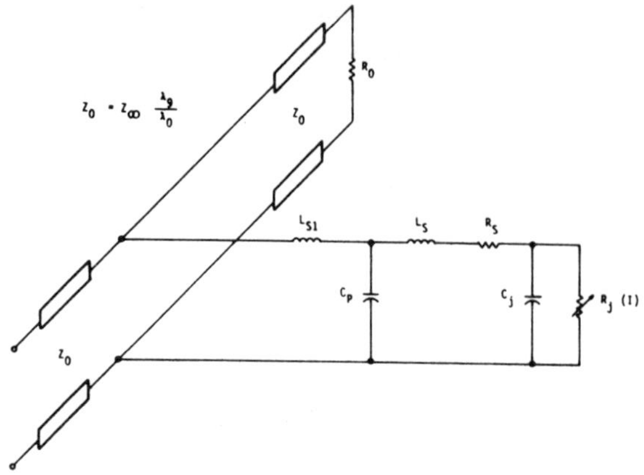

Fig. 7. Equivalent circuit for p-i-n diode.

Table II lists the values of the parasitic elements that provided the best fit (rms error equals 4 percent) with 21 SWR measurements at three equally spaced frequencies from 26.5 to 40.0 GHz and seven logarithmically spaced bias currents from 0 to 1.5 mA. This simple technique provided an unambiguous determination of the parasitics without complex impedance measurements (which prove difficult to reference at the plane of the diode). The validity of the equivalent circuit was confirmed by accurately predicting the total insertion loss of the diode which is generally greater than that due to the observed SWR alone.

The key parasitic element, which limits the high-frequency performance in the existing test fixture, is the series inductance L_s. The inductive reactance at 40 GHz is 35 Ω, which is significant relative to the forward-biased junction resistance and the characteristic impedance of the line ($Z_0 = 73.5$ Ω). This accounts for the reduced attenuation under forward bias. For reverse bias, the series inductance tends to resonate with the junction capacitance, thereby introducing a reflection which increases the minimum insertion loss.

The high-frequency response of the p-i-n attenuator may be improved by several methods. The series inductance can be lowered by decreasing the gap between the fins; however, this further reduces Z_0 and compounds the problem under forward bias. A better approach is to raise Z_0 by constricting the width of the fin-line housing and thereby increasing the ratios λ_g/λ_0 and b/a. This minimizes the effect of the series inductance under forward bias and increases the maximum attenuation. Raising Z_0 will also

TABLE II

P-I-N EQUIVALENT CIRCUIT PARAMETERS

L_s	= 0.14 nH
L_{s1}	≈ 0
C_p	= 0.002 pF
C_j	= 0.037 pF
Z_x	= 80.6 ohms[a]

[a] Power current definition.

Fig. 8. Filter components.

increase the reflection of a single diode under reverse bias, but this can be offset over a moderate bandwidth by spacing two diodes such that their reflections cancel. The upper frequency limitation then becomes the series resonance between L_s and C_j which is 70 GHz for the parasitics listed in Table II. Substituting a diode with a smaller junction capacitance (such as the D5840E) can move this series resonance well above 100 GHz. Further increases in the operating frequency and bandwidth are feasible with a more complex fin-line mount which tunes out the inductance L_s.

A p-i-n attenuator has been selected for measurements and analysis to demonstrate the potential of a broad class of semiconductor fin-line components. Other solid-state millimeter components which are amenable to fin-line construction include mixers, oscillators, amplifiers, limiters, switches, and phase shifters. A fin-line Gunn oscillator, which provides a power output of 45 mW in the 9-mm band, has been described elsewhere [14]. In addition to the existing examples of one- and two-port components, fin-line is well suited to the fabrication of E-plane Y junctions. Consequently, three-port networks, such as parametric amplifiers, phase modulators, and switching matrices, are feasible. For the reasons presented previously, integrated fin-line is preferred to microstrip for the construction of solid-state components and systems throughout the millimeter spectrum.

APPLICATION TO PASSIVE COMPONENTS

Although fin-line configurations with insulated fins and small gaps are best suited to solid-state applications, lower loss can be obtained with grounded fins and larger gaps. It has been shown, in connection with the discussion of Fig. 2, that an unloaded Q of 700–850 is possible for grounded fins with moderated loading ($d/b > 0.6$). Although another planar configuration [15] will provide higher values of Q, many millimeter applications exist where fin-line should prove to be adequate.

To demonstrate the capability of fin-line components fabricated according to the cross section of Fig. 1(a), various filter components were printed on 0.020-in Duroid. Test samples include the one- and four-pole inductively coupled filters shown at the center of Fig. 8, surrounded by four substrates, each printed with a single inductive element. By measuring the insertion loss of these elements across the 26.5–40-GHz band, families of design curves were generated to present the shunt susceptance as a function of strip width, with free-space wavelength as a parameter.

Fig. 9 shows how the shunt susceptance (normalized to the characteristic admittance of the fin-line) varies with respect to the strip width (normalized with respect to the wide inner dimension of the housing) for the special case of $d/b = 1.0$. Larger values of B/Y_0 can be obtained for a given w/a by reducing d/b but only at the expense of the unloaded resonator Q. Additional one-pole resonator measurements have been performed to determine the unloaded resonator Q and the reference plane at which the inductive strip can be modeled as a pure shunt susceptance. Based upon the characterization of fin-line filter elements and published design curves [16], a four-pole equal-element filter has been constructed and tested. Fig. 10 compares the calculated response with measurements. The calculation was performed by a computer-aided technique which solves for the overall $ABCD$ matrix of the entire filter network including:

1) shunt susceptance of each filter element and its variation with respect to frequency;

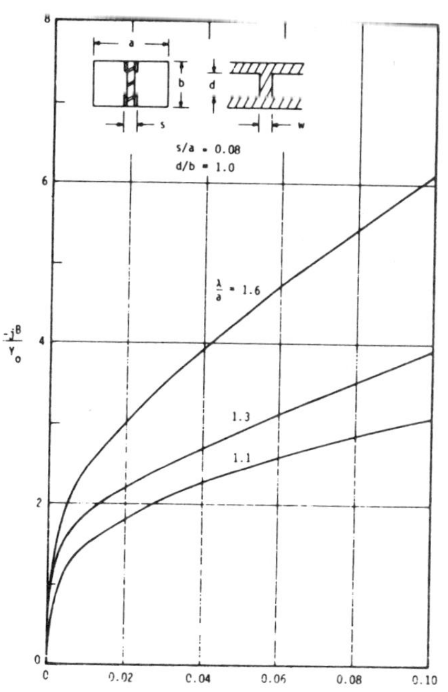

Fig. 9. Susceptance of inductive elements.

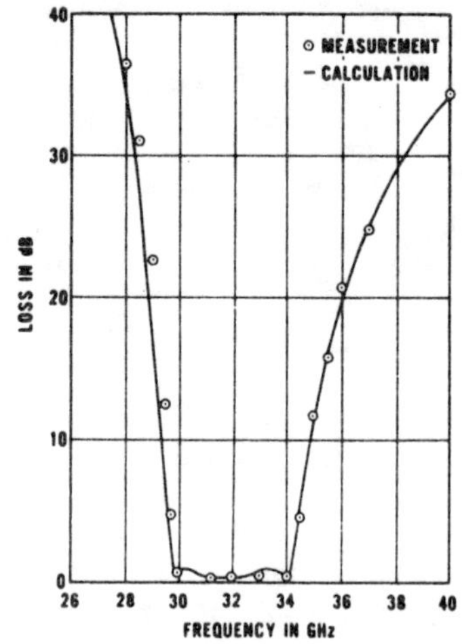

Fig. 10. Response of four-pole filter.

2) offset between the electrical and mechanical centerlines of the inductive elements;

3) variation of guide wavelength with respect to frequency, which is known accurately for $d/b = 1.0$ [5], [10].

4) unloaded resonator Q of 350 for center sections (based upon one-pole measurement).

The good agreement that has been obtained between measurements and theory serves to illustrate the straightforward application of fin-line techniques to the design of millimeter filters.

To explore the possible advantages of fin-line filters with $d \neq b$, the susceptance of a more generalized fin-line inductor has been measured. Table III lists the susceptance measured at three frequencies across the instrumentation band for two inductor configurations. For the first configuration, the fins do not load the waveguide and a fairly wide strip ($w/a = 0.2$) is required to establish a high susceptance suitable for narrow-band filters. In the second configuration, the fins provide heavy loading to the waveguide and a higher susceptance can be obtained with a moderately narrow strip ($w/a = 0.1$). Heavy fin loading not only provides a larger susceptance but one whose magnitude is less dependent upon frequency. Consequently, inductors in heavily loaded fin-line should prove to be of great value and deserve further study.

CONCLUSIONS

The application of standard microstrip techniques at millimeter wavelengths can result in problems which include excessive radiation loss, spurious coupling, dispersion, and higher mode propagation. These problems can be controlled by choosing progressively thinner substrates as the operating frequency is raised but this only serves to degrade the Q, increase tolerance problems, restrict the range over which Z_0 can be varied, and complicate the design of transitions to waveguide instrumentation. Integrated fin-line is superior to microstrip at millimeter wavelengths as the former provides eased production tolerances, compatibility with chip and beam-lead devices throughout the millimeter spectrum, and single-mode octave bandwidths, combined with the ability to construct simple transitions to waveguide. Fin-line avoids excessive miniaturization and offers low loss combined with the potential for low-cost production through batch-processing techniques.

The basic characteristics of integrated fin-line have been reviewed previously. It has been demonstrated that practical fin-line structures can provide an ur 'oaded Q of 260–850 and that the guide wavelength ca be accurately predicted by a simple equation.

Examples of solid-state and passive components have been presented which serve to illustrate the potential of integrated fin-line at millimeter wavelengths. These examples have included a p-i-n attenuator which has demonstrated the capability of constructing low-loss semiconductor mounts in fin-line. An analysis has shown that existing beam-lead devices in a simple fin-line mount can be useful well beyond 100 GHz.

Based upon the analysis and examples presented, it is clear that fin-line construction techniques can be applied to a broad class of millimeter components including mixers, oscillators, amplifiers, and phase shifters. Fin-line is also suited to the fabrication of E-plane Y junctions and is therefore compatible with three-port networks such as parametric amplifiers, phase modulators, and switching matrices.

Since integrated fin-line can provide high performance combined with low-cost batch processing, it is ideally suited to a wide range of millimeter applications including communication systems, high-resolution imaging, collision-avoidance radars, and environmental monitoring equipment.

TABLE III

SUSCEPTANCE OF FIN-LINE INDUCTORS

Frequency (GHz)	λ_{g}/a	Normalized Susceptance (B/Y_0): No Fin Loading $d/b = 1.0$, $w/a = 0.2$	Normalized Susceptance (B/Y_0): Heavy Fin Loading $d/b = 0.1$, $w/a = 0.1$
26.5	1.59	13.2	18.1
33.0	1.28	6.8	14.7
40.0	1.05	4.9	12.0

ACKNOWLEDGMENT

This program was sponsored by AIL under the direction of K. S. Packard, M. T. Lebenbaum, and J. J. Whelehan. D. Fleri and J. J. Taub formulated the initial program plan and made technical contributions throughout the program. Technical assistance was provided by R. Chew, R. Gibbs, L. Hernandez, A. Kunze, and W. Reinheimer.

REFERENCES

[1] D. Dobramysl, "Integrated mixer for 18 and 26 GHz," in *IEEE 1971 G-MTT Symp. Digest* (Washington, D. C.), May 16–19, 1971, pp. 18–19.

[2] T. H. Oxley, K. J. Ming, G. H. Swallow, B. J. Climer, and M. J. Sisson, "Hybrid microwave integrated circuits for millimeter wavelengths," in *IEEE 1972 G-MTT Symp. Digest* (Arlington Heights, Ill.), May 22–24, 1972, pp. 224–226.

[3] J. J. Taub, "The status of quasi-optical waveguide components for millimeter and submillimeter wavelengths," *Microwave J.*, pp. 57–62, Nov. 1970.
[4] P. J. Meier, "Two new integrated-circuit media with special advantages at millimeter wavelengths," in *IEEE 1972 G-MTT Symp. Digest* (Arlington Heights, Ill.), May 22–24, 1972, pp. 221–223.
[5] ——, "Equivalent relative permittivity and unloaded Q factor of integrated fin-line," *Electron. Lett.*, vol. 9, pp. 162–163, Apr. 1973.
[6] ——, "Microwave transmission line," U. S. Patent 3 825 863, July 1974.
[7] S. B. Cohn, "Properties of ridge wave guide," *Proc. IRE*, vol. 35, pp. 783–788, Aug. 1947.
[8] S. Hopfer, "The design of ridged waveguides," *IRE Trans. Microwave Theory Tech.*, vol. MTT-3, pp. 20–29, Oct. 1955.
[9] F. E. Gardiol, "Higher-order modes in dielectrically loaded rectangular waveguides," *IEEE Trans. Microwave Theory Tech.*, vol. MTT-16, pp. 919–924, Nov. 1968.
[10] P. H. Vartanian, W. P. Ayres, and A. L. Helgesson, "Propagation in dielectric slab loaded rectangular waveguide," *IRE Trans. Microwave Theory Tech.*, vol. MTT-6, pp. 215–222, Apr. 1958.
[11] D. Fleri, "AIL project R035 status report," Internal Memo, Oct. 1973.
[12] R. A. Pucel, D. J. Massé, and C. P. Hartwig, "Losses in mircostrip," *IEEE Trans. Microwave Theory Tech.*, vol. MTT-16, pp. 342–350, June 1968.
[13] G. C. Southworth, *Principles and Applications of Waveguide Transmission*. Princeton, N. J.: Van Nostrand, 1950, pp. 104–105.
[14] D. A. Fleri *et al.*, "A low cost X-band MIC paramp," presented at the IEEE Int. Microwave Symp., Atlanta, Ga., June 1974.
[15] Y. Konishi, K. Uenakada, N. Yazawa, N. Hoshino, and T. Takahashi, "Simplified 12-GHz low-noise converter with mounted planar circuit in waveguide," *IEEE Trans. Microwave Theory Tech.* (Short Paper), vol. MTT-22, pp. 451–454, Apr. 1974.
[16] J. J. Taub, "Design of minimum loss band-pass filters," *Microwave J.*, Nov. 1963.

5.1.2

Equivalent Reactance of a Shorting Septum in a Fin-Line: Theory and Experiment

JEFFREY B. KNORR, MEMBER, IEEE

Abstract—This paper presents the results of theoretical and experimental investigations of the equivalent reactance of a shorting septum in a fin-line. Numerical and experimental data showing septum reactance for several different fin-lines are presented and compared. Some design curves are included for millimeter-wave fin-lines with this type of discontinuity.

I. Introduction

THE USE of fin-line as a millimeter-wave circuit medium has been discussed rather widely in the microwave literature during the past several years. Various fin-line components and circuits have been described and a number of analyses have been published. The analytical work which has been presented thus far has dealt with methods for determining the wavelength and characteristic impedance of a fin-line. While this is very necessary, it alone is not sufficient to permit the practical application of fin-line in millimeter-wave circuits. It is also necessary to analyze the various types of discontinuities that are required in order to realize the impedances needed in a practical circuit of finite extent.

One fin-line discontinuity which is of practical interest is the shorting septum. Located in the plane of the fins, the ideal, lossless septum is a perfectly reflecting termination which may be produced conveniently by printed circuit methods. Fig. 1 shows a fin-line cavity with a shorting septum at each end. A large current flows on the septum, and behind a reference plane coincident with the end of the septum the net stored energy is magnetic. (It is assumed here that the fin-line is operated in the dominant mode at a frequency below the cutoff for the next highest order mode.) The septum is not an ideal short, therefore, but behaves like a perfectly reflecting inductive reactance at the specified reference plane. The equivalent circuit for the cavity of Fig. 1 is shown in Fig. 2. It is a transmission line terminated with an inductor which represents the reactance of the septum as seen at the reference plane.

Theoretically, the equivalent reactance of the septum may be found by first determining the fin-line wavelength, λ', at a given frequency and then determining the shortest resonant length of the structure at that same frequency. The reactance is then found from the difference between the resonant length and $\lambda'/2$. Experimentally, the equivalent reactance may be measured using a suitably constructed test fixture with which a perfect short may be established at the desired reference plane. By measuring the difference between the angle of the reflection coefficient of the septum and that of the perfect short the reactance may be determined.

Manuscript received February 23, 1981; revised July 13, 1981. This work was supported by the Naval Postgraduate School Foundation Research Program.

The author is with the Department of Electrical Engineering, Naval Postgraduate School, Monterey, CA 93940.

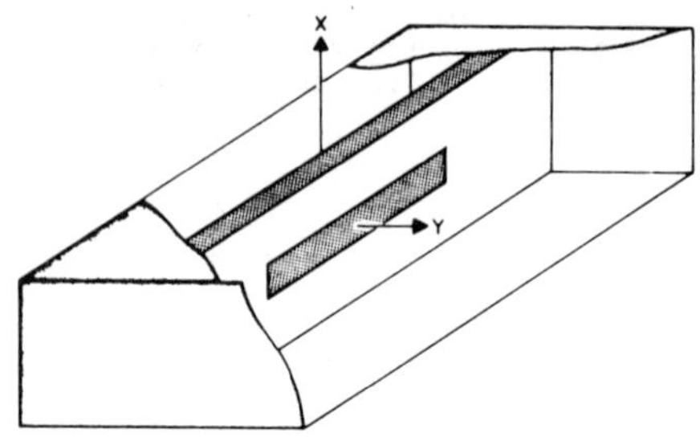

Fig. 1. Cutaway view of a fin-line cavity with a shorting septum at each end.

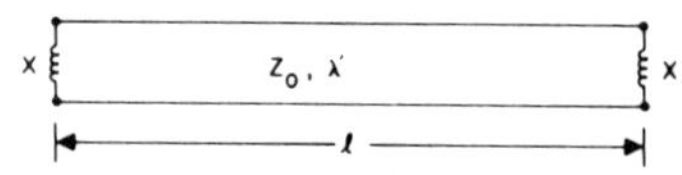

Fig. 2. Equivalent circuit representation of a fin-line with a shorting septum.

The purpose of this paper is to present the results of an investigation of the equivalent reactance of a septum in a fin-line. The septum was investigated both theoretically and experimentally. Section II describes the theoretical and experimental methods employed in the investigation and Section III presents the numerical and experimental data. Section IV contains some useful millimeter-wave fin-line circuit design data.

II. Theoretical and Experimental Methods

A. Theoretical Method

The theoretical analysis of the fin-line resonator may be carried out using the spectral domain method which has been discussed in several earlier papers by the author and co-workers [1]–[3]. A microstrip resonator has also been analyzed by Itoh using this method [4]. It will be shown below that the analysis of the fin-line resonator may be carried out by making relatively minor modifications to the equations developed for fin-line transmission [3]. The fields in the fin-line cavity may be written in the form [5]

$$E_{zi} = \left(\frac{\partial^2}{\partial z^2} + k^2\right)\phi_i^e \tag{1a}$$

J.B. Knorr, *IEEE Transactions on Microwave Theory and Techniques*, Vol. MTT-29, No. 11, Nov. 1981, pp. 1196–1202.

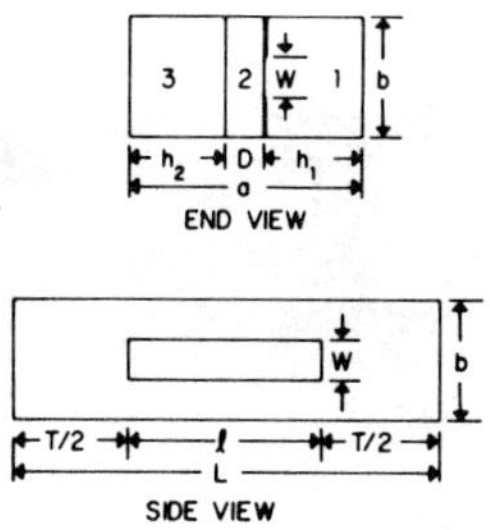

Fig. 3. End and side views of fin-line cavity showing dimensions and regions.

$$H_{zi}=\left(\frac{\partial}{\partial z^2}+k^2\right)\phi_i^h \tag{1b}$$

$$E_{xi}=\frac{\partial^2\phi_i^e}{\partial x\partial z}-j\omega\mu_i\frac{\partial\phi_i^h}{\partial y} \tag{1c}$$

$$E_{yi}=\frac{\partial^2\phi_i^e}{\partial y\partial z}+j\omega\mu_i\frac{\partial\phi_i^h}{\partial x} \tag{1d}$$

$$H_{xi}=j\omega\epsilon_i\frac{\partial\phi_i^e}{\partial y}+\frac{\partial^2\phi_i^h}{\partial x\partial z} \tag{1e}$$

$$H_{yi}=-j\omega\epsilon_i\frac{\partial\phi_i^e}{\partial x}+\frac{\partial^2\phi_i^h}{\partial y\partial z} \tag{1f}$$

where the scalar potential functions $\phi_i^e(x,y,z),\phi_i^h(x,y,z)$ satisfy the Helmholtz equation

$$\nabla^2\phi_i+k_i^2\phi_i=0 \tag{2}$$

and

$$k_i^2=\omega^2\mu_i\epsilon_i$$

for each of the three regions $i=1,2,3$ as shown in Fig. 3. For this problem we define the two-dimensional Fourier transform

$$\Phi_i(\alpha_n,y,\xi_k)=\int_{-L/2}^{+L/2}\int_{-b/2}^{+b/2}\phi_i(x,y,z)e^{j\alpha_n x}e^{j\xi_k z}\,dz\,dx \tag{3}$$

where

$$\alpha_n=\begin{cases}2n\dfrac{\pi}{b}, & \phi^h \text{ even in } x \quad (4a)\\ (2n-1)\dfrac{\pi}{b}, & \phi^h \text{ odd in } x \quad (4b)\end{cases}$$

$$\kappa_k=\begin{cases}2k\dfrac{\pi}{L}, & \phi^h \text{ odd in } z \quad (5a)\\ (2k-1)\dfrac{\pi}{L}, & \phi^h \text{ even in } z. \quad (5b)\end{cases}$$

With this definition the Helmholtz equation is transformed to obtain

$$\frac{\partial^2\Phi_i}{\partial y^2}(\alpha_n,y,\xi_k)=\gamma_i^2\Phi(\alpha_n,y,\xi_k) \tag{6}$$

with

$$\gamma_i^2=\alpha_n^2-(k_i^2-\xi_k^2). \tag{7}$$

Next, the solutions for the Φ_i are obtained. Equation (6) and boundary conditions are applied to obtain the set of transformed boundary equations. When these equations are compared with those for fin-line transmission [3], it is found they are identical if we choose

$$\xi_i=-\beta \tag{8}$$

where

$$\beta=\frac{2\pi}{\lambda'} \tag{9}$$

is the propagation constant which appears in the fin-line transmission analysis. Thus the final form of the equations,

$$\mathcal{G}_{11}(\alpha_n,\xi_k,\lambda)\mathcal{E}_x(\alpha_n,\xi_k)+\mathcal{G}_{12}(\alpha_n,\xi_k,\lambda)\mathcal{E}_z(\alpha_n,\xi_k)=\mathcal{J}_x(\alpha_n,\xi_k) \tag{10a}$$

$$\mathcal{G}_{21}(\alpha_n,\xi_k,\lambda)\mathcal{E}_x(\alpha_n,\xi_k)+\mathcal{G}_{22}(\alpha_n,\xi_k,\lambda)\mathcal{E}_z(\alpha_n,\xi_k)=\mathcal{J}_z(\alpha_n,\xi_k) \tag{10b}$$

is exactly the same as for fin-line transmission. The $\mathcal{G}_{pq}$ are the Fourier transformed components of the dyadic Green's function, $\mathcal{E}_x$ and $\mathcal{E}_z$ are the transformed components of the electric field in the plane of the fins at $y=D$ and $\mathcal{J}_x$ and $\mathcal{J}_z$ are the transformed components of the surface current on the fins.

Equations (10) may be further manipulated to eliminate $\mathcal{J}_x$ and $\mathcal{J}_z$ through the application of Galerkin's Method. Define the inner product as

$$\langle A(\alpha_n,\xi_k),B(\alpha_n,\xi_k)\rangle=\sum_{n=-\infty}^{+\infty}\sum_{k=-\infty}^{+\infty}A(\alpha_n,\xi_k)B^*(\alpha_n,\xi_k). \tag{11}$$

Then taking the inner product of (10a) with $\mathcal{E}_x$ and (10b) with $\mathcal{E}_z$ and applying Parseval's theorem, we obtain

$$\langle\mathcal{G}_{11}\mathcal{E}_x,\mathcal{E}_x\rangle+\langle\mathcal{G}_{12}\mathcal{E}_z,\mathcal{E}_x\rangle=0 \tag{12a}$$

$$\langle\mathcal{G}_{21}\mathcal{E}_x,\mathcal{E}_z\rangle+\langle\mathcal{G}_{22}\mathcal{E}_z,\mathcal{E}_z\rangle=0 \tag{12b}$$

where the right-hand side of (12) is zero because the current on the fins is orthogonal to the field between the fins. Explicit expressions for the $\mathcal{G}_{pq}$ exist and the field components $\mathcal{E}_x$ and $\mathcal{E}_z$ may be expanded in a suitable set of basis functions. The resonant frequency of the cavity is then found from (12) by numerical computation. It is interesting to note that the fin geometry enters the problem only through $\mathcal{E}_x$ and $\mathcal{E}_z$.

The accuracy of the solution obtained using (12) depends upon the quality of the basis set used to expand $\mathcal{E}_x$ and $\mathcal{E}_z$. In practice, the accuracy must be traded off against computation time, memory space and program complexity. Previous work has shown that accurate results may be obtained under the assumption $\mathcal{E}_z\approx 0$ in which case (12a) simplifies to

$$\sum_n\sum_k\mathcal{G}_{11}(\alpha_n,\xi_k,\lambda)|\mathcal{E}_x(\alpha_n,\xi_k,W,b,l,L)|^2=0.$$

It is also known from previous work that accurate results are obtained when the field between the fins is assumed constant with respect to x. The main problem in the present case is thus to expand $\mathcal{E}_x$ in such a manner that the amplitude variation of the x directed electric field (between the fins) with respect to z is properly represented.

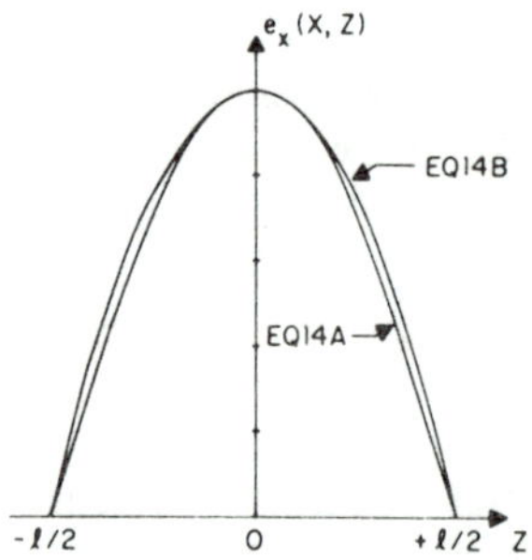

Fig. 4. One term approximations to e_x for the case $l=0.4\lambda'$.

Two different single-term expansions of $\mathcal{E}_x$ were investigated. These were obtained from Fourier transformation of assumed electric field distributions, $e_x(x,z)$ between the fins given by

$$e_x(x,z)=A\cos\left(\frac{\pi z}{l}\right), \qquad |x|\leq W/2,\ |z|\leq l/2 \quad (14a)$$

and

$$e_x(x,z)=A\cos\left(\frac{2\pi z}{\lambda'}\right)-A\left[\frac{\cos\frac{\pi l}{\lambda'}}{\cosh\frac{Cl}{2}}\right]\cosh\left(\frac{Cl}{2}\frac{2z}{l}\right), \qquad |x|\leq W/2,\ |z|\leq l/2 \quad (14b)$$

where in (14b) C is a parameter which will be discussed. Both 14(a) and 14(b) assume the field is independent of x. Equation 14(a) assumes a sinusoidal variation of field with z with the amplitude going to zero at $z=\pm l/2$. This is exactly correct for the case where the septum is absent ($T=0$). However, as the septum protrudes further into the cavity and the equivalent reactance terminating the cavity increases, the correctness of this representation deteriorates. This deterioration was reflected in poor accuracy of the numerical results obtained using (14a). The representation of (14b) corrects this difficulty. The first term of (14b) gives the correct amplitude variation for the dominant mode. When the cavity is reactively terminated, $l<\lambda'/2$, this term is nonzero for $z=\pm l/2$. The total field, e_x, is zero, however, due to the presence of evanescent modes. The presence of the evanescent modes is accounted for approximately by the second term of (14b). The rate of decay is adjusted by the parameter $Cl/2$. The results obtained using (14b) appeared to be quite satisfactory and were not extremely sensitive to the choice of $Cl/2$. The value $Cl/2=10$ was used in the computations reported here. The basis function of (14b) will be referred to as the end corrected basis function. The basis functions of (14) are plotted in Fig. 4 for the case $l=0.4\lambda'$ which corresponds to an equivalent normalized septum reactance, $x=0.158$.

The theoretical method described above provides a straightforward procedure for calculating the equivalent reactance of a fin-line septum. For a specified frequency and structure geometry the first step is to calculate the fin-line wavelength, λ', [3]. Next, the shortest resonant length, l, is calculated for the same frequency and geometry. The difference between the length of a perfectly shorted

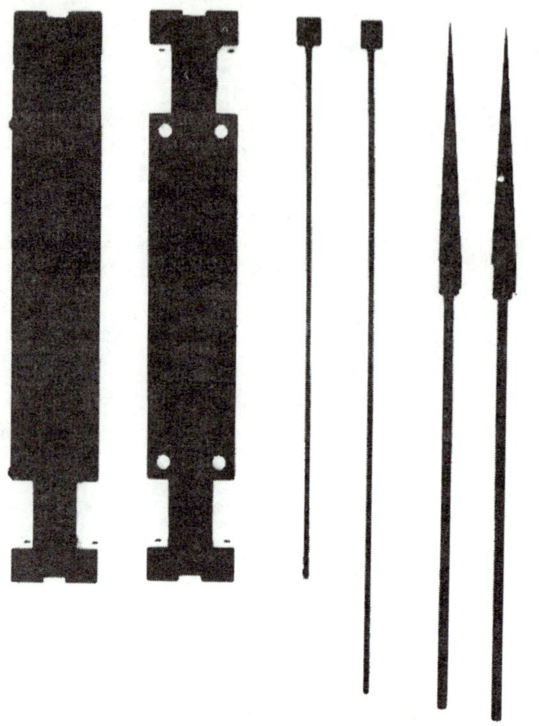

Fig. 5. Photograph of waveguide test fixture, a fin, shorting blocks, and wedge terminations.

cavity and the length of a reactively terminated cavity is

$$2\Delta l=(\lambda'/2)-l \quad (15)$$

which may be used to calculate the equivalent reactance

$$x=\tan(2\pi\Delta l/\lambda'). \quad (16)$$

This relation holds for any septum length, $T/2$. The equivalent reactance of a long septum is found simply by allowing $T/2$ to become large which effectively eliminates interaction between the fields and the end wall of the cavity.

B. Experimental Method

Experimental measurements of septum reactance were made using a test fixture which was machined specifically for that purpose. The measurements were carried out in the 8–12-GHz frequency range for reasons related to availability of test instrumentation and ease of fabrication of the test fixture. The test fixture consisted of a 10-in length of X-band guide ($a=0.9$ in, $b=0.4$ in) machined in two identical longitudinal sections which were bolted together. The necessary fin patterns were etched from 0.002-in thick beryllium copper sheet and inserted between the two halves of the shielding guide. No dielectric material was present. Shorting blocks and wedge terminations were constructed in such a manner that they could be positioned on both sides of the fin. A photograph of all the test apparatus is shown in Fig. 5.

The experimental measurements reported here were all obtained with $\epsilon_{r2}=1$ (no dielectric) as described above. The decision not to print the fins on dielectric was made because in practice the dielectric must be positioned and secured in the shield. This is done by machining a groove or a slot in the wall of the shield. In this case the actual fin-line geometry differs from that shown in Fig. 3 and this introduces a source of measurement uncertainty that is difficult to quantify. Thus the use of fins but no dielectric resulted, we believe, in the closest possible correspondence between the experimental structure and its theoretical model.

Measurements were made on fin-lines with $W/b=0.1$, $W/b=0.5$, and $W/b=1.0$. When $W/b=1.0$, we have a special case, that of a bifurcated guide. In this case the septum is preceeded by empty, unloaded rectangular waveguide and it is convenient to measure the reactance of the septum by observing the standing wave pattern with a slotted line. It is a simple procedure to place a short at the end of the slotted line to establish the load position and then to replace the short with the test fixture. A metal septum is placed in the test fixture and the leading edge of the septum is aligned with the flange on the test fixture so that it is in the same plane as was the short. This method was employed here for $W/b=1.0$ and is believed to yield experimental results with uncertainty limited to several percent.

For the fin-lines with $W/b<1$, the septum was printed at one end of the fin-line while at the other end was printed a tapered transition to the empty waveguide from which the fin-line was excited. The test fixture was connected to a HP 8410S microwave network analyzer through a slide screw tuner and precision waveguide to coax transition. For each measurement frequency, sliding matched terminations were first placed along the finned section of line and the slide screw tuner was used to eliminate reflections. The sliding terminations were then removed and replaced by the shorting blocks which were aligned with the end of the septum to establish the reference plane for calibration of the network analyzer. The shorting blocks were then removed and the normalized reactance of the septum was measured.

Although considerable care was taken to obtain accurate experimental data, it was not possible to eliminate all sources of experimental error. The two most significant sources of error appear to be due to positioning and to matching as discussed below:

1) Positioning Error: The reference plane is established by positioning sliding shorting blocks to the plane coincident with the leading edge of the septum. This must be accomplished after the fixture is assembled with the fins in place and some phase error may result from imprecise positioning. The phase error $\Delta\phi$ due to a positioning error Δl is given by

$$\Delta\phi/\Delta l = 720/\lambda' \text{ degrees/m} \tag{17}$$

where λ' is the fin-line wavelength. Suppose, for example, $\lambda'=30$ mm, a typical value for the X-band lines used in this study. Then the error is 24°/mm. If it is assumed the

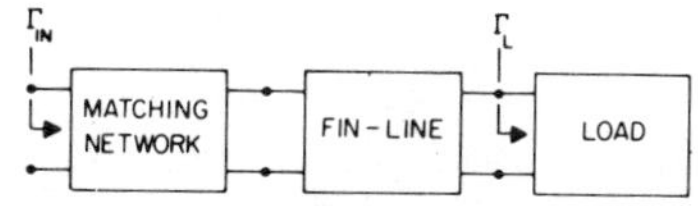

Fig. 6. Error model for the determination of phase uncertainty due to a matching network.

shorting blocks can be positioned with an accuracy of ±.1 mm, then the phase error is $\Delta\phi=\pm 2.4$ degrees. If the measured reactance is $x=0.190$ ($\Gamma=1\angle 159°$) the uncertainty in the reactance will be $\Delta x=\pm.023$ and the error is ±12 percent. This clearly illustrates the fact that a very small positioning error will introduce a significant uncertainty in the measured reactance.

2) Matching Error: For network analyzer measurements the analyzer was connected to the fin-line with its septum through a coax-waveguide transition, slide screw tuner and tapered transition. The slide screw tuner was adjusted to obtain the best possible match (smallest reflection) between the analyzer and the fin-line when the sliding matched terminations were in place along the fin-line just ahead of the septum. The combination of transition-tuner-taper may be modeled as a two-port network through which the reflection from the septum must be viewed. This is illustrated in Fig. 6.

The reflection coefficient, Γ_{in}, measured at the input to the matching network is

$$\Gamma_{in} = S_{11} + \Gamma'_L \tag{18}$$

where

$$\Gamma'_L = \frac{S_{12}S_{21}\Gamma_L}{1-S_{22}\Gamma_L} \tag{19}$$

is the load reflection coefficient, Γ_L, transferred to the input of the matching network. If the matching network is assumed lossless and reciprocal with $|S_{11}|=|S_{22}|<0.1$ the difference between the angle of Γ_L and the angle of Γ'_L can be shown to be

$$\phi'_L - \phi_L = 2\phi_{12} + |S_{11}|\sin(\phi_{22}+\phi_L). \tag{20}$$

The matching network, therefore, introduces a phase uncertainty, $\Delta\phi \leq \pm|S_{11}|$ radians when either the reference short or the septum is measured. In the worst case the total error would be less than $2|S_{11}|$ if the errors for short and septum add together. Thus the phase uncertainty introduced by the matching network is bounded by

$$\Delta\phi \leq \pm 2|S_{11}| \text{ rad}. \tag{21}$$

When the fin-line is terminated in a perfect matched load, $\Gamma_L=0$ and the reflection coefficient seen at the input of the matching network is $\Gamma_{in}=S_{11}$. The matching procedure described here typically resulted in $|\Gamma_{in}|\approx 0.01$. Assuming a perfect load, the corresponding phase uncertainty due to the matching network was therefore ±1.1 degrees. If the matched load is not perfect then there will be some instances where the phase error due to matching will be greater.

The uncertainty due to matching error must be added to the uncertainty due to positioning. For the conditions cited in the example discussed under positioning error it was shown that an uncertainty of ± 12 percent resulted from a positioning error of ± 0.1 mm. If we assume the phase error due to matching is ± 1.1 degrees (which is probably optimistic) then the total phase uncertainty in this example is ± 3.5 deg and the resultant error in reactance may be as large as ± 17 percent.

Clearly, although the measurement of septum reactance is straightforward in principle, it is not an easy task to achieve high experimental accuracy in practice. This is particularly true when $W/b<1$ and the problem becomes more serious as W/b is made smaller because the fin-line wavelength, λ', decreases.

III. Numerical and Experimental Results

The numerical data obtained using the theoretical method described in Section II of this paper have been compared with other available theoretical results and with experimental data. For those cases where such comparisons were made, the various results were found to agree quite well as described below.

A. Comparison with Other Theory

When $W/b=1$ and $\epsilon_i=\epsilon_0$ for all three fin-line regions, we obtain the special case of a bifurcated rectangular waveguide when T is large. Results for this special case have been presented by Marcuvitz [6]. Fig. 7 shows the normalized septum reactance versus frequency for a bifurcated X-band waveguide. The dashed curves were computed using the spectral-domain method with (14a) and the solid curves were computed using the end corrected basis function of (14b). The discrete points (x) were plotted using data from [6]. Clearly, there is good agreement between the two theories when the basis function of (14b) is used. There is considerable disagreement when (14a) is used. The experimental results were also in close agreement when (14b) was used and it was, therefore, concluded that the quality of this approximation was good.

B. Comparison with Experiment

Measurements of septum reactance were made in X-band using fin-lines with $W/b=1.0$, $W/b=0.5$, and $W/b=0.1$. In all cases the fins were centered ($h_1=a/2$) and no dielectric was present ($\epsilon_i=\epsilon_o$ for all three regions).

Fig. 8 shows the normalized reactance calculated using (14b) (solid curves) and the measured reactance (discrete points). The experimental data for $W/b=1$ were obtained from slotted line measurements and the uncertainty is believed to be small. The difference between the calculated and measured reactance values is better than 5 percent in all cases. The experimental data for $W/b=0.5$ and $W/b=0.1$ were obtained using a network analyzer. This required the use of transitions and a more complex tuning and calibration procedure as described earlier. The errors which resulted introduced a greater experimental uncertainty than in the case $W/b=1$. This is seen in Fig. 8 where the discrepancy between numerical and experimental results is somewhat greater for $W/b=0.5$ and $W/b=0.1$.

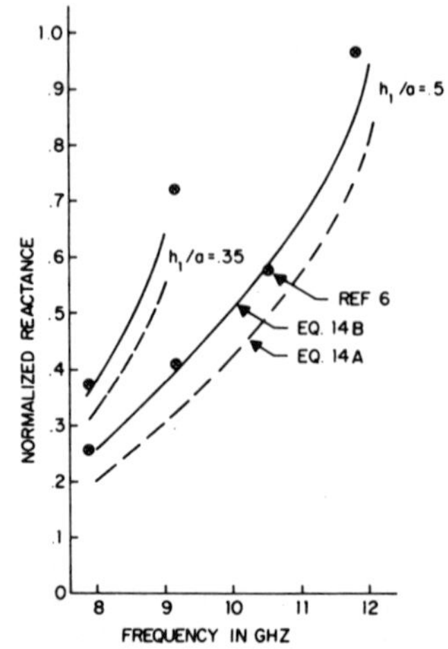

Fig. 7. Normalized septum reactance versus frequency for a bifurcated WR (90) X-band waveguide with the septum in several different positions $\epsilon_{r2}=1$, $T\rightarrow\infty$.

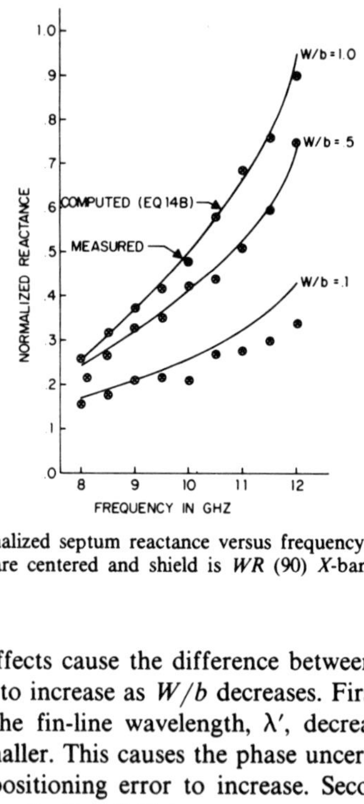

Fig. 8. Normalized septum reactance versus frequency for several fin-lines. Fins are centered and shield is WR (90) X-band guide $\epsilon_{r2}=1$, $T\rightarrow\infty$.

Several effects cause the difference between theory and experiment to increase as W/b decreases. First, at a given frequency the fin-line wavelength, λ', decreases as W/b becomes smaller. This causes the phase uncertainty associated with positioning error to increase. Second, as W/b decreases the fin-line field becomes more concentrated in the region of the fins which causes the termination to behave less ideally. Finally, although the fin thickness was only 0.002 in, the effect of the finite metal thickness was unknown. It is to be expected, however, that any error due to fin thickness will increase with decreasing W/b since the field becomes more concentrated in the vicinity of the fins. This deduction follows from perturbation theory.

C. Effect of Septum Length

The effect of septum length was investigated numerically by calculating the septum reactance for various values of

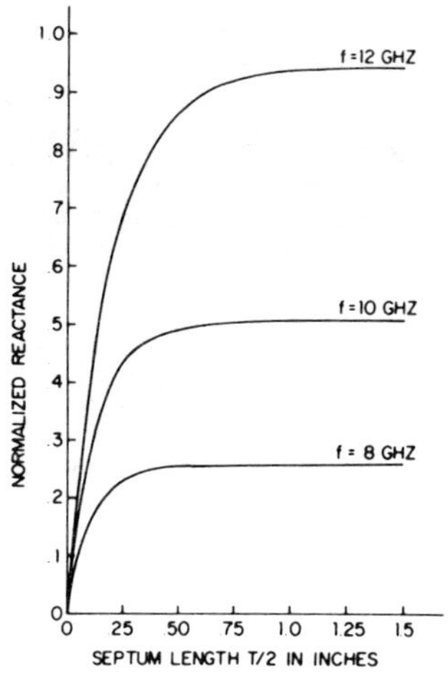

Fig. 9. Normalized septum reactance versus septum length for several frequencies. Fins are centered, $W/b=1$, $\epsilon_{r2}=1$ and shield is WR (90) X-band guide.

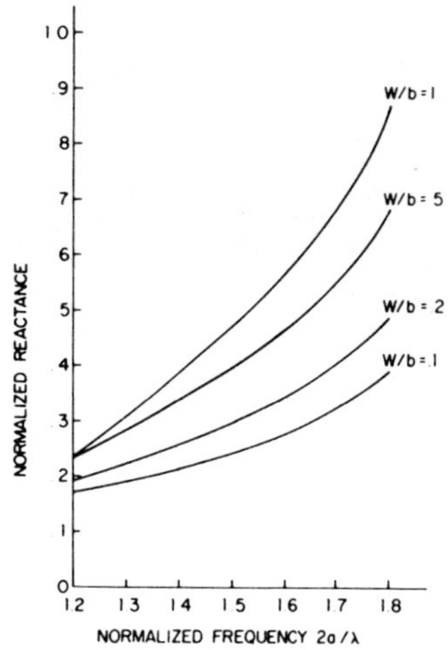

Fig. 10. Normalized septum reactance versus normalized frequency for a fin-line with $\epsilon_{r2}=2.2$, $b/a=0.5$, $h_1/a=0.5$, $D/a=0$, and $T\to\infty$.

distance from the leading edge of the septum to the end of the cavity. This is shown in Fig. 9. For all frequencies, the reactance increases rapidly as the length of the septum is increased and then saturates. Once the distance to the end wall of the cavity has become sufficiently large there is negligible interaction between the field and the end wall. Beyond this point the reactance does not change with further increases in septum length. It can also be seen from Fig. 9 that the septum length required to reach maximum reactance increases with frequency. This is a result of the fact that the evanescent field behind the leading edge of the septum decays more slowly as the frequency increases.

Some experimental measurements were made to verify the computed curves shown in Fig. 9. The results were quite satisfactory. These data are not shown here, however, since the test fixture was not designed in such a way that highly accurate results could be obtained.

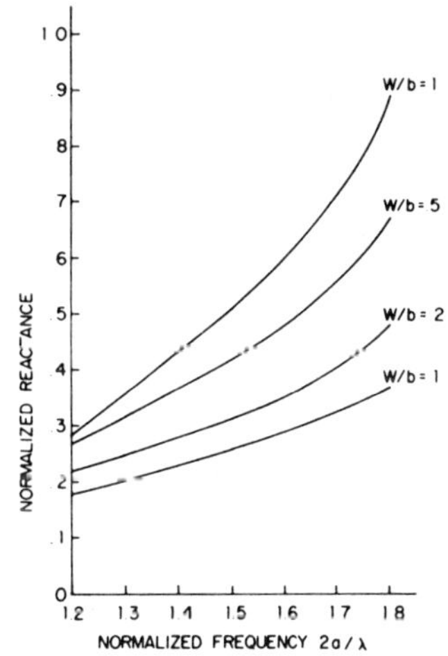

Fig. 11. Normalized septum reactance versus normalized frequency for a fin-line with $\epsilon_{r2}=2.2$, $b/a=0.5$, $h_1/a=0.5$, $D/a=0.05$, and $T\to\infty$.

IV. Design Curves

The structure which has been discussed here is described by a number of independently variable parameters. This precludes the possibility of producing any finite set of universal design curves. However, in practice fin-lines are normally constructed with the fins at or near the center of the shield because of bandwidth considerations. Further, if dielectric is used, it is normally thin (typically, $D=0.005$ in) and one popular choice of material has $\epsilon_r=2.2$. Thus the design curves shown in Figs. 10–12 should prove useful in many cases of practical interest. These curves cover the frequency range over which the empty shield would normally be operated in the TE_{10} waveguide mode.

An inspection of the curves in Figs. 10–12 shows that the reactance tends to increase as the thickness of the dielectric increases. The increase in reactance results from a relatively higher proportion of stored magnetic energy beyond the leading edge of the septum. This in turn is caused by a slower decay of the field on the dielectric loaded side of the septum. It is not universally true that the dielectric

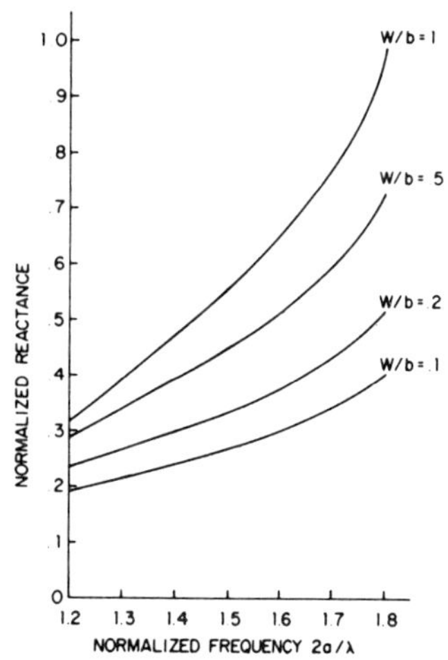

Fig. 12. Normalized septum reactance versus normalized frequency for a fin-line with $\epsilon_{r2}=2.2$, $b/a=0.5$, $h_1/a=0.5$, $D/a=0.10$, and $T\to\infty$.

will cause an increase in reactance, however. For 2 $a/\lambda=1.8$, for example, the reactance first decreases slightly and then increases as the dielectric thickness is increased from $D/a=0$ to $D/a=0.1$.

V. Conclusions

The application of the spectral domain method to the analysis of a fin-line resonator has been described. It has been shown the equivalent reactance of a bifurcating septum may be found by this method. Numerical and experimental data for septum reactance have been presented and have been shown to be in quite good agreement. Since the fin-line wavelength, λ', is used in the calculation of the septum reactance, the results presented here also provide an indirect verification of the accuracy of the wavelength calculations described in [3]. Selected design curves have been presented which should cover many cases encountered in practice.

The successful use of the spectral-domain approach in solving the septum discontinuity problem treated here suggests that the method should prove useful in solving a variety of planar line discontinuity problems.

Acknowledgment

The author wishes to acknowlege the contribution of LT. Guy Miller, U.S. Navy who supervised the design of the experimental apparatus and provided some of the experimental data reported here.

References

[1] J. B. Knorr and K-D. Kuchler, "Analysis of coupled slots and coplanar strips on dielectric substrate," *IEEE Trans. Microwave Theory Tech.*, vol. MTT-23, pp. 541–548, July 1975.

[2] J. B. Knorr and A. Tufekcioglu, "Spectral-domain calculation of microstrip characteristics impedance," *IEEE Trans. Microwave Theory Tech.*, vol. MTT-23, pp. 725–728, Sept. 1975.

[3] J. B. Knorr and P. M. Shayda, "Millimeter-wave fin-line characteristics," *IEEE Trans. Microwave Theory Tech.*, vol. MTT-28, pp. 737–743, July 1980.

[4] T. Itoh, "Analysis of microstrip resonators," *IEEE Trans. Microwave Theory Tech.*, vol. MTT-22, pp. 946–962, Nov. 1974.

[5] R. Harrington, *Time Harmonic Electromagnetic Fields*. New York: McGraw-Hill, 1961.

[6] N. Marcuvitz, *Waveguide Handbook*. New York: McGraw-Hill, 1951.

5.1.3

Scattering Coefficients of an Inductive Strip in a Finline: Theory and Experiment

JEFFREY B. KNORR, SENIOR MEMBER, IEEE, AND JOHN C. DEAL

Abstract—This paper describes the application of the spectral-domain method to the computation of the scattering coefficients of an inductive strip in a millimeter-wave finline. Measured scattering data are compared with numerical data to establish the accuracy of the results. The predicted and measured responses of several finline resonators are also compared.

I. Introduction

THE USE OF finline as a transmission structure for millimeter-wave integrated circuits was first proposed by Meier in 1974 [1]. Since that time, the practical utility of finline has been discussed in the literature by numerous authors and various theoretical analyses have been published. Recently, a comprehensive review, "The status of printed millimeter-wave E-plane circuits," was published by Solbach [2]. That review contains an exhaustive list of references to the existing literature so that effort will not be duplicated here. Finline has become established as a viable structure and, therefore, analyses contributing to its effective utilization are of widespread interest.

The spectral-domain method has proven to be a particularly versatile technique for analyzing the behavior of planar transmission lines in general and finline in particular. Knorr and Shayda have applied this method to determine the transmission characteristics of finline [3] and Jansen [4] and Knorr [5] have used it to investigate the end effect in a shorted finline. The natural extension of the spectral-domain technique to various discontinuity problems has been pointed out by Knorr [5] and by Koster and Jansen [6], [7].

One discontinuity of considerable practical importance is the inductive strip in finline. It is used to construct simple resonators as well as more complicated multiresonator filters. The geometry for an inductive strip is illustrated in Fig. 1, and a cross-sectional view of the finline in which the strip resides is shown in Fig. 2. The electric field in the plane containing the fins is predominantly in the x-direction. A wave incident on the strip induces a current on the strip and the associated stored magnetic energy of the evanescent field gives the strip its inductive property. If the strip is very long, the field decays with distance and the strip behaves like a shorting septum [5]. If the strip is sufficiently short, the incident wave is partially backscattered and partially forward scattered. The scattering coefficients s_{11} and s_{21} are functions of strip width, fin separation, dielectric constant, and shield size. Theoretically, the problem may be solved by computing the odd- and even-mode resonant lengths of finline cavities which are end coupled by the strip. These lengths are then used to compute the complex scattering coefficients of the strip at the specified frequency.

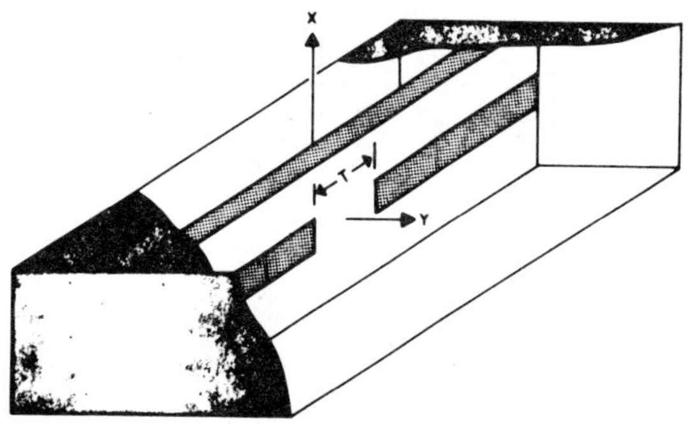

Fig. 1. View of finline inductive strip of axial length T centered in a finline cavity.

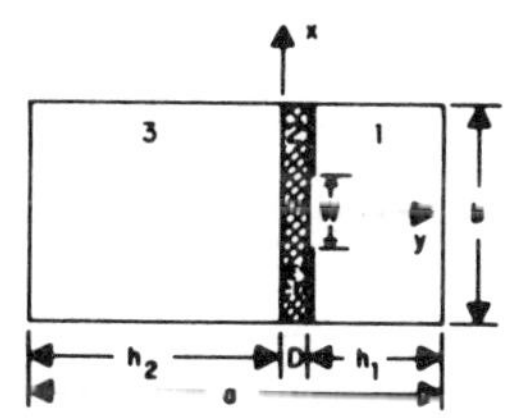

Fig. 2. Cross-sectional view of a finline.

The purpose of this paper is to present the results of both theoretical and experimental investigations of an inductive strip in a unilateral finline. The problem is treated theoretically by applying the spectral-domain method to finline resonators which are end-coupled by a thin, lossless strip. This permits the scattering matrix of the strip to be computed. Experimentally, the strips were investigated in two ways. First, the scattering coefficients for various strips were measured directly. When the strip resides on a dielectric substrate, however, direct measurement of scattering coefficients presents serious practical difficulties since the reference planes for the measurements cannot be easily established. Thus, a second method, measurement of resonator response, was also employed. Two identical strips were used to form a resonator, and the strip scattering

Manuscript received February 20, 1985; revised May 14, 1985.
J. B. Knorr is with the Department of Electrical and Computer Engineering, Naval Postgraduate School, Monterey, CA 93943.
J. C. Deal is with the Department of Electrical Engineering, U.S. Military Academy, West Point, NY 10996.

J.B. Knorr and J.C. Deal, *IEEE Transactions on Microwave Theory and Techniques*, Vol. MTT-33, No. 10, Oct. 1985, pp. 1011–1017.

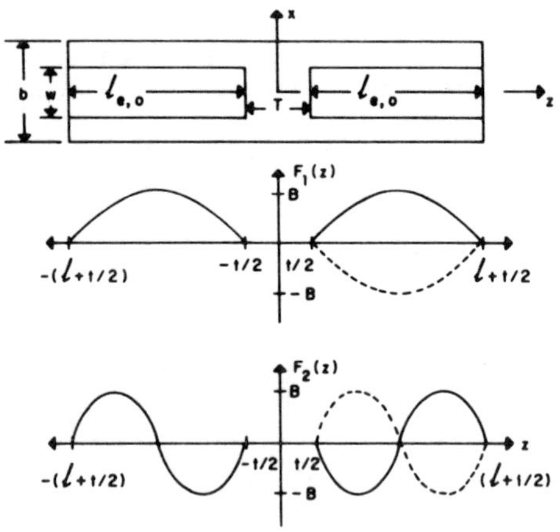

Fig. 3. Coupled finline cavities and basis functions used to expand the even- (solid) and odd- (dashed) mode electric field.

coefficients were extracted from the resonator response. Excellent agreement between numerical and experimental data is demonstrated, thereby establishing the validity of the computational results. Section II of this paper describes the theoretical and experimental methods, and Section III contains the numerical and experimental results. Conclusions appear in Section IV.

II. Theoretical and Experimental Methods

A. *Theoretical Method*

The spectral-domain solution of the finline cavity problem proceeds as described by Knorr [5]. The particular configuration of interest here, end-coupled resonators, is accommodated by proper choice of modes and a basis set with which the fields of the modes can be expanded. The z-component of the electric field between the fins is assumed to be negligibly small. The geometric structure of the coupled cavities and field distributions of the selected basis functions for the z-dependence of the x-directed electric field e_x^z for both odd and even modes are described in Fig. 3. The end walls are perfect shorts, and the inductive discontinuity is in the center of the structure. The solid curve in Fig. 3 represents the field distribution of the even mode and the dashed curve represents that of the odd mode.

The electric field between the fins may be expanded in the basis set depicted in Fig. 3 as

$$e_x(x,z) = A\sum_{q=1}^{Q} B_q F_q(z), \qquad (|x|<W/2,\ t/2\leqslant |z|\leqslant l+t/2) \quad (1)$$

where the $F_q(z)$ are defined as

$$F_q(z)=\begin{cases} \sin(q\pi/l)[z+(l+t)/2] & (z<0, q \text{ even})\\ \pm\sin(q\pi/l)[z-(l+t)/2] & (z>0, q \text{ even})\\ \cos(q\pi/l)[z+(l+t)/2] & (z<0, q \text{ odd})\\ \pm\cos(q\pi/l)[z-(l+t)/2] & (z>0, q \text{ odd})\end{cases} \quad (2)$$

For the sine function, plus (+) applies for the odd mode and minus (−) for the even mode, while for the cosine function, plus (+) applies for the even mode and minus (−) for the odd mode.

In the transform domain,

$$F[e_x(x,z)] = E_x^x(\alpha_n)\sum_{q=1}^{Q} B_q E_{xq}^z(\xi_k) \quad (3)$$

where α_n and ξ_k are the x and z transform variables, respectively, as defined by [5, eqs. (4) and (5)]. For q even

$$E_{xq}^{zs}(\xi_k)=\begin{cases} (-1)^q(q\pi l_e/2)\dfrac{\sin\theta}{\theta^2-\left(\frac{q\pi}{2}\right)^2}2\sin\xi_k\left(\dfrac{l_e+t}{2}\right) & \text{(even mode)}\\ (-1)^q(q\pi l_o/2)\dfrac{\sin\theta}{\theta^2-\left(\frac{q\pi}{2}\right)^2}2j\cos\xi_k\left(\dfrac{l_o+t}{2}\right) & \text{(odd mode)}\end{cases} \quad (4a)$$

and for q odd

$$E_{xq}^{zc}(\xi_k)=\begin{cases} (-1)^q(q\pi l_e/2)\dfrac{\cos\theta}{\theta^2-\left(\frac{q\pi}{2}\right)^2}2\cos\xi_k\left(\dfrac{l_e+t}{2}\right) & \text{(even mode)}\\ (-1)^q(q\pi l_o/2)\dfrac{\cos\theta}{\theta^2-\left(\frac{q\pi}{2}\right)^2}2j\sin\xi_k\left(\dfrac{l_o+t}{2}\right) & \text{(odd mode)}\end{cases} \quad (4b)$$

where $\theta=(\xi_k l)/2$. Using Galerkin's Method with the inner product defined as given in [5], we now obtain

$$\langle G_{11}E_x, E_x\rangle = 0 \quad (5)$$

where $E_z=0$ is assumed and where G_{11} is the Fourier transformed dyadic Green's function component. Assuming $E_x=E_x^xE_x^z$ yields

$$\left[\langle\{E_x^x(\alpha_n)\}^2 G_{11}E_{xi}^z(\xi_k), E_{xj}^z(\xi_k)\rangle\right]\left[B_j\right]=0 \quad (6)$$

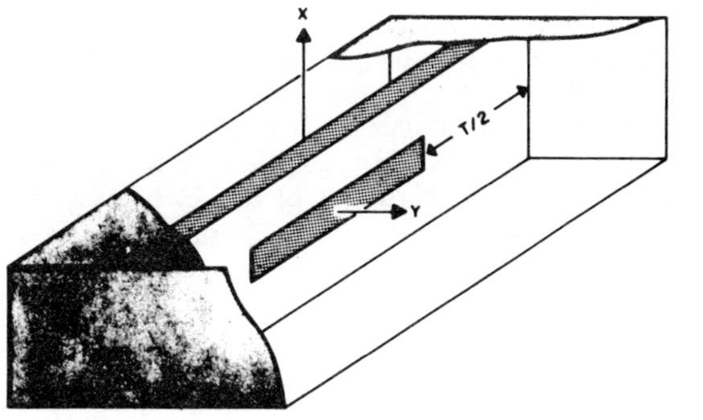

Fig. 4. View of a finline cavity terminated with inductive septae having half the length of the inductive strip between the coupled cavities shown in Fig. 1. The septae are backed by perfect electric walls.

where i and j range from 1 to Q and the E^z_{xi} and E^z_{xj} are sine functions for even index and cosine functions for odd index. The matrix represented by the bracketed term on the left in (6) is $Q \times Q$ and contains the elements of the inner product defined over α_n and ξ_k [5, eq. (11)].

Computationally, the first step is to determine the guide wavelength (see [3]) and then to find the odd- and even-mode resonant lengths of the two coupled cavities. The resonant lengths are found by computing the roots of the characteristic equation

$$\det\left[\langle\{E^x_x(\alpha_n)\}^2 G_{11} E^z_{xi}(\xi_k), E^z_{xj}(\xi_k)\rangle\right] = 0. \quad (7)$$

It should be noted that a single-term expansion leads to inaccurate results as discussed in [5].

A compromise must be made between numerical accuracy and computation time, which are both affected by the size of the search interval, limits of summation for the inner product, iteration accuracy (answer residue), and number of expansion terms used (order of the matrix). It was determined through convergence testing that a matrix of order 8 would permit odd- and even-mode resonant lengths to be computed with an accuracy of better than 2 percent when $W/b = 0.5$ or $W/b = 0.25$. For $W/b = 1$, a 12-term expansion of e_x was used.

Several limiting cases were investigated to check the correctness of the computer program. First, as the length T of the inductive strip increases to the point where the two cavities are uncoupled, the odd- and even-mode resonant lengths become equal. The end effect is the same as for a semi-infinite septum [5]. Second, as the length of the strip becomes infinitesimally small, the odd-mode resonant length approaches $\lambda'/2$. Lastly, consider a single cavity terminated at both ends by inductive septae having half the length of the strip coupling the two cavities and backed by perfect shorting walls as illustrated in Fig. 4. The resonant length of the odd mode of the coupled cavities is related to the resonant length of the single cavity by

$$l_0/\lambda' = l/\lambda' + l_{eq}/\lambda' \quad (8a)$$

where

$$l_{eq}/\lambda' = (1/2\pi)\tan^{-1}(x_{eq}) \quad (8b)$$

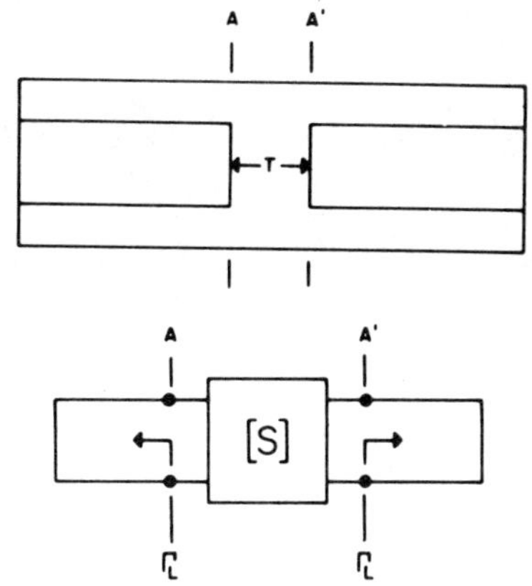

Fig. 5. Representation of the inductive strip by a two-port network characterized by the scattering matrix $[S]$.

and λ' is the guide wavelength. This last check is useful since the equivalent reactance x_{eq} of a variable length septum in a single cavity has been computed previously [5].

All these limiting cases were checked and the results were found to be correct to within 0.5 percent. This helped to develop confidence in the computer program.

The scattering coefficients of the inductive strip can be computed from the odd- and even-mode resonant lengths. To accomplish this, the strip is represented by a two-port network as depicted in Fig. 5. The inductive strip is lossless, reciprocal, and symmetric so its scattering matrix is unitary with $s_{12} = s_{21}$ and $s_{11} = s_{22}$. The unitary property leads to

$$|s_{21}| = \left[1 - |s_{11}|^2\right]^{1/2} \quad (9a)$$

$$\theta_{21} = \theta_{11} \pm \pi/2 \quad (9b)$$

where, in (9b), the minus $(-)$ sign is chosen because the strip is inductive. With reference to Fig. 5, Γ_L is the reflection coefficient looking into the perfectly shorted finline at the reference plane A or A' and may be written as

$$\Gamma_L = -1e^{-j2\beta l}. \quad (10)$$

Using (10), the scattering equations may be solved to find

$$s_{11} = \frac{1}{2}\left[\frac{\Gamma_{Le} + \Gamma_{Lo}}{\Gamma_{Le}\Gamma_{Lo}}\right] \quad (11a)$$

$$s_{21} = \frac{1}{2}\left[\frac{\Gamma_{Le} - \Gamma_{Lo}}{\Gamma_{Le}\Gamma_{Lo}}\right] \quad (11b)$$

where

$$\Gamma_{Le} = -e^{-j4\pi l_e/\lambda'} \quad (12a)$$

and

$$\Gamma_{Lo} = -e^{-j4\pi l_o/\lambda'}. \quad (12b)$$

From (11) and (12), we obtain the final result

$$s_{11} = \exp\left[j2\pi(l_e/\lambda' + l_o/\lambda')\right]\cos 2\pi(l_e/\lambda' - l_o/\lambda') \quad (13a)$$

$$s_{21} = j\exp\left[j2\pi(l_e/\lambda' + l_o/\lambda')\right]\sin 2\pi(l_e/\lambda' - l_o/\lambda'). \quad (13b)$$

Throughout the above equations, the subscripts e and o denote the even and odd modes of the coupled cavities.

To summarize, the field equations are first solved to find the odd- and even-mode resonant lengths of the coupled finline cavities at a specified frequency. These lengths are then used to compute the scattering coefficients of the inductive strip coupling and cavities. It is evident from (13) that the magnitudes of the scattering coefficients depend upon the difference between the odd- and even-mode resonant lengths. It is therefore necessary to compute these lengths very accurately to obtain accurate values for the scattering coefficients. This implies that e_x must be expanded in a basis set that can approximate the true field distribution with a reasonable number of terms.

B. *Experimental Method*

The scattering coefficients of inductive strips in finlines have been measured in two ways. Direct measurements were made for strips etched from 2-mil-thick beryllium copper and centered in a WR (90) waveguide shield. The test fixture (shield) was 15 in long. For those cases where $W/b<1.0$, the finline was matched to the WR (90) empty guide with a linearly tapered section. Except for the length, the test fixture, shorting blocks, and load are as described in [5] (see [5] for photographs). An HP 8409B Automatic Microwave Network Analyzer was used to measure $|s_{11}|$ and $|s_{21}|$ and θ_{11} was measured using a slotted line which seemed to produce better results.

Indirect measurement of strip scattering coefficients is also possible. In general, a reciprocal, symmetric two-port network is characterized by a scattering matrix which has four independent parameters: $|s_{11}|$, θ_{11}, $|s_{21}|$, and θ_{21}. Two identical strips can be used to form a high-Q finline resonator. It can be shown that the resonant frequency (f_0), Q, insertion loss (at f_0), and return loss (at f_0) are determined by the four independent parameters of the scattering matrix. If the strips (two-port networks) are lossless, then the situation is even simpler. The resonator properties are determined by $|s_{11}|$ and θ_{11} (see (9)). Conversely, the scattering coefficients (at f_0) may be expressed in terms of the resonator parameters. Measurement of the resonator response therefore permits the scattering coefficients to be computed. This approach has the advantage of requiring no phase information from the measurements. A scalar analyzer can be used. The approach is particularly attractive when it is difficult to establish reference planes for measurement of phase as it is when a strip is printed on a dielectric substrate. It is also a useful approach if a vector analyzer is unavailable, as is often the case at millimeter wavelengths. The disadvantage of this approach, of course, is that one resonator provides data for the computation of scattering coefficients at only a single frequency, f_0. Identical strips must be used to build resonators of several different lengths (different f_0's) to obtain values of scattering coefficients over a range of frequencies.

The derivation of the relations between resonator characteristics and strip scattering coefficients is lengthy. Therefore, a simpler and perhaps more satisfying approach will be used here. Numerical and experimental resonator insertion loss data will be presented and compared. Computed strip scattering coefficients have been used in the MICRO-Compact® program [8] to predict the resonator response numerically. The agreement between the predicted and measured response of the resonator is thus a good indication of the accuracy of the computed strip scattering coefficients. Since strips are used to build practical components such as filters, this type of comparison would ultimately be desired anyway and is therefore quite satisfying.

III. Numerical and Experimental Results

The spectral-domain analysis described in the previous section was implemented in a computer program, FINCAV. Numerical data generated using FINCAV were compared with experimental data to verify the accuracy of the computational results. To ease measurement difficulties, all measurements were made using X-band test structures. Shield size was 0.4 in$\times$0.9 in (WR (90)), and for those cases where no dielectric was present, the fins were centered in the shield. For $W/b<1.0$, a linear taper was employed and where dielectric was present, a dielectric taper was used for matching purposes. Return loss for back to back tapers was measured and found to be greater than 20 dB. Direct measurement of scattering coefficients was accomplished using both a slotted line and an HP 8409B Automatic Vector Microwave Network Analyzer. In those cases where two inductive strips were used to form a resonator, the response was measured using an HP 8756 Scalar Network Analyzer with an HP 8350 Sweep Generator and plotter. Since a considerable amount of experimental data was collected, only representative results will be presented here.

A. *Scattering Coefficient Measurements*

Direct measurement of inductive strip scattering coefficients was made for several cases where no dielectric was present ($\epsilon_{r2}=1$). With no dielectric, it is easy to establish the reference planes required for the measurement. Fig. 6 and 7 show $|s_{11}|$ and θ_{11} for several strips in finlines with $W/b=1$. In the worst case, the difference between numerical and experimental data is about 7 percent. The rms error is much less. Similar data for $W/b=0.25$ is shown in Figs. 8 and 9. The worst-case difference there is 5 percent. Data for s_{21} is not presented here since similar agreement was obtained. For lossless strips, s_{21} can be found from s_{11} using (9).

B. *Resonator Measurements*

A number of resonators were constructed using two identical inductive strips. Analysis shows that the resonant frequency is determined by

$$\theta_{11}=2\pi l/\lambda' \tag{14}$$

where l is the distance between the strips. The shape of the transmission response curve over the band is determined by both $|s_{11}|$ and θ_{11}. As mentioned earlier, s_{11} completely defines the behavior of a lossless strip (see (9)). The reso-

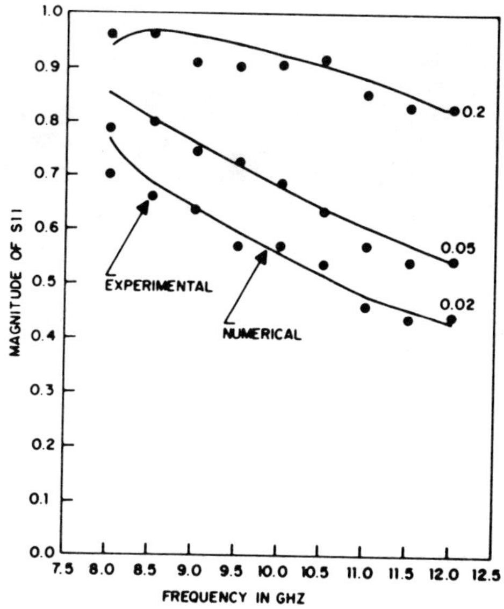

Fig. 6. $|s_{11}|$ versus frequency for $T = 0.02$-, 0.05-, and 0.2-in inductive strips centered in a WR (90) finline with $W/b = 1.0$ and $\epsilon_{r2} = 1.0$.

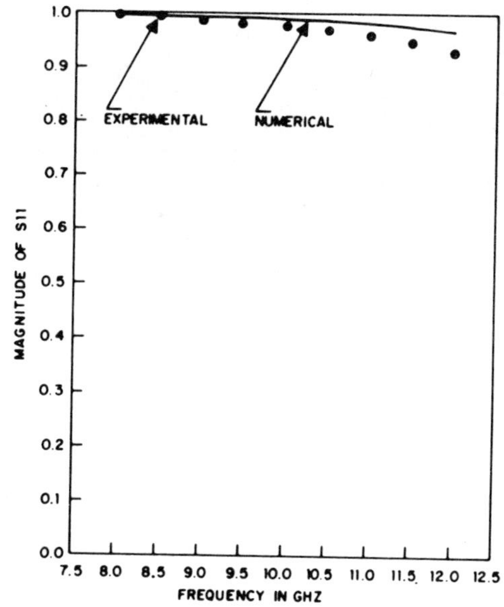

Fig. 8. $|s_{11}|$ versus frequency for a $T = 0.2$-in inductive strip centered in a WR (90) finline with $W/b = 0.25$ and $\epsilon_{r2} = 1.0$.

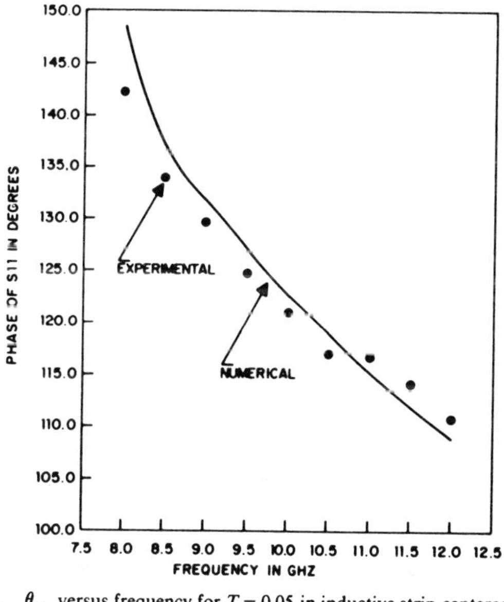

Fig. 7. θ_{11} versus frequency for $T = 0.05$-in inductive strip centered in a WR (90) finline with $W/b = 1.0$ and $\epsilon_{r2} = 1.0$.

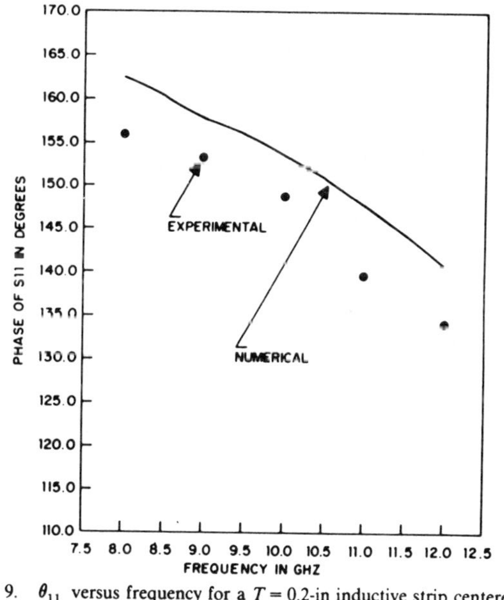

Fig. 9. θ_{11} versus frequency for a $T = 0.2$-in inductive strip centered in a WR (90) finline with $W/b = 0.25$ and $\epsilon_{r2} = 1.0$.

nator approach is particularly attractive when $\epsilon_{r2} > 1$ because it is difficult to establish reference planes for vector measurements.

The FINCAV program was used to compute scattering coefficients for the resonator inductive strips. The strips were modeled as two-port black boxes in MICRO-Compact®, and the computed values of the scattering coefficients were supplied as data for the 8–12-GHz band in 1-GHz steps. The resonator response predicted by MICRO-Compact® was compared with the measured response. Results for two resonators, one with dielectric and one without dielectric, will be presented here.

The first resonator had $W/b = 1.0$ and consisted of strips with length $T = 0.22$ in separated by a distance $l = 0.53$ in and centered in the WR (90) shield. The measured transmission response (solid curve) is shown in Fig. 10 along with the response predicted using MICRO-Compact® (circled points). The agreement is quite good. The predicted and measured resonant frequencies differed by only 0.5 percent (50 MHz) and the difference in the two transmission curves is only about 2 dB. The ripple in the measured response is due to the low VSWR coax-waveguide transitions which were used on each end of the finline test fixture.

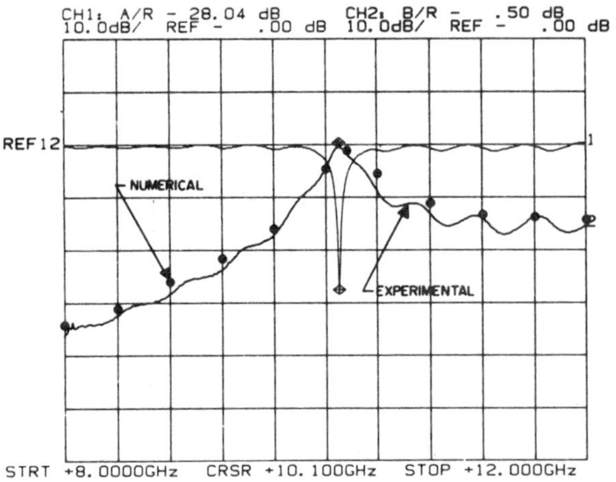

Fig. 10. Measured (solid curve) and predicted (circled points) transmission loss and return loss for a finline resonator. Inductive strips with $T = 0.22$ in are separated by 0.53 in and centered in the WR (90) shield with $\epsilon_{r2} = 1.0$.

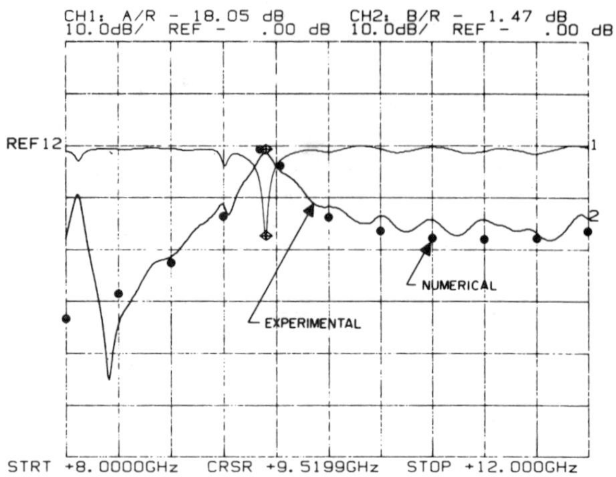

Fig. 11. Measured (solid curve) and predicted (circled points) transmission loss and return loss for a finline resonator. Inductive strips with $T = 0.22$ in are separated by 0.53 in and reside on a 1/32-in dielectric substrate with $\epsilon_{r2} = 2.5$, which is centered in the WR (90) shield.

The second resonator was identical to the first (within manufacturing tolerances) except the strips were printed on a 1/32 in dielectric substrate with $\epsilon_r = 2.5$. The substrate was centered in the WR (90) shield so the fins were 0.015-in off center. Slots extending through the shield walls were used to support the substrate, and the wall thickness was chosen equal to one quarter of a dielectric wavelength at band center, 10 GHz. Ideally, this causes an RF short circuit to appear at the inside wall but clearly represents a difference between the computer model and the actual test fixture. The measured transmission loss (solid curve) of this filter is shown in Fig. 11 along with the response predicted using MICRO-Compact® (circled points). The predicted and measured resonant frequencies differ by 0.2 percent (20 MHz). The difference between the measured and predicted response is generally 2–3 dB, although there are clearly several anomalies in the measured response below resonance. These were due to the slot which extended almost the full length of the 10-in test fixture. This suggests that the slots should not be any longer than necessary to support the dielectric substrate. It should be mentioned that the finline section between the two inductive strips in this resonator was also modeled as a two-port black box. Transmission phase data was derived from the computed finline dispersion. The overall response of this resonator is almost identical to that of the first resonator except that the resonant frequency is shifted down about 600 MHz (approx. 6 percent) by the presence of the dielectric substrate.

The agreement between predicted and measured resonator response was quite satisfying since strip scattering coefficients would most often be used in this manner.

C. Errors

Although good agreement between numerical and experimental data has been demonstrated, some discussion of the discrepancies is warranted. There are a number of sources of error which contribute to the differences seen in Figs. 6–11. Numerical computations were carried out with a goal of obtaining better than 2-percent accuracy for the values of odd- and even-mode resonant lengths. However, (13) shows that the scattering coefficients depend upon the difference, $(\lambda_e/\lambda' - \lambda_o/\lambda')$, so the s_{ij} may be in error by more than 2 percent. Another source of numerical error arises in the prediction of resonator response and is due to interpolation at frequencies between those for which scattering coefficient data were entered in the MICRO-Compact® program. Experimentally, there are a number of sources of error. Error in the determination of the reference plane for vector measurements and error due to transitions are discussed in [5]. Other sources of experimental error include losses, manufacturing tolerances, and instrumentation error (principally, sweeper frequency accuracy). Finally, when dielectric is used, any slot or groove used to support the substrate results in a physical structure which is different from the numerical model. These factors easily account for the observed differences between numerical and experimental data. Considerable effort would be required to further resolve the observed difference.

IV. Conclusions

This paper treats the general case of an inductive strip in a finline. The strip may be printed on a dielectric substrate and any value of $0 < W/b \leq 1.0$ is permitted. Previous papers have been restricted to special cases and have generally used other mathematical methods of analysis. This paper extends analysis to the most general case.

Numerical and experimental scattering coefficient data are presented for strips, and good agreement is demonstrated. The transmission response for several resonators has also been measured and compared with the predicted response based on use of the computed strip scattering coefficients. Results agree within the limits imposed by the various sources of numerical and experimental error.

Lastly, the data presented here demonstrates the applicability of the spectral-domain technique to this discontinuity problem. This is particularly significant. Since the spectral-domain technique has previously been shown to work successfully for various configurations of strips and slots, the possibility exists that a single program could be created to analyze the behavior of various planar structures using this technique.

References

[1] P. J. Meier, "Integrated fin-line millimeter components," *IEEE Trans. Microwave Theory Tech.*, vol. MTT-22, pp. 1209–1216, Dec. 1974.

[2] K. Solbach, "The status of printed millimeter-wave *E*-plane circuits," *IEEE Trans. Microwave Theory Tech.*, vol. MTT-31, pp. 107–121, Feb. 1983.

[3] J. B. Knorr and P. M. Shayda, "Millimeter wave fin-line characteristics," *IEEE Trans. Microwave Theory Tech.*, vol. MTT-28, pp. 737–743, July 1980.

[4] R. H. Jansen, "Hybrid mode analysis of end effects of planar microwave and millimetrewave transmission lines," *Proc. Inst. Elec. Eng.*, vol. 128, pt. H, pp. 77–86, Apr. 1981.

[5] J. B. Knorr, "Equivalent reactance of a shorting septum in a finline: theory and experiment," *IEEE Trans. Microwave Theory Tech.*, vol. MTT-29, pp. 1196–1202, Nov. 1981.

[6] N. H. L. Koster and R. H. Jansen, "Some new results on the equivalent circuit parameters of the inductive strip discontinuity in unilateral fin-lines," *Arch. Elek. Übertragung.*, vol. 35, no. 12, pp. 497–499, 1981.

[7] R. H. Jansen and N. H. L. Koster, "A unified CAD basis for the frequency dependent characterization of strip, slot and coplanar MIC components," in *11th Eur. Microwave Conf. Proc.* (Amsterdam), 1981, pp. 682–687.

[8] MICRO-Compact®, available from Compact Software, 1131 San Antonio Road, Palo Alto, CA 94303.

5.1.4

Theoretical and Experimental Investigation of Finline Discontinuities

MARYLINE HELARD, JACQUES CITERNE, ODILE PICON, AND VICTOR FOUAD HANNA

Abstract—The dominant and the first-five higher order modes in a unilateral finline are precisely described from a thorough spectral-domain approach. Then, using the modal analysis, coupling coefficients between eigenmodes at a discontinuity that have to be introduced into the scattering matrix formulation are directly computed in the spectral-domain, and, consequently, the equivalent circuit parameters of the discontinuity are determined. Finally, finline discontinuities often used for impedance transformation are investigated and a good agreement between theoretical and experimental results is reported.

I. Introduction

The field theoretical solution of finline discontinuities presents a complex problem. This explains the small number of concise and rigorous attempts of analyses reported in the recent literature [1].

In fact, the major complexity for efficient and accurate analysis of single or complex discontinuities in finline structures lies in the necessary treatment of hybrid eigenmodes to which there is no closed-form solution.

Firstly, the modal analysis [2] has been applied [3] to single and double steps in the finline slot width, while the characterization of uniform finline structures was obtained by the moment mode-matching technique [4].

This pioneering work provided an insight into CAD potentialities of integrated millimeter-wave finline circuits [5] that such a field theoretical solution could provide. However, a hasty model in terms of equivalent circuits prevents an unbiased evaluation of computed intermediate field parameters.

A. Beyer and I. Wolff [6] used a nearly similar approach of the finline treatment and solved the discontinuity problem by a flexible and elegant combination of the moment mode-matching technique and the modal analysis.

In solving the problem of the rectangular waveguide finline tapers, use was made, for the first time, of the powerful method of coupled modes [7] accompanied by the spectral-domain approach. It was shown that such a combination offers a very efficient computational scheme of the return and insertion losses of the "back-to-back" tapering transition. Unfortunately, experiments were unable to validate the presumed exact fields computations.

Manuscript received February 12, 1985; revised May 31, 1985.

M. Helard is with the Centre Commun d'Etudes des Télécommunications et Télédivision, 35510 Cesson Sevigne, France.

J. Citerne is with the Laboratories Structures Rayonnantes, Department Genie Electrique, INSA, 35031 Rennes Cedex, France.

O. Picon and V. Fouad Hanna are with the Division Espace et Transmission Radioélectrique, Centre National d'Etudes des Télécommunications, 92131 Issy-Les-Moulineaux, France.

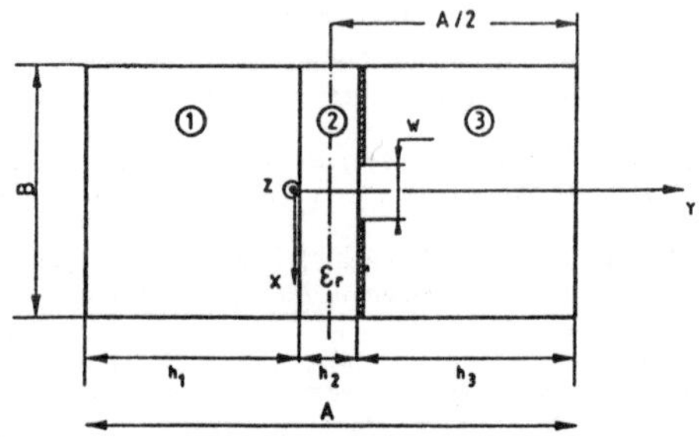

Fig. 1. Cross-section parameters of a unilateral finline. In the *Ku*-band (12–18 GHz), $A = 15.8$ mm, $B = 7.9$ mm, $h_2 = 0.635$ mm, and $\epsilon_r = 9.6$ (Alumina). In the *Ka*-band (26–40 GHz), $A = 7.112$ mm, $B = 3.556$ mm, $h_2 = 0.254$ mm, and $\epsilon_r = 2.22$ (Duroid 5880).

Recently, the transverse resonance techniques [8] were used to compute the resonant frequencies of a resonant structure containing the finline discontinuity and in consequence to determine the equivalent-circuit parameters of the discontinuity.

This paper presents a precise field theoretical solution of finline discontinuities using a combination of the spectral-domain approach and the modal analysis.

The aim of the paper is threefold: 1) to give an accurate evaluation of six eigenmodes of the unilateral finline via a thorough spectral-domain approach; 2) to perform calculations of the generalized scattering matrix of a single-step slot width discontinuity by combining the direct modal analysis and the spectral-domain approach [9] and, consequently, to determine the equivalent-circuit parameters of this discontinuity; 3) to compare the theory and experiments in the *Ku*- and *Ka*-bands on complex discontinuities [10], as this is the only method that enables the direct comparison of measured and calculated scattering parameters of either the simple or the complex discontinuity.

II. Spectral-Domain Approach of a Unilateral Finline

The axial field components $E_{z,i}(x,y)$ and $H_{z,i}(x,y)$ in the ith region ($i = 1,2,3$) are expanded in Fourier series (Fig. 1) within their domain

$$-B/2 \leqslant x \leqslant B/2.$$

M. Helard, J. Citerne, O. Picon, and V. Fouad Hanna, *IEEE Transactions on Microwave Theory and Techniques*, Vol. MTT-33, No. 10, Oct. 1985, pp. 994–1003.

For odd waves, the following expansions of $E_{z,i}(x,y)$ and $H_{z,i}(x,y)$ are valid:

$$E_{zi}(x,y)=\sum_{m=1}^{\infty}\tilde{E}_{zi}(m,y)\sin\alpha_m x$$

$$H_{zi}(x,y)=\sum_{m=0}^{\infty}\tilde{H}_{zi}(m,y)\cos\alpha_m x \quad (1)$$

where quantities with the sign (~) designate the line amplitude (i.e., the mth term of the Fourier series) associated with the space harmonic $\alpha_m = 2m\pi/B$.

The partial differential equations for the axial field components $E_{z,i}(x,y)$ and $H_{z,i}(x,y)$ are also Fourier expanded with respect to x; ordinary differential equations are derived for the mth line amplitudes $\tilde{E}_{z,i}(m,y)$ and $\tilde{H}_{z,i}(m,y)$, respectively.

For a unilateral finline (Fig. 1), these mth line amplitudes are given in Appendix I.

Through the application of boundary conditions at $y=0$ and $y=h_2$, which are also Fourier expanded, the spectral coefficients are related to each other, to the mth line amplitudes of fin surface current components denoted $\tilde{J}_x(m,h_2)$ and $J_z(m,h_2)$, and to the mth line amplitude of the slot aperture field components denoted $\tilde{E}_x(m,h_2)$ and $\tilde{E}_z(m,h_2)$. Extensive algebraic manipulations of these boundary conditions then yield functional equations relating the mth line amplitude of the fin surface current components to the mth line amplitude of the slot aperture field components. For a unilateral finline, the standard computational scheme uses the following admittance matrix representations of these functional equations, namely:

$$\begin{bmatrix} G_{11} & G_{12} \\ G_{21} & G_{22} \end{bmatrix}\begin{bmatrix} \tilde{E}_x(m,h_2) \\ \tilde{E}_z(m,h_2) \end{bmatrix}=\begin{bmatrix} \tilde{J}_x(m,h_2) \\ \tilde{J}_z(m,h_2) \end{bmatrix}. \quad (2)$$

Closed-form expressions of the $\bar{\bar{G}}$ matrix elements are listed in Appendix II.

Now it is the moment to start the numerical part through the application of the Galerkin's form of the general method of moments to the functional equations (2). The slot aperture field components $E_x(x,h_2)$ and $E_z(x,h_2)$ are expanded in terms of two complete sets of R and S basis functions denoted $\mathscr{E}_{x,r}(x,h_2)(r=1,\cdots,R)$ and $\mathscr{E}_{z,s}(x,h_2)(s=1,\cdots,S)$

$$E_x(x,h_2)=\sum_{r=1}^{R}a_r\mathscr{E}_{x,r}(x,h_2)$$

$$E_z(x,h_2)=\sum_{s=1}^{S}b_s\mathscr{E}_{z,s}(x,h_2). \quad (3)$$

Obviously, the basis functions in (3) have nonzero values in the interval $-W/2 \leqslant x \leqslant W/2$ of the cell $-B/2 \leqslant x \leqslant B/2$.

The expansions (3) are displayed in Fourier series to bring the mth line amplitude of the aperture field components and the mth line amplitude of basis functions into relationship denoted by

$$\tilde{E}_x(m,h_2)=\sum_{r=1}^{R}a_r\tilde{\mathscr{E}}_{x,r}(m,h_2)$$

$$\tilde{E}_z(m,h_2)=\sum_{s=1}^{S}b_s\tilde{\mathscr{E}}_{z,s}(m,h_2). \quad (4)$$

The coefficients a_r and b_s involved in (3) and (4) are the first true constants of the waveguide problem.

By using an inner product consistent with the Parseval theorem, the Galerkin's procedure is directly applied to the matrix form (2) in the Fourier domain. A set of $R+S$ homogeneous and linear equations, for which the $R+S$ unknowns are precisely the constants a_r and b_s, is obtained.

Nontrivial solutions of this set of equations occur for zero values of its matrix determinant. The real roots (i.e., $\beta^2>0$) determine the propagating eigenmodes, whereas the imaginary roots (i.e., $\beta^2<0$) determine evanescent ones.

For any given root β, the associated mth line amplitudes of the aperture field components are expressed in terms of the $R+S$ coefficients a_r and b_s involved in (4). By a substitution process of $\tilde{E}_x(m,h_2)$ and $\tilde{E}_z(m,h_2)$ in boundary conditions, the coefficients $A(m)$ through $H(m)$ are determined, as well as the mth line amplitude of the axial field components. The summation of Fourier series gives finally the axial field components of eigenmodes anywhere in the waveguide cross section.

Eigenmode normalization is not a necessary task as far as the waveguide treatment is concerned. However, this operation can be found relevant in the further discontinuity problem. So, applying again the Parseval's theorem, the orthogonality of eigenmodes in lossless waveguides can be expressed directly in the spectral domain as

$$\frac{B}{2}\sum_{m=0}^{\infty}\int_{-h_1}^{h_2+h_3}\left[\tilde{E}_{x,p}(m,y)\tilde{H}^*_{y,p'}(m,y) - \tilde{E}_{y,p}(m,y)\tilde{H}^*_{x,p'}(m,y)\right]dy = s\frac{\beta_p}{|\beta_p|}P\delta_{pp'} \quad (5)$$

where δ_{pp} represents the Kroneker delta and the coefficient s can take the values $+1$ or -1 depending on the propagating or evanescent nature of the eigenmode labelled p. Line amplitudes of transverse field components involved in (5) are derived from longitudinal components given by (1). The eigenmode normalization is carried out if the $R+S$ unknowns a_r and b_r in (3) and (4) associated with a given root β are reevaluated to ensure a power flow value P equal to 1 W in (5).

III. Modal Analysis of Single- and Multiple-Step Slot Width Discontinuities

The problem to be treated is how an incident accessible power entering each side of the junction is apportioned between the various scattered eigenmodes. The solution lies in the derivation of the generalized scattering matrix $[S]$ of the junction.

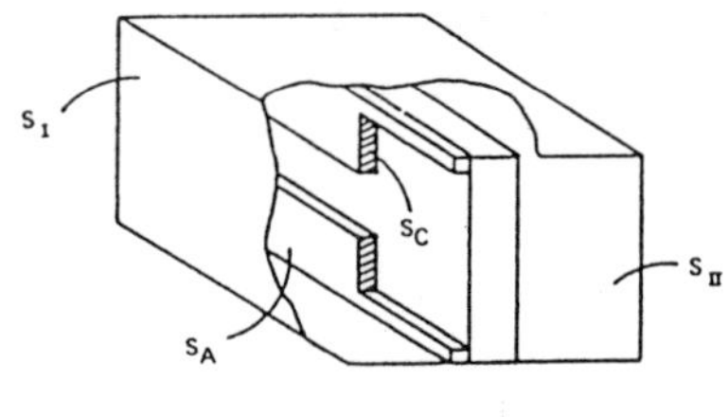

(a)

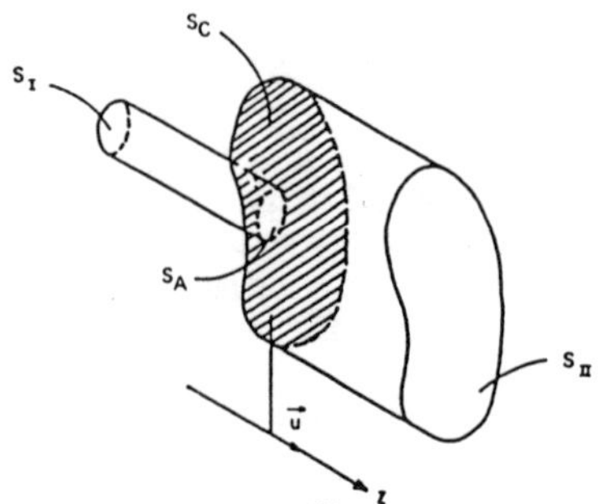

(b)

Fig. 2. (a) A single-step slot width junction between two unilateral finlines. (b) Model for modal analysis.

A single slot width discontinuity as that shown in Fig. 2(a) can be modeled by a junction between two cylindrical closed waveguides as shown in Fig. 2(b). We have to notice that the conducting fins in Fig. 2(a) have negligible thickness. They have been drawn so for a better comprehension of the model of Fig. 2(b).

The transverse electric and magnetic field to the left of the junction plane $z=0$ (waveguide I side) can be expressed as

$$\vec{E}_T^{\mathrm{I}} = \sum_{p=1}^{P} \left(A_p^{\mathrm{I}} + B_p^{\mathrm{I}}\right)\vec{e}_{p,T}^{\,\mathrm{I}}$$

$$\vec{H}_T^{\mathrm{I}} = \sum_{p=1}^{P} \left(A_p^{\mathrm{I}} - B_p^{\mathrm{I}}\right)\vec{h}_{p,T}^{\,\mathrm{I}}. \tag{6}$$

In a similar manner, to the right of the junction plane $z=0$, the expansion of the transverse electric and magnetic fields is written as

$$\vec{E}_T^{\mathrm{II}} = \sum_{q=1}^{Q} \left(A_q^{\mathrm{II}} + B_q^{\mathrm{II}}\right)\vec{e}_{q,T}^{\,\mathrm{II}}$$

$$\vec{H}_T^{\mathrm{II}} = \sum_{q=1}^{Q} \left(B_q^{\mathrm{II}} - A_q^{\mathrm{II}}\right)\vec{h}_{q,T}^{\,\mathrm{II}}. \tag{7}$$

The above expansions involve normalized transverse electric and magnetic fields $\vec{e}_{p,T}^{\,\mathrm{I}}, \vec{h}_{p,T}^{\,\mathrm{I}}$ (resp.: $\vec{e}_{q,T}^{\,\mathrm{II}}, \vec{h}_{q,T}^{\,\mathrm{II}}$) associated with the eigenmode p (resp.: q) of the waveguide I (resp.: II). Eigenmodes p and q are forward traveling waves in the two waveguides I and II, while the modal amplitudes $A_p^{\mathrm{I}}, B_p^{\mathrm{I}}$ (resp. $A_q^{\mathrm{II}}, B_q^{\mathrm{II}}$) are referred to incident and reflected waves in the waveguide I (resp. II) at the junction plane.

If $S_{\mathrm{I}} < S_{\mathrm{II}}$ (this is the case outlined in Fig. 2), the boundary conditions at the junction plane $z=0$ can be expressed as

$$\vec{E}_T^{\mathrm{I}} = \vec{E}_T^{\mathrm{II}} \tag{8a}$$

$$\vec{H}_T^{\mathrm{I}} = \vec{H}_T^{\mathrm{II}} \tag{8b}$$

on the aperture surface S_A and as

$$\vec{E}_T^{\mathrm{II}} = 0 \tag{8c}$$

$$\vec{H}_T^{\mathrm{II}} + \vec{J}_T^{\mathrm{I}} \Lambda \vec{u} = 0 \tag{8d}$$

on the transverse conducting wall S_C.

The junction surfaces S_A and S_C are related to waveguide cross sections S_{I} and S_{II} by $S_A = S_{\mathrm{I}}$ and $S_A + S_C = S_{\mathrm{II}}$.

The surface current $\vec{J}_T^{\mathrm{I}}$ on the conducting wall S_C is labeled with a subscript I to indicate a relationship with waveguide I.

The next step is to transform pairs of boundary functional equations (8a) and (8b) into an equivalent set of linear equations involving the $2(P+Q)$ modal amplitudes A_p^{I}, B_p^{I}, A_q^{II}, and B_q^{II} to be determined.

There is a unique procedure to derive such an equivalent linear set of $2(P+Q)$ equations that is closely related to the basic assumption $S_{\mathrm{I}} < S_{\mathrm{II}}$. This point has never been clarified enough in the literature, especially according to the uniqueness of the solution; therefore, it is summarized briefly in Appendix III.

Thus, the generalized scattering matrix of the simple discontinuity can be constructed from this system of equivalent linear equations which can be written in a matrix form as

$$\begin{aligned} {}^{T}[P]([A^{\mathrm{I}}]+[B^{\mathrm{I}}]) &= [L]([A^{\mathrm{II}}]+[B^{\mathrm{II}}]) \\ [K]([A^{\mathrm{I}}]-[B^{\mathrm{I}}]) &= [N]([B^{\mathrm{II}}]-[A^{\mathrm{II}}]) \end{aligned} \tag{9}$$

with

$$[A^{\mathrm{I}}] = \begin{bmatrix} A_1^{\mathrm{I}} \\ A_2^{\mathrm{I}} \\ \vdots \\ A_p^{I} \\ \vdots \\ A_P^{\mathrm{I}} \end{bmatrix} \quad [B^{\mathrm{I}}] = \begin{bmatrix} B_1^{\mathrm{I}} \\ B_2^{\mathrm{I}} \\ \vdots \\ B_p^{\mathrm{I}} \\ \vdots \\ B_P^{\mathrm{I}} \end{bmatrix}$$

$$[A^{\mathrm{II}}] = \begin{bmatrix} A_1^{\mathrm{II}} \\ A_2^{\mathrm{II}} \\ \vdots \\ A_q^{\mathrm{II}} \\ \vdots \\ A_Q^{\mathrm{II}} \end{bmatrix} \quad [B^{\mathrm{II}}] = \begin{bmatrix} B_1^{\mathrm{II}} \\ B_2^{\mathrm{II}} \\ \vdots \\ B_q^{\mathrm{II}} \\ \vdots \\ B_Q^{\mathrm{II}} \end{bmatrix} \tag{10}$$

where

$$[P] \text{ matrix } P \times Q;\ P_{p,q} = \int_{S_{\mathrm{I}}} \vec{e}^{\,\mathrm{I}}_{p,T} \Lambda \vec{h}^{\,\mathrm{II}*}_{q,T} \cdot \vec{u}\, dS$$

$$[L] \text{ diagonal matrix } Q \times Q;\ L_{q,q} = S_q \frac{\beta_q}{|\beta_q|}$$

$$[K] \text{ diagonal matrix } P \times P;\ K_{p,p} = S_p \frac{\beta_p}{|\beta_p|}$$

$$[N] \text{ matrix } Q \times P;\ N_{q,p} = \int_{S_{\mathrm{I}}} \vec{e}^{\,\mathrm{I}*}_{p,T} \Lambda \vec{h}^{\,\mathrm{II}}_{q,T} \cdot \vec{u}\, dS. \tag{11}$$

From (9), the generalized scattering matrix can be written in the form

$$[B] = [S] \cdot [A] \tag{12}$$

where

$$[B] = \begin{bmatrix} [B^{\mathrm{I}}] \\ [B^{\mathrm{II}}] \end{bmatrix} \quad [A] = \begin{bmatrix} [A^{\mathrm{I}}] \\ [A^{\mathrm{II}}] \end{bmatrix}. \tag{13}$$

The generalized scattering matrix of the junction assembles two reflection blocks denoted $[S_{11}]$ and $[S_{22}]$ and two transmission blocks denoted $[S_{21}]$ and $[S_{12}]$ arranged as

$$[S]_{S_{\mathrm{I}} < S_{\mathrm{II}}} = \begin{bmatrix} [S_{11}]_{S_{\mathrm{I}} < S_{\mathrm{II}}} & [S_{12}]_{S_{\mathrm{I}} < S_{\mathrm{II}}} \\ [S_{21}]_{S_{\mathrm{I}} < S_{\mathrm{II}}} & [S_{22}]_{S_{\mathrm{I}} < S_{\mathrm{II}}} \end{bmatrix}. \tag{14}$$

Each block has a size $(P+Q) \times (P+Q)$ and can be determined separately from matrices $[P]$, $[L]$, $[K]$, and $[N]$ after extensive algebraic manipulation of (9). Results are listed below:

$$\begin{aligned}
[S_{11}]_{S_{\mathrm{I}} < S_{\mathrm{II}}} &= \left([K] + [N] \cdot [L]^{-1} \cdot {}^{T}[P]\right)^{-1} \\
&\quad \cdot \left([K] - [N] \cdot [L]^{-1} \cdot {}^{T}[P]\right) \\
[S_{12}]_{S_{\mathrm{I}} < S_{\mathrm{II}}} &= 2\left([K] + [N] \cdot [L]^{-1} \cdot {}^{T}[P]\right)^{-1} \cdot [N] \\
[S_{21}]_{S_{\mathrm{I}} < S_{\mathrm{II}}} &= 2\left([L] + {}^{T}[P] \cdot [K]^{-1} \cdot [N]\right)^{-1} \cdot {}^{T}[P] \\
[S_{22}]_{S_{\mathrm{I}} < S_{\mathrm{II}}} &= -\left([L] + {}^{T}[P] \cdot [K]^{-1} \cdot [N]\right)^{-1} \\
&\quad \cdot \left([L] - {}^{T}[P] \cdot [K]^{-1} \cdot [N]\right).
\end{aligned} \tag{15}$$

The above $[S]$ matrix is labeled with subscripts $S_{\mathrm{I}} < S_{\mathrm{II}}$ to recall that it is addressed only to the case $S_{\mathrm{I}} < S_{\mathrm{II}}$. As mentioned in Appendix III, the $[S]$ matrix addressed to the alternative case $S_{\mathrm{II}} < S_{\mathrm{I}}$ is derived in a quite different way. However, the results can be related to the previously studied case: $S_{\mathrm{I}} < S_{\mathrm{II}}$. This relation can be written as

$$[S]_{S_{\mathrm{I}} > S_{\mathrm{II}}} = \begin{bmatrix} [S_{22}]_{S_{\mathrm{I}} < S_{\mathrm{II}}} & [S_{21}]_{S_{\mathrm{I}} < S_{\mathrm{II}}} \\ [S_{12}]_{S_{\mathrm{I}} < S_{\mathrm{II}}} & [S_{11}]_{S_{\mathrm{I}} < S_{\mathrm{II}}} \end{bmatrix}. \tag{16}$$

IV. Combination of the Spectral-Domain Approach and the Modal Analysis

The spectral-domain approach evaluates both the phase constants and the associated line amplitudes of eigenmode field components.

TABLE I
Basis Functions and Associated Line Amplitudes of the Aperture Field Corresponding to Odd Modes

$E_x(x, D)$	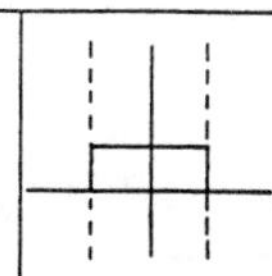	$f(x) = \begin{cases} 1. & ;\lvert x\rvert < W/2 \\ 0 & ;\lvert x\rvert > W/2 \end{cases}$	$\tilde{f}(\alpha_n) = \frac{2}{\alpha B} \sin \frac{\alpha_n W}{2}$
	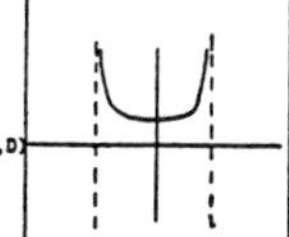	$f(x) = \begin{cases} \frac{1.}{\left(1. - (2x/W)^2\right)^{1/2}} & ;\lvert x\rvert < W/2 \\ 0 & ;\lvert x\rvert > W/2 \end{cases}$	$\tilde{f}(\alpha_n) = \frac{\pi W}{2B} J_0\left(\frac{\alpha_n W}{2}\right)$
	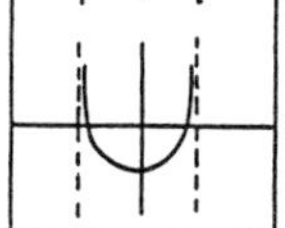	$f(x) = \begin{cases} -\frac{\cos(2\pi x/W)}{\left(1. - (2x/W)^2\right)^{1/2}} & ;\lvert x\rvert < W/2 \\ 0 & ;\lvert x\rvert > W/2 \end{cases}$	$\tilde{f}(\alpha_n) = -\frac{\pi W}{4B}\left[J_0\left(\pi + \frac{\alpha_n W}{2}\right) + J_0\left(\pi - \frac{\alpha_n W}{2}\right)\right]$
$E_x(x, D)$	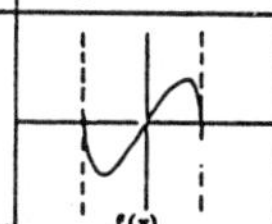	$f(x) = \begin{cases} 2x/W \cdot \left(1. - (2x/W)^2\right)^{1/2} & ;\lvert x\rvert < W/2 \\ 0 & ;\lvert x\rvert > W/2 \end{cases}$	$\tilde{f}(\alpha_n) = -\frac{j\pi}{\alpha B} J_2\left(\frac{\alpha_n W}{2}\right)$
	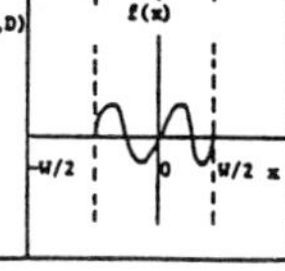	$f(x) = \begin{cases} \frac{\sin(4\pi x/W)}{\left(1. - (2x/W)^2\right)^{1/2}} & ;\lvert x\rvert < W/2 \\ 0 & ;\lvert x\rvert > W/2 \end{cases}$	$\tilde{f}(\alpha_n) = -\frac{j\pi W}{4B}\left[J_0\left(2\pi - \frac{\alpha_n W}{2}\right) - J_0\left(2\pi + \frac{\alpha_n W}{2}\right)\right]$

As a result of the eigenmode normalization, the calculation of elements L_{qq} and K_{pp} of matrices $[L]$ and $[K]$ is reduced to the determination of the propagating or evanescent nature of eigenmodes labeled p and q in waveguides I and II, respectively.

The computation of elements $P_{p,q}$ and $N_{q,p}$ of matrices $[P]$ and $[Q]$ can be quickly accomplished as it is done directly in the spectral domain. For example, the $P_{p,q}$ coefficient can be expressed from (11) as

$$P_{p,q} = B \sum_{m=0}^{\infty} \int_{-h_1}^{h_2+h_3} \left[\tilde{e}^{\,\mathrm{I}}_{p,x}(m, y) \tilde{h}^{\,\mathrm{II}*}_{q,y}(m, y) - \tilde{e}^{\,\mathrm{I}}_{p,y}(m, y) \tilde{h}^{\,\mathrm{II}*}_{q,x}(m, y)\right] dy. \tag{17}$$

V. Computed Results

A. Unilateral Finline Analysis

The sets of basis functions selected to describe the aperture field are represented in Table I. As shown in [11], the dominant mode can be described precisely by means of a single $E_x(x, D)$ basis function: the unit rectangular pulse denoted $f_1(x)$ in Table I. However, to describe both the dominant and the higher order modes, the aperture field expanded with the basis functions denoted $f_1(x)$, $f_2(x)$, $f_3(x)$, and $f_4(x)$ in Table I appears as a more judicious choice. Checks of this aperture field have been made in two limit cases for the unilateral finline: the standard rectangular waveguide ($\epsilon_r \to 1$; $W/B \to 1$) and the rectangular waveguide loaded symmetrically by a dielectric slab ($W/B \to 1$). They allowed conclusions to be made about the completeness of at least the first six eigenmodes.

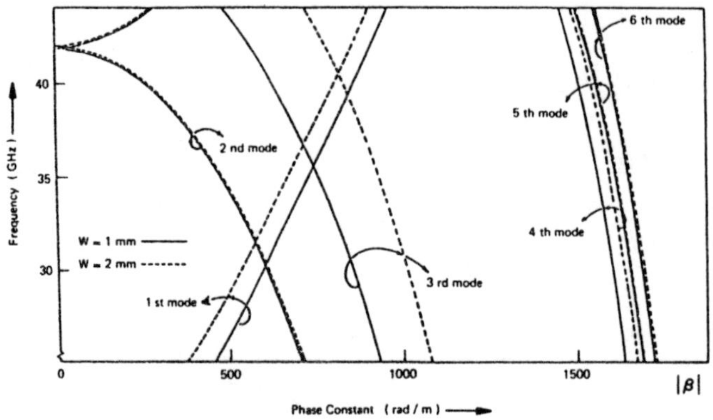

Fig. 3. Dispersion characteristics of eigenmodes in a unilateral finline in the *Ku*-band ($\epsilon_r = 2.22, h_2 = 0.254$ mm).

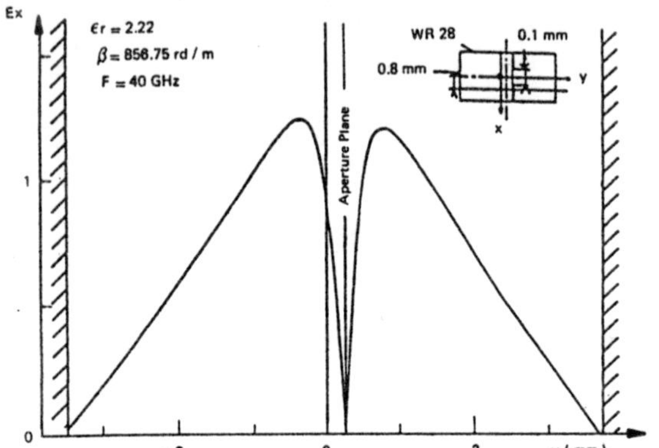

Fig. 4. E_x field component of the fundamental eigenmode for the shown finline.

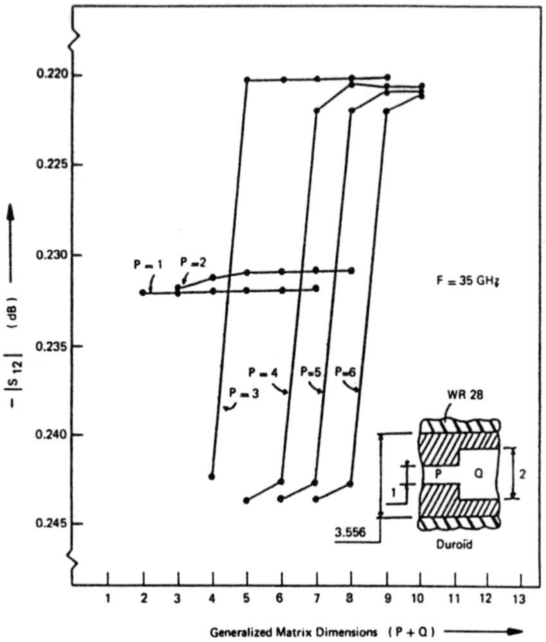

Fig. 5. Convergence test of the amplitude of the transmission coefficient for a step slot width discontinuity between two unilateral finlines (dimensions are in millimeters).

Dispersion characteristics of the first six eigenmodes in a unilateral finline in the *Ka*-band are plotted in Fig. 3 for two values of the slot width. They show that the frequency band for single-mode operation in a unilateral finline is quite identical to those of the standard WR28 rectangular waveguide. Another check of the aperture field is to compute the eigenmode distribution to be sure of the boundary conditions as shown in Fig. 4 for the E_x component of the fundamental mode as a function of y at $x = 0.8$ mm.

B. Scattering Parameters of Single-Step Slot Discontinuity in a Unilateral Finline

The most critical factor in the modal analysis is the convergence of reflection and transmission coefficients as a function of the number of modes taken into consideration when writing the boundary condition equations (8) at the junction plane. A relative convergence towards wrong values may be obtained if the number of modes is not sufficiently high.

Convergence tests on the moduli and the phases of reflection and transmission coefficients are performed making use of the description possibilities that are offered by the spectral-domain approach for the finline. The results of these tests applied on an abrupt junction between two finlines, as well as on an abrupt junction between a finline and a rectangular waveguide, are respectively given in Figs. 5 and 6. Here, the total number of modes did not exceed 14. The highest level of the curves drawn in Figs. 5 and 6 is supposed to be the real convergence level. Fig. 7 shows the computed scattering parameters of a unilateral finline discontinuity as a function of frequency compared with these computed by Schmidt [8] using the mode-matching procedure and with these computed by Sorrentino and Itoh [8] using the transverse resonance technique. Our results are in excellent agreement with Schmidt's results and in good agreement with Sorrentino's.

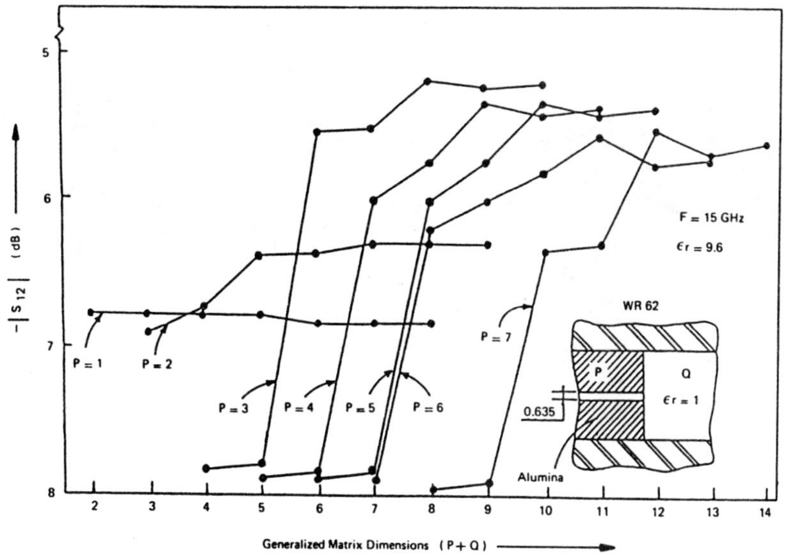

Fig. 6. Convergence test of the amplitude of the transmission coefficient for an abrupt junction between an empty rectangular waveguide and a unilateral finline (dimensions are in millimeters).

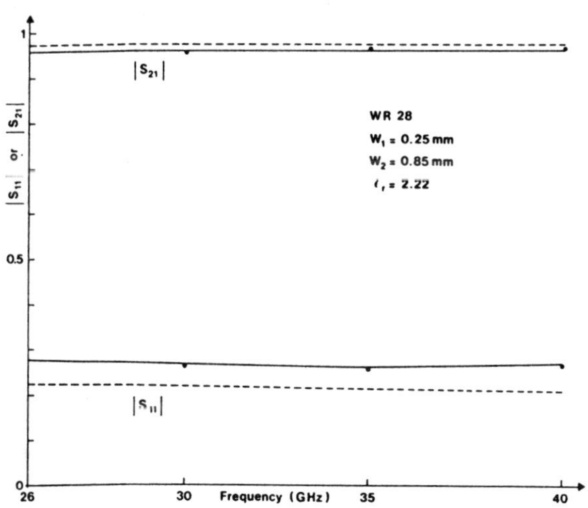

Fig. 7. Scattering parameters of a unilateral finline step discontinuity. —— Our theory, ----- Sorrentino's [8] results, and ∘ Schmidt's results.

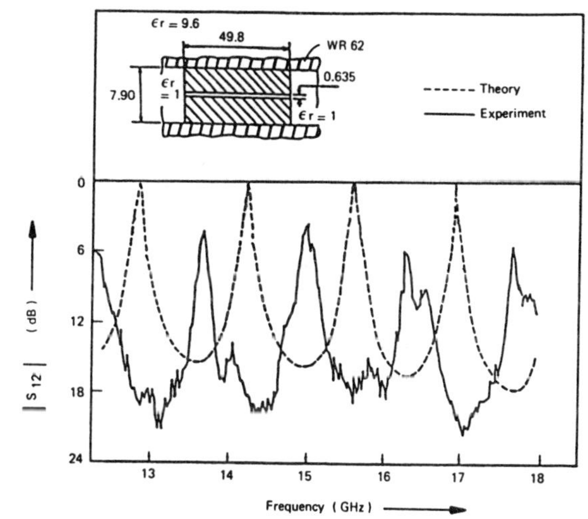

Fig. 8. Back-to-back arrangement of two abrupt junctions between an empty rectangular waveguide and a unilateral finline (dimensions are in millimeters).

C. *Comparison Between Theory and Experiments*

In order to evaluate objectively the effectiveness of the direct modal analysis for computer-aided design of finline circuits containing different discontinuities, the following three circuits have been fabricated, their scattering parameters have been calculated, and finally measured. The calculation of each circuit includes the effect of the discontinuity created by the narrow face of the substrate. The first circuit is that of two simple rectangular waveguide-unilateral finline junctions as shown in Fig. 8. The agreement can be judged satisfactory for the module of the transmission coefficient. The frequency shift of about 500 MHz can be explained by the mechanical constraints in the fabrication process (e.g., positioning grooves). The second circuit represents a pair of complex transitions each composed of three single simple transitions as shown in Fig. 9. The agreement between theory and experiment is still good in spite of the systematic frequency shift like that observed in the results of the first circuit. The third circuit is that given in Fig. 10, which represents a gradual transition operating in the *Ku*-band. The calculations are performed after dividing the transition into nine single simple discontinuities (Fig. 11). The agreement between theory and experiment is considered definitively satisfactory.

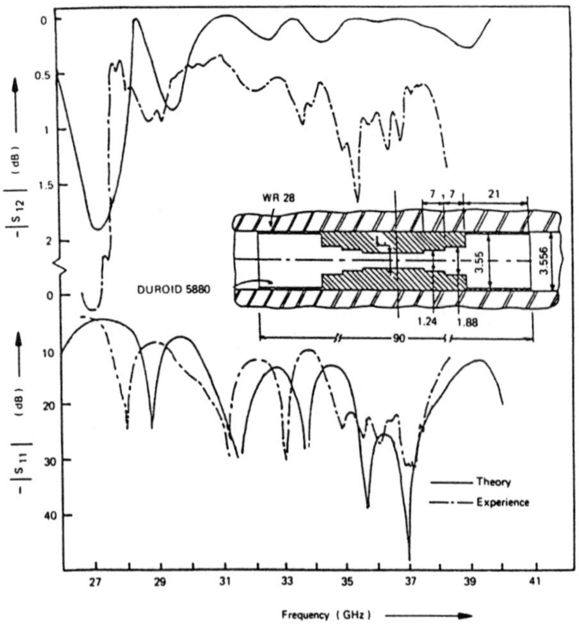

Fig. 9. Measurements and simulation results of a waveguide to finline complex transition in the *Ka*-band (dimensions are in millimeters).

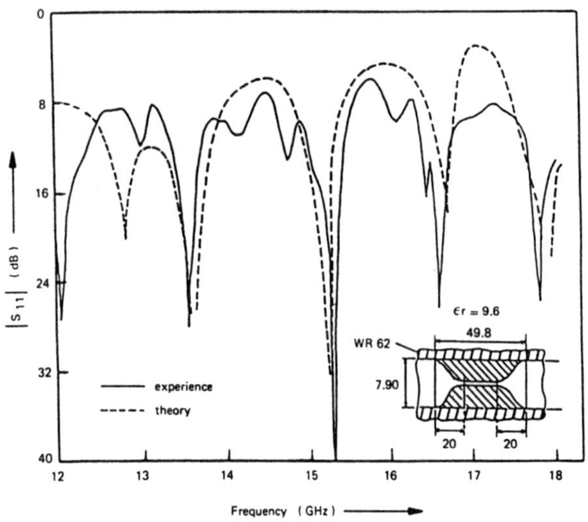

Fig. 10. Comparison between theory and experiment for the shown tapered transition (dimensions are in millimeters).

VI. Equivalent-Circuit Parameters of Simple Finline Discontinuities

The knowledge of both the reflection coefficient S_{11} and the transmission coefficient S_{21} in a given frequency band allows one to construct the equivalent circuit of the junction.

Each line having an access to the junction is considered as a lossless transmission line for which the effective dielectric constant and the characteristic impedance are those corresponding to the fundamental mode. The spectral-domain approach of a finline allows one to calculate these parameters directly in the spectral domain with excellent

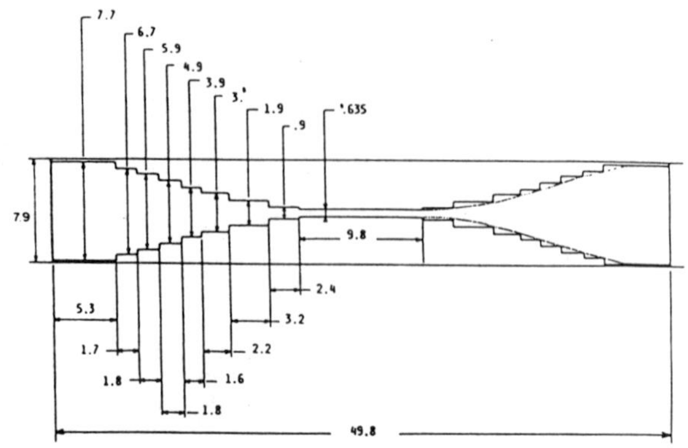

Fig. 11. Theoretical cutting used in the simulation of the transition reported in Fig. 10 (dimensions are in millimeters).

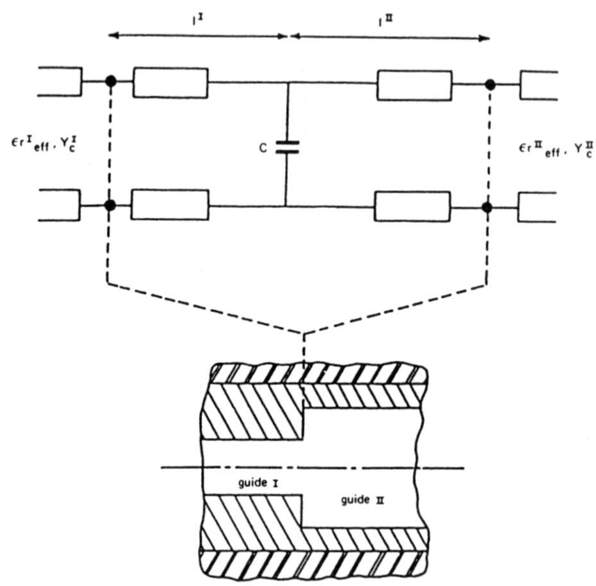

Fig. 12. Equivalent-circuit parameters of an elementary discontinuity.

accuracy. It is worth mentioning that the characteristic impedance is calculated according to a definition that relates the power flow of the fundamental mode and the potential that is induced between the slot edges.

As far as the single-step slot discontinuity is concerned, an equivalent circuit like that shown in Fig. 12 can be selected. The parameters C, l^{I}, and l^{II} of this equivalent circuit can be calculated by comparison of its scattering matrix to that of the single simple discontinuity restricted to four coefficients which represent the reflection and the transmission on only the fundamental eigenmodes of the two finlines having an access on the junction.

Examples of the results are reported in Figs. 13 and 14.

VII. Conclusion

The spectral-domain approach combined with the direct modal analysis appears as a very promising technique for calculation of scattering matrix elements of finline discontinuities.

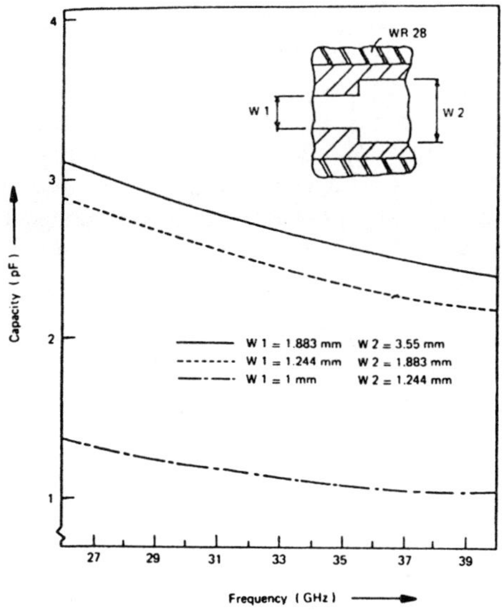

Fig. 13. Values of the lumped capacitance of the equivalent circuit of an elementary discontinuity as a function of frequency ($\epsilon_r = 2.22$, $h_2 = 0.254$ mm).

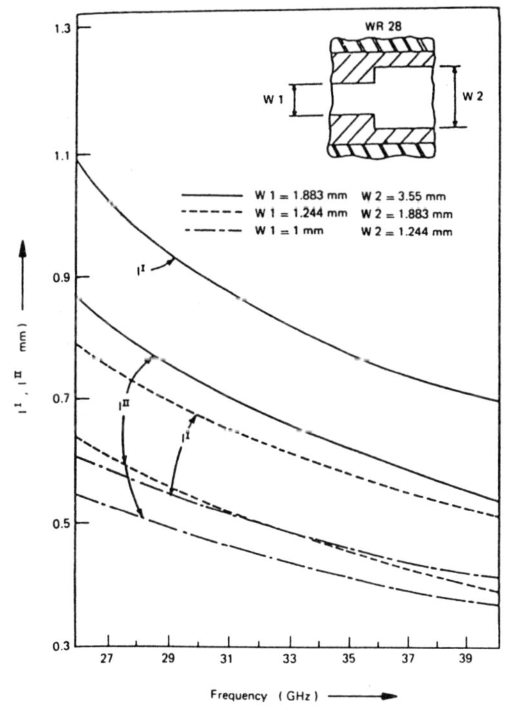

Fig. 14. Reference plane localization as a function of the frequency for the given elementary discontinuity ($\epsilon_r = 2.22$, $h_2 = 0.254$ mm).

Convergence is indeed necessary both in the waveguide eigenmode evaluation and in the waveguide discontinuity problem. In the case of a unilateral configuration, results show that a relatively simple aperture field allowed an unambiguous identification of six eigenmodes. Moreover, they show that, within such an identification, a satisfactory convergence on scattering parameters of a step slot width discontinuity can be achieved.

The agreement between theory and measurements performed on three different complex finline discontinuities is quite satisfactory.

Appendix I

The mth line amplitudes $\tilde{E}_{z,i}(m, y)$ and $\tilde{H}_{z,i}(m, y)$ can be written in the case of unilateral finline in the following form:

$$\begin{aligned} \tilde{E}_{z,1}(m, y) &= A(m)\sinh\gamma_1(y + h_1) \\ \tilde{H}_{z,1}(m, y) &= B(m)\cosh\gamma_1(y + h_1) \end{aligned} \tag{A1a}$$

in region (2)

$$\begin{aligned} \tilde{E}_{z,2}(m, y) &= C(m)\sinh\gamma_2 y + D(m)\cosh\gamma_2 y \\ \tilde{H}_{z,2}(m, y) &= E(m)\sinh\gamma_2 y + F(m)\cosh\gamma_2 y \end{aligned} \tag{A1b}$$

and in region (3)

$$\begin{aligned} \tilde{E}_{z,3}(m, y) &= G(m)\sinh\gamma_3(h_2 + h_3 - y) \\ \tilde{H}_{z,3}(m, y) &= H(m)\cosh\gamma_3(h_2 + h_3 - y) \end{aligned} \tag{A1c}$$

where coefficients γ_i ($i = 1$, 2, and 3) are defined as

$$\begin{aligned} \gamma_1^2 &= \gamma_3^2 = \alpha_m^2 - k_1^2 \\ \gamma_2^2 &= \alpha_m^2 - k_2^2 \end{aligned} \tag{A2a}$$

with

$$\begin{aligned} k_1^2 &= \omega^2\mu_0\epsilon_0 - \beta^2 \\ k_2^2 &= \omega^2\mu_0\epsilon_0\epsilon_r - \beta^2 \end{aligned} \tag{A2b}$$

represent the transverse wavenumbers of the mth line amplitude of the guided wave inside each region.

In (A2b), β denotes the phase constant of the guided wave to be determined at any given angular frequency ω.

Appendix II

The elements of the admittance matrix representation $[G]$ given in (2) can be written as

$$\begin{aligned} G_{11} = -j\Bigg[& k_2^2\frac{\tanh(\gamma_2 h_2)}{\mu_0\omega\gamma_2} - \epsilon_0 F\left(k_2^2\frac{\gamma_1}{\tanh(\gamma_1 h_1)}\right. \\ & \left. + k_1^2\frac{\epsilon_r\gamma_2}{\tanh(\gamma_2 h_2)}\right) + k_1^2\frac{\coth(\gamma_1 h_3)}{\mu_0\omega\gamma_1}\Bigg] \end{aligned} \tag{A3}$$

$$\begin{aligned} G_{12} = G_{21} = j\alpha\beta\Bigg[& \frac{\tanh(\gamma_2 h_2)}{\mu_0\omega\gamma_2} - \epsilon_0 F\left(\frac{\gamma_1}{\tanh(\gamma_1 h_1)}\right. \\ & \left. + \frac{\epsilon_r\gamma_2}{\tanh(\gamma_2 h_2)}\right) + \frac{\coth(\gamma_1 h_3)}{\mu_0\omega\gamma_1}\Bigg] \end{aligned} \tag{A4}$$

$$\begin{aligned} G_{22} = j\Bigg[& B_2\left(\frac{\tanh(\gamma_2 h_2)}{\mu_0\omega\gamma_2} - F\frac{\epsilon_0\gamma_1}{\tanh(\gamma_1 h_1)}\right) \\ & + B_1\left(\frac{\coth(\gamma_1 h_3)}{\mu_0\omega\gamma_1} - F\frac{\epsilon_0\epsilon_r\gamma_2}{\tanh(\gamma_2 h_2)}\right)\Bigg] \end{aligned} \tag{A5}$$

where

$$k_2^2 = \omega^2 \mu_0 \epsilon_0 \epsilon_r - \beta^2$$

$$k_1^2 = \omega^2 \mu_0 \epsilon_0 - \beta^2$$

$$B_2 = \alpha^2 - \mu_0 \epsilon_0 \epsilon_r \omega^2$$

$$B_1 = \alpha^2 - \mu_0 \epsilon_0 \omega^2$$

$$F = \frac{k_1^2 k_2^2}{\cosh^2(\gamma_2 h_2)} \cdot \frac{\omega}{\delta}$$

$$\delta = (\alpha\beta)^2 (k_1^2 - k_2^2)^2 - \epsilon_0 \mu_0 \omega^2 \Bigg[k_1^4 \epsilon_r \gamma_2^2 + k_2^4 \gamma_1^2 + (k_1 k_2)^2 \gamma_1 \gamma_2 \left(\frac{\tanh(\gamma_2 h_2)}{\tanh(\gamma_1 h_1)} + \epsilon_r \frac{\tanh(\gamma_1 h_1)}{\tanh(\gamma_2 h_2)} \right) \Bigg].$$

Appendix III

Equations (8a) and (8c) are scalarly multiplied by $\vec{h}^{\text{II}*}_{q,T} \Lambda \vec{u}$ and $e^{\text{I}*}_{p,T} \Lambda \vec{u}$, respectively, and then integrated over S_A. Since $S_{\text{I}} < S_{\text{II}}$, $S_A = S_{\text{I}}$, thus the orthogonality properties of eigenmodes in waveguide I can be used to obtain the following set of $P+Q$ linear equations:

$$\sum_{p=1}^{P} \left(A_p^{\text{I}} + B_p^{\text{I}}\right) \int_{S_{\text{I}}} \vec{e}^{\,\text{I}}_{p,T} \Lambda \vec{h}^{\text{II}*}_{q',T} \vec{u}\, dS =$$

$$\sum_{q=1}^{Q} \left(A_q^{\text{II}} + B_q^{\text{II}}\right) \int_{S_{\text{I}}} \vec{e}^{\,\text{II}}_{q,T} \Lambda \vec{h}^{\text{II}*}_{q',T} \vec{u}\, dS, \qquad q' = 1, \cdots, Q \quad \text{(A6)}$$

$$S_{p'} \frac{\beta^{\text{I}}_{p'}}{|\beta_{p'}|} \left(A^{\text{I}}_{p'} - B^{\text{I}}_{p'}\right) = \sum_{q=1}^{Q} \left(B_q^{\text{II}} - A_q^{\text{II}}\right) \int_{S_{\text{I}}} \vec{e}^{\,\text{I}*}_{p',T} \Lambda \vec{h}^{\text{II}}_{q,T} \vec{u}\, dS, \qquad p' = 1, \cdots, P \quad \text{(A7)}$$

which are found quite equivalent to boundary functional equations (8).

Similarly, the scalar multiplication of (8b) by $h^{\text{II}*}_{q,T} \Lambda \vec{u}$ and (8d) by $e^{\text{I}*}_{p,T} \Lambda \vec{u}$ and the surface integration over S_C provides the set of $P+Q$ equations

$$\sum_{q=1}^{Q} \left(A_q^{\text{II}} + B_q^{\text{II}}\right) \int_{S_C} \vec{e}^{\,\text{II}}_{q,T} \Lambda h^{\text{II}*}_{q,T} \cdot \vec{u}\, dS = 0, \qquad q' = 1, \cdots, Q \quad \text{(A8)}$$

$$\sum_{q=1}^{Q} \left(B_q^{\text{II}} - A_q^{\text{II}}\right) \int_{S_C} \vec{e}^{\,\text{I}*}_{p',T} \Lambda \vec{h}^{\text{II}}_{q,T} \cdot \vec{u}\, dS = 0, \qquad p' = 1, \cdots, P. \quad \text{(A9)}$$

During derivation, use has been made of the following equation:

$$\int_{S_C} (\vec{e}_{p',T} \Lambda \vec{u}) \cdot (\vec{J}^{\,\text{I}}_T \Lambda \vec{u})\, dS = \int_{S_C} \vec{J}^{\,\text{I}}_T \cdot e^{\text{I}*}_{p'T}\, dS = 0$$

due to the above-mentioned connection of the surface current $\vec{J}^{\,\text{I}}_T$ with waveguide I.

Clearly, the set (A8) can be imbedded in set (A6) after enlarging the surface integration at its right-hand side from S_{I} to $S_{\text{II}} = S_A + S_C$. Such a widening is done without altering the left-hand side of set (A6), and, moreover, use can then be made of the orthogonality in waveguide II.

Now, as far as the set (A7) is concerned, another look at the set (A9) shows that its right-hand side already includes it.

Considering now the case $S^{\text{I}} > S^{\text{II}}$, the functional boundary conditions of (8) still hold except those over S_C ((8c) and (8d)), which must be replaced by

$$E^{\text{I}}_T = 0 \quad \text{(A10)}$$

$$\vec{H}^{\text{I}}_T = \vec{J}^{\,\text{II}}_T \Lambda \vec{u} \quad \text{(A11)}$$

where the superscript II of the surface current notation $\vec{J}^{\,\text{II}}_T$ indicates now a relationship with the waveguide II. There, another unique procedure must be employed to derive the linear set of $P+Q$ equations equivalent to the boundary functional equations (8), (A9), and (A10).

This alternative procedure starts from the scalar multiplication of (8a) and (A6) by $h^{\text{I}*}_{p',T} \Lambda \vec{u}$ and of (8b) and (A7) by $e^{\text{II}*}_{q'T} \Lambda \vec{u}$, followed by surface integration where use must be made of the basic assumptions $S_A = S_{\text{II}}$ and $S_A + S_C = S_{\text{I}}$.

References

[1] K. Solbach, "The status of printed millimeter-wave *E*-plane circuits," *IEEE Trans. Microwave Theory Tech.*, vol. MTT-31, pp. 107–121, Feb. 1983.

[2] A. Wexler, "Solution of waveguide discontinuities by modal analysis," *IEEE Trans. Microwave Theory Tech.*, vol. MTT-15, pp. 508–517, Sept. 1967.

[3] H. El Hennaway and K. Schunemann, "Analysis of finline discontinuities," in *Proc. 9th Eur. Microwave Conf.* (Brighton, England), 1979, pp. 448–452.

[4] H. Hofmann, "Dispersion of planar waveguides for millimeter-wave application," *Arch. Elek. Übertragung.*, vol. 31, pp. 40–44, Jan. 1977.

[5] H. El Hennaway and K. Schunemann, "Computer-aided design of finline detectors, modulators and switches," *Arch. Elek. Übertragung.*, vol. 36, pp. 49–56, Feb. 1982.

[6] A. Beyer and I. Wolff, "Calculation of transmission properties of inhomogeneous finlines," in *Proc. 10th Eur. Microwave Conf.* (Warsaw, Poland), 1980, pp. 322–326.

[7] D. Mirshekar-Syahkal and J. B. Davies, "Accurate analysis of tapered planar transmission lines for microwave integrated circuits," *IEEE Trans. Microwave Theory Tech.*, vol. MTT-29, pp. 123–128, Feb. 1981.

[8] R. Sorrentino and I. Itoh, "Transverse resonance analysis of finline discontinuities," *IEEE Trans. Microwave Theory Tech.*, vol. MTT-32, pp. 1633–1638, Dec. 1984.

[9] M. Helard, J. Citern, O. Picon, and V. Fouad Hanna, "Exact calculation of scattering parameters of a step slot width discontinuity in a unilateral finline," *Electron. Lett.*, vol. 14, pp. 537–539, July 1983.

[10] M. Helard, J. Citern, O. Picon, and V. Fouad Hanna, "Solution of finline discontinuities through the identification of its first four higher order modes," in *IEEE MTT-S Int. Microwave Symp. Dig.*, May 1983, pp. 387–389.

[11] J. B. Knorr and P. M. Shayda, "Millimeter wave finline characteristics," *IEEE Trans. Microwave Theory Tech.*, vol. MTT-28, pp. 737–742, July 1980.

5.2.1

A COMPUTER-CONTROLLED INTEGRATED MICROWAVE MEASUREMENT SYSTEM IN FINLINE TECHNIQUE FOR AUTOMATIC MATERIAL PARAMETER MEASUREMENTS

Adalbert Beyer, Wolfhardt Müller-Gronau
and Ingo Wolff

Duisburg University, Department of Electrical
Engineering, Bismarckstr. 81, D-4100 Duisburg
Fed. Rep. of Germany

Abstract

Integrated circuit elements in finline technique have been developed for the application in an automatic measurement system which determines the complex material parameters of ferrites and dielectrics. These elements render possible a very fast tuning of the measurement setup in comparison with mechanically controlled systems. They can also be produced much cheaper and more easily than waveguide elements. The design of fully integrated microwave measurement systems in finline technique will be discussed.

Introduction

The realization of automatic measurement systems for measuring e.g. the complex material parameters of ferrites and dielectrics requires the use of electronically controllable components, such as attenuators, phase shifters, switches etc.. The consideration of possible concepts for the construction of these components yields that integrated circuits in finline technique show certain advantages with respect to other principles like conventional diode-waveguide attenuators or electromechanically controlled rotary vane devices.

Such advantages are, besides others, that in generalfinline components can be produced much cheaper and more easily than waveguide elements.Furthermore, the use of pin-diodes renders possible a very fast tuning in comparison with mechanically controlled systems employing servo motors; in contrast to conventional diode-waveguide attenuators, the pin-diodes can be easily integrated into the finline.

Details of system design

In a first version, we have designed a computer-controlled integrated microwave unit in finline technique, which uses a finline attenuator and a finline isolator. Basically, a finline attenuator consists of three sections: a homogeneous center section,where one or more diodes are located, and two taper sections, which provide the necessary matching between the impedances of the waveguide and the diodes. The resulting performance of the attenuator largely depends on the properties of the taper sections. Therefore, extensive research with respect to the characteristics of these tapers was done, the results will be discussed in the oral presentation. Since a single diode does not meet the required specifications with regard to bandwidth and attenuation behaviour, the use of two or more diodes is necessary. Investigations show that a configuration according to Fig. 1 yields the best results.

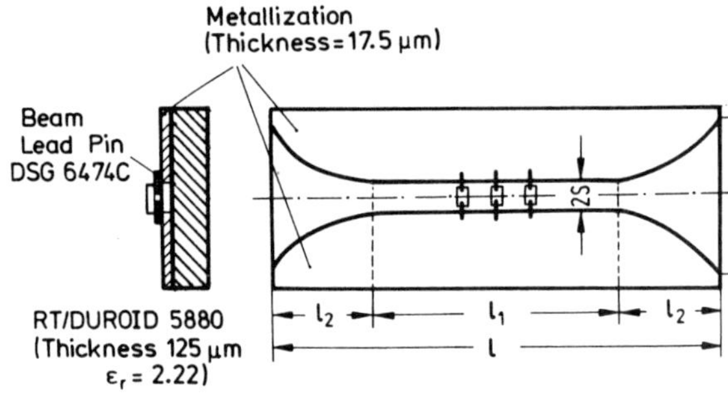

Fig. 1: The finline attenuator.

In Figs. 2 and 3, the measured results for the insertion loss and the reflection loss of the attenuator are plotted versus the frequency with the overall bias current I_D of the three diodes as a parameter. It can be seen from Fig. 2 that the attenuation for a constant bias current is frequency-dependent; therefore the application of the attenuator for precise wide-band measurements requires the use of a correction mechanism for the bias current in order to achieve a frequency-independent attenuation.
In Fig. 4, the bias current I_D, which is frequency-dependent for a constant value of the attenuation, is plotted versus the frequency with the insertion loss as a parameter.

A. Beyer, W. Muller-Gronau, and I. Wolff, *IEEE MTT-S International Microwave Symposium Digest*, 1984, pp. 345–347. Copyright © 1984 IEEE. Reprinted with permission.

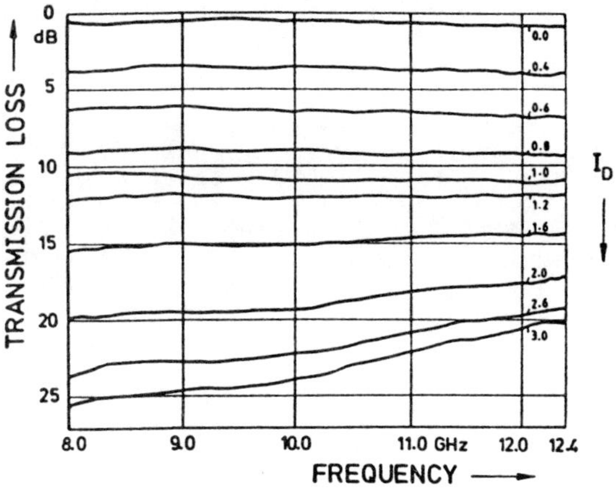

Fig. 2: Insertion loss of the finline attenuator.

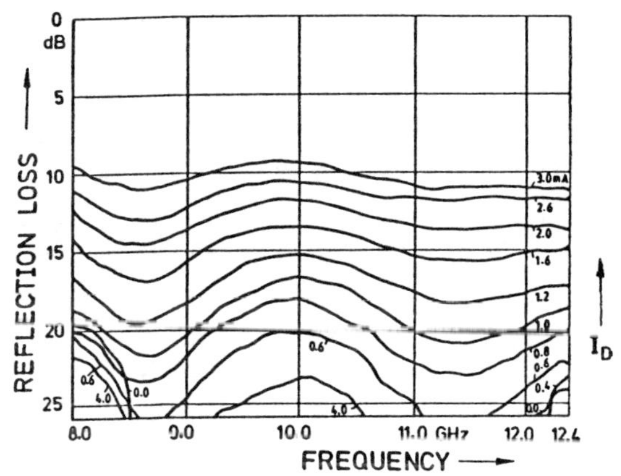

Fig. 3: Reflection loss of the finline attenuator.

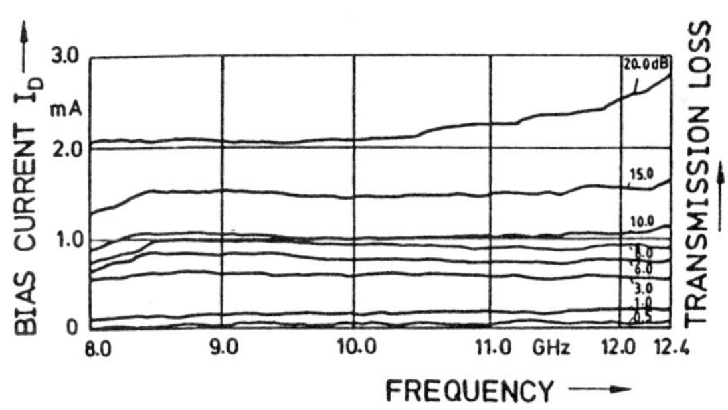

Fig. 4: The bias current of the finline attenuator.

The above-described attenuator was developed as a part of an automatic system for measuring the complex magnetic and dielectric material parameters of microwave ferrites or dielectrics. Employing a recently published method /3/ (and its just completed extension for measuring the imaginary parts of the material parameters), the complex material parameters can be computed from the measured resonant frequencies and Q-factors of resonant modes that occur in a cavity partly filled with the material sample to be measured.

The same desk-top computer that controlls the automatic measurement system shown in Fig. 5 is also used for the computation of the material parameters. The determination of the Q-factor requires an exact 3 dB variation of the microwave power that is available at the input of the transmission-type cavity. The correction mechanism consists of the desk-top computer (microcomputer) in combination with a digital-to-analog converter (DAC). Depending on the frequency, the computer provides the adequate bias current I_D for the desired attenuation. By means of a second DAC, the computer also provides the sweep voltage for the microwave generator. The output signal of the crystal detector is digitized by an analog-to-digital converter (ADC) and, like the output signal of the frequency counter, is then transferred to the computer.

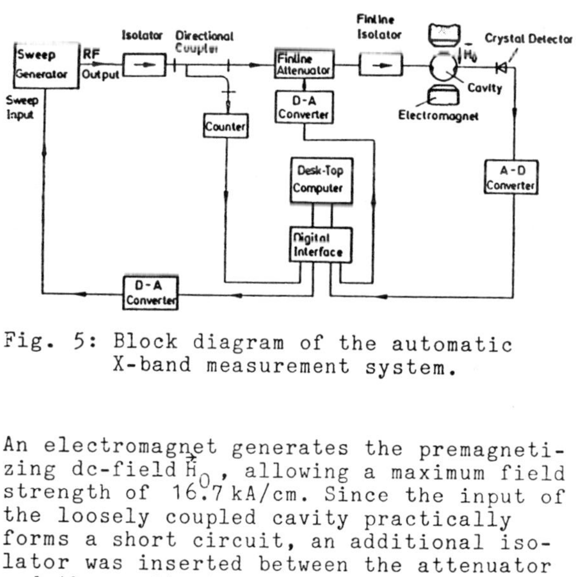

Fig. 5: Block diagram of the automatic X-band measurement system.

An electromagnet generates the premagnetizing dc-field $\vec{H}_0$, allowing a maximum field strength of 16.7 kA/cm. Since the input of the loosely coupled cavity practically forms a short circuit, an additional isolator was inserted between the attenuator and the cavity in order to prevent the occurrence of reflections which could deteriorate the accuracy of the attenuator. The isolator /5/ was also realized in finline technique. Fig. 6 shows a cross-sectional view of the isolator.

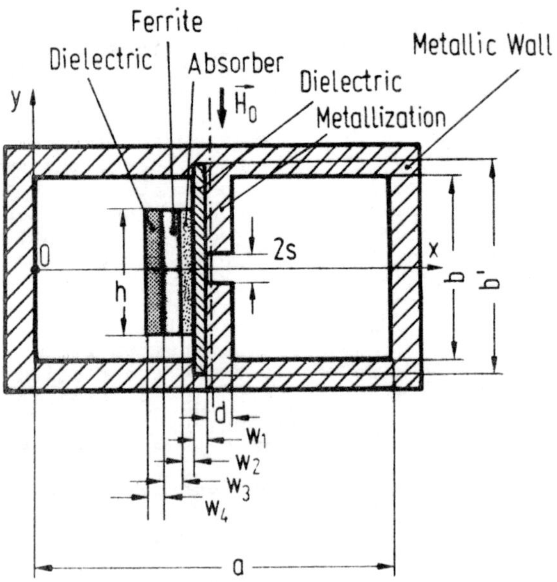

Fig. 6: The cross-sectional view of the isolator in finline technique.

In Fig. 7, measured results for the insertion loss in forward (transmission loss, Fig. 7) and backward (reverse loss, Fig. 7) direction are plotted as a function of the frequency.

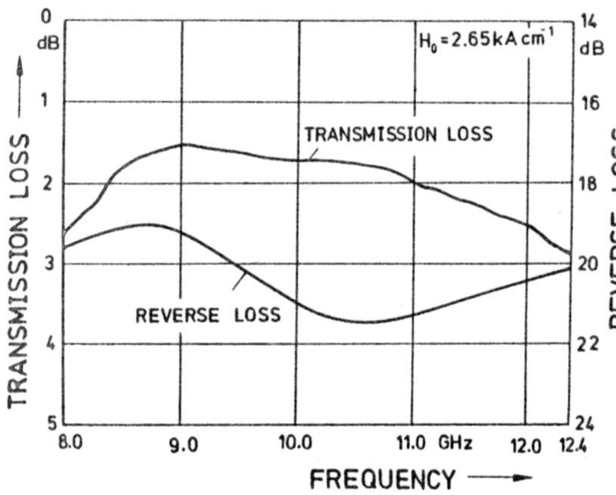

Fig. 7: Characteristics of the finline isolator.

The application of the automatic measurement system yields that the difference between automatically and manually measured resonant frequencies was less than 0.5 MHz, while the automatically and manually obtained Q-factors showed a typical deviation between each other of about 3 %; tests proved that the measurement accuracy for the Q-factor can still be improved.

At present, a second computer-controlled fully integrated microwave unit in finline technique, which includes several of the - up to now - seperate microwave components shown in Fig. 5, is under development: the attenuator, two isolators and two directional couplers (one for levelling purposes) are combined in a compact finline circuit.

Conclusion

For the application in an automatic measurement system, integrated elements in finline technique have been developed, which allow a simpler and more efficient system design than conventional waveguide elements. A wide-band precision attenuator is described, whose bias current is controlled by a microcomputer in order to achieve a frequency-independent attenuation Besides controlling the measurement system, the same computer is also used for the calculation of the material parameters from the measurement data. Furthermore, an isolator also realized in finline technique is described, which improves the accuracy of the attenuator. At present, a fully integrated microwave unit in finline technique is under development, which includes most of the microwave elements of the measurement system.

References

/1/ Beyer, A. and Wolff, I.: Calculation of the transmission properties of inhomogeneous finlines. Proc. of the 10th Eu. M.C., Warsaw (Poland), 322 - 326 (1980)

/2/ Beyer, A.: Analysis of the characteristics of an earthed finline. IEEE Trans. vol. MTT-29, No. 7, 676 - 680 (1981)

/3/ Müller-Gronau, W. and Wolff, I.: A microwave method for the determination of the real parts of the magnetic and dielectric material parameters of premagnetized microwave ferrites. IEEE Trans. vol. IM-32, No. 2, 377 - 382 (1983)

/4/ Meier, P.J.: Integrated finline millimeter-wave components. IEEE Trans. vol. MTT-22, No. 12, 1209 - 1216 (1974)

/5/ Beyer, A. and Solbach, K.: A new finline ferrite isolator for integrated millimeter-wave circuits. IEEE Trans. vol. MTT 29, No. 12, 1344 - 1348 (1981)

5.3.1

INTEGRATED K_A-BAND FIN-LINE SIX-PORT

M. Malkomes, R. Walsdorf

ABSTRACT

The design and performance of an integrated mm-wave 6-port is presented. The design uses Fin-Line technology and beam-lead diode detectors in a Ka-full-band application. The high precision measurement capability of the 6-port device has been verified in an automated 6-port network analyzer setup.

INTRODUCTION

Network analyzers (NWA) on the basis of the 6-port measurement scheme offer an attractive low cost alternative to the well known fundamental/harmonic-mixer equipted NWA's. Two important advantages qualify the 6-port NWA: The controller and data acquisition unit operate (except of some minor software adaptions) at any frequency band. Merely by changing the 6-port units the measurement system is completed for measurement in the frequency band associated with these 6-ports. Despite it's lack of real-time measurement capabilities a 6-port NWA can be constructed to form a relatively low-cost, light weight, compact, precise measurement setup with "quasi real-time" performance under e. g. 16-bit-microprocessor control.

The second important advantage of the 6-port measurement scheme is that the expences in microwave NWA equipment can be reduced merely to the components or circuits needed for the 6-ports including four power-detection devices such as Schottky-diodes or Thermistor detectors.

In the past these 6-ports have been constructed by use of stripline and microstrip technology with over one octave bandwidth in the frequency range below 18 GHz [1, 2]. For millimeter-wave and/or high precision measurements dielectric [3] and various standard waveguide 6-port setups have been reported [e. g. 4]. No compact, integrated 6-port realization above 18 GHz is actually known to the authors. To reduce system size and cost and to achieve premises for industrial application in cm- and mm-wave ranges a planar, integrated 6-port reflectometer design has been evaluated.

THE 6-PORT DESIGN

This paper reports the design and application (Ka-band) of an integrated 6-port detecting device using E-plane Fin-Line technology in a split-block housing.

Institute for High Frequency Techniques, Technical University of Aachen, Aachen, West Germany

M. Malkomes and R. Walsdorf, *Proceedings of 15th European Microwave Conference*, 1985, pp. 323–328.

The Fin-Line design (fig. 1) consits of a symmetrical ring 5-port in conjunction with a 10 dB distributed parallel-line directional coupler.

As power sensors beam-lead silicon Schottky-diodes are implemented either in series configuration (at the 5-port arms) to achieve low reflective detection or in parallel (at the coupler arm D) for high sensitivity power detection.

By our design concept a dense concentration of all high frequency circuitry is achieved resulting in a compact 6-port detection device with a footprint of 10 cm by 15 cm.

It is well known that a 6-port with power detecting points "q_i-points" [5] equally distributed in the complex reflection plane (fig. 2) represents an optimum configuration for 6-port measurement application. The Fin-line ring 5-port presented here in conjunction with a directional coupler gives a practical realization according to the above requirements: Matching at all ports and 120 degree phase condition between adjacent arms of the 5-port. The coupler, the ring-5-port and the detector circuits have been designed and optimized using computer simulation.

Measured performance of the coupler with a beam-lead diode at one arm and the ring with three detector diodes are displayed in fig. 3 and fig. 4 respectively. Input reflection under -10 dB can be observed, which is sufficient when an appropriate error correction and calibration scheme [6] is applied in the measurement procedure.
It can be deduced from figs. 4 and 5 that good matching under -15 dB and transmission of -6 dB has been achieved with our Fin-line design.

The simple detector circuits with Fin-line matched loads operated with a Tss= -50 dBm (zero-bias, Siemens diode), conversion k = 200 ÷ 400 mV/mW and an input reflection under -15 dB in Ka-band. The figures 6 and 7 demonstrate the 6-port operation with two calibration standards (matched waveguide load, waveguide short) connected to the test-port 3. The measurement of the short (fig. 7) visualizes the 120 degree phase difference between the three power readings at the 5-port, which thus are ideally conditioned for the 6-port calibration and measurement scheme.

6-PORT NWA APPLICATION AND PERFORMANCE

A dual 6-port Ka-band NWA set-up operating with our new Fin-line integrated devices has been tested in Ka-band. The performance data which have been achieved by measuring some known standards as "thru" ($|S_{ik}|$ = 0dB) and shorts ($|S_{ii}|$ = 0 dB) after calibration are summarized in the following table:

Transmission Measurement: (S_{ik})			
System Directivity	> 40 dB		
relative accuracy, $\Delta\|S_{ik}\|$	< 0,1 dB	at	$\|S_{ik}\|$ = 0 dB
	< 0,5 dB	at	$\|S_{ik}\|$ = -10dB
Δ arc S_{ik}	< 3 deg		
Reflection Measurement: (S_{ii})			
Sensitivity	<-35 dB		
relative accuracy, $\Delta\|S_{ii}\|$	< 0,1 dB	at	$\|S_{ii}\|$ = 0 dB
Δ arc S_{ii}	< 1 deg		
(Generator power level: 0 dBm)			

Tabel I 6-port NWA performance data

These data reveal the same degree of measurement accuracy compared to 6-port set-ups constructed by bulky and expensive standard waveguide components.

CONCLUSION

With our design concept presented and the successful practical realization an integrated 6 port reflectometer for mm-wave Network Analyzer applications has been achieved. The design goal has been met with a compact, cheap planar Fin-Line circuit. The measurcment accuracy in an automated, error corrected 6-port NWA is equal to results achieved with NWA's utilizing waveguide 6-ports.

REFERENCES

[1] F. C. DeRonde, "Octave-wide matched symmetrical, reciprocal, 4- and 5-ports", 1982 MTT-S Symposium Digest, June 1982, pp. 521 - 523

[2] M. Malkomes, G. Kadisch, H. J. Schmitt, "Optimized microstrip ring-star 5-ports for broadband 6-port measurement application", 1984 MTT-S Symposium Digest, June 1984, pp. 472 - 474

[3] D. Radovich, "Coupler for 94 GHz network analyzer", Microwave Journal, Dec. 1982, pp. 90 - 97

[4] H. M. Cronson, R. A. Fong-Tom, "A 94 GHz diode-based single six-port reflectometer", IEEE MTT-30, No. 8, Aug. 1982, pp. 1260 - 1264

[5] G. F. Engen, "The Six-Port Reflectometer: An Alternative Network Analyzer", IEEE MTT-25, No. 12, Dec. 1977, pp. 1075 - 1079

[6] M. Malkomes, R. Meyer, "A self-calibration technique for dual six-port network analyzers", 14. European Microwave Conf. Digest, Sept. 1984, pp. 796 - 800

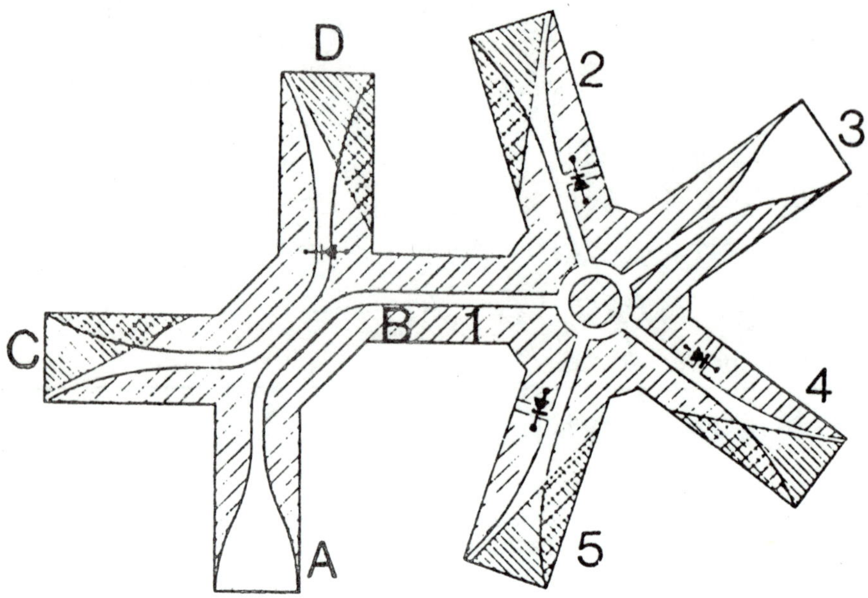

Fig. 1 Schematical layout of the integrated Fin-Line 6-port

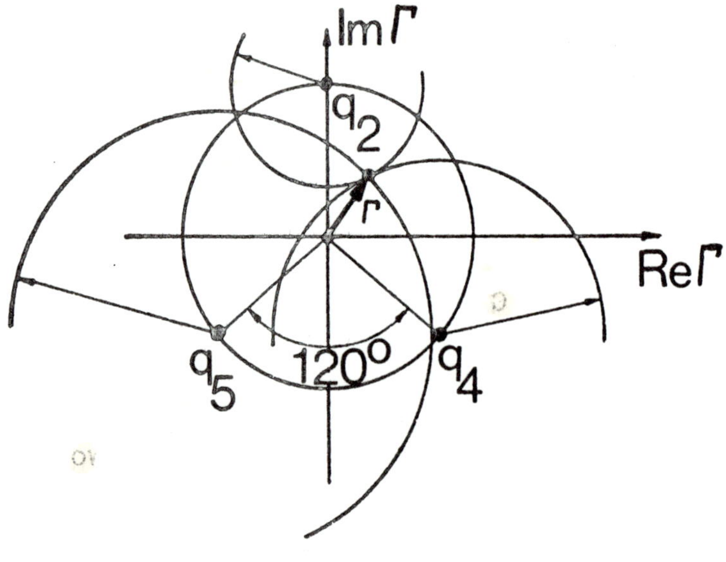

Fig. 2 6-port measurement scheme: Determination of reflection Γ from the intersection of circles in a 6-port NWA

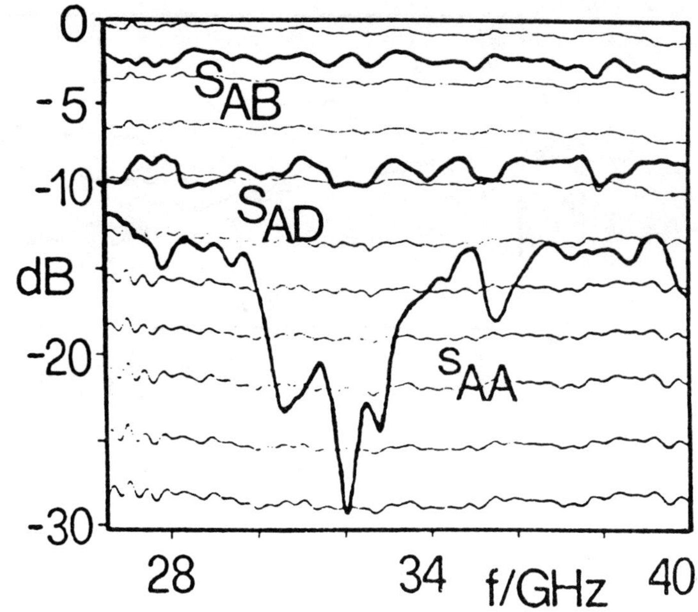

Fig. 3 Input reflection of 10 dB coupler with detector diode

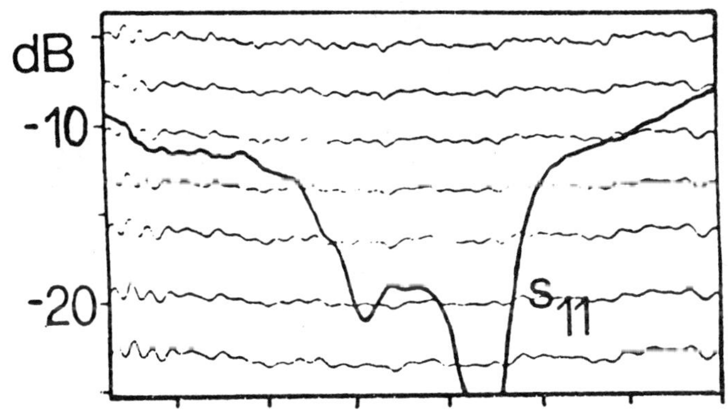

Fig. 4 Input reflection of 5-port with three detector diodes

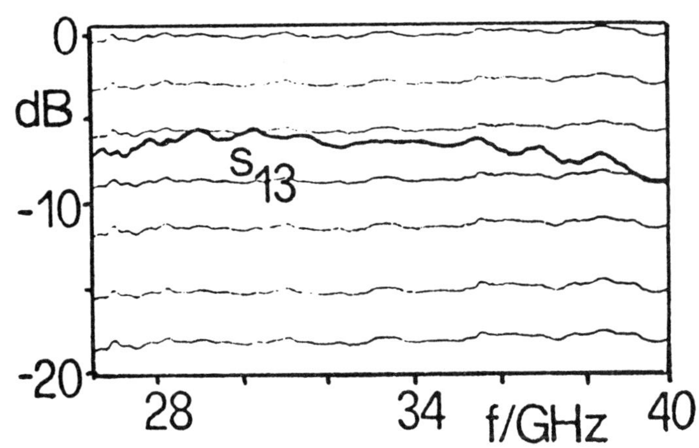

Fig. 5 Transmission of the ring 5-port

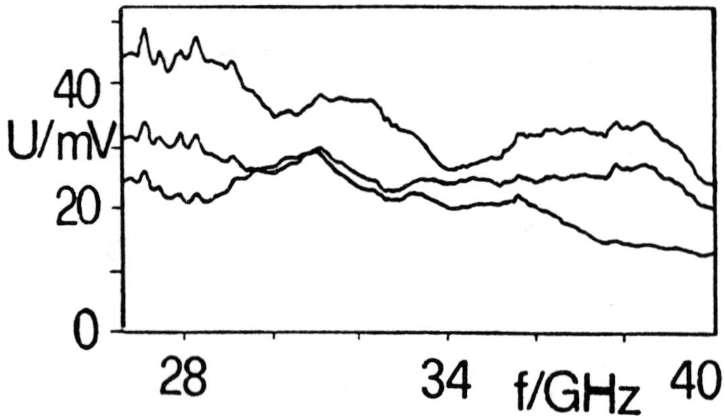

Fig. 6 Detector voltages, matched load at port 3

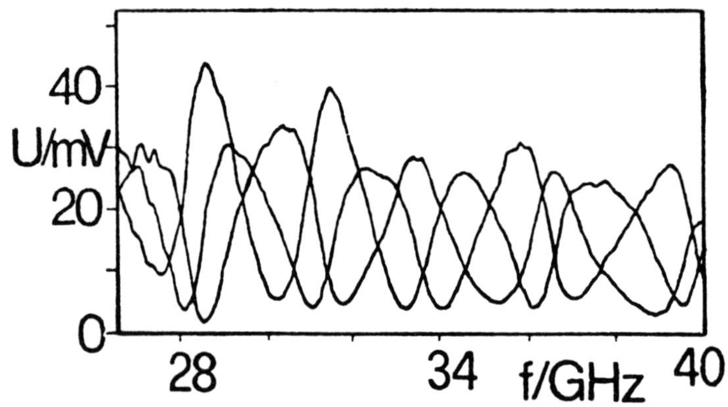

Fig. 7 Detector voltages, short at port 3

Chapter 6
Passive Components

6.1 TAPERS AND MATCHING NETWORKS

Most practical fin-line circuit designs include one or several types of transitions, either between various types of fin lines, or between fin lines and different types of transmission lines. The transition between a fin line and its corresponding waveguide is particularly important because of the need to connect fin-line circuits with conventional components and systems. One of the ways of realizing such transitions is with the use of tapered fin lines.

Because a fin line is dispersive, the phase constant varies along the length of a tapered fin line. Therefore, a rigorous mathematical treatment of a tapered fin line leads to a nonlinear differential equation, for which a closed-form solution may not exist. In any case, it is impossible to define clearly the local characteristic impedance of a tapered section. In addition, the characteristic impedance of a dispersive line is not uniquely defined. Although fin-line taper design has many drawbacks, we observe one advantage: the taper can be as long as three wavelengths for certain applications in which the insertion loss is not critical.

Besides the problem of matching two different fin gaps, or matching fin lines and other transmission lines, we must contend with important problems of matching a two-terminal device across a fin line. Unlike in the previous case, a correct value of the fin-gap impedance must be accurately known.

In this section, we will include some of the most significant publications in fin-line tapers and matching network designs.

6.1.1 "Simplified Analysis and Synthesis of Fin-Line Tapers" by Hinken

This paper deduces a formula for the local-mode coupling coefficient in a fin-line taper with an assumed homogeneous dielectric over the cross section. The coupling coefficient is subsequently defined as a function of the operating frequency and the derivative with respect to the axial coordinate. The solution is then extended to fin-line structures. We develop a synthesis method based on the local coupling coefficient and known taper design methods. This method is applicable to any fin-line structure with a low substrate dielectric constant and small substrate thickness.

6.1.2 "Synthesis of Optimum Finline Tapers Using Dispersion Formulas for Arbitrary Slot Widths and Locations" by Schieblich, Piotrowski, and Hinken

This paper describes the design of a Dolph-Chebyschev taper in a unilateral fin line at Ka-band. Although the method is simple, Fig. 8 does not show true Chebyschev behavior because the response is not equally rippled.

6.1.3 "Finline Taper Design Made Easy" by Beyer and Wolff

This unique paper presents an unconventional method for the quick and exact design of fin-line tapers in the form of circular arcs. Figure 6 shows that a circular arc taper performs better than a near optimum one. This technique is very simple to implement on a programmable calculator.

6.1.4 "Analysis and Synthesis of Tapered Finlines" by Pramanick and Bhartia

This paper presents a simple but rigorous solution of the associated, nonlinear, differential equations of fin-line tapers for an assumed taper profile. The method is very fast because we use closed-form equations for uniform fin lines in the computing process. The synthesis does not require solutions for the differential equations.

6.1.5 "Hybrid Fin-Line Matching Structures" by El Hennawy and Schunemann

This paper exploits the unique feature of a back-to-back transition between a unilateral and bilateral fin line for impedance matching of a device. The capacitive series reactance of the equivalent T-network is used to design a broadband switch with two *pin* diodes and 20 dB isolation over the full Ka-band. A modal analysis is used to analyze the transition.

6.1.6 "Miniaturisation in E-Plane Technology" by de Ronde

This paper discusses various inexpensive ways for full-band matching of T-junctions, corners, magic tees, crossovers, *et cetera.* The methods used in achieving such matchings are quite different from conventional ones, but are superior to conventional tapers and transitions in many respects. The usefulness of the techniques has been proven at X-band. Also, these techniques look promising at millimeter-wave frequencies.

6.1.7 "A Waveguide-to-Microstrip Transition" by Piotrowski and Wojtasiak

A waveguide-to-microstrip transition is usually realized by a ridged waveguide or an antipodal fin line. This paper presents a new approach, using a bilateral fin line. The transition operates over the entire fundamental mode bandwidth of the waveguide. Such an X-band transition offers 20 dB return loss and about 0.6 dB insertion loss. The structure is very simple to design and less complicated than other types of waveguide-to-microstrip transitions.

6.2 FILTERS AND MULTIPLEXERS

E-plane filters were first realized by Paul Meier in 1974 by experimental investigation of inductive strips bridging the broad walls of a rectangular waveguide housing. To demonstrate the feasibility of various fin-line filters with bilateral fin lines, different filter components were printed on 0.02-inch Duroid™ substrate. The test samples included one- and four-pole inductively coupled filters. By measuring the insertion loss of the elements across the entire Ka-band, Meier generated families of design curves to present the shunt susceptances as a function of strip width, with frequency as the parameter.

Today, E-plane filter design has changed considerably. Several software packages exist, which are based on the rigorous solution of Maxwell's equations in an E-plane filter structure. Each of the methods has its own merits and faults. This section will include papers presenting the methods, as well as papers on multiplexer designs using E-plane filters.

6.2.1 "Millimetre-Wave Low-Loss Finline Lowpass Filters" by Nguyen and Chang

This paper shows how an E-plane, low-loss, low-pass filter can be realized by converting the conventional, corrugated waffle-iron filter into fin-line form. The filter consists of a sequence of notches periodically spaced along a single-ridge fin line. This filter can be easily designed using the distributed low-pass prototype and Burton and Hoefer's design equations for unilateral fin-line stubs (Sec. 4.2.2).

6.2.2 "Computer-Aided Design of Millimeter-Wave E-Plane Filters" by Shih, Itoh, and Bui

This paper presents one of the most efficient computer-aided design algorithms for a class of fin-line bandpass filters. The design program consists of two algorithms, one for analysis and the other for optimization. There are three steps to the analysis. The first step analyzes a limiting case of a single junction created by a semi-infinite septum in a waveguide, using the residue calculus technique. The next step calculates the S-parameters of a finite length septum by placing two junctions back to back and using the generalized S-matrix concept. The third step uses the generalized S-matrix concept again to obtain the overall S-matrix of the filter. Finally, the performance of the filter is optimized by the quasi-newton method. This method can analyze filters with zero metalization thickness only, but has lately been modified for finite metalization thickness.

6.2.3 "Material and Circuit Evaluation for Millimetre-Wave Applications: High-*Q* Fin-Line Filters for Millimetre Waves," by Arndt, Bornemann, Grauerholz, and Vahldieck

This is one of the first papers to describe a design theory of E-plane filters, which is based on rigorous and accurate mode-matching techniques. In this approach, the field is expanded into eigenmodes within each subregion of the filter. By matching various tangential field components, the authors derive linear equation systems for the scattering coefficients. Finally, an optimization procedure is used to obtain the desired response.

6.2.4 "Broad-Band Millimeter-Wave E-Plane Bandpass Filters" by Bui, Ball, and Itoh

In this paper, the authors synthesize the E-plane bandpass filter from an optimum Chebyschev distributed low-pass prototype. Their method saves a great deal of computation time by eliminating eventual optimization of the filter response.

6.2.5 "Realizations and Design of Fin-Line Bandstop Filters" by Omar and Schunemann

This paper presents an experimental and theoretical study of potential slot patterns in fin lines, for the realization of E-plane bandstop filters. It reports on filters using a double-slot fin line and hybrid structure. This is the only paper currently available on E-plane bandstop filters.

6.2.6 "Millimeter-Wave Diplexers with Printed Circuit Elements" by Shih, Bui, and Itoh

The authors of this paper present a new idea. Two E-plane filters are fed by the transition between a suspended microstrip line and a rectangular waveguide, thus producing an inexpensive W-band diplexer with wide channel bandwidths. This novel structure is printed on a single substrate and assembled in a split block housing. The diplexer has 1 dB of loss at W-band. This structure can be easily reproduced at lower millimeter-wave bands with lower loss.

6.2.7 "Diplexer Arrangement Utilizing an E-Plane Metal Insert" by Reiter and Kolumban

The authors present a waveguide diplexer with waveguide input and coaxial line outputs. The waveguide has a metal plate that runs parallel to the broader waveguide walls, which separate the waveguide into two parts. Each part contains a simple, E-plane, metal insert filter tuned to the center frequency of the channels. The diplexer is realized at X-band, where tuning screws are used for tuning the diplexer to achieve the desired performance. However, such tuning and coaxial outputs may make it impossible to realize such a diplexer at millimeter-wave frequencies.

6.3 COUPLED FIN LINES AND DIRECTIONAL COUPLERS

E-plane directional couplers are mostly of the distributed type, and are used for many functions such as mixing, *pin* diode phase modulating, and multiplexing. In a distributed fin-line coupler, the coupling effect takes place between two fin gaps that are in close proximity to each other in the same housing.

In addition to the distributed type of coupler, printed probe couplers are also very popular in many E-plane integrated circuits. The present section will include the most significant publications on the analysis and design of distributed and printer probe couplers, and on coupling between a fin line and dielectric resonator.

6.3.1 "Propagation in Coupled Unilateral and Bilateral Finlines" by Sharma and Hoefer

This important paper provides design data on the even- and odd-mode phase constants, and the characteristic impedances of coupled unilateral and bilateral fin lines over a wide range of structural parameters for the most commonly used RT-Duroid-5880™ substrate material. The data were generated by using the spectral-domain approach. Similar results for quartz substrate ($\epsilon_r = 3.80$) would be useful.

6.3.2 "Broadband Finline Directional Couplers" by Callsen, Schmidt, and Solbach

This paper, first published in German, gives a complete account of E-plane couplers, from analysis to complete fabrication and measured results. It discusses every aspect of the distributed type of coupler and the de Ronde (fin line-suspended microstrip) coupler.

6.3.3 "Development of a Coupler in Finline Technique" by Beyer, Kother, and Wolff

This brief paper focuses attention on a generalized solution of coupled fin gaps in the same housing, including the finite thickness of the strips. The analysis uses the symmetry in the odd and the even modes, and decomposes the fields into the fundamental LSE and LSM modes. Measured results have been presented, together with a discussion on the influence of various structural parameters on the performance of a 3 dB coupler for the Ka-band.

6.3.4 "Material and Circuit Evaluation for Millimetre-Wave Applications: 3 dB Fin-Line Printed Probe Coupler for 31 GHz" by Arndt, Bornemann, Grauerholz, and Vahldieck

This experimental paper presents a 3 dB coupler based on the printed probe principle. Figure 2, which shows the coupling coefficient as a function of probe length, will be useful in designing printed probe couplers for coupling values ranging up to 40 dB.

6.3.5 "Analysis of the Coupling Coefficient between a Cylindrical Dielectric Resonator and a Fin-Line" by Hernandez-Gil and Perez

Dielectric resonators play an important role in today's microwave and millimeter-wave integrated circuits. This paper investigates the coupling between a circular cylindrical dielectric resonator and a unilateral fin line. The analysis uses the finite element method, and is supported by experimental results. We must investigate this particular topic further, and generate more accurate design data.

6.4 NONRECIPROCAL DEVICES

Nonreciprocal devices is of interest in millimeter-wave E-plane integrated circuits because they are more compact and integrable than the nonreciprocal devices built with conventional ferrite-loaded waveguides.

Unilateral fin lines on magnetized ferrite substrate have been analyzed. It has been found that a single ferrite does not show adequate nonreciprocal properties. Therefore, it is necessary to incorporate a dielectric spacer between the fins and the ferrite substrate in order to obtain an adequate nonreciprocal property. The following papers expand on some of these concepts and devices.

6.4.1 "18–26 GHz Finline Resonance Isolators Using Hexagonal Ferrites" by Helszajn and Thorpe

This paper offers a practical solution to the design of millimeter-wave fin-line isolators requiring large, direct magnetic fields. The solution is a hexagonal ferrite material with large uniaxial anisotropy. Experimental data over 18–26 GHz in WR42 housing show that stacking different substrate materials gives a better arrangement than cascading different substrate materials for full waveguide band isolators.

6.4.2 "Power Density Distribution Analysis of Ferrite Loaded Finlines for the Development of Integrated Nonreciprocal Millimeter Wave Elements" by Beyer and Wolff

This paper develops a field displacement isolator. The design is based on a rigorous-mode field expansion method of analyzing the propagation characteristics of a unilateral fin line. Figures 2, 3, and 4 of the paper show the power density distribution in the fin line in different directions. The experimental isolator at 28.5 GHz has 1.2 dB loss over a 2 GHz bandwidth.

6.4.3 "Millimetric Nonreciprocal Coupled-Slot Finline Components" by Davis and Sillars

This paper describes novel two-port and four-port Ka-band isolators. The design is based on the nonreciprocal effects of a ferrite magnetized parallel to the direction of propagation, and similar effects of the even- and odd-modes of coupled fin gaps. A 3.6 GHz bandwidth is obtained for a 20 dB isolation. This design does not necessarily need hexagonal ferrites for low applied fields, and offers a large bandwidth. The authors meticulously study various ferrite arrangements on coupled fin gaps.

6.4.4 "Progress on Finline Non Reciprocal Phase-Shifters" by Tedjini, Saguet, and Rauly

This paper proposes a nonreciprocal phase shifter based on the gyroelectrical property of a semiconducting material under an applied magnetic field. The structure consists of a fin line loaded in its slot by a thin slab of high mobility, semiconducting material, which is transversally magnetized. A $2 \times 1 \mu m$ InSb is used as the semiconductor, and a differential phase shift of 90° is obtained for a length of 1.65 cm at 36 GHz. Optimization of the bandwidth and insertion loss will result in a useful nonreciprocal phase shifter.

6.4.5 "A Fin Line Ferrite Isolator and Circulator for the R-Band" by Beyer and Wolff

The last paper of this section describes the first approach in fin-line isolator design. An antipodal fin-line Y-junction is used in conjunction with three impedance transformers and a ferrite sphere to realize the isolator. A cosine-squared taper is found to be best suited for impedance transformation in antipodal fin line. We obtain a transmission loss better than 3 dB and an isolation better than 10 dB.

6.5 ANTENNAS

So far, most of the studies on the feasibility of fin-line antennas have been experimental. The fin line, because of its low loss and great compatibility with solid-state devices, shows much promise for millimeter-wave antenna applications. In this section, we include a unique publication on the experimental study of unilateral fin-line antenna design for imaging array applications.

6.5.1 "Studies in Fin-Line Antenna Design for Imaging Array Applications" by Farr, Webb, and Mittra

This paper presents an experimental study on the feasibility of various fin-line antenna designs for a 220 GHz imaging radar system. The study has produced two possible antenna designs. One design is a transition from fin line to rectangular waveguide followed by a horn or open-ended waveguide antenna. The other design is an end-fire antenna with a dielectric rod protruding from the front end. Thus, this paper shows the potential of fin-line antennas.

BIBLIOGRAPHY

Tapers and Matching Networks

1. A.M.K. Saad and K. Schunemann, "Design of Finline Tapers, Transitions, and Couplers," *Proc. 11th European Microwave Conf.*, 1981, pp. 205–308.
2. D. Mirshekar-Syahkal and J.B. Davis, "Accurate Analysis of Tapered Planar Transmission Lines for Microwave Integrated Circuits," *IEEE Trans. Microwave Theory and Techniques*, Vol. MTT-29, No. 2, Feb. 1981, pp. 123–128.
3. A. Beyer, "Analysis of Transmission Characteristics of Inhomogeneous Grounded Finline," *IEEE Trans. Microwave Theory and Techniques*, Vol. MTT-33, No. 2, Feb. 1985, pp. 145–149.
4. P. Pramanick and P. Bhartia, "A Generalized Theory of Tapered Transmission Line Matching Transformers and Asymmetric 180° Coupler," *IEEE MTT-S Symposium Digest*, 1987.

Filters and Multiplexers

1. Y. Konishi and K. Uenakada, "The Design of a Bandpass Filter with Inductive Strip-Planar Circuit Mounted in a Waveguide," *IEEE Trans. Microwave Theory and Techniques*, Vol. MTT-22, No. 10, Oct. 1974, pp. 869–873.
2. Y. Tajima and Y. Sawayama, "Design and Analysis of a Waveguide — Sandwich Microwave Filter," *IEEE Trans. Microwave Theory and Techniques*, Vol. MTT-22, No. 9, Sept. 1974, pp. 839–841.
3. A.M.K. Saad and K. Schunemann, "Design and Performance of Finline Bandpass Filters," *Proc. 9th European Microwave Conf.*, 1979, pp. 397–401.
4. F. Arndt et al., "Low-Insertion-Loss Finline Filters for Millimeter-Wave Applications," *Proc. 11th European Microwave Conf.*, 1981, pp. 309–314.
5. R.R. Monsour and R.H. Macphie, "An Improved Transmission Line Matrix Formulation of Cascaded Discontinuities and Its Application to E-Plane Circuits," *IEEE Trans. Microwave Theory and Techniques*, Vol. MTT-34, No. 12, Dec. 1986, pp. 1490–1498.
6. Y.C. Shih, "Design of Waveguide E-Plane Filters with All-Metal Inserts," *IEEE Trans. Microwave Theory and Techniques*, Vol. MTT-32, No. 7, July 1984, pp. 695–704.
7. Y.C. Shih and T. Itoh, "E-Plane Filters with Finite Thickness Septa," *IEEE Trans. Microwave Theory and Techniques*, Vol. MTT-31, No. 12, Dec. 1983, pp. 1009–1012.
8. I.D. Cohen, N. Worontzoff, J. Levy, and A. Harvey, "Millimeter-Wave Multiplexer with Printed Circuit Elements for the 88 to 100 GHz Frequency Range," *IEEE MTT-S Symposium Digest*, 1984, pp. 233–235.
9. C. Nguyen and K. Chang, "Design and Performance of a W-Band Broadband Finline Diplexer with over 20 GHz Bandwidth," *IEEE MTT-S Symposium Digest*, 1985, pp. 349–352.

Coupled Fin Lines and Directional Couplers

1. A.M.K. Saad and K. Schunemann, "Design of Finline Tapers, Transitions and Couplers," *Proc. 11th European Microwave Conf.*, 1981, pp. 305–308.
2. H. El Hennawy, R. Knoechel, and K. Schunemann, "Octave-Band Finline Hybrid," *Proc. 11th European Microwave Conf.*, 1981, pp. 301–303.
3. P.J. Meier, "Printed Probe Hybrid Coupler for the 3 mm Band," *Proc. 9th European Microwave Conf.*, 1979, pp. 443–447.
4. F.C. de Ronde, "A New Class of Microstrip Directional Coupler," *IEEE MTT-S Symposium Digest*, 1970, pp. 184–186.
5. H. El Hennawy, R. Knoechel, and K. Schunemann, "A Wideband Branch Line Coupler in Finline Technology," *Archiv fur Elek Ubertra*, Vol. 37, Jan. 1983, pp. 40–46.

6. R.N. Bates, S.J. Nightingale, and P.M. Ballard, "Millimeter-Wave E-Plane Components and Subsystems," *Radio and Electronic Engineers*, Vol. 52, Nov.-Dec. 1982, pp. 506–512.

Nonreciprocal Devices

1. A. Beyer and K. Solbach, "A New Finline Isolator for Integrated Millimeter-Wave Circuits," *IEEE Trans. Microwave Theory and Techniques*, Vol. MTT-29, No. 12, Dec. 1981, pp. 1344–1348.
2. Y. Hayashi and R. Mittra, "An Analytical Investigation of Finlines with Magnetized Ferrite Substrate," *IEEE Trans. Microwave Theory and Techniques*, Vol. MTT-31, No. 6, June 1983, pp. 495–498.
3. K. Solbach, "E-Plane Circulators 30 through 150 GHz for Integrated Millimeter-Wave Circuits," *Proc. 13th European Microwave Conf.*, 1983, pp. 163–167.
4. D. Guan and R. Hoffman, "An Analytical Investigation of Finline Structures Containing Magnetized Ferrite Slab Smaller than the Waveguide," *Proc. 16th European Microwave Conf.*, 1986, pp. 493–498.

Antennas

1. A.K. Sharma, R.M. Wilson, and A. Rosen, "An Experimental Investigation of Millimeter-Wave Finline Antennas," *Proc. IEEE AP-S Symp.*, 1985, pp. 97–100.
2. P.J. Meier, "Millimeter Integrated Circuits Suspended in the E-Plane of Rectangular Waveguide," *IEEE Trans. Microwave Theory and Techniques*, Vol. MTT-26, No. 10, Oct. 1978, pp. 726–733.
3. A. Beyer, "Millimeter-Wave Antenna in Finline Technique," *Proc. ICAP '83*, Part I, Norwich, England, 1983, pp. 44–46.
4. T. Theingren, E.I. Kollberg, and K.S. Yugvesson, "Vivaldi Antennas for Single Beam Integrated Receivers," *Proc. 12th European Microwave Conf.*, 1982, pp. 361–366.
5. P.J. Gibson, "The Vivaldi-Aerial," *Proc. 9th European Microwave Conf.*, 1979, pp. 101–105.

6.1.1

Simplified Analysis and Synthesis of Fin-Line Tapers

by Johann H. Hinken*

A formula is deduced which gives the local reflection coefficient in gradual fin-line tapers in terms of the frequency, the local cut-off frequency and its derivative with respect to the axial coordinate. The formula is valid for all common types of fin-lines and at least for thin substrates of low ε_r. On the basis of this formula and of known taper design methods, a procedure is described for the synthesis of near-optimum fin-line taper contours. Measurements agree well with theoretically obtained frequency responses.

Vereinfachte Analyse und Synthese von Flossenleitungs-Übergängen

Für den lokalen Reflexionskoeffizienten in einem allmählichen Flossenleitungs-Übergang wird eine Formel abgeleitet, die nur von der Frequenz, der lokalen Grenzfrequenz und ihrer Ableitung nach der axialen Koordinate abhängt. Die Formel gilt für alle gewöhnlichen Typen von Flossenleitungen und wenigstens für Substrate, die dünn sind und ein kleines ε_r haben. Basierend auf dieser Formel und auf bekannten Entwurfsmethoden für Leitungsübergänge wird ein Verfahren zur Synthese von nahezu optimalen Konturen in Flossenleitungs-Übergängen beschrieben. Messungen stimmen gut mit den theoretischen Frequenzgängen überein.

1. Introduction

Fin-line structures have gained much attention for use in microwave integrated circuits e.g. [1]. They consist principally of a slot line on a dielectric substrate plate, which is mounted in the direction of the fundamental mode electric field within a rectangular waveguide. Unilateral, bilateral, and antipodal fin-lines are the terms given to special types of these structures as shown in Fig. 1. They all have somewhat different properties. The antipodal fin-line seems especially well-suited for low impedance superconducting circuits [2].

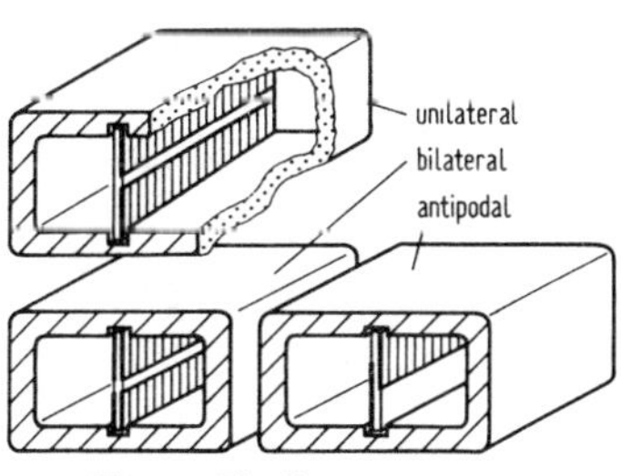

Fig. 1. Fin-line structures.

This and other applications require fin-line tapers, i.e. gradual transitions from one slot width to another, so changing the fin-line impedance. Transitions from standard waveguide to fin-line normally consist of fin-line tapers [3] – [5]. Fin-line tapers are often also a part of standard waveguide to microstrip transitions [3], [6], [7].

Designing a fin-line taper means choosing a suitable function for the length dependence of the slot width so that in a prescribed bandwidth, the return loss is never lower than a tolerable value and the physical taper length is small. This small length ensures a low insertion loss. Both demands are somewhat contradictory and a compromise has to be found.

Up to now, the design procedure has been to assume an analytical slot width function and either determine the frequency behaviour experimentally [3], or calculate it numerically [4], [5]. The function can then be slightly changed and the frequency behaviour redetermined. Finally, by trial and error, the best one of the class of functions under consideration can be obtained.

The types of calculation used up to now are rather involved. In [4], when integrating the Riccati differential equation for the reflection coefficient, surface integrals have to be computed numerically over and over again. This is the case in [8]. In [5] and [9], less involved calculation procedures are described, but the method presented in the present paper is thought to be even simpler.

In the following section, a formula will be deduced for the local fundamental mode coupling coefficient in a fin-line taper with assumed homogeneous dielectric over the cross section. The coefficient is expressed in terms of the operating frequency and of its derivative with respect to the axial coordinate. The solution is then approximately extended to dielectrically inhomogeneous cross sections as is the case of practical fin-line structures.

On the basis of this local coupling coefficient and known taper design methods [10], in the third section, a synthesis method for fin-line tapers is described, followed by an example in the fourth section.

It should be noted that in this article, it is assumed that the slot width dependence of the local fin-line cut-off frequency is known. This knowledge may, for example, be obtained from diagrams, e.g. [11].

* Dr.-Ing. J. H. Hinken, Physikalisch-Technische Bundesanstalt, Postfach 33 45, D-3300 Braunschweig.

J.H. Hinken, *Archiv fur Elektronik und Ubertragungstechnik*, Vol. 37, No. 11/12, Nov.-Dec. 1983, pp. 375–380.

2. Fundamental Mode Coupling Coefficient

For TEM lines with varying cross section, the coupling coefficient between the forward and backward travelling waves is simply [10]

$$\varkappa_{\mathrm{TEM}}^{-+} = -\frac{1}{2}\frac{1}{Z}\frac{\mathrm{d}Z}{\mathrm{d}z}, \tag{1}$$

Z wave impedance, z longitudinal coordinate.

For the analysis of non-TEM waveguides of varying cross section in the monomode frequency region, it is normally sufficient to consider only the forward and backward travelling waves of the fundamental mode. But the application of eq. (1) is problematic, because it is not known what impedance definition should be used for Z. The present paper circumvents this problem by deriving a formula for $\varkappa_{11}^{-+}$ which depends on the local cut-off frequency instead of the local impedance.

2.1. Basic equations

The following calculations are valid for the unilateral, the bilateral, and the antipodal fin-line without dielectric substrate. Fig. 2 shows the antipodal fin-line cross section as an example. This is an empty waveguide of a special cross section. This type of waveguide is considered in [10] along with others. The formulas in Section 2.1 are taken from [10].

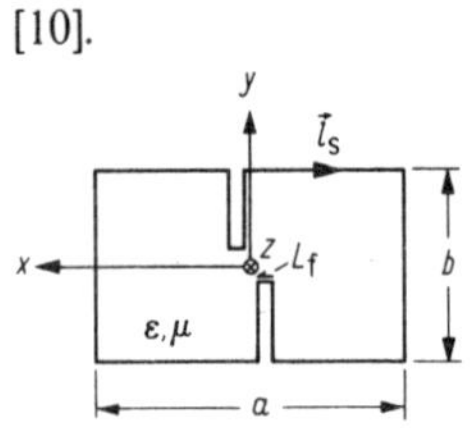

Fig. 2. Cross section of an antipodal fin-line with homogeneous dielectric.

At a varying cross section, the coupling coefficient is

$$\varkappa_{11}^{-+} = C_{11} - \frac{1}{\sqrt{Z_1}}\frac{\mathrm{d}\sqrt{Z_1}}{\mathrm{d}z} \tag{2}$$

with the field-wave impedance

$$Z_1 = \sqrt{\frac{\mu}{\varepsilon}}\,\frac{1}{\sqrt{1-(f_c/f)^2}}, \tag{3}$$

μ permeability, ε permittivity, f operating frequency, f_c local cut-off frequency.

The fundamental fin-line mode is an H-mode. The coupling factor C_{11} can therefore be calculated according to

$$C_{11} = -\frac{1}{2}\oint_L \tan\Phi\left(\frac{\partial T_1}{\partial l_s}\right)^2 \mathrm{d}l_s \tag{4}$$

from an integration along the boundary of the cross section. Φ is the angle between the unit vectors $\vec{n}_s$ and $\vec{n}_0$ which point outwards from the waveguide, and with $\vec{n}_s$ being in the cross-sectional plane and $\vec{n}_0$ being perpendicular to the waveguide wall. For the symmetrical arrangement of Fig. 3, with the angle Φ_0 at the fin edge, and with $(\partial T_1/\partial l_s)^2 = E_{tn}^2/Z_1$, where $\vec{E}_{tn}$ is the real electric field distribution of the

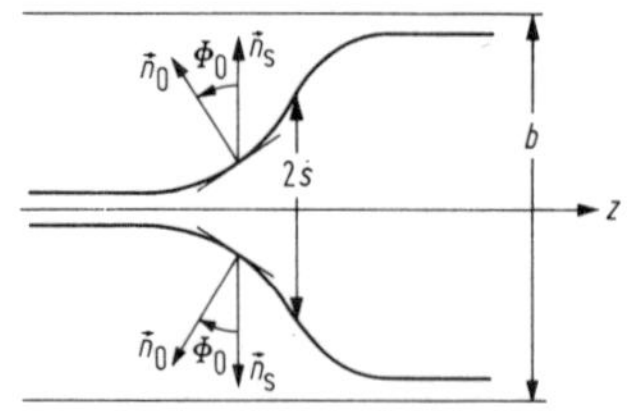

Fig. 3. Side view of a fin-line taper.

H-mode under consideration in a waveguide with nonvarying cross section and normalized to the power 1, eq. (4) simplifies to

$$C_{11} = -\frac{1}{Z_1}\tan\Phi_0 \int_{L_f} E_{tn}^2\,\mathrm{d}l_s. \tag{5}$$

2.2. Waveguide with perturbed cross section

The integral in eq. (5) also occurs in the perturbation calculation of the fin-line mode phase constant when it is changed due to a small variation of the slot width, as shown in Fig. 4. This perturbation calculation should now be performed in order to replace the integral of eq. (5) by a known term.

Fig. 4. Partial view of a waveguide cross section with z-independent wall deformation.

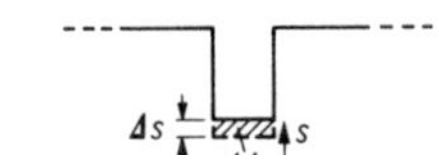

From [12] it can be seen that the relative change of the phase constant β in a waveguide with a flat, z-independent wall deformation is approximately

$$\frac{\Delta\beta}{\beta} = -\frac{1}{1-(f_c/f)^2}\,\frac{\iint_{\Delta A}(\mu|H_0|^2-\varepsilon|E_0|^2)\,\mathrm{d}A}{\iint_{A}(\mu|H_0|^2+\varepsilon|E_0|^2)\,\mathrm{d}A}, \tag{6}$$

A total waveguide cross section, ΔA perturbed area.

$|H_0|$ and $|E_0|$ are the magnitudes of the magnetic and electric field distribution before the perturbation. For H-waves, $\vec{E}_0=\vec{E}_t$ is purely transverse. $|H_0|^2$, however, is constituted by the transverse field $\vec{H}_t$ and the longitudinal component H_z.

The ratio q of their two respective contributions in the denominator of eq. (6) can be expressed through the transverse eigenfunction T_1, [12], according to

$$q \equiv \frac{\iint_A |H_z|^2\mathrm{d}A}{\iint_A |H_t|^2\mathrm{d}A} = \frac{k_c^4}{\beta^2}\,\frac{\iint_A |T_1|^2\mathrm{d}A}{\iint_A |\mathrm{grad}\,T_1|^2\mathrm{d}A} \tag{7}$$

with the cut-off wave number $k_c = 2\pi f_c\sqrt{\mu\varepsilon}$.

The second fraction on the very right hand side of eq. (7) may now, according to [12], be replaced by $1/k_c^2$, yielding $q = k_c^2/\beta^2$, or with

$$\beta = 2\pi f\sqrt{\mu\varepsilon}\,\sqrt{1-(f_c/f)^2}$$

yielding

$$1+q = \frac{1}{1-(f_c/f)^2}. \tag{8}$$

Considering, too, that $E_t/H_t = Z_1$ is valid over the whole cross section, eq. (6) is now obtained as

$$\frac{\Delta\beta}{\beta} = -\frac{1}{1-(f_c/f)^2} \cdot \frac{\iint\limits_{\Delta A} [\mu(|H_t|^2+|H_z|^2) - \varepsilon|E_t|^2]\,dA}{\left[\frac{1}{1-(f_c/f)^2}\frac{\mu}{Z_1^2}+\varepsilon\right]\iint\limits_A |E_t|^2 dA}. \tag{9}$$

In the numerator of eq. (9), the contribution of H_z will now be assumed to be negligibly small compared with the residual contributions from H_t and E_t. To the author's knowledge, there are no numerical field calculations which support this assumption, but it can be made plausible by the following considerations.

1. It is assumed that the fins are in planes near to the middle of the waveguide, i.e. near $x=0$.
2. In the unperturbed rectangular waveguide, i.e. with slot width $2s$ = waveguide height b, the longitudinal magnetic field component is exactly zero at $x=0$.
3. In the other limit, too, i.e. in a wide overlapping antipodal fin-line with small fin separation in x-direction, H_z is nearly zero, as the fin-line then approaches a TEM parallel plate transmission line.
4. These considerations in the limits can lead to the assumption that at moderate slot widths, too, the longitudinal magnetic field is negligibly small at the fin edges at L_s. Only at these edges the integration in the numerator of eq. (9) has to be performed.

With $E_t/H_z = Z_1$ eq. (9) then becomes

$$\frac{\Delta\beta}{\beta} = -\frac{1}{1-(f_c/f)^2} \frac{\left(\frac{\mu}{Z_1^2}-\varepsilon\right)\iint\limits_{\Delta A} |E_t|^2 dA}{\left[\frac{1}{1-(f_c/f)^2}\frac{\mu}{Z_1^2}+\varepsilon\right]\iint\limits_A |E_t|^2 dA}. \tag{10}$$

At this point, the field E_t which is of arbitrary amplitude should be replaced by the real distribution E_{tn}, the field which is normalized to the power 1 according to

$$1 = \iint\limits_A \vec{E}_{tn} \times \vec{H}^*_{tn}\,dA = \frac{1}{Z_1}\iint\limits_A E^2_{tn}\,dA\,. \tag{11}$$

Furthermore, it should be considered that within the area ΔA, the electric field will be constant along Δs and that the integration over the cross section gives the same value at two lines.

Eq. (11) then becomes

$$\frac{\Delta\beta}{\Delta s} = k\frac{(f_c/f)^2}{\sqrt{1-(f_c/f)^2}}\frac{1}{Z_1}\int\limits_{L_t} E^2_{tn}\,dl_s \tag{12}$$

with $k = 2\pi f\sqrt{\mu\varepsilon}$ being the wave number.

2.3. Coupling coefficient

The integral in eq. (12) can now be introduced into eq. (5). Furthermore, when it is considered that

$$\lim_{\Delta z\to 0}\frac{\Delta\beta}{\Delta s} = -\frac{d\beta}{ds} = \frac{k}{\sqrt{1-(f_c/f)^2}}\frac{f_c}{f^2}\frac{df_c}{ds} \tag{13}$$

and $\tan\Phi_0 = ds/dz$, a rather simple expression for C_{11} follows:

$$C_{11} = -\frac{1}{f_c}\frac{df_c}{dz}. \tag{14}$$

Into the second term on the right hand side of eq. (2), eq. (3) can be introduced, which gives

$$-\frac{d\sqrt{Z_1}}{dz}\frac{1}{\sqrt{Z_1}} = -\frac{1}{2}\frac{(f_c/f)^2}{1-(f_c/f)^2}\frac{1}{f_c}\frac{df_c}{dz}. \tag{15}$$

Together with eq. (14), the coupling coefficient then results as

$$\varkappa_{11}^{-+} = -\frac{1-(f_c/f)^2/2}{1-(f_c/f)^2}\frac{1}{f_c}\frac{df_c}{dz}. \tag{16}$$

Eq. (16) is an approximate expression for the local reflection coefficient of a fin-line transition with tapered slot width and with homogeneous dielectric. The approximation is the neglect of the longitudinal magnetic field at the fin edges. Eq. (16) gives the reflection coefficient for any type of fin-line taper in terms of the operating frequency, the local cut-off frequency and its derivative with respect to the axial coordinate. In contrast to the wave impedance, the cut-off frequency is well defined. It is that characteristic value of a waveguide cross section which is normally calculated before other values (e.g. phase constant, wave impedance) in its dependence on the slot width $2s$. From this dependence, the local reflection coefficient in a fin-line taper can then be directly calculated.

Eq. (16) has been deduced for fin-lines with homogeneous dielectric. In practice, however, the fins are held by thin dielectric substrates which are also inside the waveguide. These substrates lower the cut-off frequency. The introduction of this lowered cut-off frequency instead of f_c into eq. (16) was an attempt to apply the formula (16) to practical fin-line tapers. This attempt was successful, as will be shown in Section 2.4.

2.4. Theoretical and measured taper return loss

In a fin-line taper, the phase constant β depends on the longitudinal coordinate z. For such a taper of length L, the input reflection coefficient R is [13]

$$R = -\int\limits_0^L \varkappa_{11}^{-+}\exp\left(-j\,2\int\limits_0^z \beta\,dz'\right)dz\,, \tag{17}$$

provided the tapering is smooth enough to ensure that the voltage standing wave ratio is not too large all along the taper. This restriction is normally fulfilled.

In order to show the validity of eq. (16) for a double transition from a Ka-band waveguide to a 50 Ω antipodal fin-line and back to a Ka-band waveguide, the calculated and the measured return loss is compared. The transition is made on RT/duroid 5880 substrate, 0.127 mm thick and $\varepsilon_r = 2.2$. One taper length is 8.12 mm. The total length of the double transition is 30.0 mm. The exact profile is synthesized as described in Sections 3 and 4.

The calculation was made by the numerical integration of eq. (17) with $\varkappa_{11}^{-+}$ from eq. (16). The phase constant β in eq. (17) was calculated according to [11]

$$\beta = 2\pi f \sqrt{\mu_0 \varepsilon_0} \sqrt{k_e} \sqrt{1-(f_c/f)^2}\,. \tag{18}$$

During the numerical integration, the local values of f_c and the equivalent dielectric constant k_e were taken from interpolation in a value matrix which was placed at our disposal by [14]. This table gives $f_c(s)$ and $k_e(s)$ for the antipodal fin-line structure used.

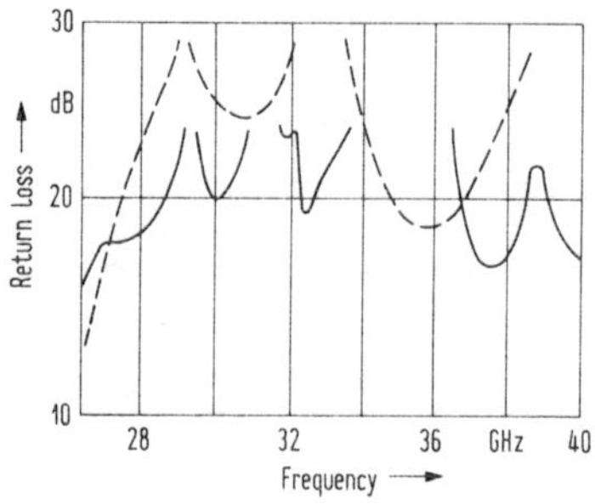

Fig. 5. Calculated (– – –) and measured (——) return loss of a Ka-band waveguide to 50 Ω antipodal fin-line double transition. Material RT/duroid 5880 with $\varepsilon_r = 2.2$ and 0.127 mm thick; one taper length 8.12 mm, total length 30.00 mm. Synthesized profile according to Sections 3 and 4. The overetching offset was compensated for in advance.

Fig. 5 shows this calculated return loss in the Ka-band, together with the measured return loss. The overall frequency responses seem very similar. The residual differences are obviously caused by failures in the realisation process, because the following observations prove the tapers very sensitive against failures:

Overetching occured and the desired curves at first were not exactly realised. Offsets in the slot width were detected and measured by microscopic measurements. Recalculation of the frequency response with this offset taken into consideration resulted in good agreement. An example is shown in Fig. 6.

Later on, and also in the mask for the double transition of Fig. 5, the profile was corrected in advance for a typical overetching offset of 70 μm perpendicular to the profile.

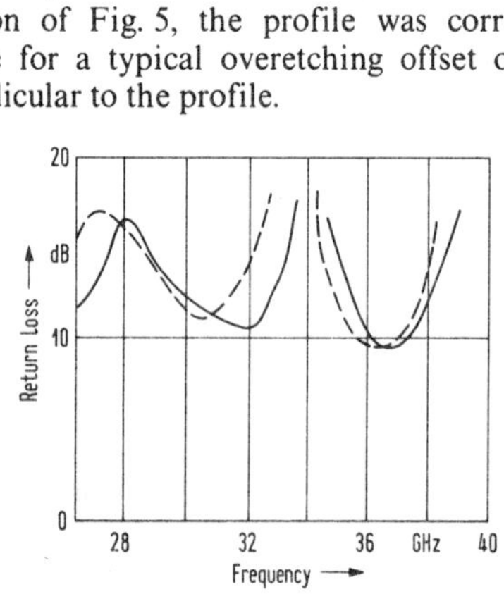

Fig. 6. As Fig. 5, but with an intended slot width offset of 2×30 μm from synthesized profile.

Smaller irregular failures, which could not be corrected in advance are assumed to be the reason for the residual differences in calculated and measured return loss.

These comparisons show that eq. (16) is not only valid for fin-lines with homogeneous cross section. It may also be applied to fin-line structures with inhomogeneous dielectric if f_c is regarded as the cut-off frequency of the dielectrically loaded fin-line. This seems to apply at least to dielectric substrates of low dielectric constant and small thickness, as in the case examined. The use of such material, however, is widespread.

3. Synthesis of Slot Width Function

In [10], the synthesis of optimum TEM tapers is described in detail. To take the advantages of that method into the design of fin-line tapers, several points must be considered in advance. If eq. (17) is applied to a fin-line taper, $\varkappa_{11}^{-+}$ and β depend on both the frequency f and the longitudinal coordinate z. In TEM tapers, eq. (17) is valid, but β depends only on frequency and $\varkappa_{11}^{-+}$ depends only on z. Because the frequency dependence of eq. (17) is much more determined by β in the exponent than by $\varkappa_{11}^{-+}$, in a first approximation $\varkappa_{11}^{-+}(f, z)$ should be replaced by its value at a fixed frequency f_0:

$$\varkappa_{11}^{-+}(f, z) \approx \varkappa_{11}^{-+}(f_0, z)\,. \tag{19}$$

In a further step, the integral in the exponent of eq. (17) should be approximated by a product of a factor which is only f-dependent and a factor which is only z-dependent according to

$$2\int_0^z \beta(f, z')\,\mathrm{d}z' \approx \eta(f)\,\zeta(z)\,. \tag{20}$$

Analogously to [10], we choose $\eta(f_0) = 1$. Then

$$\zeta(z) = 2\int_0^z \beta(f_0, z')\,\mathrm{d}z' \tag{21}$$

is valid, and

$$\mathrm{d}\zeta = 2\beta(f_0, z)\,\mathrm{d}z\,. \tag{22}$$

Analogously to [10], we denominate $\zeta(L) \equiv 2\vartheta$ where L is the taper length.

With eqs. (19) to (21), eq. (17) can be written approximately as

$$R(\eta) \approx -\int_0^{2\vartheta} \frac{\varkappa_{11}^{-+}[f_0, z(\zeta)]}{2\beta[f_0, z(\zeta)]}\,\mathrm{e}^{-\mathrm{j}\eta\zeta}\,\mathrm{d}\zeta\,. \tag{23}$$

Eq. (23) can be basis for a design procedure according to [10]. In accordance with this work we abbreviate

$$\frac{\varkappa_{11}^{-+}(f_0, z)}{2\beta(f_0, z)} = C\,K(\zeta) \tag{24}$$

with the coupling distribution function $K(\zeta)$ being normalized by

$$\int_0^{2\vartheta} K(\zeta)\,\mathrm{d}\zeta = 1\,. \tag{25}$$

From eqs. (16), (22), (24), and (25), after analytical integration, the constant C follows as

$$C = \ln\left[\frac{f_{c1}}{f_{c2}}\left(\frac{1-f_{c2}^2/f_0^2}{1-f_{c1}^2/f_0^2}\right)^{1/4}\right], \tag{26}$$

f_{c1}, f_{c2} local cut-off frequencies at the taper ends. The normalized input reflection coefficient is

$$F(\eta) = \int_0^{2\vartheta} K(\zeta)\, e^{-j\eta\zeta}\, d\zeta\,. \tag{27}$$

In [10], for $F(\eta)$ being smaller than a prescribed discrimination D at normalized frequencies $\eta \geqq 1$, such distributions $K(\zeta)$ are given, which lead to short coupling sections. Depending on the magnitude of D, different types of functions are to be chosen for $K(\zeta)$: const, sin, $\sin^2$, ... The optimum function, and this may be adjusted to all D, is the Dolph-Chebyshev distribution, which, however, leads to abrupt steps in the slot width function at the taper ends. The effect of these steps on higher order mode excitation and their influence on the fundamental mode input reflection coefficient, however, can be estimated only with difficulty.

The fin-line taper design procedure, which is based on knowledge of the dependences $f_c(s)$ and $k_e(s)$, can now be described as follows:

1. Choose values $2s_1$ and $2s_2$ for the slot width at the beginning of the taper and the taper end; choose the minimum frequency f_0 and the maximum tolerable input reflection coefficient $|R|_{max}$ for $f > f_0$.
2. Determine $f_{c1} = f_c(s_1)$ and $f_{c2} = f_c(s_2)$.
3. Determine C from eq. (26).
4. Calculate D from

$$D = |R|_{max}/C\,. \tag{28}$$

5. Choose an appropriate function $K(\zeta)$ and the value of ϑ from [10].
6. Calculate f_c according to

$$f_c(\zeta) = \tag{29}$$
$$= f_{c1}\left\{\frac{1}{2}\frac{f_{c1}^2}{f_0^2} + \left[\frac{1}{4}\frac{f_{c1}^4}{f_0^4} + \left(1-\frac{f_{c1}^2}{f_0^2}\right)e^{4CI(\zeta)}\right]^{1/2}\right\}^{-1/2}$$

with $$I(\zeta) = \int_0^{\zeta} K(\zeta')\, d\zeta'\,. \tag{30}$$

Eq. (29) is obtained from integrating eq. (24) by using eqs. (16) and (22). For standard distribution functions $K(\zeta)$, the integration in eq. (20) can be performed analytically.

7. Calculate $s(\zeta)$ from the known dependences $f_c(\zeta)$ and $f_c(s)$.
8. Calculate $z(\zeta)$ from the integrated equation (22) i.e. from

$$z(\zeta) = \frac{1}{2\pi f_0\sqrt{\mu_0\varepsilon_0}}\int_0^{\zeta}\frac{d\zeta}{\sqrt{k_e[s(\zeta)]}\sqrt{1-f_c^2(\zeta)/f_0^2}}\,. \tag{31}$$

This may be done numerically.

Steps 7 and 8 give the desired function $s(z)$ in a parametric representation.

4. Design Example

A taper from a Ka-band waveguide to an antipodal fin-line of 0.4 mm overlapping on RT/duroid 5880 substrate, 0.127 mm thick and $\varepsilon_r = 2.2$ is to be designed. The antipodal fin-line has at the end an impedance of approximately 50 Ω when calculating with microstrip formula [15]. The dependences $f_c(s)$ and $k_e(s)$ were kindly given by [14]. The preliminary maximum tolerable return loss is chosen to be $|R|_{max} = 0.1$ for frequencies $f \geqq f_0 = 26.4$ GHz.

With $2s_1 = 3.56$ mm, which is the standard height of a Ka-band waveguide, and $2s_2 = -0.4$ mm, the local cut-off frequencies at the ends result as $f_{c1} = 20.55$ GHz and $f_{c2} = 5.16$ GHz. From eqs. (26) and (28) follows $C = 1.605$ and $D = 0.0623 \triangleq 24.1$ dB. From [10], it follows that a sine-taper gives a near-optimum length for this value of D:

$$K(\zeta) = \frac{\pi}{4\vartheta}\sin\frac{\pi\zeta}{2\vartheta} \quad \text{for} \quad 0 < \zeta < 2\vartheta\,. \tag{32}$$

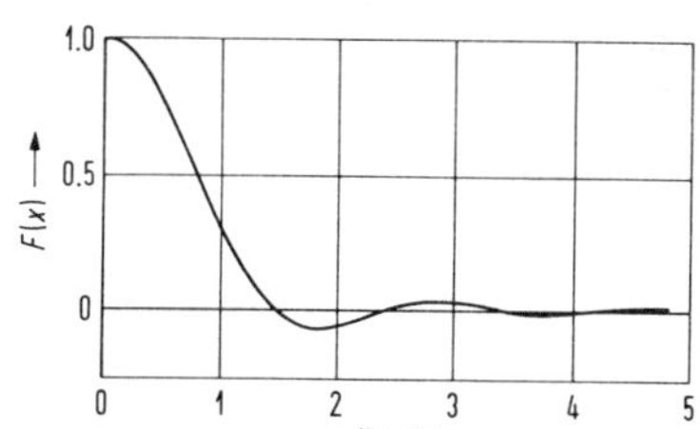

Fig. 7. Normalized input reflection coefficient $F(x)$ for a sine-type coupling distribution function [10].

The normalized input reflection coefficient is taken from [10] and shown in Fig. 7, curve a, versus

$$x = \eta\vartheta/\pi\,. \tag{33}$$

The value $x_0 = \vartheta/\pi$, which corresponds to $\eta = 1$ and $f = f_0$, must now be chosen appropriately. To fulfill the requirement of $D = 0.0623$ strictly, a value of $x_0 = 2.05$ would have to be chosen. If, however, $D = 0.071$ is sufficient, $x_0 = 1.36$ can be chosen, which considerably reduces ϑ and thus the taper length. The maximum input reflection coefficient for $f \geqq f_0$ will then be $|R|_{max} = C \cdot D = 1.605 \cdot 0.071 = 0.114$. This means, the minimum return loss reduction is only 1 dB, i.e. from 20 dB to 19 dB. D was therefore chosen $D = 0.071$, and from eq. (23) follows $\vartheta = 1.36\,\pi$.

According to eq. (30) eq. (32) must be integrated, yielding

$$I(\zeta) = \frac{1}{2}\left(1 - \cos\frac{\pi\zeta}{2\vartheta}\right). \tag{34}$$

As described in steps 6 to 8, the functions $f_c(\zeta)$, $s(\zeta)$, and $z(\zeta)$ are now obtained, with the last two functions giving the slot width function $s(z)$. This function $s(z)$ is plotted in Fig. 8. With this function known, a taper can be built using standard techniques.

The input reflection coefficient of the taper with the slot width function $s(z)$ can now be calculated using eqs. (17) and (16) which were proved valid in Section 2. Comparison of this frequency response with that expected from the normalized response in Fig. 7 shows how well the approximations (19) and (20) are fulfilled.

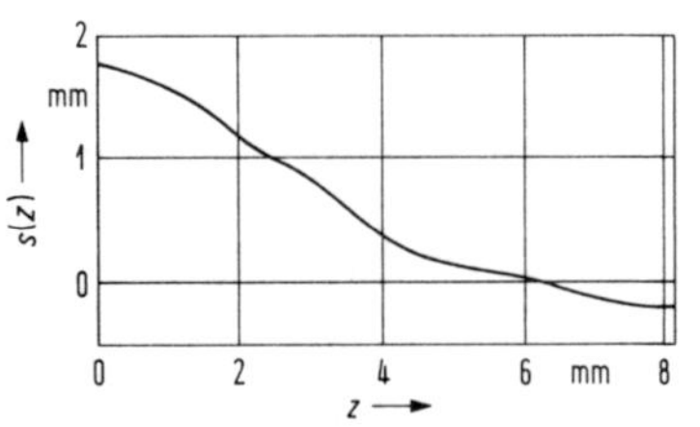

Fig. 8. Synthesized slot width function $s(z)$.

Both curves are plotted in Fig. 9. It can be seen that the approximations (19) and (20) influence the design more on the safe side. This is pleasant, but probably at the same time, the functions $s(z)$ deviated slightly even more from the optimally short function.

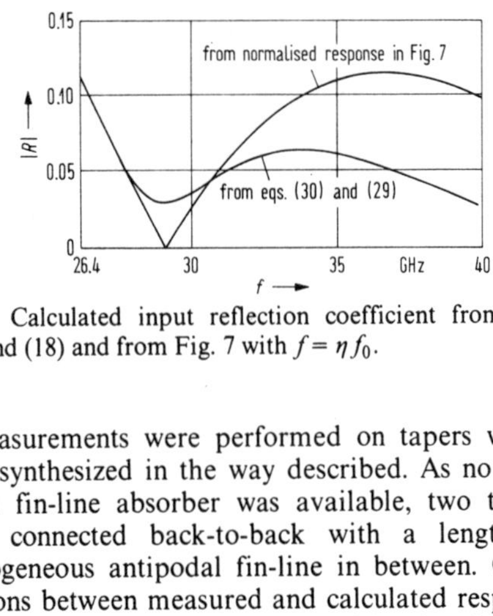

Fig. 9. Calculated input reflection coefficient from eqs. (17) and (18) and from Fig. 7 with $f = \eta f_0$.

Measurements were performed on tapers which were synthesized in the way described. As no antipodal fin-line absorber was available, two tapers were connected back-to-back with a length of homogeneous antipodal fin-line in between. Comparisons between measured and calculated response were made in Section 2.4 and considered satisfactory.

5. Conclusion

A formula has been deduced for the local reflection coefficient in fin-line tapers with homogeneous dielectric. The formula depends on the frequency, the local cut-off frequency and its z-dependence.

The formula is not only applicable to tapers in the usual type of fin-lines but to all waveguides whose cross sections vary at places with negligible longitudinal magnetic field.

The formula proved to be also applicable for fin-lines on a thin substrate with low ε_r. A quantitative limit for substrate thickness and ε_r is, however, not known.

On the basis of this formula and of the TEM taper design method presented in [10], a synthesis method for the slot width function of a nearly optimally short fin-line taper has been described. In an example, a sine-type coupling distribution function was near optimum and therefore applied. For other design goals, the Dolph-Chebyshev distribution may be necessary. With this distribution, it is probable that those reflections can be integrated into the design which occur at an abrupt transition from an empty waveguide to one which is loaded by the dielectric substrate without fins. This effect was not considered in the presented design.

The author wishes to thank Dr. K. Schünemann and his group, and Mr. R. Pöpel for fruitful discussions. He is especially indebted to Dr. A. M. K. Saad for providing a value matrix of the functions $f_c(s)$ and $k_e(s)$. The author also wishes to thank Mr. H. Flügel for performing the necessary computations and Mr. F.-T. Ullmann for fabricating the fin-line structures. Finally, the author gratefully acknowledges the financial support of the Deutsche Forschungsgemeinschaft.

(Received December 15th, 1982.)

References

[1] Adelseck, B. et al., Neue Millimeterwellenkomponenten in quasi-planarer Leitungstechnik. Frequenz **35** [1981], 118–123.

[2] Hohenwarter, G. K. and Beyer, J. B., Broadband RF-coupling to thin-film Josephson and SIS devices. Transact. IEEE MAG-**17** [1981], 826–829.

[3] Begemann, G., An X-band balanced fin-line mixer. Transact. IEEE MTT-**26** [1978], 1007–1011.

[4] Beyer, A. and Wolff, I., Calculation of the transmission properties of inhomogeneous fin lines. Proc. 10th Europ. Microwave Conf., Warsaw 1980, pp. 322–326.

[5] Saad, A. M. K., Analyse von Eigenwellen planarer Leitungen. Doctoral Thesis, Universität Braunschweig, 1982.

[6] van Heuven, J. H. C., A new integrated waveguide-microstrip transition. Transact. IEEE MTT-**24** [1976], 144–147.

[7] Lavedan, L. J., Design of waveguide-to-microstrip transitions specially suited to millimeter-wave applications. Electron. Letters **13** [1977], 604–605.

[8] Mirshekar-Syahkal, D. and Davies, J. B., Accurate analysis of tapered planar transmission lines for microwave integrated circuits. Transact. IEEE MTT-**29** [1981], 123–128.

[9] Saad, A. M. K. and Schünemann, K., Design of fin-line tapers, transitions and couplers. Proc. 11th Europ. Microwave Conf., Amsterdam 1981, pp. 305–308.

[10] Sporleder, F. and Unger, H.-G., Waveguide tapers, transitions and couplers. P. Peregrinus, Stevenage 1979.

[11] Saad, A. M. K. and Begemann, G., Electrical performance of fin-lines of various configurations. Microwaves, Optics and Acoustics **1** [1977], 81–88.

[12] Unger, H.-G., Elektromagnetische Theorie für die Hochfrequenztechnik, Vol. I and II. Dr. A Hüthig Verlag, Heidelberg 1981.

[13] Tang, C. C. H., Nonuniform waveguide high-pass filters with extremely steep cut-off. Transact. IEEE MTT-**12** [1964], 300–309.

[14] Saad, A. M. K., Private communication.

[15] Gupta, K. C. et al., Microstrip lines and slotlines. Artech, Dedham, Mass., 1979.

6.1.2

Synthesis of Optimum Finline Tapers Using Dispersion Formulas for Arbitrary Slot Widths and Locations

CHRISTIAN SCHIEBLICH, JERZY K. PIOTROWSKI, AND J. H. HINKEN, SENIOR MEMBER, IEEE

Abstract **—The theory of TEM matching sections has been modified so that it can be applied to finline tapers. A step-by-step procedure is given to calculate the taper contour for a given maximum VSWR. The taper is optimum in the sense that its length is the shortest possible for the required VSWR. To achieve fast convergence, a transversal resonance method was developed to calculate finline dispersion, which is valid for arbitrary slot widths and slot locations. The finline can be unilateral as well as bilateral, and the slot may be off-centered. The dispersion data are compared with values found in the literature, and the calculated taper performance with the authors' own measurements, both showing good agreement.**

Manuscript received April 25, 1984; revised July 27, 1984. This work was supported in part by the Deutsche Forschungsgemeinschaft (DFG) and the German Academic Exchange Service (DAAD).

C. Schieblich is with the Technische Universität Hamburg-Harburg, Arbeitsbereich Elektrotechnik III, Hochfrequenztechnik, Wallgraben 55, D-2100 Hamburg 90, West Germany.

J. K. Piotrowski is with the Institute of Electron Technology, Warsaw Technical University, Koszykowa 75, Warsaw, Poland.

J. H. Hinken is with the Physikalisch-Technische Bundesanstalt, Postfach 3345, D-3300 Braunschweig, West Germany.

I. INTRODUCTION

FINLINE COMPONENTS have attracted much attention due to their favorable properties, such as broad single-mode bandwidth, moderate attenuation, sim-

C. Schieblich, J.K. Piotrowski, and J.H. Hinken, *IEEE Transactions on Microwave Theory and Techniques*, Vol. MTT-32, No. 12, Dec. 1984, pp. 1638–1645.

ple fabrication, and integration feasibility. For the design of such components, even if they are planned for a later integration, low return loss transitions to other waveguides are essential. Transitions in the final device should be as short as possible to achieve compactness and low insertion loss. This contribution deals with tapers between different slot widths with the above-mentioned features, and particularly transitions between finline and rectangular hollow waveguide. The slot may be located arbitrarily on the substrate. The computation is very fast, because an easily evaluated transversal resonance method is sufficient to obtain the slot profile.

II. TEM Theory

The taper design for TEM structures is well-known from the literature. The input reflection coefficient is [1]

$$R(\beta) = -\int_0^l \kappa^{-+}(z')e^{-2j\beta z'}dz' \tag{1}$$

with $\kappa^{-+}(z)$ the z-dependent coupling coefficient between the backward and the forward traveling fundamental wave, l the length of the taper, and $\beta = 2\pi f\sqrt{\varepsilon\mu}$ the non-z-dependent phase constant. From (1), $\kappa^{-+}(z)$ can be deduced by Fourier transformation, and solutions are known which hold the input reflection coefficient R below a certain value $R_{\max}$ for phase constants $\beta > \beta_0$, thus achieving a high-pass performance.

III. Taper Synthesis for Non-TEM Structures

According to [2], an expression equivalent to (1) can be found for non-TEM waves if we approximate the coupling coefficient

$$\kappa^{-+}(f,z) \simeq \kappa^{-+}(f_0,z) \tag{2}$$

by its value at a fixed frequency f_0. Furthermore, we approximate the integral

$$\int_0^z 2\beta(f,z')\,dz' \simeq \eta(f)\cdot\xi(z) \tag{3}$$

as a product of a purely frequency-dependent and a purely z-dependent factor. $\eta(f)$ is normalized so that $\eta(f_0)=1$. This results in

$$R(\eta) = \int_{-\theta}^{\theta} C\cdot K(\xi)e^{-j\eta\xi}d\xi \tag{4}$$

with $\theta = \xi(l) = -\xi(0)$ and

$$C\cdot K(\xi) = -\frac{\kappa^{-+}(f_0,\xi)}{2\beta(f_0,\xi)}. \tag{5}$$

The integral in (4) is of the same type as (1), and the coupling distributions for $C\cdot K(\xi)$ known from TEM theory can be applied.

IV. Non-TEM Theory for the Coupling Coefficient in Finline

If we have a relation between $\kappa^{-+}(f_0,\xi)$ and $\beta(f_0,\xi)$, the function $\beta(f_0,\xi)$ can be evaluated from $CK(\xi)$. In fact, on certain assumptions, both values can be expressed with the local cutoff frequency $f_c(z)$ as a parameter.

According to [3], the coupling coefficient κ^{-+} in an empty waveguide with arbitrary and varying cross section is

$$\kappa^{-+} = C_{11} - \frac{1}{\sqrt{Z_1}}\frac{d\sqrt{Z_1}}{dz} \tag{6}$$

where Z_1 is the field-wave impedance.

The finline taper is modeled by a double ridge waveguide with varying gap s. With $\tan\phi_0 = ds/dz, C_{11}$ for the fundamental mode (H mode) can be written as an integral along both foreheads of the ridges

$$C_{11} = -\frac{1}{Z_1}\tan\phi_0\int_{L_1,L_2} E_{tn}^2\,dl_s. \tag{7}$$

E_{tn} is the electric-field distribution normalized to the power 1 of the fundamental mode if $\phi_0 = 0$.

According to [2], (7) simplifies to

$$C_{11} = -\frac{1}{f_c}\frac{df_c}{dz}. \tag{8}$$

With

$$Z_1 = \sqrt{\mu_0/\varepsilon_0}\Big/\sqrt{1-(f_c/f)^2}$$

(6) and (8) result in

$$\kappa^{-+} = -\frac{1-(f_c/f)^2/2}{1-(f_c/f)^2}\frac{1}{f_c}\frac{df_c}{dz}. \tag{9}$$

Equation (9) has been derived on the following assumptions:

- $|\tan\phi_0| \ll 1$, i.e., smoothly varying slot contour,
- negligible longitudinal magnetic-field components on and negligible transversal current density across the forehead of the ridge (i.e., along the thickness of the finline metallization) [2],
- TE character of the fundamental mode.

The latter assumption follows from the ridge-waveguide model. Although the field distribution is not the same, the dielectric substrate in real finlines can be considered by an effective dielectric constant k_e, which lowers the cutoff frequency f_c. With this cutoff frequency inserted in (9) and the relation

$$\beta = 2\pi f\sqrt{\varepsilon_0\mu_0}\sqrt{k_e}\sqrt{1-(f_c/f)^2} \tag{10}$$

(5) can be evaluated.

V. Distribution of the Coupling Coefficient

In (4), the integrand is split into a ξ-dependent part and the normalizing constant C, so that

$$\int_{-\theta}^{\theta} K(\xi)\,d\xi = 1. \tag{11}$$

Integrating (5) from $\xi = -\theta$ to θ with (9), (2), and $d\xi = 2\beta(f_0,z)\,dz$ from (3) yields

$$C = \frac{1}{4}\ln\left[\left(\frac{f_c(0)}{f_c(l)}\right)^4\cdot\frac{1-(f_c(l)/f_0)^2}{1-(f_c(0)/f_0)^2}\right]. \tag{12}$$

Integrating (5) from $-\theta$ to ξ results in the ξ-dependent cutoff frequency

$$f_c(\xi) = f_c(0)\cdot\left[F/2+\sqrt{F^2/4+(1-F)\cdot\exp(4CI(\xi))}\,\right]^{-1/2} \tag{13}$$

with $F=(f_c(0)/f_0)^2$ and

$$I(\xi) = \int_{-\theta}^{\xi} K(\xi')\,d\xi'.$$

The coupling distribution $K(\xi)$ along the taper must be chosen so that the reflection coefficient is below a certain value $R_{\max}$ for frequencies $f > f_0$. In [3], a procedure is given to evaluate the coefficients a_i in the Fourier series

$$K(\xi) = \sum_{i=1,3,\cdots}^{2n+1} a_i \cos\left(i\frac{\pi}{2}\frac{\xi}{\theta}\right) \tag{14}$$

so that the reflection coefficient $R(\eta)$ from (4) reaches the value $R_{\max}$ n times for $\eta > 1$ (i.e., n equal ripples for $f > f_0$). With an infinite number of terms, (14) becomes the Dolph–Chebyshev distribution.

It is characterized by

$$K(\xi) = \frac{D}{2}\left\{\frac{I_1\left(\theta\sqrt{1-(\xi/\theta)^2}\right)}{\sqrt{1-(\xi/\theta)^2}} + \delta(\xi-\theta)+\delta(\xi+\theta)\right\} \tag{15}$$

for $-\theta \leqslant \xi \leqslant \theta$ with $D = R_{\max}/C$, $\theta = \operatorname{arcosh}(1/D)$, $I_1(x)$ the modified Bessel function of the first order, and $\delta(t)$ Dirac's delta function.

The shape of the function $K(\xi)$ in (14) with $n=4$ compared with that of (15) is shown in Fig. 1 with $D=0.01$ in both cases. The functions are even in ξ, so only the branch for $\xi > 0$ is shown. The shapes of $K(\xi)$ are similar where both are nonzero. The Dirac's function at $\xi = \theta$ makes the Dolph–Chebyshev distribution the shortest possible taper for a given reflection loss.

Due to the normalization (11) and the symmetry $K(\xi) = K(-\xi)$ we can evaluate the integral in (13)

$$I(\xi) = \begin{cases} \dfrac{1}{2} + \dfrac{D\theta^2}{2}\displaystyle\int_0^{\xi/\theta} \frac{I_1\left(\theta\sqrt{1-y^2}\right)}{\theta\sqrt{1-y^2}}\,dy, & \text{for } |\xi| < \theta \\ 0, & \text{for } \xi = -\theta \\ 1, & \text{for } \xi = \theta \end{cases} \tag{16}$$

The Bessel function in (16) can be expanded in a power series [4], and the integral in (16) can be evaluated by integrating it term by term.

This expansion was previously published in [5], but because of some misprints there, we give the result once more

$$\int_0^x \frac{I_1\left(\theta\sqrt{1-y^2}\right)}{\theta\sqrt{1-y^2}}\,dy = \sum_{k=0}^{\infty} a_k b_k$$

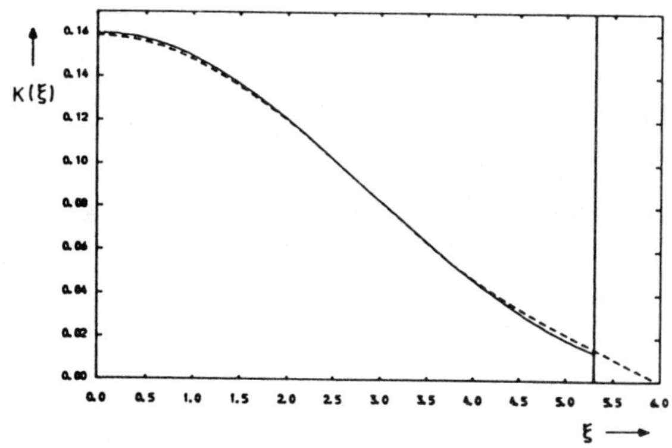

Fig.. 1. Normalized coupling distribution for $D = 0.01$. ——Dolph–Chebyshev (15). ----cosine-series $n = 4$ (14).

with

$$a_0 = 1 \qquad a_k = \frac{\theta^2}{4k(k+1)}a_{k-1}$$

$$b_0 = \frac{x}{2} \qquad b_k = \frac{\frac{x}{2}(1-x^2)^k + 2kb_{k-1}}{2k+1}. \tag{17}$$

The series converges rapidly, and taking only 12 terms makes the relative error smaller than 10^{-5}.

Dirac's function in (15) produces discontinuities at the taper ends, a fact which might cause problems due to the excitation of higher order modes. The step in cutoff frequency can be calculated from (12), (13), and (16)

$$\frac{f_c(+0)}{f_c(0)} = 1 - R_{\max}\cdot\frac{f_0^2 - f_c^2(0)}{2f_0^2 - f_c^2(0)}. \tag{18}$$

The discontinuity at the taper end with the narrow slot is even smaller. The lower the $R_{\max}$ chosen, the smaller the step. For $R_{\max} = 0.01$ ($\triangleq -40$ dB) and $f_0 = 1.4\ f_c(0)$, the ratio in (18) is 0.997. The Fourier series distribution (14) does not have this discontinuity. Comparative measurements with both taper types showed that the influence of the step is negligible.

VI. Synthesizing the Slot Contour

Up to now, we have been able to synthesize the function of the cutoff frequency f_c along the taper. It should be noted that no wave impedances are required for this synthesis, thus avoiding the problem of finding the appropriate impedance definition.

To obtain the slot width, we need a relation between slot width and cutoff frequency. This relation should be easy to evaluate, because it is called at every knot on the z axis. It is inconvenient to apply spectral-domain techniques [6]–[9], which are very precise, but time-consuming. The simple formulas of [10] are limited to special values for the permittivity of the substrate and are not valid for large slot widths, [11] gives no data for small slot widths, and the accuracy of the results in [12] seems too poor.

VII. Transversal Resonance Condition

A new attempt was therefore made to evaluate cutoff frequency and effective permittivity by a transversal resonance method. The method is valid for arbitrary slot widths and slot locations. The computed results are compared with data from [8], [10], [11], and [13].

The analysis is based on the following simplifications:

- isotropic, homogeneous, and lossless dielectric layer,
- zero-thickness metallization with infinite conductivity,
- thickness of substrate small compared with waveguide width,
- symmetrically located substrate.

In the dispersion relation (10) for the phase constant β, which underlies our taper synthesis, the effective dielectric constant k_e may be approximated as frequency-independent [14]

$$k_e \approx (f_{c0}/f_c)^2 \tag{19}$$

with f_{c0} the cutoff frequency of a finline of the same dimensions and a substrate's permittivity $\varepsilon_r = 1$. Equation (10) is then the dispersion of a homogeneously filled waveguide. In the following, our task will be to find the cutoff frequencies f_c and f_{c0}.

A. Bilateral Finline

The cross section of a bilateral finline and its equivalent transverse network at cutoff are shown in Fig. 2. The finline is symmetrical with respect to the $x = a/2$ plane, where the admittance in the x-direction at cutoff is zero. The equivalent transverse network for the dominant mode consists of a capacitive susceptance shunting the TEM transmission line with short-circuit termination. The cutoff wavenumber k_x is determined by the resonance condition at reference plane T

$$-\cot(k_x \cdot l) + \frac{B_b}{Y_b} = 0 \tag{20}$$

with

$$\frac{B_b}{Y_b} = \frac{b}{\pi} \cdot k_x \cdot (P_w + \varepsilon_r \cdot P_d) \tag{21}$$

where

$$P_w = \ln(\csc(\alpha_w) \cdot \csc(\beta_w)) \tag{22}$$

$$P_d = r_d \cdot \arctan\left(\frac{1}{r_d}\right) + \ln\sqrt{1 + r_d^2} \tag{23}$$

and

$$\alpha_w = \frac{\pi}{2} \cdot \frac{s}{b} \qquad r_d = \frac{d}{s}$$

$$\beta_w = \frac{\pi}{2}\left(1 - 2\frac{e}{b}\right).$$

The field distortions by the metal fins have been modeled by the susceptance jB_b which is composed of two parts. The first part models the field distortions left of reference plane T. It is taken from Marcuvitz's *Waveguide Handbook* [15, p. 218], where the susceptance of a window of zero thickness in a rectangular waveguide has been derived. In fact, it is sufficient to take just the first term in P_w into account (22) because the other terms influence the results by less than 0.2 percent.

The second part models the field distortion between plane T and the symmetry plane $x = a/2$. It has been taken from the equivalent circuit of an open E-plane T junction (see [15, p. 337]). Due to the symmetry with respect to $x = a/2$, this part of the susceptance is characterized only by the susceptance jB_a of the equivalent circuit in [15].

The first zero of k_x in (20) is the cutoff wavenumber for the fundamental mode. The cutoff frequencies f_{c0} and f_c are obtained from k_x with $\varepsilon_r = 1$ and $\varepsilon_r \neq 1$, respectively. Numerical results have been compared to those taken from [10] and a disagreement of less than 1 percent has been found. (Setting $\varepsilon_r = 2.22$, $\left(f_c\sqrt{\varepsilon_0\mu_0} \cdot b\right)$ varies between 0.1 and 0.2 for d/b between 1/32 and 3/4. In all cases, the deviations are smaller than 0.001.) Fig. 3 shows dispersion in a bilateral finline of relatively small slot width. The agreement with results taken from [13] is very good.

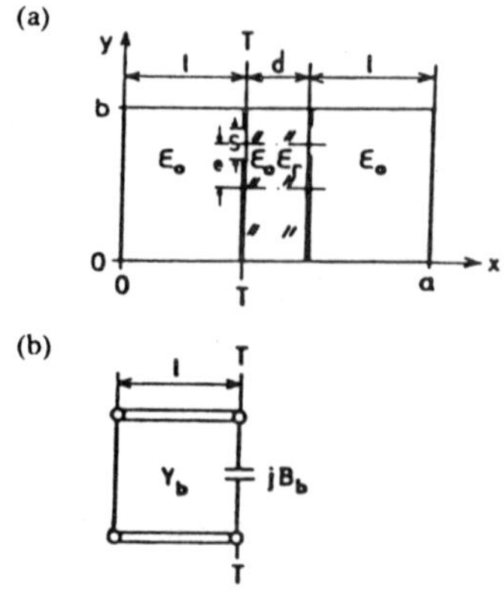

Fig. 2. The cross section of (a) a bilateral finline and (b) its equivalent transverse network.

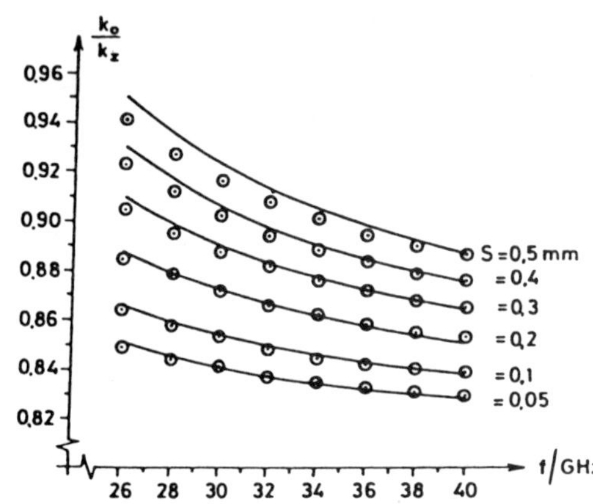

Fig. 3. Dispersion of a bilateral finline with slot centered in waveguide. ——Simons and Tech [13]. ⊙ Our approximation.

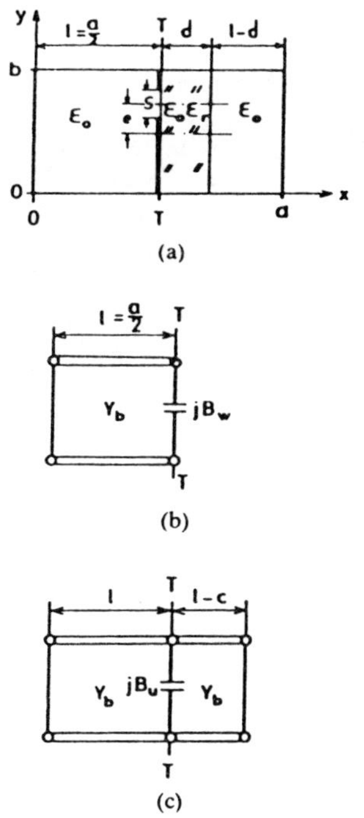

Fig. 4. The cross section of (a) a unilateral finline, (b) its equivalent transverse network for $\varepsilon_r = 1$, and (c) for $\varepsilon_r \neq 1$.

B. Unilateral Finline

Unilateral finline is often preferred because it is easy to fabricate and semiconductor devices are simple to mount. The cross section of this finline is shown in Fig. 4. The thin metal layer is placed at the $x = a/2$ plane, so that the structure is symmetrical for $\varepsilon_r = 1$. It is useful to construct two equivalent transverse networks at cutoff for the dominant mode: one for $\varepsilon_r = 1$ and another for $\varepsilon_r \neq 1$ (Fig. 4). The equivalent network for $\varepsilon_r = 1$ consists simply of a transmission line of length $a/2$ which is short-circuited at one end and shunted by jB_w at the other. The shunt susceptance can be taken from (21) and (22).

The cutoff wavenumber k_{x0} in the air-filled ridged waveguide is determined by the resonance condition at reference plane T

$$-\cot(k_{x0}\cdot l) + \frac{B_w}{Y_b} = 0 \tag{24}$$

with

$$\frac{B_w}{X_b} = \frac{b}{\pi}\cdot k_{x0}\cdot P_w. \tag{25}$$

P_w is given by (22). The equivalent network in Fig. 4(c) is formed by a susceptance jB_u shunted by two TEM transmission lines with short-circuit terminations. The relation governing cutoff of the H_{10} mode is given by

$$-\cot(k_x\cdot l) - \cot[k_x\cdot(l-d)] + \frac{B_u}{Y_b} = 0 \tag{26}$$

with

$$\frac{B_u}{Y_b} = \frac{b}{\pi}\cdot k_x\cdot[2P_w + \varepsilon_r(P_d + P_b)] \tag{27}$$

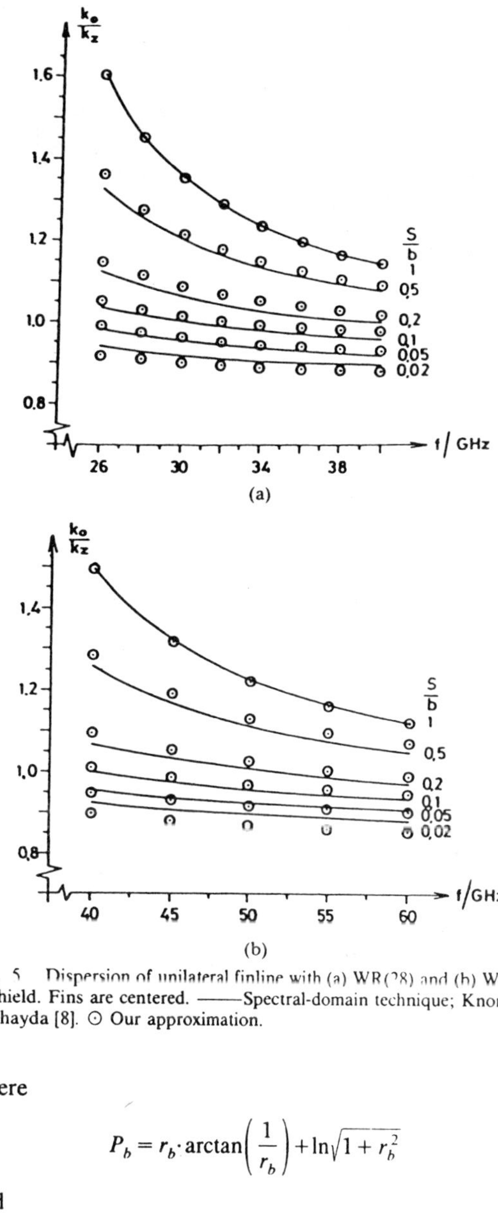

Fig. 5. Dispersion of unilateral finline with (a) WR(28) and (b) WR(19) shield. Fins are centered. ——Spectral-domain technique; Knorr and Shayda [8]. ⊙ Our approximation.

where

$$P_b = r_b\cdot\arctan\left(\frac{1}{r_b}\right) + \ln\sqrt{1 + r_b^2} \tag{28}$$

and

$$r_b = \frac{d}{b}.$$

The susceptance jB_u has been constructed by superposing the susceptance of the window (the term proportional to $2P_w$) and another susceptance (the term proportional to $\varepsilon_r(P_d + P_b)$) representing the influence of the dielectric and the transformation through the layer of length d. While P_w (22) and P_d (23) have already been given, P_b in (28) has been taken from [15, p. 337].

The unilateral finline with metal fins centered in the waveguide (Fig. 4) has been analyzed in [8], [10], and [13]. Other authors have considered a unilateral finline with the dielectric layer symmetrically located in the waveguide [6], [9]. The modifications of (24) and (26) for this case are easy to carry out.

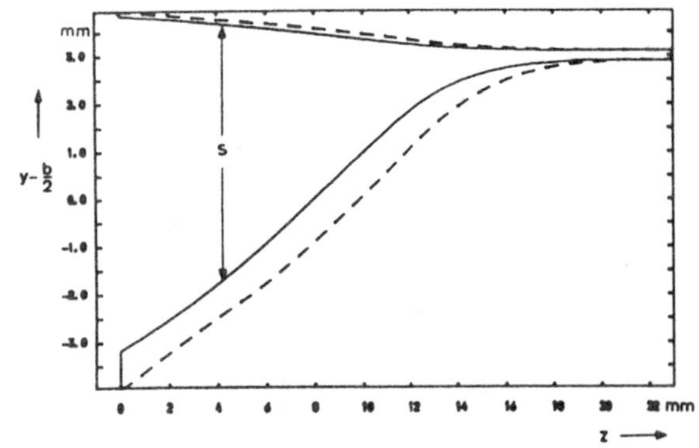

Fig. 6. Taper contour. Design data: $f_0 = 12.3$ GHz, $R_{max} = -30$ dB. ——Dolph–Chebyshev. ----cosine-series $n = 4$.

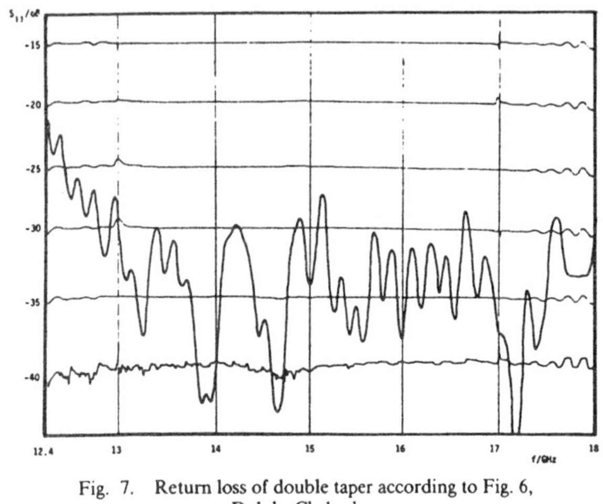

Fig. 7. Return loss of double taper according to Fig. 6, Dolph–Chebyshev.

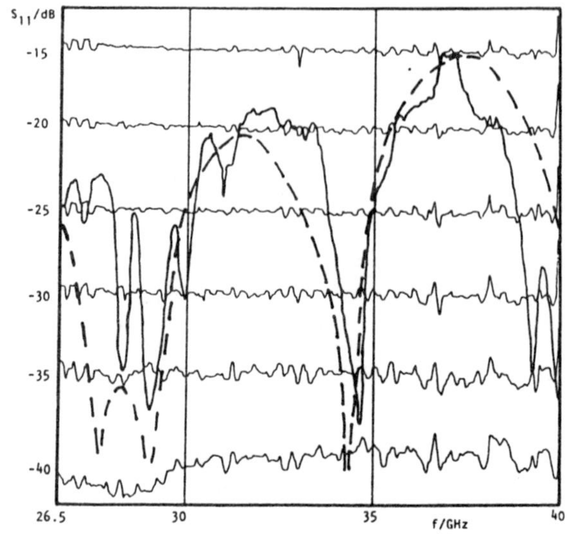

Fig. 8. Return loss of double taper $R_{max} = -20$ dB, Dolph–Chebyshev. ——Measured, ----calculated.

Comparing the cutoff wavenumber of the unilateral finline with results published in [10] yields deviations of less than 1.5 percent if $d/a < 1/8$. This restriction is, however, usually fulfilled for practical finlines. Some dispersion curves are presented in Fig. 5 showing good agreement (better than 3 percent) with published results. The authors' results deviate from those calculated with Hoefer's diagram for the slot capacitance [11] less than 0.5 percent. This diagram, however, gives only data for a normalized slot width $s/b \geqslant 0.1$.

VIII. Synthesis Procedure

The synthesis procedure can now be summarized as follows.

1) Choose the cutoff frequency f_0 of the taper and the maximum input reflection coefficient R_{max} for $f > f_0$. Choose an appropriate step width Δz for the longitudinal coordinate. If the taper is eccentric, select a relation between slot width s and eccentricity e (Figs. 2(a) and 4(a)) to achieve a unique function $s(f_c)$ from the characteristic equation. A linear function

$$e(s(z)) = e(0) + \frac{e(l) - e(0)}{s(l) - s(0)} \cdot (s(z) - s(0)) \quad (29)$$

guarantees that the slot edges never exceed the waveguide height.

2) Determine $f_c(0)$ and $f_c(l)$ from $s(0)$ and $s(l)$ and the transverse resonance condition.

3) Determine the normalization factor C from (12).

4) Set initial values $z = 0$, $\xi = -\theta$, and $s = s(0)$.

5) Next take $z{:} = z + \Delta z$. Evaluate the related $\xi(z)$ from (3) as $\xi{:} = \xi + 2\beta(f_0, z)\cdot\Delta z$, f_c from (13), and s from the transversal resonance condition with $k_x = 2\pi f_c \sqrt{k_e}\sqrt{\mu_0\varepsilon_0}$.

6) Repeat 5) until the final slot width is reached.

IX. Experimental Results

Several double-tapers were fabricated on RT/Duroid 5880 in WR-62 (and WR-28) housing. They lead from the empty waveguide to a slot width of 0.2 mm and back again to the empty waveguide. The slot was located either in the center or 3 mm (1.5 mm) below, R_{max} was chosen to be -20, -30, -35, -40, and -60 dB. To estimate the influence of the steps in the Dolph–Chebyshev profile, a set of tapers with the cosine-series coupling distribution (14) was also made.

The dynamic range of the measurement equipment was -40 dB (-35 dB) for the reflection coefficient, restricted by the waveguide termination behind the tapers. No double-taper (except the 20-dB type) turned out to be worse than -25 dB in the return loss from 12.4 to 18 GHz (26.5 to 40 GHz), and most of them were better than -30 dB. The taper length for a taper designed for -30 dB is only 18 mm (8.2 mm), which is 0.75 free-space wavelengths at the lower frequency limit.

The results of the measurements can be summarized as follows.

1) There is no significant difference in the performance of cosine-series and Dolph–Chebyshev profile tapers. The steps in the profile apparently have no influence.

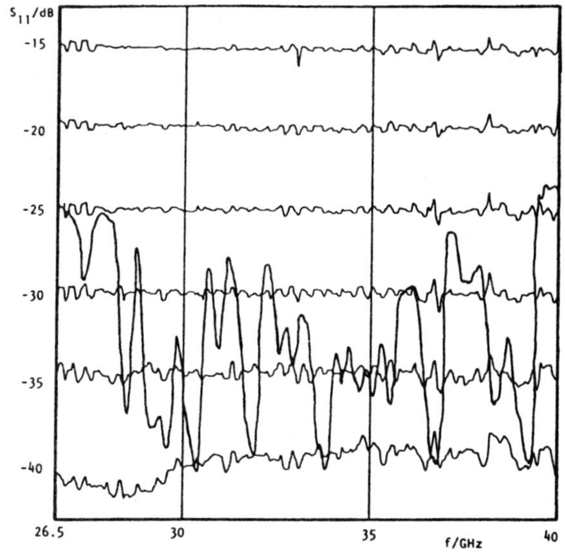

Fig. 9. Return loss of double taper. Design data: $f_0 = 26$ GHz, $R_{max} = -30$ dB, length 8.2 mm, centered slot.

2) Designing the taper for reflection loss better than -35 dB does not improve the performance. The taper may even become worse. This probably means that the accuracy limit of the transversal resonance method has been reached.

3) Tapers with a return loss of -30 dB or better should be fabricated within ± 5 percent in slot width.

For lower requirements, however, tolerances are not so critical: For a maximum reflection coefficient of -20 dB, a profile shift of ± 20 percent is admissible.

Fig. 6 shows the slot profile of a -30-dB *Ku*-band eccentric taper with cosine-series and Dolph–Chebyshev profile. Fig. 7 shows the input reflection of the corresponding double taper. Fig. 8 compares the measured input reflection of a 20-dB double taper compared with values calculated by direct integration of the coupled wave equations [3] with phase constant $\beta(f, z)$ and coupling coefficient $\kappa^{-+}(f, z)$. Finally, Fig. 9 shows the performance of a -30-dB taper in *Ka*-band. Insertion loss was about 0.3 dB for any double taper.

X. Conclusion

A fast and precise method has been presented for the design of optimum finline tapers with arbitrary slot widths and locations. The method can be applied for tapers down to -35-dB return loss. An even better performance may be obtained with a more exact relation between slot width and cutoff frequency, involving more computer time. The practical results would be doubtful, however, because requirements for mechanical tolerances become very stringent. The design procedure can also be applied to antipodal finlines, provided that a relation between slot width and cutoff frequency is known. The tapers have already been successfully applied to finline circulators, phase modulators, mixers, and oscillators.

Acknowledgment

The authors wish to thank Prof. Dr. K. Schünemann for helpful discussions.

References

[1] R. W. Klopfenstein, "A transmission line taper of improved design," *Proc. IRE*, vol. 44, pp. 31–35, Jan. 1956.

[2] J. H. Hinken, "Simplified analysis and synthesis of fin-line tapers," *Arch. Elek. Übertragung.*, vol. 37, pp. 375–380, Nov./Dec. 1983.

[3] F. Sporleder and H.-G. Unger, *Waveguide Tapers Transitions and Couplers*. Stevenage: Peregrinus, 1979.

[4] M. Abramowitz and I. A. Stegun, *Handbook of Mathematical Functions*. New York: Dover, 1965, sec. 9.6.10.

[5] A. M. Khilla, "Computer aided design of an optimum Chebyshev microstrip taper," *Arch. Elek. Übertragung.*, vol. 35, pp. 133–135, Mar. 1981.

[6] H. Hofmann, "Dispersion of planar waveguides for millimeter wave application," *Arch. Elek. Übertragung.*, vol. 31, pp. 40–44, Jan. 1977.

[7] L. P. Schmidt and T. Itoh, "Spectral domain analysis of dominant and higher order modes in fin-lines," *IEEE Trans. Microwave Theory Tech.*, vol. MTT-28, pp. 981–985, Sept. 1980.

[8] J. B. Knorr and P. M. Shayda, "Millimeter-wave fin-line characteristics," *IEEE Trans. Microwave Theory Tech.*, vol. MTT-28, pp. 737–743, July 1980.

[9] D. Mirshekar-Syahkal and J. B. Davies, "An accurate, unified solution to various fin-line structures, of phase constant, characteristic impedance and attenuation," *IEEE Trans. Microwave Theory Tech.*, vol. MTT-30, pp. 1854–1861, Nov. 1982.

[10] A. K. Sharma and W. J. R. Hoefer, "Empirical expressions for fin-line design," *IEEE Trans. Microwave Theory Tech.*, vol. MTT-31, pp. 350–356, Apr. 1983.

[11] W. J. R. Hoefer, "Fin-line design made easy," in *Proc. Int. Microwave Symp.*, (Ottawa), 1978, p. 471, and report distributed at the Symposium.

[12] A. M. K. Saad and G. Begemann, "Electrical performance of fin-lines of various configurations," *Microwaves Opt. Acoust.*, vol. 1, pp. 81–88, Jan. 1977.

[13] R. N. Simons and M. Tech, "Analysis of millimeter-wave integrated fin-line," *Inst. Elec. Eng. Proc., Part H*, vol. 130, pp. 166–169, Mar. 1983.

[14] P. J. Meier, "Integrated fin-line millimeter components," *IEEE Trans. Microwave Theory Tech.*, vol. MTT-22, pp. 1209–1216, Dec. 1974.

[15] N. Marcuvitz, *Waveguide Handbook*. New York: McGraw-Hill, 1951.

6.1.3

FINLINE TAPER DESIGN MADE EASY

Adalbert Beyer and Ingo Wolff

Duisburg University, Department of Electrical Engineering,
Bismarckstrasse 81, D-4100 Duisburg, FRG

Abstract

In this paper a method is presented which permits the quick and exact fabrication of tapers in finline technique. This method uses tapers in the form of circular arcs which are matched to the slotwidth as a function of the characteristic impedance. For the application of the described method the field distributions and the characteristic impedances of finlines must be known.

Introduction

In finline technique optimized bent structures are used in many components and systems. Examples of such structures are tapers which match different geometrical and electrical parameters of two waveguide sections or bent coupling sections which realize the desired coupling ratios of two slot structures.

For an easy fabrication of millimeter wave circuits a computer aided production of the layout masks is needed. The complexity of the circuit structures and especially the arbitrarily bent contours of e.g. tapers on the one hand render the fabrication difficult, on the other hand they effect the accuracy of the desired layouts. In the following a method is presented which allows to fabricate the layouts faster and with a higher degree of accuracy.

Circular Arc Finline Tapers

Some papers have presented calculation methods for inhomogeneous finlines /1,4/. The optimization of the inhomogeneous sections of finlines yields arbitrary contour functions which for practical applications are approximated by polynomials. Plotters and mask cutters normally can produce circular arcs quickly and with a high degree of accuracy. For this reason the optimized tapers instead of using polynomials are approximated by means of circular arc sections in this paper.

Fig.1 shows the slotwidth s of a finline taper (2) which is used to match a rectangular waveguide (1) and a homogeneous finline (3). This taper is approximated by means of two circular arcs (Fig.2); its taper function is not only continuous but it is also characterized by a taper function which is continuously differentiable in the interconnection points P_H and P_F to the homogeneous waveguides (1) and (3). From these conditions of the taper function it follows that at the points P_H, P_S, P_F:

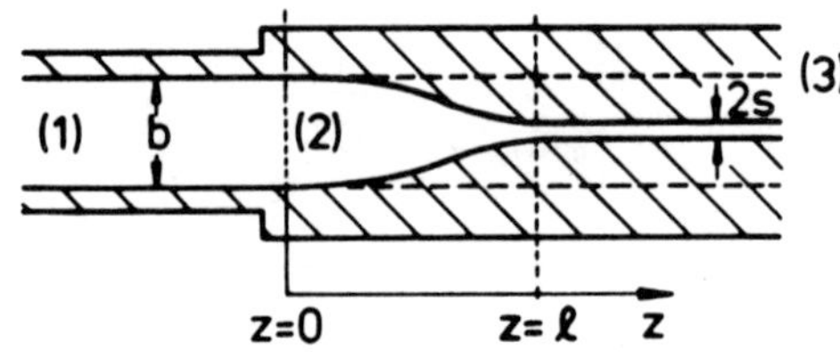

Fig.1: Finline taper.

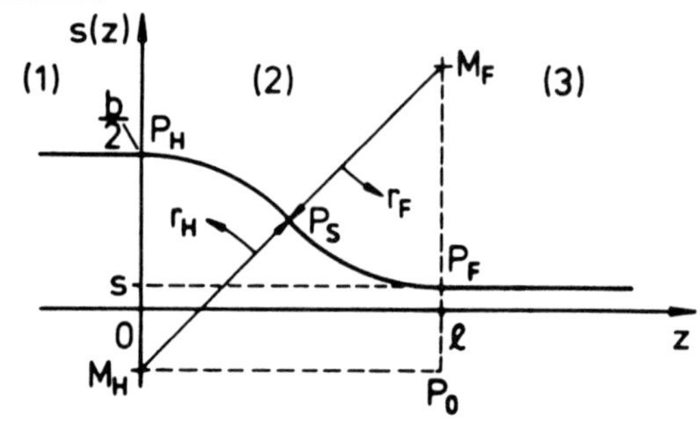

Fig.2: Geometry of a finline taper composed by two circular arcs.

- the center M_H of the circular arc which issues into the waveguide section is a point of the line z=0,
- the center M_F of the circular arc which issues into the finline section is a point of the line z=l
- the intersection point of the circular arcs lies on the connecting line between M_H and M_F.

The Pythagorean theorem can be applied to the triangle (M_H, M_F, P_0) and the sum r_F+r_H can be determined to:

$$r_F + r_H = \frac{l^2+(b/2-s)^2}{2(b/2-s)} \quad . \tag{1}$$

The geometry of the taper (l,s,b) defines a quantity of circle radii which permit to find an overall continuous and continuously differentiable

A. Beyer and I. Wolff, *IEEE MTT-S International Microwave Symposium Digest*, 1985, pp. 493–496.

taper function. Moreover (1) still contains a degree of freedom which can be used with regard to other than geometrical aspects. Two cases shall be considered which a) use equal radii and b) unequal radii of the two circular arcs.

a) Equal radii:

$$r = r_F = r_H = \frac{l^2+(b/2-s)^2}{4(b/2-s)} \quad . \qquad (2)$$

The solution given in (2) is the easiest and most obvious one. In this case only one adjustment of the radii is needed so that the structural function can be produced with a high degree of accuracy (only the intersection point P_S is critical in this context).

Fig.3 shows the qualitative dependences of the slot widths s(z) and of the characteristic impedances $Z_L(z)$ for the cases a) and b). The

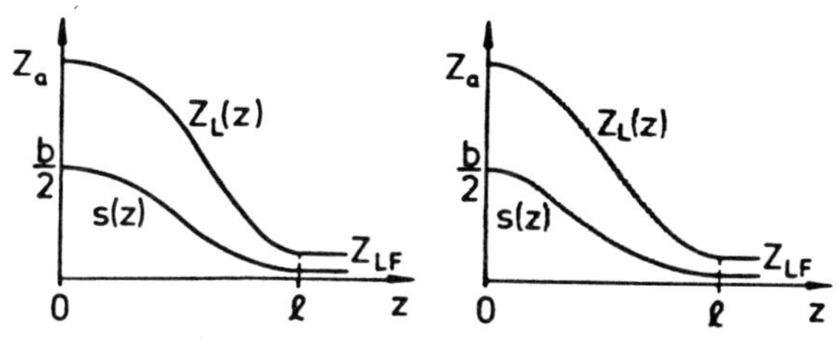

Fig.3: Plot of the functions $Z_L(z)$ and s(z)
a) for identical radii $r_H=r_F$,
b) for different radii $r_H \neq r_F$.

field theoretical background for the determination of these quantities is described in /1/.

b) Unequal radii:

The structural function according to a) implies that

$$s(z=l/2) = \frac{b/2+s}{2} \quad , \qquad (3)$$

i.e. at z=l/2 the slotwidth s is the mean value between the height of the waveguide and the slot width of the homogeneous finline. It clearly can be seen that the structural function s(z) in case a) and $Z_L(z)$ in case b) have more smooth taper functions. However, the reflection properties of the tapers are largely determined by the dependence $Z_L(z)$. Consequently, it is more reasonable to match the dependences of the characteristic impedances at z=l/2:

$$Z_L(z=l/2) = \frac{Z_{LF}+Z_a}{2} = Z_M \qquad (4)$$

and to require:

$$s(z=l/2) = s(Z_M), \qquad (5)$$

where $s(Z_M)$ is the slotwidth which is adjoint to the impedance Z_M.

The slotwidth at z=l/2 (half taper length) must be determined so that it is the algebraic mean value of the characteristic impedances Z_{LF} and Z_a at the ends of the taper (i.e. at z=0 and z=l). The function $Z_L(z)$ is known from a taper analysis program, e.g. /1/. Equ.(4) then directly can be determined from $Z_L(z)$ and (5) from its inverse function. Thus from the pairs of circular arcs which fit (1) that pair which fulfills the condition (5) can be found.

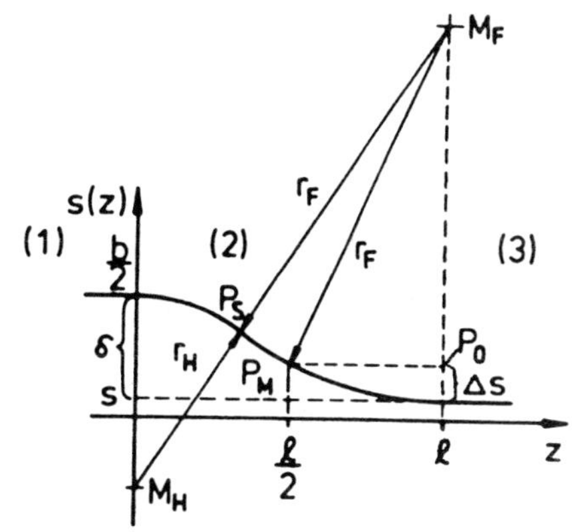

Fig.4: Geometry of the matched radii of a circular arc taper.

If Δs is the extension of the slot width s at z=l/2 compared to the slotwidth of the homogeneous finline, in the case a) of equal radii this value is: $\Delta s = \delta/2$. For all practical applications of the structural function which is defined by the characteristic impedance up to now only cases with $\Delta s < \delta/2$ have been found. Then it follows from Fig.4 that $r_F \to r_H$, i.e. that the circular arc around M_F with the radius r_F intersects the point P_M. The radius r_F can be calculated from the triangle (M_F, P_M, P_0):

$$r_F = \frac{(l/2)^2+\Delta s^2}{2\Delta s} \qquad (6)$$

If (considering Fig.4) the case $\Delta s \to 0$ is regarded, it becomes clear that below a lower bound of Δs no pair of circles can be found, because for such values of Δs negative radii r_H result. From (6) it follows that in the limiting case $r_H \to 0$ the condition

$$\frac{l^2+\delta^2}{2\delta} - r_F > 0 \qquad (7)$$

results for the existence of the structural function. If (7) is inserted into (6) and solved with respect to Δs, the following condition for Δs must be fullfilled:

$$\Delta s > \frac{\delta}{2}\{1+(1/\delta)^2(1 \mp \sqrt{1+(\delta/1)^2+(\delta/1)^4})\} \ . \quad (8)$$

In the case of $r_H \to 0$ the negative sign of the root has to be used. The positive sign of the root holds for analogous considerations in the case $r_F \to 0$. For a K-band finline mount the following geometrical dimensions are used: $b/2 = 2.159$ mm, $s = 100...200$um and $1 > 15$ mm or $1/\delta > 7$. Under the condition $1/\delta \gg 1$ the fourth power of $\delta/1$ is neglected and because

$$\sqrt{1+x} \cong 1 + 0.5x \ , \ |x| \ll 1 \ , \quad (9)$$

the following condition results from (8):

$$\Delta s > \frac{\delta}{4} \ . \quad (10)$$

For all substrate materials for which the characteristic impedance fulfills the following conditions:

$$Z_L(s+\frac{\delta}{4}) < \frac{Z_a+Z_{LF}}{2} < Z_L(s+\frac{\delta}{2}) \quad (11)$$

(with Z_{LF} the characteristic impedance of the homogeneous finline and Z_a the characteristic impedance of the waveguide) a matched circular arc taper can be found using equs. (6) and (7).

A similar condition can be found for the slot-width s(z):

$$s + \frac{\delta}{4} < s(z=1/2) < s + \frac{\delta}{2} \ . \quad (12)$$

All tapers discussed in this paper meet these conditions.

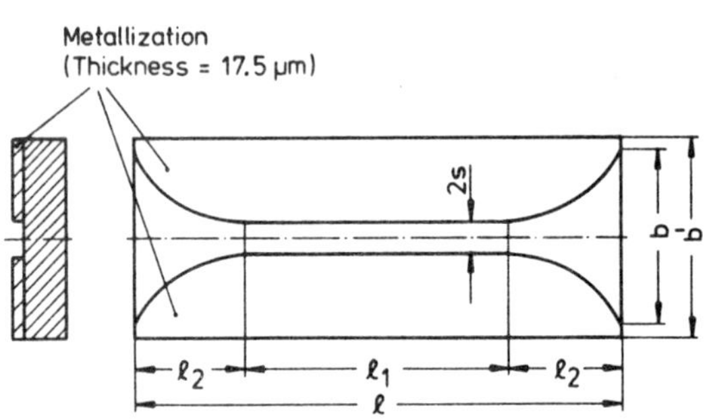

Fig.5: Geometry of a two-taper section. Material: RT/Duroid 5880, height 125 um, $\varepsilon_r = 2.22$.

Experimental Verification

In order to examine the efficiency of this method a double circular arc taper designed for the K-band (18 GHz - 26.5 GHz) is compared to a linear and a polynomial taper. The structure of each taper corresponds to that shown in Fig.5, which also specifies the common geometrical dimensions and electrical parameters. Fig.6a) shows the return loss of the linear two-taper section.

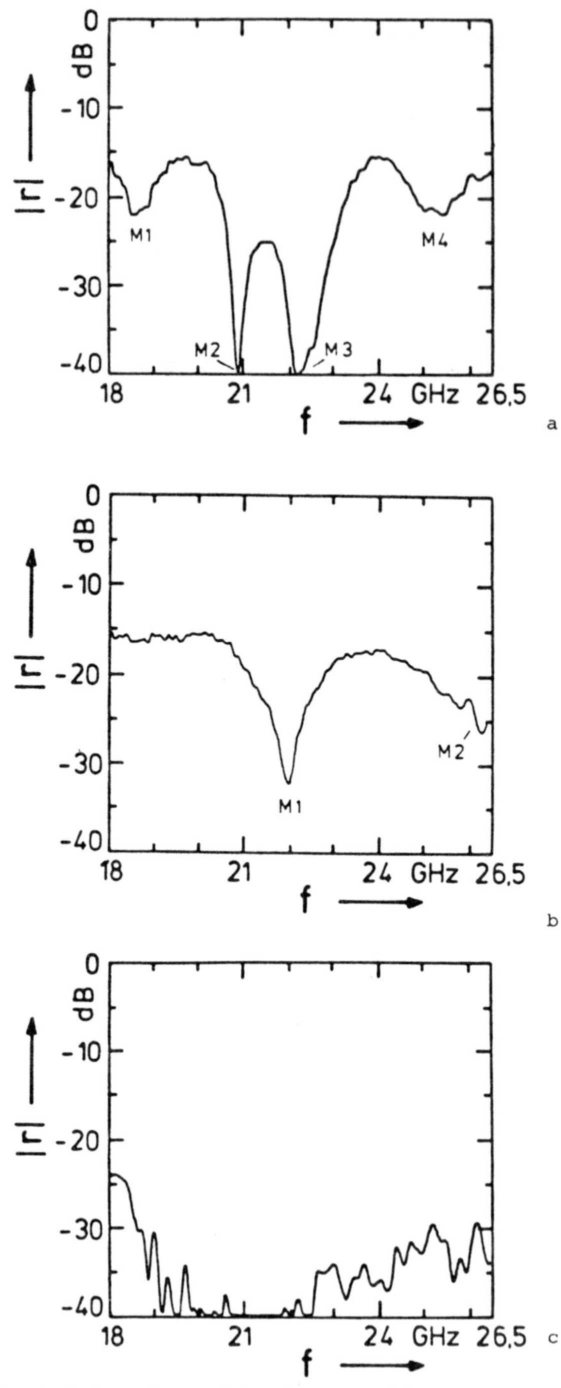

Fig.6: Return loss of two-taper sections versus the frequency at K-band:
a) for a linear taper,
b) for a near optimum taper,
c) for a circular arc taper.

This figure shows four minima M_1, M_2, M_3 and M_4; two of them are in the order of -40 dB and the other two only nearly -20 dB. The four minima result from reflexions at the beginnings and the ends of the tapers. The frequency spacing between the minima is inversely proportional to the geometrical spacing between these reflecting planes (consider: The phase constant β varies along the taper). In the case shown in Fig.6a) the frequency spacings between the minima are relatively small because the geometrical spacing l_1 is large. From the figure it can be concluded that the reflexion at the taper input where the empty waveguide changes into the the finline section is large (-20 dB) compared to the reflexion at the end of the taper and the beginning of the homogeneous finline.

Fig.6b) shows the return loss of an optimized two-taper section with a polynomial slot function. The bent contour sections have been produced using a plotter and thus they surely do not exhibit the desired smooth curve. In the case of this taper M_1 and M_2 indicate the reflection minima. A comparison with the preceding structure shows that the frequency spacing between the minima is more than three times larger as in the case of the linear taper (Fig.6a)) because the geometrical spacing between the reflecting planes is about one third compared to that of the linear taper.

It should be mentioned that the tolerances for producing the slot at the taper end, where the taper changes into the homogeneous finline, are much more critical than at the interconnection between the empty waveguide and the taper, because at the end of the taper the slotwidth is 15 times smaller than the waveguide height.

The best results for the return loss have been measured for the circular arc two-section taper (Fig.6c)). No dominant reflection can be observed in this case. The minimum return loss is improved by -8 dB; the mean value of the return loss is even -14 dB lower in the case of the circular arc taper compared to the other tapers.

Conclusions

From the results presented in this paper it can be concluded that in a practical finline taper design it is much more efficient to realize a smooth taper function than to put high effort into a numerical optimization of other structural functions. This conclusion is made under the assumption that the layout masks of the tapers are produced with a plotter of low accuracy as it often is done in the laboratory. The results presented here do not mean that the theoretically optimized taper functions could not deliver even better properties if they can be produced with a very high accuracy. So the arguments given in this paper could be a little bit debilibitated if a high accuracy mask cutter is used. In any case the production of the circular arc tapers is much simpler than that of the other tapers.

References

/1/ A.Beyer and I.Wolff: Calculation of the transmission properties of inhomogeneous finlines. Proc. 10th European Microwave Conference, Warsaw (Poland), 1980, pp. 322 -326.

/2/ R.P.Hecken: A near-optimum taper matching section without discontinuities. IEEE Trans. MTT-20, No. 11, pp. 734-739, Nov. 1972.

/3/ J.R.W.Hoefer: Fin line design made easy. 1978 IEEE MTT-S Internat. Microwave Symp. Digest, Ottawa (Canada), pp. 471-474.

/4/ A.M.Saad, K.Schünemann: Design of fin-line tapers, transistions, and couplers. Proc. 11th European Microwave Conference, Amsterdam (Netherlands), (1981), pp. 305-308.

/5/ L.Solymar: Spurious modes in nonuniform waveguides. IRE Trans. MTT-7, No.7, pp. 379-383, 1959.

/6/ Ch.P.Womack: The use of exponential transmission lines in microwave components. IRE Trans. MTT-10, No.3, pp. 124-132, 1962.

6.1.4

DESIGN TAPERED FINLINES USING A CALCULATOR

The approach is mathematically rigorous, but there's no need to solve complex equations for boundary conditions.

FINLINE-taper transmission lines are indispensable in any millimeter-wave E-plane circuit that uses circulators, couplers, mixers, oscillators, waveguide-to-finline transitions, matching networks, and filters.

But the design of finline tapers has traditionally required solving complex eigenvalue boundary-condition equations. Solutions to these equations have been virtually impossible without the use of time-consuming computer techniques.

PROTAP PRAMANICK, Senior Member, Technical Staff, COM DEV Ltd., 115 Sheldon Dr., Cambridge, Ontario, Canada N1R 7H6, (519) 622-2300; and **PRAKASH BHARTIA**, Director, R&D, Communication and Space, Dept. of National Defense, Ottawa, Ontario, Canada

There is, however, another approach that uses only a programmable calculator. While mathematically rigorous, this approach does not require solutions to boundary condition equations.

The technique is based on analysis of the closed-form expression for the power-voltage definition of the characteristic impedance. With this approach, reflection coefficients have been successfully generated for exponential-, parabolic-, and cosine-squared tapers, matching 200- and 300-Ω fingaps, in unilateral finline, in WR-19 housing, and on 150μm RT/duroid substrate.

The value of this accomplishment is best understood by assessing the weaknesses of existing design methods. Most E-plane circuit designs require at least one type of transition between various types of finlines, or between finlines and different types of transmission lines or microstrip. Of particular importance is the waveguide-to-finline transition because of the need to interface finline circuits with conventional components and systems. One way to accomplish such transitions is with the use of a tapered finline.

There have been quite a few methods reported during the last decade for designing a tapered finline. Some methods carefully avoid the controversial definition of the characteristic impedance of the non-transverse electromagnetic (TEM) transmission line by computing the finline taper in terms of the coupling coefficient between the forward and backward traveling waves. The computational method assumes that

(continued on next page)

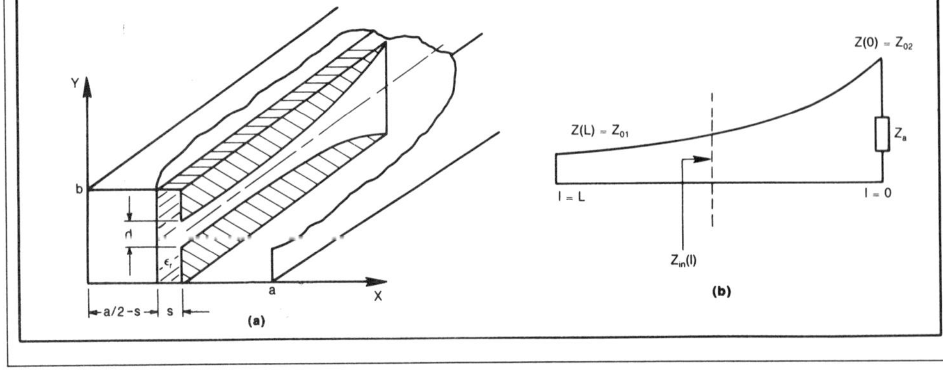

1. Computing tapered-finline contours (a) requires the impedance schematic representation of the finline (b) in order to determine the change in impedance due to an incremental change in distance along the finline.

P. Pramanick and P. Bhartia, *Microwaves and RF*, Vol. 26, No. 6, June 1987, pp. 111–114.

the fingap contour is in the form of circular axes, which are matched to the fingap as a function of the characteristic impedance. An optimization process is then used to obtain the best return loss of these finlines.

In order to apply this method, however, the field distributions and the characteristic impedances of the finlines must be known. A third method assumes a wave-impedance profile, described mathematically by a polynomial, for the finline taper. The exact solution of this profile is obtained by adjusting the coefficients of the polynomial for the best return loss and the maximum bandwidth.

MODIFIED APPROACH

In general, all of the above methods result in approximate solutions. Obtaining an exact solution requires, first of all, a proper understanding of a non-TEM-mode, tapered transmission line. In particular, the propagation constant along the taper must be analyzed. Since the propagation constant varies along the length of a finline taper, the existing analyses used for pure TEM lines are inadequate in dealing with finline tapers to take into account the variation of the propagation constant.

This analysis is modified by evaluating the equivalent circuit of the finline taper (Fig. 1). The values of the input impedances Z_{in} at a distance, l, from the load end, and the value of $Z_{in} + dZ_{in}$ at l + dl must be calculated. If characteristic impedance Z(l) and the propagation constant β(l) are assumed to remain constant over the incremental length "dl", then the following equations result:

$$z_{in} + dz_{in} = z(l) \times \frac{z_{in} + jz(l)\tan\{d[l\beta(l)]\}}{z(l) + jz_{in}\tan\{d[l\beta(l)]\}} \quad (1)$$

For small d [l β(l)], the value of tan {d[l β(l)]} is approximately equal to β (l) × dl. Therefore, Eq. 1 becomes

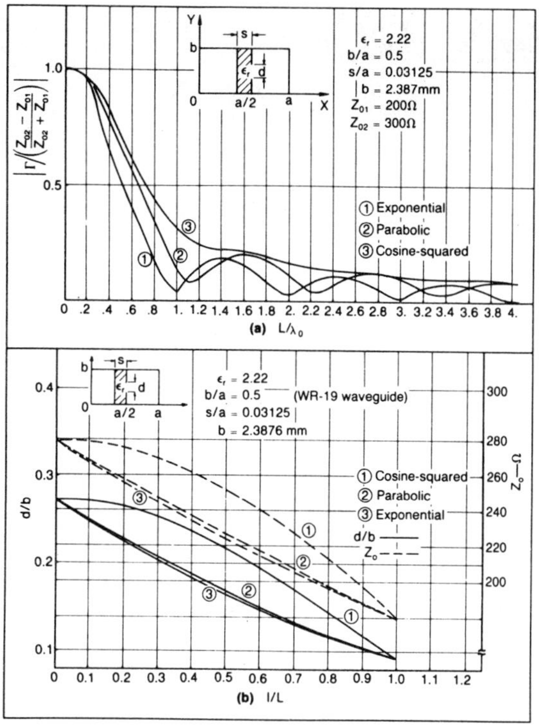

2. Values for the reflection coefficient (a) and impedance (b) for a given finline configuration and length are easily obtained from the above design curves. These curves are based on the closed-form characteristic impedance definition.

$$\frac{dz_{in}}{dl} = ja(l) + jb(l)z_{in}^2 \quad (2)$$

where

$$a(l) = z(l)\beta(l) \quad (2a)$$
$$b(l) = a(l)/[z(l)]^2$$

Equation 2 is called the Riccati equation, and can be solved numerically.

The reflection coefficient, Γ, and Z_{in} at a given point, l, on the line are related through the equation

$$z_{in} = \frac{1 + \Gamma}{1 - \Gamma} z(l) \quad (3)$$

Combining Eqs. 2 and 3 gives

$$\frac{d\Gamma}{dl} = \frac{1}{z(l)}\left[ja(l)(1 - \Gamma)^2 + jb(l)(1 + \Gamma^2)z^2(l) + (\Gamma^2 - 1)\frac{dz(l)}{dl}\right] \quad (4)$$

Equation 4 can be solved numerically. Solutions of Eqs. 3 and 4 can be obtained very easily using a programmable calculator if the finline parameters Z(l) and β(l) are known in closed-form in terms of the structural parameters of the finline.

ANALYSIS EQUATIONS

The characteristic impedance Z_o and the propagation constant β (Fig. 1) of the unilateral finline are given by

$$z_o = \frac{240\pi^2(px + q)(b/a)}{(0.385x + 1.762)^2\sqrt{\epsilon_e(f)}} \quad (5)$$

$$\beta = (2\pi/\lambda)\sqrt{\epsilon_e(f)} \quad (6)$$

$$p = -0.763(b/\lambda)^2 + 0.58(b/\lambda) + 0.0775[\ln(a/s)]^2 - 0.668[\ln(a/s)] + 1.262 \quad (7)$$

$$q = 0.372(b/\lambda) + 0.914 \quad (8)$$

for d/b > 0.3

$$p = 0.17(b/\lambda) + 0.0098 \quad (9)$$

$$q = 0.138(b/\lambda) + 0.873 \quad (10)$$

for d/b ≤ 0.3

TAPERED FINLINES

$$x = \ln \operatorname{cosec}(0.5\pi\, d/b) \quad (11)$$

where:
a,b, and d are as defined in Fig. 1.

The effective dielectric constant $\epsilon_e(f)$ is given by

$$\epsilon_e(f) = k_e - \left(\frac{\lambda}{\lambda_{ca}}\right)^2 \quad (12)$$

where
f = the operating frequency,
λ = the wavelength,
λ_{ca} = the cutoff wavelength of the finline for $\epsilon_r = 1$,
ϵ_r = the substrate dielectric constant, and
k_e = the equivalent dielectric constant of the finline at frequency f, and is given by the following equations.

$$k_e = k_c + \left(\frac{k_1 - k_c}{\frac{b}{\lambda_1} - \frac{b}{\lambda_{cf}}}\right) \times \left(\frac{b}{\lambda} - \frac{b}{\lambda_{cf}}\right) \quad (13)$$

$$k_c = 1 + \frac{s}{a} \times (a_1 x + b_1)(\epsilon_r - 1) \quad (14)$$

where

$$a_1 = -0.5148[\ln(a/s)]^2 + 4.2145[\ln(a/s)] - 5.09 \quad (15)$$

$$b_1 = 0.0452[\ln(a/s)] + 1.8257, \quad (16)$$

for $d/b > 0.5$

and

$$a_1 = 0.4021[\ln(a/s)]^2 - 0.7685[\ln(a/s)] + 0.3921 \quad (17)$$

$$b_1 = 2.42 \sin[0.556 \ln(a/s)] \quad (18)$$

for $d/b \leq 0.5$

$$\lambda_{cf} = \lambda_{ca} / \sqrt{k_c} \quad (19)$$

$$\lambda_{ca} = 2a\left[1 + \left(\frac{4}{\pi}\right)\left(\frac{b}{a}\right) \times \left(1 + 0.2\sqrt{\frac{b}{a}}\right) x\right]^{1/2} \quad (20)$$

$$k_1 = 1 + \left(\frac{\lambda_1}{\lambda_{ca}}\right)^2 \quad (21)$$

$$\frac{\lambda_1}{\lambda_{ca}} = \pi\left\{\bar{x}\frac{(3 - 2\bar{x})(\epsilon_r - 1)}{12}\right\}^{1/2} \quad (22)$$

where

$$\bar{x} = s_1 + q_r/3 - (s_1^3 - 2q_r/\pi^2)^{1/3} \quad (23)$$

$$s = \{(0.0666 + 0.0466q_r^2 + 0.015q_r^4 - 0.000137q_r^6)^{1/2} + q_r/\pi^2\}^{1/3} \quad (24)$$

and

$$q_r = \frac{k_c - 1}{\epsilon_r - 1} \quad (25)$$

SYNTHESIS EQUATION

If Z(l) is known at any point of the taper, the fingap is obtained from

$$d/b = (2/\pi)\sin^{-1}[\exp(-x)] \quad (26)$$

where
x = the solution

$$x^4 + C_z x^3 + C_2 x^2 + C_1 x + C_o = 0 \quad (27)$$

where

$$C_z = H/G + 9.156 \quad (27a)$$

$$C_2 = I/G + 9.156(H/G) + 20.95 - 6.728pF/G\bar{Z} \quad (27b)$$

$$C_1 = 9.156(I/G) + 20.95(H/G) - 6.748(pE + qF)/G\bar{Z} \quad (27c)$$

$$C_0 = 20.95(I/G) - 6.748(qE/G\bar{Z}) \quad (27d)$$

$$\bar{Z} = Z(l)/\left[(240\pi^2)(b/a)\right] \quad (27e)$$

$$E = 8\left[1 + \frac{s}{a}b_1(\epsilon_r - 1)\right]^{1/2} \quad (27f)$$

$$F = \left(\frac{4}{\pi}\right)\left(b/a\right) \times \left(1 + 0.2\sqrt{b/a}\right)E \quad (27g)$$

$$G = 4(s/a)a_1(\epsilon_r - 1)F/E \quad (27h)$$

$$H = E(F/8 + G/F) \quad (27i)$$

$$I = E^2/8 - (b/a)^2(\lambda/b)^2 \quad (27j)$$

The above synthesis equation assumes that $k_e \approx k_c$ at all frequencies, which is true only if $s/a \ll 1$, (where "s" is as defined in Fig. 1) and $\epsilon_r < 2.5$. Otherwise the value of d/b obtained from Eq. 26 has to be corrected using the analysis equations.

Although the above procedure is applicable to any kind of taper profile that can actually be fabricated, it is especially applicable to the following profiles:

(a) *Exponential:* The exponential taper profile (Fig. 1b) is defined by

$$z(l) = \exp[\ln(z_{02}) + (l/L)\ln(z_{01}/z_{02})] \quad (28)$$

(b) *Parabolic:* The parabolic profile is defined by

$$z(l) = (a + bl)^2 \quad (29)$$

where

$$a = \sqrt{z_{02}} \quad (29a)$$

$$b = (\pm\sqrt{z_{02}} \mp \sqrt{z_{01}})/L \quad (29b)$$

(c) Cosine-squared: The cosine-squared taper profile is defined by

$$z(l) = a\cos^2 bl \quad (30)$$

where

$$a = z_{02} \quad (30a)$$

$$b = (1/L)\cos^{-1}(z_{01}/z_{02})^{1/2} \quad (30b)$$

COMPUTED RESULTS

Using the above analysis and synthesis equations, reflection coefficient curves and design data curves were generated (Fig. 2). These curves show that, although the exponential taper has the smallest fingap profile of the three tapers, it gives the best performance. ••

6.1.5

Hybrid Fin-Line Matching Structures

HADIA EL HENNAWY, STUDENT MEMBER, IEEE, AND KLAUS SCHÜNEMANN, MEMBER, IEEE

***Abstract*—Two transitions between unilateral and bilateral fin-line mounted back to back show unique features for impedance transformation. The series reactances of the equivalent T-circuit are theoretically shown to be capacitive. This is exploited by designing a broad-band switch with two p-i-n diodes. Its isolation is about 20 dB throughout the *Ka*-band.**

I. INTRODUCTION

THE BASIC building blocks of fin-line circuits are various discontinuities in the slot width. Impedance transformation is usually performed with either one or two steps in the slot width [1]. With two cascaded steps one can generate either a notch or a strip. Almost all known components are realized in this way. These structures can be analyzed by combining an eigenmode with a modal analysis [2]. The procedure has been carried through in [1]. We will apply this method here to new configurations.

The structures for impedance transformation to be described show both electrical and practical advantages over the known ones. Their slot patterns are sketched in Fig. 1. These structures consist of two cascaded transitions between unilateral and bilateral fin-lines of equal slot widths. The slot may be located either symmetrically or unsymmetrically with respect to the waveguide axis. A large range of impedances can be generated by varying two geometrical parameters: the common slot width $2s$ and the length $2l$ of the middle section. Such a line section can therefore be used in either of two ways: as an impedance transformer or as a semiconductor device mount. In the latter application, the circuit patterns show a practical advantage over conventional ones. While the circuit at the front side which contains the semiconductor devices is protected against damage, one can conveniently alter the transforming section on the back side of the substrate in order to optimize performance. In addition, there are, however, even electrical advantages over the usual notch and strip patterns, which will be derived in the following.

Manuscript received March 10, 1982; revised June 4, 1982.

K. Schünemann is with the Institut für Hochfrequenztechnik, Technische Universität Braunschweig, D-3300 Braunschweig, West Germany.
H. El Hennawy is with Ain Shams University, Cairo, Egypt.

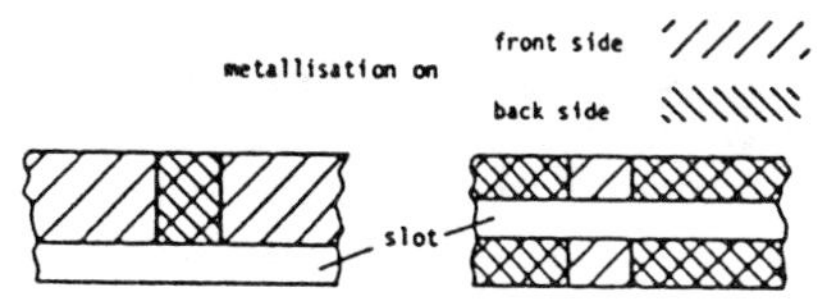

Fig. 1. Slot patterns of transitions between unilateral and bilateral fin-lines.

II. MODAL ANALYSIS

In order to analyze the structures shown in Fig. 1, one must know both the propagation constants and the field distributions of the hybrid eigenmodes of unilateral and bilateral fin-lines. The problem has been solved by a number of authors. We have adopted the spectral-domain technique presented, e.g., in [3] and modified it to determine the eigenmodes of both unilateral and bilateral fin-line in a unified form. The unilateral fin-line case is treated in Appendix I, while the results for the bilateral fin-line will be presented elsewhere [4]. The notation is the same for both cases. The key to an efficient eigenmode evaluation is a suitable choice of the system of basis functions into which the slot fields must be expanded. This has been discussed in [3]. We have used a fifth-order polynomial modified by a square-root term in order to take the edge condition correctly into account. Thus it was possible to calculate up to 30 eigenmodes with sufficient accuracy.

The modal analysis for computing the characteristics of an abrupt transition between a bilateral and a unilateral fin-line shall be briefly described (compare also to [5]). As

H. El Hennawy and K. Schunemann, *IEEE Transactions on Microwave Theory and Techniques*, Vol. MTT-30, No. 12, Dec. 1982, pp. 2132–2138.

shown in the slot patterns of Fig. 1, both the slot widths and the location of the slot above the broad wall of the waveguide are assumed to stay constant at both sides of the junction. The electric and magnetic fields of the ith eigenmode are written as

$$e_i(x, y, z) = a_i \bar{e}_i(x, y) e^{\pm \gamma_i z}$$
$$h_i(x, y, z) = a_i \bar{h}_i(x, y) e^{\pm \gamma_i z} \tag{1}$$

with $\bar{e}_i$ and $\bar{h}_i$ the transverse vector functions of the electric and magnetic field, respectively. Denoting the transverse field distribution of the left waveguide a at the junction ($z = 0$) by $\bar{E}_a$, $\bar{H}_a$, and expanding it in terms of the eigenmodes of that waveguide, reads

$$\bar{E}_a = (1+\rho) a_1 \bar{e}_{a1} + \sum_{i=2}^{\infty} a_i \bar{e}_{ai}$$
$$\bar{H}_a = (1-\rho) a_1 \bar{h}_{a1} - \sum_{i=2}^{\infty} a_i \bar{h}_{ai}. \tag{2}$$

Similarly, one writes for waveguide b

$$\bar{E}_b = \sum_{j=1}^{\infty} b_j \left(\bar{e}_{bj} + \sum_{k=1}^{\infty} s_{jk} \bar{e}_{bk} \right)$$
$$\bar{H}_b = \sum_{j=1}^{\infty} b_j \left(\bar{h}_{bj} - \sum_{k=1}^{\infty} s_{jk} \bar{h}_{bk} \right). \tag{3}$$

ρ means reflection coefficient of the incident mode ($i = 1$), s_{jk} are the scattering coefficients of the next discontinuity located at $z > 0$ in waveguide b.

In the following, we will address the so-called 'boundary reduction problem' (terminology taken from [2]), i.e., the cross section of waveguide a is larger than that of waveguide b. In our case, the former is a unilateral and the latter a bilateral fin-line. The boundary enlargement problem can be treated in the same way as described below, if the subscripts a and b in (2) and (3) are interchanged.

Equations (2) and (3) are used to formulate the boundary conditions at the junction, which are manipulated in the following way. The cross product of the electric field in (2) with $\bar{h}_{am}$ is taken and integrated over the cross section of waveguide a. For the unknown field on the left-hand side, $\bar{E}_b$ is inserted from (3). Similarly, the cross product of the magnetic field in (2) with $\bar{e}_{bn}$ is taken and integrated over the cross section of waveguide b. For the unknown field on the left-hand side, $\bar{H}_b$ is inserted from (3). One obtains

$$(1+\rho) a_1 \int_{(a)} \bar{e}_{a1} \times \bar{h}_{am} \cdot u_z \, dx \, dy + \sum_{i=2}^{\infty} a_i \int_{(a)} \bar{e}_{ai} \times \bar{h}_{am} \cdot u_z \, dx \, dy$$
$$= \sum_{j=1}^{\infty} \left[b_j \left(\int_{(b)} \bar{e}_{bj} \times \bar{h}_{am} \cdot u_z \, dx \, dy + \sum_{k=1}^{\infty} s_{jk} \int_{(b)} \bar{e}_{bk} \times \bar{h}_{am} \cdot u_z \, dx \, dy \right) \right] \tag{4a}$$

$$(1-\rho) a_1 \int_{(b)} \bar{e}_{bn} \times \bar{h}_{a1} \cdot u_z \, dx \, dy - \sum_{i=2}^{\infty} a_i \int_{(b)} \bar{e}_{bn} \times \bar{h}_{ai} \cdot u_z \, dx \, dy$$
$$= \sum_{j=1}^{\infty} \left[b_j \left(\int_{(b)} \bar{e}_{bn} \times \bar{h}_{bj} \cdot u_z \, dx \, dy - \sum_{k=1}^{\infty} s_{jk} \int_{(b)} \bar{e}_{bn} \times \bar{h}_{bk} \cdot u_z \, dx \, dy \right) \right]. \tag{4b}$$

u_z means unit vector in the z-direction. Because $\bar{E}_a$ exists on the common aperture only and vanishes elsewhere, the integrals on the right-hand side of (4a) must only be taken over the cross section of waveguide 'b'. These integrals can be solved once the eigenmodes have been formulated. The solutions are listed in Appendix II for the general case of a junction between two unilateral fin-lines. If a junction between a unilateral and a bilateral fin-line is to be treated, one must only omit some of the expansion coefficients. ($A_n^{(2)} = S_n^{(2)} = B_n^{(2)} = M_n^{(2)} = 0$ for an unsymmetrical mode in bilateral fin-line, $\bar{A}_n^{(2)} = \bar{S}_n^{(2)} = \bar{B}_n^{(2)} = \bar{M}_n^{(2)} = 0$ for a symmetrical mode.)

In order to solve (4) for the amplitudes of the excited modes, one must first determine the scattering coefficients s_{jk}. For a single transition, waveguide b may be assumed to be terminated in a matched load: $s_{jk} = 0$. For two transitions in cascade, we can take advantage of the symmetry of the structure with respect to the plane $z = l$. The equivalent circuit can then be found from both an even and an odd excitation of both ports, so that $s_{jk} = 0$ for $j \neq k$ and

$$s_{jj} = (1 - y_{bj})/(1 + y_{bj}) = \pm \exp(-2 j \beta_{bj} l). \tag{5}$$

y_{bj} is the normalized input admittance of the jth mode in waveguide b measured at $z = 0$, and β_{bj} means propagation constant of this mode.

We are now in a position to solve the system of equations (4). It is convenient to normalize the wave amplitudes on a_1. Setting

$$a_i' = a_i / a_1, \quad b_j' = b_j / a_1, \quad I'_{ambj} = I_{ambj} / (F_{am} F_{bj}) \tag{6}$$

with the I's and F's taken from Appendix II we obtain for the case that the exciting wave a_1 impinges on waveguide b from waveguide a

$$\rho I'_{a1bn} + \sum_{i=2}^{L} a_i' I'_{aibn} - \sum_{j=1}^{L} b_j' (1 + s_{jj}) I'_{bjbn} = - I'_{a1bn}$$
$$\rho I'_{ama1} + \sum_{i=2}^{L} a_i' I'_{amai} + \sum_{j=1}^{L} b_j' (1 - s_{jj}) I'_{ambj} = I'_{ama1}. \tag{7}$$

These relations are valid for the boundary enlargement problem, i.e., that waveguide a is a bilateral and waveguide b a unilateral fin-line. If a wave a_1 impinges from b on a,

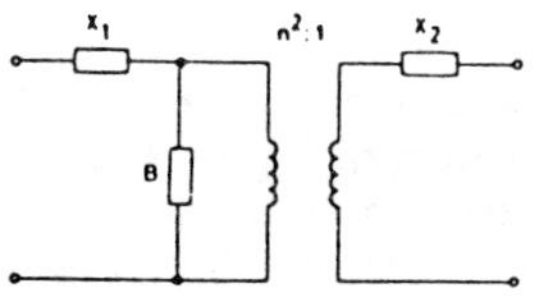

Fig. 2. Equivalent circuit of a single transition (port 1 = bilateral, port 2 = unilateral fin-line).

one obtains for the boundary reduction problem

$$\rho I'_{b1bm} + \sum_{i=2}^{L} a'_i I'_{bibm} - \sum_{j=1}^{L} b'_j (1 + s_{jj}) I'_{ajbm} = - I'_{b1bm}$$

$$\rho I'_{anbl} + \sum_{i=2}^{L} a'_i I'_{anbi} + \sum_{j=1}^{L} b'_j (1 - s_{jj}) I'_{anaj} = I'_{anbl}. \quad (8)$$

The infinite series are truncated at $i = j = L$, thus taking L modes in each waveguide into account. Then (7) or (8) generate $2L$ linear equations for $1+(L-1)+L$ unknowns.

A single transition from unilateral to bilateral fin-line is characterized by an equivalent circuit shown in Fig. 2 whose elements can be computed from the input admittances at both ports. Two cascaded transitions are represented by a T-equivalent circuit with series reactances X_{sc} and shunt reactances $(X_{oc} - X_{sc})/2$. (Subscripts *sc* denote a short circuit, subscripts *oc* an open circuit at the symmetry plane $z = l$.)

III. Performance of Transitions

The modal analysis has been applied to a single transition between unilateral and bilateral fin-line and to the cascaded transitions sketched in Fig. 1. Some results for the elements of the equivalent circuit of a single transition (Fig. 2) are shown in Fig. 3. (The turns ratio of the ideal impedance transformer has been omitted because it is very close to unity.) The general shape of the three remaining elements (note that subscript 1 refers to the bilateral fin-line port) can be explained by assuming an electric field directed perpendicularly through the dielectric substrate from front to back fin of the bilateral fin-line. This field can of course exist only in the immediate vicinity of the transition between the two fin-lines. For larger slot widths ($2s$), the substrate thickness ($2d$) is much smaller than ($2s$). Hence, a capacitive series reactance X_1 appears in the equivalent circuit. The shunt susceptance B is small because the slot width approaches the waveguide height ($2b$). The electric field concentration in the slot increases with decreasing ($2s$). Hence, B must increase monotonically and X_1 must increase, too. Finally, X_1 becomes inductive because of an increasing magnetic field concentration which is due to a surface current filament on the transversely oriented fin edge of the bilateral fin-line.

The most important result is that there may exist a capacitive series reactance. It will be shown below that the potentials for impedance transformation in fin-lines will be largely amplified by this.

The equivalent circuit elements of two cascaded transitions are shown in Fig. 4 versus the length of the middle section. The slot width is relatively small: $s = 0.2$ mm. The shape of the curves can be explained by cascading the equivalent circuits of two single transitions. (They have, however, been computed by taking higher order mode

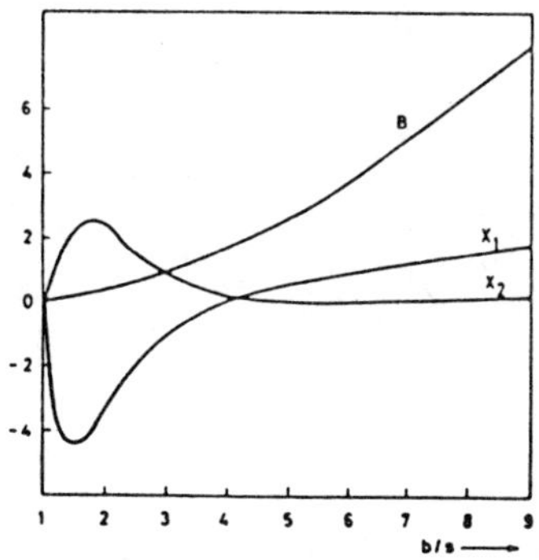

Fig. 3. Elements of the equivalent circuit of a single transition (Fig. 2) versus slot width ($2s$) normalized against waveguide height ($2b$). The elements are normalized against the wave impedance of the adjacent port. (Frequency 30 GHz, WR-28 waveguide with $2a = 7.112$ mm, $2b = 3.556$ mm, symmetrically located slot of width $2s$, RT-duroid 5880 substrate of thickness $2d = 0.254$ mm and relative dielectric constant $\epsilon_r = 2.22$.)

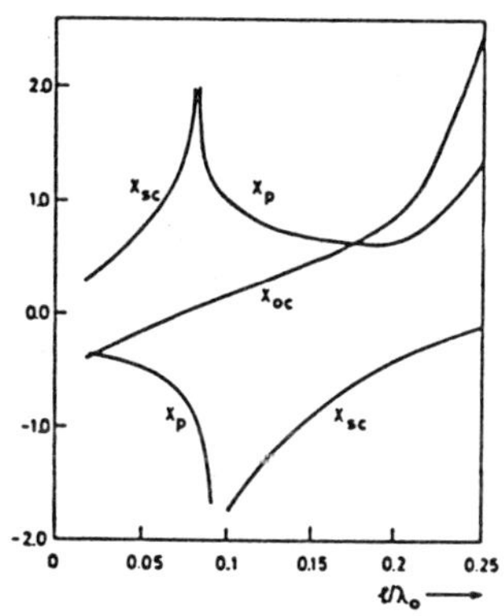

(a)

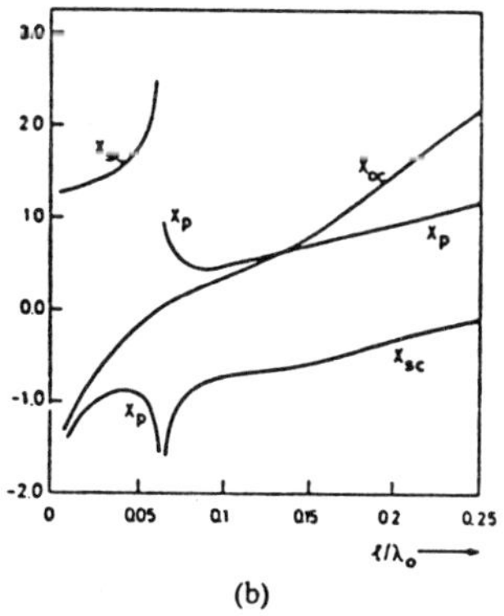

(b)

Fig. 4. (a) Elements of the equivalent T-circuit of a section of bilateral fin-line (length $2l$) embedded into a unilateral fin-line. The elements are normalized against the wave impedance of the unilateral fin-line. λ_0 means free-space wavelength. (Parameters as in Fig. 3 with $2s = 0.4$ mm.) (b) Elements of the equivalent T-circuit of a section of unilateral fin-line embedded into a bilateral fin-line. (Normalization against bilateral fin-line impedance; parameters as in Fig. 4(a).)

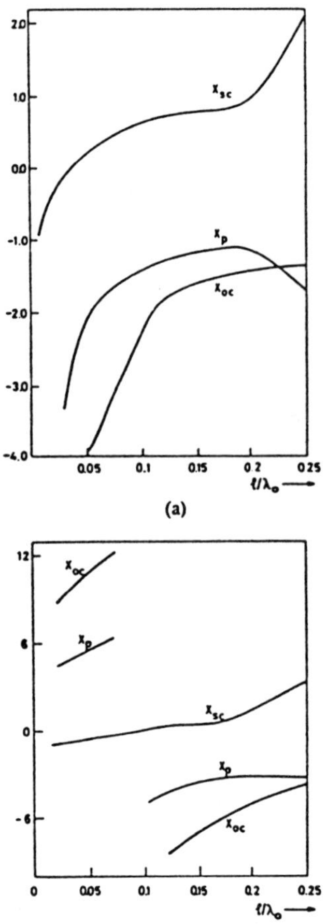

Fig. 5. (a) T-circuit elements as in Fig. 4(a) with $2s = 1.0$ mm. (b) T-circuit elements as in Fig. 4(b) with $2s = 1.0$ mm.

coupling into account.) A significant feature is the parallel resonant behavior of the series reactance for small lengths l, which can be explained if the two cascaded transitions are separately regarded (i.e., if higher order mode coupling between them is neglected). The series reactance is calculated for odd excitation so that one port of the equivalent circuit of Fig. 2 is terminated in a short-circuited transmission line of length l. Hence, the parallel resonant behavior of X_{sc} is due to B resonating with X_1 and the input reactance of the transmission line.

The series reactance is capacitive for larger lengths l. This is of great importance for circuit design because there is no other discontinuity known showing a capacitive series reactance. (Both notch and strip show an inductive series reactance [1].) This unique feature of a double transition between unilateral and bilateral fin-line offers an interesting alternative in the design of p-i-n-diode switches as will be illustrated below. It can, however, be observed only for small slot widths. The series reactance is inductive, if s exceeds a certain limit. This is illustrated by the results shown in Fig. 5. Instead of a parallel resonance, X_{sc} now shows series-resonant behavior. This can again be explained by means of the simplified equivalent circuit. The shunt susceptance B of a single transition can now be neglected (compare with Fig. 3!), so that X_1 is in series with the short-circuited transmission line of length l (odd excitation). Hence, X_{sc} models a series circuit. The shunt element X_p is, however, always capacitive.

IV. Design of a Broad-Band Switch

The unique characteristic of a section of bilateral fin-line embedded into a unilateral fin-line shall now be utilized to design a broad-band switch with two p-i-n diodes. The diodes are usually bonded across a narrow fin-line slot [7] within a distance of about a quarter of a wavelength in order to obtain broad-band performance. Then one needs three diodes in order to guarantee 20-dB isolation over a waveguide band [7]. Using the matching structures discussed above will, however, reduce the number of diodes for nearly the same performance to 2.

A p-i-n diode can approximately be characterized by its package inductance in on-state and by its i-layer capacitance shunted by the package capacitance in off-state. The on-state of the diode normally corresponds to the isolation state of the switch (diode mounted in shunt to the transmission line). Hence, the package inductance must be carefully tuned in order to enhance isolation. Guidelines for designing a broad-band switch with two p-i-n diodes in shunt are derived in detail in [8]. Both diodes should be about a quarter of a wavelength spaced. Broad-band performance demands to compensate for the package inductances at two different frequencies f_a and f_b, which are related to the upper and lower corner frequencies f_1, f_2 via [8]

$$f_{a,b} = \left[\left(\frac{f_1 + f_2}{2} \right)^2 \pm \sqrt{\left(\frac{f_1 + f_2}{2} \right)^4 - (f_1 f_2)^2} \right]^{1/2}. \quad (9)$$

The two sections of bilateral fin-line which are used to tune the diodes in their on-state must, hence, show different lengths. The slot pattern of the switch and its equivalent circuit are sketched in Fig. 6. One recognizes that a capacitive series reactance X_{sc} is necessary in order to fulfill the said tuning condition. (Hence, neither a notch nor a strip can be used instead.) Following the design rules given in [8], the equivalent circuit should present a bandpass characteristic with, e.g., a maximally flat response in the transmission state. This is achieved for $X = 2B$ with X and B defined in Fig. 6. Thus, the length of the unilateral fin-line between both diode mounts is fixed. In the isolation state, series resonance is claimed at f_a and f_b, respectively. This determines the lengths of the two mounting sections. The common slot width should correspond to the physical size of the diode package.

We have realized a Ka-band switch with two p-i-n diodes 4701 (Microwave Associates). The slot width has been chosen to $2s = 0.4$ mm. Measuring the diode impedance in a bilateral fin-line of equal slot width yields the input data of the design procedure referred to the diode plane.

$Z_{on} = 0.19 + j0.15$, $Z_{off} = -j6.53$ at $f_b = 28$ GHz, $Z_{on} =$

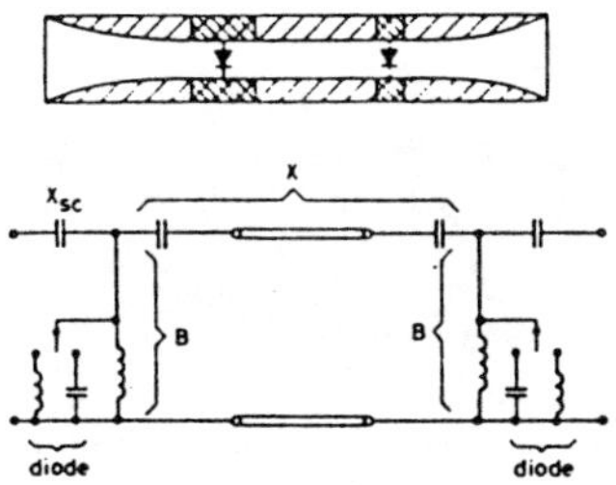

Fig. 6. Slot pattern and equivalent circuit of a switch with two p-i-n diodes 7401 (Microwave Associates).

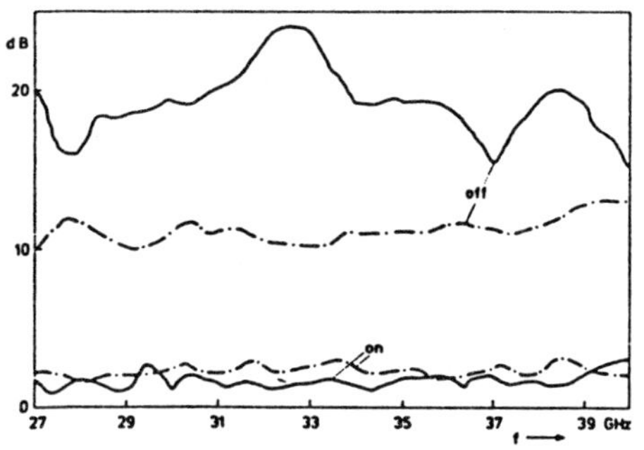

Fig. 7. Performance of the switch from Fig. 6. Solid lines: hybrid fin-line circuit. Dashed-dotted lines: all unilateral fin-line, diodes quarter of a wavelength apart.

$0.31 + j0.89$, $Z_{off} = 0.45 - j1.02$ at $f_a = 38$ GHz. Neglecting the real parts, the length of each section has to tune out Z_{on}. One recognizes from Fig. 4(a) that the shunt reactance X_p of the unloaded section shows a broad minimum versus the length l. It is hence reasonable to start with a guessed $X_p = 0.7$ taken from the minimum range. The total shunt reactance of the loaded section being known, the condition for series resonance can be evaluated for X_{sc}. This fixes l so that the choice for X_p can be checked. Repeating this cycle one obtains $l_1 = 2.2$ mm; $l_2 = 1.6$ mm. To simplify the following calculations, the frequency dependence of the equivalent circuit elements will be neglected. They are calculated at the center frequency 32 GHz: $jX_{sc1} = -0.25$, $jX_{p1} = 0.8$, $jX_{sc2} = -0.784$, $jX_{p2} = 0.75$. Then the condition for maximally flat response in the transmission state can be evaluated yielding a net length for the unilateral fin-line between the mounting sections of 4.0 mm.

The performance of the realized switch is shown in Fig. 7. These results have been obtained without any modification of the calculated slot pattern and without any adjustment. The isolation is much better than for a simple switch with the two diodes mounted a quarter of a wavelength apart in a unilateral fin-line. The isolation is about 19 dB, the insertion loss less than 2 dB. No efforts have, however, been undertaken in order to minimize the dissipative losses. Hence, it can be concluded that hybrid fin-line circuits offer new potentials to circuit design.

V. Conclusions

Using a modal analysis, cascaded transitions between unilateral and bilateral fin-lines have been investigated. These structures show a unique feature: a capactive instead of an inductive series reactance of the equivalent T-circuit. This cannot be obtained with any other known discontinuity in fin-lines. It is illustrated that this new degree of freedom can advantageously be utilized in the design of broad-band switches. The isolation obtained with two p-i-n diodes is nearly the same as has as yet been achieved with three diodes. Moreover, there are practical advantages in that a semiconductor device mount can easily be modified without affecting the device itself. Hence, the combination of unilateral and bilateral fin-lines seems to be an interesting alternative to conventional notch and strip patterns. Other applications like filters and FET mounts are also imaginable.

Appendix I

The unified calculation of the hybrid eigenmodes will only briefly be sketched because it has recently been described elsewhere [4] for the bilateral fin-line. Hence, we will complete those derivations by treating the unilateral fin-line case. The notation is the same as in [4].

The fin-line cross section is shown in Fig. 8. The slot width is $2s = d_1 - d_2$. Because of symmetry an electric wall (e.w.) may be assumed on the x-axis. The fields are derived from the E_x- and H_x-components which must satisfy the Helmholtz equation.

The transverse electric field components are written

$$\begin{aligned}
E_{ti}^{(1)} &= \sum_{n=1}^{\infty} A_n^{(1)} \cos \bar{\alpha}_n (x-a) \sin \alpha_n y \cdot u_x \\
&\quad + \sum_{n=0}^{\infty} S_n^{(1)} \sin \bar{\alpha}_n (x-a) \cos \alpha_n y \cdot u_y \\
E_{ti}^{(3)} &= \sum_{n=1}^{\infty} A_n^{(3)} \cos \bar{\alpha}_n (x+a) \sin \alpha_n y \cdot u_x \\
&\quad + \sum_{n=0}^{\infty} S_n^{(3)} \sin \bar{\gamma}_n (x+a) \cos \alpha_n y \cdot u_y \\
E_{ti}^{(2)} &= \sum_{n=1}^{\infty} \left(A_n^{(2)} \sin \tilde{\alpha}_n x \sin \alpha_n y + \bar{A}_n^{(2)} \cos \tilde{\alpha}_n x \sin \alpha_n y \right) \cdot u_x \\
&\quad + \sum_{n=0}^{\infty} \left(S_n^{(2)} \cos \tilde{\alpha}_n x \cos \alpha_n y + \bar{S}_n^{(2)} \sin \tilde{\alpha}_n x \cos \alpha_n y \right) \cdot u_y
\end{aligned} \tag{A1}$$

with

$$\bar{\alpha}_n = \sqrt{k_0^2 - \beta^2 - \alpha_n^2}, \qquad \tilde{\alpha}_n = \sqrt{\epsilon_r k_0^2 - \beta^2 - \alpha_n^2}, \quad \alpha_n = n\pi/b. \tag{A2}$$

k_0 means free-space wavenumber, β propagation constant. u_x and u_y are unit vectors in the x- and y-directions, respectively. For the magnetic field components, one has similarly

$$\begin{aligned}
H_{ti}^{(1)} &= \sum_{n=0}^{\infty} \left(B_n^{(1)} \sin \bar{\alpha}_n (x-a) \cos \alpha_n y \right) \cdot u_x \\
&\quad + \sum_{n=1}^{\infty} \left(M_n^{(1)} \cos \bar{\alpha}_n (x-a) \sin \alpha_n y \right) \cdot u_y
\end{aligned}$$

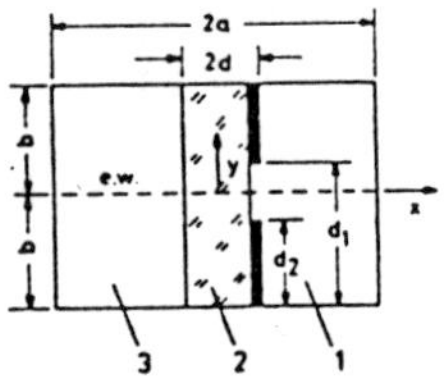

Fig. 8. Cross section of a unilateral fin-line (e.w. means electric wall).

$$H_{ti}^{(3)} = \sum_{n=0}^{\infty} \left(B_n^{(3)} \sin \bar{\alpha}_n (x+a) \cos \alpha_n y \right) \cdot u_x + \sum_{n=1}^{\infty} \left(M_n^{(3)} \cos \bar{\alpha}_n (x+a) \sin \alpha_n y \right) \cdot u_y$$

$$H_{ti}^{(2)} = \sum_{n=0}^{\infty} \left(B_n^{(2)} \cos \tilde{\alpha}_n x \cos \alpha_n y + \bar{B}_n^{(2)} \sin \tilde{\alpha}_n x \cos \alpha_n y \right) \cdot u_x + \sum_{n=1}^{\infty} \left(M_n^{(2)} \sin \tilde{\alpha}_n x \sin \alpha_n y + \bar{M}_n^{(2)} \cos \tilde{\alpha}_n x \sin \alpha_n y \right) \cdot u_y . \tag{A3}$$

The expansion coefficients are related via [4, eqs. (B4)]. In addition,

$$\begin{aligned}
\bar{S}_n^{(2)} &= \left(-\bar{A}_n^{(2)} \alpha_n \tilde{\alpha}_n - jk_0 \beta \bar{B}_n^{(2)} \right) / \left(\beta^2 + \alpha_n^2 \right) \\
S_n^{(3)} &= \left(-A_n^{(3)} \alpha_n \bar{\alpha}_n - jk_0 \beta B_n^{(3)} \right) / \left(\beta^2 + \alpha_n^2 \right) \\
\bar{M}_n^{(2)} &= \left(-\bar{B}_n^{(2)} \alpha_n \tilde{\alpha}_n - jk_0 \beta \epsilon_r \bar{A}_n^{(2)} \right) / \left(\beta^2 + \alpha_n^2 \right) \\
M_n^{(3)} &= \left(-B_n^{(3)} \alpha_n \bar{\alpha}_n - jk_0 \beta A_n^{(3)} \right) / \left(\beta^2 + \alpha_n^2 \right).
\end{aligned} \tag{A4}$$

Matching the tangential field components at the various interfaces yields equations (B5) of [4] and

$$A_n^{(2)} = A_n^{(3)} \left(\frac{-\bar{\alpha}_n}{\tilde{\alpha}_n} \cos \tilde{\alpha}_n d \sin \bar{\alpha}_n (a-d) - \frac{\sin \tilde{\alpha}_n d \cos \bar{\alpha}_n (a-d)}{\epsilon_r} \right)$$

$$\bar{A}_n^{(2)} = A_n^{(3)} \left(\frac{\cos \tilde{\alpha}_n d \cos \bar{\alpha}_n (a-d)}{\epsilon_r} - \frac{\bar{\alpha}_n}{\tilde{\alpha}_n} \sin \tilde{\alpha}_n d \sin \bar{\alpha}_n (a-d) \right)$$

$$B_n^{(2)} = B_n^{(3)} \left(\sin \bar{\alpha}_n (a-d) \cos \tilde{\alpha}_n d + \frac{\bar{\alpha}_n}{\tilde{\alpha}_n} \cos \bar{\alpha}_n (a-d) \sin \tilde{\alpha}_n d \right)$$

$$\bar{B}_n^{(2)} = B_n^{(3)} \left(\frac{\bar{\alpha}_n}{\tilde{\alpha}_n} \cos \tilde{\alpha}_n d \cos \bar{\alpha}_n (a-d) - \sin \bar{\alpha}_n (a-d) \sin \tilde{\alpha}_n d \right)$$

$$A_n^{(3)} = -A_n^{(1)} \sin \bar{\alpha}_n (a-d) \Big/ \left(\cos 2\tilde{\alpha}_n d \sin \bar{\alpha}_n (a-d) + \frac{\tilde{\alpha}_n}{\epsilon_r \bar{\alpha}_n} \sin 2\tilde{\alpha}_n d \cos \bar{\alpha}_n (a-d) \right)$$

$$B_n^{(3)} = -B_n^{(1)} \sin \bar{\alpha}_n (a-d) \Big/ \left(\cos 2\tilde{\alpha}_n d \sin \bar{\alpha}_n (a-d) + \frac{\bar{\alpha}_n}{\tilde{\alpha}_n} \sin 2\tilde{\alpha}_n d \cos \bar{\alpha}_n (a-d) \right). \tag{A5}$$

The remaining expansion coefficients are expressed in terms of the slot field components $E_y(y)$ and $E_z(y)$ after Fourier transforming all field components and boundary conditions with respect to y. The results are given in [4, eqs. (B6) and (B7)]. The problem of finding the EM field distribution has thus been reduced to determining the slot fields. This is solved by applying Ritz–Galerkin's method. The procedure has been carried through in [3] and in [4, appendix B] to which the reader is referred.

Appendix II

The integrals appearing in (4) are solved by inserting the transverse field configurations given in Appendix I. One obtains

$$\begin{aligned}
I_{aibj} &= \int_{(a,b)} \bar{e}_{ai} \times \bar{h}_{bj} \cdot u_z \, dx \, dy \\
&= \frac{b}{2} \sum_{n=0}^{\infty} \Bigg\{ \left[\delta_n \left(A_{na}^{(1)} M_{nb}^{(1)} + A_{na}^{(3)} M_{nb}^{(3)} \right) - \eta_1 \epsilon_n \left(B_{nb}^{(1)} S_{na}^{(1)} - B_{nb}^{(3)} S_{na}^{(3)} \right) \right] \cdot \frac{\sin (\bar{\alpha}_{nai} - \bar{\alpha}_{nbj})(a-d)}{2(\bar{\alpha}_{nai} - \bar{\alpha}_{nbj})} \\
&\quad + \left[\delta_n \left(A_{na}^{(1)} M_{nb}^{(1)} + A_{na}^{(3)} M_{nb}^{(3)} \right) + \eta_1 \epsilon_n \left(B_{nb}^{(1)} S_{na}^{(1)} + B_{nb}^{(3)} S_{na}^{(3)} \right) \right] \cdot \frac{\sin (\bar{\alpha}_{nai} + \bar{\alpha}_{nbj})(a-d)}{2(\bar{\alpha}_{nai} + \bar{\alpha}_{nbj})} \\
&\quad + \left[\eta_2 \delta_n \left(A_{na}^{(2)} M_{nb}^{(2)} + \bar{A}_{na}^{(2)} \bar{M}_{nb}^{(2)} \right) + \epsilon_n \left(B_{nb}^{(2)} S_{na}^{(2)} + \bar{B}_{nb}^{(2)} \bar{S}_{na}^{(2)} \right) \right] \frac{\sin (\tilde{\alpha}_{nai} - \tilde{\alpha}_{nbj}) d}{(\tilde{\alpha}_{nai} - \tilde{\alpha}_{nbj})} \\
&\quad - \left[\eta_2 \delta_n \left(A_{na}^{(2)} M_{nb}^{(2)} - \bar{a}_{na}^{(2)} \bar{M}_{nb}^{(2)} \right) + \epsilon_n \left(B_{nb}^{(2)} S_{na}^{(2)} - \bar{B}_{nb}^{(2)} \bar{S}_{na}^{(2)} \right) \right] \cdot \frac{\sin (\tilde{\alpha}_{nai} + \tilde{\alpha}_{nbj}) d}{(\tilde{\alpha}_{nai} + \tilde{\alpha}_{nbj})} \Bigg\}
\end{aligned} \tag{A6}$$

$$\eta_1 = \begin{cases} 1, & \text{for } \bar{\alpha}_{nbj} \text{ real} \\ -1, & \text{for } \bar{\alpha}_{nbj} \text{ imaginary}, \end{cases} \qquad \delta_n = \begin{cases} 0, & \text{for } n = 0 \\ 1, & \text{for } n \neq 0 \end{cases}$$

$$\eta_2 = \begin{cases} 1, & \text{for } \tilde{\alpha}_{nbj} \text{ real} \\ -1, & \text{for } \tilde{\alpha}_{nbj} \text{ imaginary}, \end{cases} \qquad \epsilon_n \begin{cases} 1, & \text{for } n \neq 0 \\ 2, & \text{for } n = 0 \end{cases}. \tag{A7}$$

Also

$$F_{ai}^2=\int_{(a,b)} \bar{e}_{ai}\,\bar{e}^*_{ai}\,dx\,dy$$

$$=\frac{b}{2}\sum_{n=0}^{\infty}\Big\{\big[\delta_n\big(|A_n^{(1)}|^2+|A_n^{(3)}|^2\big)+\eta_1\epsilon_n\big(|S_n^{(1)}|^2+|S_n^{(3)}|^2\big)\big]\cdot\frac{a-d}{2}+\big[\delta_n\big(|A_n^{(1)}|^2+|A_n^{(3)}|^2\big)-\eta_1\epsilon_n\cdot\big(|S_n^{(1)}|^2+|S_n^{(3)}|^2\big)\big]\cdot\frac{\sin 2\bar{\alpha}_n(a-d)}{4\bar{\alpha}_n}$$
$$+\big[\eta_2\delta_n\big(|A_n^{(2)}|^2+|\bar{A}_n^{(2)}|^2\big)+\epsilon_n\big(|S_n^{(2)}|^2+|\bar{S}_n^{(2)}|^2\big)\big]\cdot d-\big[\eta_2\delta_n\big(|A_n^{(2)}|^2-|\bar{A}_n^{(2)}|^2\big)-\epsilon_n\big(|S_n^{(2)}|^2-|\bar{S}_n^{(2)}|^2\big)\big]\cdot\frac{\sin 2\tilde{\alpha}_n d}{2\tilde{\alpha}_n}\Big\} \tag{A8}$$

$$I_{aiai}=\int_{(a,b)}\bar{e}_{ai}\times\bar{h}_{ai}\cdot u_z\,dx\,dy$$

$$=\frac{b}{2}\sum_{n=0}^{\infty}\Big\{\big[\delta_n\big(A_{na}^{(1)}M_{na}^{(1)}+A_{na}^{(3)}M_{na}^{(3)}\big)-\eta_1\epsilon_n\big(S_{na}^{(1)}B_{na}^{(1)}+S_{na}^{(3)}B_{na}^{(3)}\big)\big]\cdot\frac{a-d}{2}+\big[\delta_n\big(A_{na}^{(1)}M_{na}^{(1)}+A_{na}^{(3)}M_{na}^{(3)}\big)+\eta_1\epsilon_n\big(B_{na}^{(1)}S_{na}^{(1)}+B_{na}^{(3)}S_{na}^{(3)}\big)\big]\cdot\frac{\sin 2\bar{\alpha}_n(a-d)}{4\bar{\alpha}_n}$$
$$+\big[\eta_2\delta_n\big(A_{na}^{(2)}M_{na}^{(2)}+\bar{A}_{na}^{(2)}\bar{M}_{na}^{(2)}\big)-\epsilon_n\big(B_{na}^{(2)}S_{na}^{(2)}+\bar{B}_{na}^{(2)}\bar{S}_{na}^{(2)}\big)\big]\cdot d-\big[\eta_2\delta_n\big(A_{na}^{(2)}M_{na}^{(2)}-\bar{A}_{na}^{(2)}\bar{M}_{na}^{(2)}\big)+\epsilon_n\big(B_{na}^{(2)}S_{na}^{(2)}-\bar{B}_{na}^{(2)}\bar{S}_{na}^{(2)}\big)\big]\cdot\frac{\sin 2\tilde{\alpha}_n d}{2\tilde{\alpha}_n}\Big\} \tag{A9}$$

$$I_{aiaj}=\int_{(a,b)}\bar{e}_{ai}\times\bar{h}_{aj}\cdot u_z\,dx\,dy$$

$$=\frac{b}{2}\sum_{n=0}^{\infty}\Big\{\big[\delta_n\big(A_{nai}^{(1)}M_{naj}^{(1)}+A_{naj}^{(3)}M_{naj}^{(3)}\big)-\epsilon_n\big(B_{naj}^{(1)}S_{nai}^{(1)}+B_{naj}^{(3)}S_{nai}^{(3)}\big)\big]\cdot\frac{\sin(\bar{\alpha}_{ni}-\bar{\alpha}_{nj})(a-d)}{2(\bar{\alpha}_{ni}-\bar{\alpha}_{nj})}$$
$$+\big[\delta_n\big(A_{nai}^{(1)}M_{naj}^{(1)}+A_{nai}^{(3)}M_{naj}^{(3)}\big)+\epsilon_n\big(B_{naj}^{(1)}S_{nai}^{(1)}+B_{naj}^{(3)}S_{nai}^{(3)}\big)\big]\cdot\frac{\sin(\bar{\alpha}_{ni}+\bar{\alpha}_{nj})(a-d)}{2(\bar{\alpha}_{ni}+\bar{\alpha}_{nj})}$$
$$+\big[\delta_n\big(A_{nai}^{(2)}M_{naj}^{(2)}+\bar{A}_{nai}^{(2)}\bar{M}_{naj}^{(2)}\big)-\epsilon_n\big(B_{naj}^{(2)}S_{nai}^{(2)}+\bar{B}_{naj}^{(2)}\bar{S}_{nai}^{(2)}\big)\big]\cdot\frac{\sin(\tilde{\alpha}_{ni}-\tilde{\alpha}_{nj})\,d}{(\tilde{\alpha}_{ni}-\tilde{\alpha}_{nj})}-\big[\delta_n\big(A_{nai}^{(2)}M_{naj}^{(2)}-\bar{A}_{nai}^{(2)}\bar{M}_{naj}^{(2)}\big)+\epsilon_n\big(B_{naj}^{(2)}S_{nai}^{(2)}-\bar{B}_{naj}^{(2)}\bar{S}_{nai}^{(2)}\big)\big]\cdot\frac{\sin(\tilde{\alpha}_{ni}+\tilde{\alpha}_{nj})\,d}{(\tilde{\alpha}_{ni}+\tilde{\alpha}_{nj})}\Big\}. \tag{A10}$$

Some of the subscripts in (A6)–(A10) have been omitted if they were not necessary for a unique characterization. The integral F_{ai}^2 is used later for normalization. The infinite series in the above formulas should be truncated at $N = Mb/s$ in order to take the edge condition accurately into account [6]. M is the number of components used in the system of basis functions. We have chosen $M = 4$.

Acknowledgment

The authors are indebted to the Deutsche Forschungsgemeinschaft for financial support.

References

[1] H. El Hennawy and K. Schünemann, "Analysis of fin-line discontinuities," in *Proc. 9th Euro. Microwave Conf.*, (Brighton), 1979, pp. 448–452.

[2] A. Wexler, "Solution of waveguide discontinuities by modal analysis," *IEEE Trans. Microwave Theory Tech.*, vol. MTT-15, pp. 508–517, 1967.

[3] L. P. Schmidt and T. Itoh, "Spectral domain analysis of dominant and higher order modes in fin-lines," *IEEE Trans. Microwave Theory Tech.*, vol. MTT-28, pp. 981–985, 1980.

[4] H. El Hennawy and K. Schünemann, "Impedance transformation in fin-lines," *Proc. Inst. Elec. Eng.*, part H: MOA, 1982.

[5] R. F. Harrington, *Time-Harmonic Electromagnetic Fields*. New York: McGraw-Hill, 1961.

[6] R. Mittra, T. Itoh, and T. Li, "Analytical and numerical studies of the relative convergence phenomenon arising in the solution of an integral equation by the moment method," *IEEE Trans. Microwave Theory Tech.*, vol. MTT-20, pp. 96–104, 1972.

[7] H. Meinel, B. Adelseck, and H. Callsen, "A survey of planar integrated millimeter-wave components," in *Military Microwaves Symp. Dig.*, (London, England), 1980, pp. 82–87.

[8] J. L. White, *Semiconductor Control*. New York: Artech, sec. 4, 1977.

6.1.6

MINIATURISATION IN E-PLANE TECHNOLOGY

Frank C. de Ronde *

ABSTRACT

In a simple and cheap way, a planar full-band matching technique allows for miniaturisation of E-plane components, such as corners, shifts, T junctions,Magic Tees, cross-over and quadrature directional couplers a.o. A finline taper has been made half the length of the shortest computer optimized one, having similar performance. A transition to suspended stripline offers possibilities for this line instead of the finline. Although realised for X-band, scaling down to mm waves is our aim as well as application for devices.Additional technical details will be given at the oral presentation.

INTRODUCTION

In E-plane technology finline and to a lesser extent the co-planar line are very popular:their conductors are close together, so semiconductors can easily be inserted (1,2).In general they are used if really needed because they are rather lossy.Therefore waveguides have been used in between, not only as straight sections, but also as bends, junctions and even couplers.Now transitions are needed from finline to waveguide and back again, introducing losses,due to their rather long tapers in order to prevent reflection,so it is very important to minimize their lengths. Consequently this makes a complete circuit large compared to the wavelength (3,4).We realized almost ideal compact as well as unconventional waveguide components,transitions to finline or even to suspended stripline, being less lossy and compatible with S.S.L.filters and -couplers. In order to achieve this our full-band matching technique (5) has been adapted to E-plane technology : the matching elements (M.E.) have been made from copper (20 micron) on polyester film (100 micron), a cheap metallized film used for printed circuits .

I.WAVEGUIDE COMPONENTS

The size of a w.g. bend can be reduced considerable by using a corner and compensating its reflection (Fig. 1b) by a vane as matching element, located symmetrically in the corner (Fig. 1a) and clamped in position by the two w.g.halves,automatically providing the contact with its walls as applied throughout.In case of contact problems or insulation, serrated chokes might be used.The length l proved to be almost constant for angles between 0 and 90°. The vane acts as a double taper of which the reflection becomes maximum for $l/_2 = \lambda/_4$. If we choose λ just above our frequency band,then a similar reflection behaviour occurs as given in Fig.1b, however,of opposite sign so one may cancel the other and a perfect match could be obtained. Instead of a triangle, also a square vane works as well, even with a hole in it. Because currents pass over the edge of the vane, the original reflection of the corner returns, if they are cut by making a slot s or even more than one for broader bandwidth - almost completely if $s = \lambda/_4$ (Fig.1c)

* Bath University,School of Electrical Engineering,Bath BA2 7AY,England.

Chosing this above the band,waves are heavily reflected,whereas in the band it forms a series inductance L_s, which can easily be full-band matched with the proper parallel capacitance C_p in the form of a planar stub.This way a low-pass-or even band reject filter can be realized.

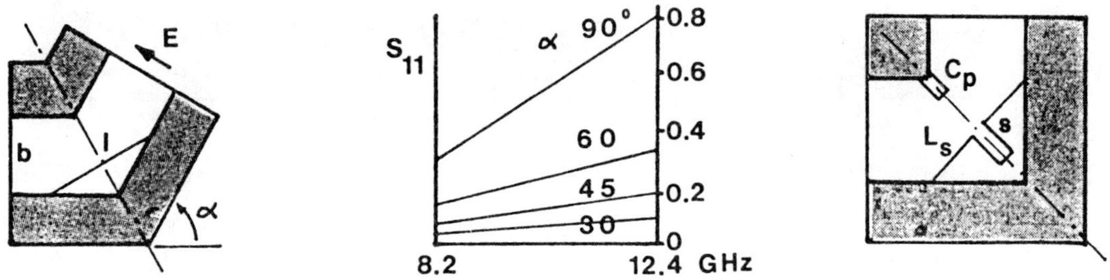

Fig. 1a F.B.M.corner α = 0-90°,$l \simeq b\sqrt{2}$ b. plain corners.   c. L.P.filter.
(shaded) w.g.wall   (blank) M.E.

An S-bend can be reduced in size by replacing it by a shift: a pair of two identical matched corners,close together (Fig.2a).Due to interference of the matching elements, their form has to be slightly modified; less if the shift becomes greater.The electrical symmetry must always be respected. The matching of a Tee, as seen from the symmetry arm(1 in Fig.2b) becomes easy if the side arms (2 and 3) have half their height ($b/_2$).Then we are dealing with 2 - 90° corners, back-to-back; a double triangular (or-square) vane on the symmetry axis will do the matching.In order to go to normal height again a taper has been used.With a circular one, tangential at $b/_2$ from the centre, the matching becomes quite simple.It is more or less the dual case of matching a microstrip taper (6).
In both cases the M.E!s are at the high impedance ends,being less critical at the same time. For arrays, a folded version of the Tee (Fig. 2c) is often needed in the feeder; here we are dealing with shifts and impedance-transformers, so steps have been chosen, such as to compensate automatically for the reflection at the common wall. This way a compact configuration is obtained with a total axial length of about $\frac{1}{4}\lambda_{g\ max}$.

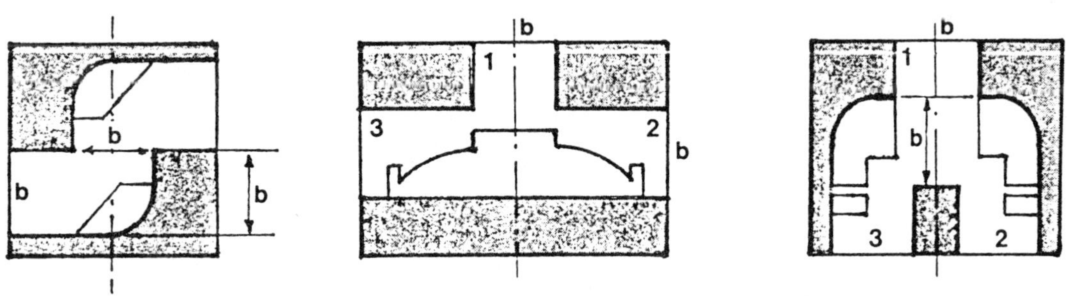

Fig.2 a.Shift $S_{11} \leqslant 0.01$ b. Tee $S_{11} \leqslant 0.01$ c.Folded Tee $S_{11} \leqslant 0.02$

With these E-Tees it is comparatively easy to realise Magic Tees (Fig.3a,b) We only have to add an H-Tee (shunt Tee), which can be done by adding a narrow strip as a probe in the symmetry plane, opposite port 1, together with a groove in the lower wall as the outer conductor of a suspended stripline (S.S.L.)forming the fourth port.In Fig.5c we will see, how a simple transition to waveguidecan be made if wanted. With the strip protuding into the waveguide over about $\lambda/_4$ a good match can be obtained with the rightZ_{os} of the S.S.L.As is well known: a hybrid tee becomes really magic if ports 1 and 4 are matched.

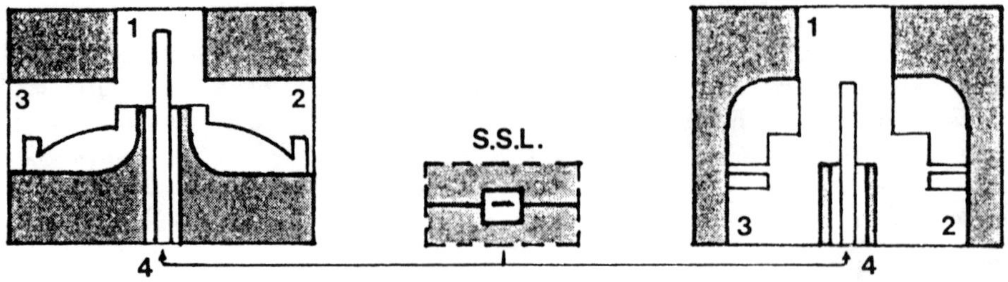

Fig.3 a. Magic Tee b. Folded Magic Tee

If a coupling weaker than 3dB is wanted,asymmetrical M.E.'s must be used. The matched probe must become asymmetrical as well and perpendicular everywhere to the electric field coming from port 1 in order to remain decoupled from this port. This might result in a curved probe,which is no problem here,unless this curve becomes frequently dependant.

Quadrature directional couplers are not very popular in finline (1) because they are rather long and lossy.Therefore waveguide principles have been applied in E-plane technology,like branchguide couplers (4).Most popular are the multiprobe couplers (1),although they are rather long,due to the tapered probes at both ends.

The most compact type is the dual of my strip-slot version (1),however,a severe drawback is the separation of the coupled ports by the isolated one. Therefore we have developed a new type of quadrature D.C.which will be explained by starting with its extreme : the cross-over (Fig. 4a).

A passive,lossless and reciprocal 4-port is a directional coupler if matched and vice versa.If it has two planes of symmetry in addition, it becomes a quadrature coupler. If the 4-port is completely symmetrical then it can be shown that the coupler becomes degenerate: there is no coupling at all and the isolation is infinite : one obtains a cross-over. This might be a useful component in E-plane technology.The problem is,however,either to match the 4-port or to cancel the side-arm coupling without perturbing the complete symmetry.

A plain symmetrical w.g.cross has a lot of reflection ($S_{ii} \leqslant 0.5$) and strong coupling to the side arms as well.Excitation of port 1 corresponds to equal even- and odd mode excitation simultaneously at ports 1 and 3 (see Fig.) Only the odd mode is responsible for side arm coupling, whatever symmetrical matching element (M.E.)will be applied, so the even mode may be neglected here.If a ring has been inserted in the cross,currents are induced clockwise and may cancel the original- anti-clockwise,electric fields $\propto E_o$ in the side arms- completely by increasing the induced currents by capacitive coupling (c) with the waveguide corners and by using a square "ring" (Fig.4a).

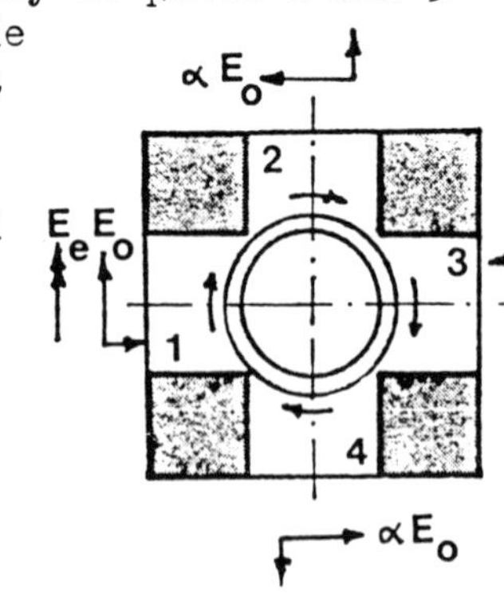

Now we have obtained a degenerate coupler: a cross-over without side-arm coupling. The performance is not too frequency dependent because the "resonant"ring is tightly coupled to the ports.

Directional coupling to the side arms can be obtained if two probes are placed diagonally (Fig.4b) or by variations of the width of the ring in diagonally located corners(less effective).A wave entering port 1 couples to port 4 because the contributions of these two probes are in phase; to port 2 they cancel,because they are anti phase due to their $\lambda/_4$ separation.

There is great similarity between this cross-over and coupler and a microstrip version(7); they are more or less duals.In both cases we are dealing with a cross (E- resp.H-plane) of higher char.impedance, giving rise to side-arm coupling due to the odd- resp.even mode and compensated by a cap.coupled "ring" by way of a $\lambda/2$ different, so anti phase,coupling. For the couplers the rings are modified in similar ways.

The E-plane coupler can also be regarded as a "planar" (and dual)version of the wellknown crossguide coupler.

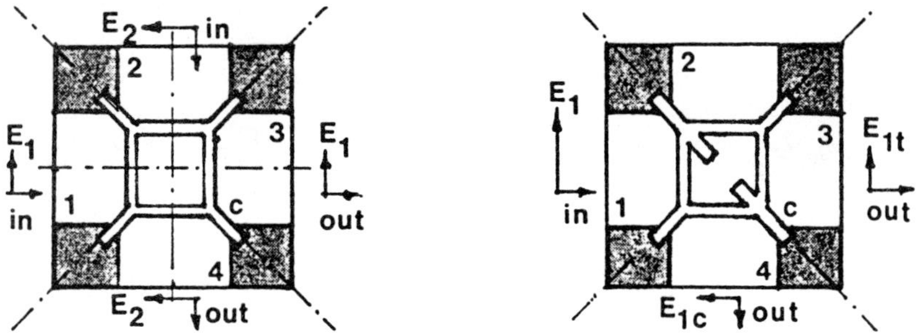

Fig.4 a. Cross-over b.0-90° Directional coupler

II. TRANSITIONS

Although a synthesis for optimum design of a finline taper has been developed recently (8), even shorter tapers with similar performance can be made by applying our full-band matching technique.Apart from being a mode transducer, the finline taper is mainly an impedance transformer and a solution for its dual, a microstrip taper, has already been given (6). In the E-Tee described above (Fig.2b) we applied already asymmetrical circular tapers,matched at the right place by capacitive stubs.For the finline taper symmetrical circular tapers have been used, which together with the M.E.'s were never longer than $3/_8$ λmax for $s/_b \geqslant 0.025$ (Fig.5a).

Instead of the finline the suspended stripline (S.S.L.)is often more favourable: lower losses,higher isolation due to higher cut-off frequency; couplers,filters,switches etc.can easily be realized. Cheap one sided metallised film can also be used.Until now the available transitions were rather long because similar principles were applied as for the finline: a taper (1,9).Probe types are much shorter but, due to the combination of mode transducer and impedance transformer, are in general not full band (10).We have separated both functions because often the transformer is not needed e.g.when used as the coupling medium between waveguides.

The new w.g.-S.S.L.transition will now be described (Fig. 5b). For a characteristic impedance of the S.S.L.above 100 Ω ,we found a position $x_o$of the probe, where the matching was merely a matter of the length of the probe: for longer wavelength a longer probe than for shorter ones.

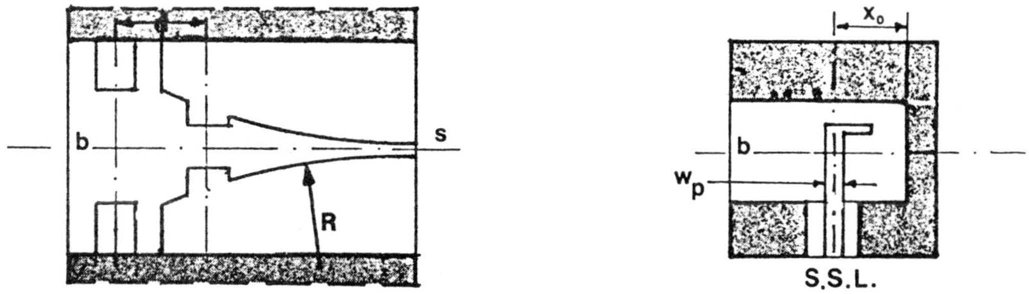

Fig.5 Transitions a.Finline taper $S_{11} \leqslant 0.03$ b. w.g - S.S.L $S_{11} \leqslant 0.02$

This could be solved full band either by using a broad probe (w_p) or by toploading the probe by a short,narrow strip in the x direction.The latter configuration proved to be the least critical.

The end-launch type (Fig.5c) is a variation of the right-angle type. With the aid of the probe parameters,l_p,w_p,w_g and g optimum match to the S.S.L.with char.impedance Z_{os} could be obtained.

The symmetrical type (Fig.5d)is even more simple to match,because there is no $\lambda/4$ distance to the plunger,so only a certain width (w_p)of the probe will be sufficient,together with the right Z_{os} for the S.S.L.This type has been used as the shunt tee in the magic tee (Fig.3)

From the S.S.L.it is very easy to go to coplanar line, coax or micro-strip.If 50 Ω is needed,an impedance transformer (6)can easily be applied: even in the waveguide to gain length for the end-launch type.

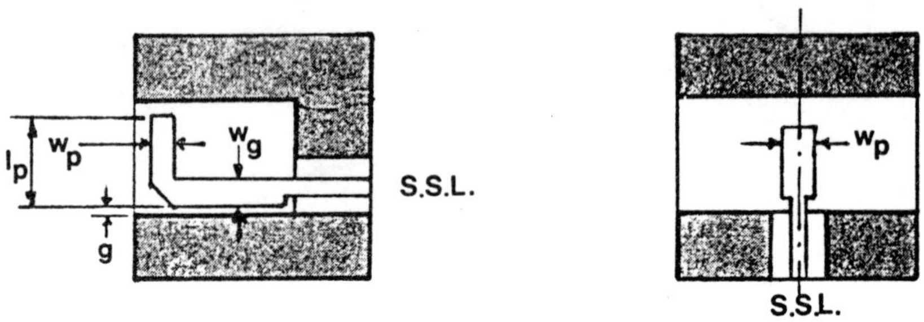

Fig.5 Transitions c.End-launch $S_{11} \leqslant 0.02$ d. Symmetrical $S_{11} \leqslant 0.01$

CONCLUSION

The planar full-band matching technique not only allows for more compact circuitry but at the same time offers possibilities for high performance mm waveguide components at an even lower price because there is no need for a good dielectric substrate; the e.m.fields are almost in air: in the waveguide and the suspended stripline.

ACKNOWLEDGEMENT

The author likes to thank Prof T.E.Rozzi for giving him the opportunity to do this research.

6.1.7

A WAVEGUIDE-TO-MICROSTRIP TRANSITION

Jerzy K. Piotrowski and Wojciech Wojtasiak

ABSTRACT

The design and performance of a waveguide-to-microstrip transition is presented. The optimum bilateral finline taper has been used to match the fundamental mode waveguide characteristic impedance to a 50-ohm microstrip mounted on a low dielectric substrate. The transverse resonance method has been applied to calculate characteristic impedance of bilateral finline taking into consideration the finite metallization thickness and substrate holding grooves. The experimental results of X-band transition are also given.

INTRODUCTION

The state of the art of microwave semiconductor devices technology and development of soft, lowloss dielectric substrates allow to realize the microstrip components in millimeter-wave bands [1, 2]. On the other hand, measurement systems for frequencies above 20 GHz are mainly produced in waveguide technique. Therefore, there is the need for wide-band waveguide-to-microstrip transition.

Former designs of such transitions, reported in literature, have been based on ridge-waveguide [3], antipodal finline [4] and waveguide below cutoff [5] transformers or on antenna-type transitions [6]. A disadvantage of the constructions described in [5, 6] is the small bandwidth so far obtained. The results of the transitions with ridge-waveguide transformer depend critically on the dimensions, what is undesirably in millimeter-wave region. The transition with antipodal finline is fully integrated on the substrate but till now to achieve desired wide-band performance the cut-and-try technique has to be employed.

This work presents the fullband waveguide-to-microstrip transition with bilateral finline transformer. The transition is shown at Fig.1.

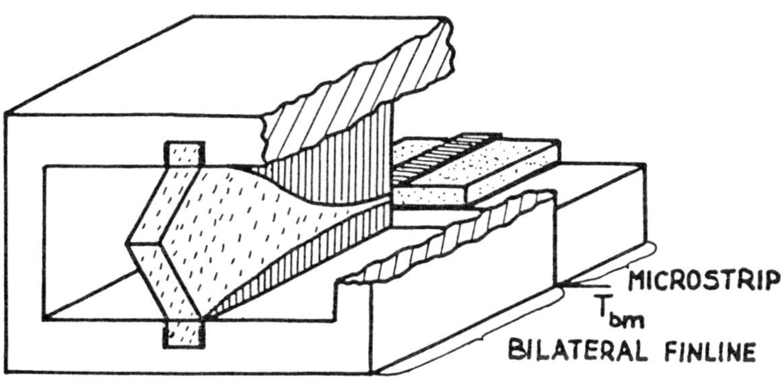

Fig.1. The view of a waveguide-to-microstrip transition.

Warsaw University of Technology , Institute of Electron Technology
Koszykowa 75, 00-662 Warsaw, Poland.

The design of the proposed transition is essentially the synthesis of a optimum finline taper when the parameters of bilateral finline structure at plane T_{bm} (Fig.1) are properly chosen. The bilateral finline at this plane should comply with the following requirements:

- its characteristic impedance is equal to characteristic impedance of microstrip (both impedances are on a power-voltage basis);
- slot width is appropiately less than substrate thickness of microstrip;
- substrate thickness is suitably less than strip width of microstrip.

The above simple conditions should allow to get the minor field distortions at plane T_{bm} and as a result of this to obtain satisfactory parameters of the transition.

BILATERAL FINLINE CHARACTERISTICS

The cross section of a bilateral finline is shown in Fig.2. The transverse resonance method (TRM), which was recently developed for unilateral finline and presented in [7], has been modified to analyse this lossless transmission line taking into consideration the finite metallization thickness (t) and the substrate holding grooves (w). The essence of the method is shortly described below.

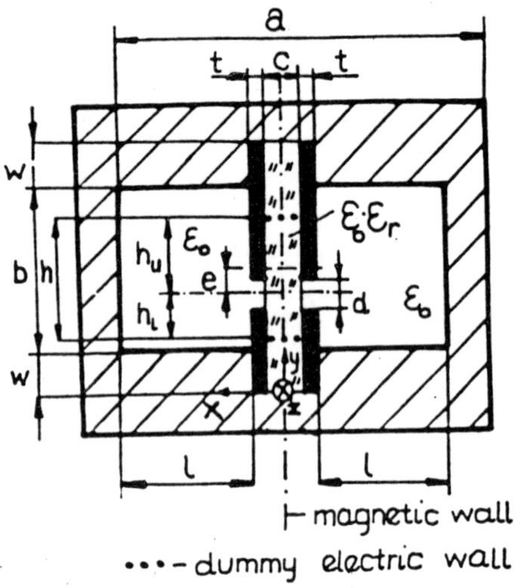

Fig.2. Bilateral finline.

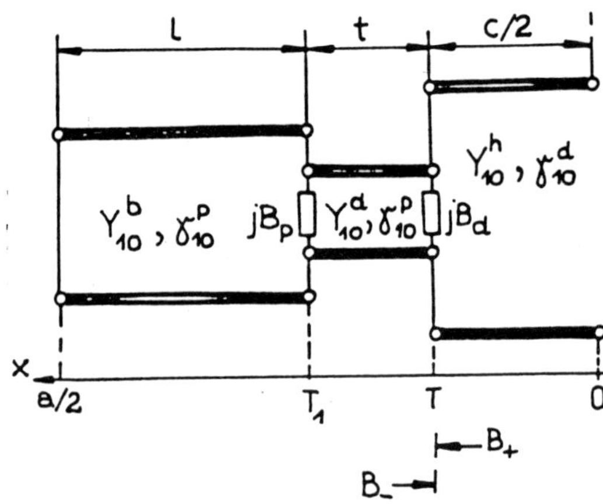

Fig.3. Equivalent circuit of bilateral finline resonator.

The half-wavelenght bilateral finline resonator (in the z direction) is considered. The transverse structure of the resonator is symmetrical in relation to x=0 plane, in which the magnetic wall occurs for the fundamental mode. Hence, the equivalent network for transverse section of this resonator is as in Fig.3. The network consists of three segments of H_{10}-mode (to x-axis) waveguides. Two of them, namely l and t, are air-filled, and described by propagation constant γ^p_{10} and characteristic admittances Y^b_{10}, Y^d_{10}, respectively. The segment c/2 is filled by dielectric with relative permittivity ε_r. This waveguide is characterised by propagation constant γ^d_{10} and characteristic admittance Y^h_{10}. All three characteristic admittances are based on the power-voltage definition.

The susceptance B_a represents in plane T_1 all higher order modes (H_{1n}, E_{1n} for n>1) from waveguide segment of lenght l. Similarly, the susceptance B_d represents in plane T all higher order modes from dielectric-filled waveguide.

The phase constant of the fundamental mode in bilateral finline is the first root of transverse resonance condition:

$$B_+ + B_- = 0 \qquad (1)$$

where B_+ and B_- are susceptances in the plane T (Fig.3) looking into the +x and -x directions, respectively.

The characteristic impedance of this transmission line should be calculated by using the procedure given in [8].

The analysis is simplified by assumption that only the H_{10}-mode is present in the slot (segment t in Fig.3). This assumption together with the short circuit in plane x=a/2 and the open circuit in plane x=0 cause, that the electromagnetic field in the resonator is H to z-axis and can be expressed as the superposition of Hz_{1n}-modes. Hence, both equivalent susceptances B_a and B_d have been obtained by using the variational technique as in [7]. The components B_+ and B_- of resonance equation (1) have been easily derived basing on the network in Fig.3.

The described method is useful to analyse the finline characteristics in dependence on location of dummy electric walls, which are shown in Fig.2. That helps to choose the proper slot location in waveguide housing. The effective dielectric constant (ε_{eff}) and characteristic impedance (Z_0) of bilateral finline at frequency 10 GHz in function of normalized (to slot width) distance between the dummy electric walls for e=0 and e=(b-d)/2 are shown at Fig.4.a and Fig.4.b, respectively. The calculations have been carried out for: ε_r=2.44, c=0.373mm and t=35μm. There has been found, that finline characteristics are less sensitive to location of the dummy electric walls when the slot is placed at the bottom of waveguide housing.

Knowing the parameters of bilateral finline structure at plane T_{bm} the finline transformer can be designed.

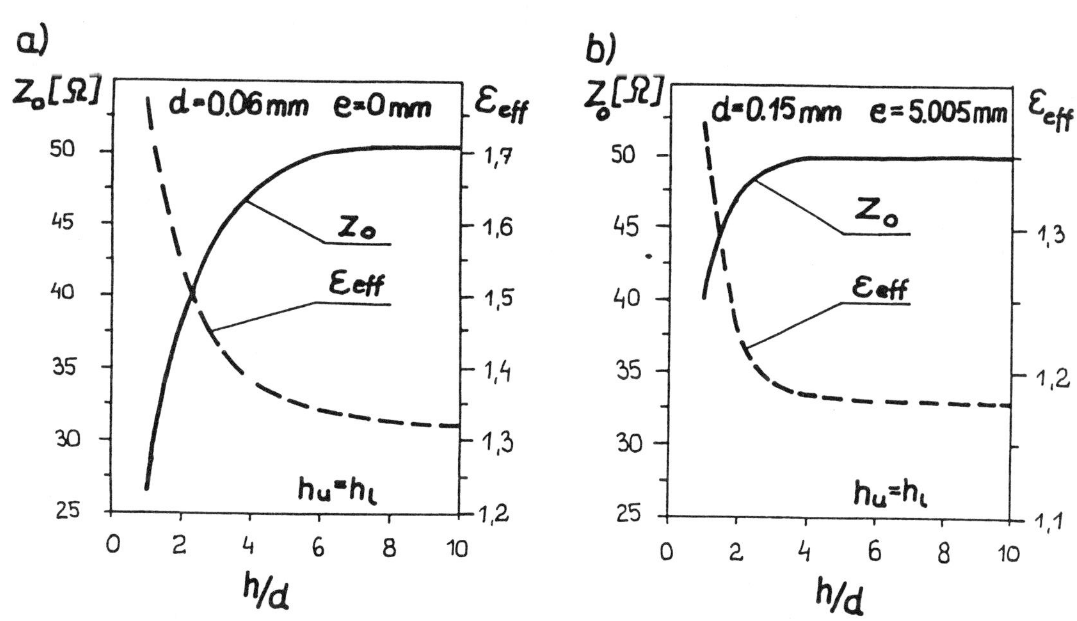

Fig.4. Bilateral finline characteristics in function of normalized distance between the dummy electric walls for e=0 (Fig.4.a) and e=(b-d)/2 (Fig.4.b).

BILATERAL FINLINE TAPER

Design of optimum finline taper resolves itself into determinetion of such a slot contour , that the input VSWR will be below a certain value R_{max} , and the length of the taper will be the shortest possible one. The synthesis procedure of finline taper is given in [9]. This approach is based on the evaluation of the cutoff frequencies along the taper. No characteristic impedances are required and that leads to considerable simplification of the calculations. The relation between the slot width and the cutoff frequency has been obtained by using transverse resonance method as in [9], but the method has been slightly modyfied and the finite metallization thickness has been taken into consideration.

The exemplary X-band taper contour has been shown in Fig.5.a. At the taper ends (planes: z=0, $z=l_t$) exist step discontinuities, which are ty-

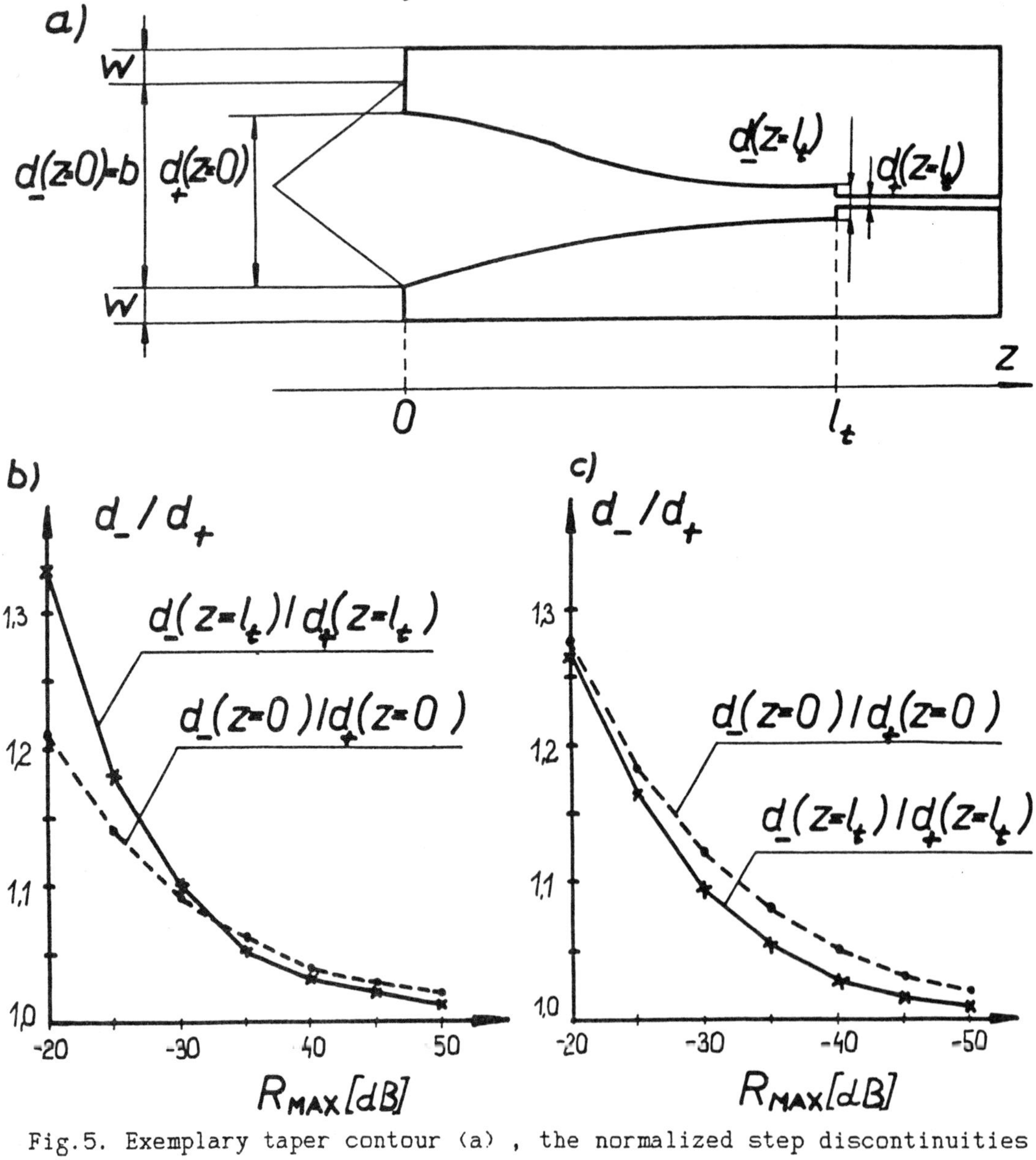

Fig.5. Exemplary taper contour (a) , the normalized step discontinuities in depedence on R_{max} for e=5.005 mm (b) and for e=0.0 mm (c).

pical of Dolph-Chebyshew transformers. These steps cause the exitation of higher order modes, but the lenght of the taper grows shorter. The normalized step discontinuities in depedence on R_{max} for e=(b-d)/2 and e=0 have been presented in Fig.5.b and Fig.5.c, respectively. The curves have been calculated for substrate parameters as in previous section and for d=0.15mm. The step at the plane z=0 is smaller for e=5.005mm than for e=0, but the transformer is longer. The discontinuity at the plane $z=l_t$ is practically independent of the slot location.

Generally, the lower the R_{max} chosen, the smaller the step, but the taper length is longer. The taper length l_t is linear function of the R_{max}, when R_{max} is in dB. Hence, this simple relation between l_t and R_{max} with the knowledge of the step discontinuities are important for design of finline tapers.

PRACTICAL PERFORMANCE

The model of double waveguide-to-microstrip transition for X-band has been realized. The Cu Clad (ε_r=2.44) with dielectric thickness 0.373mm and metallization thickness 35μm has been used for bilateral finline and 30mm long microstrip.

The performance of the transition depends greatly on the quality of electrical contact between the fin and microstrip.

Some higher order modes may be generated at the T_{bm} plane (Fig.1). To prevent them from being transmitted - which would increase the loss - a small rectangular cover with cutoff waveguide dimensions was placed over the microstrip.

The measured insertion losses and return losses for a double transition are shown in Fig.6.

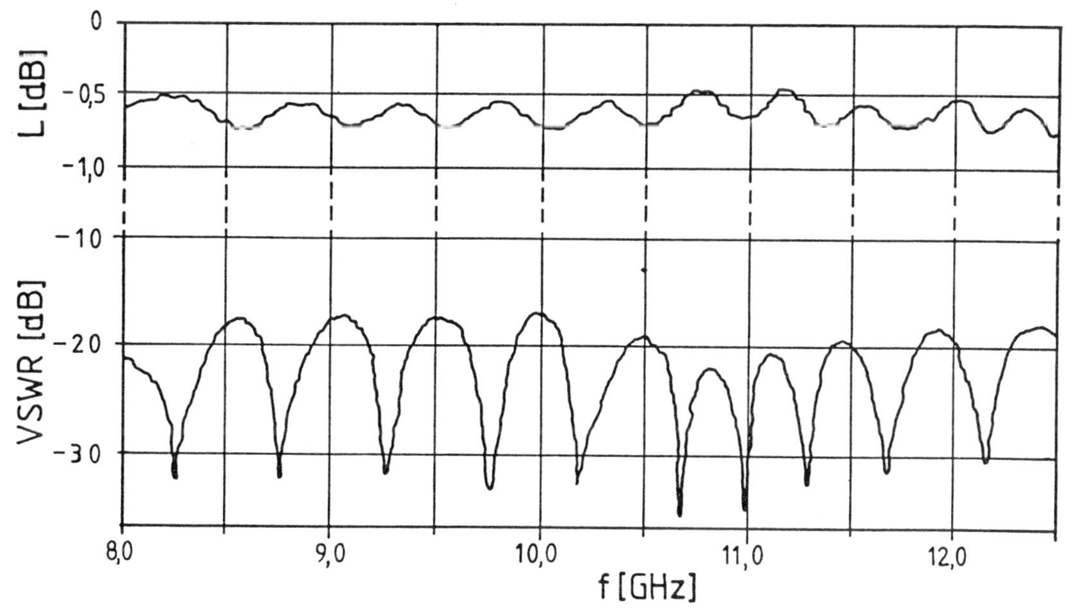

Fig.6. Return losses (VSWR) and insertion losses (L) of the X-band waveguide-microstrip-waveguide transition.

REFERENCES

[1] K. Chang, R. Tahim, D. English, A. Grote, T. Phan, C. Sun, G. Hayashibara, P. Yen and W. Piotrowski, "W-band (75 to 110 GHz) microstrip components," IEEE 1985 MTT-S Int'l Microwave Symposium Digest, pp.371-374.

[2] W. Brunner, F. Garcia, M. Mlinar, W. Piotrowski, J. Rabel, J. Raue and S. Schell, "100-300 GHz transmission media evaluation and component development," Proc. 14-th European Microwave Conf., 1984, pp.587-592.

[3] S.S. Moochalla and Chae An, "Ridge waveguide used in microstrip transition," Microwaves & RF, vol.23, no.3, 1984, pp.149-152.

[4] J.H.C.van Heuven, "A new integrated waveguide-microstrip transition," IEEE Trans. Microwave Theory Tech., vol.MTT-24, 1976, pp.144-147.

[5] S. Bharj and S. Mak, "Waveguide-to-microstrip transition uses evanescent mode," Microwaves & RF, vol.23, no.1, 1984, pp.99, 100 and 134.

[6] B. Glance, R. Trumbaruto, "A waveguide to suspended stripline transition," IEEE Trans. Microwave Theory Tech., vol.MTT-21, no.2, 1973, pp.117-118.

[7] J.K. Piotrowski, "Efficient analysis of finline with finite metallization thickness," IEEE 1986 MTT-S Int'l Microwave Symposium Digest, pp.213-216.

[8] S.B. Cohn, "Slot line on a dielectric substrate," IEEE Trans. Microwave Theory Tech., vol.MTT-17, no.10, 1969, pp.768-778.

[9] Ch. Schieblich, J.K. Piotrowski and J.H. Hinken, "Synthesis of optimum finline tapers using dispersion formulas for arbitrary slot widths and locations," ibid., vol.MTT-32, no.12, 1984, pp.1638-1645.

6.2.1

MILLIMETRE-WAVE LOW-LOSS FINLINE LOWPASS FILTERS

Indexing terms: Filters, Millimetre-wave devices

Millimetre-wave low-loss finline filters, with very sharp cutoff, consisting of a number of notches periodically spaced along the single-ridge finline are described. Measured pass-band insertion losses ranged 0·5 to 1 dB for Q-band (33 to 50 GHz) filters have been achieved. These results demonstrate the technological advance of millimetre-wave filters using finline integrated circuits.

Introduction: The increasing demand in the development of wideband low-cost millimetre-wave systems has created an urgent need for broadband low-cost low-loss filters. Because of the similarity between finline and conventional waveguides, various components have been realised using the finline techniques. Among these components are the E-plane bandpass filters proposed recently.[1–6] These bandpass filters have been successfully designed to achieve very low loss in the 18 to 170 GHz range; however, there has been little finline lowpass filter development.[7] The low-cost finline lowpass filter is an important component with many applications in receivers, multipliers and multiplexers. This letter reports the performance of a Q-band (33 to 50 GHz) finline lowpass filter (Fig. 1), which is similar to the classical corrugated waveguide or waffle-iron filters introduced by Cohn.[8] It is actually a very broadband bandpass filter in the sense that it will not propagate energy below a waveguide cutoff frequency. A maximum insertion loss of 1 dB and a stopband attenuation of 40 dB at 43 GHz have been achieved for a filter designed at the cutoff frequency of 40 MHz. The same filter can be scaled up to other frequency bands.

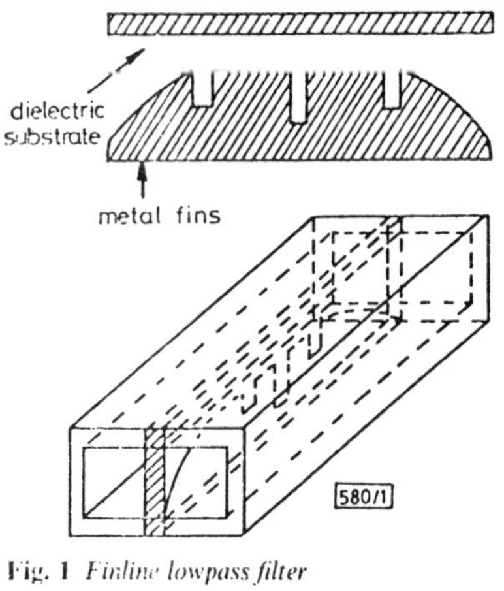

Fig. 1 *Finline lowpass filter*

Filter design and performance: The filter consists of a sequence of notches periodically located along the single-ridge finline to form a series of directly coupled apertures. The spacing between consecutive irises is approximately one-quarter of λ_g,

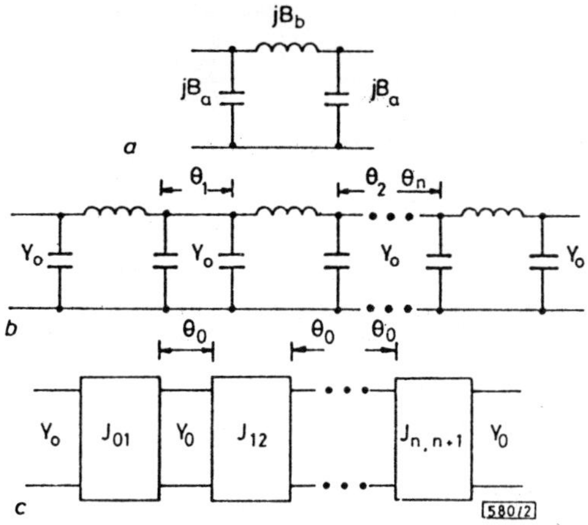

Fig. 2

a Equivalent circuit of a notch (B_a and B_b are susceptive values)
b, c Equivalent networks of a filter using notch equivalent circuits and admittance inverters, respectively

the wavelength of a single-ridge finline at the cutoff frequency of the filter. The equivalent circuit of a very short notch can be expressed as a π-network[9] as shown in Fig. 2*a*. The equivalent circuit of the filter is given in Fig. 2*b*, and in exactly equivalent

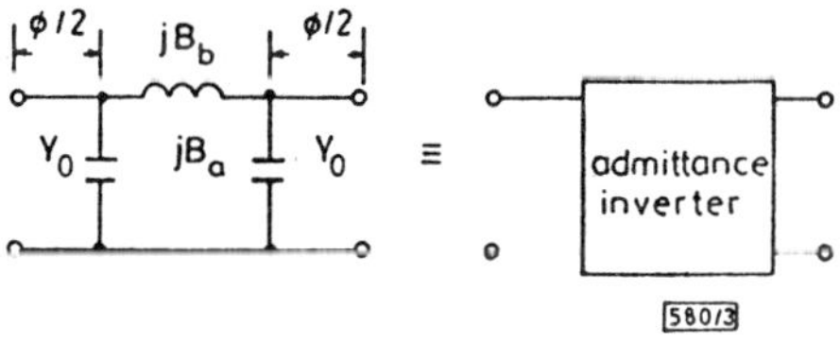

Fig. 3 *Admittance inverter*

$$\frac{J}{Y_0} = \left|\tan\left(\frac{\phi}{2} + \tan^{-1}\frac{B_a}{Y_0}\right)\right|, \text{S}$$

$$\phi = -\tan^{-1}\left(\frac{2B_b}{Y_0} + \frac{B_a}{Y_0}\right) - \tan^{-1}\frac{B_a}{Y_0}, \text{rad}$$

form using admittance inverters[10] (Fig. 3) in Fig. 2*c*. Y_0 is the characteristic admittance of finline between irises. The electrical length between irises j and $j + 1$ is given by

$$\theta_j = \theta_0 + \tfrac{1}{2}[\phi_{j-1,j} + \phi_{j,j+1}] \quad (1)$$

where θ_0 equals $\pi/2$ at the cutoff frequency of the filter. A design procedure has been developed. The notch discontinuity is characterised taking into account the dominant and evanescent modes using the rigorous modal analysis.[9] Empirical closed form expressions for the scattering parameters of this notch have been derived recently by curve-fitting to the

C. Nguyen and K. Chang, *Electronics Letters*, Vol. 20, No. 24, Nov. 22, 1984, pp. 1010–1011.

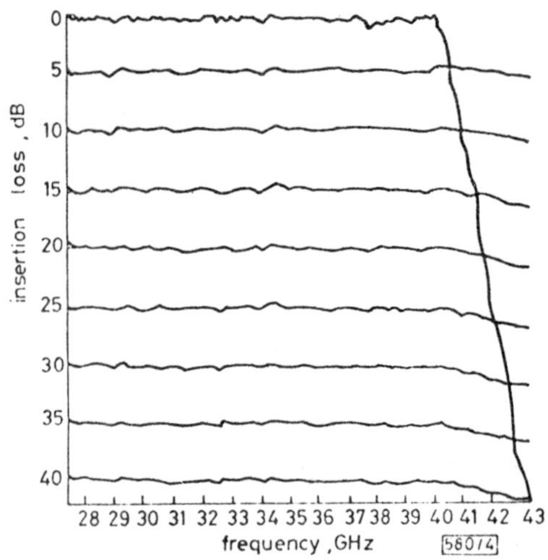

Fig. 4 *Performance of finline lowpass filter*

experimental results;[11] however, these expressions are limited to a narrow range of structural parameters. The spacing between irises can then be calculated from eqn. 1. Assuming that no evanescent-mode interaction occurs between irises, an ABCD matrix manipulation is used to determine the overall network electrical response. Owing to the low impedance level at the input and output ports of the filter, impedance transformers are used to match these ports to the terminating waveguides.

Based on the method outlined, several filters have been realised on 0·127 mm (0·005 in)-thick Duroid substrate. A maximum insertion loss of 1 dB below 40 GHz has been achieved for a seven-section bilateral finline filter. The results are shown in Fig. 4. These losses include the losses due to two cosine-taper transitions and a 1·5 in-long waveguide. The spurious responses or spikes occur near 52·7 GHz due to the propagation of the TE_{02} mode are rejected by at least 30 dB. The photograph of the 40 GHz cutoff filter and its test fixture is given in Fig. 5.

Conclusions: Finline lowpass filters with very sharp cutoff, consisting of several notches alternating with a number of single-ridge finline sections, have been reported. A typical maximum insertion loss of 1 dB has been achieved in the passband. The results demonstrated the feasibility of finline lowpass filters and should have many applications in broadband high-performance low-cost millimetre-wave systems.

Acknowledgment: The authors wish to thank Dr. C. Sun for his encouragement and helpful suggestions.

C. NGUYEN
K. CHANG

TRW Electronics & Defense Sector
One Space Park
Mail Station 138/4934
Redondo Beach, CA 90278, USA

9th October 1984

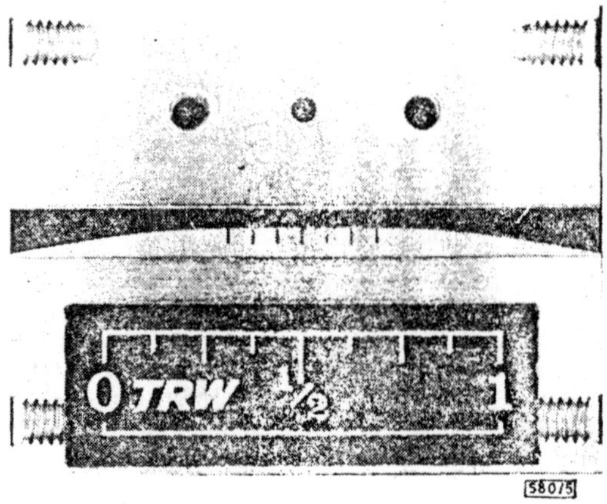

Fig. 5 *Q-band finline lowpass filter*

References

1 KONISHI, Y., and UENAKADA, K.: 'The design of a bandpass filter with inductive strip-planar circuit mounted in waveguide', *IEEE Trans.*, 1974, **MTT-22**, pp. 869–873

2 TAJIMA, Y., and SAWAYAMA, Y.: 'Design and analysis of a waveguide-sandwich microwave filter', *ibid.*, 1974, **MTT-22**, pp. 839–841

3 MEIER, P. J.: 'Integrated finline millimeter components', *ibid.*, 1974, **MTT-22**, pp. 1209–1216

4 SAAD, A. M. K., and SCHÜNEMANN, K.: 'Design and performance of finline bandpass filters'. 10th European Microwave Conference, Warsaw, Sept. 1980, pp. 397–401

5 ARNDT, F., *et al.*: 'Low-insertion-loss finline filters for millimeter wave applications'. 11th European Microwave Conference, Amsterdam, Netherlands, Sept. 1981, pp. 309–314

6 SHIH, Y. C., ITOH, T., and BUI, L. Q.: 'Computer-aided design of millimeter-wave E-plane filters', *IEEE Trans.*, 1983, **MTT-31**, pp. 135–141

7 REINDEL, J.: 'Printed WG circuits trim component costs', *Microwaves*, 1980, pp. 60–63

8 COHN, S. B.: 'Analysis of a wide band wave-guide filter', *Proc. IRE*, 1949, **37**, pp. 651–656

9 HENNAWY, H. E., and SCHÜNEMANN, K.: 'Impedance transformation in fin-lines', *IEEE Proc. H, Microwaves, Opt. & Antennas*, 1982, **129**, pp. 342–350

10 MATTHAEI, G. L., YOUNG, L., and JONES, E. M. T.: 'Microwave filters, impedance-matching network, and coupling structures' (McGraw-Hill, New York, 1964), p. 438

11 BURTON, M., and HOEFER, W.: 'An improved model for short and open-circuited series stubs in finlines'. 1984 IEEE MTT-S International Microwave Symposium Digest, pp. 330–332

6.2.2

Computer-Aided Design of Millimeter-Wave *E*-Plane Filters

YI-CHI SHIH, STUDENT MEMBER, IEEE, TATSUO ITOH, FELLOW, IEEE, AND L. Q. BUI

Abstract—**A computer-aided design (CAD) algorithm has been developed for a class of *E*-plane bandpass filters. The analysis portion of the algorithm is based on the residue-calculus technique and a generalized scattering parameter method. It is mathematically exact and numerically very efficient. Filters designed with this method have been fabricated and tested in *Ka*-band. Good agreement with design has been obtained.**

I. Introduction

SEVERAL FILTER CIRCUITS that are composed of a conductive sheet which is properly designed and sandwiched between the waveguide shells parallel to the *E*-plane have been proposed recently [1]–[5]. The circuits are developed either on a metal sheet [1], [2] or on a bilateral fin-line [3]–[5]. In these filters, the slot widths are equal to the waveguide height; thereby the structure really consists of several resonators separated by printed inductive strips. Since the structures, especially those in forms of fin-lines, are amenable to photolithographic techniques and only straight line shapes are involved, they are highly suitable for mass production. Some design procedures have been reported based on a number of methods such as mode-matching techniques and variational techniques. These methods have been applied only to metallic *E*-plane circuits or bilateral fin-line structures.

This paper presents a new efficient computer-aided design (CAD) algorithm developed for a class of *E*-plane bandpass filters. A unified treatment is introduced which applies not only to the purely metallic *E* plane circuits and the bilateral fin-line circuits, but also to the unilateral and insulated fin-line circuits (Fig. 1). Since the analysis portion of the algorithm is based on the residue-calculus technique [6] and a generalized scattering parameter method [7], it is mathematically exact and numerically very efficient, as the convergence is guaranteed. Filters thus designed have been fabricated and tested in *Ka*-band with good agreement with the design.

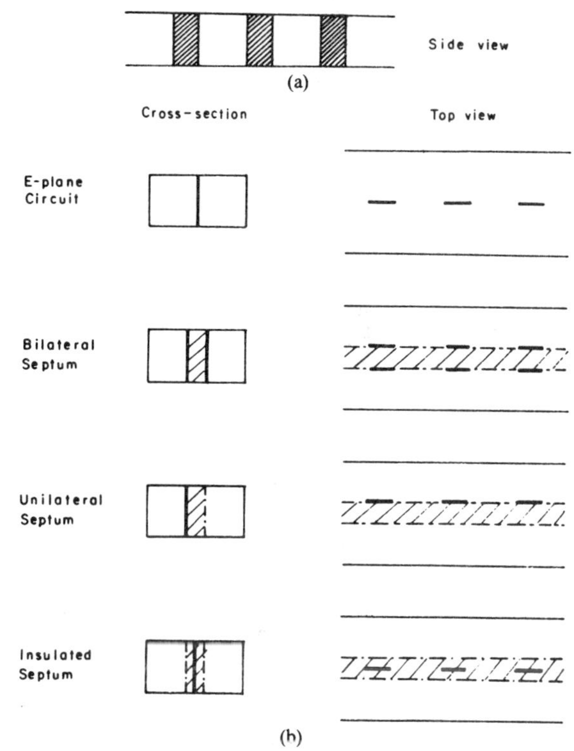

Fig. 1. Class of *E*-plane filter structures. (a) Side view. (b) Cross section and top view.

II. Theory

The CAD program consists of an optimization routine and an analysis routine. We describe the analysis routine first. The analysis is carried out in three steps. First, a limiting case of a single junction created by a semi-infinite septum in a waveguide is analyzed. This structure may be analyzed exactly using the residue-calculus technique [6] to obtain a closed-form expression for the scattering matrix. The second step is to calculate the scattering parameters for a finite-length septum. This is done by placing two junctions back to back, and utilizing the concept of the generalized scattering matrix [7], which takes into account the interaction between junctions by not only the propagating modes but also all the evanescent modes. Finally, to combine several septums into a filter circuit, the generalized *S*-matrix technique is applied again to obtain the total scattering parameters of the composite structure.

A. Semi-Infinite Septum

Let us consider an infinitesimally thin semi-infinite septum in a waveguide inhomogeneously filled with four

Manuscript received March 5, 1982; revised April 20, 1982. This work was supported in part by the U.S. Army Research Office Contract DAAG29-81-K-0053.

Y. C. Shih and T. Itoh are with the Department of Electrical Engineering, University of Texas, Austin, TX 78712.

L. Q. Bui is with the Hughes Aircraft Company, Electron Dynamics Division, Torrance, CA 90509.

Y.C. Shih, T. Itoh, and L.Q. Bui, *IEEE Transactions on Microwave Theory and Techniques*, Vol. MTT-31, No. 2, Feb. 1983, pp. 135–142. Copyright © IEEE. Reprinted with permission.

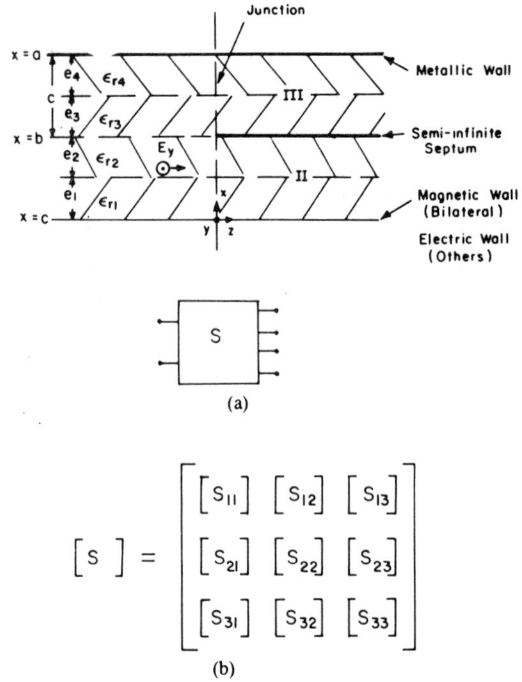

Fig. 2. Characterization of the edge of septum. (a) Geometry. (b) Matrix representation.

dielectric slabs as shown in Fig. 2(a). This is a generalized structure for which values of the thickness and dielectric constant of each layer can be chosen to represent the structures in Fig. 1(b). (Note that for the case of bilateral fin-lines, a magnetic wall is introduced at the center of the waveguide and only the region above the magnetic wall is considered. Such a region represents the upper half of the bilateral fin-line due to the symmetry of the structure.) It is then subdivided into three regions at the junction $z = 0$, with each region consisting of a uniform waveguide which possesses known eigenmodes and corresponding propagation constants.

Assuming a TE_{10}-type incident wave, the excited fields are composed of the TE_{no}-type waves since there are no structural variations in the y-direction. Therefore, the existing modal fields consist of only E_y, H_x, and H_z components. The total fields in each region are then expanded in terms of the eigenmodes. After matching the tangential field components at $z = 0$, and making use of the orthogonality of the eigenmodes, an infinite set of equations results. Since the existence of the stratified dielectric layer does not affect the edge condition of the septum, the equations are of the same form as that for a bifurcated waveguide described in [6] and the scattering parameters of the junction are obtainable by the residue-calculus technique.

The resulting scattering matrix has nine elements, each of which is also a matrix of infinite dimension. The terms in each element are of the form of an infinite product multiplied by some constant coefficient. They are summarized in the following:

$$S_{11}(m,p) = -\frac{H_p}{H_m} e^{-L(\gamma'_m+\gamma'_p)} \frac{(\gamma''_m-\gamma'_m)(\gamma'''_m-\gamma'_m)}{(\gamma''_m+\gamma'_p)(\gamma'''_m+\gamma'_p)} \cdot f(m,-\gamma'_m,\gamma'_p) \quad (1a)$$

$$S_{21}(m,p) = \frac{H_pF_m}{2\gamma''_m} e^{L(\gamma''_m-\gamma'_p)} \frac{1}{(\gamma''_m-\gamma'_p)} \cdot f(0,\gamma''_m,\gamma'_p) \quad (1b)$$

$$S_{31}(m,p) = \frac{H_pG_m}{2\gamma'''_m} e^{L(\gamma'''_m-\gamma'_p)} \frac{1}{(\gamma'''_m-\gamma'_p)} \cdot f(0,\gamma'''_m,\gamma'_p) \quad (1c)$$

$$S_{12}(m,p) = -\frac{H_mF_p}{2\gamma'_m} e^{L(\gamma''_p-\gamma'_m)} \frac{1}{(\gamma'_m-\gamma''_p)} \cdot f(0,\gamma''_p,\gamma'_m) \quad (1d)$$

$$S_{22}(m,p) = -\frac{F_p}{F_m} e^{L(\gamma''_m+\gamma''_p)} \frac{(\gamma'''_m+\gamma''_p)(\gamma'_m-\gamma''_m)}{(\gamma'''_m-\gamma''_m)(\gamma'_m+\gamma''_p)} \cdot f(m,\gamma''_p,-\gamma''_m) \quad (1e)$$

$$S_{32}(m,p) = -\frac{F_p}{G_m} e^{L(\gamma'''_m+\gamma''_p)} \frac{(\gamma''_m+\gamma''_p)(\gamma'_m-\gamma'''_m)}{(\gamma''_m-\gamma'''_m)(\gamma'_m+\gamma''_p)} \cdot f(m,\gamma''_p,-\gamma'''_m) \quad (1f)$$

$$S_{13}(m,p) = -\frac{H_mG_p}{2\gamma'_m} e^{L(\gamma'''_p-\gamma'_m)} \frac{1}{(\gamma'_m-\gamma'''_p)} \cdot f(0,\gamma'''_p,\gamma'_m) \quad (1g)$$

$$S_{23}(m,p) = -\frac{G_p}{F_m} e^{L(\gamma''_m+\gamma'''_p)} \frac{(\gamma'''_m+\gamma'''_p)(\gamma'_m-\gamma''_m)}{(\gamma'''_m-\gamma''_m)(\gamma'_m+\gamma'''_p)} \cdot f(m,\gamma'''_p,-\gamma''_m) \quad (1h)$$

$$S_{33}(m,p) = -\frac{G_p}{G_m} e^{L(\gamma'''_m+\gamma'''_p)} \frac{(\gamma''_m+\gamma'''_p)(\gamma'_m-\gamma'''_m)}{(\gamma''_m-\gamma'''_m)(\gamma'_m-\gamma'''_p)} \cdot f(m,\gamma'''_p,-\gamma'''_m) \quad (1i)$$

where

$$L = \frac{b}{\pi}\ln\left(\frac{a}{b}\right) + \frac{c}{\pi}\ln\left(\frac{a}{c}\right)$$

$$f(m,\gamma_1,\gamma_2) = \prod_{\substack{n=1 \\ n\neq m}}^{\infty} \frac{(\gamma'_n+\gamma_2)(\gamma''_n+\gamma_1)(\gamma'''_n+\gamma_1)}{(\gamma'_n+\gamma_1)(\gamma''_n+\gamma_2)(\gamma'''_n+\gamma_2)}$$

where γ'_i, γ''_i, and γ'''_i are the propagation constants of the ith mode in Regions I, II, and III, respectively. F_i, G_i, and H_i are parameters determined readily for a given structure (see Appendix I). The notation $S_{ij}(m,p)$ stands for the scattering coefficient of the mth order mode.

B. Septum with Finite Length

With the knowledge of the scattering parameters for a single junction, the generalized scattering matrix technique is applied to obtain a two-port scattering matrix as

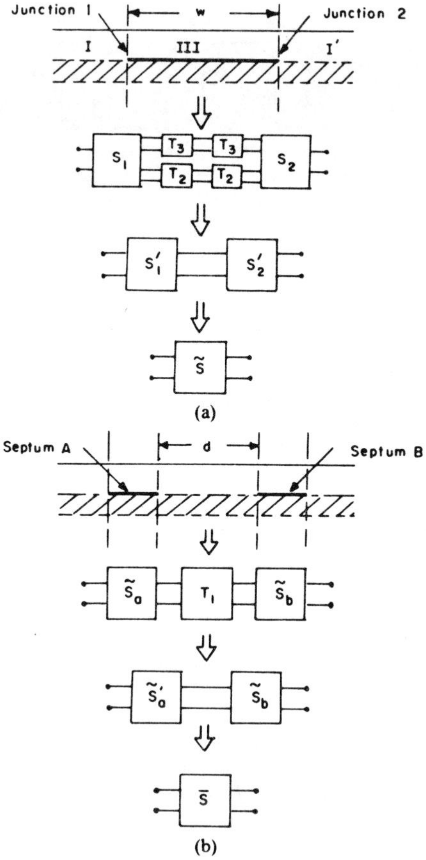

Fig. 3. S-parameter derivation. (a) For a single septum. (b) Cascaded septum.

shown in Fig. 3(a). Suppose now that the TE-type wave from Region I is incident upon Junction 1. At this junction, fields are reflected back into Region I, and transmitted into Regions II and III. After traveling a distance w, part of the wave transmitted into Regions II and III is reflected back and part is transmitted into Region I′ at Junction 2. This process continues until the intensity of the reflected wave dies out. The multiple-reflection phenomenon between Junctions 1 and 2 is implied in a matrix manipulation which yields the scattering matrix for a finite length septum.

Referring to Fig. 3(a), S_1 and S_2 represent the scattering matrices for isolated Junctions 1 and 2 (i.e., for semi-infinite septums), respectively. From (1), we obtain the value of the elements in both matrices. Since the characteristics of these two junctions are essentially the same except for the opposite orientation, we have

$$S_2 = S_1^T. \tag{2}$$

We now define the transmission matrix

$$T = \begin{bmatrix} [I] & [0] & [0] \\ [0] & [T_2] & [0] \\ [0] & [0] & [T_3] \end{bmatrix} \tag{3}$$

where $[I]$ and $[0]$ are the identity matrix and the zero matrix, respectively, and the matrices $[T_2]$ and $[T_3]$ are of infinite size and represent the wave propagating (for propagation modes) or attenuating (for evanescent modes) for a distance of $w/2$ in guided Regions II and III, respectively. The combination of the transmission matrix along with S_1 and S_2 results in S'_1 and S'_2

$$S'_1 = TS_1T \tag{4}$$

$$S'_2 = TS_2T = S_1'^T. \tag{5}$$

A further combination of S'_1 and S'_2 then yields the composite scattering matrix $\tilde{S}$ for the septum of length w, viewed from I at $z = 0$ and from I' at $z = w$.

C. Filter Structures

For the final step of the analysis, several finite length septums are cascaded with appropriate separations to form a filter circuit. This is done by using the same techniques used in Section II-B described above. The procedure depicted in Fig. 3(b) shows the cascading of two septums represented by $\tilde{S}_a$ and $\tilde{S}_b$. The combination of $\tilde{S}_a$ and the transmission matrix T_1 results in matrix $\tilde{S}'_a$ which is further combined with $\tilde{S}_b$ to yield $\bar{S}$, the total scattering matrix of two septums with a separation of d. This procedure may be repeated to obtain the scattering parameters of a filter consisting of any number of septums.

III. Design

Based on the above analysis, a computer program is developed to calculate the total scattering parameters for a filter structure with given design parameters. Although the analysis is formally exact, in practice the infinite product in (1) and the infinite dimension of the elements in the scattering matrices need to be truncated. To obtain a very accurate result for (1), at least 300 terms should be retained. The calculation of the products here is efficient, since only readily known propagation constants are involved. Furthermore, this equation needs to be evaluated only once for a design assignment at each frequency point, since the scattering parameters of the semi-infinite septum depend only on the waveguide size and dielectric properties which are constant factors in a design.

For most applications, the fundamental mode is below cutoff in the narrow waveguide section and all of the higher order modes are evanescent in the wide waveguide. Therefore, only the first few accessible modes [8] are essentially responsible for the coupling between resonators. For most cases, only two or three modes are required to describe accurate scattering behavior. Notice that the number of modes considered here indicates the dimension of the matrices to be manipulated. The calculation includes a number of complex matrix multiplications and inversions. This analysis is efficient mainly due to the fact that only two or three modes are required.

Based on the analysis program, an optimization computer program applying the quasi-Newton method [9] is used which varies the input parameters until a desired value of the insertion loss for a given bandwidth is obtained. To do this, an error function $F(\bar{x})$ to be minimized is defined

$$F(\bar{x}) = \sum_{N_s} \left(\frac{L_s}{|S_{21}|} \right)^2 + \sum_{N_p} \left(\frac{|S_{21}|}{L_s} \right)^2 \tag{6}$$

where N_s and N_p are the number of frequency sampling points in the stopband and passband, respectively, L_s and L_p are the minimum stopband and maximum passband attenuation levels, respectively, and $|S_{21}|$ is the calculated value of the filter attenuation. For a given thickness of the dielectric, the parameters $\bar{x}$ to be optimized are the septum lengths and resonator lengths.

To prevent the program from approaching nonrealizable or nonphysical results, upper and lower bounds of each parameter are specified

$$x_{li} < x_i < x_{ui}. \tag{7}$$

The constrained parameters are then reduced into essentially unconstrained ones by the transformation [10]

$$x_i = x_{li} + \frac{1}{\pi}(x_{ui} - x_{li})\cot^{-1} x_i' \tag{8}$$

where $-\infty < x_i' < \infty$, but only solutions within the range

$$0 < \cot^{-1} x_i' < \pi \tag{9}$$

are allowed. This transformation has a penalizing effect upon the parameters in the vicinity of the upper and lower bounds.

For the optimization procedure, the first two modes in each subregion are considered; the final design values are calculated with three modes. The total computing time for optimizing a five-resonator filter was about 2 min on a CDC Dual Cyber 170/750.

IV. Results and Discussion

In an earlier publication [11], it has been pointed out that the Q-factor of insulated fin-lines is inferior to that of bilateral fin-lines. Therefore, in this study, bandpass filters have been designed using only bilateral and unilateral fin-line configurations. Some typical designs for Ka-band applications are shown in Figs. 4, 6, and 7 fabricated on RT/Duroid substrates. Notice that the resonators are very weakly coupled by wide septums. In such cases, the septum cannot be represented by simple shunt-inductance equivalent circuits. Instead, it has to be represented by an equivalent T network [1]. Since the equivalent circuits are useful in conventional circuit design and are helpful for good initial guesses in the optimization procedure, this topic is discussed briefly in Appendix II.

The five-resonator filter design in Fig. 4 is a bilateral fin-line in which the hatched portion indicates metal on both sides of the substrate. The center frequency is 38.85 GHz, the 0.1-dB ripple bandwidth is 1.1 GHz, and the skirt selectivity at 500 MHz away from either edge of the passband is −40 dB. Measured characteristics of this filter are shown in Fig. 5. Computed responses are also plotted for comparison. Shapes of the calculated and measured

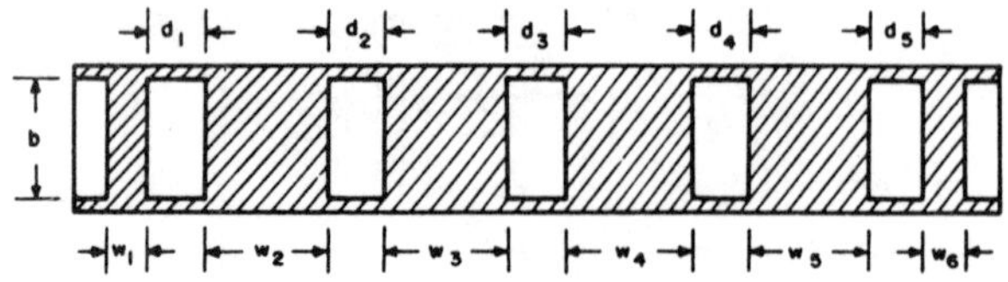

Resonator lengths: Septum widths:

$d_1 = d_5 = 2.6594$ mm $w_1 = w_6 = 1.6749$ mm

$d_2 = d_4 = 2.6205$ mm $w_2 = w_5 = 5.2786$ mm

$d_3 = 2.6195$ mm $w_3 = w_4 = 5.8584$ mm

Substrate: RT/Duroid Waveguide: Height b = 3.556 mm

$\varepsilon_r = 2.2$ Width a = 7.112 mm

Thickness = 10 mil

Fig. 4. Design example (five-resonator Ka-band bilateral fin-line filter).

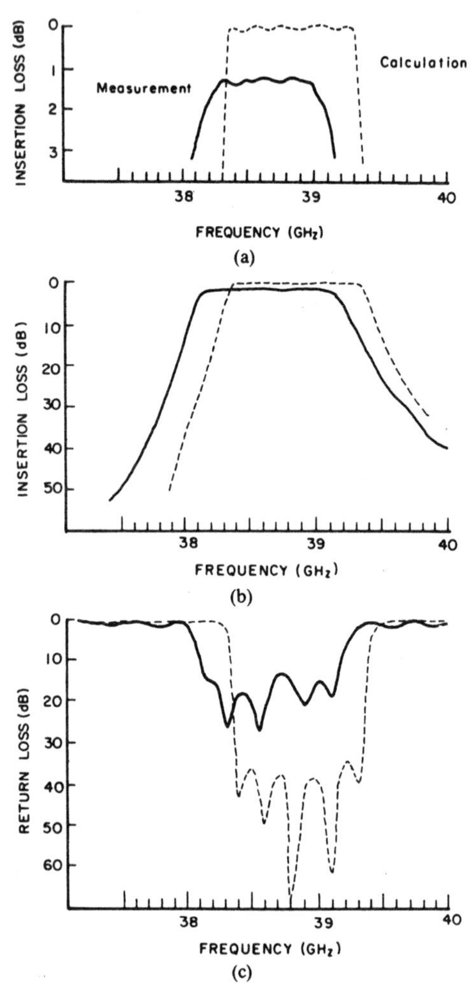

Fig. 5. Frequency responses for the filter in Fig. 4. (a) Expanded view of insertion loss. (b) Insertion loss characteristics. (c) Return loss characteristics.

response curves agree extremely well. Measured insertion loss in the passband is less than 1.3 dB of which 0.3 dB is associated with the test fixture. The measured center

Fig. 6. Three-resonator *Ka*-band bilateral fin-line filter. Resonator length: $d_1 = d_3 = 2.6559$ mm; $d_2 = 2.6187$ mm. Septum width: $w_1 = w_4 = 1.7230$ mm; $w_2 = w_3 = 5.3628$ mm.

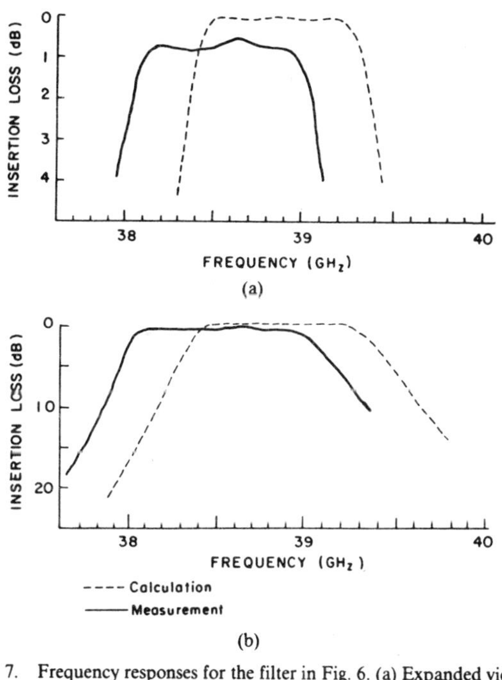

Fig. 7. Frequency responses for the filter in Fig. 6. (a) Expanded view of insertion loss. (b) Insertion loss characteristics. --- Calculation. —— Measurement.

frequency is shifted to the lower side by about 170 MHz which is less than 0.5 percent of the center frequency.

Fig. 6 shows a photograph of a three-resonator *Ka*-band bilateral fin-line filter printed on a 10-mil-thick RT/Duroid substrate. The design specifications for this filter were the same as those for the last one except for a decreased skirt selectivity of −20 dB. The measured results in Fig. 7 show that the insertion loss is less than 0.8 dB in the passband, including the fixture loss. The calculated and measured frequency responses again agree well; however, the measured center frequency is shifted down by about 300 MHz.

Bandpass filters have also been developed in unilateral fin-line configurations. A typical design for a three-resona-

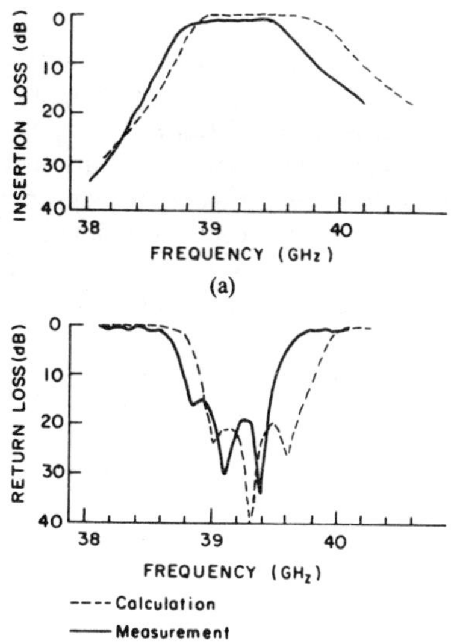

Fig. 8. Three-resonator *Ka*-band unilateral fin-line filter. --- Calculation. —— Measurement.

tor filter has the following specifications and dimensions:

Resonator lengths	$d_1 = d_3 = 2.1568$ mm, $d_2 = 2.0879$ mm;
Septum widths	$w_1 = w_4 = 2.3325$ mm, $w_2 = w_3 = 6.8718$ mm;
Passband	39.0 ~ 39.7 GHz, Chebyshev response with 0.1-dB ripple; and
Skirt selectivity	−20 dB at 500 MHz away from passband edge.

The measured results in Fig. 8 show about 2-dB insertion loss in the passband and about 200-MHz downward shift for the center frequency.

In the experiments performed, the measured center frequencies are always lower than the calculated values. The deviations range from 170 MHz to 300 MHz. One cause of this shift is due to the way the filter is assembled. To support the substrate in the waveguide, 10-mil-deep grooves are machined into the top and bottom walls. An experimental study on this effect shows that within the quarter-wavelength range, the deeper the groove is made, the more the center frequency will shift down. In addition, another possible source of the discrepancy is the finite thickness of the metal fin. The thickness effect is minor at this frequency since the thickness of the copper cladding on the substrate is about 0.5 mil. However, it will become an important factor for higher frequency applications, such as *W*-band circuits. These effects are currently under study.

From the experiments, it has been observed that the insertion loss of a unilateral fin-line filter is always greater than that of a bilateral filter for similar design specifications. One possible explanation is that the dielectric loss of the unilateral filter is greater, while the conductor loss is about the same in both cases. In the case of bilateral fin-lines, little energy can penetrate into the very narrow space between the strips. Therefore, the energy is mainly coupled through the wider empty waveguide section. In the unilateral case, the energy is coupled about evenly through the empty guide and the partially dielectric-filled guide region. Obviously, more dielectric loss is encountered in this case. By the same reasoning we can surmise that the conductor loss is about the same for both cases. Although the bilateral structure has copper cladding on both sides of the substrate, there is essentially no current flowing on the inner surface facing the dielectric. It is thus equivalent to a conducting strip with current flowing on the outer surface of both sides as in the unilateral case. Of course, these statements are strictly conjectural; further study is needed to verify them. This conjecture would also be applicable to the case of insulated fin-lines versus bilateral fin-lines.

Another point worth mentioning is the sensitivity of the filter structure to mechanical tolerances. Calculations show that for both cases, the center frequency shifts about 100 MHz if the fabrication error in resonator length is 1 mil. Also, if the dielectric substrate is shifted in the broadwall direction (x-direction) from its centered position, the filter's center frequency falls. This effect is much more noticeable with the unilateral structure. Therefore, the bilateral structure shows more promise from a constructional viewpoint.

V. Conclusions

We presented an efficient CAD algorithm based on an exact analysis for a class of E-plane bandpass filters. Filters for Ka-band applications have been fabricated based on the present design. Measured characteristics agree well with theoretical data. Extensive experiments show that the bilateral construction is more promising than the unilateral structure.

Appendix I
Expressions for H_n, F_n, and G_n

The coefficients H_n, F_n, and G_n are related to the structure geometry and the field distribution in Regions I, II, and III, respectively. They are derived from the overlapping integrals

$$\int_0^b \phi_m^{\mathrm{I}}(x)\phi_n^{\mathrm{II}}(x)\,dx = \frac{H_m F_n}{\gamma_n''^2 - \gamma_m'^2} \tag{A1}$$

$$\int_b^a \phi_m^{\mathrm{I}}(x)\phi_n^{\mathrm{III}}(x)\,dx = \frac{H_m G_n}{\gamma_n'''^2 - \gamma_m'^2} \tag{A2}$$

where $\phi_i^{\mathrm{I}}(x)$, $\phi_i^{\mathrm{II}}(x)$, and $\phi_i^{\mathrm{III}}(x)$ are the ith order orthonormal eigenfunctions for the E_y fields in Regions I, II, and III, respectively (refer to Fig. 2). The expressions for H_i, F_i, and G_i of different geometries are given in the following.

Metallic E-Plane Structure

$$H_n = \sqrt{\frac{2}{a}}\sin\frac{n\pi b}{a} \tag{A3a}$$

$$F_n = (-1)^n\sqrt{\frac{2}{b}}\,\frac{n\pi}{b} \tag{A3b}$$

$$G_n = (-1)^n\sqrt{\frac{2}{c}}\,\frac{n\pi}{c}. \tag{A3c}$$

Bilateral Fin-Line Structures

$$H_n = A_n\cos\xi_n b \tag{A4a}$$

$$F_n = (-1)^n\sqrt{\frac{2}{b}}\,\frac{\left(n-\frac{1}{2}\right)\pi}{b} \tag{A4b}$$

$$G_n = (-1)^n\sqrt{\frac{2}{c}}\,\frac{n\pi}{c} \tag{A4c}$$

where

$$A_n = \sqrt{2}\left[b + \frac{\sin\xi_n b\cos\xi_n b}{\xi_n} + \left(c - \frac{\sin\eta_n c\cos\eta_n c}{\eta_n}\right)\left(\frac{\cos\xi_n b}{\sin\eta_n c}\right)^2\right]^{-1/2}$$

and where ξ_n and η_n are x-components of nth order wavenumber in the dielectric and the air region, respectively.

Unilateral Fin-Line Structures

$$H_n = A_n C_n \sin\eta_n c \tag{A5a}$$

$$F_n = -B_n\xi_n'\frac{\sin\eta_n' e_1}{\sin\xi_n' e_2} \tag{A5b}$$

$$G_n = (-1)^n\sqrt{\frac{2}{c}}\cdot\frac{n\pi}{c} \tag{A5c}$$

where

$$A_n = \sqrt{2}\left[d - \frac{\sin\eta_n e_1\cos\eta_n e_1}{\eta_n} + \eta_n^2 e_2\left(\frac{\sin^2\eta_n e_1}{\eta_n^2} + \frac{\cos^2\eta_n e_1}{\xi_n^2}\right) + C_n^2\left(c - \frac{\sin\eta_n c\cos\eta_n c}{\eta_n}\right) + C_n\frac{\eta_n^2\sin\xi_n e_2}{\xi_n}\cdot\left(\frac{\sin\eta_n e_1\sin\eta_n c}{\eta_n^2} + \frac{\cos\eta_n e_1\cos\eta_n c}{\xi_n^2}\right)\right]^{-1/2}$$

$$C_n = \frac{\eta_n}{\sin\eta_n c}\left(\frac{\sin\eta_n e_1\cos\xi_n e_2}{\eta_n} + \frac{\sin\xi_n e_2\cos\eta_n e_1}{\xi_n}\right)$$

$$B_n = \sqrt{2}\left[e_1 - \frac{\sin\eta_n' e_1\cos\eta_n' e_1}{\eta_n'} + \left(e_2 - \frac{\sin\xi_n' e_2\cos\xi_n' e_2}{\xi_n'}\right)\left(\frac{\sin\eta_n' e_1}{\sin\xi_n' e_2}\right)^2\right]^{-1/2}$$

and where $\xi_n(\xi_n')$ and $\eta_n(\eta_n')$ are the nth order wavenumber of dielectric (air) part in Regions I and II, respectively.

Appendix II
Equivalent Circuit of a Strip

The equivalent T network for a finite-length septum is shown in Fig. 9. The values of the normalized reactance x_s and x_p are to be calculated from the scattering parameters

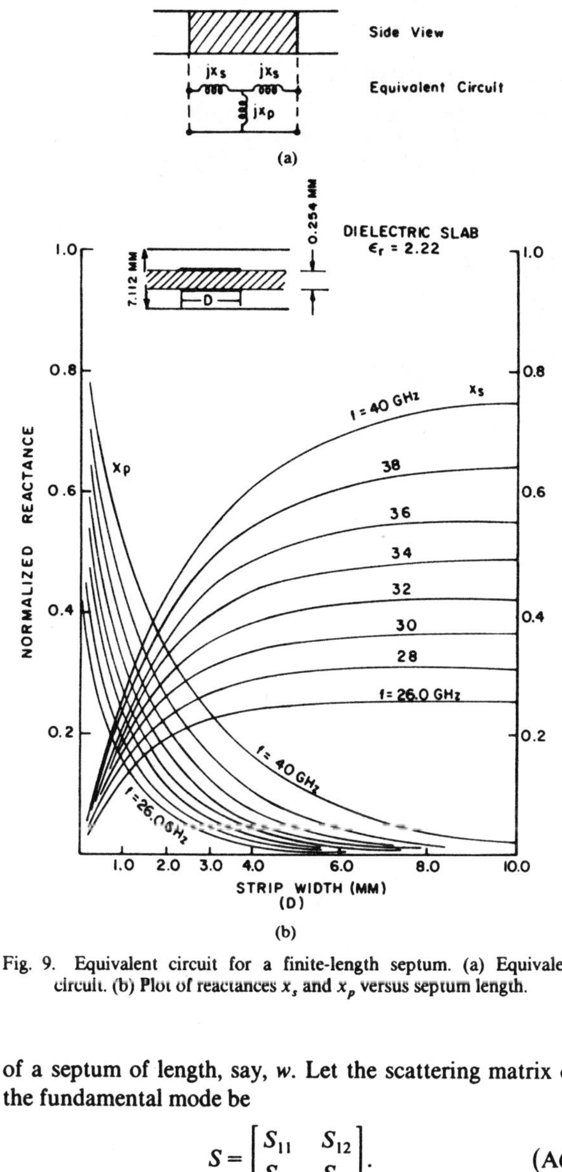

Fig. 9. Equivalent circuit for a finite-length septum. (a) Equivalent circuit. (b) Plot of reactances x_s and x_p versus septum length.

of a septum of length, say, w. Let the scattering matrix of the fundamental mode be

$$S=\begin{bmatrix} S_{11} & S_{12} \\ S_{21} & S_{22} \end{bmatrix}. \tag{A6}$$

For this reciprocal lossless two-port device, the following relations hold:

$$S_{12}=S_{21} \tag{A7}$$

$$S_{11}=S_{22} \tag{A8}$$

$$|S_{11}|^2+|S_{22}|^2=1. \tag{A9}$$

These relations provide useful error-checking criteria for the numerical calculation.

If we convert the scattering matrix into impedance matrix and equate it to the impedance matrix of the equivalent T network, we have

$$[\bar{Z}]=[I+S][I-S]^{-1}=\begin{bmatrix} jx_s+jx_p & jx_p \\ jx_p & jx_s+jx_p \end{bmatrix} \tag{A10}$$

where I is the identity matrix. By equating the elements in the matrix on both sides, x_s and x_p are uniquely determined

$$jx_p=\frac{2S_{12}}{(1-S_{11})^2-S_{12}^2} \tag{A11}$$

$$jx_s=\frac{1-S_{12}+S_{11}}{1-S_{11}+S_{12}}. \tag{A12}$$

Fig. 9 shows the calculated values for x_s and x_p as a function of septum length w with frequency as parameters.

References

[1] Y. Konishi and K. Uenakada, "The design of a bandpass filter with inductive strip-planar circuit mounted in waveguide," *IEEE Trans. Microwave Theory Tech.*, vol. MTT-22, pp. 869–873, Oct. 1974.

[2] Y. Tajima and Y. Sawayama, "Design and analysis of a waveguide-sandwich microwave filter," *IEEE Trans. Microwave Theory Tech.*, vol. MTT-22, pp. 839–841, Sept. 1974.

[3] P. J. Meier, "Integrated fin-line millimeter components," *IEEE Trans. Microwave Theory Tech.*, vol. MTT-22, pp. 1209–1216, Dec. 1974.

[4] A. M. K. Saad and K. Schünemann, "Design and performance of fin-line bandpass filters," *9th Eur. Microwave Conf.*, (Brighton, England), Sept. 17–20, 1979.

[5] F. Arndt *et al.*, "Low-insertion-loss fin-line filters for millimeter-wave applications," *11th Eur. Microwave Conf.* (Amsterdam, Netherlands), Sept. 7–10, 1981, pp. 309–314.

[6] R. Mittra and S. W. Lee, *Analytical Techniques in the Theory of Guided Waves.* New York: Macmillan, 1971, pp. 45–54.

[7] C. F. Vanblaricum, Jr. and R. Mittra, "A modified residue-calculus technique for solving a class of boundary value problems—Part II: Waveguide phased arrays, modulated surfaces, and diffraction gratings," *IEEE Trans. Microwave Theory Tech.*, vol. MTT-17, pp. 310–319, June 1969.

[8] T. E. Rozzi and W. F. G. Mecklenbräuker, "Wide-band network modeling of interacting inductive irises and steps," *IEEE Trans. Microwave Theory Tech.*, vol. MTT-23, pp. 235–245, Feb. 1975.

[9] Routin ZXMIN, IMSL Library 3 (for CDC Cyber 70/6000/7000 series). See IMSL Library Manual.

[10] J. W. Bandler, "Optimization method for computer-aided design," *IEEE Trans. Microwave Theory Tech.*, vol. MTT 17, pp. 533–562, Aug. 1969.

[11] P. J. Meier, "Equivalent relative permittivity and unloaded Q factor of integrated fin-line," *Electron. Lett.*, vol. 9, no. 7, pp. 162–163, Apr. 1973.

6.2.3

F. Arndt, J. Bornemann, D. Grauerholz & R. Vahldieck
Microwave Department, University of Bremen, Germany

Material and Circuit Evaluation for Millimetre-Wave Applications: High-Q Fin-Line Filters for Millimetre Waves*

Abstract A new design theory for high-*Q* fin-line filters is described, including higher order mode propagation. These filters are well suited for use in millimetre-wave integrated circuits. The design data for two fin-line filters are given for mid-band frequencies of approximately 15 and 34 GHz. The measured insertion losses in the passband are only 0.25 and 0.5 dB, respectively, for these two frequencies.

Résumé Cet article expose une nouvelle théorie pour le calcul des filtres à ligne nageoire à coefficient Q élevé, prenant en compte la propagation des modes d'ordre supérieur. Ces filtres conviennent bien aux circuits intégrés en onde millimétrique. On présente les caractéristiques de conception de deux filtres pour des fréquences centrées aux environs de 15 et 34 GHz; les pertes par insertion respectives, mesurées dans la bande passante, ne dépassent pas 0.25 et 0.5 dB aux deux fréquences en question.

*This article, provided via ERNO Raumfahrttechnik GmbH, is based on work conducted under ESA ESTEC Contract No. 4044/79/NL DG(SC).

F. Arndt, J. Bornemann, D. Grauerholz, and R. Vahldieck, *European Space Agency Journal*, Vol. 5, No. 1, 1981, pp. 33–42.

Introduction

In a fin-line structure[1], metal inserts ('fins') are printed on a dielectric substrate which bridges the broad walls of a rectangular waveguide (Fig. 1). This permits circuit elements to be photoetched by low-cost techniques. Because the dimensions of practical fin-line circuits remain compatible with chip and beam-lead devices, fin-line structures are very attractive when designing millimetre-wave integrated circuits.

Very many components, both active and passive, have already been made using this technique[2–5], but there still exists a paucity of suitable theoretical research on fin-line structures. So far only experimental design data[2–5] and first-order design theories[6–7] are available, based on the assumption of fundamental modes in the fin-lines only. For discontinuities in the slot width g (Fig. 1), as is the case in filter structures, however, this first-order theory leads to incorrect results because, as has been shown by our own measurements, higher order modes have an important influence on the transmission and reflection coefficients.

The measured Q (including the effect of losses within the irises) of a grounded fin-line increases with the width g of the gap between the fins. The optimal value is reached[2] when the normalised gap width g/b is 1. High-Q fin-line filters can therefore be designed (Fig. 2) consisting of a grounded fin-line structure with alternating metallic bridges and gaps over the total height of the rectangular waveguide. No design theory has been available until now for such high-Q filters.

The purpose of this paper is therefore to introduce such a theory, which will take into account the higher order mode propagation. The measured frequency response of the fin-line filters that have been designed and operated at about 15 and 34 GHz shows good agreement between theory and practical results. The high-Q design leads to losses of typically only 0.25 and 0.5 dB, respectively, at these two frequencies.

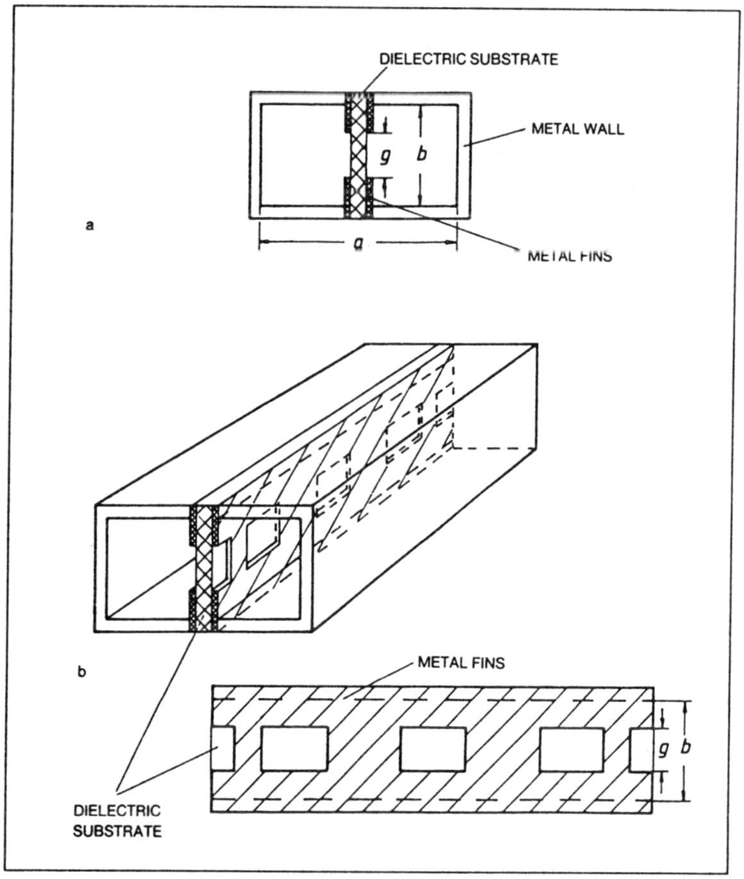

Figure 1. General fin-line filter configuration
a. grounded fins
b. bandpass structure

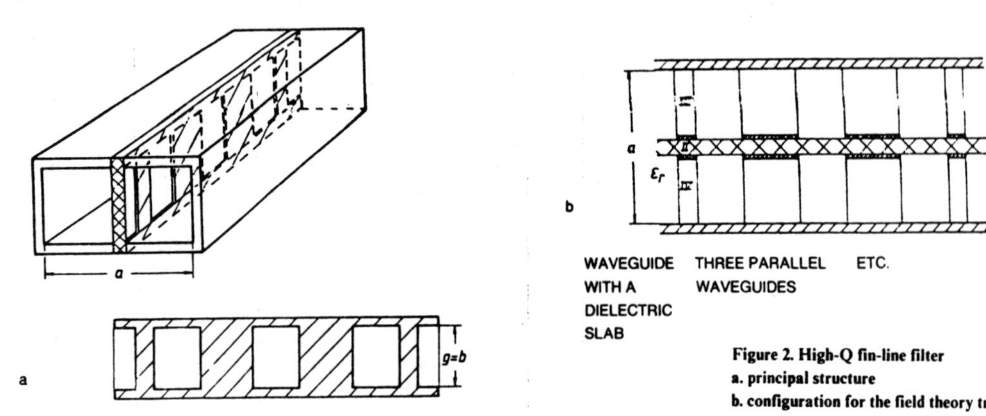

Figure 2. High-Q fin-line filter
a. principal structure
b. configuration for the field theory treatment

Design theory

Since the theory is based on the exact-field treatment of the structure shown in Figure 2, the high-Q fin-line filter is regarded as consisting of alternating waveguide structures (Fig. 2b): a waveguide with a dielectric slab perpendicular to the ground wall and three parallel waveguides, the middle one of which is filled with dielectric. This implies the calculation of the scattering coefficients of the two waveguide transitions shown in Figure 3. The scattering matrix of the total fin-line filter structure can then be obtained by suitably combining the transitions, the length of waveguide I being reduced to zero if structures a and b (or inverse) are joined together directly.

Waveguide transition to three waveguides

The wave incident in I (Fig. 3a) is assumed to consist of h_{mo}-modes; there is no discontinuity in the y-direction, so that the only field components in the v regions (v = I, II, III, IV) are E_y^v, H_x^v, H_z^v. The waves are calculated by the method of field expansion into eigenmodes:

$$\begin{bmatrix} E_y^v \\ \\ H_x^v \\ \\ H_z^v \end{bmatrix} = \begin{bmatrix} -\omega\mu \sum\limits_m k_{zm}^v A_m^v \sin \dfrac{m\pi}{p^v} \cdot f^v \\ \\ \sum\limits_m k_{zm}^{v\,2} A_m^v \sin \dfrac{m\pi}{p^v} \cdot f^v \\ \\ -j \sum\limits_m k_{zm}^{v\,2} A_m^v \cos \dfrac{m\pi}{p^v} \cdot f^v \end{bmatrix} \cdot e^{-jk_{zm}^v \cdot z} \tag{1}$$

Figure 3. Waveguide transitions
a. transition to three parallel waveguides
b. transition to a waveguide with a dielectric slab
c. wave amplitude vectors of the corresponding structures of finite length

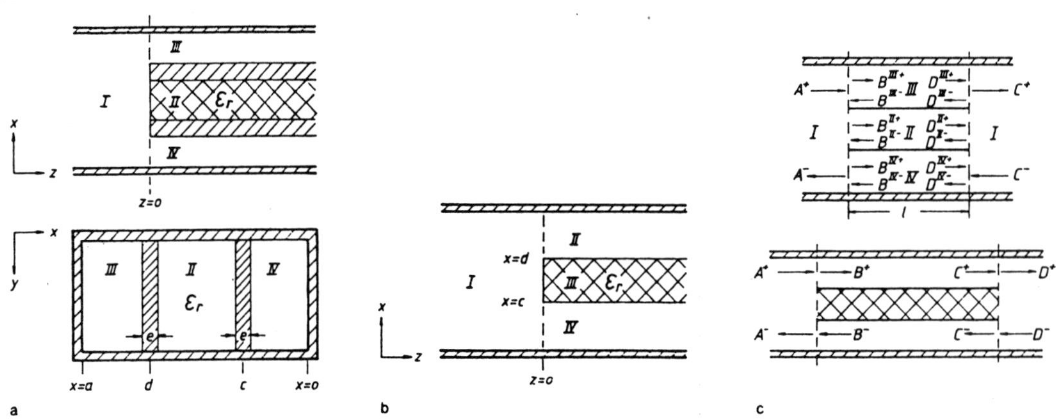

with the abbreviations (Fig. 3a)

$$\begin{bmatrix} f^{I} \\ f^{II} \\ f^{III} \\ f^{IV} \end{bmatrix} = \begin{bmatrix} x \\ d - \frac{e}{2} - x \\ a - x \\ c - \frac{e}{2} - x \end{bmatrix}$$

$$\begin{bmatrix} p^{I} \\ p^{II} \\ p^{III} \\ p^{IV} \end{bmatrix} = \begin{bmatrix} a \\ d - e - c \\ a - d + \frac{e}{2} \\ c - \frac{e}{2} \end{bmatrix}$$

$$\begin{bmatrix} k_{zm}^{I^2} \\ k_{zm}^{II^2} \\ k_{zm}^{III^2} \\ k_{zm}^{IV^2} \end{bmatrix} = \begin{bmatrix} k_0^2 - \left(\frac{m\pi}{a}\right)^2 \\ k_0^2 \varepsilon_r - \left(\frac{m\pi}{p^{II}}\right)^2 \\ k_0^2 - \left(\frac{m\pi}{p^{III}}\right)^2 \\ k_0^2 - \left(\frac{m\pi}{p^{IV}}\right)^2 \end{bmatrix}$$

$$k_0^2 = \mu_0 \varepsilon_0$$

At $z = 0$ the waves are coupled by matching the various tangential field components at the interface. This leads to two equations for the amplitudes $(A)^+$, $(B)^+$ of the forward waves and $(A)^-$, $(B)^-$ of the backward waves (Fig. 3c):

$$\begin{aligned} &(A)^+ + (A)^- = (N)^v((B)^{v+} + (B)^{v-}) \qquad v = \text{II, III, IV} \\ &(A)^+ - (A)^- = \sum_{v=\text{II}}^{\text{IV}} (L)^v((B)^{v+} - (B)^{v-}) \end{aligned} \tag{2}$$

The terms $(N)^v$ and $(L)^v$ are elucidated in Appendix 1.

The scattering matrix (S) of the structure with finite length l (Fig. 3c) is given by

$$\begin{pmatrix} (A)^- \\ (C)^+ \end{pmatrix} = \begin{pmatrix} (S_{11}) & (S_{12}) \\ (S_{21}) & (S_{22}) \end{pmatrix} \begin{pmatrix} (A)^+ \\ (C)^- \end{pmatrix} \tag{3}$$

with

$$\begin{aligned} (S_{11}) = (S_{22}) = (W)^{-1} \{ &[(\pi) + (E)]^{-1}(\phi)[(\pi)+(E)]^{-1}(\phi) - [(\pi)+(E)]^{-1} \\ &[(\pi)-(E)]\} \\ (S_{12}) = (S_{21}) = (W)^{-1} \{ &[(\pi)+(E)]^{-1}(\phi) - [(\pi)+(E)]^{-1}(\phi)[(\pi)+(E)]^{-1} \\ &[(\pi)-(E)]\} \end{aligned} \tag{4}$$

The abbreviations are explained in Appendix 2.

Transition to a waveguide with a dielectric slab

For this structure (Fig. 3b), the waves in regions II, III and IV propagate with a common propagation factor k_{zm} which is determined by the boundary conditions along the dielectric slab:

$$\left. \begin{array}{l} E_y^{II} = E_y^{IV} \\ H_z^{II} = H_z^{IV} \end{array} \right|_{x=c} \qquad \left. \begin{array}{l} E_y^{II} = E_y^{III} \\ H_z^{II} = H_z^{III} \end{array} \right|_{x=d} \tag{5}$$

With Equation 1, a system of linear equations is obtained in which the determinant is required to be zero. This leads to a transcendent equation for the propagation factor k_{zm} which is solved numerically.

The scattering matrix (S) of the structure of finite length l (Fig. 3c) is given by

$$\begin{pmatrix} (A)^- \\ (D)^+ \end{pmatrix} = \begin{pmatrix} (S_{11}) & (S_{12}) \\ (S_{21}) & (S_{22}) \end{pmatrix} \begin{pmatrix} (A)^+ \\ (D)^- \end{pmatrix} \tag{6}$$

with

$$(S_{11}) = (S_{22}) = (Q_{11}) + (Q_{12})(R)(W)(Q_{22})(R)(Q_{21})$$

$$(S_{12}) = (S_{21}) = (Q_{12})(R)(W)(Q_{21})$$

The abbreviations are explained in Appendix 3.

For the computer calculation of the total scattering matrix, the expansion into twelve eigenmodes in each region (see Equation 1) has proved to be sufficient.

Results

Figure 4 shows the calculated insertion loss (scattering coefficient S_{21}) in dB as a function of frequency for a nine-step fin-line filter (four metal inserts), designed for a mid-band frequency of 14.26 GHz, mounted into a P-band waveguide ($a = 15.699$ mm, $b = 7.899$ mm). The dielectric substrate is RT/duroid 5880 with an $\varepsilon_r = 2.22$.

The design process is started with filter dimensions given by Saad & Schünemann[6,8] for the narrow slot filter ($g \ll b$, cf. Fig. 1), which exhibits a very poor measured insertion loss of more than about 2 dB. The total scattering matrix of the fin-line filter can be calculated for given parameters. An optimising computer program is used[9], which varies the input parameters until the desired

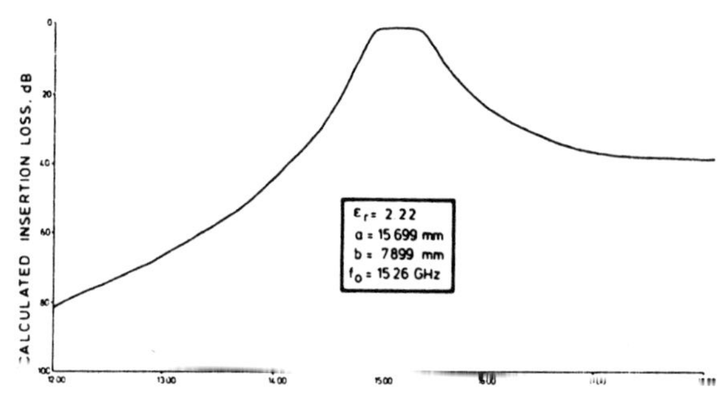

Figure 4. P-band fin-line filter. Calculated insertion loss as a function of frequency

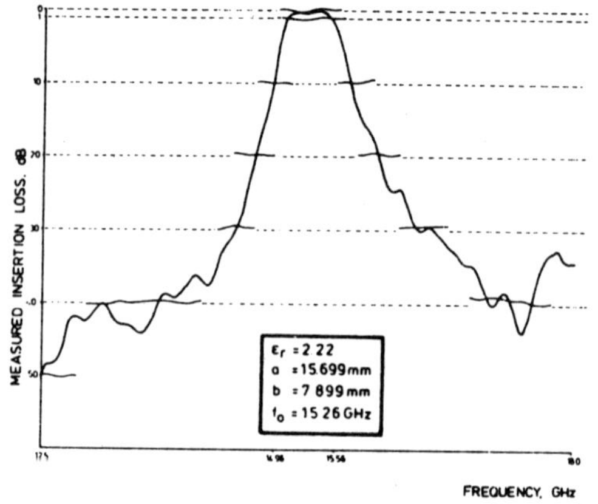

Figure 5. P-band fin-line filter. Measured insertion loss as a function of frequency

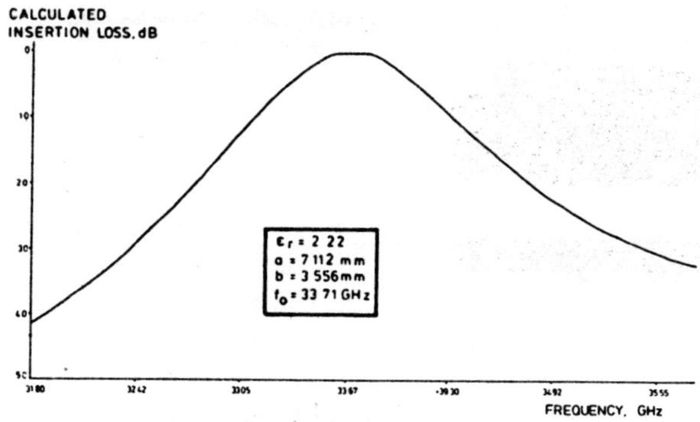

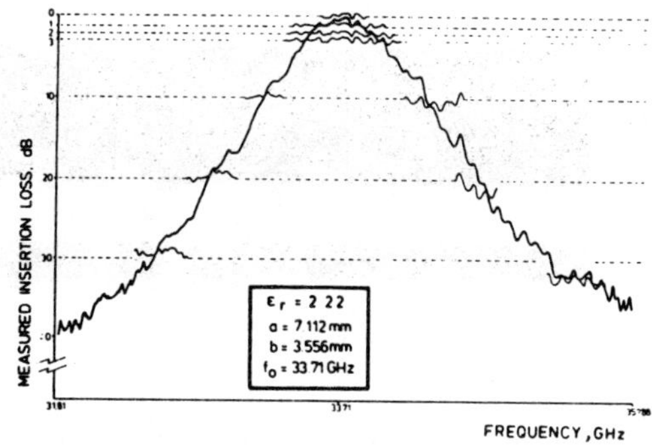

Figure 6. R-band fin-line filter. Calculated insertion loss as a function of frequency

Figure 7. R-band fin-line filter. Measured insertion loss as a function of frequency

insertion loss is obtained for a given bandwidth. The theoretical insertion losses for the 15.26 GHz and 33.71 GHz fin-line filters are 0.05 dB, and 0.10 dB, respectively, at mid-band frequency. The dimensions of the fin-line steps are given in Table 1. Figure 5 shows the measured insertion loss to be in good agreement with the calculated curve. The typical measured insertion loss in the passband is only about 0.25 dB.

Figure 6 shows the calculated insertion loss for a corresponding fin-line filter designed for 33.71 GHz (R-band waveguide, $a = 7.112$ mm, $b = 3.556$ mm). The dimensions of the fin-line steps are given in Table 1. The calculation also agrees well with the measured curve (Fig. 7). The typical measured insertion loss in the passband is only about 0.5 dB. The two photoetched fin-line structures are shown in Figure 8.

Table 1

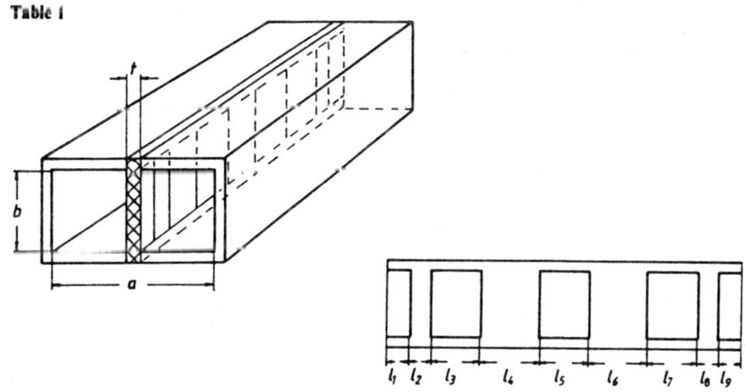

Fin-Line Filter	Dimensions* in mm a	b	$l_1 = l_9$	$l_2 = l_8$	$l_3 = l_7$	$l_4 = l_6$	l_5
P-Band f_0 = 15.26 GHz	15.699	7.899	27.432	2.1	8.376	7.8	8.384
R-Band f_0 = 33.71 GHz	7.112	3.556	39.26	0.72	3.734	3.918	3.736

Dielectric Material: RT/Duroid 5880 $\varepsilon_r = 2.22$
Thickness t = 0.254 mm

* The l-dimensions do not include the etching error. After etching l_2, l_4, l_6 and l_8 are reduced by 0.06 mm.

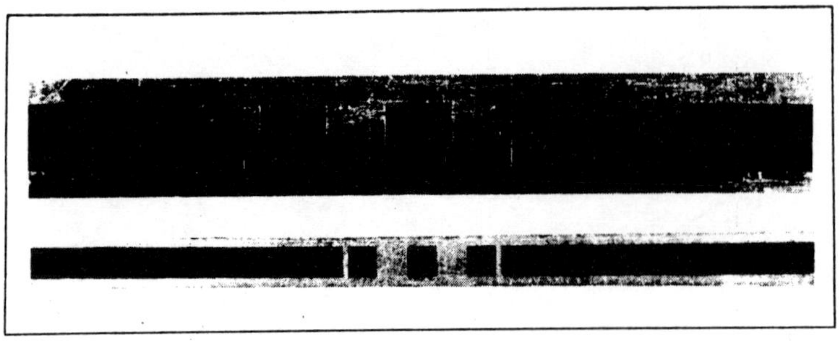

Figure 8. Two photoetched fin-line structures (for 15 GHz and 33.7 GHz)

Conclusion

A novel design theory has been shown to lead to high-Q fin-line filters suitable for millimetre waves. They consist of alternating metallic bridges and gaps on a dielectric slab mounted between the broad walls of a rectangular waveguide. The theory is based on field expansion into eigenmodes within the subregion. By matching the various tangential field components, linear-equation systems for the scattering coefficients have been derived which can be solved numerically. The calculated and measured insertion-loss curves for the two nine-step filters that have been constructed agree very well. The theoretical insertion losses for the filter designs are 0.05 dB (for 15.26 GHz) and 0.10 dB (for 33.71 GHz). The typical measured insertion losses, which are only about 0.25 dB and 0.5 dB, respectively, show the excellent suitability of these fin-line filters for use in the design of high-Q millimetre-wave integrated circuits.

References

1. Meier P J 1972, Two new integrated-circuit media with special advantages at millimeter wavelengths, IEEE 1972 G-MTT Symp. Digest (Arlington Heights, III) 22-24 May, 1972, pp. 221-223
2. Meier P J 1974, Integrated fin-line millimeter components, *IEEE Trans.* **MTT-22**, pp. 1209-1216
3. Meier P J 1978, Millimeter integrated circuits suspended in the E-plane of rectangular waveguide, *IEEE Trans.* **MTT-26**, pp. 726-733
4. Begemann G 1978, An X-band balanced fin-line mixer, *IEEE Trans.* **MTT-26**, pp. 1007-1011
5. Reindel J 1980, Printed WG circuits trim component costs, *Microwaves*, Oct. 1980, pp. 60-63
6. Saad A M K & Schünemann K 1978. *IEEE Trans.* **MTT-26**. pp. 1002-1007
7. Hennawy H E & Schünemann K 1980, Analysis of fin-line discontinuities, Proc. 10th European Microwave Conference, Warsaw. pp. 448-452
8. Saad A M K & Schünemann K 1979, Design and performance of fin-line bandpass filters, Proc. 9th European Microwave Conference, Brighton, pp. 397-401
9. Hock A & Rinderle J 1980, Zur Anwendung der Evolutionsstrategie auf Schaltungen der Nachrichtentechnik, *Frequenz* **34**, pp. 208-214

Appendix 1

Abbreviations in the scattering matrix of the waveguide transition to three waveguides (Equation 2)

$(N)^{r}$: $\quad (D_{IImn})^{-1}(D_{IInn}) = (N)^{II}$

$$(D_{Hmk})^{-1}(D_{Hkk}) = (N)^{III}$$

$$(D_{Hml})^{-1}(D_{Hll}) = (N)^{IV}$$

with the matrix coefficients

$$D_{IImn} = k_{zm}^{I^2} A_m^{I} H_{mn} \qquad D_{IInn} = \frac{d-c}{2} k_{zn}^{II^2} A_n^{II}$$

$$D_{Hmk} = k_{zm}^{I^2} A_m^I H_{mk} \qquad D_{Hkk} = \frac{a-d}{2} k_{zk}^{III^2} A_k^{III}$$

$$D_{Hml} = k_{zm}^{I^2} A_m^I H_{ml} \qquad D_{Hll} = \frac{c}{2} k_{zl}^{IV^2} A_l^{IV}$$

and coupling integrals

$$H_{mn} = \int_{x=c}^{d} \sin\frac{m\pi}{a} x \sin\frac{n\pi}{d-c}(d-x)\mathrm{d}x$$

$$H_{mk} = \int_{x=d}^{a} \sin\frac{m\pi}{a} x \sin\frac{k\pi}{a-d}(a-x)\mathrm{d}x$$

$$H_{ml} = \int_{x=0}^{c} \sin\frac{m\pi}{a} x \sin\frac{l\pi}{c} x \, \mathrm{d}x$$

$(L)^{\nu}$: $(D_{Emm})^{-1}(D_{Emn}) = (L)^{II}$

$(D_{Emm})^{-1}(D_{Emk}) = (L)^{III}$

$(D_{Emm})^{-1}(D_{Eml}) = (L)^{IV}$

with the matrix coefficients

$$D_{Emm} = \frac{a}{2} k_{zm}^{I} A_m^{I} \qquad D_{Emk} = k_{zk}^{III} A_k^{III} H_{mk}^{T}$$

$$D_{Emn} = k_{zn}^{II} A_n^{II} H_{mn}^{T} \qquad D_{Eml} = k_{zl}^{IV} A_l^{IV} H_{ml}^{T}$$

and coupling integrals

$$H_{nm} = \int_{x=c}^{d} \sin\frac{n\pi}{d-c}(d-x) \sin\frac{m\pi}{a} x \, \mathrm{d}x$$

$$H_{km} = \int_{x=d}^{a} \sin\frac{k\pi}{a-d}(a-x) \sin\frac{m\pi}{a} x \, \mathrm{d}x$$

$$H_{lm} = \int_{x=0}^{c} \sin\frac{l\pi}{c} x \sin\frac{m\pi}{a} x \, \mathrm{d}x$$

$$H_{nm}^{T} = H_{mn}$$

Appendix 2 Abbreviations in the scattering matrix of the waveguide transition to three waveguide structures of finite length (Equation 4)

$$(W) = (E) - ((\pi)+(E))^{-1}(\phi)((\pi)+(E))^{-1}(\phi)$$

(E) = unit matrix

$$(\pi) = \sum_v \left[2(L)^v(\delta)^v - (L)^v(N)^{v-1}\right] \qquad (\delta)^v = ((E)-(R)^v(R)^v)^{-1}(N)^{v-1}$$

$$(\phi) = \sum_v 2(L)^v(\varepsilon)^v \qquad (\varepsilon)^v = ((E)-(R)^v(R)^v)^{-1}(R)^v(N)^{v-1}$$

with the diagonal matrix of the waveguides with finite length l

$$(R)^v = \begin{bmatrix} e^{-jk'_{z1}\cdot l} & \cdot & \cdot & 0 \\ \cdot & \cdot & & \\ \cdot & & \cdot & \\ \cdot & & & \cdot \\ 0 & & & e^{-jk'_{zm}\cdot l} \end{bmatrix}$$

Appendix 3

Abbreviations in the scattering matrix of the waveguide transition to a waveguide with a dielectric slab of finite width and length (Equation 6)

$$(W) = ((E)-Q_{22})(R)(Q_{22})(R))^{-1} \qquad R_{mm} = e^{-jk_{zm}l}$$

The scattering coefficients of the single step are:

$$(Q_{11}) = 2(D_H)^{-1}(L_H)(M)-(E) \qquad (Q_{21}) = 2(M)$$

$$(Q_{12}) = 2(D_H)^{-1}(L_H)(M)(D_E)^{-1}(L_E) \qquad (Q_{22}) = 2(M)(D_E)^{-1}(L_E)-(E)$$

$$(M) = ((D_E)^{-1}(L_E) + (D_H)^{-1}(L_M))^{-1}$$

$$D_{Emm} = \frac{a}{2}k^{I}_{zm}A^{I}_{m} \qquad D_{Hmm} = \frac{a}{2}k^{I^2}_{zm}A^{I}_{m}$$

The coefficients of the matrices (L_E) and (L_H) are given by:

$$k_{zk}(A^{II}_k I^{II}_{mk} + B^{II}_k I^{II}_{mk} + A^{III}_k(-T_{1k}I^{III}_{mk} + I^{III}_{mk}) + A^{IV}_k I^{IV}_{mk}) = L_{Emk}$$

$$k^2_{zk}(A^{II}_k I^{II}_{mk} + B^{II}_k I^{II}_{mk} + A^{III}_k(-T_{1k}I^{III}_{mk} + I^{III}_{mk}) + A^{IV}_k I^{IV}_{mk}) = L_{Hmk}$$

with the abbreviation $T_{1k} = \tan(k^{III}_{xk}\cdot a)$.

Coupling integrals:

$$I^{II}_{mk} = \int_{x=c}^{d} \cos(k^{II}_{xk}x)\sin\left(\frac{m\pi}{a}x\right)dx$$

$$I^{II}_{mk} = \int_{x=c}^{d} \sin(k^{II}_{xk}x)\sin\left(\frac{m\pi}{a}x\right)dx$$

$$I^{III}_{mk} = \int_{x=d}^{a} \cos(k^{III}_{xk}x)\sin\left(\frac{m\pi}{a}x\right)dx$$

$$I_{mk}^{\mathrm{III}} = \int_{x=d}^{a} \sin(k_{xk}^{\mathrm{III}} x) \sin\left(\frac{m\pi}{a} x\right) dx$$

$$I_{mk}^{\mathrm{IV}} = \int_{x=0}^{c} \sin(k_{xk}^{\mathrm{IV}} x) \sin\left(\frac{m\pi}{a} x\right) dx$$

The common propagation factor k_{zm} of the regions II, III, IV is determined by:

$$\begin{bmatrix} k_{xm}^{\mathrm{II}^2} \\ k_{xm}^{\mathrm{III}^2} \\ k_{xm}^{\mathrm{IV}^2} \end{bmatrix} = \begin{bmatrix} k_0^2 \cdot \varepsilon_r \\ k_0^2 \\ k_0^2 \end{bmatrix} - k_{zm}^2$$

$$\frac{1}{k_{xm}^{\mathrm{II}}} \cdot \tan(k_{xm}^{\mathrm{II}}(d-c)) + \frac{1}{k_{xm}^{\mathrm{III}}} \cdot \tan(k_{xm}^{\mathrm{III}}(a-d)) + \frac{1}{k_{xm}^{\mathrm{IV}}} \cdot \tan(k_{xm}^{\mathrm{IV}} \cdot c)$$

$$- \frac{k_{xm}^{\mathrm{II}}}{k_{xm}^{\mathrm{III}} \cdot k_{xm}^{\mathrm{IV}}} \cdot \tan(k_{xm}^{\mathrm{II}}(\mathrm{d}-\mathrm{c})) \cdot \tan(k_{xm}^{\mathrm{III}}(a-d)) \cdot \tan(k_{xm}^{\mathrm{IV}} \cdot c) = 0$$

The waves in the regions I, II, III, IV are expanded into the eigenmodes:

$$E_y^{\mathrm{I}} = -\omega\mu \sum_{m=1} k_{zm}^{\mathrm{I}} A_m^{\mathrm{I}} \sin\frac{m\pi}{a} x \cdot e^{-jk_{zm}^{\mathrm{I}} \cdot z}$$

$$H_z^{\mathrm{I}} = -j \sum_{m=1} \frac{m\pi}{a} k_{zm}^{\mathrm{I}} A_m^{\mathrm{I}} \cos\frac{m\pi}{a} x \cdot e^{-jk_{zm}^{\mathrm{I}} \cdot z}$$

$$H_x^{\mathrm{I}} = \sum_{m=1} \left(k_0^2 - \left(\frac{m\pi}{a}\right)^2\right) A_m^{\mathrm{I}} \sin\frac{m\pi}{a} x \cdot e^{-jk_{zm}^{\mathrm{I}} \cdot z}$$

$$E_y^{\mathrm{II}} = -\omega\mu \sum_{m=1} k_{zm} (A_m^{\mathrm{II}} \cos(k_{xm}^{\mathrm{II}} \cdot x) + B_m^{\mathrm{II}} \sin(k_{xm}^{\mathrm{II}} \cdot x)) \cdot e^{-jk_{zm} \cdot z}$$

$$H_z^{\mathrm{II}} = -j \sum_{m=1} k_{zm} k_{xm}^{\mathrm{II}} (-A_m^{\mathrm{II}} \sin(k_{xm}^{\mathrm{II}} \cdot x) + B_m^{\mathrm{II}} \cos(k_{xm}^{\mathrm{II}} \cdot x)) \cdot e^{-jk_{zm} \cdot z}$$

$$H_x^{\mathrm{II}} = \sum_{m=1} (\varepsilon_r k_0^2 - k_{xm}^{\mathrm{II}^2})(A_m^{\mathrm{II}} \cos(k_{xm}^{\mathrm{II}} \cdot x) + B_m^{\mathrm{II}} \sin(k_{xm}^{\mathrm{II}} \cdot x)) \cdot e^{-jk_{zm} \cdot z}$$

$$E_y^{\mathrm{III}} = -\omega\mu \sum_{m=1} k_{zm} A_m^{\mathrm{III}} (-\tan(k_{xm}^{\mathrm{III}} \cdot a) \cos(k_{xm}^{\mathrm{III}} \cdot x) + \sin(k_{xm}^{\mathrm{III}} \cdot x)) \cdot e^{-jk_{zm} \cdot z}$$

$$H_z^{\mathrm{III}} = -j \sum_{m=1} k_{zm} k_{xm}^{\mathrm{III}} A_m^{\mathrm{III}} (\tan(k_{xm}^{\mathrm{III}} \cdot a) \sin(k_{xm}^{\mathrm{III}} \cdot x) + \cos(k_{xm}^{\mathrm{III}} \cdot x) e^{-jk_{zm} \cdot z}$$

$$H_x^{\mathrm{III}} = \sum_{m=1} (k_0^2 - k_{xm}^{\mathrm{III}^2}) A_m^{\mathrm{III}} (-\tan(k_{xm}^{\mathrm{III}} \cdot a) \cos(k_{xm}^{\mathrm{III}} \cdot x) + \sin(k_{xm}^{\mathrm{III}} \cdot x)) \cdot e^{-jk_{zm} \cdot z}$$

$$E_y^{\mathrm{IV}} = -\omega\mu \sum_{m=1} k_{zm} A_m^{\mathrm{IV}} \sin(k_{xm} \cdot x) \cdot e^{-jk_{zm} \cdot z}$$

$$H_z^{\mathrm{IV}} = -j \sum_{m=1} k_{zm} k_{xm}^{\mathrm{IV}} A_m \cos(k_{xm}^{\mathrm{IV}} \cdot x) \cdot e^{-jk_{zm} \cdot z}$$

$$H_x^{\mathrm{IV}} = \sum_{m=1} (k_0^2 - k_{xm}^{\mathrm{IV}^2}) A_m^{\mathrm{IV}} \sin(k_{xm}^{\mathrm{IV}} \cdot x) \cdot e^{-jk_{zm} \cdot z}$$

Manuscript received 6 February 1981

6.2.4

Broad-Band Millimeter-Wave *E*-Plane Bandpass Filters

L. Q. BUI, D. BALL, MEMBER, IEEE, AND T. ITOH, FELLOW, IEEE

***Abstract*—An accurate method to obtain starting estimates for an *E*-plane bandpass filter CAD program recently developed by Shih, Itoh, and Bui is presented. Results agree very well with exact values for filter design with relative bandwidths exceeding 10 percent at *W*-band and *D*-band and 20 percent at lower frequencies.**

I. Introduction

With increasing activity in millimeter-wave integrated circuit development, the *E*-plane filter shown in Fig. 1 recently has drawn considerable attention [1]–[5]. Such a structure consists of metal strips inserted along the *E*-plane of a rectangular waveguide. The strips may be fabricated on dielectric substrates [3]–[5], in which case this filter is quite compatible with finline technology.

To date, many *E*-plane filter designs reported are limited to narrow bandwidths of less than a few percent. For wide-band designs, as all the resonators are strongly coupled, an accurate analysis procedure, one which takes into account the interaction not only of the dominant mode but also of the higher order modes generated at the edge of the strips, is necessary.

We recently developed a computerized (CAD) algorithm for a class of *E*-plane filters [6]. For narrow-band designs, the algorithm converges rapidly with the prescribed response by use of available optimization routines [7]. However, since no systematic scheme was incorporated for providing starting values of the optimization routine, convergence of the CAD program was rather slow; furthermore, the cost of generating acceptable design parameters often was prohibitively expensive. This deficiency is especially noticeable in design of relative bandwidth greater than 5 percent. Since *E*-plane filters are inductively direct-coupled cavity filters, the conventional design procedure given by Levy using an impedance inverter and low-pass prototypes is available as [8]. Depending on the bandwidth equipment, either a lumped-element or a distributed prototype is used to calculate the inverter values. In general, a distributed prototype is not favorable due to difficulties in

Manuscript received April 26, 1984; revised August 1, 1984.

L. Q. Bui and D. Ball are with the Hughes Aircraft Company, Microwave Products Division, Torrence, CA 90509.

T. Itoh is with the Department of Electrical Engineering, the University of Texas at Austin, Austin, TX 78712.

L.Q. Bui, D. Ball, and T. Itoh, *IEEE Transactions on Microwave Theory and Techniques*, Vol. MTT-32, No. 12, Dec. 1984, pp. 1655–1658.

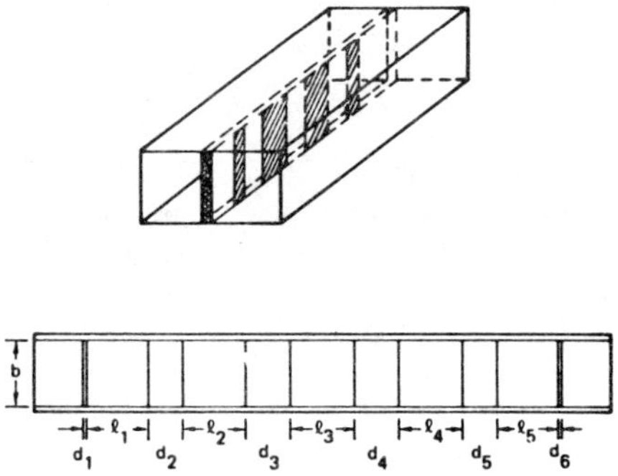

Fig. 1. *E*-plane filter structure.

obtaining either impedance values using network synthesis technique or element values for untabulated design, even if such prototype can be used for wide-band design [9].

This paper presents an accurate method that uses an explicit formula to obtain the element values of a distributed low-pass prototype element derived by Rhodes [10]. The equivalent circuit of an *E*-plane inductive strip [6] is used to obtain design parameters such as septum widths and spacing. Results agree with exact values for filter designs of up to 30-percent bandwidth extremely well. The computation cost was reduced substantially since no optimization is required. Using this modified CAD program, designs of *E*-plane bandpass filters with more than 10-percent bandwidth at *D*-band and *W*-bands and more than 20 percent at lower frequency ranges have been accomplished. Results are unprecedented to our best knowledge.

II. Procedures

As described in [6], our original CAD program consists of one optimization routine and one analysis algorithm. The analysis consists of three steps. In the first step, the generalized scattering matrix at a strip edge is calculated. For instance, $S_{11}(m, p)$ is the amplitude of the reflected mth mode when the pth mode is incident on that edge. Because each strip extends from the lower broad wall to the upper broad wall of the waveguide and is uniform, these generalized scattering parameters can be calculated exactly by residue calculus techniques [11]. The second step computes the generalized S matrix of a strip by combining the S parameters derived in step 1 for both edges of the strip. The last step combines generalized S matrices of several strips. This algorithm allows us to obtain the generalized S matrix of the entire filter structure (Fig. 1). Note that the use of the generalized S matrix enables us to find the interactions between junctions due to the dominant mode and to all higher order modes. This is quite important for accurate design of wide-band filters. The effect of higher order mode interactions becomes even more important for filters in which the strips are narrow and the resonators are coupled strongly.

The procedure is based on a formulation proposed by Rhodes for a distributed stepped-impedance low-pass prototype [10]. When the passband ripple ϵ, the lower and upper frequencies f_L and f_H, and the out-of-band rejection L (decibels) are specified, the procedure is as follows.

1) Determine the midband guide wavelength λ_{go} by solving

$$\lambda_{gL}\sin(\pi\lambda_{go}/\lambda_{gL})+\lambda_{gH}\sin(\pi\lambda_{go}/\lambda_{gH})=0 \quad (1)$$

where λ_{gL} and λ_{gH} are the guide wavelengths in the resonator section at f_L and f_H. For a narrow-band case $(\lambda_{gL}\approx\lambda_{gH})$

$$\lambda_{go}=\frac{\lambda_{gL}+\lambda_{gH}}{2}.$$

However, for the broad-band case, (1) generally must be solved numerically.

2) Determine the scaling parameter α by

$$\alpha=\lambda_{go}/\left[\lambda_{gL}\sin(\pi\lambda_{go}/\lambda_{gL})\right]. \quad (2)$$

3) Determine the number of resonators N. This is accomplished by finding the minimum value of N for which the most severe constraints on the rejection L satisfies

$$L=10\log\left\{1+\epsilon^2T_N^2\left[\alpha\frac{\lambda_g}{\lambda_{go}}\sin\left(\frac{\pi\lambda_{go}}{\lambda_g}\right)\right]\right\} \quad (3)$$

at the designated stopband frequency f_s. In (3), T_N is the Chebyshev polynomial of degree N, and λ_g is the guide wavelength at f_s.

4) Calculate the impedances of the distributed element and impedance inverter values

$$Z_n=2\alpha\sin\left[\frac{(2n-1)\pi}{2N}\right]\Big/y-\frac{1}{4y\alpha}\left\{\frac{y^2+\sin^2(n\pi/N)}{\sin\frac{(2n+1)\pi}{2N}}+\frac{y^2+\sin^2[(n-1)\pi/N]}{\sin\frac{(2n-3)\pi}{2N}}\right\},\quad n=1,\cdots,N \quad (4)$$

$$k'_{n,n+1}=\frac{\sqrt{y^2+\sin^2(n\pi/N)}}{y},\quad n=0,1,\cdots,N \quad (5)$$

where

$$y=\sinh\left[\frac{1}{N}\sinh^{-1}\frac{1}{\epsilon}\right]. \quad (6)$$

5) Recognizing that the characteristic impedances of the resonator sections are identical, scale the impedance Z_n to unity and

$$K_{n,n+1}=k'_{n,n+1}/\sqrt{Z_nZ_{n+1}},\quad n=0,\cdots,N \quad (7)$$

and

$$Z_0=Z_{n+1}=1.$$

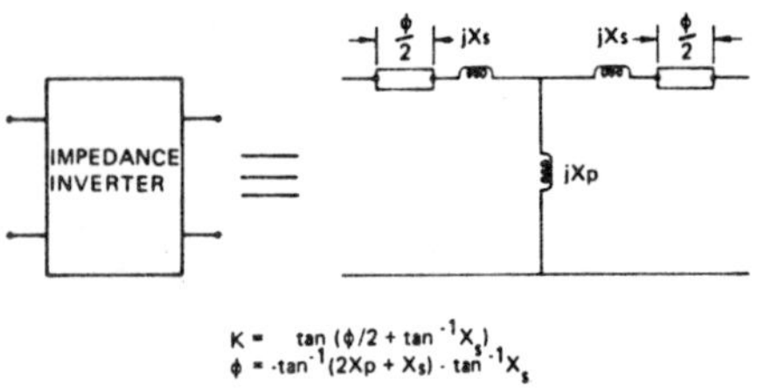

Fig. 2. Impedance inverter.

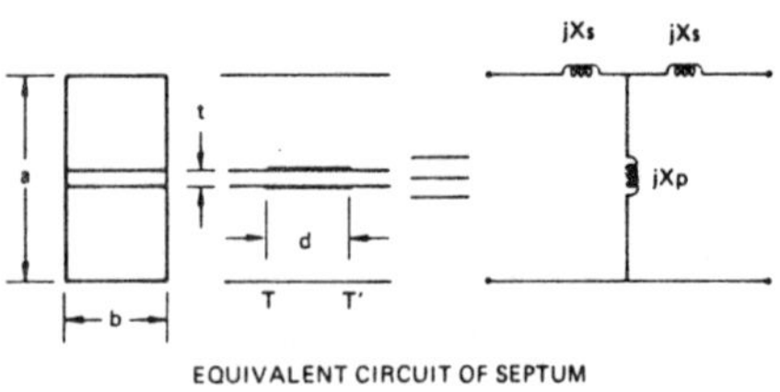

Fig. 3. Equivalent circuit of a bilateral fin.

6) Determine the widths of E-plane strips so that the required impedance inverter is realized. Use the T equivalent circuit of the strip for the inverter (Fig. 2)

$$K = \left| \tan\left(\frac{\phi}{2} + \tan^{-1} X_s\right) \right| \tag{8}$$

$$\phi = -\tan^{-1}(2X_p + X_s) - \tan^{-1} X_s. \tag{9}$$

The normalized reactances X_p and X_s in Fig. 3 are given in [10]

$$jX_s = \frac{1 - S_{12} + S_{11}}{1 - S_{11} + S_{12}} \tag{10a}$$

$$jX_p = \frac{2S_{12}}{(1 - S_{11})^2 - S_{12}^2} \tag{10b}$$

where S_{11}, etc., are the scattering coefficients of the dominant mode. Note that, in calculating X_s and X_p, effects of all of the higher order modes are incorporated. Therefore, the results are accurate.

X_s and X_p in (10) are functions of the strip width d as described in [6]. Since these functions are not available explicitly, we implemented a root-seeking routine to find the value of d that provided the required impedance value K and the angle ϕ for each impedance inverter.

7) Finally, the length of the jth resonator formed by the jth and $(j+1)$th strips is given by

$$l_j = \frac{\lambda_{go}}{2\pi}\left[\pi + \frac{1}{2}(\phi_j + \phi_{j+1})\right] \tag{11}$$

III. Design and Measurement

The method described above was applied to the design of E-plane filters for use in various frequency regions, ranging from Ka (26.5–40 GHz) to D (110–175 GHz) band. In this paper, we discuss three examples, one each for D-, W-, and U-bands. Note that the procedure is valid for E-plane filters of bilateral, and unilateral, and metal-only construction.

TABLE I
D-Band Filter Design

Ripple:	0.10 dB
Lower Band-Edge Frequency:	139.0 GHz
Upper Band-Edge Frequency:	151.0 GHz
Lower Rejection :	20.0 dB at 132.0 GHz
Upper Rejection :	20.0 dB at 160.0 GHz
Relative Dielectric Consistant:	2.1
Substrate Thickness/Inches:	0.003
Waveguide Width/Inches:	0.065
No. of Resonators:	5
Strips $d_1 = d_6 = 1.25$, $d_2 = d_5 = 13.25$, $d_3 = d_4 = 17$ mils	
Resonators $l_1 = l_5 = 34.93$, $l_2 = l_4 = 35.39$, $l_3 = 35.40$ mils	

TABLE II
D-Band Filter Response

Frequency (GHz)	Return Loss (db)	Insertion (db)
130.000	0.000	51.480
131.000	0.000	47.463
132.000	0.001	43.125
133.000	0.001	38.385
134.000	0.003	33.160
135.000	0.009	27.267
136.000	0.037	20.486
137.000	0.242	12.499
138.000	2.292	3.797
139.000	17.047	0.054
140.000	23.726	0.001
141.000	53.904	0.015
142.000	22.584	0.015
143.000	23.270	0.014
144.000	59.235	0.007
145.000	22.635	0.018
146.000	19.720	0.040
147.000	23.958	0.010
148.000	30.072	0.003
149.000	18.466	0.058
150.000	20.014	0.038
151.000	15.733	0.105
152.000	3.823	2.310
153.000	0.887	7.325
154.000	0.250	12.550
155.000	0.087	17.086
156.000	0.039	20.946
157.000	0.020	24.286
158.000	0.012	27.198
159.000	0.008	29.765
160.000	0.007	32.038

Table I shows specifications for a D-band filter and the initial guess of dimensions for the strips (d_j) and resonators (l_j) derived by procedures 1) through 7) described above. When these values are used in the analysis portion of the CAD program, the response is as shown in Table II. Since these results satisfy the specifications, no optimization was required.

A bilateral D-band filter (Table I) was fabricated and tested. A 3-mil-thick cuflon substrate is used for this filter. Fig. 4 shows the measured insertion loss of this filter. Since two sweep generators were used for this measurement, the two printouts are placed side by side to generate Fig. 4. The measured results indicate that the entire passband is shifted upward by about 3.5 GHz. However, this shift constitutes only about 2.4 percent of the center frequency. Such a shift is caused by the thickness of the strip metallization and by the grooves holding the E-plane substrate. The groove dimension is 5 mils wide by 3 mils deep, and Fig. 5 is a photograph of the D-band filter.

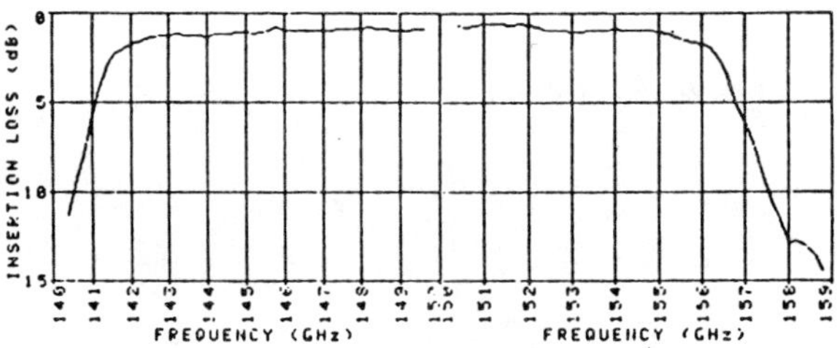

Fig. 4. Measured characteristics of a *D*-band bilateral filter on a cuflon substrate.

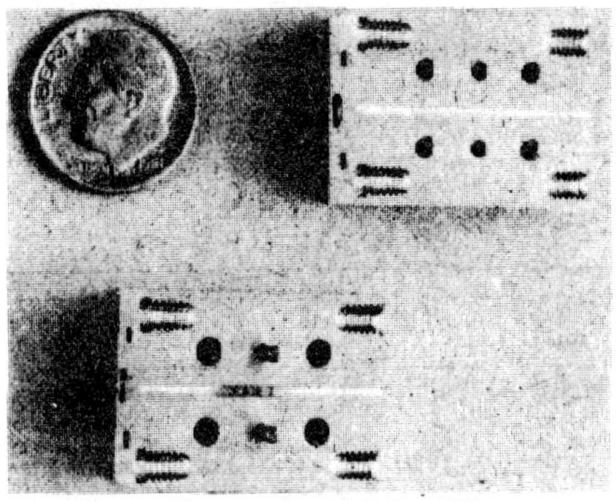

Fig. 5. Photograph of a *D*-band filter.

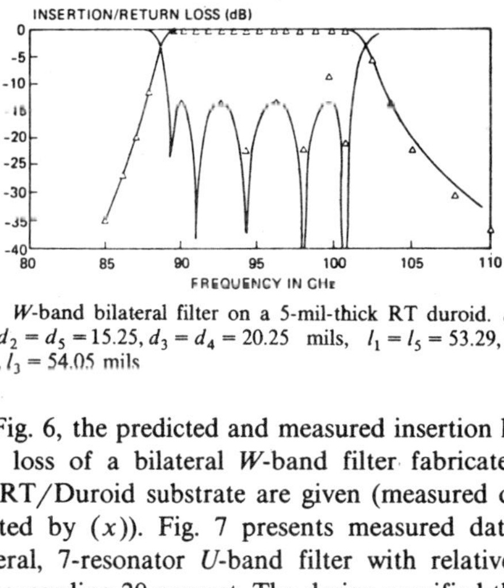

Fig. 6. *W*-band bilateral filter on a 5-mil-thick RT duroid. $d_1 = d_6 = 1.25, d_2 = d_5 = 15.25, d_3 = d_4 = 20.25$ mils, $l_1 = l_5 = 53.29, l_2 = l_4 = 54.02, l_3 = 54.05$ mils.

In Fig. 6, the predicted and measured insertion loss and return loss of a bilateral *W*-band filter fabricated on a 5-mil RT/Duroid substrate are given (measured data are indicated by (x)). Fig. 7 presents measured data for a unilateral, 7-resonator *U*-band filter with relative bandwidth exceeding 20 percent. The design specified the passband from 50 to 60 GHz. In the measurement, a *V*-band sweeper (50 to 75 GHz) was used via a tapered transition to the *U*-band setup.

IV. Conclusions

An efficient method for estimating the starting values of a broad-band *E*-plane filter was developed using a conventional direct-coupled filter and the exact equivalent circuit of an *E*-plane inductive strip. Compared to our original CAD procedure, this modified version's CPU time is reduced by a factor of 30. Excellent agreement with exact analysis and measurements has been obtained for filter designs of up to 160 GHz. Performance of these filters, particularly those at *D*-band, is unmatched to date.

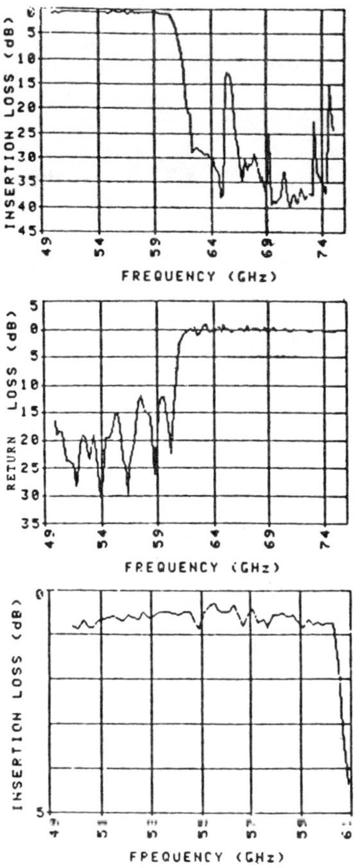

Fig. 7. Measured responses of a *U*-band, unilateral 7-resonator filter. $d_1 = d_8 = 1.5, d_2 = d_7 = 20.25, d_3 = d_6 = 29.75, d_4 = d_5 = 32.25$ mils, $l_1 = l_7 = 85.73, l_2 = l_6 = 85.53, l_3 = l_5 = 85.25, l_4 = 85.20$ mils.

References

[1] Y. Konishi and K. Uenakada, "The design of a bandpass filter with inductive strip-planar circuit mounted in waveguide," *IEEE Trans. Microwave Theory Tech.*, vol. MTT-22, pp. 1209–1216, 1974.

[2] Y. Tajima and Y. Sawayama, "Design and analysis of a waveguide-sandwich microwave filter," *IEEE Trans. Microwave Theory Tech.*, vol. MTT-22, pp. 839–841, 1974.

[3] P. Meier, "Integrated fin-line millimeter components," *IEEE Trans. Microwave Theory Tech.*, vol. MTT-22, pp. 1209–1216, 1974.

[4] F. Arndt *et al.*, "Theory and design of low-insertion loss fin-line filters," *IEEE Trans. Microwave Theory Tech.*, vol. MTT-30, pp. 155–162, 1982.

[5] A. M. K. Saad and K. Schunemann, "Design and performance of fin-line bandpass filters," in *Proc. 10th Eur. Microwave Conf.*, (Warsaw), Sept. 1980, pp. 397–401

[6] Y. C. Shih, T. Itoh, and L. Q. Bui, "Computer-aided design of millimeter-wave *E*-plane filters," *IEEE Trans. Microwave Theory Tech.*, vol. MTT-31, pp. 135–142, 1983.

[7] Subroutine ZXMIN, ZXSSQ of IMSL Library.

[8] R. Levey, "Theory of direct-coupled cavity filters," *IEEE Trans. Microwave Theory Tech.*, vol. MTT-15, pp. 340–348, 1967.

[9] L. Young, "Direct-coupled cavity filters for wide and narrow bandwidths," *IEEE Trans. Microwave Theory Tech.*, vol. MTT-11, pp. 162–178, 1963.

[10] J. D. Rhodes, "Microwave circuit realizations," in *Microwave Solid State Devices and Applications*, D. V. Morgan and M. J. Haves, Eds. England: Peregrinus, 1980, pp. 49–57.

[11] C. F. Vanblaricum, Jr. and R. Mittra, "A modified residue-calculus technique for solving a class of boundary value problems—Part II: Waveguide phase arrays, modulated surfaces, and diffraction gratings," *IEEE Trans. Microwave Theory Tech.*, vol. MTT-17, pp. 310–319, 1969.

6.2.5

REALIZATIONS AND DESIGN OF FIN-LINE BANDSTOP FILTERS

A.S. Omar and K. Schünemann*

ABSTRACT

The potentials of various slot patterns for realizing fin-line bandstop filters are discussed. Filters utilizing a double-slot fin-line, and a hybrid structure are designed and realized.

POTENTIALS OF SLOT PATTERNS FOR FILTER REALIZATIONS

Whereas a lot of papers deal with the design of bandpass filters in fin-line technology (e.g. /1/, /2/), almost nothing has as yet been reported concerning fin-line bandstop filters. We have undertaken combined theoretical and practical studies in order to realize such circuits by computer-aided design.

The first step is to investigate the potentials of possible slot patterns for realizing bandstop filters. The discontinuities which are usually used in fin-line circuits are shown in Fig.1: a conductive strip, a notch, and a post. Their equivalent circuits have been derived in the past /3/, /4/. They show a common feature: The series reactance of the T-circuit is always inductive. Hence it is not possible to realize a resonance circuit in series to the transmission line in this way. On the other hand, the shunt reactance is either purely inductive (for a strip) or purely capacitive (for a notch and a post). Hence resonance circuits in shunt to the transmission line can also not be realized in this way. This class of discontinuities is actually used for implementing coupling sections between half-wave resonators in a single-slot fin-line. Thus only bandpass filters can be realized /1/, /2/.

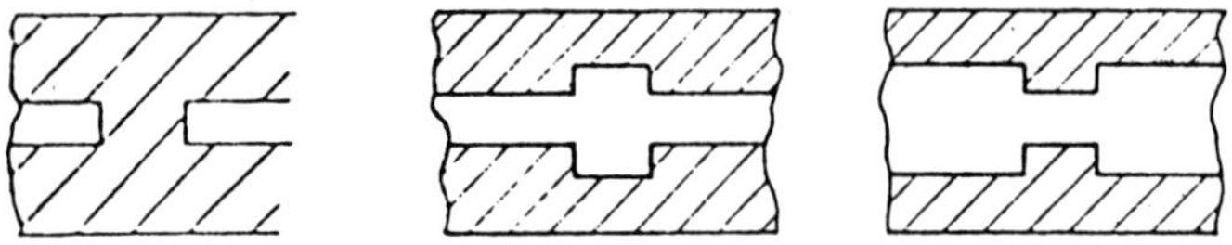

Fig.1

Slot pattern of conductive strip, notch and post

There still remain two circuits for realizing a bandstop filter, whose slot patterns have been sketched in Fig.2. The first is a section of double-slot fin-line which is approximately half a wavelength long. Then both an even and an odd mode may exist which have been calculated according to /5/. The second slot now behaves like a reaction resonator, if it is about half a wavelength long. Its coupling to the main line depends on both slot

Institut für Hochfrequenztechnik, Technische Universität Braunschweig, Postfach 3329, D-3300 Braunschweig, West-Germany
*Now with Technical University Hamburg-Harburg, Arbeitsbereich Hochfrequenztechnik, Harburger Schloßstr. 20, D-2100 Hamburg 90, West-Germany

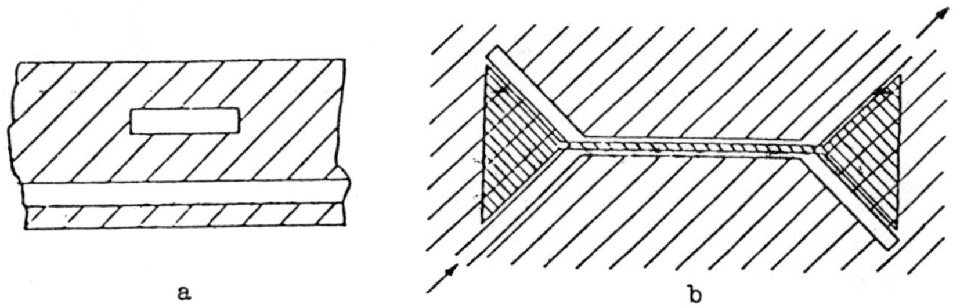

Fig.2

Slot patterns for bandstop filters

widths and the separation of the slots. Alternatively, the second slot can be loaded capacitively by means of a post in order to reduce the overall length.

Quite a different approach to a bandstop filter is shown in Fig.2b. It consists of a planar hybrid with two ports terminated by short circuits. Thus a narrow-band bandstop filter can be realized with a maximum attenuation of about 40 dB. The design procedure is straightforward (see /6/) and needs no further explanation here. The hybrid can easily be designed, too, as is described in /7/.

All types of filters except that shown in Fig.2b are designed in the following way: The first step is to characterize the discontinuity by its equivalent circuit. The procedure (known as 'modal analysis' /8/) is to expand the fields at either side of the junction between two fin-lines of different sections into their eigenmodes whose amplitudes are then determined from matching the tangential field components at the interface. Two cascaded junctions mounted back to back can thus be described as is explained elsewhere /4/. In the second step, the well-known filter design method of /6/ is applied after prescribing the response curve and the number of resonators. A K-inverter network is extracted from the equivalent circuit of the coupling section (almost quarter-wave transmission line plus additional phase shifts at the junctions of the discontinuities) and cascaded with the reaction-type cavity etc.. The design is straightforward. It neglects, however, higher-order mode coupling between the individual resonators. This is taken into account in the final step, where an optimization computer program is used which varies the dimensions of all discontinuities until the prescribed response curve is best approximated. For this step, the discontinuity (reaction cavity plus coupling junctions) is described by its scattering matrix. The procedure is similar to those used in /1/, /2/. Due to the preceding step by which good initial values can be obtained, its convergence is fast.

The fin-line eigenmodes have been obtained from a spectral domain approach according to /9/ for a single-slot and according to /5/ for a double-slot fin-line.

RESULTS

Bandpass filters are built by cascading half-wave fin-line resonators with coupling fourpoles which are realized by the equivalent circuit elements of a discontinuity (strip, notch, or post). Bandstop filters need, however, a different design as we have seen above. The reaction-type resonator is composed of a second slot section embedded into a normal fin-line. In the following we will present some numerical results for illustrating the essential features of this filter building block.

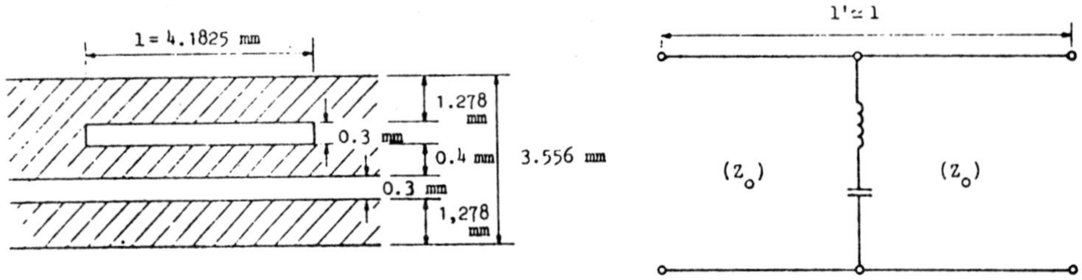

Fig.3

Slot pattern and equivalent circuit of a double-slot fin-line section

Fig.3 shows the slot pattern and the equivalent circuit used. The frequency dependence of its shunt reactance has been plotted in Fig.4. The resonant frequency is 30 GHz, the reactance slope $\omega_o L/Z_o = 22.5$. The design of bandstop filters now is to cascade such resonators with impedance inverters of any odd multiple of a quarter-wave single-slot fin-line.

Fig.4

Frequency dependence of the normalized shunt reactance X; slot width = 0.3 mm, slot separation = 0.4 mm, physical length l = 4.1825

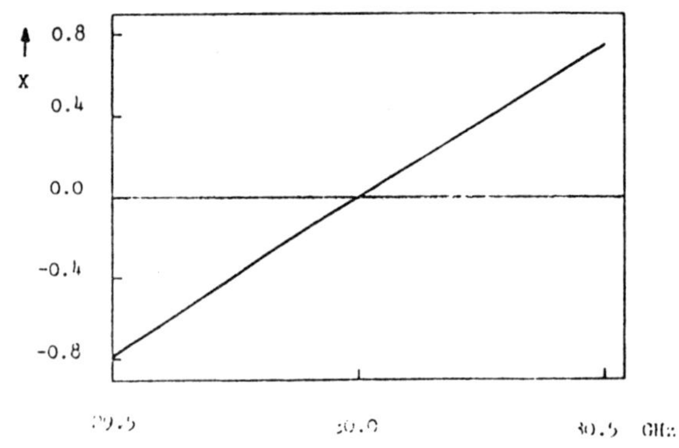

The frequency dependence of both the normalized equivalent line length l'/l and the normalized shunt susceptance B is shown in Fig.5. The following conclusions can be drawn:

- The length l'/l varies by less than 0.035 over the whole Ka-band. In addition it may be assumed to be constant and in particular equal to 1.
- There is only one pole in B(f). Hence there will not be any parasitic stop bands within the Ka-band.

The variations of l' and X with physical length l are depicted in Fig.6 showing that

- l'/l is nearly unity and only slightly depending on l/λ_o;
- the resonant length of the double-slot section is nearly independent of

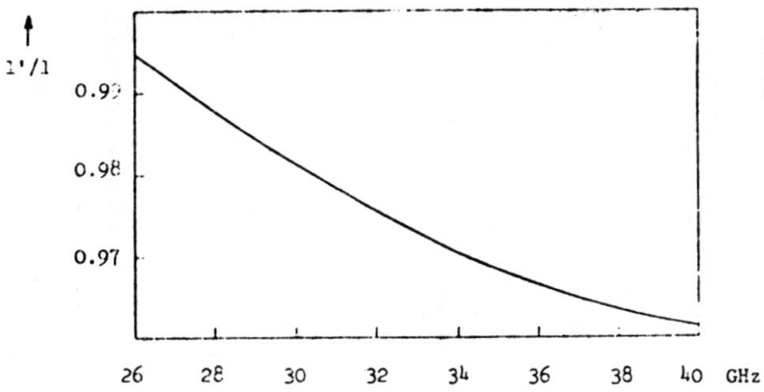

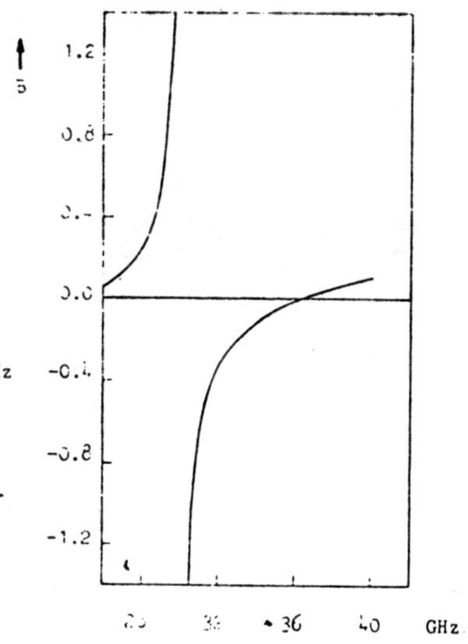

Fig.5

Frequency dependence of the normalized equivalent length l'/l and the normalized shunt susceptance B; slot width = 0.3 mm, slot separation = 0.4 mm, physical length l = 4.1825 mm

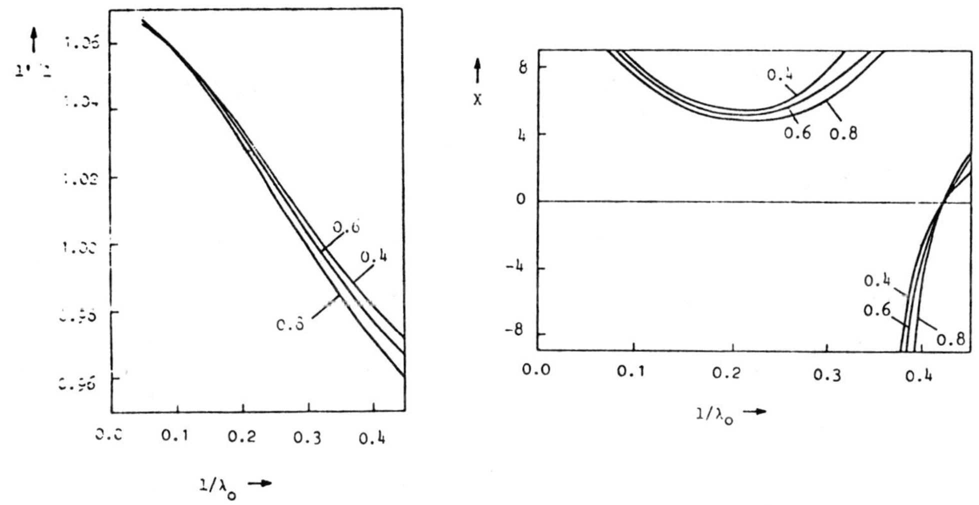

Fig.6

Normalized equivalent length l'/l and normalized shunt susceptance X versus normalized physical length l/λ_o; slot width = 0.4 mm, frequency = 30 GHz
parameter: slot separation in mm

the distance between the slots, so that it only depends on slot width;

- the normalized shunt reactance for line lengths below resonance decreases if the distance between the slots increases.

The variations of the equivalent circuit elements with slot width are depicted in Fig.7. This shows

- l'/l being nearly unity
- the strong influence of the slot width on these parameters.

Several bandstop filters have been designed and realized in Ka-band. Response curves are shown in Fig.8. The distance between the individual

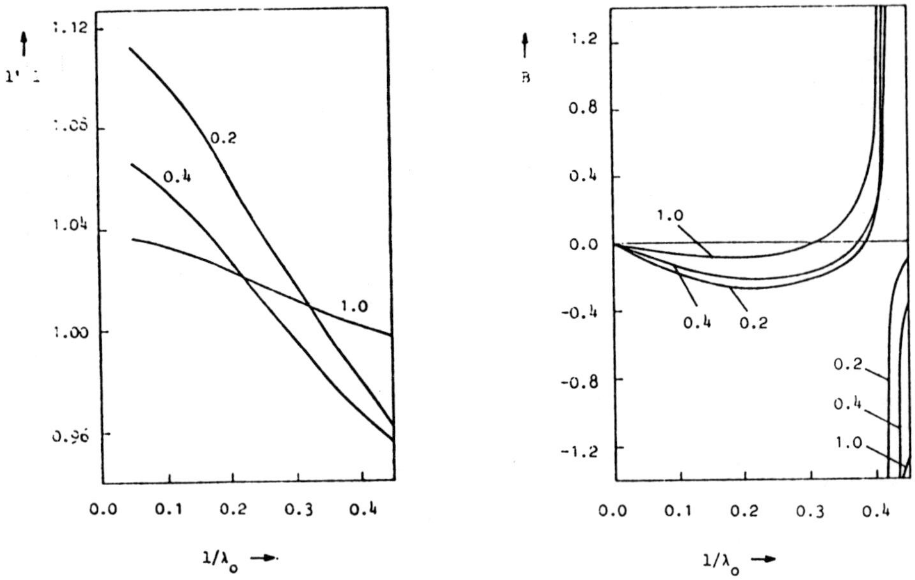

Fig.7

Normalized equivalent length l'/l and normalized shunt susceptance B versus normalized physical length l/λ_o; slot separation = 1.0 mm, frequency = 30 GHz;
parameter: slot width

reaction cavities has been chosen to 3 quarters of a wavelength in order to minimize higher-order mode coupling. The agreement to measurements is satisfying although the center frequencies differ by about 550 MHz. We attribute this effect mainly to the influence of the slit in the split waveguide housing which has been used for mounting the substrate. It has been shown in /10/ that even a short slit decreases the cutoff frequency of the fundamental mode. We have estimated by replacing the slit in the split housing by an effective reactance that the fundamental eigenvalue changes by about 1 per cent. (The reactance has been taken from /11/). This explains most of the observed difference between measured and calculated center frequencies.

Finally the measured response curve of a bandstop filter realized with a short-circuited hybrid (Fig.2b) is shown in Fig.9.

ACKNOWLEGEMENT

The authors are indebted to the Deutsche Forschungsgemeinschaft for financial support.

REFERENCES

/1/ Arndt, et al.: 11th Europ. Microwave Conf., 1981, Conf. Proc. 309-314.

/2/ Shih, Itoh, Bui: Internat. Microwave Symp., Dallas 1982, MTT-S Digest 471-473.

/3/ Jansen, Koster: 11th Europ. Microwave Conf., 1981, Conf. Proc. 682-687.

/4/ El Hennawy, Schünemann: IEE Proc. pt.H. (MOA), vol. 129, 1982, 342-350.

/5/ Schmidt, Itoh, Hofmann: Internat. Microwave Symp., Washington 1980, MTT-S Digest 255-257.

/6/ Matthaei, Young, Jones: Microwave filters, impedance-matching networks, and coupling structures. New York, 1964.

/7/ El Hennawy, Knöchel, Schünemann: AEÜ vol. 37, 1983, 40-46.

/8/ Wexler: IEEE Trans., MTT-15, 1967, 508-517.

/9/ Schmidt, Itoh: IEEE Trans., MTT-28, 1980, 981-985.

/10/ Beyer, Wolff: Internat. Microwave Symp., Washington 1980, MTT-S Digest 258-260.

/11/ Marcuvitz: Waveguide Handbook. New York, 1951.

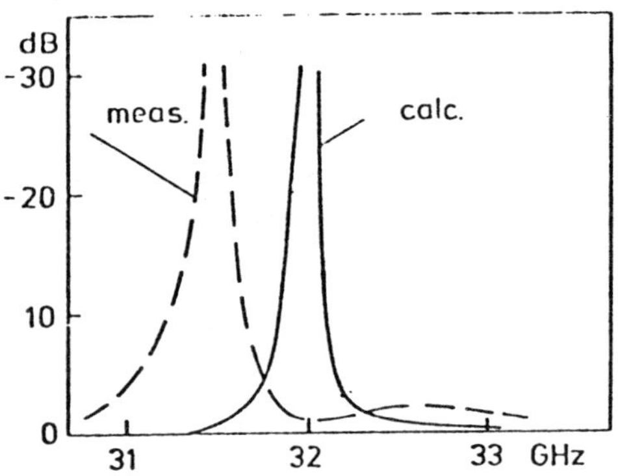

Fig.8

Response of the bandstop filter corresponding to Fig.2a

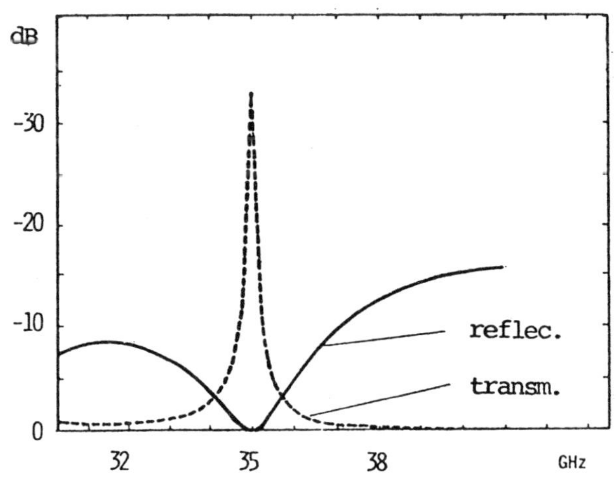

Fig.9

Response of the bandstop filter corresponding to Fig.2b

6.2.6

Millimeter-Wave Diplexers with Printed Circuit Elements

YI-CHI SHIH, MEMBER, IEEE, LONG Q. BUI, AND TATSUO ITOH, FELLOW, IEEE

Abstract—A novel design of *W*-band (75–110 GHz) noncontiguous diplexers is described. The common port of the diplexer is fed by suspended probe transitions which are suitable for wide-band applications. The circuit is printed on a single substrate and easily assembled in a split-block housing. The measured insertion loss at the passbands is about 1 dB. The calculated frequency response of a diplexer is in good agreement with the measurement.

I. Introduction

Recent advances in the design of millimeter-wave channelized receivers have created a need for small, integrated high-performance diplexers and multiplexers [1]–[4]. However, the design information on these components is very limited. In [2], the authors have designed a diplexer consisting of two *E*-plane filters and a waveguide T-junction. Since the reactance of the T-junction varies relatively rapidly as a function of frequency, it is difficult to compensate for the reactance in order to design diplexers with wide-band performance using this scheme. In this paper, we will describe a novel design of a millimeter-wave noncontiguous diplexer that is capable of wide-band applications. Channel bandwidths as wide as 10 GHz can now be achieved with an insertion loss of about 1 dB at *W* band. In Section II, the diplexer structure is described and the design principle is discussed. Section III provides a brief discussion on the theoretical analysis for the diplexer. Section IV summarizes the results of both the theory and measurement. Good agreements are obtained.

II. Diplexer Design

Fig. 1 shows a typical layout of the diplexer. It consists of two band-select filters and two waveguide-to-suspended-stripline probe transitions. The filters are *E*-plane filters [5]–[7] which can be bilateral, unilateral, or a combination of both. The complete circuit is fabricated on a single substrate, which is then cut to size and placed in a split-block housing.

The use of waveguide-to-suspended-stripline transitions at the common port of the diplexer makes is possible for wide-band applications. The idea stems from our experience in the design of suspended-stripline circuits at millimeter-wave frequencies, where a good transition has been designed to cover the full *W*-band (75–110 GHz). The design is a refinement of the work done by the other designers in this field [8].

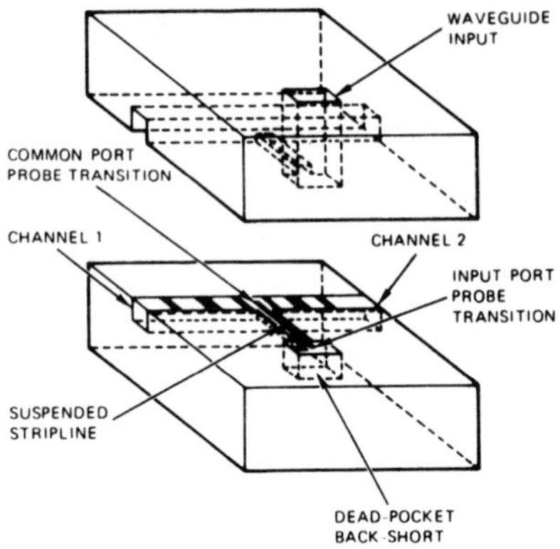

Fig. 1. Physical configuration of a diplexer.

There are two probe transitions, i.e., the input-port probe and the common-port probe. The input-port probe transition is very similar to the conventional coaxial-to-waveguide adaptor. One end of the suspended-strip probe is inserted into the broadside of the waveguide about one quarter-wavelength deep, and a waveguide backshort is located behind the probe to optimize the coupling. The other end of the probe is the common-port input to the diplexer and works in a similar way; however, this coupling structure is quite different. There are no waveguide backshorts but the bandpass filters. The transition works because at the passband of one channel, the other filter will reflect almost all the energy, and thus serves as a good short circuit. In this case, the loading effects of each filter at the short-circuit reference planes are required for the design of the transition and must be included in the net design of the diplexer to ensure good response. This requirement makes the *E*-plane filters an ideal candidate because of the existing accurate analytical modeling. High-performance *E*-plane filters have been designed up to 160 GHz using our modeling technique, and we have found excellent correlation between theory and measured data [7].

The filters may have any *E*-plane configuration, as defined earlier; however, because of the suspended-strip transition, we found it more convenient to make the filters and transitions on a single substrate. Furthermore, we have found that for the same filter specifications, the bilateral

Manuscript received March 15, 1985; revised July 1, 1985.
Y.-C. Shih and L. Q. Bui are with the Microwave Product Division, Hughes Aircraft Company, Torrance, CA 90509.
T. Itoh is with the Department of Electrical Engineering, University of Texas at Austin, Austin, TX 78712.

Y.C. Shih, L.Q. Bui, and T. Itoh, *IEEE Transactions on Microwave Theory and Techniques*, Vol. MTT-33, Dec. 1985, pp. 1465–1469.

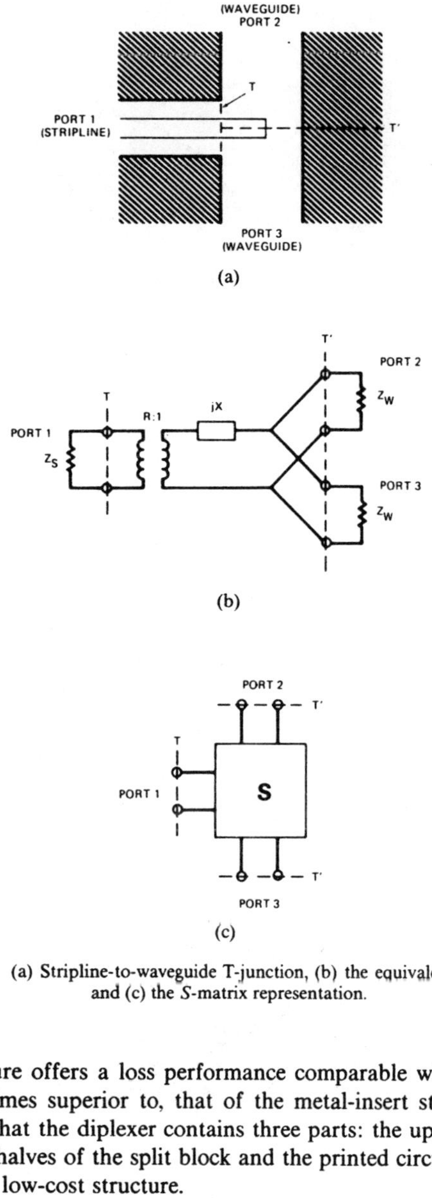

Fig. 2. (a) Stripline-to-waveguide T-junction, (b) the equivalent circuit, and (c) the S-matrix representation.

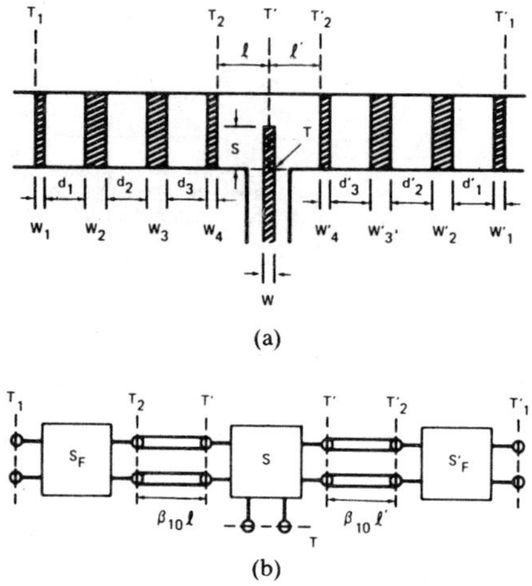

Fig. 3. (a) The printed-circuit part of the diplexer, and (b) the S-matrix network representation.

structure offers a loss performance comparable with, and sometimes superior to, that of the metal-insert structure. Note that the diplexer contains three parts: the upper and lower halves of the split block and the printed circuit. It is a very low-cost structure.

III. Analysis

In a recent study [9], we analyzed a microstrip-to-waveguide transition using an approximate theory. The probe transition is considered as a monopole antenna radiating into the waveguide. Based on an assumed current distribution on the probe antenna, a variational expression has been derived in the spectral domain for the input impedance at the feed point. An equivalent circuit is then derived as shown in Fig. 2(b) for the transition in Fig. 2(a). The expressions for the turn ratio R and the reactance jx will be presented in [9].

The three-port transition junction can then be represented by a 3×3 scattering matrix. The S-parameters are derived from the equivalent circuit as

$$S_{11}=\frac{R^2(jx+Zw/2)-Zs}{R^2(jx+Zw/2)+Zs}$$

$$S_{22}=S_{33}=\frac{-R^2Zw/2}{R^2(jx+Zw/2)+Zs}$$

$$S_{21}=S_{31}=S_{12}=S_{13}$$

$$=\frac{R\sqrt{ZwZs}}{R^2(jx+Zw/2)+Zs}$$

$$S_{23}=S_{32}=\frac{jxR^2+Zs}{R^2(jx+Zw/2)+Zs}.$$

We can now proceed to analyze the diplexer structure as shown in Fig. 3(a), where two channel filters have been connected to a common probe transition. The S-parameters for the E-plane filters are readily available from [5]. Taking into account the distance between the filters and the transition, we obtain the network representation in Fig. 3(b). The overall S-parameters for the diplexer are then obtained using the network combining technique described in the Appendix.

IV. Results

The waveguide-to-suspended-strip transition is essential for the diplexer design. Therefore, we have tested numerous W-band transition circuits to find a good combination. Fig. 4 depicts the geometry and the dimensions of an operating W-band transition. The circuit is designed on an RT-5890 Duroid substrate whose dielectric constant is 2.0. We have analyzed, fabricated, and tested two such transitions connected back-to-back by a 1-in-long suspended stripline. The results are shown in Fig. 5 with good agreement between the theory and measurement. The return loss is better than 15 dB, and the measured insertion loss is

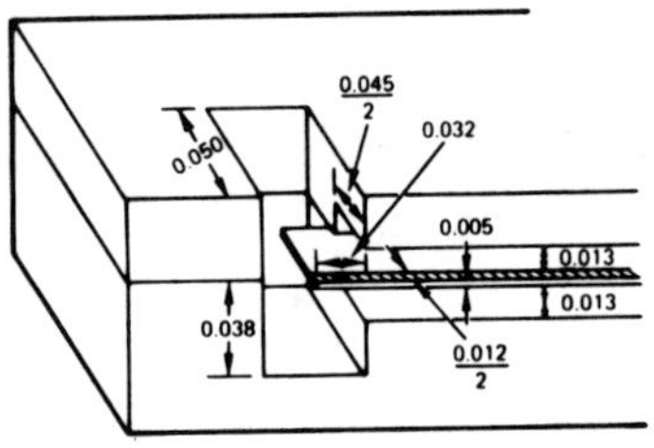

Fig. 4. Cross-section view of a *W*-band waveguide-to-suspended-strip transition in WR-10 waveguide.

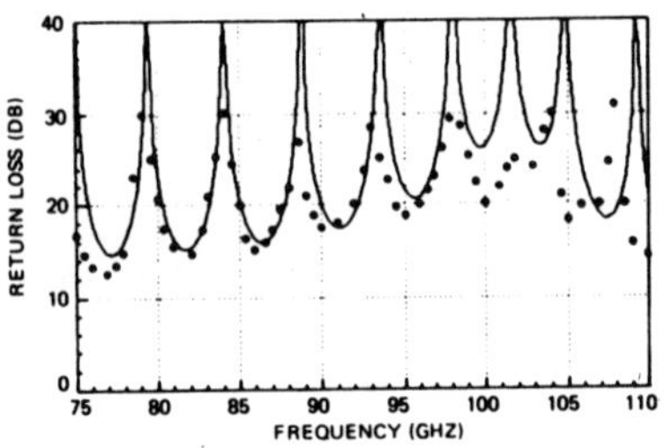

Fig. 5. Return loss of the probe transition at *W*-band (—— theory, • measurement).

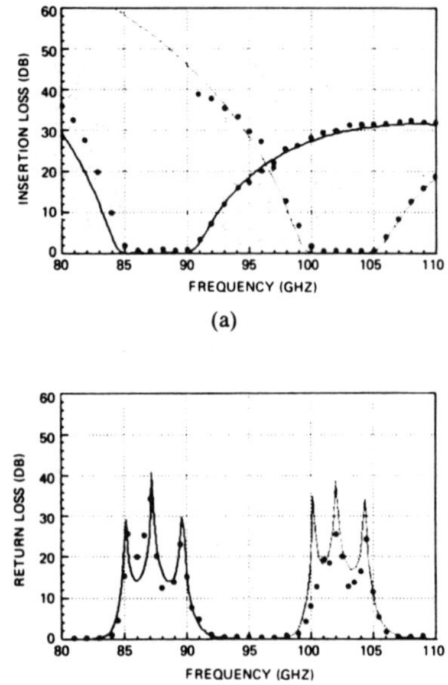

(a)

(b)

Fig. 6. (a) Insertion loss, and (b) return loss of *E*-plane filters (—— theory, • measurement).

TABLE I
DIMENSIONS FOR A *W*-BAND DIPLEXER (ALL UNITS IN MILS)

	$W_1 = W_4$	$W_2 = W_3$	$d_1 = d_3$	d_2	ℓ	W	S
Filter 1	3.0	20.9	63.4	64.4	27		
Filter 2	8.7	37.4	45.2	44.9	27		
Probe						12	32

better than 0.7 dB (not shown) over the *W*-band frequencies.

The *E*-plane filters are designed using the procedures described in [5]–[7]. As an example, we have designed two 3-cavity bilateral *E*-plane filters having 5-GHz bandwidths centered at 87.5 and 102.5 GHz, with passband ripples of 0.2 and 0.1 dB, respectively. The dimensions of the filters are listed in Table I for a dielectric substrate of 5-mil thick. Both the theoretical and experimental results are shown in Fig. 6 for comparison. The measured insertion loss is less than 1 dB in the passband of both filters. In the analyses, we have also obtained the short-circuit reference location for each filter at the passband frequencies of the other filter. Based on this information, the relative locations of the filters with respect to the probe are then adjusted to yield a complete diplexer.

Fig. 7 is the picture of a complete diplexer. The *E*-plane filters and the transitions are fabricated on a single 5-mil substrate using chemical etching. The circuit is then cut and placed in the split-block housing. The diplexer was evaluated over the frequencies of interest and its performance is shown in Fig. 8. A close agreement is found between the theoretical and experimental data. In the passband of the channels, the return loss is more than 10 dB, and more typically 15 dB. The measured insertion loss is typically 1 dB. Notice that each channel very much preserves the characteristics of the corresponding filter, except that the upper-end rejection of the lower channel has a slight improvement.

Fig. 7. Photograph of a completed *W*-band diplexer.

Diplexers having wider channel passbands can also be built in a similar manner. For example, Fig. 9 shows the performance of a diplexer having a 10-GHz passband for both channels. A 10-GHz guard band is placed between the two channels. The input return loss in the passband of the channels is more than 12 dB, and the insertion loss about 1 dB.

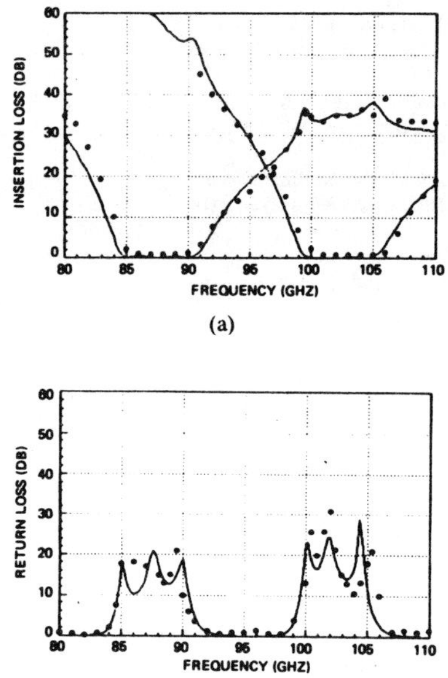

(a)

(b)

Fig. 8. The (a) insertion loss and (b) return loss of a diplexer shown in Fig. 7 (—— theory, • measurement).

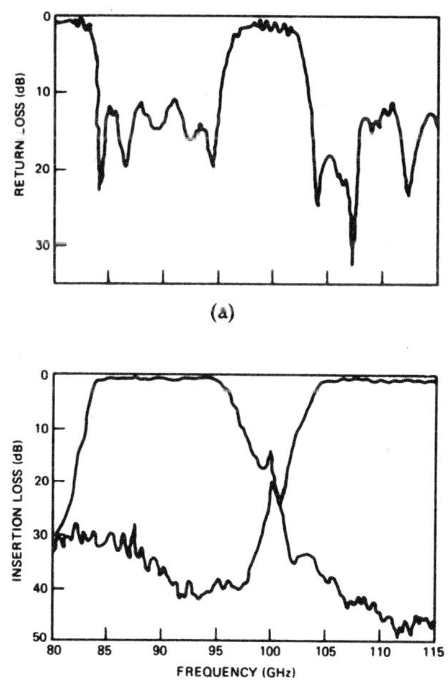

(a)

(b)

Fig. 9. The measured (a) return loss (b) insertion loss of a diplexer with 10-GHz channel bandwidths.

V. Conclusion

The combination of the E-plane filters and a unique suspended probe transition imbedded at the common port has resulted in the design of high-performance and low-cost diplexers with wide channel bandwidths. The complete circuit is printed on a single substrate and is easily assembled in a split-block housing. This makes it highly reproducible using photolithographic techniques. Typical W-band diplexers exhibit about 1-dB insertion loss over the channel passbands.

Combining the probe-transition analysis program with the E-plane filter analysis program, we have calculated the frequency response of a diplexer in good agreement with the measurement.

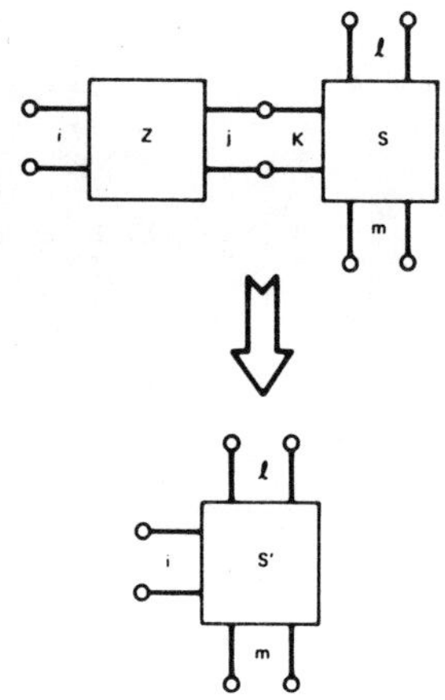

Fig. 10. The network combination scheme for a two-port and a three-port network.

Appendix

Refer to Fig. 10. Port j of a 2-port network Z is connected to port k of a 3-port network S. The result is a 3-port network S' with the following elements:

$$S'_{ii} = Z_{ii} F Z_{ij} S_{kk} Z_{ji}$$

$$S'_{ll} = S_{ll} + F S_{lk} Z_{jj} S_{kl}$$

$$S'_{mm} = S_{mm} + F S_{mk} Z_{jj} S_{km}$$

$$S'_{ml} = S_{ml} + F S_{mk} Z_{jj} S_{kl}$$

$$S'_{lm} = S_{lm} + F S_{lk} Z_{jj} S_{ml}$$

$$S'_{il} = F Z_{ij} S_{kl}$$

$$S'_{im} = F Z_{ij} S_{km}$$

$$S'_{li} = F S_{lk} Z_{ji}$$

$$S'_{mi} = F S_{mk} Z_{ji}$$

where

$$F = \frac{1}{1 - S_{kk} Z_{jj}}.$$

References

[1] P. H. Meier *et al.*, "26 to 60 GHz channelized receiver meets surveillance requirements," *Microwave Systems News*, vol. 11, pp. 60–80, Dec. 1981.

[2] L. D. Cohen *et al.*, "Millimeter-wave multiplexers with printed circuit elements for the 88 to 100 GHz frequency range," in *1984 IEEE MTT-S Dig.*, pp. 233–235.

[3] K. Breuer and N. Worontzoff, "A low cost multiplexer for channelized receiver front ends at millimeter-waves," in *1980 IEEE MTT-S Dig.*, pp. 150–152.

[4] A. Hislop and D. Rubin, "Suspended substrate *Ka*-band multiplexer," *Microwave J.*, vol. 24, pp. 73–77, June 1981.

[5] Y.-C. Shih, T. Itoh, and L. Q. Bui, "Computer-aided design of millimeter-wave *E*-plane filters," *IEEE Trans. Microwave Theory Tech.*, vol. MTT-31, pp. 135–152, Feb. 1983.
[6] Y.-C. Shih, "Design of waveguide *E*-plane filters with all-metal inserts," *IEEE Trans. Microwave Theory Tech.*, vol. MTT-32, pp. 695–704, July 1984.
[7] L. Q. Bui, D. Ball, and T. Itoh, "Broad-band millimeter-wave *E*-plane bandpass filters," *IEEE Trans. Microwave Theory Tech.*, vol. MTT-32, Dec. 1984.
[8] A. Hislop and D. Rubin, private communication.
[9] Y.-C. Shih, L. Q. Bui, and T. Itoh, "Analysis and design of microstrip-waveguide transitions," to be published.

6.2.7

DIPLEXER ARRANGEMENT UTILIZING AN E-PLANE METAL INSERT

G. Reiter, Mrs. T. Kolumbán

ABSTRACT

The design method and experimental results obtained with a realized 8-GHz metal insert diplexer are presented.

INTRODUCTION

In a recent publication, a waveguide band-pass filter utilizing an E-plane metal insert has been described [1.]. According to this filter realization method, the waveguide is split into two equal parts along the E-plane, and a thin metal insert having rectangular openings is placed between the two halves. The waveguide cavities are thus formed by empty waveguide sections between the strips of the metal insert, all reactance elements of the filter being formed by strips of a single metallic plate. This type of filter is easily reproducible, and can be realized accurately and economically by microwave integrated circuit technology.

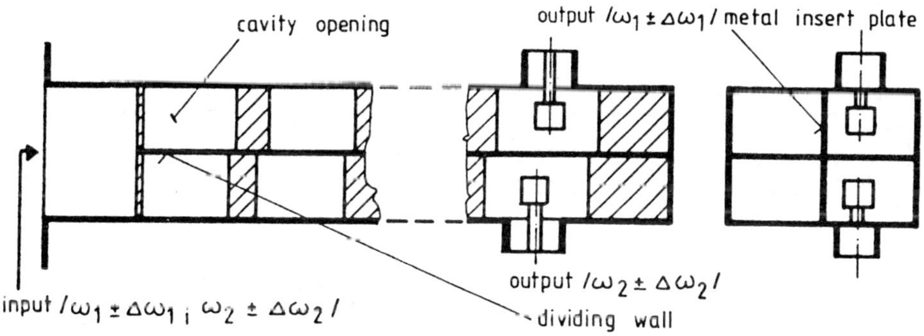

Fig. 1. Diplexer arrangement

This paper describes the application of the metal insert concept to a waveguide diplexer. The waveguide assembly has an inner wall parallel with the wider waveguide walls, dividing the waveguide into two equal parts (Fig. 1.). The filter cavities are realized by a single thin metal insert plate (see Fig. 2.) perpendicular to the dividing wall with two rows of rectangular opennings, thus forming cavities in both waveguide halves. The first two metal strips of the insert are at the beginning of the dividing wall, near the input port realized in the original wavegiude. The two similar output ports are realized in coaxial lines at the far ends of the half-heigth waveguide section, the enlarged inner conductors of these lines protruding into the last filter cavities. The frequency bands separated by the diplexer are $\omega_1 \pm \Delta\omega_1$ and $\omega_2 \pm \Delta\omega_2$. It is assumed that $\Delta\omega_1 \geqslant \Delta\omega_2$ and thus $|\omega_1 - \omega_2| \geqslant 3\Delta\omega_1$

Research Institute for Telecommunication "TKI", Budapest, Hungary

G. Reiter and T. Kolumban, *Proceedings of 15th European Microwave Conference*, 1985, pp. 859–864.

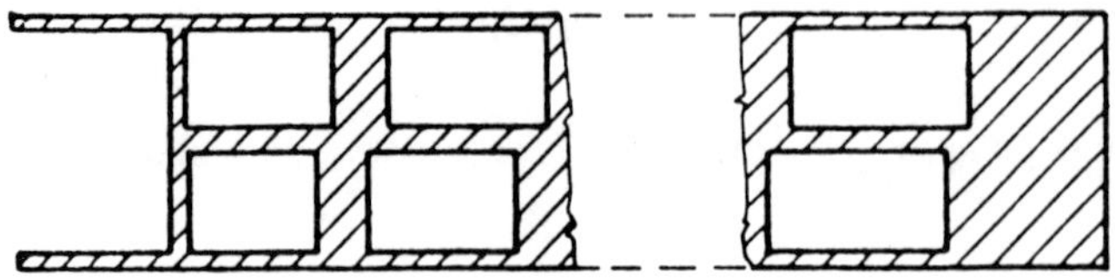

Fig. 2. Metal insert plate

DESIGN METHOD

The equivalent circuit of the diplexer arrangement shown in Fig. 1. is made up of inverters and series resonant circuits having loaded Q's of q and resonant frequencies of $\omega_{1k}(k=1,2,\ldots m)$ and $\omega_{2\ell}(\ell=1,2\ldots n)$, respectively Fig. 3.), where

$$q = \frac{\pi}{4}\left(\frac{\lambda_g}{\lambda_0}\right)^2 \qquad (1)$$

The first step of the design procedure is to calculate the impedances K_{1k} and $K_{2\ell}$ of the inverters making up the two waveguide branches and the resonant frequencies ω_{1k} and $\omega_{2\ell}$. As the second step, these impedance and resonant frequency values are used to calculate the following geometrical dimensions of the diplexer:

width and height of the original and halved waveguides,
length of the cavity openings and of the metal strips,
protruding enlarged inner conductor dimensions of the output coaxial line sections.

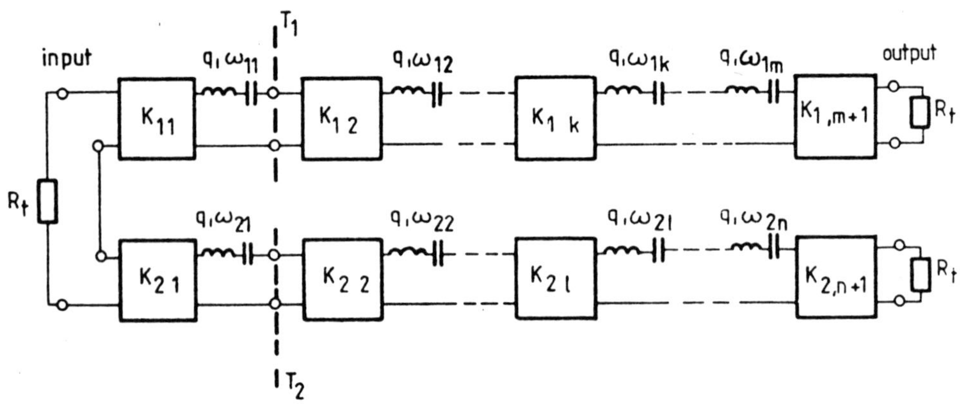

Fig. 3. Equivalent circuit of the diplexer

A circuit optimization procedure is used to calculate the impedances and resonant frequencies of the equivalent circuit in Fig. 3. It is feasible to choose the initial values of the optimizataion close to the final values in order to shorten the calculations.

This is promoted by determining two prototype filters with transmission characteristics specified for the two diplexer branches. These prototype filters are ladder networks comprising series and parallel resonant circuits. The well known method of waveguide filter design allows a transformation of the prototype filters into a network comprising inverters of impedance K^{o}_{1K} and $K^{o}_{2\ell}$ and series resonant circuits of loaded Q-factors of q and resonant frequencies of ω_1 and ω_2 (Fig. 4.). The numbers m and n are giving from this procedure. Let us divide the equivalent circuits in Fig.3. and 4. along the dashed lines at ports T_1 and T_2 into right and left parts. Suitable intitial values for impedances K_{1K} and $K_{2\ell}$ are obtained by assuming identical inverters and series resonant circuits for the right part of the equivalent circuits:

$$K_{1k} = K^{o}_{1k} \quad ; \quad \omega_{1k} = \omega_1 \qquad (k = 2, \dots m+1) \tag{2.}$$

$$K_{2\ell} = K^{o}_{2\ell} \quad ; \quad \omega_{2\ell} = \omega_2 \qquad (\ell = 2, \dots n+1) \tag{3.}$$

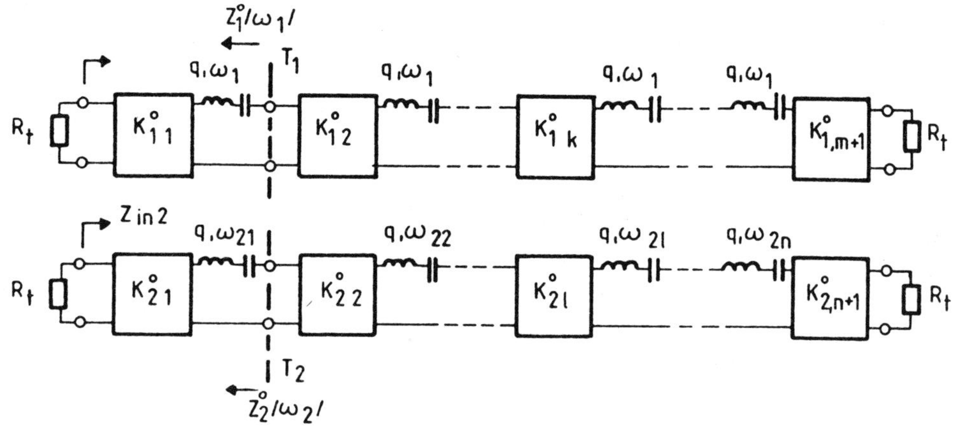

Fig. 4. Transformed prototype filters

Finally, the left part of the network in Fig. 3. is converted into a more simple network at frequencies ω_1 and ω_2 (Fig. 5) in order to determine the initial values of K_{11}, K_{21}, ω_{11} and ω_{21}, respectively. The reactances X_1 and X_2 represent the input reactance components of the networks in Fig 4.
$X_1 = \mathrm{Im}\, Z_{in2}(\omega_1)$; $X_2 = \mathrm{Im}\, Z_{in1}(\omega_2)$.
Let us assume identical input impedances for the left part networks Fig. 4. and 5. $Z_1(\omega_1) = Z^{o}_1(\omega_1)$; $Z_2(\omega_2) = Z^{o}_2(\omega_2)$. This assumption allows the calculation of the initial values from the following expressions:

$$K_{11} = K^{o}_{11}\sqrt{\frac{R_t^2 + X_1^2}{R_t^2}} \; ; \qquad \frac{\omega_{11}}{\omega_1} - \frac{\omega_1}{\omega_{11}} = \frac{X_1 (K^{o}_{11})^2}{2q\, R_t^2} \; ;$$

$$K_{21} = K^{o}_{21}\sqrt{\frac{R_t^2 + X_2^2}{R_t^2}} \; ; \qquad \frac{\omega_{21}}{\omega_2} - \frac{\omega_2}{\omega_{21}} = \frac{X_2 (K^{o}_{22})^2}{2q\, R_t^2} \; ; \tag{4.}$$

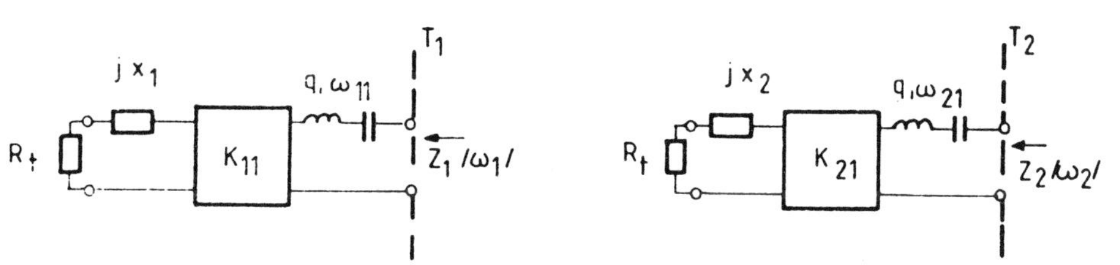

Fig. 5. Simplified equivalent circuits

From this initial values, the optimization procedure can be carried out. It is also possible to make the final values equal the initial values and to adjust the specified transmission characteristics with the tuning elements of the realized diplexer.

The knowledge of the circuit elements in Fig. 3. allows the following determination of the diplexer geometrical dimensions. The dimensions of the full height waveguide part are taken as the standardized IEC dimensions in the diplexer transmission bands. The width of the divided waveguide parts have to be reduced so as to obtain a specified attenuation of the TE_{o2} mode attenuated in the divided waveguide parts. This should be higher than the highest specified stop-band attenuation. The above consideration results in a more symmetrical transmission characteristics of the diplexer branches.

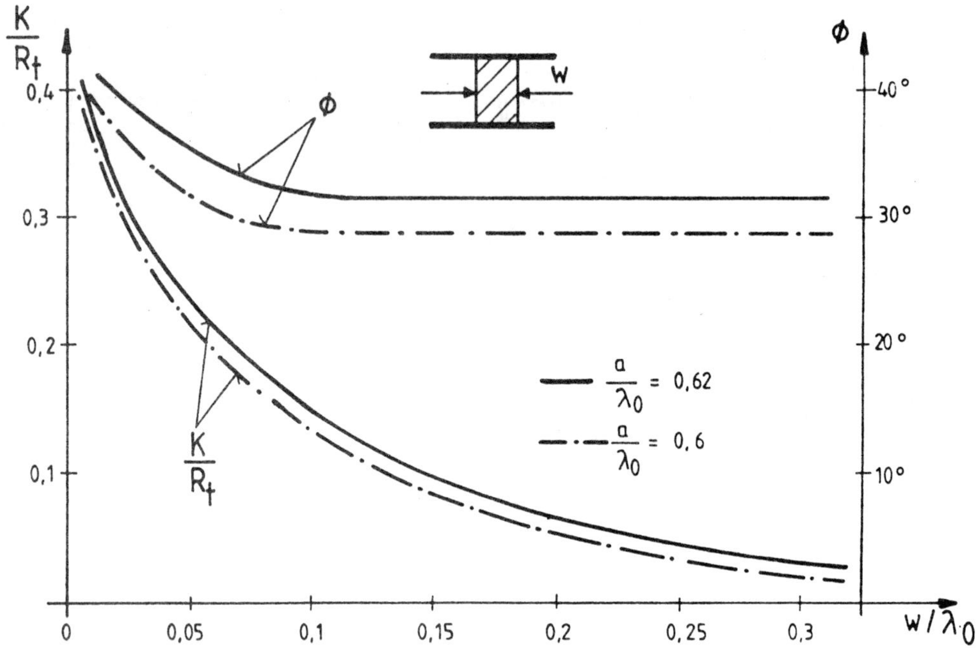

Fig. 6. Calculated values of K and ϕ

The output inverters $K_{1,m+1}$ and $K_{2,n+1}$ are realized by the protruding enlarged inner conductors of the coaxial line section, while all other inverters are made up of the metal strips. Apart of the first two metal strips at the beginning of the dividing wall, all other strip lengths can be calculated from the K_{1k} and $K_{2\ell}$ inverter impedances by the method given in Ref. [2]. According to this method, T-sections made up of inductive elements are assigned as equivalent circuits to the metal strips, and these T-sections are further developed by two transmission line sections of electrical length $\phi/2$ to form inverters. This method can also be applied for the flat waveguide parts of the divided section. Fig. 6. shows a diagram based on Ref. [2], showing the relation between the inverter impedance K, the electrical length ϕ and the length w of the metal plate.

The series resonant circuits in the equivalent circuit are realized by the cavity openings betwen the metal strips. This results in a cavity opening length wich is equal to the sum of the half waveguide wavelength at the resonant frequencies and the length of the transmission line sections pertaining to the inverters.

A cut-and-try method has been used to determine the lengths ℓ_1 and ℓ_2 of the metal strips realizing inverters K_{11} and K_{21}, further the diameter D and protruding length L of the inner conductor realizing the inverter $K_{1,m+1}$. The diplexer parts comprising the elements to be determined are separately prepared (Fig. 7a. and 8a.). Assume that the values of ℓ_1, ℓ_2, D and L are correct; the equivalent one-ports of these parts are then as shown in Figs. 7b. and 8b. Figs. 7c. and 8c. show the reflection factor phase-frequency functions of the one-ports. In these Figures, the frequency ranges pertaining to phase values $\pm 90^0$, calculated from the equivalent one-port inverter impedance and resonant frequencies of series resonant circuits from expressions (5.), are also indicated. These phase-frequency functions can be measured by a network analyzer to wich the diplexer parts shown in Figs. 7a. and 8a. are connected. The dimensions ℓ_1, ℓ_2, D and L should be adjusted to obtain the frequency ranges calculated from (5.) on the network analyzer display. The resonant frequencies are adjusted by a suitable choice of the cavity opening lengths.

$$\frac{\delta\omega_1}{\omega_1} = \frac{K_{11}^2}{2q} \quad ; \quad \frac{\delta\omega_2}{\omega_2} = \frac{K_{21}^2}{2q} \quad ; \quad \frac{\delta\omega}{\omega_1} = \frac{K_{1,m+1}^2}{2q} \qquad (5)$$

The wanted geometrical dimensions have thus been experimentaly determined. The geometrical dimensions of the output ω_2 can similarly be determined.

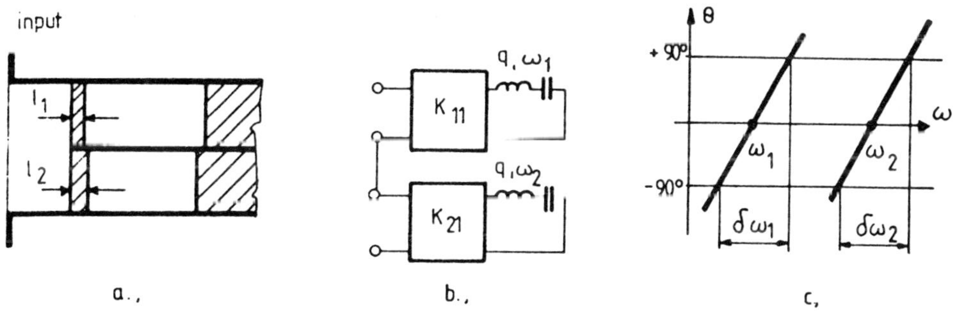

Fig. 7. Diplexer part at the input

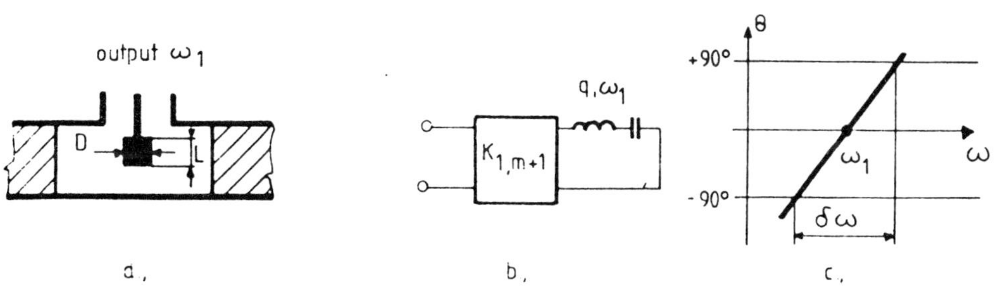

Fig. 8. Diplexer part at the output ω_1

MEASUREMENT RESULTS

According to the design method outlined, a diplexer operating in the 8 GHz frequency band intended for realizing a transmit-receive coupler has been realized. Six cavities are comprised in each of the two filter branches . Tuning screws for exact adjustment of the transmission characteristics have been placed into the cavity openings and waveguide sections between the metal strips. The measurement results obtained by a network analyzer are as follows:

at the output port	: pass-band	7,7 - 8,0 GHz
	rejection-band :	8,2 - 8,5 GHz
at the outpu port	: pass-band	: 8,2 - 8,5 GHz
	rejection-band :	7,7 - 8,0 GHz

attenuation in the pass-bands : $\leq$1 dB
attenuation in the rejection-bands: $\geq$45 dB

The transmission characteristics of the diplexer branches are practically indentical with the prototype filter characteristics. However, in the frequency range between the two pass-bands, the characteristics slopes are higher than that of the prototype filters.

REFERENCES

1. R. Vahldieck, J. Bornemann, F. Arndt, D. Grauerholz: "Optimized Waveguide E-plane Metal Insert Filters for Millimeter Wave Applications" IEEE Trans., MTT-31, pp. 65-69, January 1983.

2. Y. Konishi, K. Uenakada: "The Design of a Bandpass Filter with Inductive Strip-Planar Circuit Mounted in Waveguide" IEEE Trans., MTT-22, pp. 869-873, October 1974.

6.3.1

Propagation in Coupled Unilateral and Bilateral Finlines

ARVIND K. SHARMA, MEMBER, IEEE, AND
WOLFGANG J. R. HOEFER, SENIOR MEMBER, IEEE

Abstract —**The propagation characteristics of coupled unilateral and bilateral finlines in the even and odd modes are evaluated with the hybrid-mode formulation in the spectral domain. The frequency-dependent guide wavelength is obtained from the solution of the characteristic equation. The characteristic impedance based on the power–voltage definition is also evaluated. Numerical results are presented to study the influence of various structural parameters on the characteristics of finlines.**

I. INTRODUCTION

Among various forms of transmission media, finlines have demonstrated potential for their use in millimeter-wave integrated circuits [1], [2]. In a typical finline configuration, a planar circuit is placed in the E-plane of a rectangular waveguide. This combines the advantages of both planar circuits and waveguides. Its wide single-mode bandwidth, low dispersion, moderate attenuation and compatibility with semiconductor devices are attractive features at millimeter wavelengths.

The propagation in finlines has been a subject of considerable interest recently. An accurate description of the dispersion can be obtained with the hybrid-mode formulations [3]–[5]. This was first presented by Hofmann [6] who used the space-domain formulation. Subsequently, the spectral-domain formulation was utilized by Knorr and Shayda [7], Schmidt and Itoh [8], as well as by the authors [9]–[12], to determine the propagation characteristics of unilateral and bilateral finlines.

With the increased interest in finlines, there is a need to

Manuscript received September 29, 1982; revised January 25, 1983.
A. K. Sharma is with RCA Laboratories, David Sarnoff Research Center, Princeton, NJ 08540.
W. J. R. Hoefer is with the Department of Electrical Engineering, University of Ottawa, Ottawa, Ontario K1N 6N5, Canada.

A.K. Sharma and W.J.R. Hoefer, *IEEE Transactions on Microwave Theory and Techniques*, Vol. MTT-31, No. 6, June 1983, pp. 498–502.

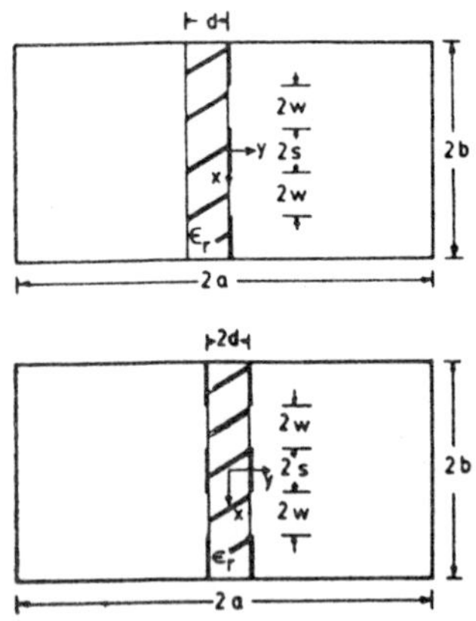

Fig. 1. Coupled unilateral and bilateral finlines.

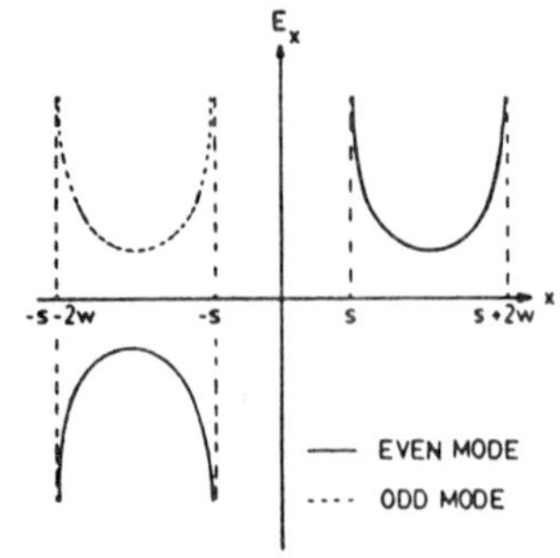

Fig. 2. Variation of the transverse electric field in the even and odd modes.

analyze coupled unilateral and bilateral finlines which are useful in the realization of various millimeter-wave components such as directional couplers and filters. An approximate solution for the propagation in the coupled unilateral finline, presented by Simons and Arora [13], is based on Cohn's procedure for the slot-line on a dielectric substrate. This procedure must take into account a large number of orthogonal modes to achieve reasonable accuracy. Schmidt, Itoh, and Hofmann [14] have recently proposed a full-wave analysis for the symmetric and asymmetric coupled unilateral finlines in the spectral domain.

In this paper, the propagation characteristics of the symmetric coupled unilateral and bilateral finlines [11] are investigated in detail. The full-wave formulation of the spectral-domain technique is utilized to evaluate the guide wavelength. The characteristic impedance is computed on the basis of the power–voltage definition.

II. Analysis

The coupled unilateral and bilateral finline configurations, as shown in Fig. 1, comprise a substrate of relative permittivity ϵ_r in the E-plane of a rectangular waveguide with dimensions $2a$ and $2b$. The thickness of the substrate is d and $2d$ for the coupled unilateral and bilateral finlines, respectively. The slots of width $2w$ are separated by a distance $2s$.

In the spectral-domain analysis of the structure, the Fourier transform of the Green's functions are related to the transforms of the current densities on the conductors and the electric fields in the region of the interface complementary to the conductors, via the equation

$$\begin{bmatrix} \tilde{H}_{11}(\alpha_n,\beta,k_0) & \tilde{H}_{12}(\alpha_n,\beta,k_0) \\ \tilde{H}_{21}(\alpha_n,\beta,k_0) & \tilde{H}_{22}(\alpha_n,\beta,k_0) \end{bmatrix} \begin{bmatrix} \tilde{E}_x(\alpha_n) \\ \tilde{E}_z(\alpha_n) \end{bmatrix} = \begin{bmatrix} \tilde{J}_x(\alpha_n) \\ \tilde{J}_z(\alpha_n) \end{bmatrix} \tag{1}$$

where α_n is the Fourier transform variable, β is the propagation constant, and k_0 is the free-space wavenumber. $\tilde{E}_x$, $\tilde{E}_z$, $\tilde{J}_x$, and $\tilde{J}_z$ are the electric fields in the aperture and the current densities on the conductors, respectively. With the application of Galerkin's procedure and Parseval's theorem, we obtain a set of algebraic equations in terms of the unknown constants of the basis functions. At a given frequency, a nontrivial solution for the propagation constant β is obtained by setting the determinant of the coefficient matrix equal to zero and finding the root of the equation. The guide wavelength is then obtained from the propagation constant.

The characteristic impedance can be evaluated in the spectral domain with the definition

$$Z_0 = \frac{V_0^2}{2P_{avg}} \tag{2}$$

where V_0 is the voltage between the fins along the air–dielectric interface and is given by

$$V_0 = \int_s^{s+2w} E_x(z,d)\,dx. \tag{3}$$

The time-averaged power P_{avg} is evaluated as follows:

$$P_{avg} = 1/2\,\mathrm{Re}\int_{-a}^{a}\int_{-b}^{b}(E_x H_y^* - E_y H_x^*)\,dx\,dy. \tag{4}$$

III. Basis Functions

Following the mathematical formulation presented above, the longitudinal and transverse electric fields are expressed in terms of known basis functions with unknown coefficients. The electric fields are appropriately chosen as follows:

$$E_x = \begin{cases} (w^2 - x^2)^{-1/2}, & s<|x|<s+2w \\ 0, & \text{elsewhere} \end{cases} \tag{5}$$

$$E_z = \begin{cases} x(w^2 - x^2)^{1/2}, & s<|x|<s+2w \\ 0, & \text{elsewhere.} \end{cases} \tag{6}$$

The Fourier transform of these functions can be easily evaluated with the definition of the discrete Fourier transform.

The propagation in a coupled finline is described in terms of two independent modes of excitation, known as even and odd modes, which corresponds to even E_z (or odd E_x) and odd E_z (or even E_x), respectively. The variation of the transverse electric field E_x in the even and odd modes is shown in Fig. 2.

IV. Numerical Results

We have developed computer programs which evaluate the guide wavelength and the characteristic impedance at a given frequency of the unilateral and bilateral finlines. We have verified these computer programs by comparing the results with those already reported by Knorr and Shayda [7] for the unilateral finline, and Schmidt and Itoh [8] for the bilateral finline. Since there was good agreement with them for various structural parameters, the numerical results presented in this paper are fairly accurate. Both the longitudinal and transverse electric fields have been taken into account to obtain these results.

The guide wavelength and characteristic impedance have been evaluated for various structural parameters of the finlines on substrates of relative permittivity $\epsilon_r = 2.22$ (RT-Duroid). The present analysis is based on the symmetries of the waveguide structures, and therefore the propagating modes correspond to the electric and magnetic wall symmetry at $x = 0$. In the case of a

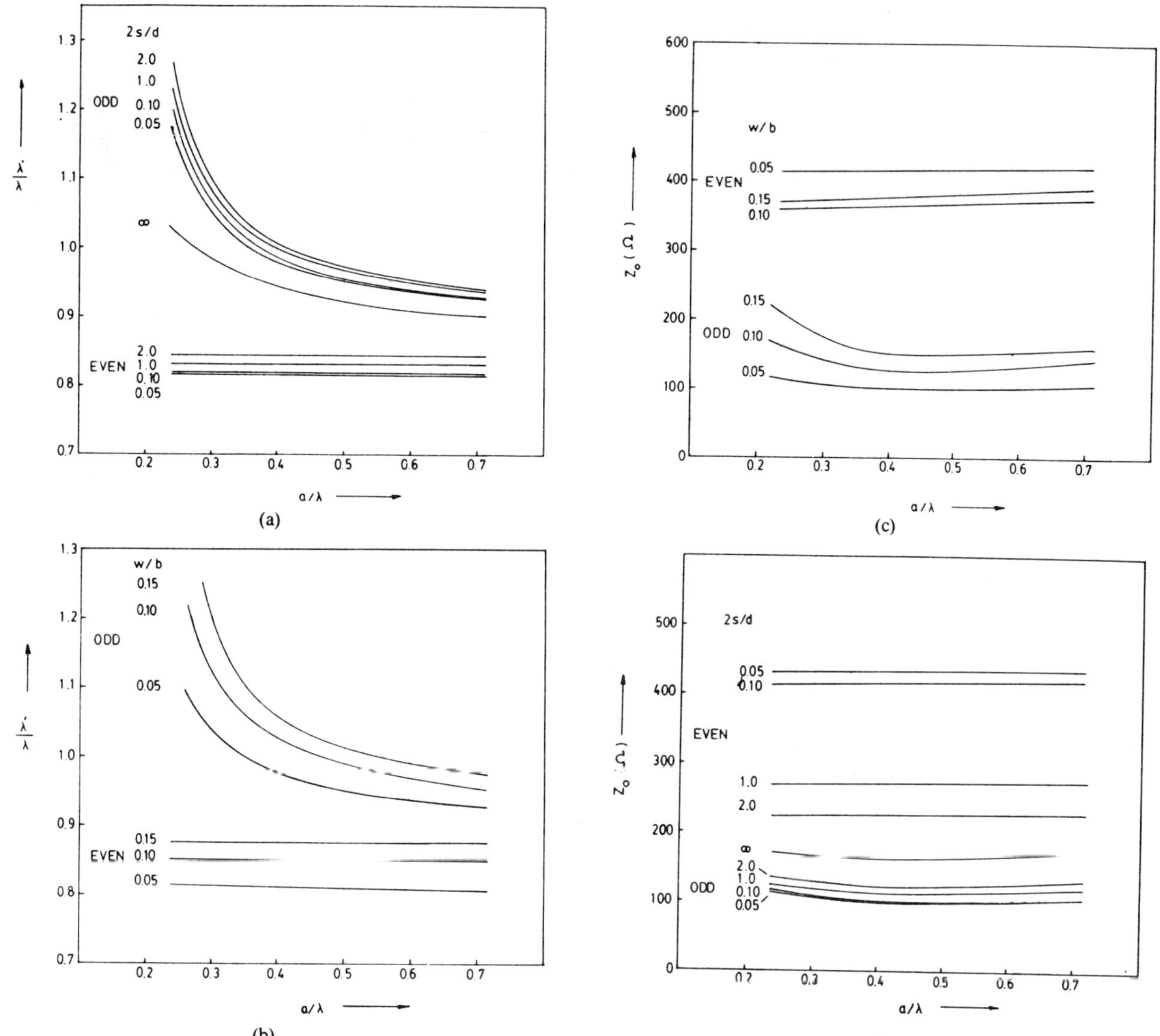

Fig. 3. Characteristics of coupled unilateral finlines in the even and odd modes as a function of normalized frequency a/λ. Normalized guide wavelength (a) with $w/b = 0.05$ and $2s/d = 0.05, 0.1, 1.0, 2.0$; (b) with $2s/d = 0.1$ and $w/b = 0.05, 0.1, 0.15$. Characteristic impedance (c) with $w/b = 0.05$ and $2s/d = 0.05, 0.1, 1.0, 2.0$; (d) with $2s/d = 0.1$ and $w/b = 0.05, 0.1, 0.15$. $d/a = 0.03515$, $a = 0.3556$ cm, $b = 0.1778$ cm, and $\epsilon_r = 2.22$.

coupled bilateral finline, there is an additional magnetic wall symmetry at $y = 0$.

The numerical results for the normalized guide wavelength (λ'/λ) and characteristic impedance (Z_0) in the even and odd modes have been evaluated as a function of the normalized frequency (a/λ). For the coupled unilateral finline, the quantities λ'/λ and Z_0 are plotted in Fig. 3 as a function of normalized frequency (a/λ) for values of normalized slot separation $(2s/d)$ and normalized slot width (w/b). Fig. 3(a) and (b) shows the variation in λ'/λ for various values of $2s/d$ and w/b, respectively. For a fixed value of $2s/d$, we observe that λ'/λ increases slightly in the even mode, whereas in the odd mode it decreases as a function of a/λ. When $2s/d$ is increased for a fixed value of a/λ, it shows an increase in the even mode, but in the odd mode it initially increases and then decreases. Both in the even and odd modes, λ'/λ increases with an increase in w/b.

The computed values of the characteristic impedance in the even and odd modes are shown in Fig. 3(c) and (d). Increasing $2s/d$ at a fixed value of a/λ decreases its value in the even mode and increases in the odd mode. The characteristic impedance in the odd mode for small values of $2s/d$ is also observed to be close to one half of the characteristic impedance of the corresponding unilateral finline with twice the slot width. With an increase in w/b, Z_0 increases in the odd mode, while in the even mode it first decreases and then increases.

For the coupled bilateral finline, the behavior of λ'/λ and Z_0 as a function of a/λ is illustrated in Fig. 4(a)–(d) for various structural parameters s/d and w/b. The variation in λ'/λ, as shown in Fig. 4(a), is similar to that observed for coupled unilateral finline. The propagating mode in the odd mode appears to behave similarly to the one on coplanar striplines. The guide wavelength in the even and odd modes increases with w/b for a fixed value of a/λ (Fig. 4(b)). With an increase in s/d at a fixed value of a/λ, the characteristic impedance, as shown in Fig. 4(c), increases in the even mode, and in the odd mode it initially increases and then decreases. However, as s/d is further

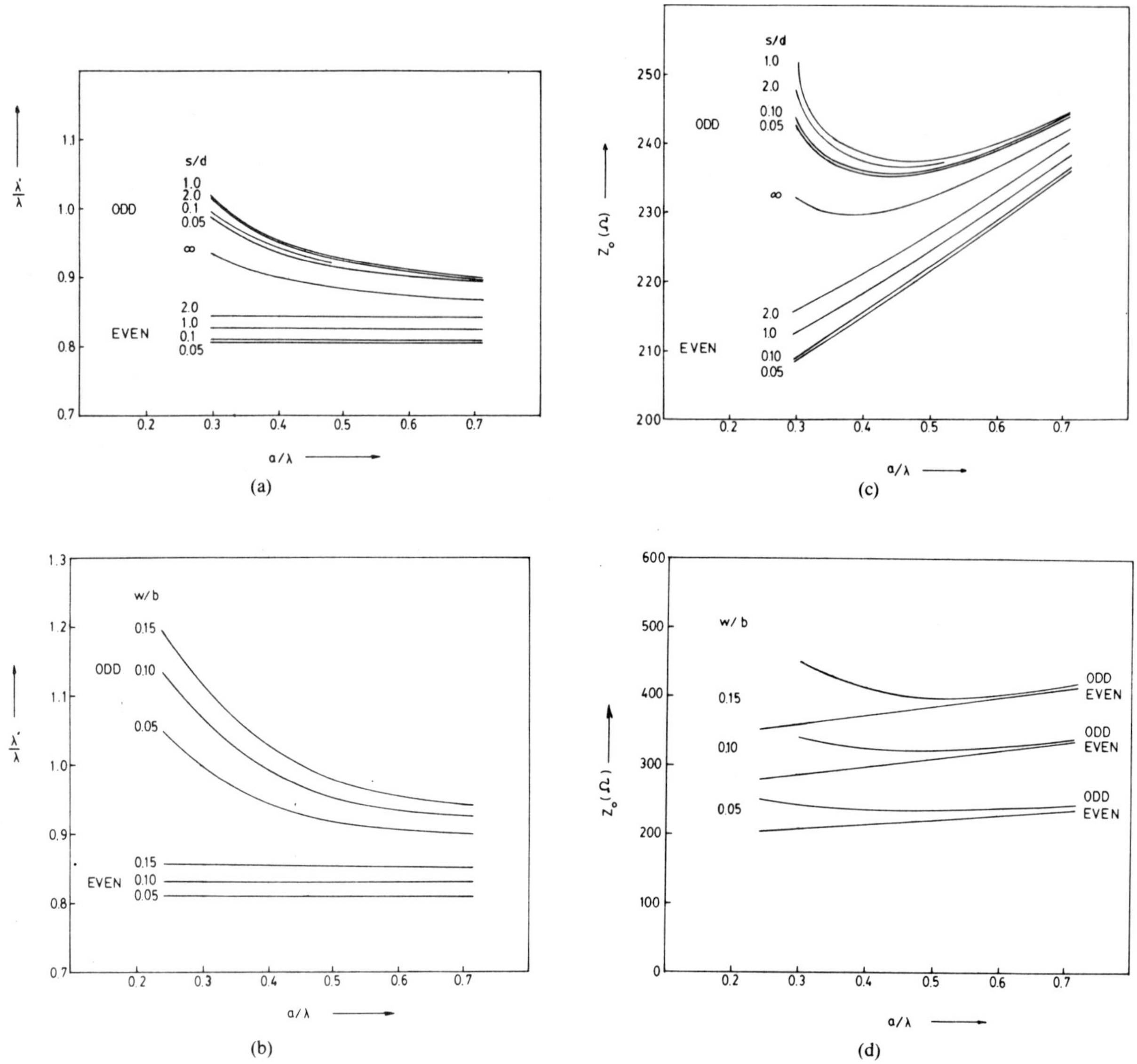

Fig. 4. Characteristics of coupled bilateral finlines in the even and odd modes as a function of normalized frequency a/λ. Normalized guide wavelength (a) with $w/b = 0.05$ and $s/d = 0.05, 0.1, 1.0, 2.0$; (b) with $s/d = 0.1$ and $w/b = 0.05, 0.1, 0.15$. Characteristic impedance (c) with $w/b = 0.05$ and $s/d = 0.05, 0.1, 1.0, 2.0$; (d) with $s/d = 0.1$ and $w/b = 0.05, 0.1, 0.15$. $d/a = 0.03515$, $a = 0.3556$ cm, $b = 0.1778$ cm, and $\epsilon_r = 2.22$.

increased, both the even- and odd-mode characteristic impedances approach to the characteristic impedance of the corresponding bilateral finline with the same slot width. This is also observed for various values of w/b (Fig. 4(d)). The coupled bilateral finline also supports quasi-TEM propagating modes which correspond to an electric wall symmetry at $y = 0$. These modes have not been considered in this investigation.

V. Conclusion

In this paper, we have presented a detailed description of the propagation in coupled unilateral and bilateral finlines using the hybrid-mode formulation in the spectral domain. The results for the normalized guide wavelength and characteristic impedance have been presented as a function of normalized frequency for some structural parameters. This technique can be utilized to obtain accurate propagation characteristics of any finline geometry.

References

[1] P. J. Meier, "Integrated finline millimeter components," *IEEE Trans. Microwave Theory Tech.*, vol. MTT-22, pp. 1209–1216, Dec. 1974.

[2] P. J. Meier, "Millimeter integrated circuits suspended in *E*-plane of rectangular waveguide," *IEEE Trans. Microwave Theory Tech.*, vol. MTT-26, pp. 726–733, Oct. 1978.

[3] J. B. Davis and D. Mirshekar-Syahkal, "Spectral domain solution of arbitrary coplanar transmission line with multilayer substrate," *IEEE Trans. Microwave Theory Tech.*, vol. MTT-25, pp. 143–146, Feb. 1977.

[4] R. H. Jansen, "Unified user-oriented computation of shielded, covered and open planar microwave and millimeter-wave transmission line characteristics," *Inst. Elec. Eng. Proc. H, Microwaves, Optics and Antennas*, vol. 3, pp. 14–22, Jan. 1979.

[5] A. Beyer and I. Wolff, "A solution of the earthed finline with finite metallization thickness," in *1980 IEEE MTT-S Int. Microwave Symp. Dig.*, (Washington, DC), pp. 258–260.

[6] H. Hofmann, "Dispersion of planar waveguides for millimeter-wave applications," *Arch. Elek. Übertragung.*, vol. 31, pp. 40–44, Jan. 1977.

[7] J. B. Knorr and P. M. Shayda, "Millimeter wave finline characteristics," *IEEE Trans. Microwave Theory Tech.*, vol. MTT-28, pp. 737–743, July 1980.

[8] L.-P. Schmidt and T. Itoh, "Spectral domain analysis of dominant and

higher order modes in finlines," *IEEE Trans. Microwave Theory Tech.*, vol. MTT-28, pp. 981–985, Sept. 1980.

[9] A. K. Sharma and W. J. R. Hoefer, "Empirical analytical expressions for finline design," in *1981 IEEE MTT-S Int. Microwave Symp. Dig.*, (Los Angeles, CA), pp. 102–104.

[10] A. K. Sharma, G. I. Costache, and W. J. R. Hoefer, "Cutoff in finlines evaluated with the spectral domain technique and with the finite element method," in *1981 IEEE AP-S Int. Antenna Propagation Symp. Dig.*, (Los Angeles, CA), pp. 308–311.

[11] A. K. Sharma and W. J. R. Hoefer, "Propagation in coupled bilateral finlines," presented at Int. Electrical, Electronics Conf. Expos., Toronto, Canada, Oct. 5–7, 1981.

[12] A. K. Sharma and W. J. R. Hoefer, "Empirical expressions for finline design," *IEEE Trans. Microwave Theory Tech.*, vol. MTT-31, pp. 350–356, Apr. 1983.

[13] R. N. Simons and R. K. Arora, "Double slot finline structure for millimeter wave applications," *Proc. IEEE*, vol. 67, pp. 1158–1160, 1979.

[14] L.-P. Schmidt, T. Itoh, and H. Hofmann, "Characteristics of unilateral finline structures with arbitrary located slots," *IEEE Trans. Microwave Theory Tech.*, vol. MTT-29, pp. 352–355, Apr. 1981.

6.3.2

Broadband finline directional couplers

Heinrich Callsen
Lorenz-Peter Schmidt
Klaus Solbach

Introduction and summary

The design of millimeter-wave circuit elements such as mixers, phase shifters or distribution networks for monopulse radar systems calls for the use of directional couplers, particularly 3-dB directional couplers. In designing such components, it stands to reason that established concepts from other conductor technology branches should be applied to the specialized field of finline technology. The **probe coupling** described by Meier [1] illustrates by way of **analogy** the carryover of multi-aperture coupling elements commonly encountered in connection with hollow waveguide applications to the field of finline technology. The disadvantage inherent in this design concept is the requirement that the two finline elements to be coupled must be fully electrically shielded relative to each other (electrical shielding walls), to the extent that the shielding must include even the probes. This technique is therefore not likely to play any significant role in the design of integrated millimeter-wave circuits.

On the other hand, the principle of coupling travelling waves to coupled lines has proved to be highly effective with respect to finline circuit applications. This concept, commonly known as **"periodic coupling,"** is used primarily in conjunction with integrated

H. Callsen, L.P. Schmidt, and K. Solbach, *Wiss Ber AEG Telefunken*, Vol. 57, No. 4/5, 1981, pp. 241–250.

optical circuits used in the design of directional couplers [2]. An equal degree of success in this regard has been achieved with slotted microstrip line couplers (hybrid couplers), a design concept taken from the field of microwave technology and adapted for finline circuit applications (Paragraph 2). Both coupling concepts lend themselves to the design of broadband directional couplers with minimal insertion losses.

1. Directional couplers with coupled finlines

In the field of microwave technology, practical results are being achieved in the application of the "periodic coupling" principle to longitudinal-slot waveguide couplers [3] as well as couplers designed for dielectric image-line applications [4]. Details of the initial applications of this coupling principle are presented in References [5] and [6].

1.1 Principle of operation

Figure 1 represents a sectional view of the structure of a finline element with coupled slots. With this kind of line structure, **two** types of waves can potentially be propagated in the operating range of the screened waveguide (mount): **even-mode waves** with symmetrical field distribution, and odd-mode waves with anti-symmetrical field distribution relative to the plane of symmetry of the structure.

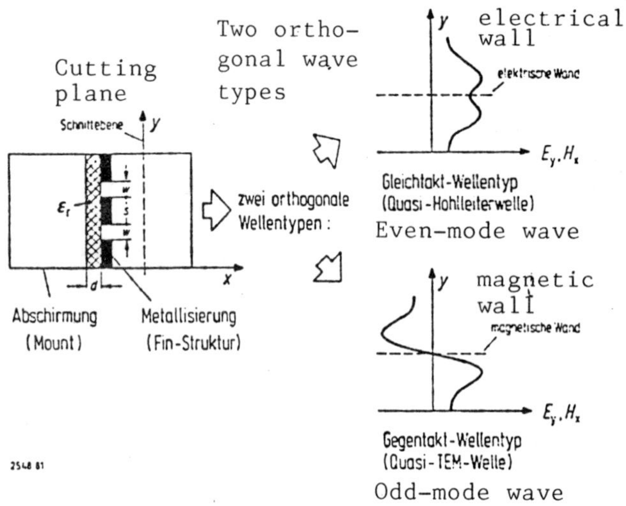

Figure 1: Sectional view of symmetrically coupled finlines, with sketches depicting the field strength distributions for even-mode and odd-mode waves.

It may be noted that the even-mode wave is characterized by identical orientation of the electrical field within the slots and therefore bears relationship to the TE_{10} mode in the waveguide. The odd-mode wave, on the other hand, is characterized by slot fields of opposite polarity, and is thus related to the quasi-TEM wave encountered with "coplanar microstrip" lines. This point is also evident in Figure 2, which by way of example shows the effective dielectric constant of the waves as a function of frequency; in this instance, the effective dielectric coefficient was completed using the method described in Reference [7]. The even-mode wave displays the dispersion characteristics of a waveguide wave (with cutoff frequencies below the frequency band shown in the illustration), while the odd-mode wave (with an effective dieletric constant that is virtually frequency-independent) displays the behaviour of a quasi-TEM wave.

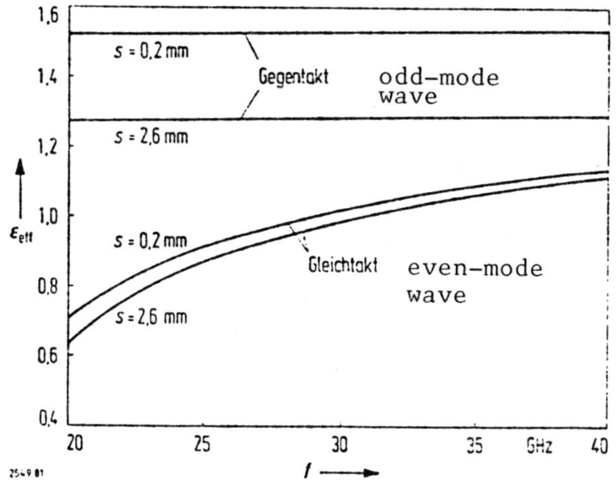

Figure 2: Effective dielectric constants for even and odd-mode waves in symmetrically coupled finlines, shown as a function of frequency.

(w = 0.2 mm; d = 0.254 mm; ε_r = 2.2, waveguide: WR-28)

Each wave condition encountered in the coupled transmission lines may be represented by an appropriate superpositioning of the two wave types. The (non-symmetrical) excitation of only one slot results in the excitation of both waves to an equal degree. The waves then propagate along the coupled transmission line structure, and their super-positioning results in new field intensity distributions on account of the different phase velocities of the waves beyond the point of excitation. This mechanism can be clarified on the basis of a simplified mathematical representation (Figure 3):

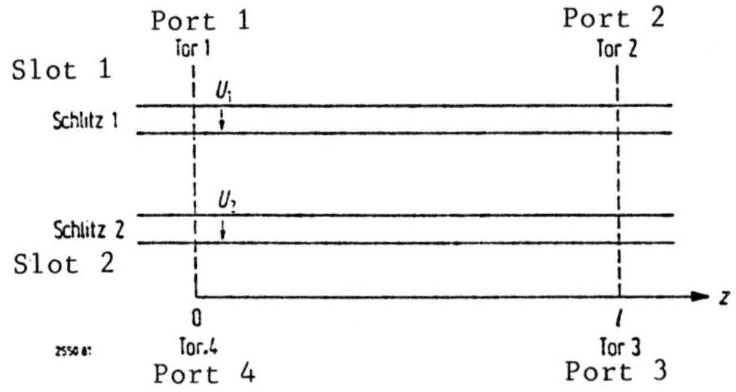

Figure 3: Longitudinal section through the coupled finlines

Let U_a represent the amplitude of the even-mode wave, and U_b the amplitude of the odd-mode wave. The waves are excited at the location $z = 0$ and propagate in the z-direction with phase constants β_a and β_b, where

$$\beta_a = \frac{2\pi}{\lambda_o} \cdot \sqrt{\varepsilon_{eff} \text{ (even-mode wave)}}$$

$$\beta_b = \frac{2\pi}{\lambda_o} \cdot \sqrt{\varepsilon_{eff} \text{ (odd-mode wave)}} \tag{1a}$$

Voltages U_1 and U_2, measured across slots ① and ②, respectively, are composed of the contributions from the individual waves:

$$U_1 = U_a + U_b$$

$$U_2 = U_a - U_b \tag{1b}$$

In the situation where only slot ① is excited with a voltage $U_1 = A$ (and a voltage $U_2 = 0$ across the second slot), the following condition is therefore obtained:

$$U_a = U_b = \frac{A}{2} \tag{1c}$$

The two wave types are thus excited with the same amplitude. The waves are out of phase behind the point of excitation:

$$U_1(z) = \frac{A}{2} \cdot (e^{-j\beta_a z} + e^{-j\beta_b z})$$

$$U_2(z) = \frac{A}{2} \cdot (e^{-j\beta_a z} - e^{-j\beta_b z}) \tag{2}$$

In view of the general condition that $\beta_a \neq \beta_b$, a point $z = L$ exists where voltage $U_1 (z)$ is nullified, and where $U_2 (z) = A$:

$$e^{-j\beta_a L} = -e^{-j\beta_b L} \curvearrowright L = \frac{\pi}{\beta_a - \beta_b} \tag{3}$$

The length L is referred to as the 0-dB coupling length, given the fact that at this point, complete coupling of the incident waves is achieved from one line slot to the next. On the basis of equation (2), the magnitude of the scattering coefficients for a section of the coupled line structure of length ℓ may be described in the following terms:

$$|s_{12}| = \cos\left(\frac{\pi}{2} \cdot \frac{\ell}{L}\right)$$

$$|s_{13}| = \sin\left(\frac{\pi}{2} \cdot \frac{\ell}{L}\right) \tag{4}$$

From equation (4), one may deduce that for a coupler length $\ell = L/2$, the power coupled in at $z = 0$ is split in **half** between the two lines **(3-dB coupler).** At port ②, the coupled wave is out of phase by 90° relative to the wave at port ③.

The designation **"periodic coupling"** for this particular coupling mechanism becomes apparent on the basis of equation (4): given an infinitely long coupled line structure, the energy of the guided waves travelling in the direction of propagation alternates back and forth from one line to the other in a periodic manner.

1.2 0-dB coupling length

On the basis of the theoretically computed [7] dispersion patterns for waves in coupled finlines, the 0-dB coupling length was determined for several cases of practical interest using equation (3). Figure 4 shows the frequency-dependent nature of the coupling length for a variety of line structures used in the Ka frequency band (26.4 to 40 GHz). From the graph, one may note that for a slot spacing of s = 0.2 mm and a slot width of w = 0.2 mm, the coupling length is virtually constant at L ≃ 22 mm over the entire band width. This

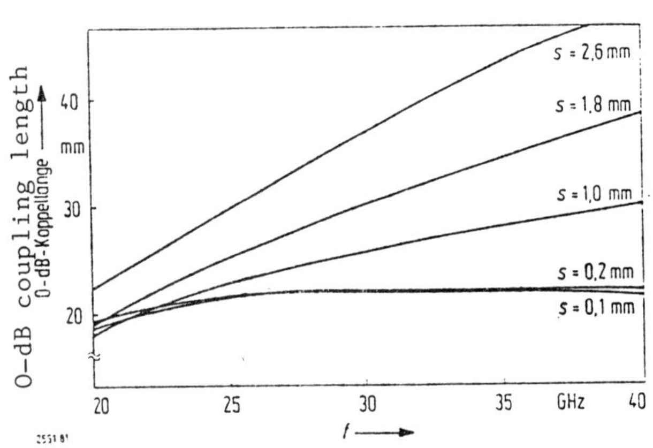

Figure 4: 0-dB coupling length for symmetrically coupled finlines, plotted **as a function of frequency**

(w = 0.2 mm; d = 0.254 mm; ε_r = 2.2; waveguide: WR-28) means that with finlines dimensioned in this particular manner, a directional coupler covering the entire bandwidth can be constructed. However, Figure 5 clearly points out the requirement for strict tolerancing in manufacturing structures for such applications. A decrease in the slot width w by 50 µm results in a 0-dB coupling length increase of approximately 5%, a condition that in the case of a 3-dB directional coupler, for example, would result in a ± 0.35 dB variation for scattering parameters s_{12} and s_{13}. An examination of Figure 6, which shows the sensitivity of the coupling length relative to changes in the slot spacing, reveals that the effects of varying this particular parameter are significantly less pronounced. For example, a 50 µm reduction in the slot length leads to a coupling length variation of only approximately 1%, and a variation of ± 0.07 dB in the case of the scattering parameters for a 3-dB coupler.

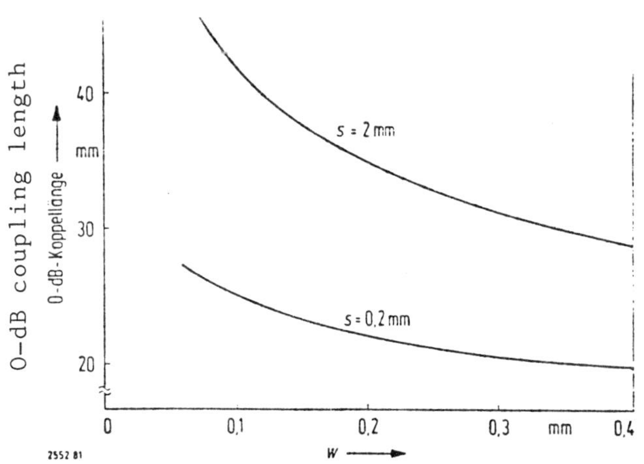

Figure 5: 0-dB coupling length for symmetrically coupled finlines, shown **as a function of slot width** for two different slot spacings.

(d = 0.254 mm; ε_r = 2.2; f = 33 GHz; waveguide: WR-28)

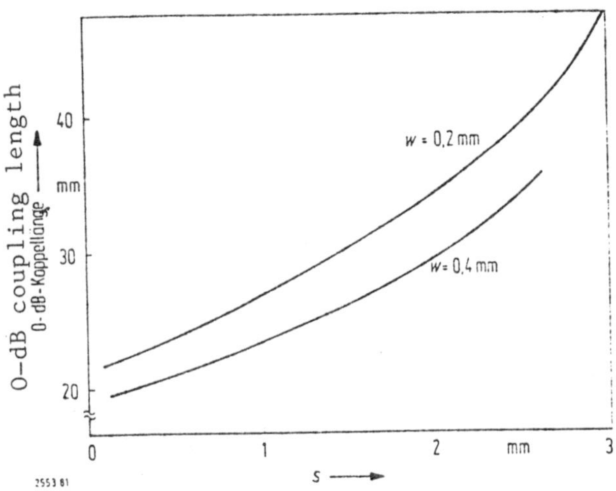

Figure 6: 0-dB coupling length for symmetrically coupled finlines, shown **as a function of the slot spacing** for two different slot widths.

(d = 0.254 mm: ε_r = 2.2; f = 33 GHz; waveguide: WR-28)

1.3 Designing a 3-dB coupler

In designing a directional coupler based on the periodic coupling principle, the manner in which the coupled lines are coupled to the input lines is of considerable importance. In order to achieve line matching over a wide frequency band with minimal reflection, curved input coupling sections (tapers) are used (Figure 7). The length of the tapers should be considerably greater than half the wavelength. As is evident from Figure 4, widely separated lines such as those encountered within the curved input coupling region yield only slightly larger 0-dB coupling lengths (i.e. only marginally less coupling) than the lines in the homogeneous coupling region.

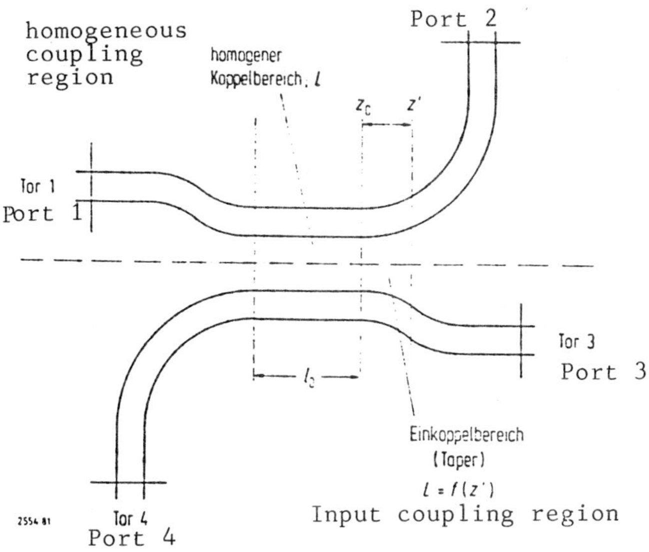

Figure 7: Finline directional coupler structure with curved input coupling lines (tapers).

This means that the line sections in the curved input coupling regions contribute significantly towards the total coupling of the line structure. The contributions of the input coupling regions towards the total coupling can be taken into account approximately by assuming that these regions are composed of numerous individual homogeneous coupled line sections, whose coupling effect can be added to that of the inner coupling region. This means that the effective coupling length l_{eff} can be represented as follows:

$$\ell_{eff} = \ell_o + \frac{2L}{\pi} \int_{z_o}^{z} (\beta_a(z) - \beta_b(z))\, dz \qquad (5)$$

The integration in this instance must be performed along the axis of the directional coupler from the coupling point of the input lines at z_o to a point z' where the coupling effect becomes negligible.

In dimensioning a 3-dB directional coupler for the 26.4 to 40 GHz frequency band, the optimal slot dimensions w = s = 0.2 mm were selected, as illustrated in Figure 4 for broadband applications. The 0-dB coupling length in this instance amounts to L ≃ 22 mm; for the given requirement, $\ell_{eff} = L/2 \simeq 11$ mm. By applying equation (5) in the case of an input coupling structure consisting of circular or elliptically-shaped input coupling lines, as depicted in Figure 7, the effective coupling length for the input coupling region is determined to be approximately 3 mm. This means that the length of the homogeneous coupling region amounts to $\ell_o = 8$ mm.

In order to achieve improved matching of the input coupling lines to the inner coupling structure, part of the originally assumed homogeneous coupling region was additionally incorporated in the input coupling zone (Figure 8) through additional tapering. As a result, wave coupling occurs primarily in the inhomogeneous regions of the coupled lines. In the case of the finline "mount," which is required for grounding and shielding the coupling structure, the distance between the axes of the coupling lines (ports) ② and ④ amounts to $\ell = 14$ mm. Due to the nature of equation (5), this distance is considerably greater than the effective coupling length ℓ_{eff}.

Figure 8: Photograph of the 3-dB directional coupler designed for application in the 26.4 to 40 GHz frequency band.

In addition, holes for short-circuiting probes were drilled at the branching points (T-junctions) in order to suppress any undesirable cavity resonance within the operating frequency range of the coupler [9]. In order to couple the finline circuit to a standard waveguide system, the finline ports were terminated by means of finline-waveguide transition elements with a length of 16 mm and a reflection factor of less than -25 dB.

1.4 Experimental results

In order to manufacture the finline structure, photomasks were first produced with the aid of a computer-based layout system. With appropriate programming instructions, this method even proved suitable for designing and producing elliptically-shaped bends for the input coupling lines. The photomasks, which were produced by means of photoscribing and subsequently reduced in size, were used in order to photo-chemically etch the line structures onto RT/Duroid 5880 with a thickness of 254 µm.

The experimentally determined scattering coefficients for the coupler shown in Figure 8 are plotted in Figure 9. It can be seen that the expected broadband nature of the coupling has been achieved to an almost perfect degree. At the midband frequency, the coupler insertion losses amount to approximately 0.5 dB, as may be deduced from the points of intersection of the s_{12} and s_{13} curves at -3.5 dB. This value corresponds approximately to a line attenuation of 0.1 dB/wavelength over the coupled lines, an amount that in turn corresponds approximately to the known values for simple finlines within this frequency range.

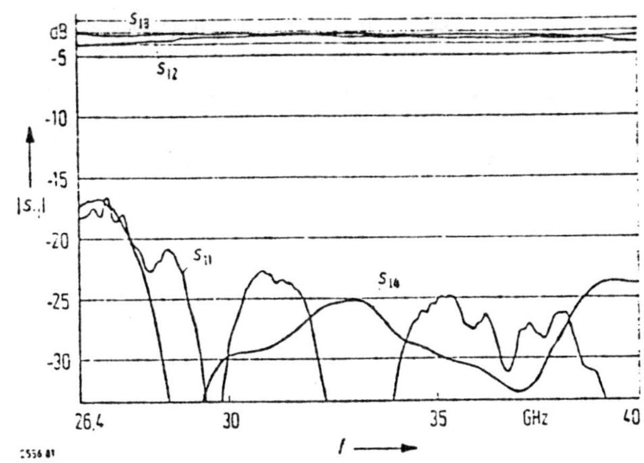

Figure 9: The experimentally measured scattering coefficients for a Ka-band 3-dB directional coupler, plotted as a function of frequency.

Given the approximate nature of the method for calculating the effective coupling length (equation 5), and taking into account the effect of the short-circuiting probes at the junction points, the characteristics of the coupling structure could be predicted only to an approximate degree. In order to achieve exact power splitting (as in this example), it was necessary to produce and examine **several** different structures whose homogeneous coupling regions were then varied relative to the original design.

Figure 9 furthermore illustrates the fact that the directivity $|s_{14}|$ and the input reflection constant of approximately -25 dB are significantly worse than for standard multi-aperture waveguide couplers. This condition does not stem from the coupling principle employed, as evidenced by the values of approximately -35 dB measured for dielectric image line couplers [10], but is essentially attributable to the discontinuities in the area of the input coupling lines (waveguide junctions) as well as the fact that the length of the input coupling taper is still insufficient. The fact that a very high degree of finline coupling still occurs in spite of increasing the distance between slots as much as possible is disadvantageous in this instance.

In addition, several coupling structures were produced whose slot widths and slot spacing dimensions differed from those selected for the original design version. An examination of the scattering parameters measured for these coupling structures reveals that the values for coupling coefficients $|s_{12}|$ and $|s_{13}|$ deviate from the values noted for the original structure, as shown in Figure 10:

w (mm)	s (mm)	$\lvert s_{12}\rvert$
0.2	0.2	-3.5 dB
0.1	0.2	-2.5 db
0.3	0.2	-4.1 dB
0.2	0.1	-3.5 dB
0.2	0.3	-3.4 dB

Figure 10: Coupling coefficient values determined for the 3-dB directional coupling structure at a frequency of 33 GHz as a result of varying slot dimensions w and s.

It is apparent that changes involving only the slot width w result in significant alteration of the power distribution between ports ② and ③; the magnitude of the variation corresponds approximately to the sensitivity determined on the basis of Figure 4 (cf. Paragraph 1.2).

The coupling structure under consideration was then reconfigured for the 60 to 90 GHz frequency band by recomputing/rescaling the 0-dB coupling length. In the process, slot widths of w = 0.1 mm and slot spacings of s = 0.1 mm were selected, and the substrate thickness was also cut in half to d = 0.125 mm. In this frequency band, the coupler displayed an increased transmission loss of approximately 0.8 dB (at the midband point) and a deterioration of approximately 5 dB for s_{11} and s_{14} (Figure 11).

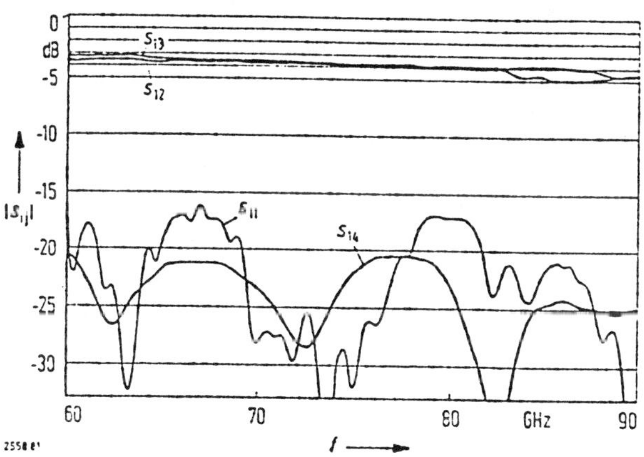

Figure 11: Measured scattering coefficients for the E-band 3-dB directional coupler, plotted as a function of frequency.

Any number of coupling ratios can be established by either increasing or decreasing the coupling length. For example, couplers with coupling losses of 10 dB and 20 dB were devised in this manner with finline structures in a cross-shaped waveguide. One of the problems encountered with this particular design configuration is the fact that the length of the taper region required to achieve favourable matching of the input lines with the coupled line structure is barely sufficient. With such structures, it was noted that the directional losses and input reflection factors were correspondingly lower than in the case of the 3-dB coupling structure already considered.

2. Hybrid directional couplers incorporating finline technology

In the field of planar and quasiplanar transmission line technology, it is common practice in designing directional couplers ("branch line" couplers) to use non-field-coupled lines instead of field-coupled lines, with the non-field-coupled lines connected to each other by means of at least two series or parallel-branched coupling lines spaced one quarter wavelength apart. The so-called hybrid directional coupler represents a special type of directional coupler, and is characterized by the fact that the coupling lines branch off from a single point as a result of using series and parallel branch elements. While directivity is determined by the propagation time differences for the partial wave components in the case of couplers with spatially distributed connecting lines, the directivity of hybrid couplers results from the differential coupling between the main line and series branch as well as between the main line and the parallel branch. Considering that both branches are connected at one point and that the coupling mechanisms are largely frequency-independent, the hybrid coupler is (in contrast to standard branch-line couplers) capable of operating over a very broad bandwidth.

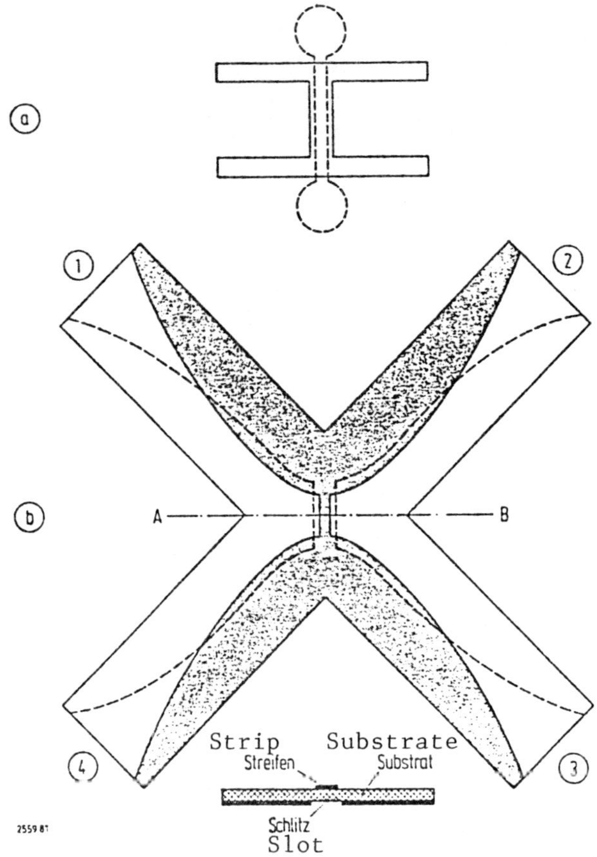

Figure 12: a) Hybrid coupler for microstrip line applications as proposed by de Ronde [11]

b) Layout of a hybrid coupler for finline applications
——————— upper substrate surface
----------- lower substrate surface.

By way of a quasiplanar circuit, de Ronde initially designed a hybrid coupler incorporating microstrip line technology ([11] and Figure 12a). The series/parallel branch was realized in the form of a slot/strip configuration. The hybrid coupler was subject to detailed examination and further development by Schiek, Köhler as well as Vogel [12 to 14]. This principle has been reapplied in this instance in order to design finline hybrid directional couplers for millimeter-wave applications.

2.1 Description of the directional coupler

The layout in principle of the coupling structure is illustrated in Figure 12b. The coupler per se is located at the centre of a cross-shaped waveguide. In each of the four arms of the waveguide, the waves in the waveguide are transferred to antipodal finlines over a broad bandwidth with the aid of continuous transition elements with minimal reflection. Respective pairs of ports are interconnected by means of circular-pattern lines that form the main lines. The coupling structure, which consists of a combination of slotted and microstrip lines, is arranged symmetrically at the centre relative to the main lines. The strip connects the oppositely situated metal layers of the antipodal finlines on the one side of the substrate, while the slot separates the common ground plane of the lines on the other side of the substrate. The detail sketch shown in Figure 12b represents a sectional view of the slot/strip configuration.

The cavity resonances produced by the finlines in the cross-shaped waveguide can be suppressed by means of an aperture along cutting plane A-B in Figure 12b, thereby simultaneously decoupling the two main lines with respect to each other [9].

2.2 Circuit analysis

The slot/strip configuration represents a quasiplanar line structure, within which two types of waves are capable of being propagated in the intended frequency band: **even-mode waves** that bear resemblance to the basic mode for microstrip lines, and **odd-mode waves** that are similar to the basic mode for unilateral finlines. These waves will therefore be designated as strip-line-type and slot-line-type waves in the ensuing text. The two wave types are orthogonal to each other; however, the line parameters for the slot line are affected by the strip line width, and the line parameters of the strip line conversely by the slot width.

The strip lines branch off from the two main lines (the antipodal finlines) in a **parallel** manner, while the slot lines branch off in **series.** Given an input applied at port ① and reflection-free closure of the remaining ports (see Figure 12b), part of the power will be directed towards port ②, and the remainder will in the case of suitable dimensioning be divided evely between the slot and strip lines of the slot/strip line. At ports ③ and ④, the two field types will in turn excite outgoing waves of equal magnitude that cancel each other out at port ③, but add constructively at port ④.

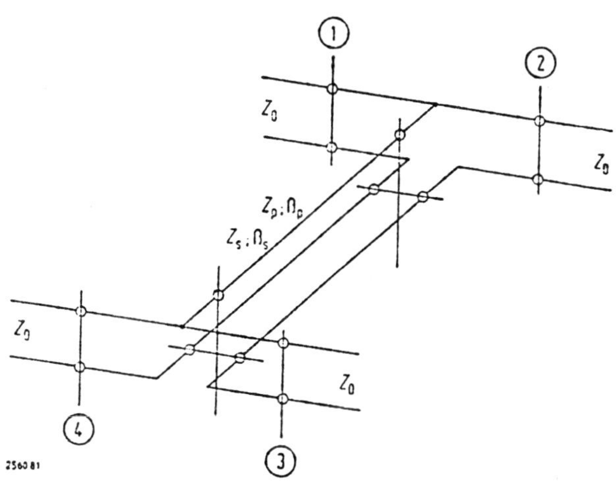

Figure 13: Equivalent circuit diagram for the hybrid directional coupler

The equivalent circuit for the coupler is shown in Figure 13. The antipodal finlines have in this instance been replaced by means of dual lines, and the slot/strip line configuration by means of a triple line. In the case of this triple-line system, the even-mode type corresponds to the strip line, and the odd-mode type to the slot line of the slot/strip configuration.

Symmetrical excitation at ports ① and ② (Figure 13) results only in even-mode excitation, with a difference in potential between the upper and the lower two lines. No difference in potential exists between the lower two lines. In considering this particular mode with wave impedance Z_p, the triple-line system forms a parallel branch relative to the main line. **Anti-symmetrical** excitation at ports ① and ②, on the other hand, results only in odd-mode excitation, with a difference in potential between the two lower lines. Due to the symmetrical configuration, there is no current flow through the third (upper) line. In considering this particular mode together with a wave impedance Z_s, the triple-line system forms a series-type branch with respect to the main line.

In order to describe the configuration in terms of network-theoretical methods, an equivalent circuit must be devised for the hybrid tee, in which the various modes are transferred from the slot/strip line to separate lines. To this end, the series/parallel branches are transferred over to so-called **hybrid circuit elements** (Figure 14), as used for telephone applications. In the case of symmetrical inputs to the hybrid circuit element at ports ① and ②, power is directed only towards port Ⓟ, whereas in the case of anti-symmetrical inputs, power is directed only towards port Ⓢ.

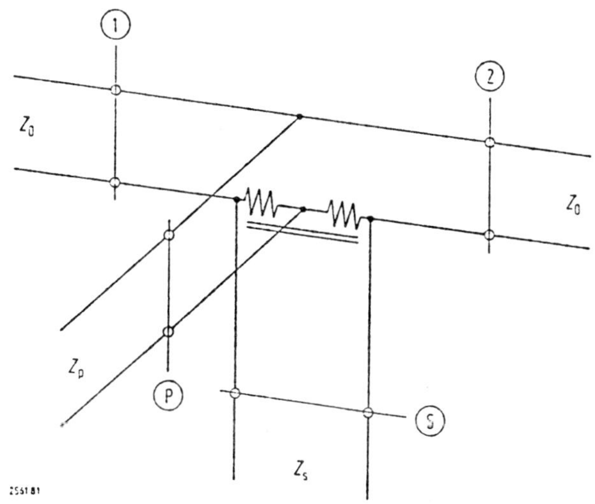

Figure 14: Equivalent circuit for a hybrid tee

One particular equivalent circuit devised for the purpose of analyzing this type of coupler has been commonly cited in a number of sources (References [12] to [14]). Specifically, the equivalent circuit in question consists of three non-coupled lines, of which one is connected in series to ports ① and ②, and the other two lines in parallel and symmetrically with respect to the first line. However, this particular equivalent circuit has been rejected following in-depth examination, because the configuration with respect to the decoupling of ports Ⓟ and Ⓢ is not correctly represented.

For reasons relating to symmetry and reciprocity, six different scattering parameters remain to be solved from the scattering matrix of the equivalent circuit for the series/parallel branch shown in Figure 14. The general scattering matrix for the hybrid tee is given by:

$$S = \begin{bmatrix} S_{11} & S_{21} & S_{p1} & S_{s1} \\ S_{21} & S_{11} & S_{p1} & -S_{s1} \\ S_{p1} & S_{p1} & S_{pp} & 0 \\ S_{s1} & -S_{s1} & 0 & S_{ss} \end{bmatrix} \qquad (6)$$

By making use of the rules for nodes and mesh networks, and by considering the relationships between currents and voltages for ideal conductors, the remaining unknown scattering parameters can be solved as functions of the impedances of the connected lines:

$$S_{11} = (Z_p Z_s - Z_o^2)/N \qquad (7a)$$

$$S_{21} = 2Z_o(Z_p - Z_s/4)/N \qquad (7b)$$

$$S_{p1} = 2\sqrt{Z_o Z_p}(Z_o + Z_s/2)/N \qquad (7c)$$

$$S_{s1} = \sqrt{Z_o Z_s}\,(Z_o + 2Z_p)/N \qquad (7d)$$

$$S_{pp} = (Z_o - 2Z_p)/(Z_o + 2Z_p) \qquad (7e)$$

$$S_{ss} = (2Z_o - Z_s)/(2Z_o + Z_s) \qquad (7f)$$

$$N = Z_o(Z_o + 2Z_p + Z_s/2) + Z_p Z_s \qquad (7g)$$

The scattering parameters for the cirectional coupler can be determined by interconnecting pairs of series/parallel branches at ports (p) and (s) as shown in Figure 14. On the basis of symmetry and reciprocity, four scattering parameters remain to be solved for the directional coupler circuit.

The scattering matrix for the directional coupler takes the following general form:

$$S = \begin{bmatrix} S^k_{11} & S^k_{21} & S^k_{31} & S^k_{41} \\ S^k_{21} & S^k_{11} & S^k_{41} & S^k_{31} \\ S^k_{31} & S^k_{41} & S^k_{11} & S^k_{21} \\ S^k_{41} & S^k_{31} & S^k_{21} & S^k_{11} \end{bmatrix} \qquad (8)$$

Given the scattering coefficients derived for the hybrid tee on the basis of equation (6), transmission lines with lengths l_s and l_p, characteristic impedances Z_s and Z_p, and phase constants β_s and β_p, the following equations are obtained for computing the scattering parameters for the directional coupler circuit:

$$S^k_{11} = S_{11} + \frac{S^2_{p1} S_{pp} \exp[-j2\beta_p l_p]}{1 - S^2_{pp} \exp[-j2\beta_p l_p]} + \frac{S^2_{s1} S_{ss} \exp[-j2\beta_s l_s]}{1 - S^2_{ss} \exp[-j2\beta_s l_s]} \qquad (9a)$$

$$S^k_{21} = S_{21} + \frac{S^2_{p1} S_{pp} \exp[-j2\beta_p l_p]}{1 - S^2_{pp} \exp[-j2\beta_p l_p]} - \frac{S^2_{s1} S_{ss} \exp[-j2\beta_s l_s]}{1 - S^2_{ss} \exp[-j2\beta_s l_s]} \qquad (9b)$$

$$S^k_{31} = \frac{S^2_{p1} \exp[-j\beta_p l_p]}{1 - S^2_{pp} \exp[-j2\beta_p l_p]} - \frac{S^2_{s1} \exp[-j\beta_s l_s]}{1 - S^2_{ss} \exp[-j2\beta_s l_s]} \qquad (9c)$$

$$S^k_{41} = \frac{S^2_{p1} \exp[-j\beta_p l_p]}{1 - S^2_{pp} \exp[-2\beta_p l_p]} + \frac{S^2_{s1} \exp[-j\beta_s l_s]}{1 - S^2_{ss} \exp[-j2\beta_s l_s]} \qquad (9d)$$

In the case of an ideal directional coupler, all ports are reflection-free in nature, and each pair of ports decoupled relative to each other. In assuming the ideal condition where the electrical lengths of the coupling lines in the hybrid coupler are equal in magnitude ($\beta_s \ell_s = \beta_p \ell_p$) , one obtains an ideal directional coupler if the criteria subsequently described are satisfied.

Ports ① and ③ are decoupled relative to each other in a frequency-independent manner ($S^k_{31} = 0$) when the following condition exists with regard to the scattering parameters for the hybrid tee:

$$S_{p1} = \pm S_{s1} \tag{10}$$

The relection factors of the coupler are cancelled out in a frequency-independent manner when the scattering parameters of the hybrid tee also fulfill the folllowing criteria:

$$S_{11} = 0 \tag{11}$$

and

$$S_{pp} = -S_{ss} \tag{12}$$

If the conditions specified in equations (10) to (12) are satisfied, the scattering parameters for the coupler can be simplified at the mid-band frequency ($\beta_s \ell_s = \beta_p \ell_p = \pi/2$) in the manner shown below:

$$S_{11}^{k} = 0 \tag{13a}$$

$$S_{21}^{k} = S_{21} + \frac{S_{pp}S_{p1}^{2}}{1 + S_{pp}^{2}} \tag{13b}$$

$$S_{31}^{k} = 0 \tag{13c}$$

$$S_{41}^{k} = -j\,\frac{2S_{p1}^{2}}{1 + S_{pp}^{2}} \tag{13d}$$

The conditions detailed in equations (10) to (12) are satisfied by the following line impedance criterion:

$$Z_o^2 = Z_s Z_p \tag{14}$$

The relationship between the line impedances and the coupling between ports ① and ② of the directional coupler can be derived from equations (7) and (13b):

$$S_{21}^{k} = \frac{\left(\dfrac{2Z_p}{Z_o}\right)^2 - 1}{\left(\dfrac{2Z_p}{Z_o}\right)^2 + 1} \tag{15}$$

The solution for the normalized impedance of the parallel-connected line can be expressed as follows:

$$\frac{Z_p}{Z_o} = \frac{1}{2}\sqrt{\frac{1 + |S_{21}^k|}{1 - |S_{21}^k|}} \qquad (16)$$

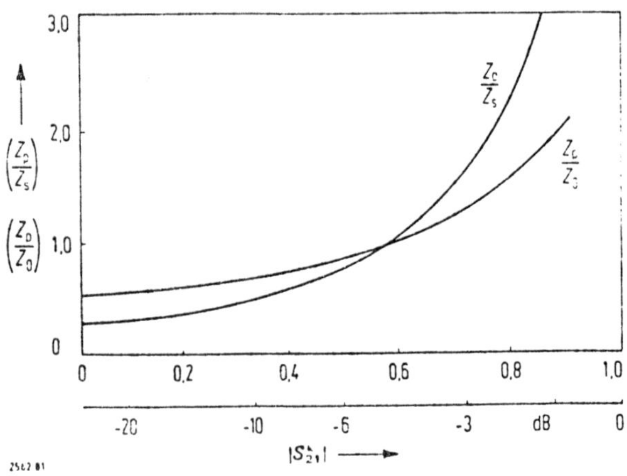

Figure 15: The relationships between even-mode and odd-mode impedances, and between even-mode and input line impedances, plotted as functions of the coupling factor $|S_{21}^k|$

The relationship between the even-mode to odd-mode impedance ratio and the coupling between ports ① and ② can be expressed in the following manner:

$$\frac{Z_p}{Z_s} = 0.25 \cdot \frac{1 - |S_{21}^k|}{1 - |S_{21}^k|} \qquad (17)$$

The dependent relationships of the line impedances relative to the coupling factor $|S_{21}^{k}|$ are shown in Figure 15 as an aid to design. With the given restrictions, the coupling between ports ① and ④ can be expressed as follows:

$$S_{41}^{k} = \frac{-j2Z_p}{Z_p^2 - 4} \qquad (18)$$

In other words the wave at port ④ is -90° out of phase relative to the wave at port ②.

2.3 Designing a 3-dB directional coupler

For the purpose of designing a 3-db directional coupler, the impedance ratios Z_p/Z_s and Z_p/Z_o can be determined from Figure 15, where Z_p represents the wave impedance of the strip line, Z_s the wave impedance of the slot line, and Z_o the wave impedance of the connecting lines (antipodal finlines):

$$Z_p/Z_o = 1.208$$

$$Z_p/Z_s = 1.49$$

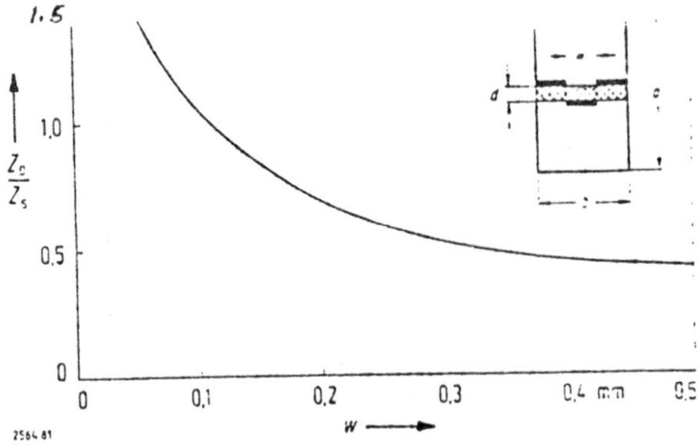

Figure 17: Relationship between the strip-line and slot-line impedances of the structure examined in Figure 15.

In order to provide a clearer overall view, the wave impedance ratio Z_p/Z_s derived from the curves in Figure 16 is plotted in Figure 17 as a function of slot width. According to Figure 17, the slot and strip widths for a 3-dB coupler amount to w = 0.055 mm. Given these dimensions, the following line parameters are obtained on the basis of Figures 15 and 16:

$Z_p = 162\,\Omega$ $\qquad \varepsilon_{eff,p} = 1.71$

$Z_s = 113.5\,\Omega$ $\qquad \varepsilon_{eff,s} = 1.33$

$Z_o = 136\,\Omega$

According to Figure 18, this impedance Z_o for the antipodal finline infers in an overlap of:

s = -0.01 mm

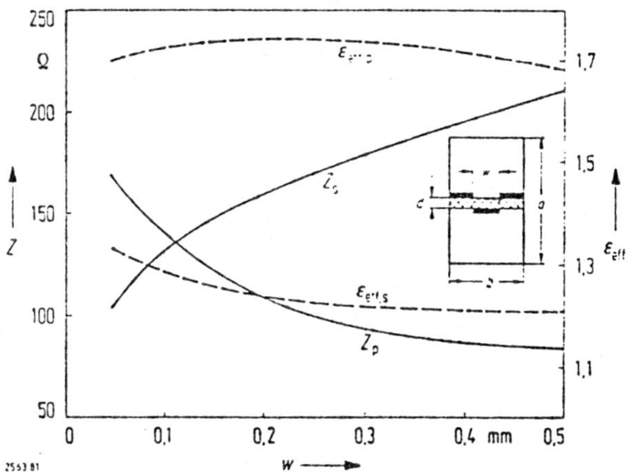

Figure 16: Line parameters for the strip and slot lines in a slot/strip configuration, plotted as a function of the slot and strip widths

(f = 35 GHz; ε_r = 2.2; d - 0.254 mm ; a = 7.11 mm; b = 3.56 mm)

Figure 16 shows the impedances and effective dielectric constants for both the slot and strip lines of a slot/strip line as respective functions of slot width and strip width on glass fibre-reinforced teflon (RT-Duroid; d = 0.254 mm; ε_r = 2.2), designed for application in Ka-band waveguides at a frequency of 35 GHz. For reasons of simplicity, only a single case involving identical slot and strip widths was considered. The line parameters for the slot/strip configuration were computed using the procedure outlined in Reference [7].

The length l of the coupling structure has been dimensioned on the basis of the larger effective dielectric constant (i.e. that of the strip-type line). The possibility also exists of incorporating a dielectric surface to increase the phase constant of the slot-type line, in order to balance the two phase constants. The geometric coupling length of the slot/strip configuration should be based on the length of the strip conductor. For a midband frequency of 35 GHz, the resultant strip conductor length is ℓ = 1.63 mm. The line parameters for the slot/stip configuration shown in Figure 16 were computed for the purpose of screening in a Ka-band waveguide. However, the line structure is positioned diagonally within the cross-shaped waveguide and features an aperture at the plane of symmetry in a configuration transverse to the coupling structure, in other words within a longitudinally inhomogeneously bounded region.

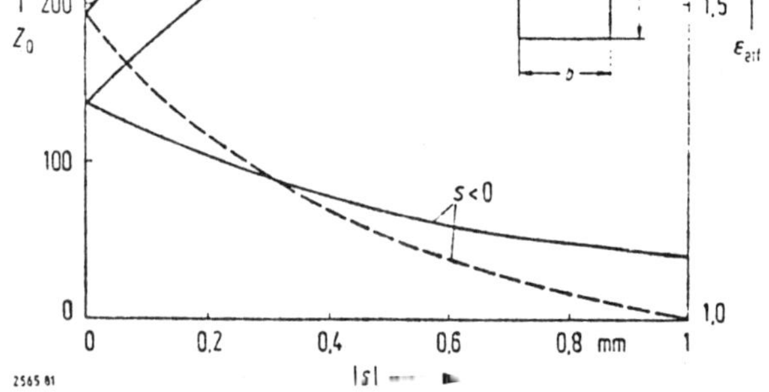

Figure 18: Line parameters of an antipodal finline, shown as a function of slot width (s 0) and overlay (s 0)

(f = 35 GHz; ε_r = 2.2; d = 0.254 mm; a = 7.11 mm; b = 3.56 mm)

As long as the waveguide/shielding dimensions are sufficiently large relative to the cross-sectional dimensions of the lines, the line parameters remain unaffected by changes in the waveguide dimensions. In the case under consideration, one may on the basis of approximation use line parameters computed for other types of waveguide shielding.

It therefore follows analogously that for the main lines (antipodal finlines), the line parameters shown in Figure 18 for the symmetrical line structure can be applied with approximately equal validity to the slightly asymmetrical lines used in the practical coupling structure.

Directional couplers with other coupling factors can only be designed using the method under consideration if the desired impedance ratios can be achieved with cross-sectional dimensions that are small in relation to the wavelength. If the line dimensions are no longer small in relation to the wavelength, the design process is no longer feasible without having to contend with the parasitic effects encountered in the branch elements.

This means that for the substrate used in the case at hand ($\varepsilon_r = 2.2$ and $d = 0.25$ mm), directional couplers with coupling attenuation values between 3 and 10 db can be achieved by allowing line structure cross-sectional dimensions of up to 0.35 mm.

2.4 Experimental results

The planar coupling structure that was produced differed with respect to the computed dimensions in the sense that the slot and strip widths amounted to $w = 0.1$ mm. The coupling length measured $\ell = 1.6$ mm, and the antipodal finline overlap amounted to 0.1 mm. Given these particular planar structure dimensions, the following line parameters were obtained:

$$Z_p = 130\ \Omega \qquad \varepsilon_{eff,p} = 1.72$$

$$Z_s = 140\ \Omega \qquad \varepsilon_{eff,s} = 1.29$$

$$Z_o = 118\ \Omega$$

Figure 19 shows both the computed and experimentally measured values for the structure of the given dimensions. The computations were performed with the aid of a computer program devised for the numerical solution of the analytical procedure detailed for the coupling structure in Paragraph 2.2.

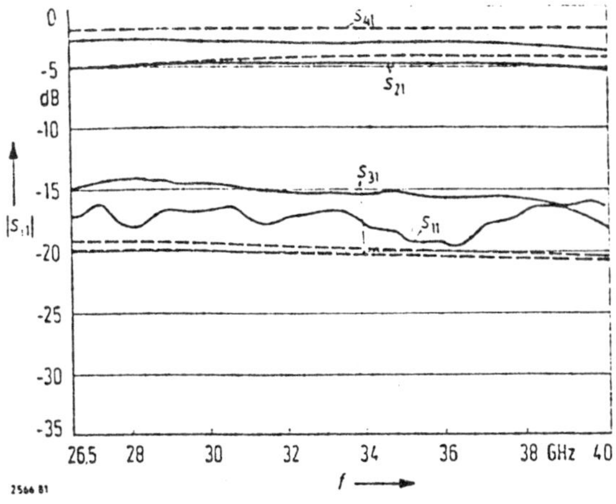

Figure 19: Scattering parameters for the hybrid coupler under consideration, plotted as a function of frequency
——————— measured values
- - - - - - - - computed values (without losses)

In comparing the theoretical and measured values, the losses experienced in practice should be taken into account; these losses correspond approximately to the taper losses (0.3 dB per taper). The computations for the distribution of power from port ① to ports ② and ④ result in values for $|S_{21}|$ and $|S_{41}|$ that are larger than the measured values. One certain explanation for this situation is the fact that the impedance ratio Zp/Zs is undervalued in the case of small slot and strip widths; this is in turn attributable to the assumption of an infinitely thin metal layer for the purpose of the computation.

A satisfactory degree of agreement exists between the computed and measured reflection factor values. The largest amount of variation between theory and practice is encounted with respect to the directional attenuation parameter $|S_{31}|$. Directional attenuation results from the destructive superpositioning of two partial waves, and is therefore particularly affected by manufacturing tolerances and deviations of the actual structure from the ideal equivalent circuit diagram.

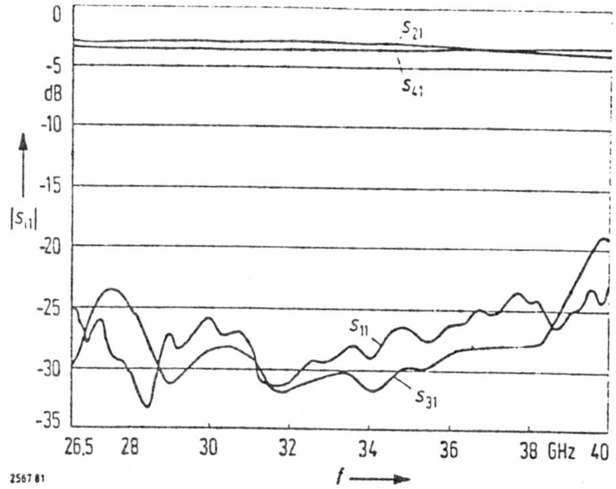

Figure 20: Measured scattering parameters for a 3-db hybrid directional coupler with quartz layer, plotted as a function of frequency.

The dielectric layer on the slotted side of the slot/strip configuration primarily affects the line parameters of the slot-type line. The phase constant for this particular mode is increased, and the wave attenuation decreased.

The dielectric layer barely has any effect on the strip-type line parameters on account of the fact that the electrical field barely extends past the slot, a phenomenon that applies particularly to narrow slots. With a quartz layer, the directivity of the coupling structure was increased by approximately 10 dB to 25 dB, while also achieving an adjustment in terms of power distribution (Figure 20). The coupling factors over the entire bandwidth consequently differ at most by 0.5 dB at a mean value of approximately 3.5 dB. The resultant insertion losses of approximately 0.5 dB in turn correspond primarily to the taper losses.

Further studies of hybrid couplers with different taper ratios have confirmed that the derived equivalent circuit diagram and computed line parameters describe the real structures in a highly representative manner.

AlE32 - Ulm, West Germany

Submitted 24 Jul 81

The research underlying this study was partially funded by the Federal Ministry of Defence. Responsibility for the material contained in this article is assumed by the authors.

REFERENCES

1. L.D. Cohen; P.J. Meier: "Advances in E-plane printed millimeter-wave circuits". 1978 IEEE International Microwave Symposium, Ottawa (Canada) Digest pp. 27-29.

2. E.A. Marcatili: "Dielectric rectangular waveguide and directional coupler for integrated optics", Bell Syst. Tech. J., vol. 48 (1969) no. 7, pp. 2071-2102.

3. X. Tanaka; X. Toshiaki: "Ridge-shaped narrow-wall coupler using TE_{10}, TE_{20} and TE_{30} modes", IEEE Trans. MTT-28 (March 1980) no. 3, pp. 239-245.

4. R.M. Knox; P.O. Toulios: "Integrated circuits for the millimeter through optical frequency range". Symposium on Submillimeter Waves, New York: Polytechnic Press (1970) pp. 497-516.

5. Neue Wellenleiter und Schaitkreise fur mm-Wellensysteme. AbschluBbericht Grundlagenentwicklung (1979) Kapitel 8.7.

6 E. Kpodzo; K. Schunemann; G. Begemann: "A quadriphase fin-line modulator", IEEE Trans. MTT-28 (July 1980) no. 7, pp. 747-752.

7. L.P. Schmidt; W. Menzel: "Berechnung der Leitungsparameter quasiplanarer Wellenleiter fur integrierte Millimeterwellen-Schaltungen". Wiss. Ber. AEG-TELEFUNKEN 54 (1981) 4-5 (dieses Heft, S. 219)

8. R. Rudokas; T. Itoh: "Passive millimeterwave IC components made of inverted strip dielectric waveguides", IEEE Trans. MTT-24 (Dec. 1976) pp. 978-981.

9. K. Solbach; H. Callsen; W. Menzel: "Spurious resonances in asymmetrical finline junctions". IEEE Trans. MTT-29, November 1981.

10. K. Solbach: "The calculation and the measurement of the coupling properties of dielectric image lines of rectangular cross section", IEEE Trans. MTT-27 (January 1979) no. 1, pp. 54-58.

11. F. de Ronde: "A New Class of Microstrip Directional Couplers", IEEE Int. Microwave Symp. (May 1970) pp. 184-186.

12. B. Schiek: "Hybrid Branchline Couplers - A Useful New Class of Directional Couplers". IEEE Trans. on MTT. Vol. 22, No. 10 (October 1974) pp. 864-869.

.../2

13. B. Schiek; J. Kohler: "Improving the Isolation of 3dB Couplers in Microstrip-Slotline Technique", IEEE Trans. on MTT, Vol. 26, No. 1 (January 1978) pp. 5-7.

14. Ryszard Vogel: "Microstrip-Slotline Components for Microwave IC's" Microwave Journal (May 1980) pp. 83-87 and p. 110.

6.3.3

DEVELOPMENT OF A COUPLER IN FINLINE TECHNIQUE

Adalbert Beyer, Dietmar Köther and Ingo Wolff

Department of Electrical Engineering, Duisburg University
Bismarckstrasse 81, D-4100 Duisburg, FRG

Abstract

Among other components directional couplers are employed in millimeterwave finline circuits. The method presented here allows a simple calculation of all modes which can exist on such couplers under consideration of the thickness of the metallization. In this way a 3-dB coupler has been developed for the Ka-band. Numerical and experimental results illustrate the applicability of this method.

Introduction

Since finlines permit a simple integration of semiconductor elements, their use as waveguides has considerably gained in importance during the last years. Today many components such as e.g. oscillators, electronically controlled attenuators and isolators can be realized in this technique and it will be possible in the future to integrate complete circuits in this way. For this purpose directional couplers are needed and, in order to design them, the propagation properties of waves on coupled finlines must be known. Therefore a field analysis of symmetrical coupled finlines must be developed first.

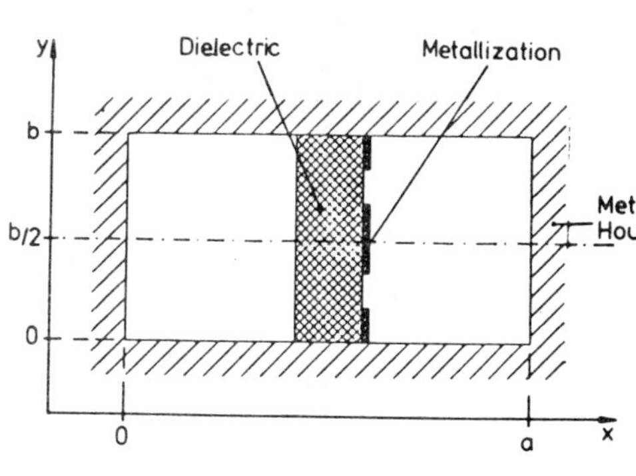

Fig.1: Cross-section of a finline coupler with two symmetrical slots.

The method presented here takes into account the finite thickness of the metallization and moreover it permits a complete solution of the boundary value problem shown in Fig.1. The methods published up to now either are approximate calculations /2/ or they did not consider the finite thickness of the metallizaton /3,4/.

Solution of the Boundary Value Problem

Making use of the symmetry, the finline coupler presented in in Fig.1 can be analysed by superposing the two fundamental modes (even and odd modes, as shown in Fig.2) which are solutions of the boundary value problem.

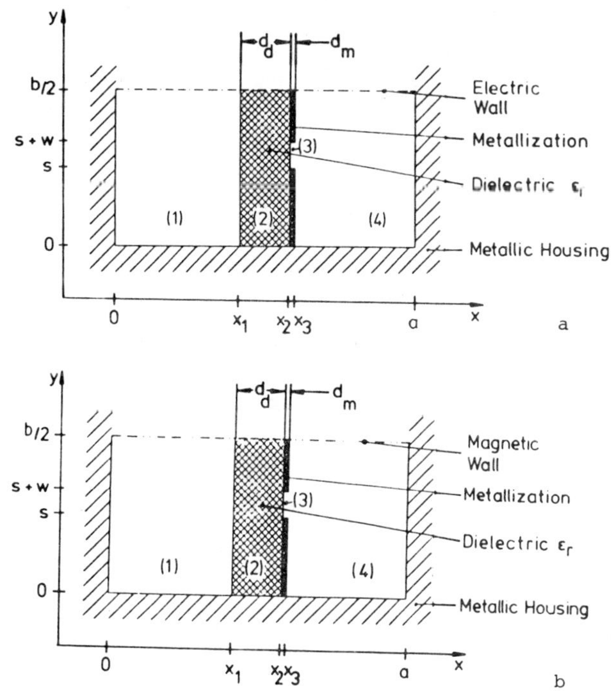

Fig.2: Geometry of the lower half of the symmetrical finline coupler,
a) for the odd mode,
b) for the even mode.

The cross sectional structures shown in Fig.2 are divided into four homogeneous subregions (1)-(4). The potential functions in the individual subregions can be described by superposing the solutions of the boundary value problems shown in Fig.3.

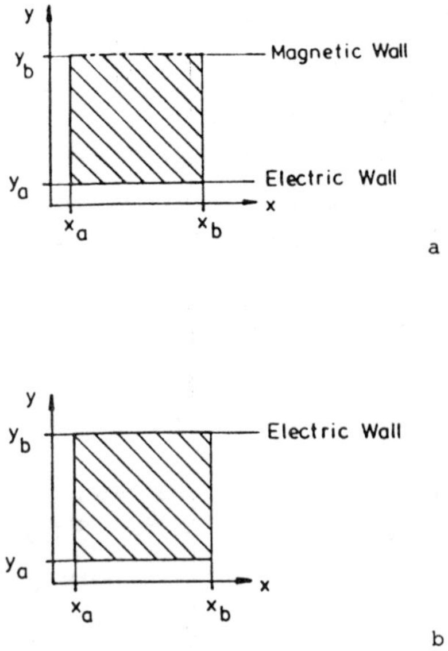

Fig.3: Geometry of the fundamental rectangular subregions which are used for solving the boundary value problem.

For a rectangular region with two electric walls, one magnetic wall and one open boundary (Fig.3a)) the potential functions are:

LSE mode:

$$\Psi_m = A_m \sin(k_{xm}(x-x_a))\cos((m-\frac{1}{2})\pi\frac{y-y_a}{y_b-y_a}) \,, \tag{1}$$

LSM mode:

$$\Phi_m = B_m \cos(k_{xm}(x-x_a))\sin((m-\frac{1}{2})\pi\frac{y-y_a}{y_b-y_a}) \,. \tag{2}$$

Analogously, for a rectangular region with three electric walls and one open boundary (Fig.3b)) the following potential functions can be found:

LSE mode:

$$\Psi_m = A_m \sin(k_{xm}(x-x_a))\cos(m\pi\frac{y-y_a}{y_b-y_a}) \,, \tag{3}$$

LSM mode:

$$\Phi_m = B_m \cos(k_{xm}(x-x_a))\sin(m\pi\frac{y-y_a}{y_b-y_a}) \,. \tag{4}$$

A coupling of the amplitude coefficients A_m, B_m results from the continuity conditions and the boundary conditions which have to be satisfied. Making use of the orthogonality of the expansion functions, a homogeneous equation system can be obtained as a function of the phase constant β, from which the phase constant can be computed by searching the zeros of the system determinant.

The employed field theoretical method has the advantage that it can be used to determine the electromagnetic field and its components directly. Fig.4 and Fig.5 show the field distributions of the even and the odd fundamental modes on two coupled finlines.

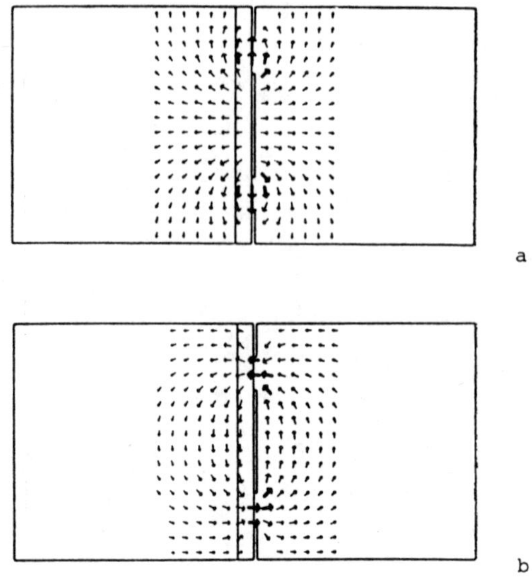

Fig.4: The field distribution of the even mode on two coupled finlines: The electric field (a) and the magnetic field (b).

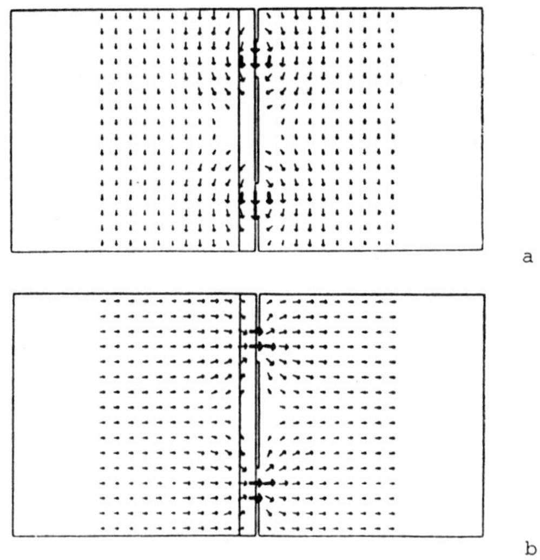

Fig.5: Field distribution of the odd mode on two coupled finlines: The electric field (a) and the magnetic field (b).

Design of a Finline Coupler

On a finline with two symmetrical slots (Fig.1), the even mode and the odd mode have different phase constants und thus a periodical coupling is obtained. The 0-dB coupling length L of such a structure is defined as /5/:

$$L = \frac{\pi}{\beta_{even} - \beta_{odd}} \tag{5}$$

For the realization of a finline coupler in the Ka-band the 0-dB coupling length L has been cal-lated in dependence on the frequency for different spacings s between the slot and the mount (Fig.6) and for different slot widths w (Fig.7).

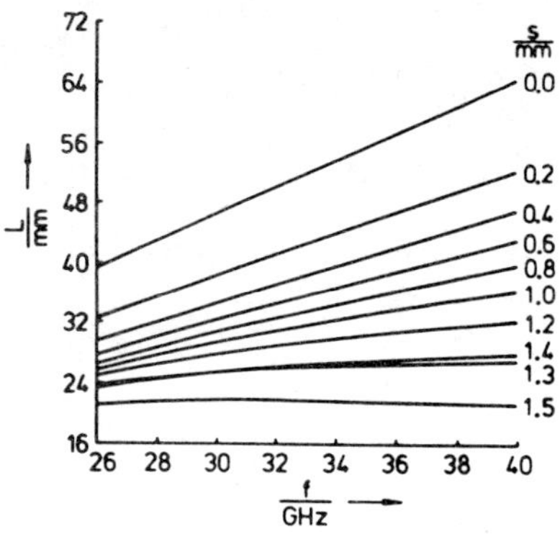

Fig.6: The 0-dB coupling length L versus the frequency f with the spacing s between the slot and the mount as a parameter. (w=0.2 mm).

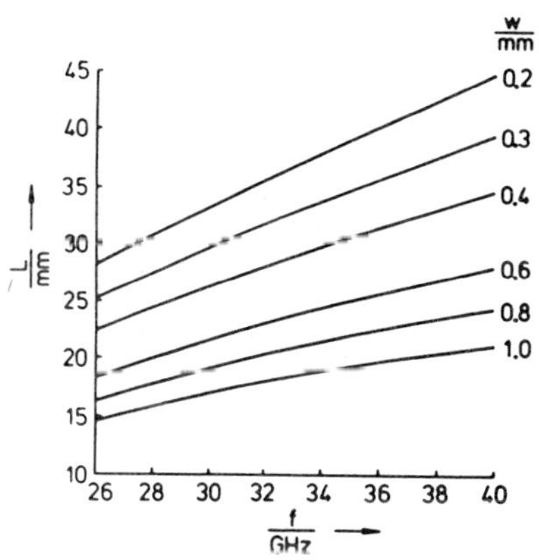

Fig.7: The 0-dB coupling length L versus the frequency f with the slot width w as a parameter. (s=0.5 mm).

For the realization of a finline coupler in the Ka-band the 0-dB coupling length L in dependence on the frequency has been calculated for different spacings s between the slot and the mount (Fig.6) and for different slot widths w (Fig.7).

From Fig.6 and Fig.7 it can be recognized that in the considered cases the frequency dependence of L is smallest for s=1.5 mm and w=0.2 mm. However, the metallic strip between the two parallel slots for technological reasons should not be too narrow; therefore in the actual coupler design s was chosen to be s=1.475 mm. In this case the slot width again is 0.2 mm. The phase constants β_{even} and β_{odd} of this configuration are shown in Fig.8.

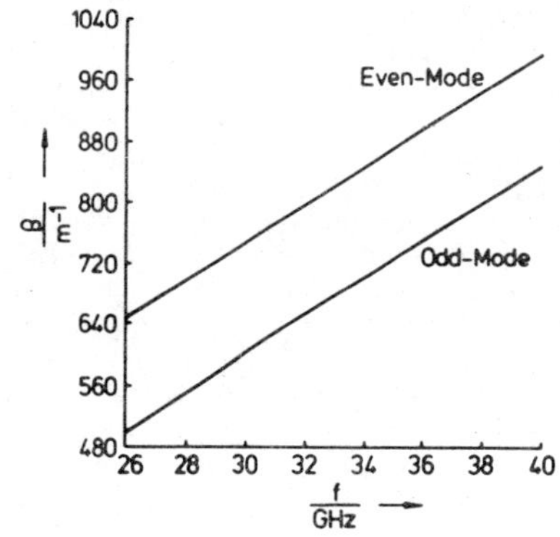

Fig.8: The phase constants of the even and the odd mode versus the frequency (s=1.475 mm, w= 0.2 mm).

The tapers which connect the coupler to the homogeneous waveguides (Fig.9) cause frequency dependent overcouplings which must be taken into account when the homogeneous coupling length L shall be determined. In order to keep this effect as small as possible, the tapers must be short. However, they should not be too steep as to avoid high reflections. In order to prevent waveguide resonances, short-circuit pins have been introduced into the two T-junction regions.

Due to the geometrical symmetry only 8 of the 16 scattering parameters are needed to describe the coupler completely. They are shown in Fig.10 for the designed coupler in the frequency range between 26 GHz and 40 GHz.

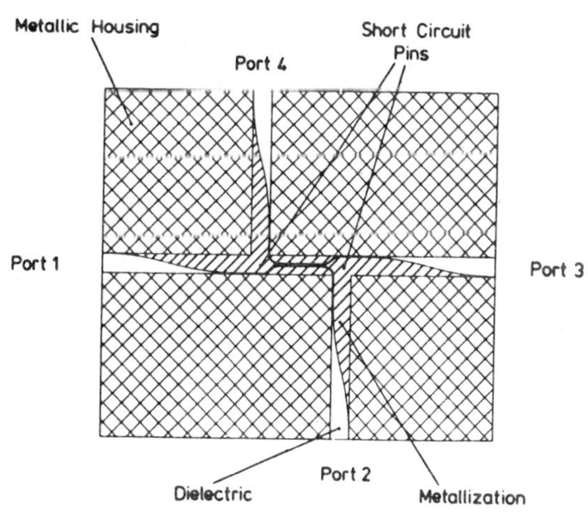

Fig 9: Top view of a four port finline coupler.

Fig.10a) and Fig.10b) show the coupling coefficients S_{31}, S_{21}, S_{12} and S_{42}. As can be recognized from these figures, the designed coupler does not posses the specified 3-dB coupling but the coupling coefficients is somewhere between 4 dB and 6 dB. There are several reasons for this deviations:

1) In the design process the bent finlines which make the connection to the tapers are not taken into account. As is known from similar couplers

in dielectric image line technique /6/ these interconnection lines have a notable influence on the coupler properties. The theoretical consideration of these lines is very difficult because they are inhomogeneously coupled finlines.
2) There is another critical aspect of the coupler shown in Fig.9. The asymmetrical tapers which are in the prototype coupler use the metal walls of the housing as one electrode of the taper finline. This construction cannot be produced with the required accuracy so that the tapers also could have an influence on the deviation of the coupling coefficients. This effect of course can be avoided by designing tapers which have a metallization on both sides of the finline slot.

The coupler has been designed for an operating frequency of 32 GHz. As Fig.10c) and Fig.10d) show, the minimum of the return loss (Fig.10d)) and the best isolation between ports 3 and 2 as well as ports 4 and 1 is reached at nearly 32.6 GHz. The deviation from the original design frequency has the same reasons as described above for the deviations of the coupling coefficients. At the frequency of 32.6 GHz a return loss of -40 dB and an isolation of 40 dB is measured for the coupler.

If a -20 dB return loss and a 20 dB isolation are required for defining the operating frequency band of the coupler, a bandwidth of about 3.5 GHz can be found for the finline coupler. Outside this band the return loss as well as the coupling between ports 3 and 2 and ports 4 and 1 are in the order of -15 dB.

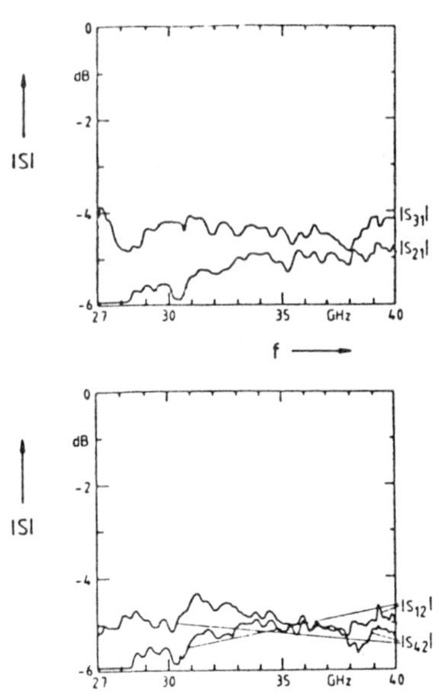

a
b

Fig.10: Measured scattering parameters of the designed finline coupler versus the frequency.

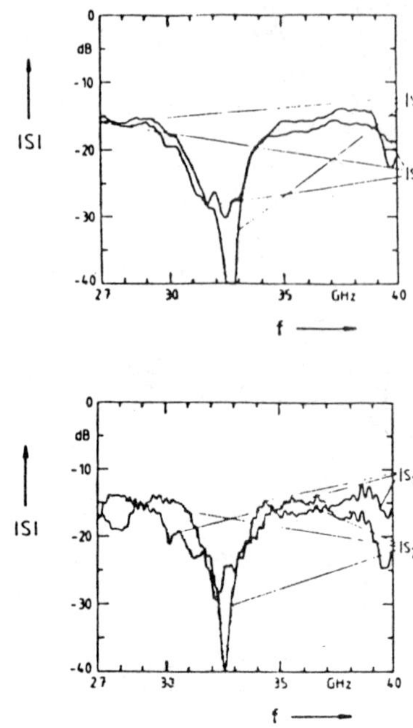

c
d

Fig.10 (cont.): Measured scattering parameters of the designed finline coupler versus the frequency.

Conclusions

The design of a finline coupler has been described. The properties of the produced coupler have to be improved e.g. by using Y-junctions instead of T-junctions in the waveguide housing. For a better understanding of the coupler, more detailed field pattern must be computed by means of the method presented here. Moreover, the problem of the inhomogeneous coupling regions and the taper between the rectangular waveguide and the finline must be considered in more detail. The described method can easily be modified for bilateral finlines or similar problems such like strip line couplers. Even isotropic magnetic materials and losses in the media could be taken into account.

References

/1/ A.Beyer and I.Wolff: 1980 IEEE MTT-S Internat. Microwave Symp. Digest, 258-260.

/2/ A.M.K.Saad and K.Schünemann: Proc. 11th European Microwave Conf., 1981, 305-308.

/3/ L.-P.Schmidt, T.Itoh and H.Hofmann: IEEE Trans. MTT-29, 1981, 352-355.

/4/ A.K.Sharma and W.J.R.Hoefer: IEEE Trans. MTT-29, 1981, 352-355.

/5/ L.Young.: Parallel coupled lines and directional couplers. Artech House, Inc., Dedham, 1972.

/6/ I.Wolff and K.Solbach: Dielectric Image Lines. Verlag H.Wolff, Aachen, 1984, 152-157.

6.3.4

F. Arndt, J. Bornemann, D. Grauerholz & R. Vahldieck
Microwave Department, University of Bremen, Germany

Material and Circuit Evaluation for Millimetre-Wave Applications: 3 dB Fin-Line Printed Probe Coupler for 31 GHz*

Abstract A 3 dB printed probe coupler has been designed for a 31.3 GHz midband frequency using a printed-circuit, seven-element, low-cost coupling structure. RT/duroid substrate material is used and an isolation of about 20 dB has been achieved.

Résumé On a conçu un coupleur de sonde de 3 dB pour une fréquence médiane de 31,3 GHz à l'aide d'une structure de couplage sur circuit imprimé à sept éléments à faible prix de revient. Avec un substrat en RT/duroid, on obtient une isolation de l'ordre de 20 dB.

* This article, provided via ERNO Raumfahrttechnik GmbH, is based on work conducted under ESA/ESTEC Contract No. 4044/79/NL/DG(SC). Companion papers have appeared in ESA Journal 81/1 and 81/2.

F. Arndt, J. Bornemann, D. Grauerholz, and R. Vahldieck, *European Space Agency Journal*, Vol. 5, No. 3, 1981, pp. 243–247. Reprinted with permission.

Introduction

A variety of coupling structures and circuits have already been developed for centimetre wavelengths (e.g. Refs. 1 and 2). For millimetre waves, however, suitable couplers, which are based on low-cost photo-etching techniques, are still very rare[3, 4].

Although common microstrip lines can also be used in the millimetre-wave region[5], some problems arise due to critical tolerances and dispersion, which may considerably reduce directivity, if other than the 3 dB hybrid ring coupling techniques are used. A very attractive new MIC coupler design principle for millimetre waves is still the printed probe coupler, suggested by Meier[3,4], in which an array of printed probes is placed in the E-plane common to parallel waveguides. The coupler in Reference 3 is designed for only 7.75 dB midband coupling at 35 GHz and there seems to be a paucity in this frequency range as far as the commonly sought 3 dB coupling is concerned.

It is therefore the purpose of this paper to provide a design suggestion for a 3 dB coupler to operate at about 31 GHz midband frequency (Fig. 1) based on the printed probe principle given in Reference 4.

The low-cost, seven-element printed probe design presented provides exact 3 dB coupling at midband frequency between ports 1 and 2, and 3, respectively. The isolation from port 4 is about 20 dB.

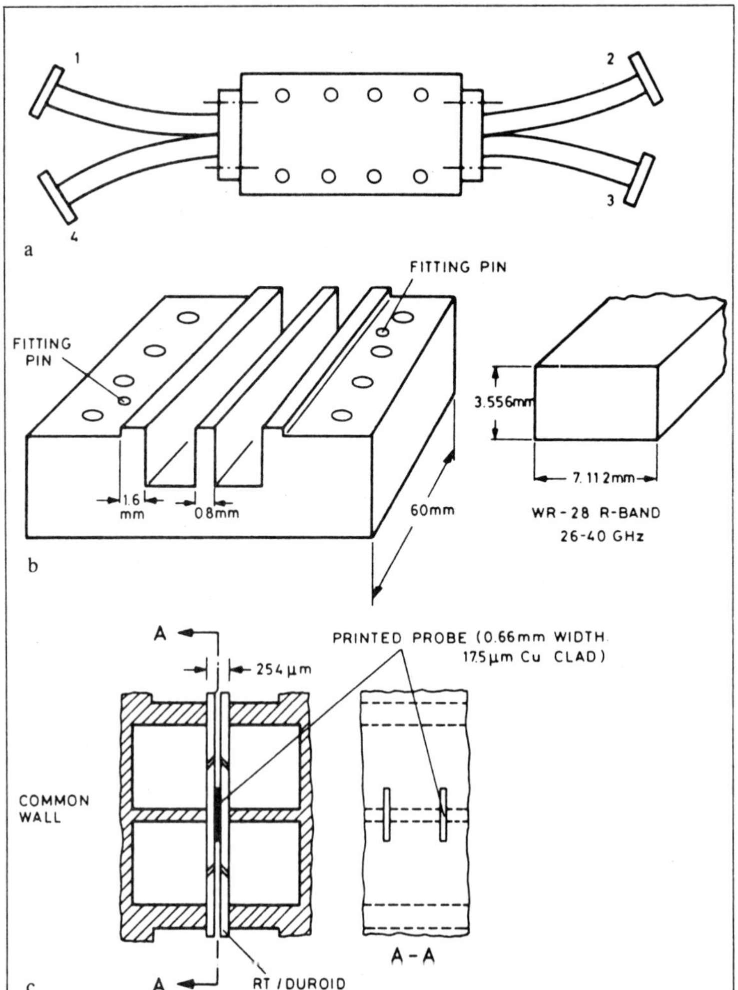

Figure 1. Fin-line coupler (printed probe) for 31.31 GHz
(a) plan view
(b) housing
(c) cross-section

Design

Figure 1 shows the printed probe fin-line coupler (plan view, housing and in cross-section) that has been designed. In this configuration, two parallel waveguides form a common housing with E-plane symmetry. Four waveguide bends are incorporated for compatibility with standard WR-28 instrumentation. The common broad wall of the two waveguides is slotted to accept a pair of dielectric RT/duroid boards. Directional coupling between the waveguides is achieved via an array of probes which is printed on one of the boards. Radiation from the outer board walls, which also contain the slot in order to simplify the manufacturing of the coupler, is negligible, because the wall thickness is chosen to be a quarter wavelength in the dielectric (about 1.6 mm, cf. Fig. 1b), providing almost a short circuit at the inner margin of the broad walls. This has been confirmed by measuring the insertion losses of the two coupled waveguides alone (with only the dielectric boards, without any probes), which gives a value of only about 0.3 dB.

After initial experiments with several probe widths, the latter was chosen to be 0.66 mm, since this value gave less variation in the coupling-response/frequency characteristic than other widths. The basic design information was also checked experimentally by measuring the midband coupling for one probe for various d/b ratios (Fig. 2). According to calculations in Reference 4 based on equivalent-circuit considerations, the midband coupling of elements 1 and 7 should be -24.1 dB, whereas the coupling of elements 2 to 6 should be -18.1 dB, if a seven-probe coupler is sought. The corresponding overall lengths l_{ges} of the probe types are shown in Figure 2.

Figure 3a shows the photo-etched, printed probe, seven-element coupling structure. The distance between the printed probe elements is $\lambda_{eff}/4 = 1.96$ mm, and this provides a midband frequency of 31.31 GHz.

Photographs of the printed probe coupler and the coupling element that have been designed are shown in Figures 3b and 3c, respectively.

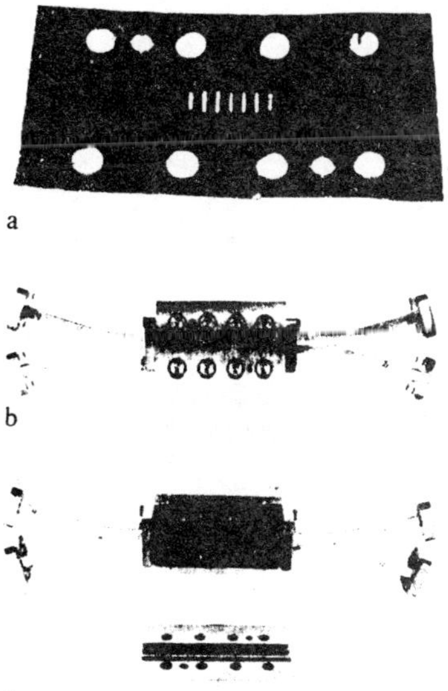

Figure 3. The coupler design (3 dB, 31.31 GHz, 20 dB isolation)
(a) seven-step printed probe coupling section (1st and 7th probe: 3.15 mm long, others: 3.65 mm long, width 0.66 mm, Cu 17.5 μm thick, substrate material RT/duroid)
(b) the complete coupler
(c) cross-section

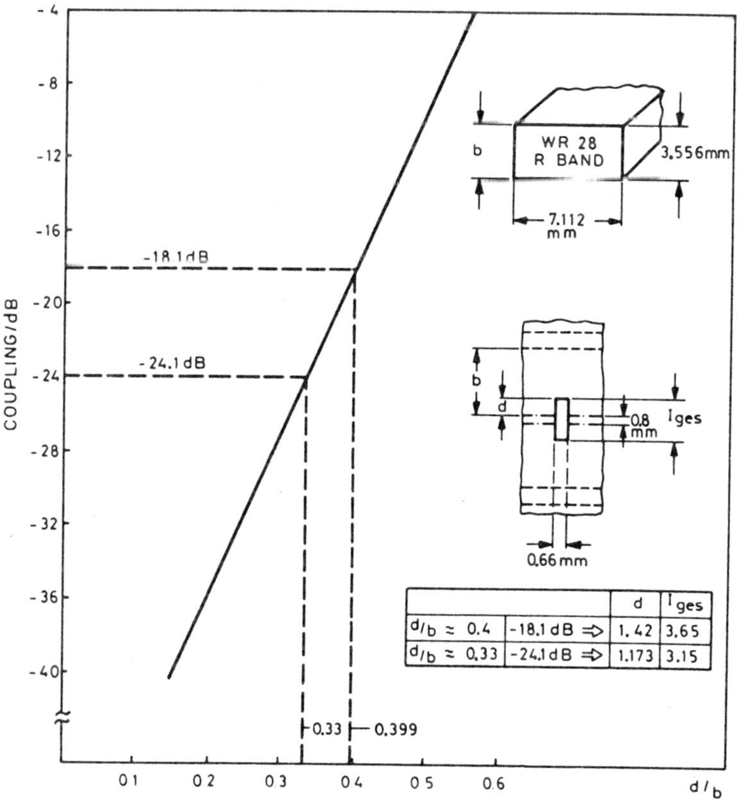

Figure 2. Measured printed probe coupling as a function of probe length (one probe only) at 31 GHz

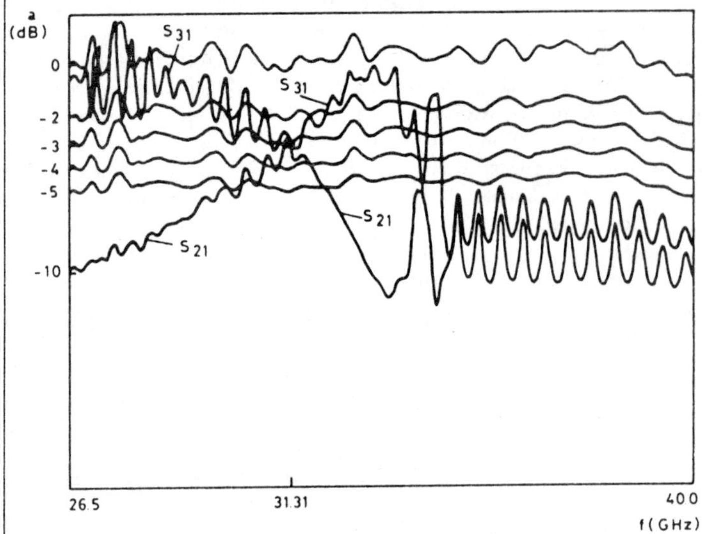

Figure 4. Measured transmission coefficients to ports 2 and 3 (S_{21}, S_{31}) of the 3 dB printed probe coupler (cf. Fig. 1a)

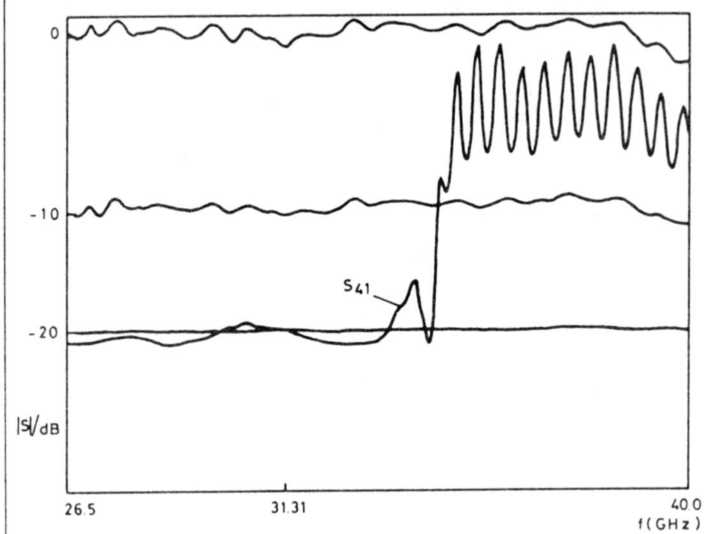

Figure 5. Measured isolation to port 4 of the 3 dB printed probe coupler

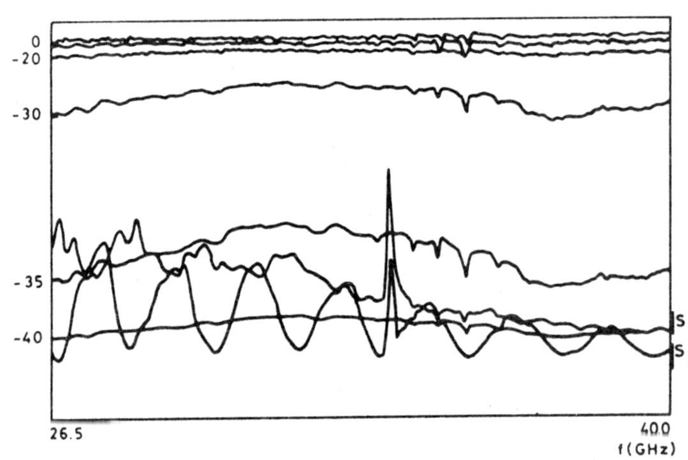

Figure 6. Measured transmission coefficients S_{31} and S_{41} of the printed probe coupler (cf. Fig. 1a) without metallic probes (with only the substrate) for comparison

Results

The measured couplings from port 1 to ports 2 and 3 (cf. Fig. 1) are shown in Figure 4. This coupler design provides exact 3 dB coupling at 31.31 GHz. If a 3 dB ± 1 dB limit is tolerated, the corresponding frequency band is about 1.48 GHz.

The isolation between ports 1 and 4 is shown in Figure 5. The measured value of about 20 dB is judged to be very good, since this measured isolation includes the effects of waveguide bends, flange discontinuities, etc.

For comparison purposes, the transmission coefficients from port 1 to ports 3 and 4 have been measured without the metallic probes, with only the dielectric boards (Fig. 6). The coupling, as well as the directivity effect, is clearly due to the printed probe arrangement, and a superimposed slot coupling effect can be neglected.

Conclusion

A seven-element printed probe coupler for 31.31 GHz has been designed based on a low-cost photo-etching technique applied on an RT/duroid substrate. This design provides exact 3 dB coupling at midband frequency, with an isolation of about 20 dB.

References

1. Matthaei G L, Young L & Jones E M T, 1964, *Microwave Filters, Impedance Matching Networks, and Coupling Structures*, McGraw Hill Book Co, New York
2. Itoh T, 1978, Generalized Spectral Domain Method for Multiconductor Printed Lines and its Application to Tunable Suspended Microstrips, *IEEE Trans*, **MTT-26**, pp 983-977
3. Meier P J, 1978, Millimeter integrated circuits suspended in the E-plane of rectangular waveguide, *IEEE Trans*, **MTT-26**, pp 726-733
4. Meier P J, 1979, Printed Probe Hybrid Coupler for the 3 mm Band, *Proc 9th European Microwave Conference*, Brighton, pp 443-447
5. Arndt F, Bornemann J, Grauerholz D, & Vahldieck R, 1981, Material and Circuit Evaluation for Millimetre-Wave Applications: A Microstrip 3 dB-Hybrid Coupler with Low-Dielectric Substrate for Millimetre Waves, *ESA Journal* **5**, No. 2, pp 127-132

Manuscript received 9 March 1981

6.3.5

ANALYSIS OF THE COUPLING COEFFICIENT BETWEEN A CYLINDRICAL DIELECTRIC RESONATOR AND A FIN-LINE

F. Hernández-Gil *, J. Pérez **

* R&D Dpt., Telefónica, Lérida 43, 28020 Madrid, Spain.
** Microwave Dpt., E.T.S.Ing. Telecomunicación, 28040 Madrid, Spain.

ABSTRACT

The external quality factor of a cylindrical dielectric resonator coupled to a fin-line is evaluated using the finite element method to analyse the transmission line and the resonator. The effects of varying the location of the resonator on the coupling are studied. Experimental results compared with the theoretical ones are presented.

INTRODUCTION

The improvement in the materials used to make dielectric resonators has made them interesting devices to be used in millimeter wave bands [1]. Most of the applications use the dielectric resonators coupled to microstrip lines to stabilize oscillators and to make filters, and several useful approaches have been reported to calculate the coupling to these lines [2]-[5]. Fin-lines have been proved to be an attractive technology to make circuits in Ka band and higher frequencies, whereas little atention has been paid to the study of the coupling to these lines.

In this paper a numerical method to calculate the coupling between a dielectric resonator and a fin-line is presented. The finite element method is used to evaluate the modes in the transmission line and in the resonator. This allows the study of different structures with the same computer program.

Several experimental and theoretical results are presented for the Ka band and the most relevant parameters which affect the coupling are studied. These results can be useful to make oscillators, filters, and others devices with cylindrical dielectric resonators.

METHOD OF ANALYSIS

The coupling to microstrip lines has been usually calculated considering the resonator field intensity in the line and calculating the energy induced by this field [2]-[4]. Other models evaluate the energy induced by the line in the resonator and model it as a magnetic loop [5]. It is more rigorous to consider both field distributions as done by Skalicky [6] to calculate the coupling to microstrip line. This last method should provide better accuracy than the others but requires the calculation of both field distributions and the evalution of a volume integral inside the resonator. This method does not require the evalution of the fin-line impedance, which is not always clearly defined.

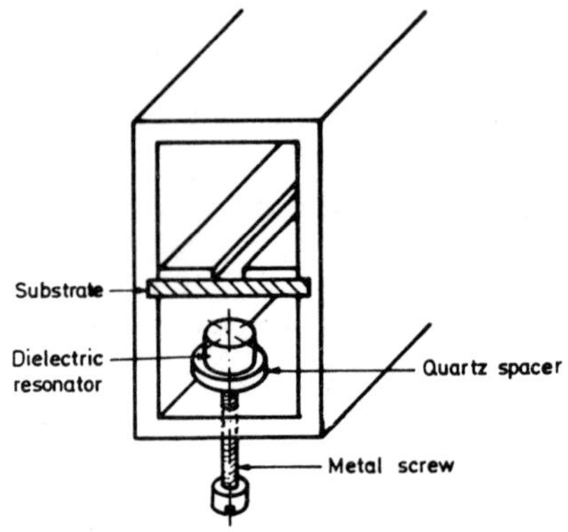

Fig.1 Cylindrical dielectric resonator coupled to a unilateral fin-line.

The structure to be analysed is shown in fig.1. A cylindrical dielectric resonator is located under the gap of a unilateral fin-line and is supported by a metal post. A spacer is under the resonator to reduce the losses. The metal post with the spacer can be substituted by a plastic screw.

External quality factor

The external quality factor of a resonator coupled to a transmission line is the ratio between the power dissipated in the line and the stored electromagnetic energy in the resonator.

$$Q_{ext} = w_o \frac{W_R}{P_d} \qquad (1)$$

F. Hernandez-Gil and J. Perez, *IEEE MTT-S International Microwave Symposium Digest*, 1986, pp. 221–224.

The power excited in the line by the resonator for the mode n can be calculated expanding the resonator field into transmission line modes according to Collin [7]. Considering the polarization current in the resonator the power dissipated in the line is

$$P_d = |C^n|^2 \; Re \; [\int_S e_n \times h_n^* \cdot \hat{z} \; dS \;] \qquad (2)$$

$$C_n = -1/P_n \int_V (e_n + e_{zn}) \cdot J_p \; e^{-\gamma_n z} \; dV \qquad (3)$$

$$P_n = 2 \int_S e_n \times h_n \cdot dS \qquad (4)$$

$$J_p = j w_0 \varepsilon_0 (\varepsilon_r - 1) E_R \qquad (5)$$

E_R : Resonator electric field

$(e_n + e_{zn})$: Electric field of the n mode of the line

The stored electromagnetic energy in the resonator at the resonant frequency is

$$W_R = 0.5 \int_V \varepsilon \; \overline{E}_R \cdot \overline{E}_R^* \; dV \qquad (6)$$

To evaluate the external quality factor using the expressions (1)-(6) the following calculations must be done.

- resonant frequency of the resonator
- electric field distribution in the resonator
- propagation constant of the line at the resonant frequency
- electric field distribution of modes above cutoff in the line.
- a volumen and a surface integral.

Analysis of the dielectric resonator

To analize the dielectric resonator the structure which contains the resonator should be considered axisymmetric to reduce the problem to two coordinates. This can be made if it is supposed that the waveguide lateral walls and the slot of the fin-line have little effect on the field distribution and on the resonant frequency.

The analysis of the dielectric resonator has been made using the finite element method decribed in [8]. This method allows to take into account the effect of the supporting structure in the resonator.

Fig.2 shows the field distribution for a dielectric resonator. The arrows length is linearly proportional to the field amplitude. As it can be seen, the field intensity decreases very quickly outside the resonator and has a small value in the lateral walls and in the fin-line

Analysis of the fin-line

The finite element method has been also used to calculate the modes of the fin-line. This allows to use the same computer code and adds flexiblity to the method.

The magnetic field variational formulation with the divergence condition imposed by the penalty method [9] was first used. However poor results were obtained in this case because the penalty term produces an important error. This effect was also observed by Hara [11] for structures with corners. Another problem of this formulation is the singularities of the field in the fin corners.

Better results were obtained using a formulation with the longitudinal components of the field [10]. In this formulation the effective permittivity is a parameter, the frequency is the eigenvalue of a linear system and the eigenfunctions are the field distributions of the modes. Third order serendipity polynomials were used to approach the field.

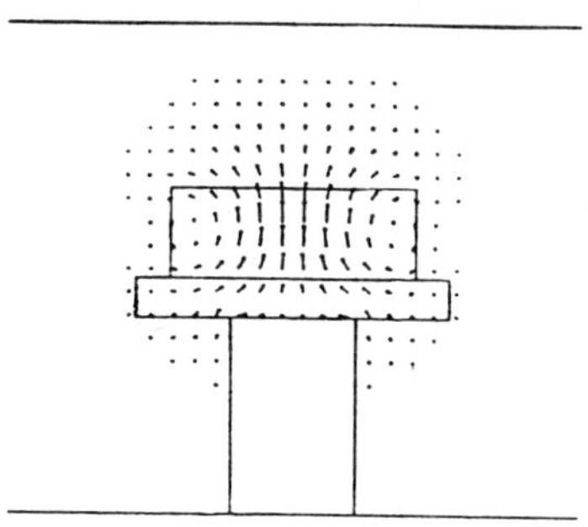

Fig.2 Magnetic field distribution for the $TE_{0\gamma\delta}$ mode in a cylindrical dielectric resonator.

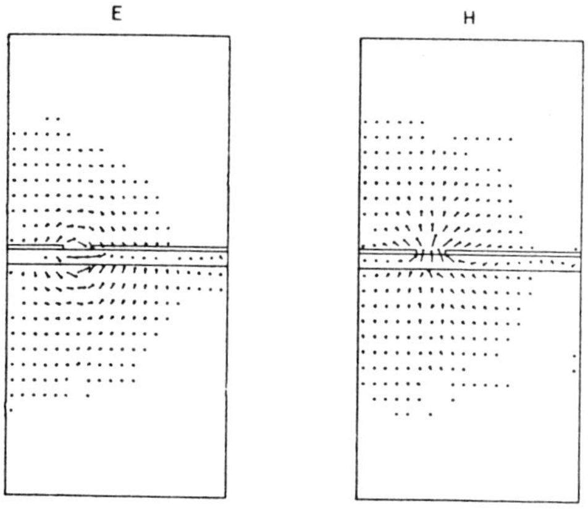

Fig.3 Transversal field distribution in a unilateral fin-line

This formulation is ill conditioned if the relative permittivity of the structure is equal to the efective permittivity of the line [10]. The modes in these points were interpolated from the whole curve.

COMPUTATION

A computer program in FORTRAN-77 to calculate the coupling was written. This program takes into account the sparsity of the matrices which arise in the finite element method. The integrals which appear in the finite element formulation and in the calculation of the coupling (1)-(6) are evaluated numerically. The geometry of the line and the resonator are entered in the program as input data files. This allows to study several kinds of lines and resonators.

The computer time required to calculate the coupling depends much on the desired precision and on the complexity of the structure to be analyzed. The total computer time for a typical case varies between 10 and 30s on a VAX11/785.

RESULTS

Experimental results

The structure used to measure the coupling between a fin-line and a dielectric resonator is shown in the fig.4a. The height of the resonator in the waveguide can be varied with a screw. The fin-line is terminated with two transitions to waveguide at both ends.

Fig.4b shows the measured and the theoretical values of the external quality factor for the $TE_{0\gamma\delta}$ mode of a resonator in Ka band varying the height of the resonator in the waveguide. As it can be seen the slope of the curve of the theoretical values is greater than that of the measured curve. Fig.4c shows the theoretical resonant frequency and the measured values. The value of the permittivity provided by the manufacturer was used to calculate the frequency.

The $TE_{0\gamma\delta}$ mode is the most coupled to the line and only two modes were observed in the 26.5-40 GHz band. The difference in the resonant frequency for the two first modes is always greater than 5 Ghz.

Variation of the coupling

The height of the resonator in the waveguide is the parameter which affects more strongly the external quality factor of the resonator. Values between 10 and 10^4 can be obtained varying this parameter. However strong variation of the resonant frequency is observed if the resonator is near the fin-line or the waveguide wall. So it is not very useful to be used as a parameter to adjust the coupling.

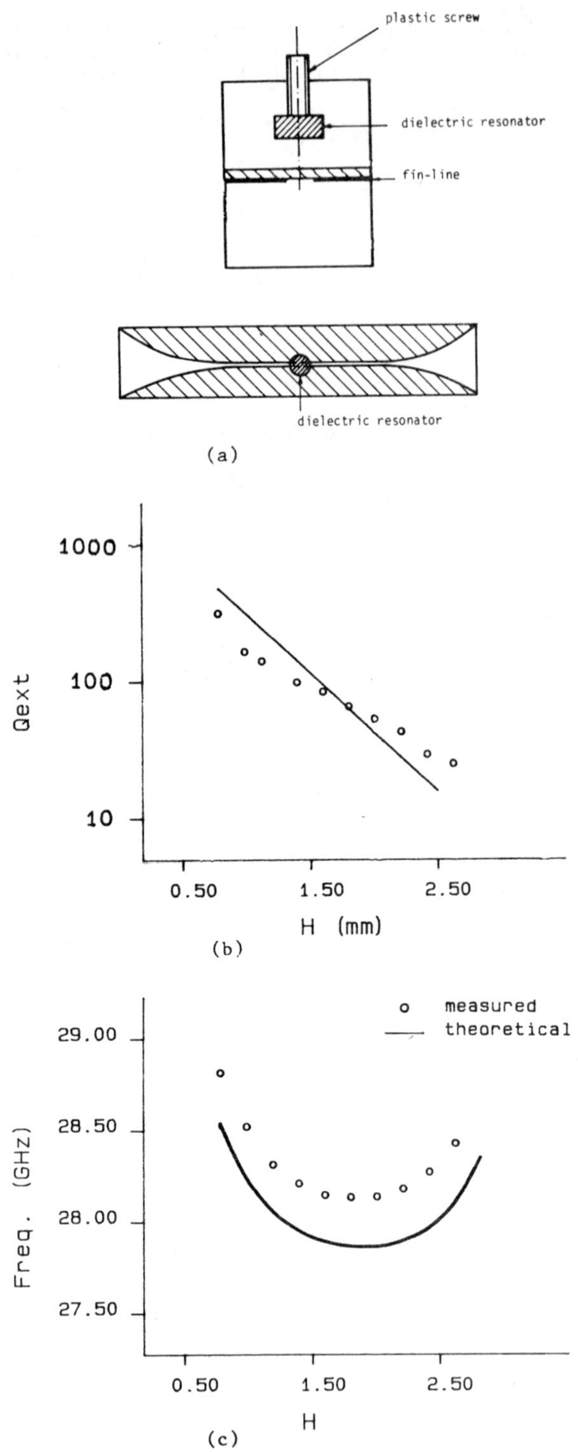

Fig.4 (a) Structure used to measure the coupling. (b) Theoretical and measured coupling. (c) Resonant frequency. A=7.11 mm, B=3.556 mm, b=0.32 mm, D_r=1.888 mm, H_r= 0.844 mm, R_t=1 mm, ε_r=37, R=1.778 mm, H_m=0.034 mm, H_s=0.254 mm, H_f=3.429 mm, $\varepsilon_{t=2.2}$

Offset of the resonator in the waveguide. This parameter allows to vary slightly the coupling with small variation in the frequency. It can be useful to adjust the coupling. The greatest coupling is obtained when the resonator is just under the fin-line gap. Fig.5 shows the variation of the coupling for a resonator in Ku band.

Width of the fin-line. The coupling increases with the width of the fin-line gap. Fig.6 shows the variation of the coupling for several widths of the gap.

CONCLUSION

A numerical method to calculate the coupling between a cylindrical dielectric resonator and a fin-line has been proposed and experimental measures were made to evaluate its precision.

The coupling of a dielectric resonator to fin-lines is generally greater than that obtained with microstrip lines. However a wide margin of values of the external quality factors can be obtained, which is useful to design oscillators and filters.

The finite element formulation used to study the fin-line should be improved in order to avoid the singularities for some values of the effective permittivity.

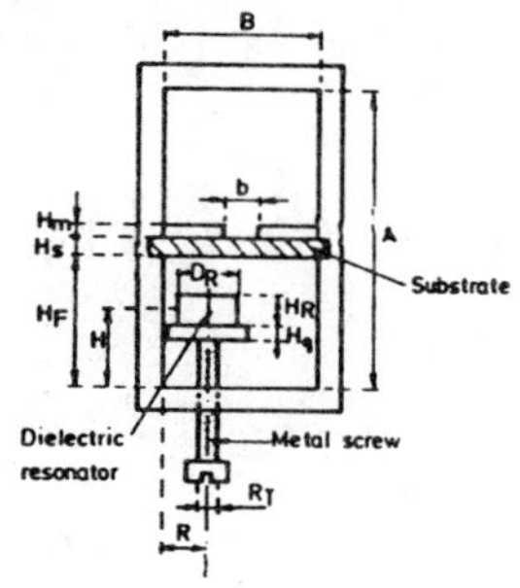

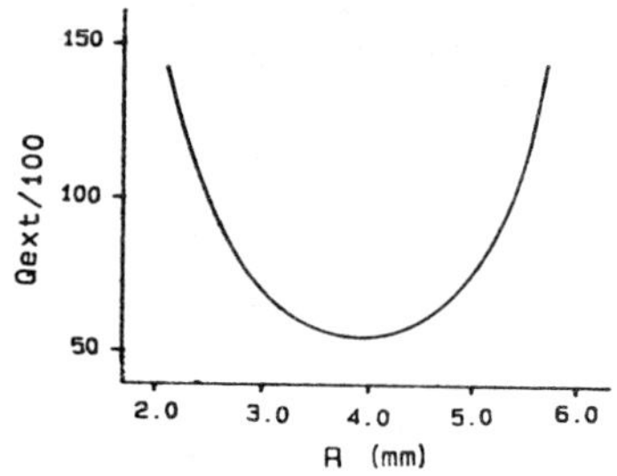

Fig.5 Effect of varying the distance to the lateral waveguide wall on the external quality factor. A=15.6 mm, B=7.8 mm, b=2 mm, D_r=3.912 mm, R_r=1.8 mm, ε_r=37, ε_q=3 mm, H_m=0.3 mm, H_s=0.5 mm,

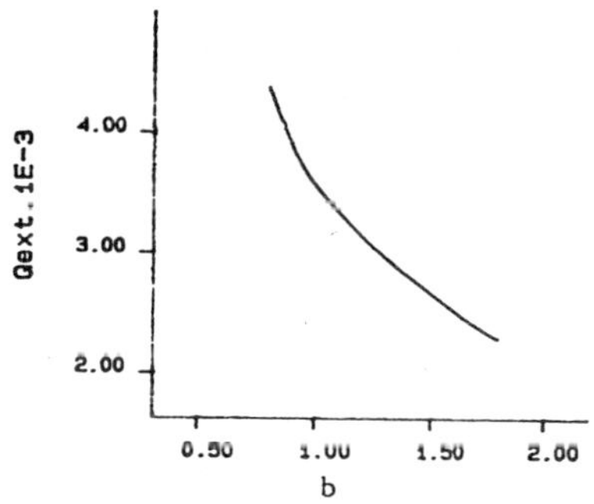

Fig.6 Effect of varying the width of the fin-line on the external quality factor. A=15.6 mm, B=7.8 mm, D_r=3.912 mm, R_r=1.8 mm, ε_r=37, ε_q=3 mm, H_m=0.3 mm, H_s=0.5 mm,

REFERENCES

[1] N.Mamodaly et al. "A fundamental-mode InP Gunn dielectric Resonator Oscillator at 94 GHz", 15 European microwave Conference Proceeding, pp.170-176, September 1985.

[2] P.Guillon, B.Byzery, M.Chaubet, "Coupling parameters between a dielectric resonator and a Microstrip line", IEEE Transactions on Microwave Theory and Techniques, Vol.MTT-33, No 3, March 1985, pp.222-226.

[3] Y.Komatsu, Y.Murakami, "Coupling Coefficient between Microstrip Line and Dielectric Resonator", IEEE Transactions on Microwave Theory and Techniques, Vol.MTT-31, No 1, January 1983, pp.34-40.

[4] R.Bonetti, A.Atia, "Analysis of Microstrip Circuits Coupled to Dielectric Resonators", IEEE Transactions on Microwave Theory and Techniques, Vol.MTT-29, No.12, pp.1333-1337, Dec. 1981.

[5] P.Guillon, S.Mekerta, Y.Garault, "Microstrip Bandstop Filter using a Dielectric Resonator", IEE Proc. Vol.128, Pt.H. No.3, June 1981.

[6] P. Skalicky, "Coupling Coefficient between Dielectric Resonator and Microstrip Line", Electronics Letters 22nd January 1981, pp.99-100.

[7] R.Collin, "Foundations for Microwave Engineering", Chapter 4, McGraw-Hill, New York 1966.

[8] F. Hernández-Gil, J. Pérez, "Analysis of Dielectric resonators with Tuning Screw and Supporting Structure", IEEE Transactions on Microwave Theory and Techniques, PP.1453-1457, Dec. 1985.

[9] B.M.Azizur Rahman, J.Brian Davies, "Penalty Function Improvement of Waveguide Solution by Finite Elements", IEEE Transactions on Microwave Theory and Techniques, Vol.MTT-32, pp.922-928, August 1984,

[10] N.Mabaya, P.E.Lagasse, P.Vandenbulcke, "Finite Element Analysis of Optical waveguides", IEEE Transactions on Microwave Theory and Techniques, Vol.MTT-29, pp.600-605, June 1985.

[11] M.Hara, T.Wada, T.Fukasa, F.Kikuchi, "A Three dimensional Analysis of the RF Electromagnetic Fields by the Finite Element Method", IEEE Transactions on Magnetic, Vol. Mag-19, No.6, pp.2417-2420, Nov. 1983.

[12] A. Jacob, C. Ansorge, "Stabilized Fin-line Oscillators", 1983 European Microwave Conference Digest, pp.303-307.

6.4.1

18–26 GHz finline resonance isolators using hexagonal ferrites

Prof. J. Helszajn, D.Sc., C.Eng., F.I.E.E., and W. Thorpe, M.Sc.

Indexing terms: *Waveguides, Microwave systems*

Abstract: A finline resonance isolator is an attractive solution to the construction of millimetre wave isolators, except for the fact that a large direct magnetic field is normally required for its operation. One way to overcome this drawback is to employ an hexagonal ferrite material with a large internal uniaxial anisotropy energy. Such devices are simple to fabricate and provide perhaps the most practical solution to the design of millimetre band isolators. The paper describes some experimental finline devices in WR 42 waveguide.

1 Introduction

One transmission line that is of practical interest at millimetre wavelengths is the finline waveguide [1, 2] (Fig. 1).

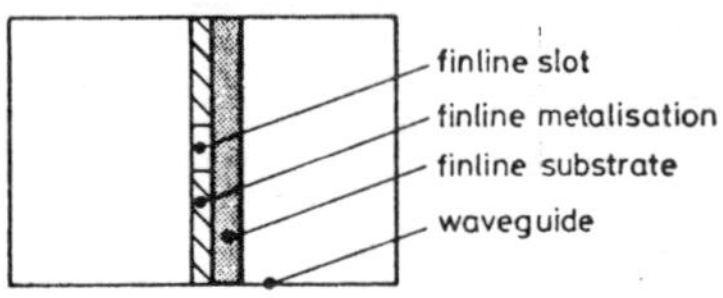

Fig. 1 *Schematic diagram of finline waveguide*

The need for nonreciprocal ferrite devices in this waveguide has been partly met in the form of field displacement isolators [3, 4], junction circulators [5, 6] and, at least in principle, a finline resonance isolator [6]. The construction of practical wideband resonance isolators is especially straightforward if the required circularly-polarised magnetic-field region is directly embodied in the finline circuit, this is achieved here by noting that such polarisation is always approximately exhibited at the interface between two different dielectric regions [12]. However, one shortcoming of the resonance isolator at millimetre wavelengths is the need for a large direct magnetic field and therefore a cumbersome magnet structure. This drawback may, of course, be alleviated by employing highly aligned polycrystalline hexagonal materials with uniaxial anisotropy energy along the C-axis of the crystal lites, as is well understood [7–10, 14]. The purpose of this paper is to briefly describe a number of finline isolators in WR 42 waveguide using hexagonal ferrite materials and circular polarisation elements. Such devices are easy to fabricate and provide perhaps the most practical solution to the design of full waveguide bandwidth isolators at very high frequencies [10]. The bandwidth of the resonance isolator is, in the first instance, mainly determined by its linewidth, but can also be widened by tapering the direct magnetic field or by either cascading or stacking ferrite substrates with different values of magnetisation and anisotropy field. Each of these techniques provides a practical means of achieving standard waveguide bandwidths, but the stacking technique described here is by far the most attractive solution, both with regards to insertion loss and mechanical layout.

Although hexagonal ferrite materials are usually employed in the design of filters and resonance devices, there is no reason for not using them for the construction of circulators, switches, phase shifters and the like.

Paper 3729H (E12), first received 25th June and in revised form 10th December 1984

Prof. Helszajn is with the Department of Electrical and Electronic Engineering, Heriot-Watt University, 31–35 Grassmarket, Edinburgh EH1 1HT, Scotland. Mr. Thorpe is with Com-Dev, Ontario, Canada

2 Finline isolator using hexagonal ferrite substrate

The operation of the ferrite resonance isolator relies on the different values of dissipation exhibited by a magnetised ferrite material under the influence of clockwise and anticlockwise polarised magnetic fields; its theory is, of course, well rehearsed and will not be repeated here [11]. Such polarisation is nearly always met at the interface between two different dielectric matcrials and is usually quite wideband [12]. Likewise, the merits of employing hexagonal ferrite materials with large internal uniaxial anisotropy energy is also well understood. Suffice to note that, with appropriate doping, these materials may exhibit effective internal uni-axial direct magnetic fields that can be varied in value between about 200 kA/m and 3000 kA/m [7], and which adds directly to the usual external or applied direct magnetic field; a feature that makes them especially useful for the construction of tunable filters and resonance devices at frequencies between 18 GHz and 90 GHz.

The schematic diagram of one possible finline isolator is illustrated in Fig. 2. It consists of a unilateral finline on a

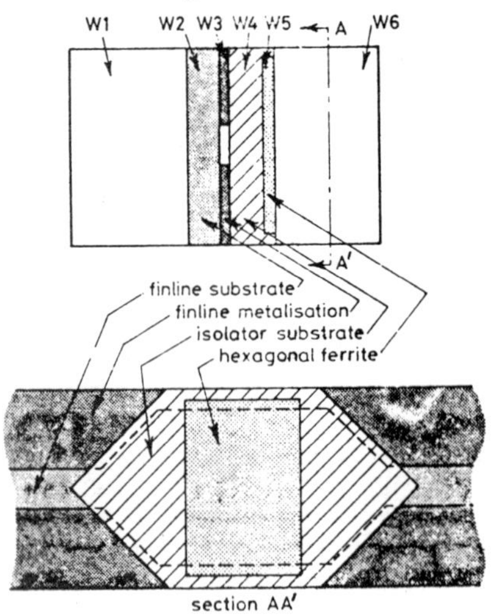

Fig. 2 *Schematic diagram of finline resonance isolator*

carrier loaded on one side by alumina and hexagonal ferrite substrates; with the circuit metallisation deposited on the finline substrate. The gap dimension of the input and output finline circuits is flared on both sides to that of the finline isolator which may, but need not, coincide with the narrow dimensions of the waveguide; the alumina is also tapered at the input and output terminals of the device for matching purposes. Contact problems between the wide dimensions of the waveguide and the finline circuit are avoided by incorporating suitable chokes in the assembly. A small external direct magnetic field is provided to orientate the internal field along the positive or negative C-axis of the uni-axial ferrite material and for tuning purpose.

The figure of merit of this class of device is defined as the ratio of its isolation and insertion loss at its midband frequency. It is in part determined by the power density in each region of the structure, the quality of the two hands of polarisation and the relative dielectric constants of the different regions, and it is in part dependent on the dielectric loss tangents of the ferrite and the other dielectric regions, and on the uniform linewidth. Except for the dielectric and magnetic dissipation parameters of the hexagonal class of ferrite compounds, none of these quantities (with proper adjustment) are likely to be very different from those met in rectangular waveguide isolators. A full theoretical model, although perhaps a standard problem, must still deal, even with the finline gap equal to that of the narrow dimension of the rectangular waveguide, with a six or more (see last Section) region boundary value problem. Furthermore, some of the regions exhibit tensor properties and magnetic and/or dielectric dissipation parameters; a situation in which the experimental approach is still second to none, provided the material resources are at hand.

3 Bandwidth and magnetic variables of resonance isolators

A knowledge of the microwave performance of the finline isolator requires a full analysis of the device which is outside the scope of this work and is indeed perhaps not an easy task. For practical purposes (at least for materials with large uniform mode linewidths) some estimate of the useful bandwidth of the device may be formed with the aid of Kittel's relationship by defining the bandedge frequencies ($f_{1,2}$) of the device at the effective direct magnetic fields ($H_{1,2}$) displaced by half a linewidth (ΔH) on either side of the anisotropy field (H_a).

$$H_{1,2} = H_a \mp \frac{\Delta H}{2} \tag{1}$$

$H_a(A/m)$ is the effective anisotropy field required to establish the main resonance line at the centre frequency (f_0) of the device, $\Delta H(A/m)$ is the linewidth of the material.

The relationships between the magnetic variables, the demagnetising factors of the ferrite shape and the microwave frequencies $f_{1,2}$ and f_0 are given in the usual way by Kittel's equation [13]. Approximating the shape demagnetising factors N_x, N_y and N_z for the geometry used here as $N_x = N_z \approx 0$ and $N_y \approx 1$ readily gives the required relationships for f_0 and $f_{1,2}$:

$$f_0 \approx \frac{\gamma}{2\pi}\sqrt{H_a\left(H_a + \frac{M_0}{\mu_0}\right)} \tag{2}$$

$$f_{1,2} \approx \frac{\gamma}{2\pi}\sqrt{\left(H_a \mp \frac{\Delta H}{2}\right)\left(H_a \mp \frac{\Delta H}{2} + \frac{M_0}{\mu_0}\right)} \tag{3}$$

γ is the gyromagnetic ratio (2.21×10^5 (rad/s)/(A/m)), μ_0 is the free space permeability ($4\pi \times 10^{-7}$ H/m), M_0 is the saturation magnetisation (T).

One typical commercial material which is suitable for use in the 18 GHz to 26 GHz band is

anisotropy field (H_a)	469 kA/m
magnetisation (M_0)	0.3400 T
linewidth (ΔH)	159 kA/m
dielectric constant (ε_f)	16
dielectric loss tangent ($\tan\delta$)	0.002

The results are

$f_1 = 17.8$ GHz
$f_0 = 20.7$ GHz
$f_2 = 23.6$ GHz

A small external direct magnetic field H_0 (A/m) is now all that is required to tune this frequency interval to operate in the WR 42 waveguide band.

The classic method of realising resonance isolators over full waveguide bands is to either employ ferrite assemblies with different values of direct magnetisation, or internal effective anisotropy fields. The former arrangement is the preferred approach at microwaves and the latter one is the more desirable choice at millimetre waves. The possibility of using tapered direct magnetic fields is also noted. The attenuation envelope formed by employing three discrete resonance lines is depicted in Fig. 3.

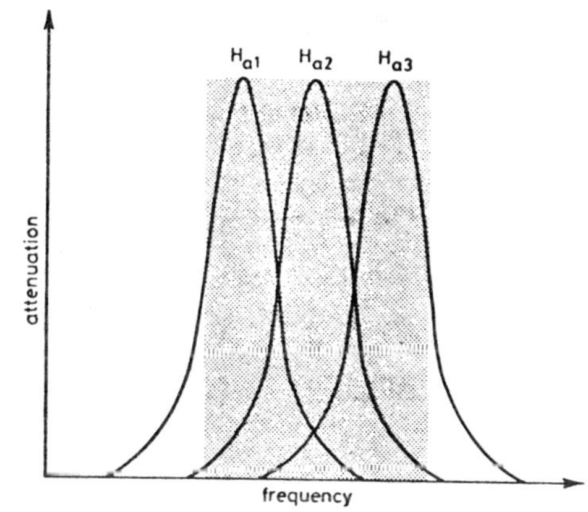

Fig. 3 *Overall response of resonance isolator using three discrete resonance lines*

The uniform mode linewidth of the polycrystalline material employed in this work is perhaps somewhat on the large side compared to that exhibited by some other families of materials, and whereas this is not a handicap in the development of devices with modest bandwidths, it is not altogether appropriate for the construction of high-quality narrow-bandwidth devices. A single crystal hexagonal ferrite material with a relatively narrow uniform linewidth has in fact been employed in the construction of a conventional waveguide resonance isolator [14].

If the shape demagnetising factors perpendicular to the easy axis of the hexagonal ferrite are unequal, as they are here, it is strictly speaking appropriate to recall that the normal modes are elliptically rather than circularly polarised. The ellipticity is given in terms of the transverse frequencies, ω_x and ω_y, by [18]

$$e = \sqrt{\frac{\omega_x}{\omega_y}} \tag{4}$$

ω_x and ω_y may be formed by noting that ω_0 in eqn. 2 is given in terms of the transverse frequencies by

$$\omega_0 = \sqrt{\omega_x \omega_y} \tag{5}$$

4 Finline isolators using single ferrite substrate

One parameter at hand in the experimental adjustment of this class of device is the finline gap. This was, in one instance, chosen to correspond to that of the input and output finline circuits, and in another case made equal to the narrow dimension of the rectangular waveguide. The circuit in the latter situation is, of course, akin to that of a standard E-plane isolator. The details of the isolator element are in this case directly taken from the work in [10]. The figure of merit of the first arrangement varied between 15 and 18 and that of the second one ranged between 22 and 24. Fig. 4 depicts the insertion loss of one of the finline transmission lines with its transitions but without the isolator structure. Figs. 5 and 6 indicate the

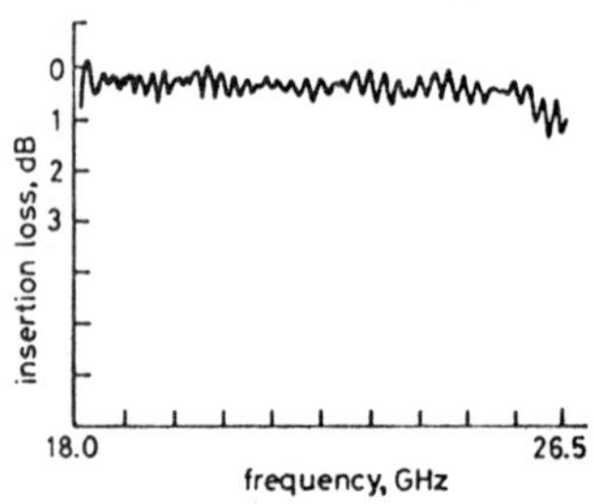

Fig. 4 *Insertion loss of finline circuit using $2\lambda_g$ exponential tapers*

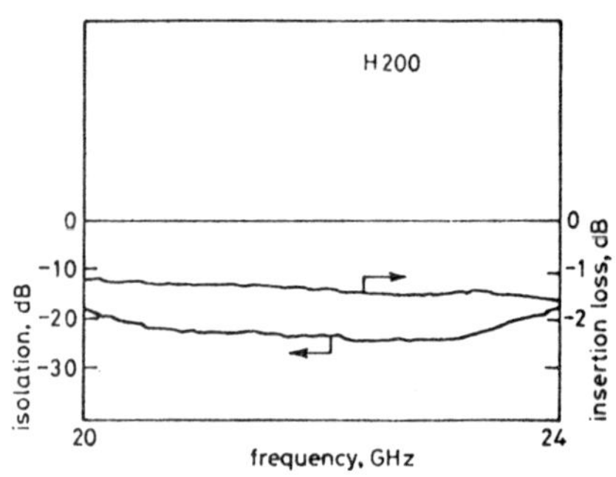

Fig. 5 *Insertion loss and isolation of finline isolator with finline gap equal to narrow dimension of WR 42 waveguide*

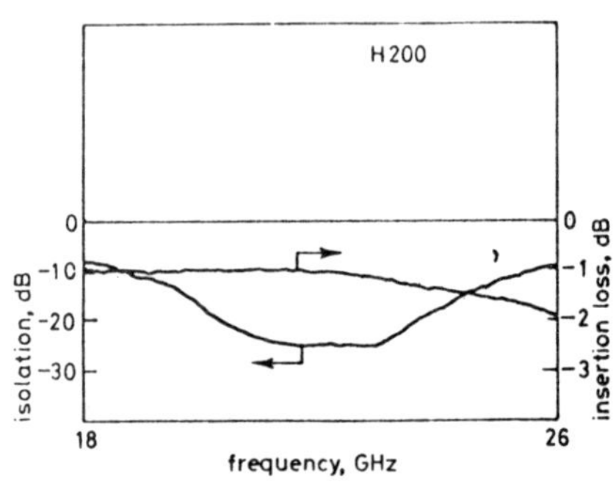

Fig. 6 *Insertion loss and isolation of finline isolator with finline gap equal to that of finline waveguide*

insertion loss and isolation of the two devices. These results were obtained with a residual direct magnetic field of about 1.8 kA/m. W_1–W_6 in Fig. 2 were chosen on the basis of experience without further ado; the length of the ferrite substrate was 45.7 mm. The overall length of the device including the tapers was of the order of six wavelengths. The return loss at each port (not shown) was better than 18 dB. However, these preliminary results are not considered optimum in any sense of the word in that the dimensions W_1–W_6 were not varied in this work, and thus the power density in each region and the polarisation of the alternate magnetic fields in the forward and backward directions of propagation have not been optimised; as mentioned previously, such polarisation is nearly always met at the interface between two different dielectric regions, provided the ratio of their dielectric constants is about 10 or larger.

5 18–26 GHz finline isolators using composite ferrite substrates

Scrutiny of the data in the section on the magnetic variables, or the prior art on conventional full waveguide bandwidth resonance isolators [10], suggests that it is necessary to employ two or more materials, with different values of effective anisotropy field, in the fabrication of a full 18–26 GHz finline isolator in WR 42 waveguide. The conventional arrangement involves cascading the substrates, but another possibility that holds some promise is obtained by stacking them (Fig. 7). Fig. 8 displays the

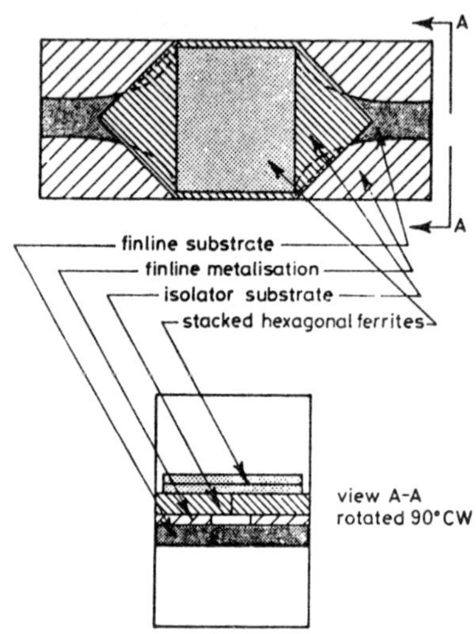

Fig. 7 *Schematic diagram of finline resonance isolator employing stacked ferrite substrates*

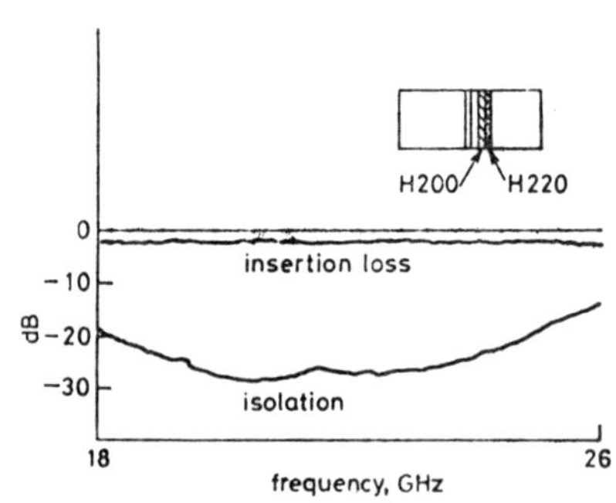

Fig. 8 *Insertion loss and isolation of finline isolator using two hexagonal materials with different internal fields in a stacked configuration*

insertion loss and isolation of one such finline isolator using two substrates over the recommended frequency range of WR 42 waveguide. Some similar preliminary results for a finline isolator obtained by cascading rather than stacking the two materials is illustrated in Fig. 9. Although there is perhaps not too much to choose between the two implementations, in so far as the isolation specification is concerned the insertion loss of the former con-

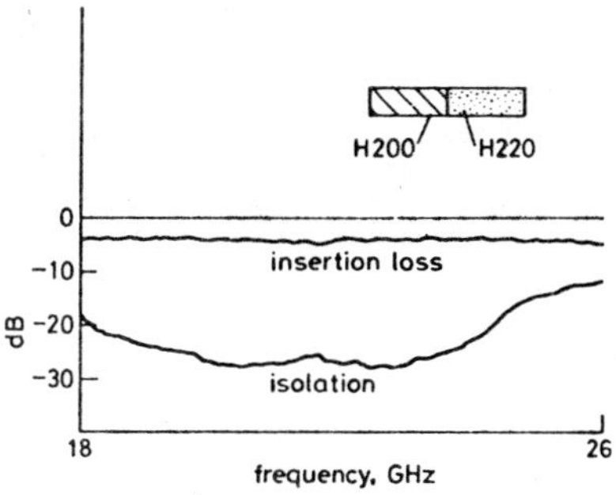

Fig. 9 *Insertion loss and isolation of finline isolator using two hexagonal materials with different internal fields in a cascade configuration*

figuration is somewhat lower as is, obviously, its overall length. The appropriate figures of merit for these two devices are readily evaluated from these two illustrations. Again, as for the single substrate assemblies, it is not suggested that the structures used here are optimum. Stacking three rather than two substrates probably also has merit, but has not been investigated here.

6 Conclusions

A practical solution to the design of millimetre wave finline isolators is the resonance isolator using hexagonal ferrites with uni-axial anisotropy energy. A number of such devices have been experimentally demonstrated here in WR 42 waveguide. One conclusion of this work is that stacking rather than cascading different substrate materials is the more attractive arrangement for the fabrication of full waveguide bandwidth isolators. The temperature behaviour of these devices has not been determined at this time.

7 Acknowledgment

One of the authors, J. Helszajn, would like to thank Trak Microwave Inc., USA, for supplying the ferrite material employed in this work.

8 References

1 MEIR, P.J.: 'Two new integrated circuit media with special advantages at millimeter wavelengths'. IEEE G-MTT Symp. Digest, 1972, pp. 221–223

2 MEIR, P.J.: 'Millimeter integrated circuits suspended in the E-plane of rectangular waveguide', *IEEE Trans.*, 1978, **MTT-26**, pp. 726–733

3 BEYER, A., and SOLBACH, K.: 'A new finline isolator for integrated millimeter wave circuits', *ibid.*, **1981**, **MTT-29**, pp. 1344–1348

4 BOCK, J.: 'New multilayered slot-line structures with high nonreciprocity', *Electron. Lett.*, 1983, **19**, (23), pp. 966–968

5 BRAAS, M., and SCHIEBLICH, C.: 'E-type circulator for finlines', *ibid.*, 1981, **17**, (19), pp. 701–702

6 BEYER, A., and WOLFF, I.: 'Finline ferrite isolator and circulator for the R-band'. 11th European Microwave Conf. Dig., Amsterdam 1981, pp. 321–326

7 DE BITETTO, D.J., DU PRE, F.K., and BROCKMAN, F.G.: 'Highly anisotropic magnetic materials for millimeter wave applications', presented at the Symposium on Millimeter Waves, Poly. Inst. of Brooklyn, March–April 1959

8 WEISS, M.T., and DUNN, F.A.: 'A 5-mm resonance isolator', *IRE Trans.*, 1958, **MTT-6**, p. 331

9 RODRIGUE, G.P., PIPPIN, J.E., and WALLACE, M.E.: 'Hexagonal ferrites for use at x to v-band frequencies', *J. Appl. Phys.*, 1962, Suppl. to Vol. **33**, pp. 1366–1368

10 TAFT, D.R., HARRISON, G.R., and HODGES, L.R.: 'Millimeter resonance isolators utilizing hexagonal ferrites', *IEEE Trans.*, 1963, **MTT-11**, pp. 346–350

11 LAX, B., and BUTTON, R.J.: 'Microwave ferrites and ferrimagnetics' (McGraw-Hill, 1962)

12 COHN, M.: 'Propagation in a dielectric-loaded parallel plane waveguide', *IRE Trans.*, 1959, MTT-7, pp. 202–208

13 KITTEL, C.: 'On the theory of ferromagnetic resonance absorption', *Phys. Rev.*, 1948, **73**, pp. 155–161

14 LEMKE, M., HOPPE, W., TOLKSDROF, W., and WEBZ, F.: 'Hexaferrites and their application at millimeter waves'. IEE colloquium *on* 'Ferrite materials, devices and technologies', London, 14th February 1985

15 SODHA, M.S., and SRIVASTAVA, N.C.: 'Differential phase shift at microwave frequencies using planar ferrites', *IEEE Trans.*, 1976, **MTT-24**, pp. 215–216

16 HAYASHI, Y., and MITTRA, R.: 'An analytical investigation of finlines with magnetized ferrite substrate', *ibid.*, 1983, **MTT-31**, pp. 495–498

17 HODGES, L.R., Jr., and HARRISON, G.R.: 'Oriented hexagonal ferrite compounds for application at millimeter wave frequencies', *J. Am. Ceram. Soc.*, 1964, **47**, (12)

18 HELSZAJN, J., and McSTAY, J.: 'External susceptibility tensor of magnetised ferrite ellipsoid in terms of uniform-mode ellipticity', *Proc. IEE*, 1969, **116**, pp. 2088–2093

6.4.2

POWER DENSITY DISTRIBUTION ANALYSIS OF FERRITE LOADED FINLINES FOR THE DEVELOPMENT OF INTEGRATED NONRECIPROCAL MILLIMETER WAVE ELEMENTS

Adalbert Beyer and Ingo Wolff

Duisburg University, Department of Electrical Engineering, Bismarckstr. 81, D-4100 Duisburg
Fed. Rep. of Germany

Abstract

A hybrid mode field analysis for a finline loaded with a ferrite- and additionally a dielectric speciem is presented.

Using this theory the power density distribution of forward and backward waves is calculated and demostrated.

The results are used to design a finline isolator for the 28.5 GHz range with a bandwidth of 2 GHz, an insertion loss smaller than 1.2 dB and an isolation higher than 39 dB.

Introduction

The development of integrated nonreciprocal millimeter wave elements in finline technique requires the use of an efficient theory, which besides the properities of a realistic finline model, takes into account the nonreciprocal behaviour of the necessary ferrite sample, too. In earlier papers /1-3/ preliminary numerical and experimental results were obtained for the development of such elements. There were also obtained /3/ preliminary results for the power density distribution at the cross-section of a field displacement isolator in finline technique.

Following, a ferrite loaded finline-structure which serves as the basis for the realization of an isolator in finline technique, is analised using a hybrid mode field analysis technique, which also provides the calculation of second order effects as e.g. the influence of the metallization thickness and substrate holding grooves.

The Field Theoretical Method

For the calculation of the electromagnetic fields which can propagate on a combined line as shown in Fig. 1, the cross-section is subdivided into five field regions (1)-(5), and a complete set of field solutions is derived for each subarea. The x and y dependences of the electromagnetic fields in subregions (1), (3)-(5) are formulated using harmonic functions, so that the boundary conditions are fulfilled on the metallic wall or metallization of the dielectric substrate. As an example, the scalar potentials in subregion (1) are given by

$$\Phi^{(1)} = \sum_{m=1}^{\infty} \{\bar{A}_m^{(1)} \sin(\bar{k}_{xm}^{(1)} x) + \bar{B}_m^{(1)} \cos(\bar{k}_{xm}^{(1)} x)\} \sin\{\bar{k}_{ym}^{(1)}(y - \frac{b}{2})\} \tag{1}$$

$$\bar{k}_{ym}^{(1)} = \frac{m\pi}{b} \tag{2}$$

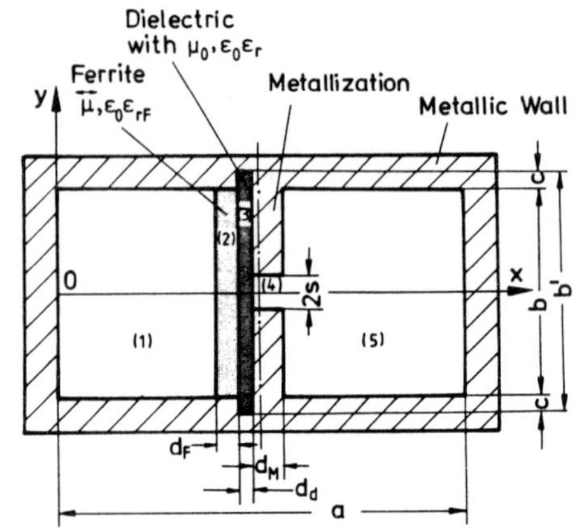

Fig. 1: Cross-section of the ferrite loaded unilateral grounded finline. (The dimensions for Ka-band are: a=7.112mm, b=3.556mm, b'=4.55mm, d_M=17.5µm, d_d=125mm, ε_{rd}=2.22, d_f=1.5mm, H_0=7.96kAcm^{-1}; ferrite material: TT2-111; the frequency: 28.5GHz.)

and

$$\Psi^{(1)} = \sum_{n=0}^{\infty} \{A_n^{(1)} \cos(k_{xn}^{(1)} x) + B_n^{(1)} \sin(k_{xn}^{(1)} x)\} \cos\{k_{yn}^{(1)}(y - \frac{b}{2})\} \tag{3}$$

$$k_{yn}^{(1)} = \frac{n\pi}{b}\ . \tag{4}$$

In region (2) the fields are expanded in following terms /4/:

$$E_{xm}^{(2)} = C_{xm}^{(2)}(x)\sin(k_{ym}^{(2)} y),\quad H_{xm}^{(2)} = K_{xm}^{(2)}(x)\cos(k_{ym}^{(2)} y), \tag{5,6}$$

$$E_{ym}^{(2)} = C_{ym}^{(2)}(x)\cos(k_{ym}^{(2)} y),\quad H_{ym}^{(2)} = K_{ym}^{(2)}(x)\sin(k_{ym}^{(2)} y), \tag{7,8}$$

$$E_{zm}^{(2)} = C_{zm}^{(2)}(x)\sin(k_{ym}^{(2)} y),\quad H_{zm}^{(2)} = K_{zm}^{(2)}(x)\cos(k_{ym}^{(2)} y) \tag{9,10}$$

with

$$k_{ym}^{(2)} = \frac{m\pi}{b}\ . \tag{11}$$

From the coupled differential equations

$$\{\frac{\partial}{\partial x^2} + (k^2 - \frac{1}{\mu_1} k_{ym}^{(2)^2} - \beta^2)\}K_{ym}^{(2)} + jkk_{ym}^{(2)} \frac{\mu_2}{\mu_1} C_{ym}^{(2)} = 0 \tag{12}$$

and

$$\{\frac{\partial}{\partial x^2} + (k^2 \mu_{eff1} - k_{ym}^{(2)} - \beta^2)\}C_{ym}^{(2)} + jkk_{ym}^{(2)}(\mu_2/\mu_1)K_{ym}^{(2)} = 0, \tag{13}$$

the unknown function $C_{ym}^{(2)}(x)$ and $K_{ym}^{(2)}(x)$ can be de-

A. Beyer and I. Wolff, *IEEE MTT-S International Microwave Symposium Digest*, 1984, pp. 342–344.

termined. The effective permittivity μ_{eff1} of the lossless ferrite is calculated from the permittivity tensor $\overleftrightarrow{\mu}$ as

$$\mu_{eff1} = \frac{\mu_1^2 - \mu_2^2}{\mu_1} \tag{14}$$

with

$$\overleftrightarrow{\mu} = \mu_0 \begin{bmatrix} \mu_1 & 0 & j\mu_2 \\ 0 & 1 & 0 \\ -j\mu_2 & 0 & \mu_1 \end{bmatrix} \tag{15}$$

$$\mu_1 = 1 + \frac{\omega_M \omega_L}{\omega_L^2 - \omega^2} \tag{16}$$

$$\mu_2 = \frac{\omega \omega_M}{\omega_L^2 - \omega^2} . \tag{17}$$

The quantities ω_M and ω_L are related to the gyromagnetic ratio γ, the saturation magnetization M_s, and the bias magnetic field H_0 as

$$\omega_M = \gamma \dot{M}_s \tag{18}$$

$$\omega_L = \gamma H_0 . \tag{19}$$

With the help of the solutions for $C_{ym}^{(2)}(x)$ and $K_{ym}^{(2)}(x)$, the other functions $C_{xm}^{(2)}(x)$, $K_{xm}^{(2)}(x)$, $C_{zm}^{(2)}(x)$ and $K_{zm}^{(2)}(x)$ - after many mathematical operations - can also be determined.

Using the given field equations, the field components of the electromagnetic field can be derived using well-known methods. The continuity conditions for the field at the boundaries between the subareas (1) - (2), (2) - (3), (3) - (4), (4) -(5) lead to a system of integral eigenvalue equations, which, using the Ritz-Galerkin method as described for unloaded finlines in /1/, is converted into a system of homogeneous equations

$$\overleftrightarrow{F} \cdot \vec{A} = \vec{0} \tag{20}$$

with $\overleftrightarrow{F}$ the system matrix and $\vec{A}$ the amplitude vector. This eigenvalue equation can be solved determining the system determinant

$$\det(\overleftrightarrow{F}) = 0 \tag{21}$$

thus yielding the phase constant and the field distributions of the waves propagating on the ferrite loaded finline. For practical calculations the size of the system was truncated to a number of about 30 equations to achieve satisfactory convergence of the results.

Numerical and Experimental Results

Using the results for the fields, the power density in a cross-section of ferrite loaded fin-

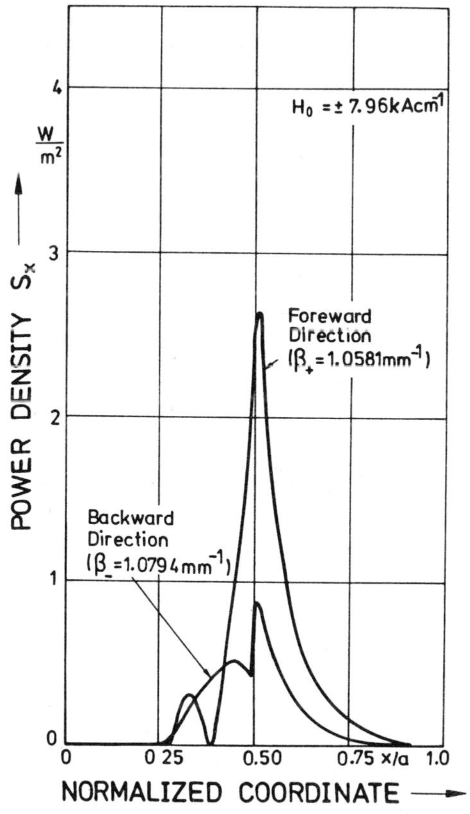

Fig. 2: Power density distribution for the finline given in Fig. 1 in x-direction versus the x-coordinate at y=0.

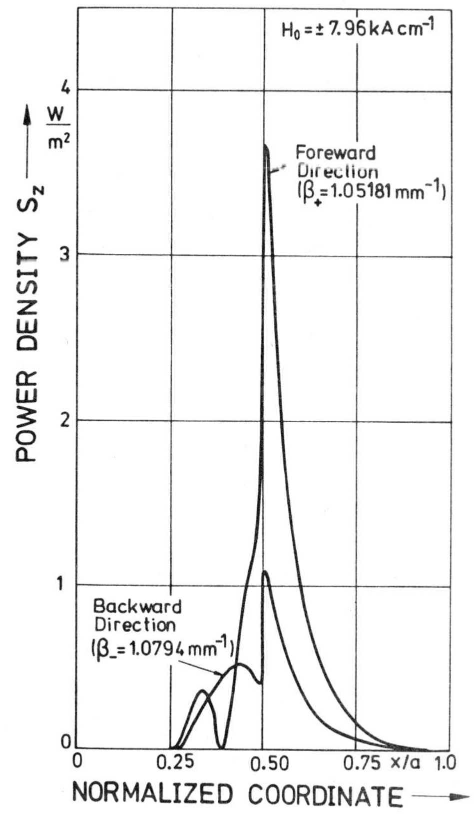

Fig. 3: Power density distribution for the finline given in Fig. 1 in z-direction versus the x-coordinate at y=0.

line can also be calculated. In Fig. 2, 3 and 4 the power density distributions are plotted at y = 0. It is seen that the ferrite sample displaces the fields of the guided waves similar to the field displacement arrangements in metal waveguide isolators.

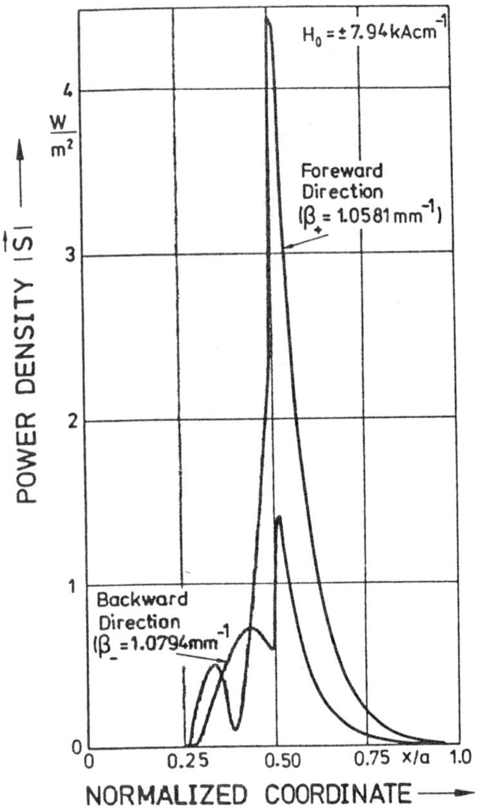

Fig. 4: Absolute values of power density for the finline given in Fig. 1 versus the x-coordinate at y=0.

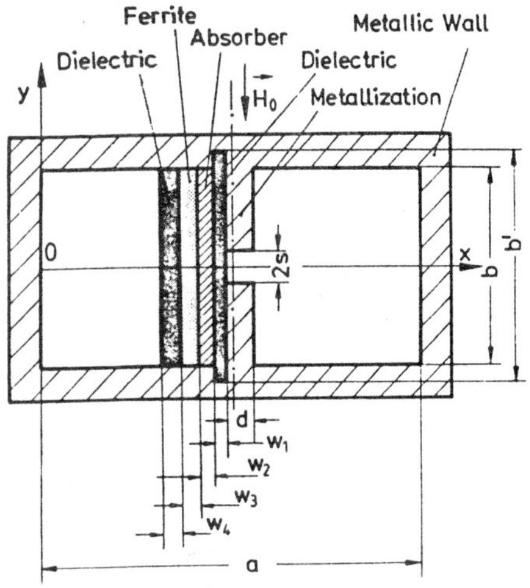

Fig. 5: Cross-section of the designed isolator in finline technique

Fig. 5 shows a composite ferrite slab-loaded finline structure designed to realize a field displacement isolator in finline technique at 28.5GHz. From the above-mentioned results, the ferrite slab was placed in the plane of circular polarization ($|H_x| = |H_z|$); its thickness was optimized, so that the backward/forward ratio is an optimum. Since so far no theoretical investigation of this structure, which is still more complicate than the structures in Fig. 1, were carried out, the determination of the optimum permittivity of the dielectric spacer and the optimum position of the absorber, were found experimentally.

In Fig. 6 the measured isolator characteristics are plotted for this isolator.

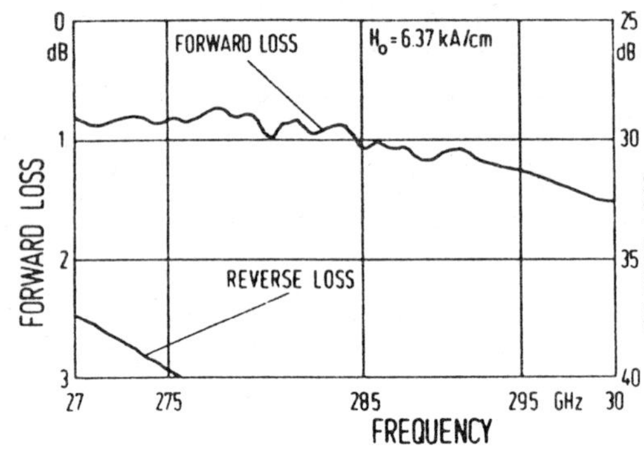

Fig. 6: The characteristics of the realized isolator in finline technique at 28.5 GHz. (w_2=0, w_3=1.6mm, w_4=1.9mm. Thin resistive sheets are placed on the left side of the ferrite sample and of the dielectric, respectively.)

It is seen that the transmission loss is better than 1.2 dB,while over the required 2 GHz bandwidth the minimum reverse loss is 39 dB. The isolator exhibits virtually the same VSWR for both forward and reverse direction, with VSWR < 1.37.

Conclusions

A rigorous expansion method for the analysis of a ferrite loaded finline has been obtained, which can be considered as a basis for the development of a field desplacement isolator in finline technique. Theoretical results fof the power density distribution show the possibilities of the method to help in the optimization of a field displacement isolator in finline technique.

This work was done in connection with a cooperation with British Telecom Research Laboratories, England.

References

/1/ Beyer, A. and Solbach, K.: 1981 IEEE MTT-S International MW Symposium Digest, 296-298

/2/ Beyer, A. and Wolff, I.: 11th Eu.M.C., Amsterdam, 321-325 (1981)

/3/ Beyer, A. and Solbach, K.: IEEE Trans. MTT-29, No. 12, 1344-1348 (1981)

/4/ Wolff, I.: Fields and waves in gyrotropic microwave structures. F. Braunschweig, Germany: Vieweg & Sohn, 1973 (In German)

6.4.3

Millimetric Nonreciprocal Coupled-Slot Finline Components

LIONEL E. DAVIS, SENIOR MEMBER, IEEE, AND DAVID B. SILLARS, MEMBER, IEEE

Abstract —**Promising preliminary results are presented for isolators and a four-port circulator in novel finline structures in the frequency range 26.5–40.0 GHz. The basic configuration is a ferrite-loaded coupled-slot finline with the ferrite magnetized parallel with the direction of propagation. The nonreciprocal effects of the odd mode propagating alone and of the odd and even modes propagating are described. All structures exhibit a 20-dB isolation bandwidth greater than 3.6 GHz. It is suggested that such structures would also be suitable for higher frequencies.**

I. Introduction

THE EASE of fabrication of finline technology together with its applicability at millimeter-wave frequencies has resulted in some effort to produce nonreciprocal devices [1], [2]. However, for traditional nonreciprocal structures (junction circulators, resonance isolators, etc.,) operating at millimeter-wave frequencies, a strong magnetic bias field or an anisotropic ferrite would normally be required together with a ferrite sample which is physically small and machined to close tolerances.

Nonreciprocal finline structures are reported here in which a) rectangular ferrite slabs are magnetized longitudinally enabling saturation to be achieved from weak magnetic fields due to the small demagnetizing factor, b) hexagonal ferrites are not required, and c) nonreciprocity is developed with a "large" slab, i.e., over several wavelengths, rather than a fraction of a wavelength as in traditional junction devices.

II. Reciprocal Structures

The finline structures considered are shown in Fig. 1(a) and (b). They consist of a ferrite slab placed either on the conductor surface (sandwich-layer) or on the substrate surface (double-layer) of the finline circuit. Two parallel slots etched from the copper conductor produce the guiding structure.

In the unmagnetized isotropic case, the double substrate coupled-slot finlines can support two propagating modes, the odd and even modes. In accordance with the convention adopted in [3], the odd mode is defined here as having the electric field E_y an even function of y with respect to the x-axis and even mode having E_y an odd function of y with respect to the x-axis. Fig. 2(a) and (b) show the electric fields of both modes in the air and dielectric

Manuscript received December 6, 1985; revised February 11, 1986.
The authors are with the Department of Electrical and Electronic Engineering, Paisley College of Technology, Paisley, PA1 2BE, Scotland.
IEEE Log Number 8608678.

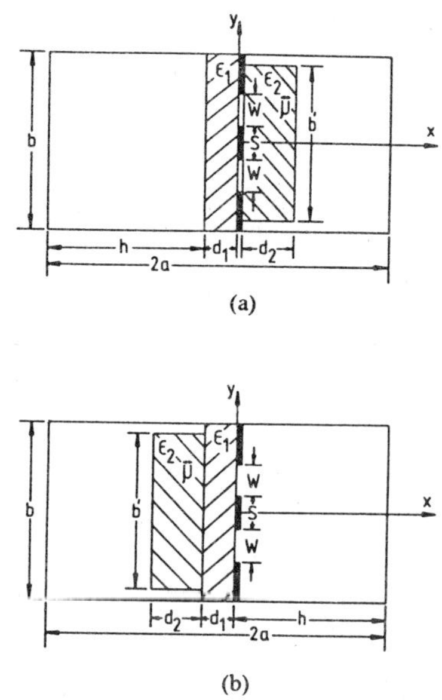

Fig. 1. Ferrite-loaded coupled-slot finline. (a) Sandwich-layer structure. (b) Double-layer structure. $a = b = 3.556$ mm, $b^1 = 3.0$ mm, $\epsilon_1 = 2.22$, $\epsilon_2 = 13.0$, $d_1 = 0.127$ mm, $d_2 = 0.5$ mm, $h = 3.4925$ mm, $M_s = 4000$ A/cm.

regions. The odd- and even-mode dispersion characteristics of a typical double-dielectric substrate coupled-slot finline are shown in Fig. 3. These results were computed using the transverse resonance technique as applied to finlines [4], [5].

The nonreciprocal properties of these structures will now be described for the cases where only the odd mode is launched alone and where both the odd and even modes are launched.

III. Nonreciprocal Behavior

Preliminary results for a coupled-slot finline isolator have recently been described [6], [7]. It was reported that when a coupled-slot finline loaded with a ferrite slab (Fig. 1(a)) and supporting only the odd mode was magnetized parallel with the direction of propagation, a displacement of the field from one slot to the other occurred. Excitation of the odd mode was achieved by a transition from a unilateral single-slot finline taper into the coupled-slot region via a tapered center conductor. The insertion loss for two back-to-back waveguide to finline tapers plus

L.E. Davis and D.B. Sillars, *IEEE Transactions on Microwave Theory and Techniques*, Vol. MTT-34, No. 7, July 1986, pp. 804–888.

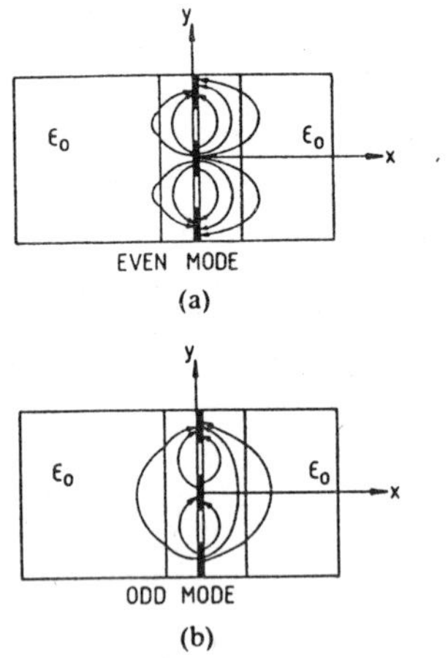

Fig. 2. Electric-field distribution of the dielectric sandwich-layer coupled-slot finline for (a) even-mode and (b) odd-mode excitation.

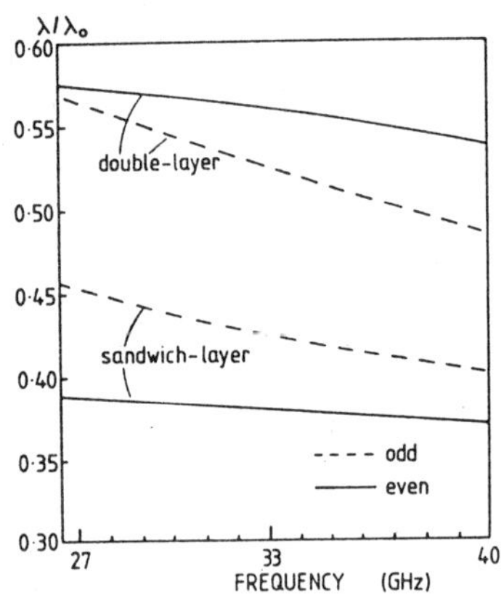

Fig. 3. Dispersion characteristics of even and odd modes in dielectric double-layer and dielectric sandwich-layer coupled-slot finline. $W = S = 0.5$ mm, $\epsilon_1 = 2.22$, $\epsilon_2 = 13.0$, $d_1 = 0.127$ mm, $d_2 = 0.5$ mm.

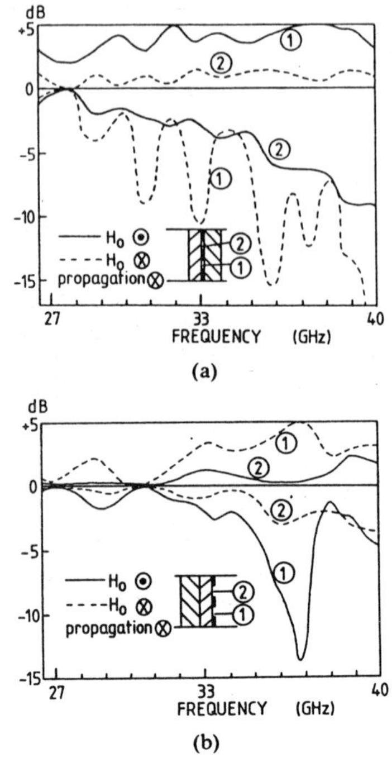

Fig. 4. Signal difference curves for each slot port and direction of H_0 for (a) ferrite-dielectric sandwich-layer structure and (b) ferrite-dielectric double-layer structure. Each characteristic is shown with respect to its own (0 dB) reference level which was obtained with the ferrite unmagnetized. $W = S = 0.5$ mm, ferrite length = 14.0 mm, $H_0 = 400$ A/cm.

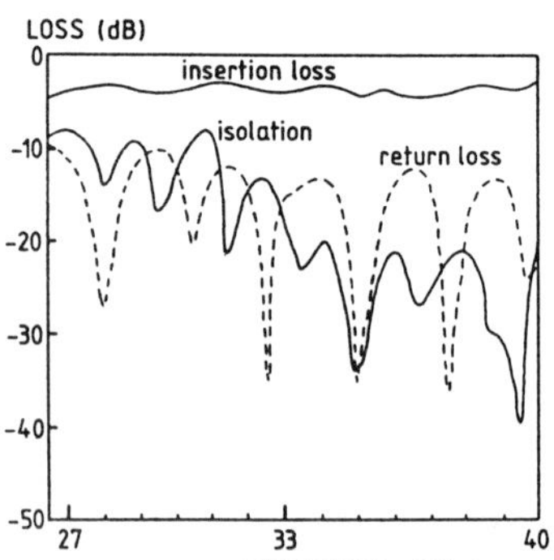

Fig. 5. Performance of single ferrite coupled-slot finline isolator. $W = S = 0.5$ mm, ferrite length = 14.0 mm, $H_0 = 400$ A/cm.

single-slot to coupled-slot transitions separated by a coupled-slot section of 28 mm in length is better than 0.4 dB over the 26.5–40.0-GHz frequency band.

The field displacement phenomenon was observed in a coupled-slot finline in which one slot incorporated a bend to permit waveguide detectors to monitor each slot separately. Fig. 4(a) and (b) show the nonreciprocal change in power that occurred in each slot for the sandwich-layer and double-layer structures for both directions of the applied magnetic field. In the diagrams, 0 dB corresponds to the output power at each slot for zero applied magnetic field. The lack of symmetry in the amount of shifting may be attributable to the bend in the finline circuit. These graphs show that, with a fixed direction of bias field, the nonreciprocity is reversed by moving the ferrite slab from one side of the finline plane to the other. They also show that in the double-layer structure the ferrite produces weaker nonreciprocity because it is in a region of weaker field [8]. It was also observed that once the ferrite was saturated there was no further change in performance as the applied field was increased, bearing in mind that it was still far below that required for ferrimagnetic resonance.

IV. Isolators

One application of this phenomenon is in an isolator, produced by placing a piece of resistive card over one slot. Fig. 5 shows results for insertion loss, return loss, and isolation for the case where a dielectrically-matched ferrite slab was placed on the conductor surface and magnetized parallel to the direction of propagation. The isolation was measured by reversing H_0, and the broad-band potential of this structure is evident.

TABLE I
CHARACTERISTIC IMPEDANCE IN Ω OF RECIPROCAL COUPLED-SLOT FINLINE STRUCTURES

Unloaded line	Sandwich-layer dielectric (ferrite) line	Matching impedance	Matching dielectric-loaded line
178	117	144	147

$W = S = 0.5$ mm, dielectric (ferrite): $\epsilon_2 = 13.0$, $d_2 = 0.5$ mm, matching dielectric: $\epsilon_2 = 3.0$, $d_2 = 1.0$ mm.

TABLE II
WORST CASE RESULTS OVER FREQUENCY RANGE 26.5–40.0 GHz

Fin-line circuit configuration	Insertion loss (dB)	Return loss (dB)
unloaded	0.4	16.0
with unmagnetised ferrite (1)	4.5	4.5
with ferrite (1) and matching-dielectric	2.2	13.0
with tapered ferrite (2)	3.0	10.0

$W = S = 0.5$ mm. Ferrite ①: 14 mm long, dielectrics: $\epsilon_r = 3.0$, 1.5 mm long. Ferrite ②: 20 mm long with 5 mm long tapers.

A. Dielectric Matching of Ferrites

The discontinuity produced at the ends of the ferrite slab was reduced by the use of low-permittivity quarter-wavelength dielectric sections. These sections were designed to match, ideally, at 33.5 GHz using the transverse resonance technique [4], [5]. The unmagnetized isotropic ferrite-loaded finline structure was simulated by a double-dielectric sandwich-layer substrate finline to permit calculation of the odd-mode characteristic impedance. Table I shows the characteristic impedance for the unloaded and dielectric (ferrite)-loaded finline and the matching section which was obtained using standard transmission-line theory. Also quoted is the characteristic impedance of the resultant matching dielectric-loaded section. This value is a function of the dielectric permittivity and its thickness as well as the finline characteristics. Therefore, the same value of characteristic impedance can be obtained for different combinations of these parameters. However, the choice of dielectrics used was limited to what was at hand resulting in a dielectric with a permittivity of 3.0 being chosen to give the nearest impedance.

Table II gives results of insertion loss and return loss for the unloaded, unmagnetized ferrite-loaded and the ferrite/dielectric finline circuits. Also given are results for the finline circuit loaded with unmagnetized tapered ferrites which were used in the devices described in the following sections.

B. Twin Ferrite Arrangement (Anti-Parallel Fields)

The investigation was extended to study the effects of placing two ferrite slabs on the coupled-slot finline separated by a length of resistive card placed over one slot. Nonreciprocity was produced by magnetizing the ferrites with anti-parallel magnetic fields, achieved by placing a bar magnet alongside each ferrite region. The small longitudinal field produced by each magnet was sufficient to saturate the ferrite slabs. Fig. 6 shows the structure with the ferrites and resistive card tapered at each end to help

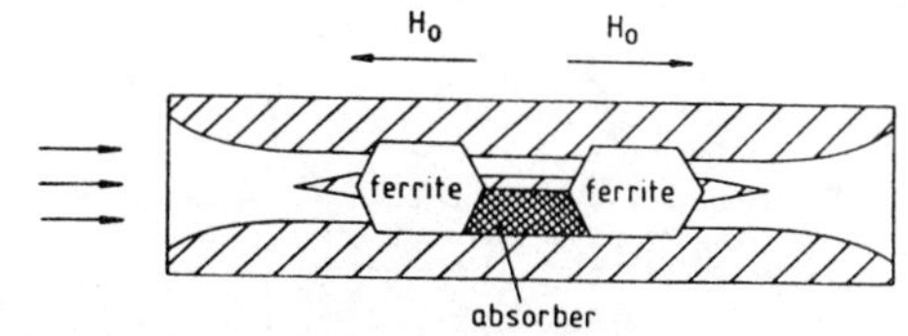

Fig. 6. Coupled-slot finline with two ferrite slabs magnetized in opposite directions and separated by a length of resistive card. With the input and the static fields directed as shown, the device exhibits isolation.

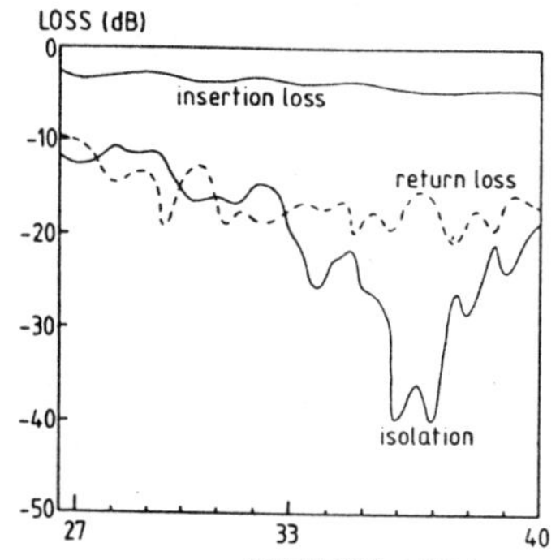

Fig. 7. Performance of twin ferrite coupled-slot finline isolator. $W = S = 0.5$ mm, ferrite length = 20 mm, $H_0 = 160$ A/cm.

reduce losses and results are shown in Fig. 7. It can be seen that a maximum isolation of 41.5 dB is obtained with a 20-dB isolation bandwidth of 6.75 GHz (18.4 percent) together with insertion losses of between 3.5–6.0 dB and a return loss better than 16 dB over the 20-dB isolation bandwidth.

For the same isolator structure, but using dielectrically-matched ferrite slabs [6], the 20-dB isolation bandwidth was measured to be 3.8 GHz (13.0 percent) with the insertion loss varying between 2.3–4.5 dB over this range. The return loss was better than 10 dB.

Although the results of Table II and for the twin ferrite isolator structures indicate that the use of dielectrically-matched ferrites compared to tapered ferrites produced better performance, it would be desirable to improve the performance of the isolator incorporating tapered ferrites as there are obvious physical problems in attaching small dielectric pieces onto the ends of the rectangular ferrite slabs. It is hoped to achieve this by optimizing the taper length of the ferrites and also investigating the effects of different taper profiles.

C. Twin Ferrite Arrangement (Unidirectional Field)

An alternative arrangement of the twin ferrite coupled-slot finline isolator is shown in Fig. 8. Here it is shown from above a plan view of the finline circuit with the ferrites placed on opposite sides of the conductor surface. Thus, one section constitutes a double-layer structure and the other a sandwich structure. A length of resistive card is positioned over the lower slot in the region between the ferrites.

From the signal difference curves of Fig. 4(a) and (b), it was shown that the field displacement effect is reversed

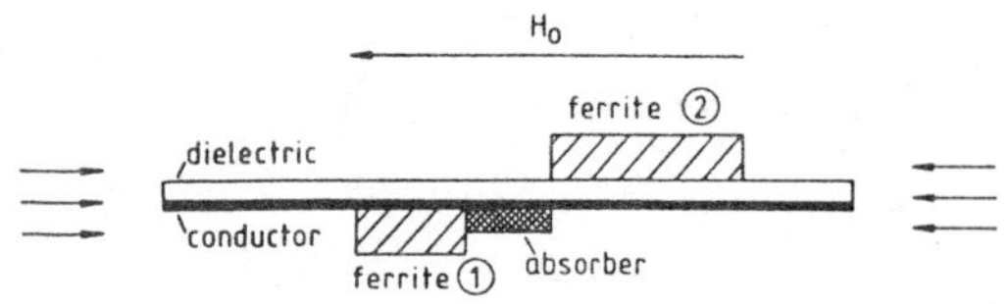

Fig. 8. Plan view of finline circuit with one ferrite slab placed on each side of the conductor and a mid-section of resistive card.

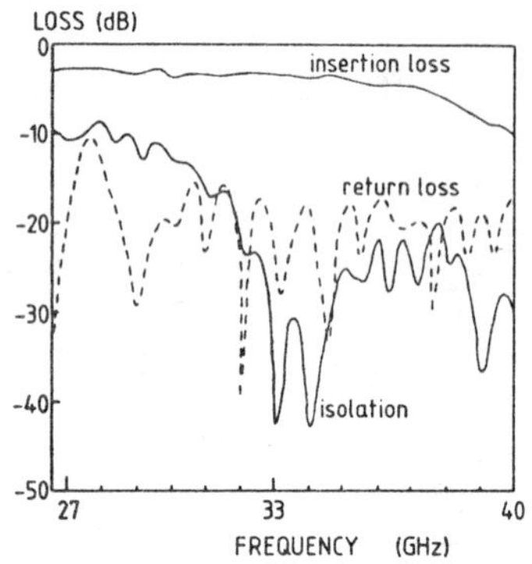

Fig. 9. Performance of the finline isolator with tapered ferrites placed on opposite sides of the finline circuit. $W = S = 0.5$ mm; ferrite ①: $20 \times 3 \times 0.5$ mm^3, ferrite ②: $25 \times 3 \times 0.5$ mm^3, $H_0 = 160$ A/cm.

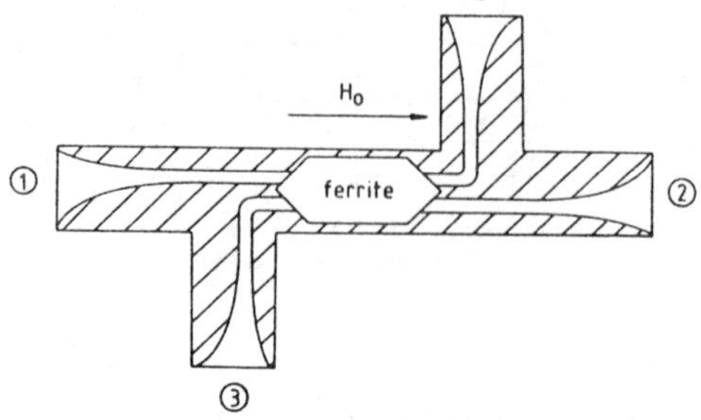

Fig. 10. Ferrite-loaded four-port finline coupler.

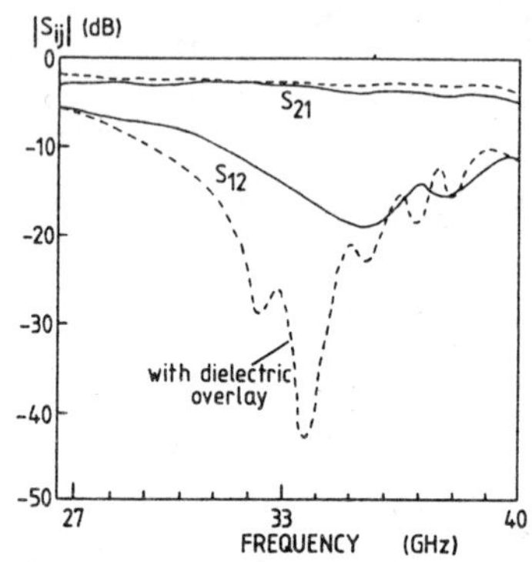

Fig. 11. Nonreciprocal coupling characteristics of ferrite-loaded four-port finline coupler with and without dielectric overlay placed on top of ferrite slab. $W = 0.2$ mm, $S = 1.0$ mm; ferrite: 30 mm long with 10 mm long tapers; dielectric overlay: $18 \times 3 \times 2$ mm^3, permittivity = 3.0, length of coupled-slot section = 35 mm, $H_0 = 160$ A/cm.

when the ferrite is moved from one side of the finline plane to the other. If the twin ferrite arrangement of Fig. 8 is magnetized by a single unidirectional magnetic field as shown, with the resistive card placed over slot ①; then, for propagation from the left-hand side into the sandwich-layer structure, the field will be perturbed into the lower slot (slot ① in Fig. 4(a)) and be absorbed. For propagation from the right-hand side into the double-layer structure, the field will be perturbed into the upper slot (slot ① in Fig. 4(b) if looking at the structure with the ferrite placed to the left of the finline) and, hence, away from the resistive card. Alternatively, isolation may be produced by reversing the applied magnetic field.

Because weaker nonreciprocity resulted when the ferrite was placed on the dielectric substrate side as was shown in Fig. 4(b) it was found that a longer ferrite slab was required to improve the nonreciprocity. Preliminary results of insertion loss, return loss, and isolation for this structure are shown in Fig. 9. A maximum isolation of 43.5 dB is observed with isolation greater than 20 dB, insertion loss less than 4.5 dB, and return loss better than 17 dB obtained over a bandwidth of 32–37 GHz (14.5 percent). However, greater than 20-dB isolation is retained over a bandwidth of 32–40 GHz (22.2 percent), but the insertion loss is prohibitively high at the higher frequencies.

V. Four-Port Circulator

Investigations have also been carried out on a nonreciprocal four-port finline coupler [7]. In the unmagnetized isotropic ferrite-loaded finline coupler, both odd and even modes propagate. With a longitudinally-magnetized ferrite slab placed on the coupled-slot section (Fig. 10), nonreciprocal coupling occurs, i.e., circulation. The circulator characteristics are shown in Fig. 11, where it is also shown that the nonreciprocal isolation (S_{12}) is improved by placing a dielectric overlay on top of the ferrite slab. This has the effect of concentrating the field within the ferrite slab [8], [9]. S-parameter curves of the nonreciprocal coupler are shown in Fig. 12(a) and (b) where four-port circulator behavior is clearly evident. Fig. 12(a) shows that a 20-dB isolation bandwidth of 3.6 GHz was produced at the coupling ports (S_{12}, S_{34}) with between 2.5–3.5-dB losses produced at the transmission ports (S_{14}, S_{32}). The return loss (S_{11}) was measured to be better than 15 dB. As shown in Fig. 12(b), the direction of circulation is reversed by reversing the applied field, as expected, but the isolation exhibits broad-band behavior, albeit at only approximately 15 dB.

It has been found from investigations on a ferrite-loaded coupled-slot finline where only the even mode propagates that for longitudinal magnetization there were no appreciable field displacement effects as has been demonstrated for the odd mode propagating only. Work is in hand to predict the odd-mode nonreciprocal propagation constants required to produce the conditions for nonreciprocal coupling. It is of interest to note that the selective behavior of two different modes has been discussed by John and Böck [10] in connection with a completely different structure, i.e., a microstrip-slot coupler.

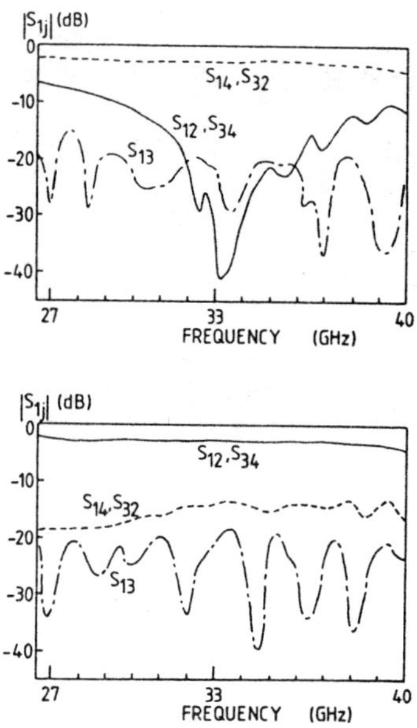

Fig. 12. Nonreciprocal transmission and coupling characteristics. Fig. 12(b) obtained by reversing H_0. Finline coupler parameters are as given in Fig. 11.

VI. Operation Above 40 GHz

These nonreciprocal coupled-slot finline configurations have a number of features that will be useful at frequencies above 40 GHz.These are

a) the closer mechanical tolerances are associated with the finline circuit (where they are more easily controlled) rather than with the ferrite slab,
b) the ferrite slab is several wavelengths long which makes grinding and handling easier,
c) the applied field is low despite the high frequency,
d) the need for hexagonal materials is less urgent,
e) broad-bandwidths appear to be possible.

VII. Conclusions

Isolator and circulator behavior have been demonstrated with coupled-slot finline structures loaded with ferrite slabs magnetized parallel to the direction of propagation.

The isolator structures consist of a single slab ferrite-loaded line and twin slab ferrite-loaded lines with anti-parallel and unidirectional applied magnetic fields. It has been shown that, in the frequency range 26.5–40 GHz, reasonable isolation is achieved with very low values of applied magnetic field. The results have implications for similar components at higher frequencies. Theoretical and experimental work is in hand to improve these promising devices.

Acknowledgment

The authors wish to acknowledge the support of the U.K. Science and Engineering Research Council.

References

[1] A. Beyer and I Wolff, "Fin-line ferrite isolator and circulator for the *R*-band," *11th Eur. Microwave Conf. Dig.* (Amsterdam), 1981, pp. 321–329.

[2] U. Goebel and C. Schieblich, "Broadband fin-line circulators," in *IEEE MTT-S Int. Microwave Symp. Dig.* (Dallas), 1982, pp. 249–251.

[3] D. Mirshekar-Syahkal and J. B. Davies, "An accurate, unified solution to various fin-line structures of phase constant, characteristic impedence, and attenuation," *IEEE Trans. Microwave Theory Tech.*, vol. MTT-30, pp. 1854–1861, Nov. 1982.

[4] R. N. Simons, "Suspended coupled slot line using double layer dielectric," *IEEE Trans. Microwave Theory Tech.*, vol. MTT-29, pp. 162–165, Feb. 1981.

[5] R. N. Simons, "Analysis of millimetre-wave integrated fin-line," *Microwaves, Optics, Antennas*, vol. 130, pp. 166–169, Mar. 1982.

[6] D. B. Sillars and L. E. Davis, "Coupled-slot fin-line isolators," *Electron. Lett.*, vol. 21, no. 3, pp. 97–98, Jan. 1985.

[7] L. E. Davis and D. B. Sillars, "Millimetric non-reciprocal coupled-slot fin-line components," in *IEEE MTT-S Int. Microwave Symp. Dig.* (St. Louis), 1985, pp. 237–238.

[8] G. Böck, "New multilayered slot-line structures with high non-reciprocity," *Electron. Lett.*, vol. 19, no. 23, pp. 966–968, Nov. 1983.

[9] K. Solbach, "*E*-plane circulators aid mm-wave design," *Microwaves and RF*, pp. 73–78, Dec. 1983.

[10] W. John and G. Böck, "A new MIC 4-port circulator," in *14th Eur. Microwave Symp. Dig.* (Liège), 1985, pp. 522–527.

6.4.4

PROGRESS ON FINLINE NON RECIPROCAL PHASE-SHIFTERS

S. TEDJINI* P. SAGUET* D. RAULY*

ABSTRACT

The aim of this communication is to propose a non reciprocal phase-shifter in fin-line technology and its theoretical analysis. The proposed device is based on semiconductor submitted to a magnetostatic field which leads to a gyroelectric material and then to a differential phase-shifter.

PRINCIPLE

It is well known that ferrite submitted to a magnetostatic field becomes a gyromagnetic media and then leads to non reciprocal effects. In similar ways, a semiconductor submitted to a magnetostatic field becomes a gyroelectric material and could lead to non reciprocal effects also. The proposed devices lies in the use to the field displacement effect in gyroelectric structures at room temperature prooved by HIROTA et al [1]. The basic structure proposed by HIROTA et al [1] is shown on figure 1. Such a structure can't produce non reciprocal effect, and generally the insertion losses are so important. To have non reciprocal effect, and reduce the insertion losses, we propose the structure figure 2. It is a fin-line loaded in its slot by a thin slab of high mobility semiconductor which is transversally magnetized.

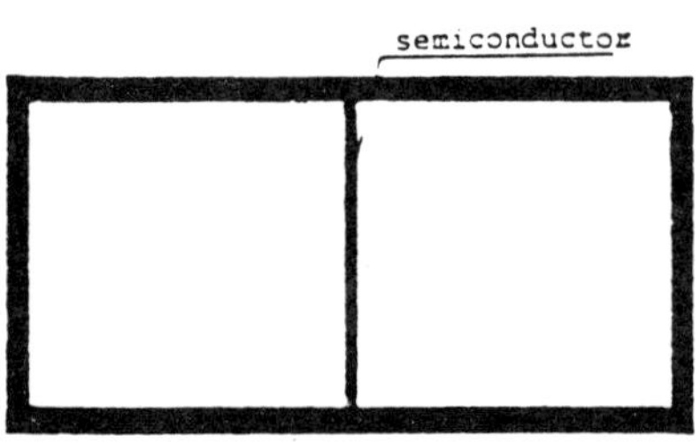

Figure 1 -
Basic structure. In the E-plane of the waveguide we have a semiconductor slab submitted to magnetostatic field B_0. The semiconductor thickness is δ . $\delta \ll a$.

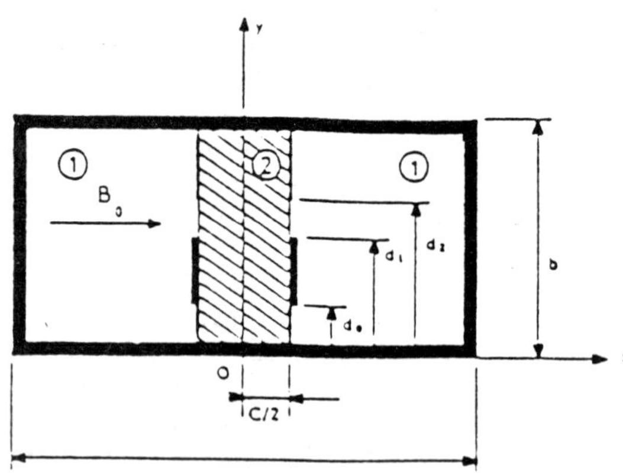

Figure 2 -
The studied structure. It is bilateral fin-line loaded by a semiconductor slab. The semiconductor lies between $y = d_0$ and $y = d_1$. The media 1 is the air, the media 2 is a dielectric ($\epsilon = \epsilon_0 \epsilon_r$).

LEMO - UA 833
23 Avenue des Martyrs 38031 GRENOBLE Cedex

S. Tedjini, P. Saguet, and D. Rauly, *Proceedings of 16th European Microwave Conference*, 1986, pp. 459–462.

Under the effect of the magnetostatic field, the semiconductor slab becomes an anisotropic media. This slab may be modelised by its conductivity tensor given by the expression 1 (and its thickness).

$$S = \sigma_o \begin{bmatrix} \frac{1}{\alpha} & 0 & 0 \\ 0 & \frac{\alpha}{\alpha^2 + (\omega_c \tau)^2} & \frac{-\omega_c \tau}{\alpha^2 + (\omega_c \tau)^2} \\ 0 & \frac{\omega_c \tau}{\alpha^2 + (\omega_c \tau)^2} & \frac{\alpha}{\alpha^2 + (\omega_c \tau)^2} \end{bmatrix} \tag{1}$$

where $\omega_c = e\, B_o/m^* =$ cyclotron frequency

$\sigma_o = Ne^2\, \tau/m^* =$ semiconductor conductivity

$\alpha = 1 + j\, \omega\, \tau$

$e =$ electron charge, $m^* =$ effective mass of the electrons

$N =$ electron density, $\tau =$ momentum relaxation time

The tangential current densities on the semiconductor layer is given by [2].

$$J_T = \delta\, \sigma\, E_T \tag{2}$$

where δ is the thickness of the semiconductor and σ is 2 x 2 matrix deduced from 1.

$$\sigma = \sigma_o \begin{bmatrix} \frac{\alpha}{\alpha^2 + (\omega_c \tau)^2} & \frac{-\omega_c \tau}{\alpha^2 + (\omega_c \tau)^2} \\ \frac{\omega_c \tau}{\alpha^2 + (\omega_c \tau)^2} & \frac{\alpha}{\alpha^2 + (\omega_c \tau)^2} \end{bmatrix} = \begin{bmatrix} \sigma_\perp & -\sigma_x \\ \sigma_x & \sigma_\perp \end{bmatrix} \tag{3}$$

Under the excitation of the fin-line fundamental mode, (quasi TE) the field displacement appears. It results from the superposition of TE_{m0} modes, excited by displacement currents and TM_{mn} modes, excited by conduction currents on the semiconductor slab. The later are generated by the σ_x term of the matrix σ. Finally, the non reciprocity is obtained because of the geometrical dissymetry in the structure (figure 2) along the oy direction. Indeed, the resulting electromagnetic field shows an asymetric distribution along oy. This leads to different propagation constants (phase constant and attenuation constant) for forward and backward waves.

ANALYSIS

To study the structure figure 2 we have used the Modified Spectral Domain Technique (MSDT) proprosed in [2,3]. The MSDT allows the calculation of attenuation and phase constants.

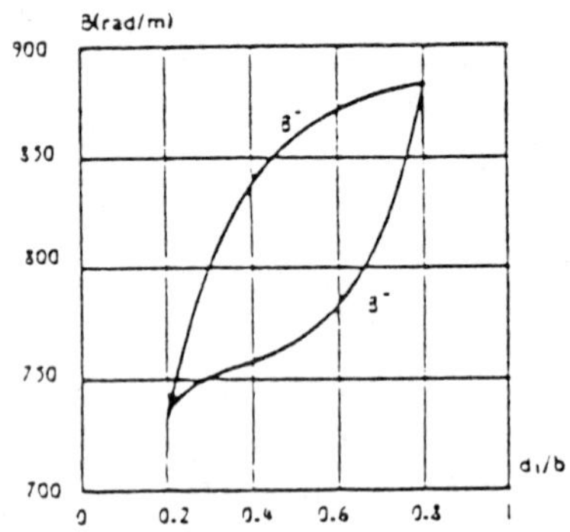

Figure 3 -
Variation of β^+ and β^- versus the semiconductor height. The InSb thickness is 2 x 1 μm.

Finline case d_0 = 0.2b, d_2 = 0.8b, f = 36 GHz, B_0 = 0.8T, c = 0.125 mm, ϵ_r = 5.5.

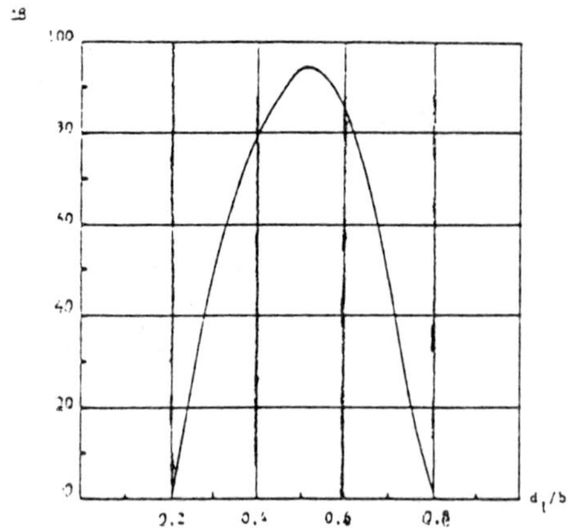

Figure 4 -
$\Delta\beta = \beta^- - \beta^+$ versus the semiconductor height.
f = 36 GHz. For $\Delta\Phi = \frac{\pi}{2} = \Delta\beta l$ we have l = 1.65 cm.

An example of computation is given on figure 3. Where the forward propagation constant β^+ and the backward one β^- are plotted versus the semiconductor height. One notices a significant difference between β^+ and β^-. The figure 4 shows the difference versus the semiconductor height. The $\Delta\beta$ variation present a maximum of 95 rad/m for a semiconductor height of 0.5 d_1/b. Which leads to a differential phase-shift of $\pi/2$ for a 1.65 cm length.

The semiconductor used for the previous calculation is the InSb. It is a high mobility one. The InSb thickness is this application is 2 x 1 μm and the magnetic field is B_0 = 0.8 T.

The performance of the structure depends mainly on the σ_x term and particularly on the product $\mu_0 \; B_0$; where μ_0 is the low electronic mobility. On the other hand the losses in this structure depends on $\sigma_\perp$ term. The ideal configuration must be : $\sigma_\perp \approx 0$.

CONCLUSION

An attempt to design a non reciprocal phase shifter in fin-line technology is proposed. It exhibits a $\frac{\pi}{2}$ differential shift for a length of 1.65 cm at 36 GHz. Other important aspects must be considered. The insertion losses due to the conductivity of the semiconductor and the bandwidth. This work is now going on.

REFERENCES

[1] R. HIROTA, T. SUZUKI
"Field distribution in a magneto-plasma loaded waveguide at room temperature"
IEEE, MTT-18, n° 4, pp. 188-195, april 1970

[2] **S. TEDJINI**
"Contribution à l'étude d'un isolateur à base de semiconducteur en Ondes milletriques"
Doctorat thesis INPG France 1982

[3] S. TEDJINI, E. PIC
"New analysis of semiconductor isolators : The Modified Spectral Domain Analysis"
IEEE, MTT-33, n° 1, pp. 59-64, jan. 1985

6.4.5

A Fin Line Ferrite Isolator and Circulator for the R-Band

A. Beyer and I. Wolff

ABSTRACT

First approaches of a fin line Y-circulator and a fin line resonance isolator are described, and experimental data of the performance are reported. The Y-circulator consist of a antipodal fin line Y-junction with three impedance transformers and a ferrite sphere as magnetically active element. The inhomogeneous fin lin tapers are mounted in the E-plane across the center of a rectangular R-band waveguide. For the optimization of the circulator tapers various antipodal fin line contour functions are calculated and a $\cos^2$-taper is chosen.

The resonance isolator in fin line technique is realized using a thin slab of ferrite which is placed in a plane of circularly polarized magnetic field in an unilateral fin line. The field distributions are discussed and first preliminary experimental results are presented.

INTRODUCTION

In the foregoing years the fin line technique has emerged as an excellent possibility to integrate millimeterwave components and subsystems. Fin lines now are used for transformers, filters, couplers, attenuators, detectors, mixers and semiconductor oscillator circuits. Therefore, non-reciprocal devices in fin line technique are of great interest in the millimeterwave range; using the results of several studies [1,3,7] which described the dispersion and characteristic impedance of homogeneous and inhomogeneous fin lines, passive non-reciprocal circuits in this technique are described here.

THE FIN LINE FERRITE CIRCULATOR

The fin line ferrite circulator described here consists of a fin line Y-junction, which contains the ferrite material of the device, connected to the three ports of the component by means of antipodal fin line impedance transformers (Fig. 1). The magnetically active region is realized using a magnetized ferrite sphere.

It is well known that the dominant mode in an inhomogeneous antipodal fin line near the slot region has a main direction of the electrical field which changes along the inhomogeneous line. It is possible to utilize this property to construct non-reciprocal millimeterwave components such as e.g. a circulator.

For the calculation of the transmission properties of inhomogeneous antipodal fin lines accurate solutions for the characteristic impedances and the effective dielectric constants of the homogeneous antipodal fin line

---

Adalbert Beyer and Ingo Wolff, Fachgebiet Allgemeine und Theoretische Elektrotechnik, Department of Electrical Engineering, Duisburg University, Bismarckstr. 81, D-4100 Duisburg 1, Fed. Rep. of Germany

A. Beyer and I. Wolff, *Proceedings of 11th European Microwave Conference*, 1981, pp. 321–326.

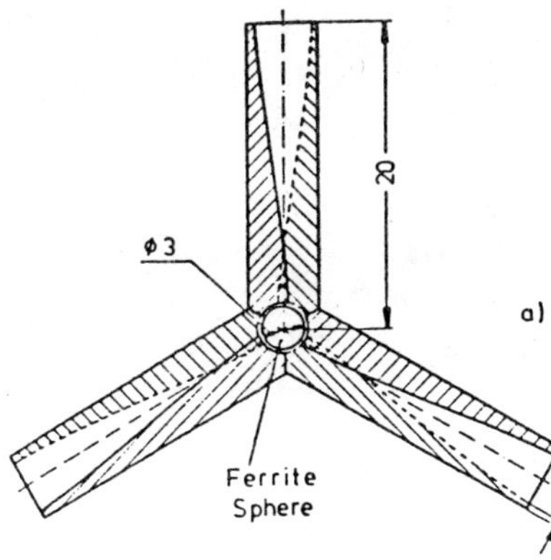

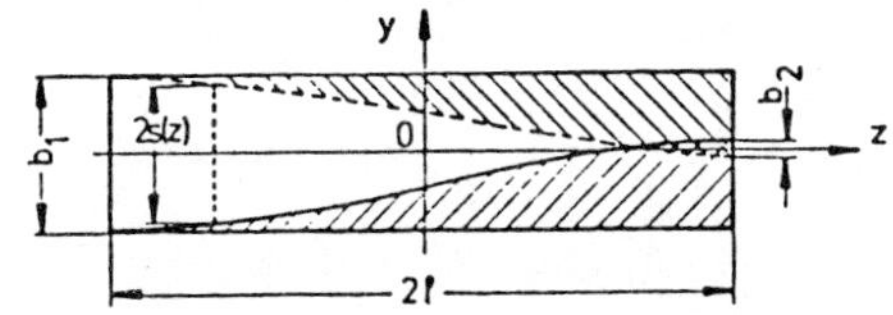

b_1 = 3.56 mm, b_2 = 0.36 mm

ε_r = 2.33, thickness = 0.25 mm

b)

Metallization on the front side of the dielectric

Metallization on the back side of the dielectric

Fig. 1: The 3-port Y-junction for the antipodal fin line ferrite circulator.
a) The Y-junction ; b) The cos²-taper.

must be known. A method for calculating this parameters has been presented in [1]. Fig. 2 shows how the effective dielectric constant and the characteristic impedance varies with respect to the slot width of the antipodal fin line. (Negative slot widths describe an overlapping of the metallization of the antipodal fin line.) Using the method given in [2] the transmission properties of the inhomogeneous antipodal fin line can be calculated.

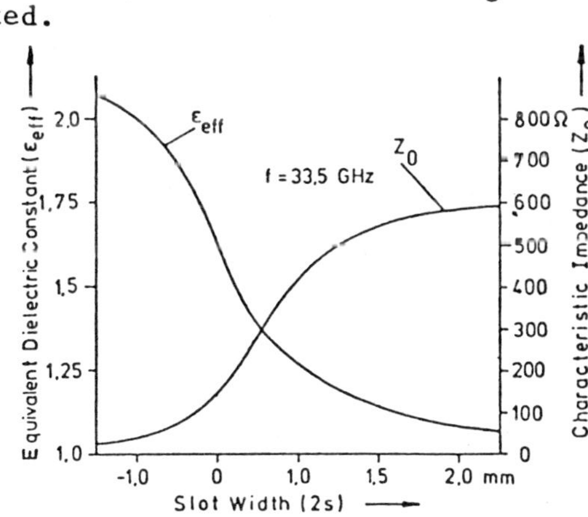

Fig. 2: Value of the effective dielectric constant and the characteristic impedance of the antipodal fin line.

With these results the impedance transformers are designed on an isotropic substrate with a thickness 0.25 mm and a dielectric constant ε_r = 2.20 (RT/Duroid 5880), the geometrical dimensions of the used cos²-taper are shown in Fig. 1.

The mode configuration of the oscillating ferrite sphere (Fig. 1) can be written using spherical coordinates (r, ϑ, φ). The ferrite medium is described by the permeability tensor in this coordinate system. It is easily shown that the components of the permeability tensor are dependent on the spherical coordinates [9], [10]. As a consequence of this dependence the oscillation modes of the sphere are coupled. It is possible to classify the electromagnetic oscillations of the sphere into coupled transverse electric modes (TE) with $E_r = 0$ and transverse magnetic modes (TM) with $H_r = 0$. The fields of the TM modes inside the sphere can be represented in terms of the first kind associated Legendre function of order n and of the first kind Bessel function of the order n + 0.5 (spherical Bessel functions). Outside the sphere, the field equations are described by the first kind associated Legendre function of order n and of the third kind cylinder (Hankel) function of the order n + 0.5. From the boundary conditions on the seperation surface the eigenvalue equations can be derived. The approximate values of the solutions of these equations deliver the value of the sphere diameter equal to about 3 mm.

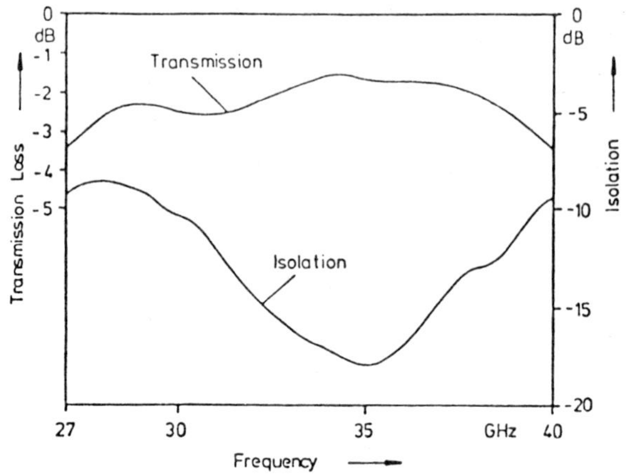

Fig. 3: The measured isolation and transmission loss of the experimental antipodal fin line circulator.

In Fig. 3, the preliminary measured characteristics of a first experimental realization of an antipodal fin line circulator are plotted. The available isolation and the bandwidth still are small and the insertion loss still is between 2 dB and 3.5 dB. It is assumed that by a proper choice of the ferrite material and the diameter these results can be improved.

THE FIN LINE RESONANCE ISOLATOR

It is well known that a magnetized ferrite slab placed in a plane of circularly polarized magnetic field results in different phase shifts and attenuations for a forward and a backward propagating wave in the slab. Fig. 4 shows the calculated normalized field components for a wave propagating in + z-direction in an unilateral fin line at 31.6 GHz. The figure illustrates the location of the plane of pure circularly polarized magnetic field, which is dependend on the operating frequency.

The field distributions shown in Fig. 4 have been calculated using the Ritz-Galerkin's method described in [1]. The fin line structure utilized has a slot width 2s = 0.2 mm, a substrate thickness w = 0.125 mm, a dielectric constant of the substrate material $\varepsilon_r = 2.20$ (RT/Duroid 5880) and a characteristic impedance $Z_0 = 180\ \Omega$. The fin line mount used was a standard R-band waveguide. Since it is no exact theory available for a ferrite-loaded fin line operating at gyromagnetic resonance, the fin line ferrite isolators has no be discussed mainly experimentally on the basis of the field distributions calculated for the inloaded fin line.

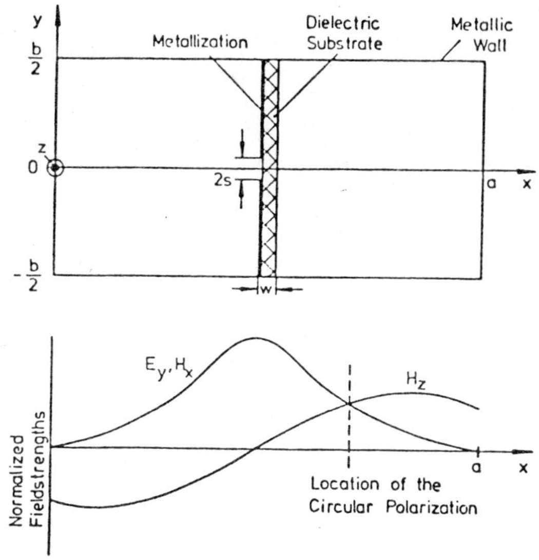

Fig. 4: Field components of the unilateral fin line in dependence on the coordinate x.

Fig. 5 shows a cross-section of the structure designed to examine the feasibility of a resonance isolator in fin line technique.

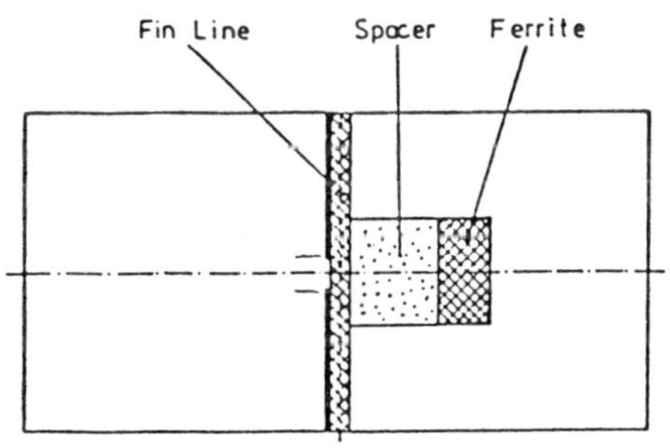

Fig. 5: Resonance isolator in fin line technique.

A fin line section with impedance transformers at both ends is used for insertion into a waveguide measurement set-up [2]. The employed optimum tapers have a maximum bandwidth for a given length.

The best compromise due to the position of the ferrite in the fin line, as to find a high backward-to-forward loss ratio, is found when ferrite is placed in a position where the forward resonant loss of the slab is a minimum. For this reason the position of the ferrite has to be chosen near the plane of circularly polarized magnetic fields. The ferrite slab is adjusted in this plane with a dielectric spacer.

Fig. 6 shows the forward and backward losses in the frequency range from 28 to 35 GHz for the examinated resonance fin line isolator. The device was an experimental laboratory model which was designed for a operating frequency f = 31.6 GHz; no attempt was made to optimize it for a broad band operation.

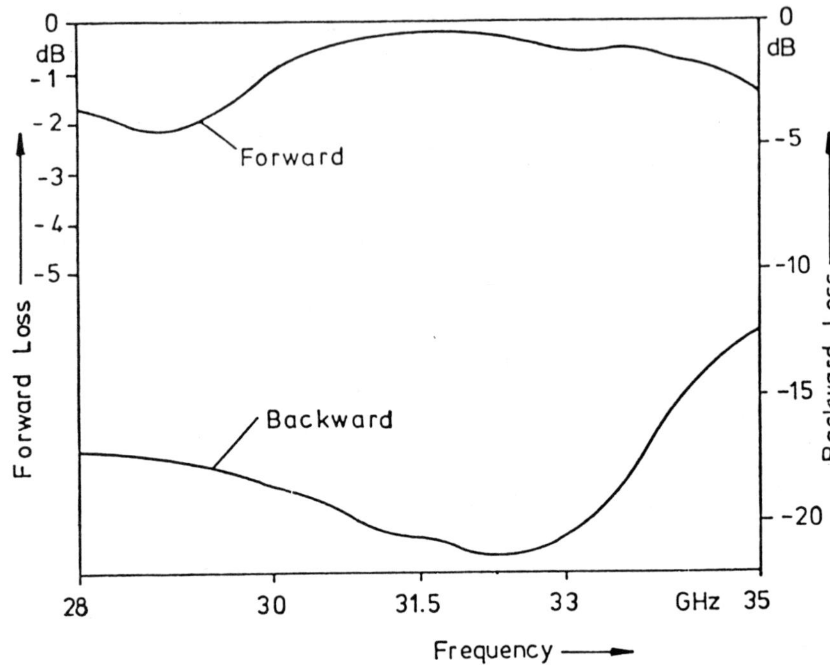

Fig. 6: Transmission properties of a 31.6 GHz fin line resonator isolator in dependence on the frequency.

The length of the isolator composite slab structure was 12.5 mm, the length of the fin line was 25 mm including 10 mm for both tapered ends. The saturation magnetisation and the applied dc magnetic field were M_S = 4000 A/cm (material TT2-111, Trans Tech. Inc., USA) and H_0 = 1800 A/cm respectively.

CONCLUSIONS

It was the purpose of this work to present first attemps for the realization of non-reciprocal devices in the fin line technique. As the results show it is principally possible to realize those devices but still much work has to be done, to get qualified devices. As the authors believe, a theoretical investigation of the field distributions of fin lines [1] with ferrite loading must be used to get better information for the design of passive non-reciprocal devices in fin line technique.

It is believed that by optimization of the ferrite location, as well as of the ferrite slab geometrical form of several other parameters, non-reciprocal fin line devices are feasible.

REFERENCES

[1] A. Beyer and I. Wolff: A Solution of the Earthed Fin Line with Finite Metallization Thickness. 1980 IEEE MTT-S Intern. Microwave Symposium Digest, Washington (USA), 258 - 260.

[2] A. Beyer and I. Wolff: Calculation of the Transmission Properties of Inhomogeneous Fin Lines. Proc. of the 10th European Microwave Conference, Warsaw (Poland), (1980), 322 - 326.

[3] Ch. Chang - T. Itoh: Spectral Domain Analysis of Dominant and Higher Order Modes in Fin Lines. 1979 IEEE MTT-S Intern. Microwave Symposium Digest, Orlando (USA), 344 - 346.

[4] L. Courtois and M. DeVecchis: A New Class of Non-reciprocal Components Using Slot Lines. IEEE Trans. MTT, vol. MTT-23, June 1975, 511 - 516.

[5] M. Gastine, L. Courtois and J.L. Dormann: Electromagnetic resonances of free dielectric spheres. IEEE Trans. MTT, vol. MTT-15, Dec. 1967, 694 - 700.

[6] R.P. Hecken: A Near-Optimum Matching Section Without Discontinuities. IEEE Trans. MTT, vol. MTT-20, November 1972, 734 - 739.

[7] H. Hofmann: Dispersion of Planar Waveguides for Millimeter-Wave Application. Arch. Elektron. Uebertragungstech., 31 (1977), 40 - 44.

[8] I. Wolff: The Lowest Order Mode and the Quasi-TEM Mode in a Ferrit-Filled Coaxial Line or Resonator. IEEE Trans. MTT, vol. MTT-20, August 1972, 558 - 559.

[9] I. Wolff: Die elektromagnetischen Eigenschwingungen einer offenen Ferritkugel. Arch. Elektron. Uebertragungstech., 23 (1969), 561 - 569.

[10] I. Wolff: Felder und Wellen in gyrotropen Mikrowellenstrukturen. Verlag Fr. Vieweg, Braunschweig, 1973.

6.5.1

Studies in Fin-Line Antenna Design for Imaging Array Applications

by Everett Farr*, Kevin Webb*, and Raj Mittra*

An experimental study was conducted concerning the development of an endfire fin-line antenna, suitable for imaging array applications at 220 GHz. Experimental models of several designs were built and tested at 80 GHz. The most successful design used a fin-line to dielectric rod transition followed by a dielectric rod antenna.

Flossenleitungsantennen für Abbildungs-Strahlergruppen

Zur Entwicklung eines Flossenleitungs-Längsstrahlers zur Anwendung in Abbildungs-Strahlergruppen für 220 GHz wurden verschiedene Versuchsmodelle für 80 GHz gebaut und geprüft. Die günstigste Ausführung besteht aus einer Flossenleitung, die auf einen dielektrischen Stielstrahler übergeht.

1. Introduction

There is, at present, a great deal of interest in imaging radar systems to be operated at 220 GHz. We are interested in this frequency because waves in this region penetrate dust and fog better than optical wavelengths, and because systems with such short wavelengths tend to be compact and lightweight. Additonally, at 220 GHz, there is a local minimum in the attenuation constant through air [1]. These characteristics enhance the desirability of this frequency range for imaging radar applications in battlefield situations, where compactness and mobility are of prime concern, and dust and fog may obstruct visibility.

It has been proposed, furthermore, that such an imaging radar system be built primarily from fin-line, a diagram of which appears in Fig. 1. Fin-line components are low loss at high frequencies and are easily built using standard printed circuit techniques [2]. In addition, fin-lines are very compatible with solid state integrated circuit technology. Thus, one might integrate a detector directly onto a GaAs substrate in order to enhance repeatability and reduce cost. The particular part of the imaging radar system we worked on was the antenna and feed mechanism. Several designs were developed, which were tested at 80 GHz. These designs and antenna patterns are presented in the following sections.

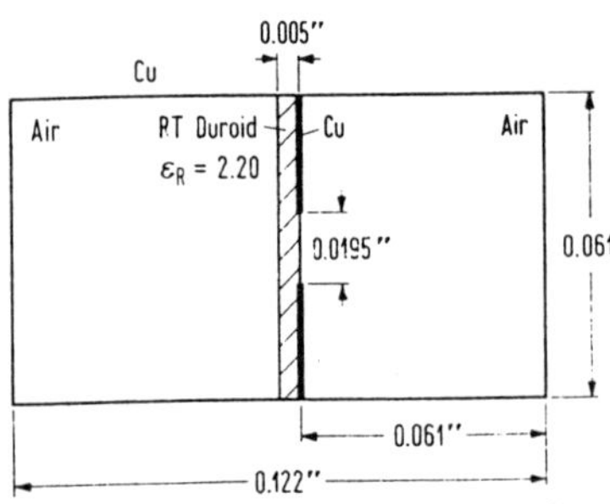

Fig. 1. Cross-section and dimensions of an E-band fin-line.

2. The Fin-Line Antenna

In order to develop an imaging radar system, a component of primary interest must be the antenna. For this application, we envisage an array of a large number antennas and feed structures arranged in a densely-packed formation. Before studying the array as a whole, we must first study the characteristics of a single antenna element.

Such a fin-line antenna must satisfy several criteria. First, it must have a relatively small cross sectional area. Next, it must have an overall design that is compatible with a densely-packed configuration. Furthermore, this antenna must have a fairly broad beamwidth. Finally, this antenna must be endfire in order to achieve a compact configuration when these antennas form an array.

3. Antenna Design and Experimental Setup

We now turn to the design of the fin-line antennas. In Fig. 2, an antenna is shown in which the dielectric portion of the fin-line extends past the end of the copper shielding. The slot gradually widens to the width of the shield, so we hoped to see a pattern roughly approximating that of an open-ended waveguide.

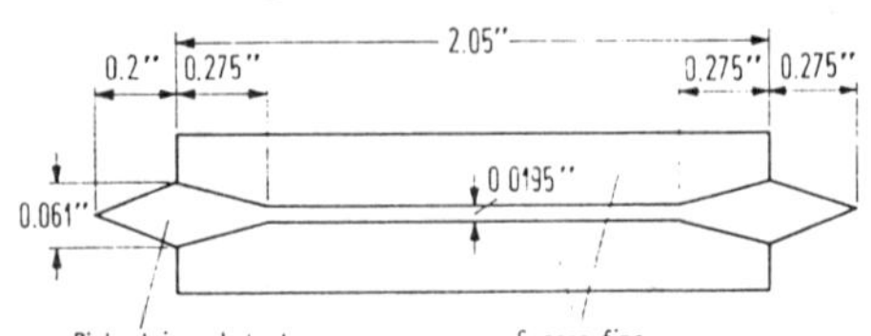

Fig. 2. Extended dielectric fin-line antenna. The rectangular waveguide couples in from the left, and the radiating element is on the right.

* Mr. E. Farr, Dr. K. Webb, Prof. Dr. R. Mittra, Electrical and Computer Engineering Department, University of Illinois, Urbana, Ill. 61801, USA.

E. Farr, K. Webb, and R. Mittra, *Archiv fur Elektronik und Ubertragungstechnik*, Vol. 39, No. 1, 1985, pp. 87–89.

In Fig. 3 we have a design where a thick dielectric rod made of Rexolite (dielectric constant = 2.53), protrudes from the front end of the shield. The intention here was to launch the wave onto a dielectric rod, and then use a standard dielectric rod antenna as a radiator. Several references exist for the design of dielectric rod antennas. See, for example [3] and [4].

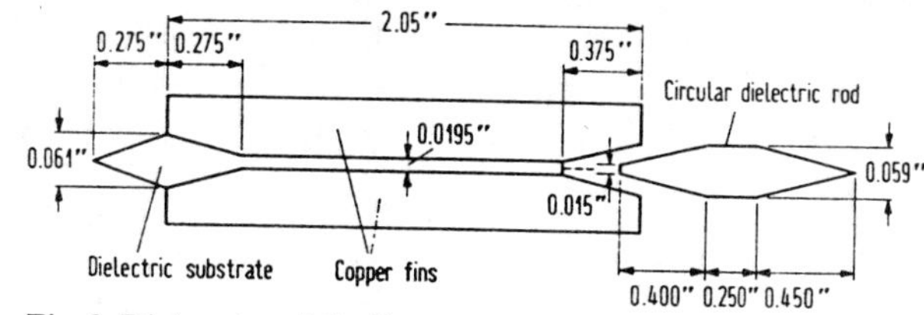

Fig. 3. Dielectric rod fin-line antenna. The rectangular waveguide couples in from the left.

Finally, we tried two printed dipole configurations. First we tried a printed dipole which was approximately a half wavelength long at 80 GHz, shown in Fig. 4. Next, we tried a printed vee dipole, shown in Fig. 5. Each of the arms of this dipole is one wavelength long, and the arms are separated by a 90 degree angle. It was hoped that these dipole configurations would act in a manner similar to the analogous thin wire configurations. For design information on wire dipoles, see for example [5].

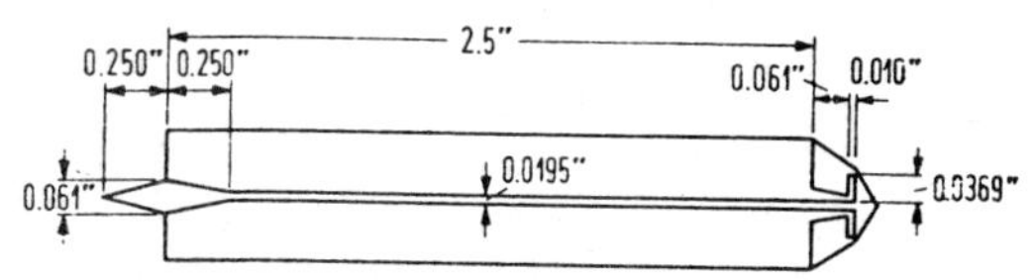

Fig. 4. Printed dipole fin-line antenna. The lengths of each half of the dipole are quarter wavelength at 80 GHz. This simulates a half wave dipole.

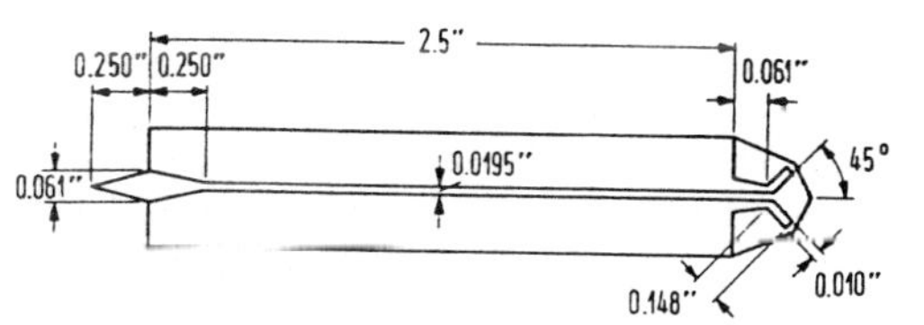

Fig. 5. Printed vee dipole fin-line antenna. The length of each vee section is approximately one wavelength at 80 GHz.

The fin-lines were mounted between two blocks of copper whose inner dimensions were those of standard E-band rectangular waveguide. A diagram of one of these copper blocks is shown in Fig. 6. Note that a recess was formed on the left side of the feed in order to allow a standard E-band metal waveguide to couple into the fin-line. The fin-lines were fabricated with standard photolithographic techniques. The substrate was 5 mil 1/2 oz. copper clad RT/duroid 5880, made by Rogers Corporation, with a dielectric constant of 2.20 at 10 GHz.

We now turn to the experimental setup used to measure these antennas, a diagram of which is shown in Fig. 7. An 80 GHz oscillator, modulated at 1 kHz, generated waves which were transmitted to a antenna under test which was 35″ away. The antenna under test was connected via a transition to a standard rectangular metal waveguide which was terminated in a diode detector. The level of the diode detector was measured by an audio-frequency voltmeter, and recorded on an X-Y plotter as a function of angle.

Now that the antenna designs and experimental apparatus have been discussed, we are ready to look at the results of our tests. These are presented in the next section.

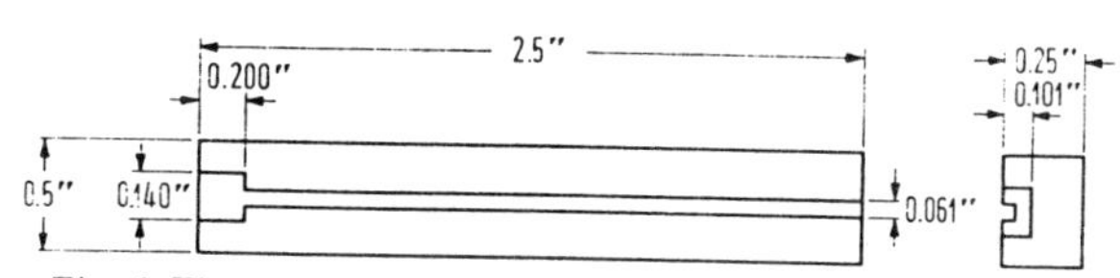

Fig. 6. The copper shield for all the fin-line antennas tested.

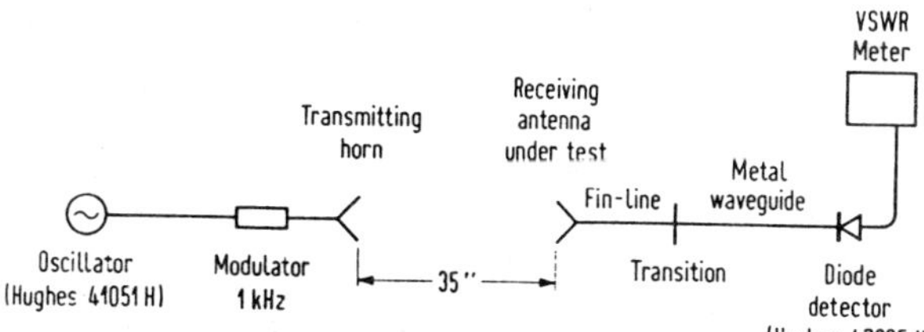

Fig. 7. Experimental setup. The antenna under test was mounted on a rotating platform.

4. Experimental Results

In order to have a pattern with which to compare our test antenna patterns, we began by plotting the *H*-plane pattern of an open-ended waveguide. This plot is shown in Fig. 8.

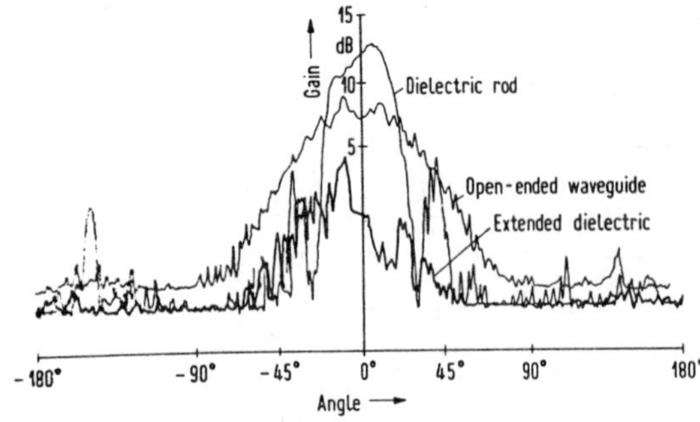

Fig. 8. *H*-plane patterns of fin-line antenna with extended dielectric (see Fig. 2), and dielectric rod (see Fig. 3). An open-ended waveguide pattern is also shown for comparison.

Next, we plotted *H*-plane patterns of fin-line antennas for the extended dielectric case (Fig. 2) and the dielectric rod case (Fig. 3). These patterns are shown in Fig. 8. Note that the gain of the dielectric rod configuration (13 dB) is approximately 4 dB higher than that of the open-ended waveguide (9 dB).

Finally, we measured the *H*-plane pattern for the printed dipole (Fig. 4) and vee dipole (Fig. 5). Neither of these two designs performed satisfactorily, so their patterns are not shown.

From these antenna patterns we can draw several conclusions. While neither the dipole nor vee dipole configurations performed satisfactorily, the extended dielectric configuration showed some promise, and the dielectric rod configuration showed a great deal of promise, with a gain of 13 dB and a 3 dB beamwidth of 40 degrees. It is likely that adjustments can be made to this design in order to further improve its performance.

5. Recommendations for Further Study

We now present our recommendations for further study of the 220 GHz imaging radar system. We have generated two distinct possibilities for the antenna array, so we will look at each individually.

The first possibility for this antenna is to have a transition from fin-line to metal waveguide, and build an antenna from metal waveguide components. A fin-line to rectangular waveguide transition is easily built, an example of which is shown in Fig. 9. Furthermore, the antenna would consist of either a horn or an open-ended waveguide. The horn would give a higher gain than the open-ended waveguide, but it would also have a larger cross-

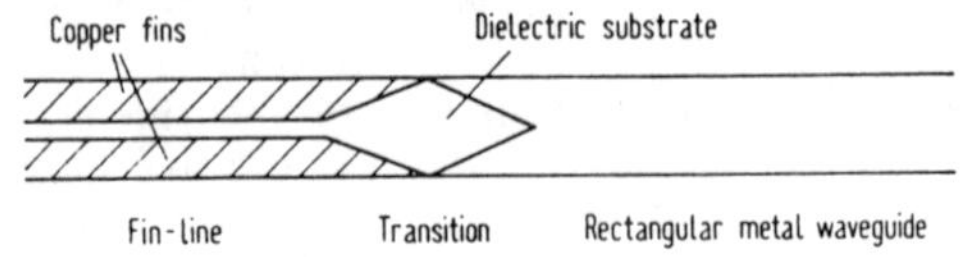

Fig. 9. An example of a fin-line to rectangular waveguide transition.

sectional area, thereby increasing the size of the array. Therefore one would use only as much flare in the waveguide as was required for sufficient gain. With this configuration, one has the option of integrating the detector directly onto the dielectric substrate, thereby eliminating the need for rectangular waveguide behind the fin-line.

The second possibility for the fin-line antenna is to use the fin-line directly in a dielectric rod configuration, shown in Fig. 3. This configuration produced the best pattern of all the new designs tested. It had a gain of 14 dB and a 3 dB beamwidth of 40 degrees. With further study, it should be possible to improve the antenna pattern by varying the parameters in the design of the antenna. As with the first configuration, one can integrate the detector directly onto the fin-line.

6. Conclusions

An experimental study was performed to determine the feasibility of several fin-line antenna designs for a 220 GHz imaging radar system. Two possible designs emerged from this study. One design involved a transition from fin-line to rectangular waveguide, followed by a horn or open-ended waveguide antenna. The other design involved an endfire fin-line antenna with a dielectric rod protruding from the front end. The initial experimental results are quite promising as this research effort continues.

The work reported in this paper was supported in part by the Army Research Office on Grant DAAG 29-82-K-0084 and in part, by the Joint Services Electronics Program, N00014-79-C-0424.

(Received April 17th, 1984.)

References

[1] Wrixon, G. T. and McMillan, R. W., Measurements of earth-space attenuation at 230 GHz. Transact. IEEE MTT-**26** [1978], 434–439.
[2] Meier, P. J., Integrated fin-line millimeter components. Transact. IEEE MTT-**22** [1974], 1209–1216.
[3] Kobayashi, S., Mittra, R., and Lampe, R., Dielectric tapered rod antennas for millimeter-wave applications. Transact. IEEE AP-**30** [1982], 54–58.
[4] Johnson, R. and Jasik, H. (Eds.), Antenna engineering handbook; 2nd ed. McGraw-Hill Book Co., New York 1984, Ch. 12.
[5] Stutzman, W. L. and Thiele, G. A., Antenna theory and design. J. Wiley & Sons, New York 1981, pp. 192–205.

Chapter 7
Active Devices

One of the main advantages of E-plane integration is the compatibility of fin lines with semiconductor devices. In particular, beam-lead devices can be very easily bonded across the small gap of a fin line, thus exposing the device to more concentrated electric fields. Today, virtually all kinds of active devices have been realized using E-plane technology, which has led to the complete integration of a receiver or millimeter-wave radar front end, up to 170 GHz.

This chapter will include some of the most significant papers on detectors, modulators, amplifiers, oscillator, switches, mixers, and front ends realized using E-plane technology.

7.1 DETECTORS AND SWITCHES

7.1.1 "High Sensitivity Millimeter Wave Detectors Using Fin-Line Technology" by Meinel and Schmidt

This paper gives a brief description of a systematic, successful approach to the realization of high sensitivity millimeter-wave detectors using E-plane technology and zero bias Schottky diodes. Several detectors were realized for different waveguide bands ranging from 18 to 170 GHz. The propagation parameters of the fin line were calculated using the spectral-domain technique and Ritz and Galerkin's procedure. This paper has served as reference by many subsequent researchers.

7.1.2 "Ka-Band Fin-Line SPDT Switch" by Koh, Williamson, Nyss, Gadoury, and Wheeler

This paper describes an E-plane SPDT switch using a fin-line series Y-junction. It presents a simple equivalent network of the junction. We analyze the switch by assuming that a forward-biased *pin* diode approximates a short circuit across the fin gap, and a reverse-biased *pin* diode approximates an open circuit across the fin gap. We also analyze the equivalent circuit to determine where a forward-biased *pin* diode should be positioned so that the two other ports may be matched. The network parameters are calculated using simple equations.

7.1.3 "Computer-Aided Design of Fin-Line Detectors, Modulators, and Switches" by El Hennawy and Schunemann

This detailed paper on computer-aided design (CAD) of E-plane detectors, modulators, and switches masters the discontinuities associated with various types of diode mounts by using a modal analysis. The equivalent network parameters are accurately predicted and adjusted for the discontinuities to match the corresponding diode equivalent circuit. This is the principal paper which promotes CAD of E-plane detectors, modulators, and switches.

7.1.4 "Integrated Microstrip and Antipodal Fin Line Leads to Enhanced Performance PIN Diode Components" by Greed

This unique paper shows how to broaden the bands of E-plane *pin* diode switches by using distributed microstrip tuning elements to tune out the junction capacitance and series inductance of beam-lead diodes. The paper successfully reports on Ka- and W-band switches. Greed did not include design details, which would have been useful.

7.2 MODULATORS

7.2.1 "A Quadriphase Fin-Line Modulator" by Kpodzo, Begemann, and Schunemann

This paper presents a quadriphase modulator realized by cascading two hybrid-coupled 180° and 90° modulators in an integrated fin-line circuit. The quadriphase modulator achieves an insertion loss of 2.2 dB at 30 GHz over 15% bandwidth. The modulator performance remains practically insensitive to housing dimensions.

7.2.2 "Millimeter-Wave QPSK Modulator in Fin Line" by Gajda and Verver

This QPSK modulator, realized for the ESA Olympus satellite at Ka-band, consists of two 180° biphase modulators, an input 3 dB quadrature hybrid, and an in-phase output combiner, all of which are integrated on a single substrate. The circuit is devoid of bond wires, and is plated through holes. Fabrication tolerance is of little concern.

The paper received the "best paper" award in the open forum session at the IEEE MTT-S Symposium in Baltimore in 1986.

7.2.3 "An E-Plane Broadband Bi-Phase Modulator for Ka-Band" by Thorpe

This paper presents the development of a biphase PSK modulator for Ka-band applications. This circuit has several good features: it does not require any post-production tuning or alignment; switching speed as high as 100 Mbit/s can be obtained; the entire design is based on a simple concept.

7.3 AMPLIFIERS

7.3.1 "A FET Amplifier in Fin-Line Technique" by l'Ecuyer, Gajda, and Hoefer

This is the only paper that describes in E-plane technique the successful realization of an unconditionally stable FET amplifier. The biasing circuits use microstrip and coplanar waveguide in a unique way to achieve the stability of the amplifier.

7.4 OSCILLATORS

7.4.1 "Design and Performance of Microwave Oscillators in Integrated Fin-Line Technique" by Knoechel

This article gives a thorough investigation of the design and performance of oscillator circuits, which are compatible with E-plane technology. It pays special attention to diode coupling structures, which maximize the efficiency and Q-factor of the circuit. The investigation generated guidelines for the design of cavity-stabilized, injection-locked, and voltage-tuned oscillators. Regardless of the new techniques developed for E-plane oscillators, this article will always remain valuable.

7.4.2 "Fin-Line Varactor-Tuned Gunn Oscillators for 35 and 60 GHz" by Solbach, Adelseck, and Sicking

This paper is an enhanced version of Solbach's paper presented at the European Microwave Conference in 1982. It presents 35 and 60 GHz E-plane oscillators with high efficiency and varactor-tuning facility. The authors doubt that the circuit can be reproduced, because of the availability of beam-lead varactor diodes and the influence of large tolerances in commercial Gunn diode characteristics. Discontinuity effects of the soldered contacts are also of serious concern.

7.4.3 "A 30 GHz FET-Oscillator Using Fin-Line Circuitry" by Meinel

This paper demonstrates the feasibility of GaAs field-effect transistor oscillators in E-plane technology. A Ka-band FET oscillator is realized by using the fin-line technique, for which we achieve 4% efficiency. Meinel proposes to optimize channel parameters for higher efficiency.

7.4.4 "Advances in E-Plane Printed Millimeter-Wave Circuits" by Cohen and Meier

Cohen and Meier also discuss printed probe four-port coupling elements for wideband applications. This paper demonstrates the simplest form of fin-line Gunn oscillator with mechanical tuning. This is one of the oldest papers on E-plane oscillator circuits.

7.5 MIXERS

7.5.1 "Material and Circuit Evaluation for Millimeter-Wave Applications: A Broadband Fin-Line Mixer for Millimetre Waves" by Arndt, Bornemann, Grauerholz, and Vahldieck

This paper presents a simple, inexpensive, E-plane broadband mixer for millimeter-wave applications. This unbalanced mixer for Ka-band offers a conversion loss of 10 dB.

7.5.2 "Wideband mm-Wave Mixers for EW Applications" by Ball and Bui

This paper is one of many published works on E-plane balanced mixers. It shows how the entire mixer can be integrated on a single substrate using different types of suspended substrate transmission lines in conjunction with fin lines.

7.5.3 "A New Type of Upconverter for Space Applications" by Delamer

While all other published works on E-plane mixers are on down converters, this paper presents an up converter in E-plane. The up converter's performance meets that of standard domestic satellites.

7.5.4 "Wide-Band Subharmonically Pumped W-Band Mixer in Single-Ridge Fin Line" by Meier, Calviello, and Bie

Until the publication of this paper, all papers presented E-plane subharmonically pumped mixers that used double-ridged fin lines, tunable waveguide backshorts, and narrow bandwidths. In contrast, this paper presents dual-function antiparallel diodes, and unique bandpass and band-reject filters in single-ridge unilateral fin line, which yields a subharmonic mixer superior to conventional E-plane subharmonic mixers in many respects. For example, this subharmonic mixer has nearly

11–14 GHz bandwidth in the W-band. This approach will be an efficient one when combined with modern E-plane filter design techniques.

7.6 RECEIVERS AND FRONT ENDS

7.6.1 "Ka-Band Radar Increases Target Resolution" by Menzel, Callsen, Solbach, and Meinel

This paper shows the potential of E-plane integration techniques in the realization of a millimeter-wave radar system. Although the system is reported for Ka-band, similar systems at higher frequency bands and with improved performance are possible. Figure 3 of the paper shows how a Ka-band pulse radar front end can be integrated on a single substrate board using different kinds of suspended-substrate transmission lines together with fin lines.

7.6.2 "Compact 94-GHz Dual-Polarization Radar Receiver Realized Using Different Integration Techniques" by Menzel

This paper combines two separate front ends, described in the previous paper, to realize a 94 GHz dual-channel receiver for a radar with dual-polarization capabilities. From circulators to mixers, every circuit is realized using E-plane techniques.

BIBLIOGRAPHY

Detectors and Switches

1. H. Meinel, "Millimeter-Wave Detectors Using Finline Technology," *Wiss. Ber. AEG Telefunken*, Vol. 54, Nos. 4–5, Sept.–Oct. 1981, pp. 233–234.

Modulators

1. C. Schieblich, U. Goebel, and V. Beres, "Broadband Reflection Type Phase Modulators," *IEEE MTT-S Symposium Digest*, 1983, pp. 570–572.
2. B. Adelseck and B. Rembold, "A 180-degree Finline Phase Modulator," *Wiss. Ber. AEG Telefunken*, Vol. 54, Nos. 4–5, Sept.–Oct. 1981, pp. 238–240.

Oscillators

1. A. Jacob and G. Ansorge, "Stabilized Finline FET Oscillators," *Proc. 13th European Microwave Conf.*, 1983, pp. 303–307.
2. B. Adelseck, F. Sicking, and H. Hofmann, "Gunn Oscillators Using Distributed Feedback Finline Structures," *Wiss. Ber. AEG Telefunken*, Vol. 54, Nos. 4–5,

Sept.–Oct. 1981, pp. 233, 235, 237.
3. H. Meinel, "A 30 GHz FET Oscillator Using Finline Circuitry," *Proc. 11th European Microwave Conf.*, 1981, pp. 297–300.

Mixers

1. T. Berceli, V. Geleji, and T. Kolumban, "Optimum Design of Single Sideband Finline Mixers with Filters," *Proc. 13th European Microwave Conf.*, 1983, pp. 573–578.
2. R. Knoechel, "A Double Balanced Finline Mixer with Broad IF Bandwidth Capability," *Proc. 13th European Microwave Conf.*, 1983, pp. 567–572.
3. W. Menzel, "A 150 GHz Balanced Mixer for Finline Integrated Circuits," *Proc. 13th European Microwave Conf.*, 1983, pp. 179–182.
4. R.H. Tahim, G.M. Hayashibara, and K. Chang, "Design and Performance of W-Band Broadband Integrated Circuit Mixers," *IEEE Trans. Microwave Theory and Techniques*, Vol. MTT-31, No. 3, March 1983, pp. 277–283.
5. W. Menzel and H. Callsen, "Ultrabroadband Balanced Finline Mixer," *Electronic Letters*, Vol. 18, No. 17, 1982, pp. 724–725.
6. P.J. Meier, "E-Plane W-Band Printed Circuit Balanced Mixer," *IEEE Trans. Microwave Theory and Techniques*, Vol. MTT-31, No. 2, Feb. 1983, pp. 227–230.
7. J.S. Wong and K.I. Chung, "A Ka-Band Orthogonal Hybrid Finline Mixer," *IEEE MTT-S Symposium Digest*, 1982, pp. 379–381.
8. R.N. Bates, R.K. Surridge, J.G. Summers, and J. Woodcock, "Millimeter-Wave Low Noise E-Plane Balanced Mixers Incorporating Planar MBE GaAs Mixer Diodes," *IEEE MTT-S Symposium Digest*, 1982, pp. 13–15.
9. G. Begemann and J.E. Raue, "A Sub-Harmonic Pumped Finline Mixer for Satellite TV Receiver Applications," *IEEE MTT-S Symposium Digest*, 1981, pp. 307–309.

Receivers and Front Ends

1. L. Szabo and E. Jensen, "A Doppler Radar Sensor in Integrated Finline Technique," *IEEE MTT-S Symposium Digest*, 1983, pp. 472–474.
2. W. Menzel, H. Callsen, and K. Solbach, "An Integrated Receiver Front End for a 94-GHz Dual Polarization Radar," *Proc. 13th European Microwave Conf.*, 1983, pp. 142–147.
3. T. Berceli, V. Galeji, G. Hammer, K. Juhasz, T. Kolumban, and G. Reiter, "A Multiple Channel 12 GHz Receiver for Satellite Television Broadcasting," *IEEE International Conf. Communications for World Progress,* Boston, Vol. 1, 1983, pp. 39–43.
4. W. Menzel, H. Callsen, K. Solbach, and H. Meinel, "A Millimeter-Wave Pulse Radar Using Integrated Finline Technique," *Mikrowellen Magazin*, Vol. 9, No. 3, 1983, pp. 322–324.

5. W. Menzel and H. Callsen, "Integrated Finline Components and Subsystems at 60 and 94 GHz," *IEEE Trans. Microwave Theory and Techniques*, Vol. MTT–31, 1983, pp. 142–146.
6. R.N. Bates, I.M. Clarke, and B.C. Barnes, "An Integrated E-Plane Millimeter-Wave 29-GHz Transreceiver," *Proc. Colloquium on Millimeter-Waves*, London, pp. 11/1-4, 1983.
7. R.N. Bates, S.J. Nightingale, and P.M. Ballard, "Millimeter-Wave E-Plane Components and Subsystems," *Radio and Electronics Engineer*, Vol. 52, Nos. 11–12, 1982, pp. 506–512.
8. W. Menzel and H. Callsen, "Millimeter-Wave Receivers in Finline Technology," *Wiss. Ber. AEG-Telefunken*, Vol. 54, Nos. 4–5, Sept.–Oct. 1981, pp. 251–253.

7.1.1

HIGH SENSITIVITY MILLIMETER WAVE DETECTORS USING FIN-LINE TECHNOLOGY

H. Meinel and L.-P. Schmidt
AEG-TELEFUNKEN
Geschäftsbereich Hochfrequenztechnik
Postfach 1730
7900 Ulm, Germany

Abstract

For the frequency range 18-170 GHz fin-line detectors have been built, covering one or even two waveguide bands. Based on accurate calculations of the propagation constants and characteristic impedances, sensitivity values of 400-200mV/mW could be achieved, using commercially available zero bias Schottky diodes.

Introduction

Commercially available mm-wave detectors for frequencies above 60 GHz normally are not broadband. They have to be DC-biased for good sensitivity, and they are very sensitive against static voltages. Thus they are not so well suited for versatile and reliable operation in laboratory environments.

Zero-bias Schottky diodes promise better performance under the described circumstances. Since two years they are commercially available as beam lead elements, the type of device, which is very well suited for fin-line application /1, 2/.

Fin-Line parameters

Matched detectors with high sensitivity can be designed using fin-line technique, provided that propagation constants and wave impedances of fin-lines are known. These frequency dependent parameters have been calculated accurately and numerically efficient by applying Galerkin's method /3/. The wave impedance definition based on the slot voltage and the fin current has been proven to be very well adapted; the implemented beam-lead diodes can be matched over a wide frequency range. Results for fin-lines on a 110 μm quartz substrate in a WR-7 waveguide housing are shown in fig. 1.

Detector design

Our fin-line detectors have been designed for Alpha Ind. type DDC 4565 D zero-bias Schottky diodes. Fig. 2 displays the construction. On the front side of the substrate the fin-line circuit, containing the semiconductor device is etched, while absorber material is painted on the back side of the substrate. The material being used is fused quartz of 220 μm and 110 μm thickness for frequencies below and above 110 GHz respectively. A good transition from waveguide to the semiconductor device could be achieved, using on optimized fin-line taper /2/. The VSWR of the entire detector unit is 1.7:1.

Results

Several units have been set up for different waveguide bands between 18 and 170 GHz, Minimum sensitivity ranges from 400 to 200 mV/mW, fig. 3, and excellent broadband behaviour could be achieved. More than octave band wide two-band units have been built. The flatness is below $\pm$ 0.8 dB over one or two entire waveguide bands.

The transfer characteristics - voltage output versus input power - are shown in fig. 4 a and b for 35 and 132 GHz respectively; good square low dependance can be testified.

Conclusion

It has been shown that fin-line technique is well suited for the construction of broadband mm-wave detectors. Commercially available beam-lead devices have been applied for frequencies as high as 170 GHz.

Acknowledgement

This work has been supported by the German Government (Grant: T/RF 31/60007/61010).

References

/1/ Adelseck, B., Barth, H., Hofmann, H., Meinel, H., Rembold, B.: Advances in mm-wave components and systems, AGARD Conf. on mm- and sub-mm-wave propagation and circuit, Munich 1978, Conf. Proc. 245, 25-1 to -17.

/2/ Hofmann, H., Meinel, H., Adelseck, B.: New integrated mm-wave components using fin-lines, IEEE Symp. Dig., MTT-S, 1978, Ottawa, 21-23.

/3/ Hofmann, H.: Dispersion of planar waveguides for millimeter-wave application, AEÜ (Electron. and Comm.) 31, 1977, 40-44.

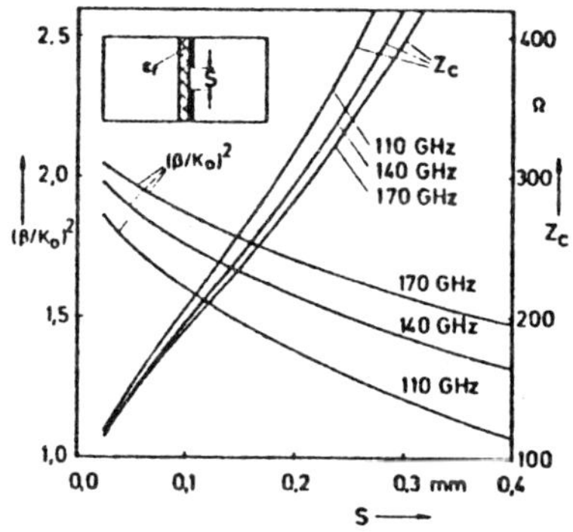

Fig. 1 Fin-line propagation constant and wave impedance

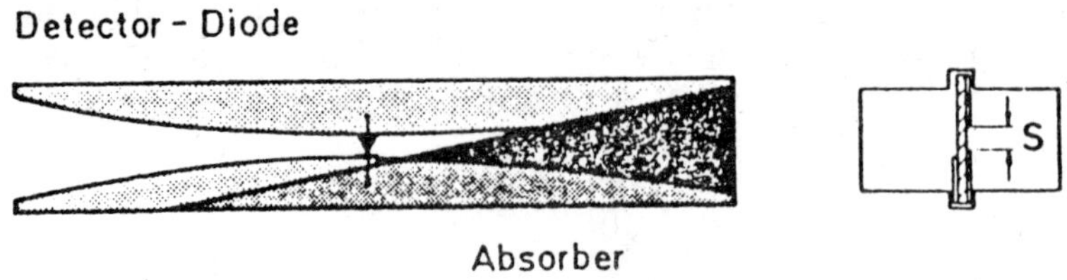

Fig. 2 General configuration of a fin-line detector

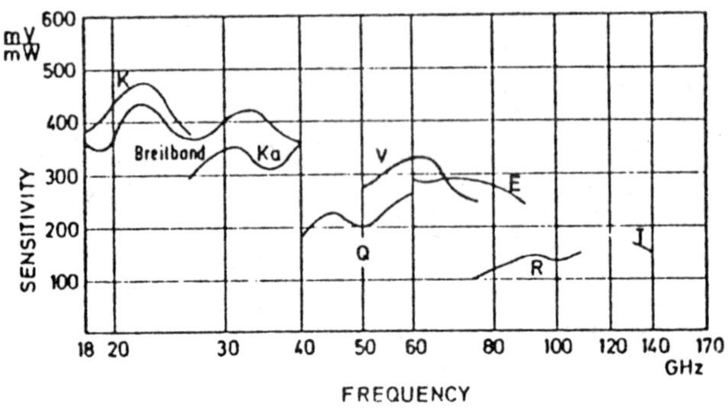

Fig. 3 Sensitivity of fin-line detectors for different waveguide bands

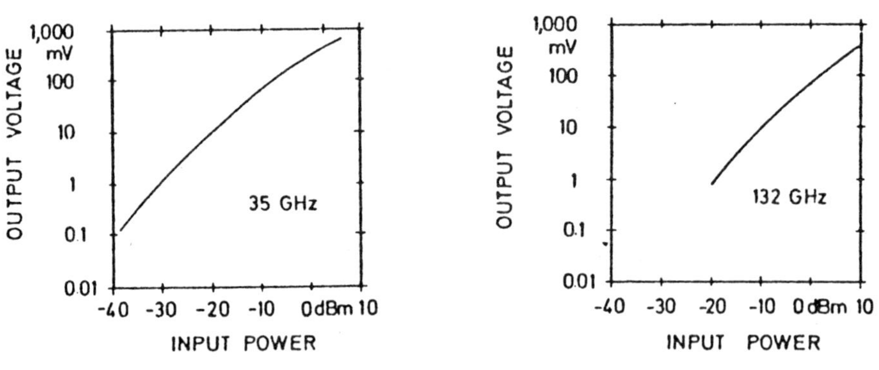

Fig. 4 Transfer characteristics of fin-line detectors

7.1.2

Technical Note

Ka-Band Fin-Line SPDT Switch

L.W. Koh, I.M.H. Williamson, M.L. Nyss
Microwave Associates, Ltd.
Dunstable, England

Omer Gadoury, D.E. Wheeler
M/A-COM Millimeter Products, Inc.
Burlington, MA

Why Fin-Line

In a fin-line structure, longitudinal metal strip and dielectric layer are suspended in the E-plane of a rectangular waveguide housing to provide capacitance loading to the quasi-TE_{10} mode. This loading widens the single mode bandwidth as in similar structure-ridge waveguide. The fin-line has a quasi-TE_{10} cut-off frequency and does not require a high k substrate to prevent radiation. The ability to choose low k substrates is advantageous at millimeter wavelengths, where excessive miniaturization can lead to tolerance problems.

Bandwidth, Cost, Easy Manufacturing

It has been demonstrated that wide-band, low loss integrated circuits can be constructed at millimeter wavelengths by inserting etched conductor sheets, with or without parallel dielectric layers, into the E-plane of a rectangular waveguide. The advantages of such circuits include a wide single-mode bandwidth, low insertion loss, production economy and compatibility with unpackaged devices and waveguide test equipment. The equivalent dielectric constant of these transmission lines can be close to unity, which avoids excessive miniaturization and thereby eases mechanical tolerances.

Design Criteria

The substrate used in the fin-line SPDT switch is cut from a 0.005″ sheet of laminate (Duroid 5880) and includes six mounting holes. The switch was designed by using a series Y junction. The substrate metalization is tapered to the desired gap which is 0.2mm between the fins. The substrate metalization also includes two RF blocking networks which are integrated with the fins.

The switches described in the article contain commercially available beam lead diodes (MA 47302; Alpha 6474-30). They are connected across the low impedance part of the fin-line.

Theoretical Analysis

The junction between one main line and the series branch is shown in Figure 1. It is similar to the E-plane junction of a rectangular waveguide and is therefore described by the same equivalent circuit shown in Figure 2.

An analysis of the switch performance was carried out by making the following assumptions, i.e. the forward biased PIN diode approximates to a short circuit across the fin-line and that the reversed biased diode approximates to an open circuit across the fin-line. The equivalent circuit of this analysis can be shown in Figure 3.

This was used to calculate the position at which the forward biased PIN diode should be placed across one port of the junction to ensure that the other port was matched. If the guide wavelength of the fundamental fin-line mode is inserted in the following expressions, the individual elements of

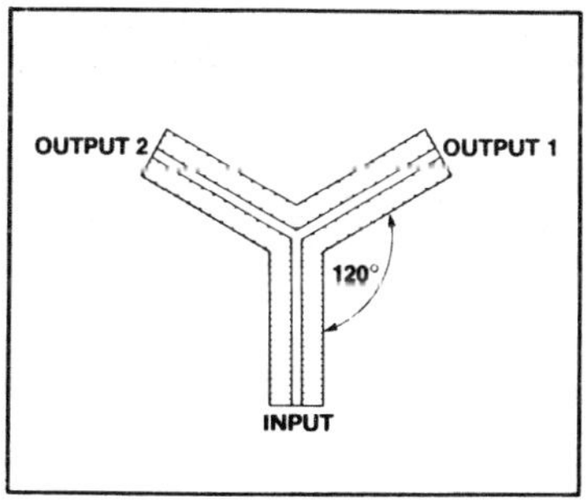

Fig. 1 Junction of main line and series branch.

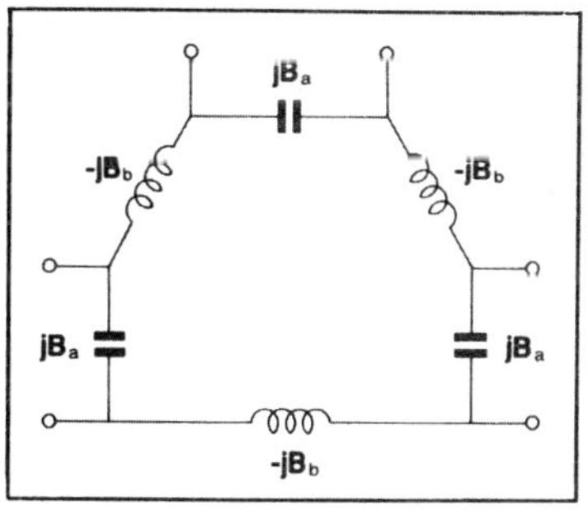

Fig. 2 Rectangular waveguide E-plane circuit.

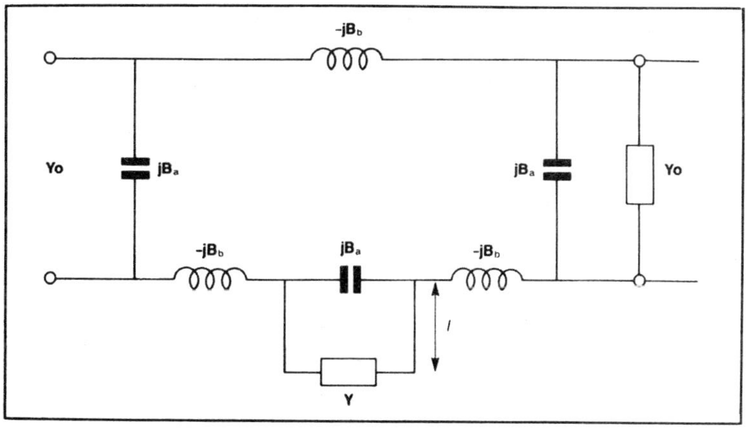

Fig. 3 Switch performance analysis equivalent circuit.

L.W. Koh, I.M.H. Williamson, M.L. Nyss, O. Gadoury, and D.E. Wheeler, *Microwave Journal*, Vol. 26, No. 6, June 1983, pp. 105–107.

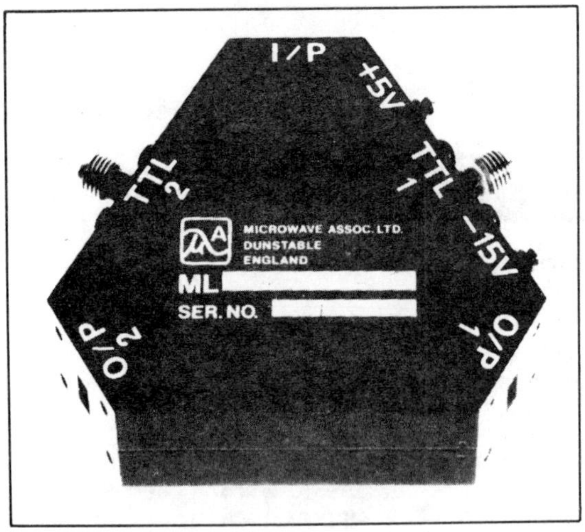

Fig. 4 Fin-line SPDT switch.

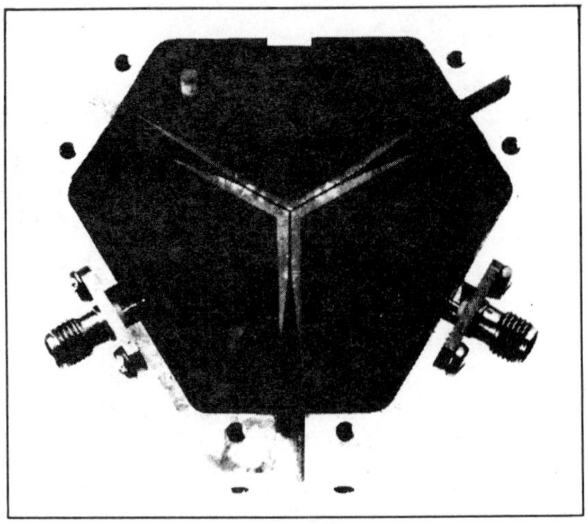

Fig. 5 Junction symmetry of fin-line SPDT switch.

the equivalent circuit can be approximately calculated.

$$\frac{Ba}{Yo} = (0.6455)\ \frac{2b}{\lambda g} \quad (1)$$

$$\frac{Bb}{Yo} = \frac{2\sqrt{3}}{\pi}\ \frac{\lambda g}{b} \quad (2)$$

The individual elements of Figure 3 can be used in conjunction with the following expression to determine the correct position of the forward biased PIN diode.

$$Z_{S.C.} = j\ Z_o \tan \beta l \quad (3)$$

The value of βl was calculated to be -0.184° or 179.816° away from the junction. It is not practical to mount the diodes at -0.184° from the junction and, thus, the first real position at which the PIN diodes could be placed, on the basis of the analysis, is a little less than $\lambda g/2$ away from the junction. At this position the bandwidth of the switch would be narrow, so this solution was not used. Therefore, the diodes were mounted as close to the junction as possible and this resulted in the performance reported in this article.

Construction

In fin-line structure, the halves of the housing are bolted together in the flange region and the dielectric board does not need to be under compression. A slot in the housing along the E-plane provides clearance for the board which is aligned by dielectric dowels. The thickness of the broad waveguide wall is chosen to be a quarter wavelength (including dielectric loading). This effectively shorts the fins to the housing at the inner edge of the broad wall and prevents radiation. The quarter-wave choke also suppresses TEM propagation by effectively connecting all conductors at the inner wall. Additional TEM suppression, where required, can be obtained by etching transverse slots into the conductor pattern within the choke region. A thick-film integral driver is included in the milled housing which accepts TTL-compatible drive pulses. The external view of the Fin-Line SPDT Switch is shown in Figure 4.

Problems in Spacing of Diodes

As expected from the theoretical analysis of the switch, the spacing of the first diode in the two output arms is critical. Furthermore, the symmetry of the junction is also critical and careful attention to the fabrication of the switch was thus essential. This is clearly shown in Figure 5. The second diode is spaced $\lambda g/4$ from the first one.

Moding

With conventional fin-line structures, it was found that higher order modes of propagation can exist and that this condition degrades the isolation performance, particularly at the top end of the frequency band. In order to suppress these modes, metal structures were inserted in the region of the switch where the active semiconductor elements are located. These effectively reduced the cutoff frequency of the trans-

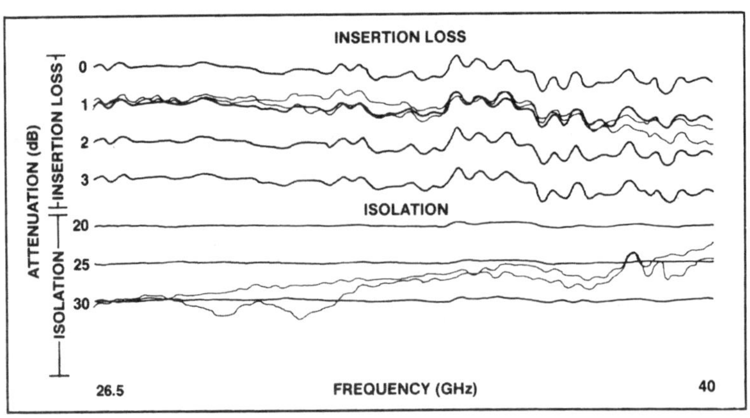

Fig. 6 SPDT fin-line switch (two diodes).

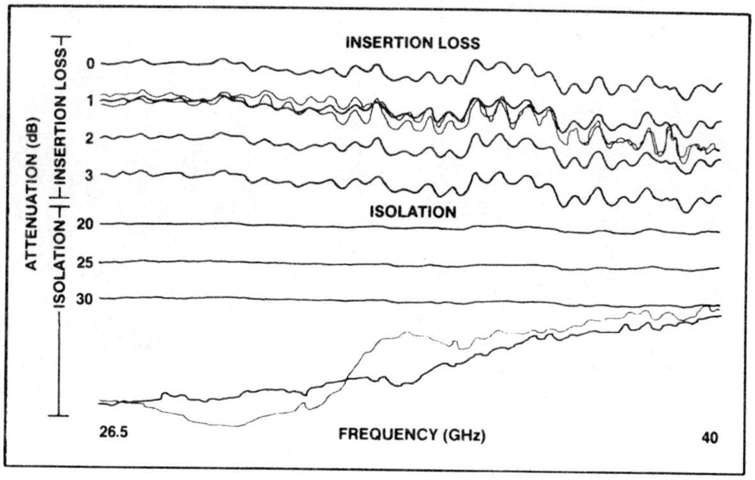

Fig. 7 SPDT fin-line switch (three diodes).

mission line and permitted a smooth isolation response, as shown in the data in Figures 6 and 7.

Switch Performance — Loss and Isolation

Figure 6 shows that the measured insertion loss of the SPDT switch, with two diodes in each of the output arms, is 1.75dB max. at the top end of the band, but is 1dB across most of the band at 26.5 to 40GHz. When forward bias is applied to the PIN diodes, the isolation is 23dB at the top end of the band, rising to 30dB at the lower end of the band. The measured return loss of the switch is greater than 15dB across the band. Figure 6 also shows the close tracking of the insertion loss between the two output arms, thus indicating that the symmetry of the design was successfully maintained in the manufactured switch.

The results obtained for the SPDT switch with three diodes in each of the output arms is shown in Figure 7. The measured insertion loss is 2dB at the top end of the band but is 1.3dB across most of the band. The isolation is 30dB across the whole band but is 35dB minimum over most of the band. From the measured results it has been clearly demonstrated that an additional diode can be introduced to obtain higher isolation with only a small increase in insertion loss. It is considered that the losses are predominantly due to the transitions from waveguide to fin-line, the fin-line circuit itself and the PIN diode closest to the junction.

Switching Speed

The switching speed, measured between 50 percent logic to 90 percent detected RF is 110nS. This is partially due to the capacitance loading of the fin-line structure and the driver delay.

Isolation — Insertion
50% logic — 90% detected RF 110nS
10% logic — 90% detected RF 85ns

Insertion — Isolation
50% logic — 10% detected RF 15ns
10% logic — 90% detected RF 6ns

The power handling capability of the switch is 100mW max.

Conclusion

Theoretical analysis has been used to predict the position of the diode in the "off" arm for optimum match of the symmetrical Y junction. A circuit was manufactured with diodes mounted as close to this position as is practical with the given physical configuration. Results are presented for a full waveguide band SPDT fin-line switch, thus confirming the analysis used in this article. Design problems such as moding and constructional difficulties were successfully overcome. The measured results indicate that higher isolation can be achieved by adding extra diodes without greatly degrading the insertion loss. ■

7.1.3

Computer-Aided Design of Fin-Line Detectors, Modulators, and Switches

by Hadia El Hennawy * and Klaus Schünemann *

Report from the Institut für Hochfrequenztechnik, Technische Universität Braunschweig

Based on both a rigorous eigenmode calculation and a modal analysis for discontinuities, a computer-aided design of some diode mounts in fin-line has been performed. Theoretical results show good agreement with measurements for a detector with a Schottky-diode, a reflection-type phase modulator with a pin-diode, and a switch with two pin-diodes. Impedance transformation has exclusively been performed with notch-type discontinuities which are capable of realizing a broad range of impedances.

Rechnergestützter Entwurf von Detektoren, Modulatoren und Schaltern mit der Flossenleitung

Ein rechnerunterstützter Entwurf einiger Diodenhalterungen in der Flossenleitung wird durchgeführt, der auf einer umfassenden Eigenwellenberechnung sowie auf einer Modenanalyse von Störstellen aufbaut. Für Detektoren mit Schottky-Diode, einen Reflexions-Phasenmodulator mit pin-Diode und einen Schalter mit zwei pin-Dioden ist die Übereinstimmung zwischen theoretischen und Meßergebnissen gut. Impedanztransformationen werden dabei ausschließlich mit Aussparungen im Flossenleitungsschlitz durchgeführt, womit Impedanzen in einem weiten Bereich realisiert werden können.

1. Introduction

The potential of fin-line technology in the realization of millimeter-wave integrated circuits has been demonstrated in the last five years. Various passive and active components have been developed including couplers, bandpass filters, tapers, detectors, switches, modulators, mixers, and oscillators [1]—[12]. Many of these components had to be designed by trial and error, because nearly all theoretical investigations published so far exclusively deal with the calculation of the propagation constants and the wave impedances. Both approximate [13], [14] and rigorous methods [15]—[19] have been developed for calculating the characteristics of unilateral, bilateral, and antipodal fin-lines. Recently, these investigations have been completed to encompass multi-slot and inhomogeneous fin-lines [20], [21]. There is, however, a lack of fundamental theoretical research concerning a description of most passive circuit elements which are used for impedance transformation and/or mounting a semiconductor device. We have recently presented an analysis of various discontinuities [22] which are often used for impedance transformation. The present work continues these investigations by using the equivalent circuits of the discontinuities in a computer-aided design of some semiconductor mounts.

The aim of this work is three-fold: 1) Typical circuit configurations will be analyzed for the particular semiconductor devices. 2) This analysis will characterize performance limits imposed by the circuit. (For practical examples we have therefore used devices which are commercially available but do not necessarily show a high performance.) 3) Design rules for mounting semiconductor devices will be outlined.

* Dipl.-Ing. H. El Hennawy, Dr.-Ing. K. Schünemann, Institut für Hochfrequenztechnik, Technische Universität, Postfach 3329, D-3300 Braunschweig.

2. Circuit Elements

Impedance transformation in fin-lines is mainly carried out by discontinuities in the slot width. A metallic strip which bridges the slot has been analyzed in [23]. It is used as a coupling element to transmission- and reflection-type resonators [6]. Nearly all other discontinuities can be composed from one or two steps in the slot width. Fig. 1 shows various steps (a, b, c), notches (d, e, f), strips (g, h, i), and transitions between bilateral and unilateral fin-lines (j, k). These discontinuities are used in bilateral and in unilateral fin-lines. Furthermore, the shapes of the upper and the lower metallic fin may or may not be congruent. Another parameter which further increases the variety of conceivable slot patterns is the distance between the slot and the broad side of the waveguide housing.

The equivalent circuits of these discontinuities have been determined from a modal analysis. According to [24] a step discontinuity in the slot width can be regarded as a junction between two different waveguides. The boundary conditions can be satisfied for two suitable infinite series of modes which are appropriate to each side of the junction. (These modes are well known from the theoretical papers cited above.) The properties of the junction can then be calculated if one finds out how power is

H. El Hennawy and K. Schunemann, *Archiv fur Elektronik und Ubertragungstechnik*, Vol. 36, No. 2, 1982, pp. 49–56.

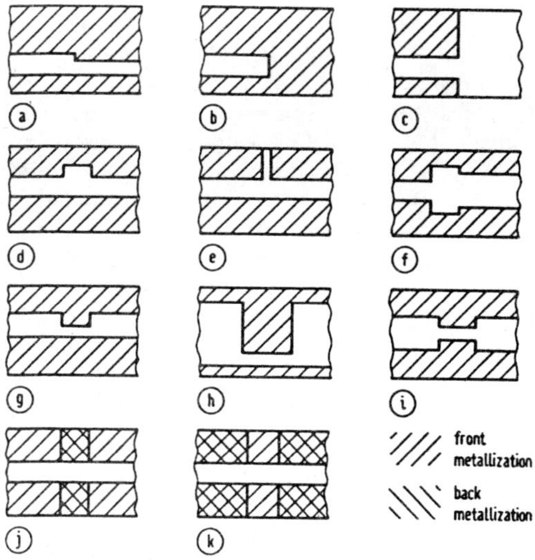

Fig. 1. Slot patterns of some fin-line discontinuities.

distributed between the various scattered modes. A solution of this problem has been developed in [24] and applied to the discontinuities of Fig. 1 in [22]. (A detailed version of this work has just been finished and will be submitted for publication [25].) Hence we can restrict ourselves here to characterizing the individual discontinuities by presenting numerical results for their equivalent circuits.

We will, in the following sections, describe the computer-aided design of those diode mounts which utilize notch-type discontinuities for impedance transformation. Applications include Schottky-diode detectors, phase modulators and switches with pin-diodes. (Components with strip mounts such as Schottky-diode mixers, VCOs with varactor

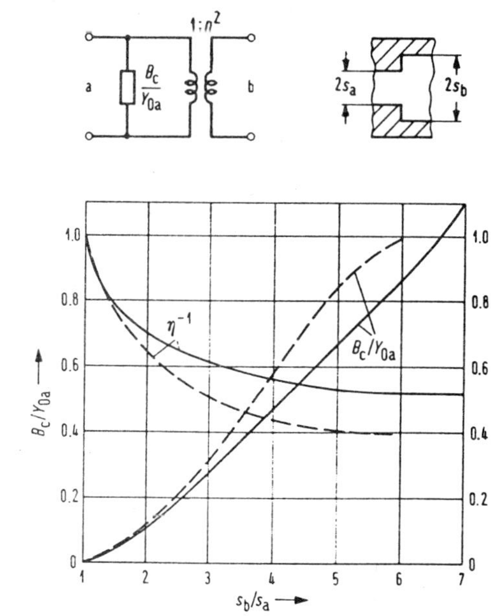

Fig. 2. Slot pattern, equivalent circuit, and performance of a step discontinuity at 30 GHz for a unilateral (solid lines) and a bilateral fin-line (dashed lines); $\eta = n^2 Y_{0b}/Y_{0a}$.

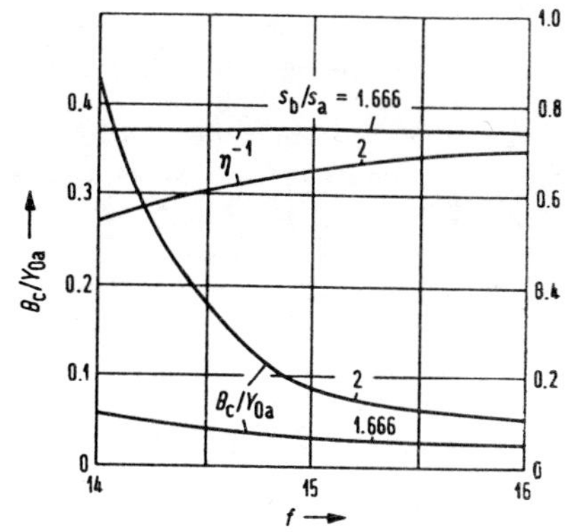

Fig. 3. Performance of a symmetrical step discontinuity in bilateral fin-line; $\eta = n^2 Y_{0b}/Y_{0a}$, $s_a = 0.2$ mm.

diode and Gunn elements, and high-power IMPATT-diode oscillators will be treated in a subsequent paper.)

If not stated otherwise, the numerical results refer to a fin-line in a standard WR-28 waveguide with cross section 7.1 mm × 3.56 mm. The dielectric substrate (RT-Duroid 5 880) is 0.254 mm thick and has a relative dielectric constant $\varepsilon_r = 2.22$. The frequency is $f = 30$ GHz. The slot pattern and the equivalent circuit of a step discontinuity are shown

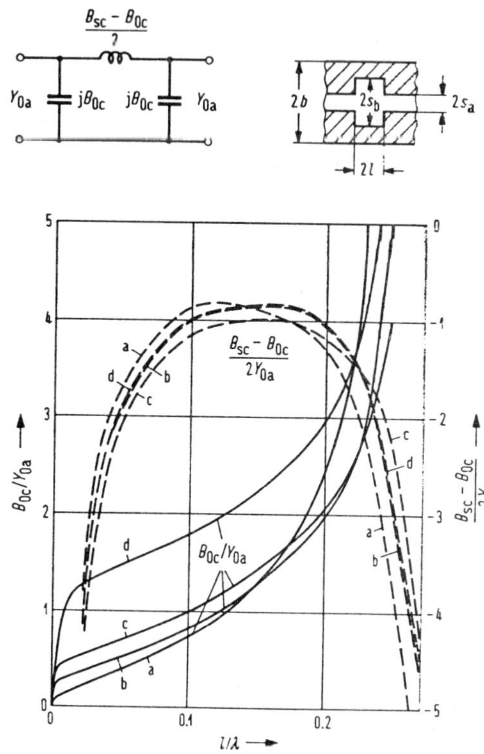

Fig. 4. Slot pattern, equivalent circuit, and performance of a notch in unilateral fin-line at 30 GHz; λ means wavelength;

(a) $s_b = 1$ mm, $s_a = 0.5$ mm,
(b) $s_b = 1.5$ mm, $s_a = 0.5$ mm,
(c) $s_b = 1.78$ mm, $s_a = 0.5$ mm,
(d) $s_b = 1.5$ mm, $s_a = 0.2$ mm.

in Fig. 2. The step can be characterized by a capacitive shunt susceptance B_c (which has been normalized to the wave admittance Y_{0a} of one of the fin-lines) and by an impedance transformer of turns ratio of n. Both elements depend strongly on the ratio of the slot widths. They are different for unilateral and bilateral fin-lines. The circuit elements depend, of course, on frequency as is illustrated in Fig. 3 for a symmetric step in the slot width (i.e., a step in both sides of the metallic fins) of a bilateral fin-line mounted in a standard WR-62 waveguide. (The detectors reported below have been realized at this frequency band.)

An equivalent Π-network has been chosen to characterize the performance of a symmetrical notch in unilateral fin-line at 30 GHz (Fig. 4). The series element turns out to be always inductive as long as the notch length is less than a half wavelength. This is plausible from a physical point of view. The slot width within the notch equals the full height of the waveguide for the curves labelled with "c". A ratio of the slot widths $s_b/s_a > 2$ mainly influences the series element. The shunt elements do not vary much with s_b/s_a; they are always capacitive. An important question is under which circum-

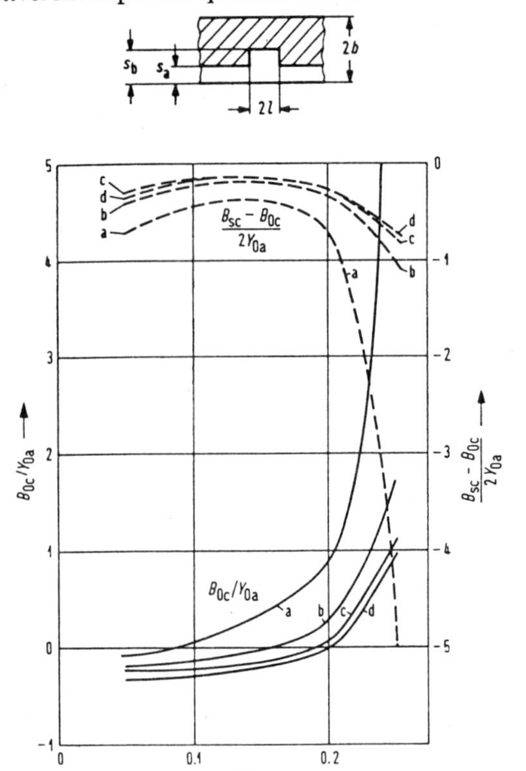

Fig. 5. Slot pattern and performance of an unsymmetrical notch in unilateral fin-line (same parameters as in Fig. 4).

stances higher-order mode coupling between the individual step discontinuities must be taken into account. Test calculations yielded as a rule of thumb that it might be neglected for $2l/\lambda > 0.1$. The thus introduced error is then less than 10 per cent.

The performance of a notch in unilateral fin-line is shown in Fig. 5. The magnitude of the shunt elements does not differ much from the corresponding values for the symmetric notch while the difference for the series elements is more pronounced. The degree of impedance transformation which can be achieved with either structure is nonetheless very similar, so that there is no preference for one or the other type of fin-line from this point of view. The frequency dependence of the equivalent circuit elements of a notch is illustrated in Fig. 6. It increases with the length of the notch.

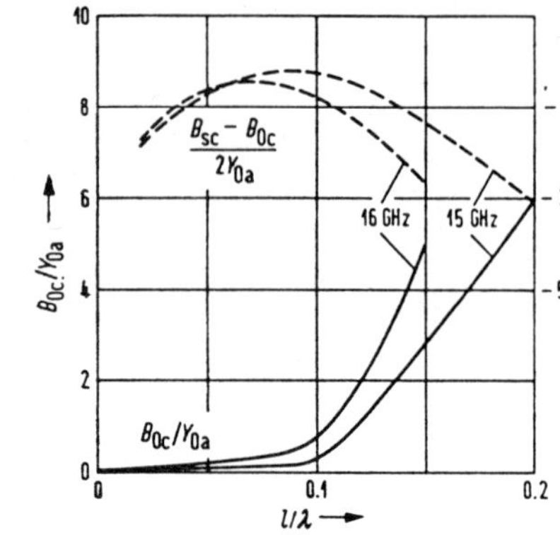

Fig. 6. Performance of a symmetrical notch in bilateral fin-line.

The last discontinuity which will be regarded here is a narrow slit in one or both of the metallic fins. It has already been mentioned that the slit is often used for dc-separation within a complex network. The analysis of extremely short notches shows, however, that resonance phenomena may occur prohibiting an application as a separating slit. This is illustrated by the results in Fig. 7. There are dimen-

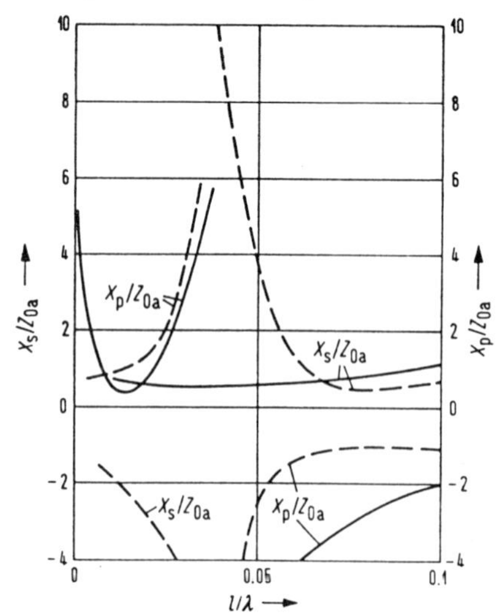

Fig. 7. Performance of a short notch (slit) in a unilateral fin-line at 30 GHz (X_s series, X_p parallel reactance in a T-circuit);

dashed lines: unsymmetrical slit, $s_a = 2.0$ mm, $s_b = 3.0$ mm;

solid lines: symmetrical slit, $s_a = 1$ mm, $s_b = 1.78$ mm.

sions for an unsymmetric notch, i.e. for a slit in only one fin, which lead to a bandstop performance, because the shunt reactance in a T-equivalent circuit shows series resonance behaviour. This effect which seems to be curious at first glance has also been observed for a post with gap mounted in a rectangular waveguide [26]. It could be attributed to a resonance between the inductance which is due to the excited TE_{m1}-modes and the capacitance which is due to the TM_{m1}-modes. (Similar effects occur, of course, also if the post is replaced by a planar axial strip [27].) The coupling to these modes becomes zero when the gap is centered in the waveguide cross section. For fin-line circuits one can conclude that the short notch should be centered in the waveguide height as well, in order not to perform an unwanted impedance transformation. Hence a dc-separating slit can only be realized with a symmetrical notch, while the corresponding unsymmetrical notch is very well suited for impedance transformation. It has in fact been used in this way [9].

3. Detectors

Schottky diodes are usually bonded across the fin-line slot in order to realize a broadband detector circuit. The fin-line is then terminated in a matched load. Full-band operation can thus be achieved at the sacrifice of minimum detectable power and tangential sensitivity. The design rule then is to match the slot width to the length of the diode package, in order not to enhance circuit parasitics. We have investigated the bandwidth restrictions which would occur due to a short-circuited or an "open-circuited" stub behind the diode (these discontinuities have been sketched in Fig. 1b and c). Then full-band operation is no longer possible, if bandwidth is defined by a 1 dB decline in i.f. signal but the bandwidth may still exceed 30 percent of a waveguide band. More important is the case of a matched detector. The stub is then used for tuning out the imaginary part of the diode impedance, while the real part is matched by placing a notch network directly in front of the diode. The bandwidth is then, however, limited to less than 10 percent.

We will in the following demonstrate a computer-aided design of matched detectors. Two mounts have been designed for a Schottky diode (HP 5082-2709). The first task is to determine the impedance of the diode as mounted in the final circuit. We have chosen a unilateral fin-line with a slot width of 0.3 mm which corresponds to the physical size of the diode, and hence minimizes the bonding wire inductance. The measurements have been performed at Ku-band in a standard WR-62 waveguide. The diode impedance has then been approximated, in a ± 1 GHz-band around 15.3 GHz, by a constant real and a linearly varying imaginary part which form the input data for the CAD-procedure. The length of the short-circuited stub behind the diode is chosen so that the diode reactance is tuned out. Two different networks have

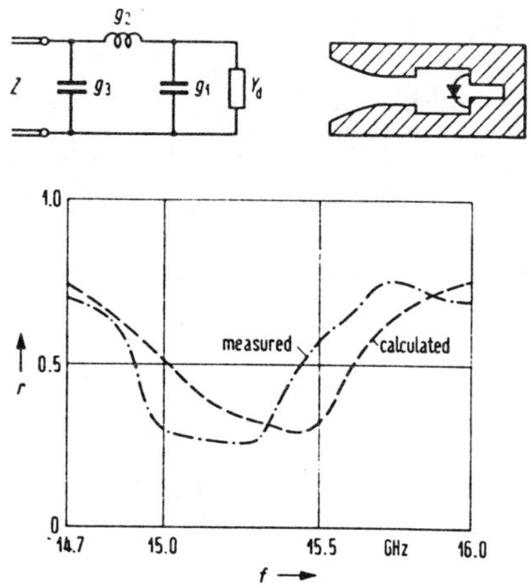

Fig. 8. Slot pattern, equivalent circuit, and input reflection coefficient of a detector in unilateral fin-line.

been designed for matching the diode resistance. Results for a notch with three different slot widths are shown in Fig. 8. The notch can be modelled by an unsymmetrical Π-network, Y_d means the sum of the diode and the stub admittances. The method of finding the g-parameters of the transforming section is described in section 4 of [28]. The problem we are concerned with here is to realize this circuit with fin-lines. The unknown parameters are the slot widths and the length of the notch. They are determined in the following way: The series parameter g_2 is directly proportional to the difference between the short- and open-circuit impedances of the notch section, which depends only on the length of the notch but not on the widths if higher-order mode coupling is neglected. The length can hence be calculated from g_2 (by using e.g. the results presented in Fig. 4). The slot width within the notch could be determined from g_1, if the propagation constant were known. This problem can be solved iteratively starting with an assumed value for the propagation constant. The whole process, which includes calculating corrected values for the notch length (taking higher-order mode coupling into account), converges within two or three steps. The slot width at the output is finally determined from g_3. A comparison between calculated and measured values of the input reflection coefficient (Fig. 8) shows a good agreement. In comparison to a non-matched mount, the minimum detectable power could be lowered by up to 8 dB. (The relatively large mismatch of the reflection coefficient is due to the strong frequency dependence of the diode admittance, which is only roughly modelled by a straight line approximation. This is illustrated by numerical results summarized in Appendix 1.)

A three-section transformer in front of the diode has been used in the second example. Slot pattern, equivalent circuit, and input reflection coefficient are shown in Fig. 9. The design proceeds in the following way: Input data are Z_d, the impedance of the diode plus that of the stub, the load R_A (i.e.

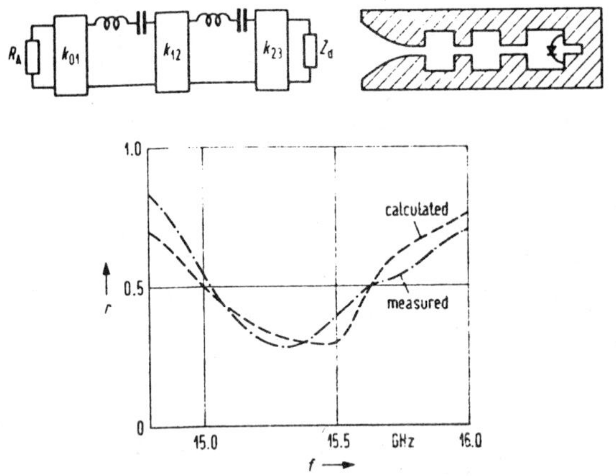

Fig. 9. Slot pattern, equivalent circuit, and input reflection coefficient of a detector in unilateral fin-line.

the wave impedance), and the number of the notches. The k-parameters of the impedance inverters are calculated in Appendix 2. Each inverter must be realized by a notch, which is represented by a T-equivalent network with series reactances X_a and shunt reactance X_b. Introducing a further parameter φ (which means the electrical length of a line section adjacent to the inverter), two relations can be established between k, φ, X_a, and X_b [28]. (The notation is the same as in [28].) They can be solved for X_a and φ provided that X_b is known. This problem has been solved iteratively by utilizing the fact that the plot of the shunt reactance X_b versus the notch length l resembles a parabola (Fig. 4). We choose now a suitable value for the slot width within the notch and assume by trial an X_b which corresponds to the vertex of the parabola. This X_b is then insensitive to variations in l. The two equations cited above can now be solved for φ and X_a which finally fixes l. The value originally chosen for X_b can now be corrected and the total cycle repeated. The procedure must be carried out for each notch. The distances between the notches are then given by combining the individual φ-parameters. One must now decide whether the dimensions of the notches are appropriate for realization, and whether the distances between the notches are large enough that the implied neglect of higher-order mode coupling was correct. If not, the total cycle must be repeated.

Regarding the performance of the optimized detector circuit, the agreement between calculations and measurements was excellent (Fig. 9).

4. Phase Modulators

Digital phase modulators are usually realized with pin-diodes which are characterized by short switching times, large power-handling capability, and almost linear steady-state impedances. Reflection-type modulators are preferred because their bandwidth is generally larger than for transmission-type modulators. Available packages for pin-diodes are similar or even identical to those for Schottky diodes. Hence both mounts for a pin-diode in a modulator and for a Schottky diode in a detector are similar. The slot width of the fin-line should correspond to the physical size of the pin-diode in order to minimize the bonding wire inductance. In fact, the similarity of both diode mounts is even more far-reaching: both mounts must match a certain diode impedance to the load. In the case of the pin-diode, this impedance is a well-defined, weighted average of the two steady-state impedances.

The task of the pin-diode embedding will now be defined taking a π-modulator as an example. This modulator is characterized by its two input reflection coefficients which show equal magnitude but opposite phase. The reflection coefficients of the pin-diode itself do not fulfill this condition due to both diode and package parasitics. Hence the diode embedding must simultaneously transform both steady-state impedances in such a way that a π-modulator is realized. It has been shown in [29] that this task is equivalent to matching the so-called "hyperbolic middle-point impedance" of the pin-diode. This particular impedance which defines the middle-point between the steady-state impedances in non-Eucledian geometry is uniquely related to these impedances. The diode embedding of a π-modulator is hence an impedance-matching network. It has been shown in [29] (and explained in detail in [10]) that this statement holds also for a phase modulator with arbitrary phase shift; the hyperbolic middle-point impedance must only be replaced by another characteristic impedance.

The guidelines for designing a phase modulator can now be formulated: One chooses a final slot width for mounting the pin-diode, measures its two steady-state impedances versus frequency, calculates the characteristic impedance according to [29], and linearizes its frequency dependence in the same way as described above for the Schottky-diode. The characteristic reactance of the pin-diode is tuned out by a short-circuited stub behind the diode, while a notch structure is designed for matching the characteristic resistance. The procedure is just the same as for the detector circuit.

Performance data of a π-modulator are presented in Fig. 10 for the pin-diode DSG 6470 C of Alpha Industries. The fin-line is unilateral with slot width 0.25 mm. A single notch has been chosen for matching the diode impedance. The fin-lines on either side of the notch have unequal impedances in order to enhance the bandwidth. A comparison between theory and experiment shows again an excellent agreement (Fig. 11). Details are given in Appendix 3. The performance of a $\pi/2$-modulator is shown in Fig. 12.

5. Switches

Another important application of pin-diodes lies in the realization of switches. A switch has at least two ports, so that its realization differs from the mounts discussed so far. Possible configurations

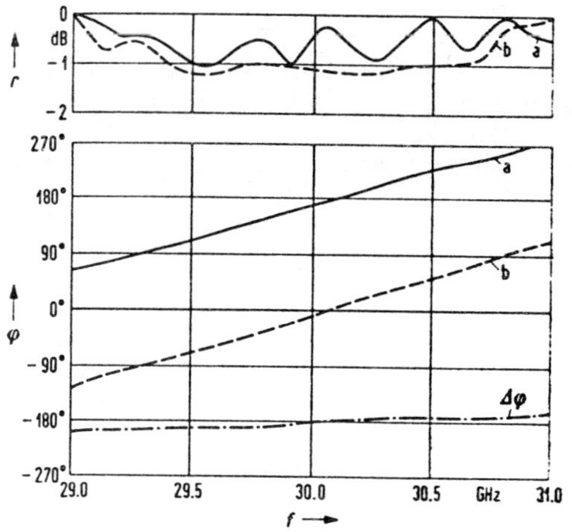

Fig. 10. Magnitudes r and phase angles φ of the input reflection coefficients (diode states a and b) and phase shift $\Delta\varphi$ of a π-modulator in unilateral fin-line.

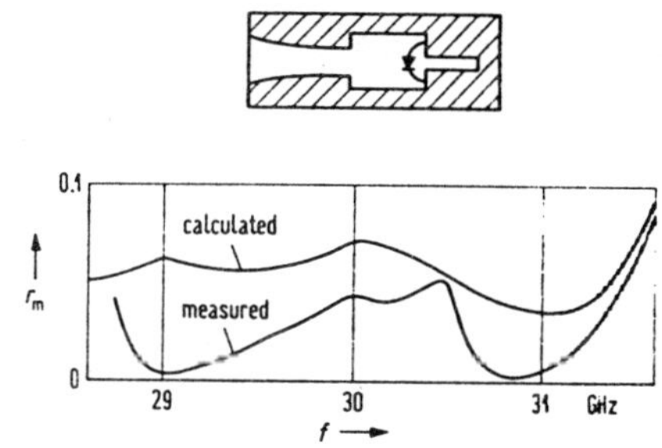

Fig. 11. Slot pattern and magnitude r_m of the middle-point reflection coefficient of the π-modulator of Fig. 10.

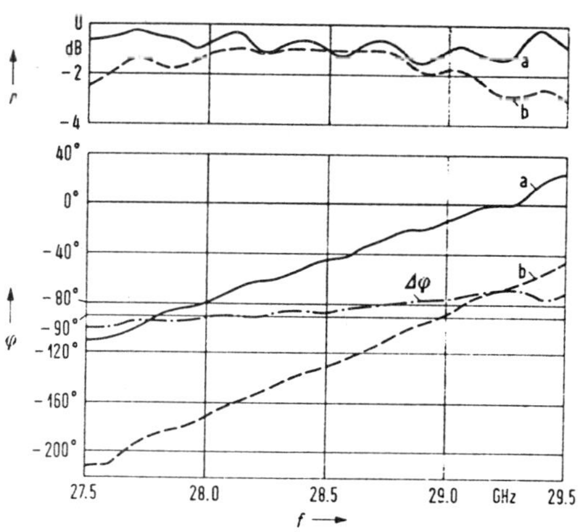

Fig. 12. Magnitude r and phase angles φ of the input reflection coefficients (diode states a and b) and phase shift $\Delta\varphi$ of a $\pi/2$-modulator in unilateral fin-line.

have recently been compared in [30]. More than one pin-diode is usually used in order to obtain high isolation. Broadband performance is achieved if the pin-diodes are mounted across the fin-line slot (shunt configuration). Impedance-matching can then be performed by strip discontinuities. This type of switch is hence beyond the scope of this paper. We will here design a configuration with diodes mounted in series with the fin-line. The slot pattern and equivalent circuit are sketched in Fig. 13. This switch is known to be a narrow-band device [30].

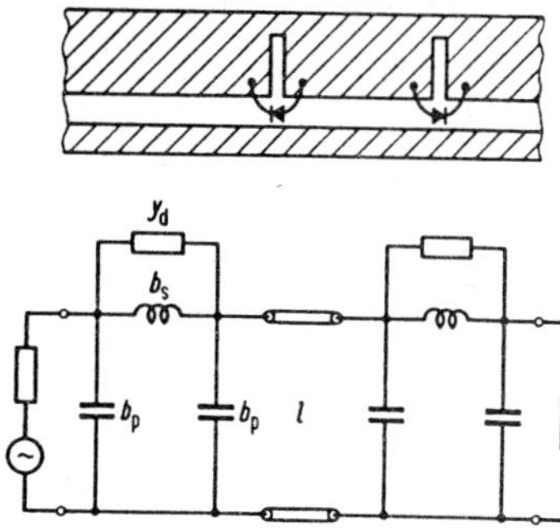

Fig. 13. Slot pattern and equivalent circuit of a switch.

The pin-diodes are directly soldered across the short notch section of length 0.25 mm, which corresponds to their physical size. The diode admittance y_d must be measured in a fin-line of equal slot width. The notch can be characterized by a Π-network with purely inductive series and capacitive shunt elements. The dimensions of the notches and their distance l are calculated for a given y_d in the following way: The slot width of the main fin-line should be as small as possible so that y_d in the forward-bias state is much larger than the wave admittance. This yields minimum insertion loss for the switch closed. According to the design procedure given in [31] y_d in the reversed-bias state should be tuned out by b_s. This determines the slot width in the short notch section (i.e. the "stub length" in the terminology of [30]). Associated with b_s are certain values for the shunt elements b_p. The two b_p which are adjacent to the line between the notches must be incorporated into the line, so that the effective electrical length of this section equals a quarter wavelength. (Details are given in Appendix 4.)

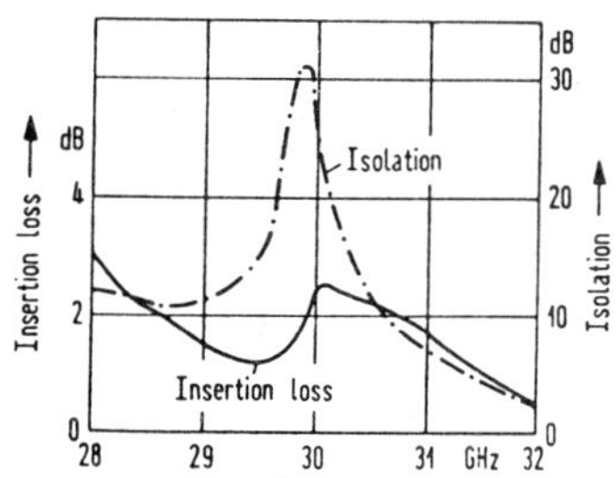

Fig. 14. Performance of the switch of Fig. 13 in a unilateral fin-line; pin-diode 47 301 of Microwave Associates.

Performance curves for a Ka-band switch are given in Fig. 14. The switch has been designed for a single frequency, 30 GHz. Theoretical values are 0.5 dB for the insertion loss and 26 dB for the isolation. One recognizes that these design goals are nearly met at a frequency which is 150 MHz lower than the desired 30 GHz. This slight shift can be attributed to the influence of the slit in the waveguide housing which has not been taken into account in the eigenmode analysis. The larger insertion loss is due to errors in the measurement of y_d.

6. Conclusions

Impedance-matching is usually performed in fin-lines by means of discontinuities in the slot width. Several discontinuities have been investigated in order to determine their potential for impedance transformation. It has been shown that the notch and its modifications (open circuit and short slit) are versatile instruments for tuning and matching semiconductor devices. Schottky diode detectors, digital phase modulators and narrow-band switches (both with pin-diodes) have been theoretically designed. Good agreement with measured performance data has been achieved. Hence a computer-aided design of fin-line circuits, which is essential for an extension of this promising technology, has been demonstrated.

Acknowledgement

The authors are indebted to the Deutsche Forschungsgemeinschaft for financial support.

(Received September 3rd, 1981.)

Appendix 1

Measuring the (normalized) diode admittance y at a constant current of 1 mA versus frequency f yielded

$$f = 14\,\text{GHz}, \qquad 15\,\text{GHz}, \qquad 16\,\text{GHz},$$
$$y = 1.24 + \text{j}1.61, \quad 2.23 + \text{j}2.67, \quad 3.56 + \text{j}4.71.$$

Compensating the imaginary part b_{15} at 15 GHz with a short-circuited stub according to

$$\cot \beta (l + \Delta l) = b_{15}$$

(Δl means distance between real and effective short-circuit plane) one finds for the load y_d (see Fig. 8)

$$y_d = 1.24 - \text{j}1.06, \quad 2.23, \quad 3.56 + \text{j}2.04.$$

Following [28], sec. 4, and using the same notation, the lowpass prototype for the bandpass matching network can be calculated. Choosing $n = 3$, one finds

$$G_A = 2.23, \quad g_0 = 1, \quad g_1 = 2.9, \quad g_2 = 0.575,$$
$$g_3 = 1.9, \quad g_4 = 0.448,$$
$$B_L = G_A/g_2 = 3.878, \quad B_{c1} = G_A g_1 = 6.467,$$
$$B_{c0} - G_A g_3 = 4.237.$$

These values are realized by a fin-line slot pattern with dimensions

$$s_a = 0.3\,\text{mm}, \quad s_b = 4.5\,\text{mm}, \quad s_c = 1.4\,\text{mm},$$
$$l_b = 5.47\,\text{mm},$$
$$l = 0.75\,\text{mm} \quad \text{and} \quad \Delta l = 0.33\,\text{mm}.$$

Appendix 2

The measured diode admittance (given in Appendix 1) shall be matched using k-inverters. According to [28], sec. 4, we find for $n = 2$, $R_A = 0.448$, $R_B = 1$, and $\delta_s = 1.727$ the following prototype elements:

$$g_0 = 1, \quad g_1 = 0.58, \quad g_2 = 0.43, \quad g_3 = 1.15,$$
$$\omega_\lambda = 0.187, \quad \lambda_{g0} = 1.89,$$
$$\omega = 0.133, \quad \omega_1' = 1, \quad k_{01} = 0.4769,$$
$$k_{12} = 0.589, \quad k_{23} = 0.7714.$$

Each k-inverter is realized by a notch with corresponding slot widths $s_b/s_a = 8$, $s_b = 2.4$ mm. The relations between the k-values, the series (jX_a) and shunt reactances (jX_b) of a T-equivalent circuit, and the contribution φ to the electrical length of the adjacent sections are given in [28]:

$$k = 1/2 \tan[-\arctan(2X_b + X_a) + \arctan X_a],$$
$$\varphi = -\arctan(2X_b + X_a) - \arctan X_a.$$

Taking k from above and assuming an initial X_b, the first relation has been solved for X_a. Then φ is specified by the second relation. The line length d_j between two notches is then given by

$$2\pi d_j/\lambda_g = \pi + (\varphi_{j-1,j} + \varphi_{j,j+1})/2.$$

The various line sections (see insert of Fig. 9) are $l = 0.75$ mm, $2l_{01} = 9.85$ mm, $\varphi_{01} = -0.89$, $d_1 = 5.87$ mm, $2l_{12} = 8.82$ mm, $\varphi_{12} = -1.065$, $d_2 = 6.51$ mm, $2l_{23} = 7.41$ mm, $\varphi_{23} = -1.314$.

Appendix 3

The two steady-state impedances $Z_1 = R_1 + \text{j}X_1$, Z_2 of the pin-diode are measured at frequencies 28, 29, and 30 GHz. The diode is bonded across a fin-line with slot width 0.25 mm. The middle point impedance Z_m is calculated from [29]

$$R_m = \sqrt{R_1 R_2}\sqrt{1 + (X_1 - X_2)^2/(R_1 + R_2)^2},$$
$$X_m = (R_1 X_2 + R_2 X_1)/(R_1 + R_2).$$

One yields for $Y_m = 1/Z_m$

$$f = 28\,\text{GHz}, \qquad 29\,\text{GHz},$$
$$Y_m = 1.087 + \text{j}0.665, \quad 1.157 + \text{j}0.428,$$
$$f = 30\,\text{GHz},$$
$$Y_m = 0.909 + \text{j}0.409.$$

A stub length l is calculated as in Appendix 1. One obtains $l = 1.95$ mm, $\Delta l = 0.1$ mm. The parameters for a lowpass prototype matching network with $n = 3$ and an average G_A are

$$G_A = 1.1, \quad g_0 = 1, \quad g_1 = 0.83, \quad g_2 = 0.95,$$
$$g_3 = 0.67, \quad g_4 = 0.909,$$
$$G_B = 1, \quad B_L = 1.158, \quad B_{c1} = 0.913,$$
$$B_{c3} = 0.737.$$

This yields the dimensions

$$s_a = 0.25\,\text{mm}, \quad s_b = 0.9\,\text{mm}, \quad s_c = 0.474\,\text{mm},$$
$$l_b = 3.924\,\text{mm}.$$

Appendix 4

The steady-state impedances in on- and off-state are measured at 30 GHz in a slot of width 0.25 mm. One obtains

$$y_d(\text{on}) = 1.675 - j4.05,$$
$$y_d(\text{off}) = 0.150 + j0.955.$$

The notch length is taken as 0.25 mm. Tuning out the susceptance in off-state requires $s_b = 1.5$ mm, which leads to $b_p = 0.2098$. The distance between the notches is then calculated from

$$\beta l = \arctan(1/b_p).$$

Using the $ABCD$-matrix, the insertion loss is given by [31]

$$\text{I.L.} = (A + B + C + D)^2/4.$$

Numerical results are

$$\text{I.L.(on)} = 0.518\,\text{dB}, \quad \text{I.L.(off)} = 25.97\,\text{dB}.$$

References

[1] Meier, P. J., Integrated fin-line millimeter components. Transact. IEEE MTT-22 [1974], 1209—1216.

[2] Cohen, L. D. and Meier, P. J., Advances in E-plane printed millimeter-wave circuits. IEEE-MTT Symp. Dig., Ottawa, 1978, pp. 27—29.

[3] Hofmann, H., Meinel, H., and Adelseck, B., New integrated millimeter-wave components using fin-lines. IEEE-MTT Symp. Dig., Ottawa, 1978, pp. 21—23.

[4] Begemann, G., An X-band balanced fin-line mixer. Transact. IEEE MTT-26 [1978], 1007—1011.

[5] Meinel, H. and Rembold, B., New millimeter-wave fin-line attenuators and switches. IEEE-MTT Symp. Dig., Orlando, 1979, pp. 249—252.

[6] Knöchel, R., Design and performance of microwave oscillators in integrated fin-line technique. IEE J. Microwaves, Opt. and Acoust. 3 [1979], 115—120.

[7] Saad, A. M. K. and Schünemann, K., Design and performance of fin-line bandpass filters. Proc. 9th Europ. Microwave Conf., Brighton, 1979, pp. 397—401.

[8] Cohen, L. D., Advances in printed millimeter-wave oscillator circuits. IEEE-MTT Symp. Dig., Washington D. C., 1980, pp. 264—266.

[9] Hofmann, H., Millimeter-wave Gunn oscillator with distributed feedback fin-line circuit. IEEE-MTT Symp. Dig., Washington D. C., 1980, pp. 59—61.

[10] Kpodzo, E., Schünemann, K., and Begemann, G., A quadriphase fin-line modulator. Transact. IEEE MTT-28 [1980], 747—752.

[11] Knöchel, R. and Schlegel, A., Octave-band double-balanced integrated fin-line mixers at millimeter-wavelengths. Proc. 10th Europ. Microwave Conf., Warszaw, 1980, pp. 722—726.

[12] Meinel, H. and Schmidt, L.-P., High sensitivity millimeter-wave detectors using fin-line technology. 5th Internat. Conf. Infrared and Millimeter-Waves, Würzburg. 1980.

[13] Saad, A. M. K. and Begemann, G., Electrical performance of fin-lines of various configurations. IEE J. Microwaves, Opt. and Acoust. 1 [1977], 81—88.

[14] Hofmann, H., Dispersion of planar waveguides for millimeter-wave applications. AEÜ 31 [1977], 40—44.

[15] Hoefer, W. J. R., Fin-line design made easy. IEEE-MTT Symp. Dig., Ottawa, 1978, p. 471.

[16] El-Sherbiny, A. and Said, M., Spectral domain calculation of the fin-line dispersion characteristics. 3rd Internat. Congress for Statistics and Computer Science, Cairo, 1978.

[17] Chang, C. and Itoh, T., Spectral domain analysis of dominant and higher-order modes in fin-lines. IEEE-MTT Symp. Dig., Orlando, 1979, pp. 344—346.

[18] Schmidt, L.-P. and Itoh, T., Spectral domain analysis of dominant and higher-order modes in fin-lines. Transact. IEEE MTT-28 [1980], 981—985.

[19] Beyer, A. and Wolff, I., A solution of the earthed fin-line with finite metallization thickness. IEEE-MTT Symp. Dig., Washington D. C., 1980, pp. 258—260.

[20] Schmidt, L.-P., Itoh, T., and Hofmann, H., Characteristics of unilateral fin-line structures with arbitrarily located slots. IEEE-MTT Symp. Dig., Washington D. C., 1980, pp. 255—257.

[21] Beyer, A. and Wolff, I., Calculation of the transmission properties of inhomogeneous fin-lines. Proc. 10th Europ. Microwave Conf., Warszaw, 1980, pp. 322—326.

[22] El Hennawy, H. and Schünemann, K., Analysis of fin-line discontinuities. Proc. 9th Europ. Microwave Conf., Brighton, 1979, pp. 448—452.

[23] Saad, A. M. K. and Schünemann, K., A simple method for analyzing fin-line structures. Transact. IEEE MTT-26 [1978], 1002—1007.

[24] Wexler, A., Solution of waveguide discontinuities by modal analysis. Transact. IEEE MTT-15 [1967], 508—517.

[25] El Hennawy, H. and Schünemann, K., Impedance transformation in fin-lines. Submitted for publication to IEE J. Microwaves, Opt. and Acoust., 1981.

[26] Eisenhart, R. L. and Khan, P. J., Some tuning characteristics and oscillation conditions of a waveguide-mounted transferred-electron diode oscillator. Transact. IEEE ED-19 [1971], 1050—1055.

[27] Chang, K. and Khan, P. J., Equivalent circuit of a narrow axial strip in waveguide. Transact. IEEE MTT-24 [1976], 611—615.

[28] Matthaei, G. L., Young, L., and Jones, E. M. T., Microwave filters, impedance-matching networks, and coupling structures. McGraw-Hill Book Co., New York 1964, Sec. 4.8.

[29] Dorschner, T. A., Characterization of reflection phase modulators using hyperbolic geometry. Proc. 3rd Europ. Microwave Conf., Bruxelles, 1973, pp. A.9.1—A.9.4.

[30] Meinel, H., Adelseck, B., and Callsen, H., A survey of planar integrated millimeter-wave components. Military Microwaves Symp. Dig., London, 1980, pp. 82—87.

[31] White, J. L., Semiconductor control. Artech House, Dedham, Mass. 1977, Sec. 4.

7.1.4

INTEGRATED MICROSTRIP AND ANTIPODAL FIN-LINE LEADS TO ENHANCED PERFORMANCE P.I.N. DIODE COMPONENTS.

R.B. GREED

ABSTRACT

Distributed microstrip line tuning elements are used in conjunction with fin-line techniques to design p.i.n. diode switches for the 26 - 110 GHz frequency band. Experimental results are presented for components utilising the forward and reverse switching modes operating at centre frequencies of 34 GHz and 94 GHz.

INTRODUCTION

A great many papers have been published in recent years which have demonstrated the versatility of fin-line techniques as applied to millimetric components and sub-systems. In a survey paper [1] no less than 94 papers were referenced which covered most millimetric circuit functions. The main advantages of the technique are that it combines the low-loss properties of waveguide with the integration capabilities of planar transmission lines, noteably hybrid microstrip. The technique also benefits from a reduced dependency upon critical housing dimensions [2]. The critically toleranced areas of the circuit are removed to the substrate which itself is fabricated, using well controlled and accurate photolithographic techniques. This feature becomes increasingly more important as frequency is increased with the corresponding decrease in linear dimensions. Furthermore, the integration of semiconductors, in particular diodes, within the fin-line can be achieved with ease.

P.I.N. DIODE SWITCHES

The beam lead packaged p.i.n. diode has been shown to be particularly useful to millimetric switch designs using the fin-line techniques. The combined attributes of small size, low series inductance, low junction capacitance and relatively low forward and reverse bias series resistance have been clearly demonstrated in past papers. Within these papers various approaches are discussed which lead to methods of overcoming the parasitic elements of currently available beam lead diodes. In one of the earliest papers [3], Meier

GEC Research Laboratories, Marconi Research Centre, Great Baddow, Essex, England.

successfully mounted beam lead diodes in shunt with the fin-line to produce an attenuator, this component achieved broadband performance. The attenuator exhibited an insertion loss of 1,5 dB and a dynamic range in excess of 30 dB over a frequency range of 36 - 42 GHz. Good input match was achieved by binomial matching an array of four diodes. An alternative approach used by [4] makes use of a series diode in conjunction with a parallel slot line stub. The combination of series diode and stub forms a parallel resonant circuit under forward bias and high attenuation results. Low attenuation is achieved in the reverse bias state. Here the stub is loaded only by the low capacitance diode junction. The technique was demonstrated in a two diode 90 GHz attenuator which has an insertion loss of 2 dB and a dynamic range in excess of 40 dB. The increased dynamic range is achieved at the expense of reduced bandwidth. A third approach [5] makes use of an array of microstrip resonators to which the beam lead diodes are mounted. The resonator-diode elements are etched on the reverse side of a unilateral fin-line. For 94 GHz operation a three element array resulted in an insertion loss of 3 dB and a dynamic range of 17 dB.

The performance demonstrated by these components can clearly satisfy many of the current applications. However, to take account of the inevitable manufacturing tolerances to hardware and device parameters, together with environmental requirements it is desirable to maintain the performance over an increased bandwidth. By doing so the tuning of individual components can be reduced or eliminated. In our work distributed line microstrip tuning elements have been used to achieve this goal. The technique has been applied to components for operation in the 26 to 110 GHz frequency band.

The microstrip elements are integrated with an antipodal form of fin-line upon a single 254 μm thick RT-Duriod substrate mounted in the E-plane of a waveguide in the conventional way. Each fin of the antipodal fin-line forms the ground plane of a microstrip structure, the tuning elements are etched on the corresponding opposite side of the substrate over this ground plane. Similarly integral microstrip low-pass bias filter networks are incorporated with each diode mount to allow individual diode biasing. Accurate alignment of the microstrip patterns and the fin-line ground planes is achieved using conventional double sided printed circuit board techniques. The design of the microstrip tuning elements follows the practices used at microwave frequencies. In particular, the width of each microstrip element is kept small to maintain the distributed line characteristics.

A design for a forward mode switch for the 30 - 38 GHz frequency range has been implemented. In this design tuning is applied only to the forward biased state to tune out the effects of the diode series inductance. An open circuit stub is used, the length of the stub is determined from standard transmission line equations and the diode inductance. Compensation for the open circuit end effect is applied. The resultant forward biased diode impedance shunting the fin-line becomes real and low valued which results in high transmission loss. In the reverse bias state, the tuning element has little effect on the overall diode circuit and the fin-line is now shunted by a high capacitive reactance resulting in low transmission loss. At the frequencies considered here it is not necessary to tune the reverse bias state. A broadband match is achieved by properly spacing the diodes. A microwave circuit analysis program in conjunction with diode equivalent circuits has been used to determine the optimum spacings. A layout of one side of the circuit is shown in Figure 1, the reverse side is similar. In Figure 2a the transmission characteristics for the switch are shown, Figure 2b shows the reverse bias return loss characteristics. It is seen that over the frequency range 30 - 38 GHz the isolation is ⩾ 50 dB with a midband insertion loss of 1 dB rising to 2 dB at the band edges. A return loss in excess of 18 dB have been achieved over most of the band. The switch has been subjected to thermal cycling between -55°C and +125°C and showed no degradation after the final test.

Similar tuning elements have been applied to diode switches for the 75 to 110 GHz band. Owing to the increased effect of the diode parasitics at these higher frequencies, it is necessary to tune both bias states to achieve good performance. The diode is tuned to present a series or parallel tuned circuit in shunt with the fin-line. Forward and reverse mode switching circuits have been realised. Both types require similar tuning elements, but are applied to opposite bias states. The capacitive tuning makes use of an open circuited line as previously used, the inductive tuning is achieved through the use of a short circuited stub. This is realised by laser drilling a small hole through the substrate, a fine gold wire is then bonded between the stub and the ground plane. Owing to the self inductance of the bondwire it has been found necessary to determine the actual length required to tune the diode empirically. Any excess stub length behind the short circuit has also to be removed for repeatable performance. Open circuited stubs whose length lies between one quarter and one half wavelength could be used to realise the required inductance, however, they are in this instance physically impractical and would lead to a more frequency sensitive solution.

As both states of the diode are tuned for optimum conditions, an iterative switch design is adopted for multiple diode arrays. The diodes are spaced one quarter wavelength apart. Two diode switches have been investigated. Both diodes are individually biased in a way to avoid using the bond wire as a bias return. Figure 3 shows the general layout of the substrate for both modes of switching. The circuits in this case are etched on 127 μm thick RT-Duroid. Figures 4 and 5 show the performance of a reverse and a forward mode two diode switch. The reverse mode switch exhibits a very broad band insertion loss characteristic. Typically a midband loss of 1,5 dB at 40 mA forward bias is achieved. The band edge loss is in the order of 3 dB. Forward bias return loss is similarly broadband being ⩾ 15 dB over a major part of the waveguide band. The reverse bias isolation exceeds 20 dB over a 7 GHz band. By comparison the forward mode switch exhibits complementary characteristics, that is a broadband isolation is achieved at the expense of insertion loss and return loss bandwidth. Typically the midband loss is 3 dB over 6 GHz, outside this range the insertion loss rises rapidly. The return loss is typically ⩾ 15 dB over a similar bandwidth. The isolation however is ⩾ 20 dB over a greater bandwidth. Figure 5 shows that further optimisation to the open and short-circuited stub lengths is required to achieve the design centre frequency of 94 GHz.

The tuning techniques used in the SPST switches can equally be applied to the SPDT configuration. Initial results on a prototype reverse mode 90 GHz switch indicates an insertion loss of 3 dB, return loss of ≥ 15 dB and an isolation of ≥ 25 dB over 6 GHz bandwidth. Work on this configuration is continuing with the aim of improving the parameters.

CONCLUSIONS

Distributed microstrip tuning elements have been used to tune out the junction capacitance and series inductance of beam lead diodes in fin-line p.i.n. switches. The use of this technique together with very low capacitance diodes leads to broadband performance. At the lower millimetric frequencies currently available diodes permit designs to be realised in which only the forward bias state is optimised with tuning elements. For higher frequency components the junction capacitance of available diodes is too high for this approach to be adopted. Iterative designs in which both diode states are optimised through tuning have to be used in this case.

REFERENCES

[1] K. Solbach. The Status of Printed Millimetric Wave E-Plane Circuits. IEEE Trans., Vol. MTT-31, No. 2, February 1983.

[2] E. Kpodzo, K. Schünemann and G. Begemann. A Quadriphase Fin-line Modulator. IEEE Trans., Vol MTT-28, No. 7, July 1980.

[3] P. Meier. New Developments with Integrated Fin-line and Related Printed Millimetric Circuit. IEEE MTT-S Int. Microwave Symp. Dig. 1975.

[4] H. Meinel and H. Callsen. Fin-line P.I.N. Diode Attenuators and Switches for 94 GHz Range. Electronics Letters Vol. 18, No. 13 1982.

[5] H. Meinel, B. Adelsech and H. Callsen. A Survey of Planar Integrated mm Wave Components. Military Microwaves 80 Symp. Dig.

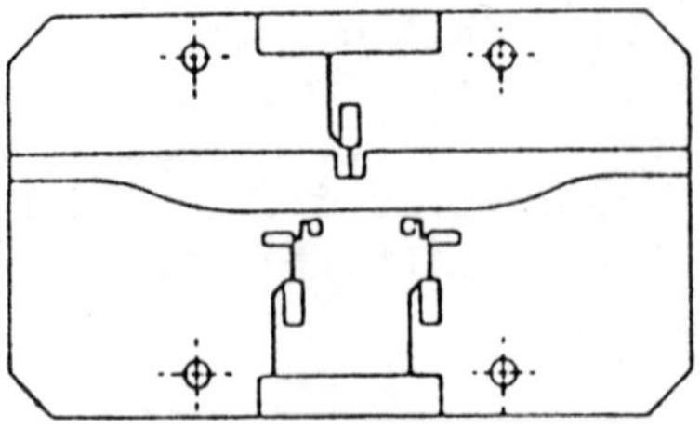

FIG. 1
Layout of Q-Band
Fin-Line Switch Circuit

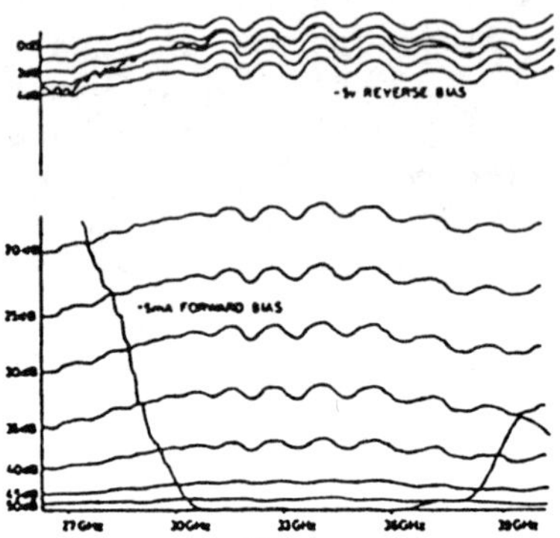

FIG. 2(a)
Transmission Loss

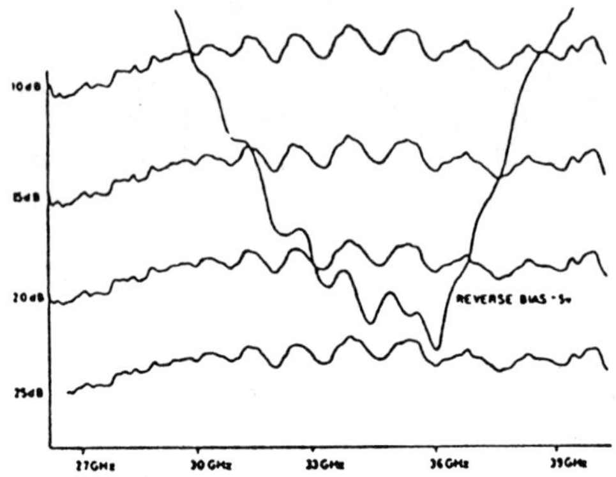

FIG. 2(b)
Return Loss Reverse Bias

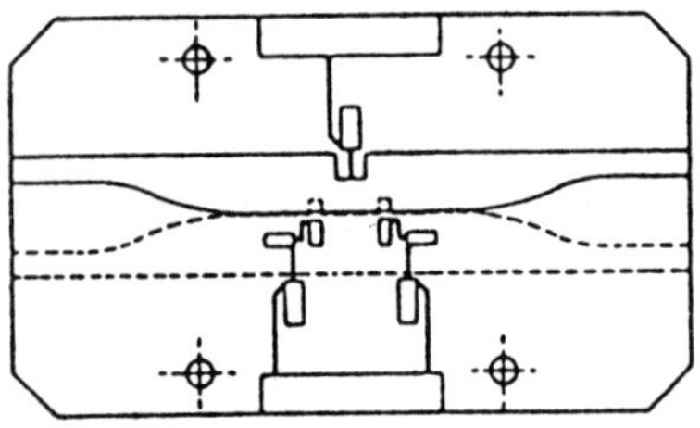

FIG. 3
General Layout of W-Band
Fin-Line Switch

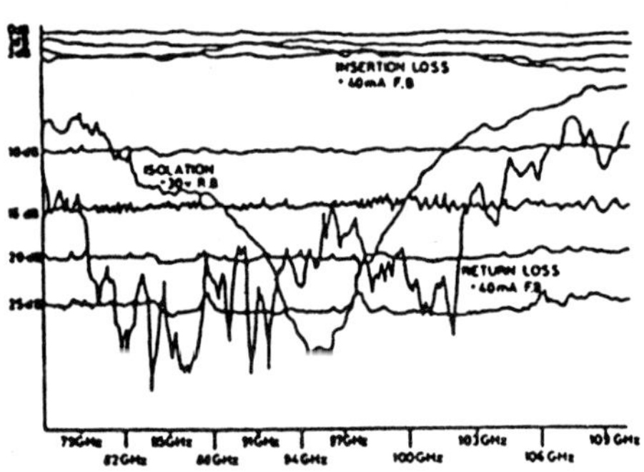

FIG. 4.
Performance of Two
Diode Reverse Mode Switch

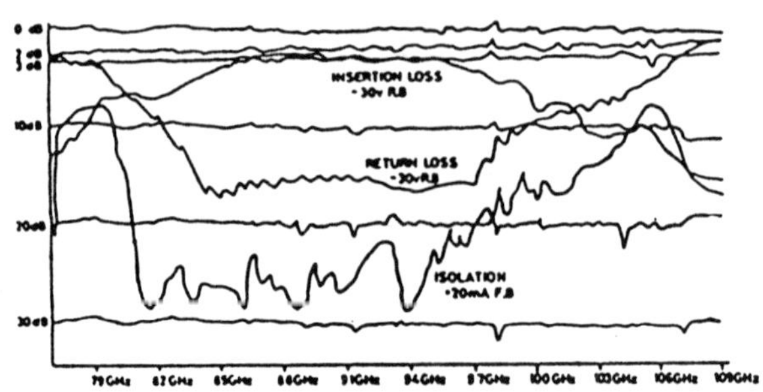

FIG. 5.
Performance of Two Diode Forward
Mode Switch

7.2.1

A Quadriphase Fin-Line Modulator

ELIAS KPODZO, STUDENT MEMBER, IEEE, KLAUS SCHÜNEMANN, MEMBER, IEEE, AND GÜNTHER BEGEMANN

Abstract—**A hybrid-coupled quadriphase modulator which has been realized in fin-line technique is described. It operates over a 15-percent frequency band around 15 GHz with an insertion loss of 1.7 dB. The circuit has been scaled to twice the operating frequency. The insertion loss then amounts to 2.2 dB. Furthermore, it is shown experimentally that the performance of the modulator does not degrade if the dimensions of the waveguide housing are altered to amounts of up to 10 percent.**

I. Introduction

DIGITAL PHASE modulators with p-i-n diodes are normally realized in rectangular waveguide techniques [1] in order to keep the insertion loss low and the modulator bandwidth wide. (The latter is defined here with respect to deviations in the desired phase shift of less than 4 deg and to an amplitude imbalance of less than 5 percent.) Insertion losses of 2 dB and a bandwidth of 1 percent of the quadriphase modulator have been achieved at millimeter wavelengths [1]. More recently, a microstrip realization has been reported [2] which shows higher losses (3 dB at *Ku*-band frequencies) but the same bandwidth. This performance could even be considerably improved at *X* band. A 4-bit phase shifter with p-i-n diodes has been described in [3] which operates over 4-percent bandwidth with an average insertion loss of 1.6 dB.

In fin-line technique, the advantages of rectangular waveguides (low loss) and of microstrip lines (low-cost production) can be combined [4]. It should hence be possible to realize a phase modulator whose performance is similar to that of a rectangular waveguide structure. This paper presents a solution to this problem.

II. Reflection Phase Modulator

The reflection phase modulator consists of a fin-line network, which must transform the two steady-state impedances of a p-i-n diode, in such a way that the phase between the two input reflection coefficients has a prescribed value. The magnitude of the reflection coefficients must be equal. For the case of a 180-deg modulator, it has been shown in [5] that the imbedding network must just match the "hyperbolic middle point" of the diode. This is illustrated in Fig. 1. In hyperbolic geometry, circles perpendicular to the periphery of the unit circle are considered to be straight lines. States ρ_1 and ρ_2 determine a segment of one such line, the hyperbolic middle point of which is denoted by ρ_m. Segments $\rho_1\rho_m$ and $\rho_2\rho_m$ have equal hyperbolic lengths and subtend an angle of 180 deg with respect to ρ_m. Since lossless bilinear transformations are congruent, thereby preserving these angle and distance relationships, any lossless network which matches ρ_m is a 180-deg modulator. ρ_m is uniquely related to the reflection coefficients ρ_1 and ρ_2 which are realized by the two impedance states of a p-i-n diode.

Manuscript received November 7, 1979; revised February 14, 1980. This work was supported by the Deutsche Forschungsgemeinschaft.

The authors are with the Institut für Hochfrequenztechnik der Technischen Universität, Postfach 3329, D-3300 Braunschweig, West Germany.

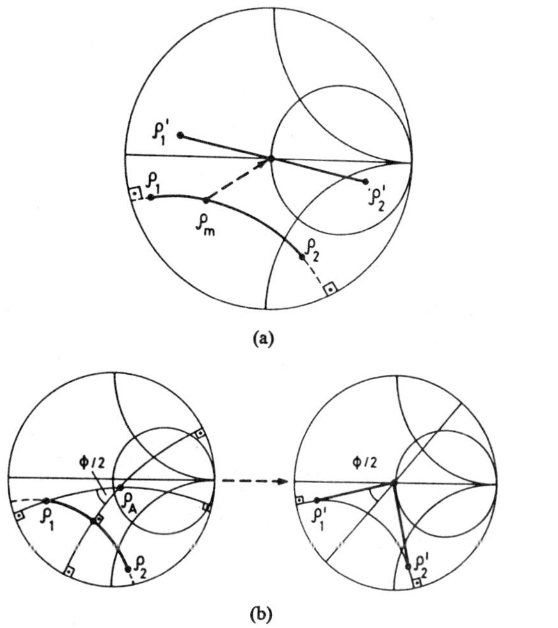

Fig. 1. Equivalence between a phase modulator and an impedance matching network. (a) 180-deg modulator. (b) 90-deg modulator. $\rho_1, \rho_2 \cdots$ reflection coefficients of the switching diode; $\rho_1', \rho_2' \cdots$ reflection coefficients at the modulator input port; $\rho_m \cdots$ hyperbolic middle point; $\rho_A \cdots$ matching point.

This result can be generalized. A two-state switching network is equivalent to an impedance-matching network. The impedance to be matched varies according to the requirements (e.g., according to the desired phase shift ϕ), but it is always uniquely defined by the states of the switching element and these requirements. It has been stated in [6] that the state to be matched for an equal-loss ϕ modulator is the apex of a hyperbolic isosceles triangle having segment $\rho_1\rho_2$ as base and opposite angle ϕ. As an example, Fig. 1(b) shows the state ρ_A to be matched for the case of a 90-deg modulator.

E. Kpodzo, K. Schunemann, and G. Begemann, *IEEE Transactions on Microwave Theory and Techniques*, Vol. MTT-28, No. 7, July 1980, pp. 747–752.

The realization of a (phase) modulator is hence reduced to that of an impedance-matching network. The requirements of a phase modulator, i.e., linearity with respect to RF signals, short switching times, large power handling capability, and low insertion loss, are best met by using a p-i-n diode as switching element. Its equivalent circuit is given in Fig. 2 together with typical values of the various circuit elements for high frequency applications. Calculating the hyperbolic middle point impedance $Z_m = R_m + jX_m$ (which corresponds to ρ_m) from the formulas [6]

$$R_m = \sqrt{R_1 R_2}\sqrt{1+\left(\frac{X_1 - X_2}{R_1 + R_2}\right)^2}$$

$$X_m = \frac{R_1 X_2 + R_2 X_1}{R_1 + R_2}$$

with $Z_1 = R_1 + jX_1$, $Z_2 = R_2 + jX_2$, the two impedance states of the p-i-n diode yield values in the range of 10–100 Ω. It is hence desirable to choose the wave impedance of the fin line in the same range, in order to facilitate the matching procedure. The fin-line circuit has been realized on a dielectric substrate (RT-duroid 5880) of thickness 0.25 mm and relative permittivity $\epsilon_r = 2.22$. The waveguide housing is a split block rectangular waveguide of standard cross section 15.8 mm×7.9 mm (WR 62). A unilateral fin line has been chosen (which yields lower circuit losses than a bilateral fin line) with a slot width of 0.2 mm. The wave impedance of the fundamental mode can be calculated from the formulas in [7] to be 125 Ω at 15 GHz. The slot width has not been reduced further, in order to facilitate scaling of the modulator into the millimeter-wave region. (A slot width of 0.1 mm gives a wave impedance of about 110 Ω in *Ku* band.)

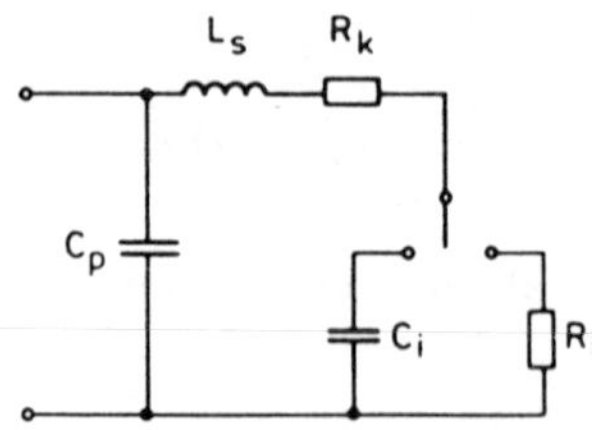

Fig. 2. Equivalent circuit of a p-i-n diode. Typical data for applications up to 20 GHz: $R_i = 1$-Ω *i*-layer resistance; $C_i = 0.1$-pF *i*-layer capacitance; $R_k = 1.5$-Ω contact resistance; $L_s = 0.2$-nH case inductance; $C_p = 0.1$-pF case capacitance.

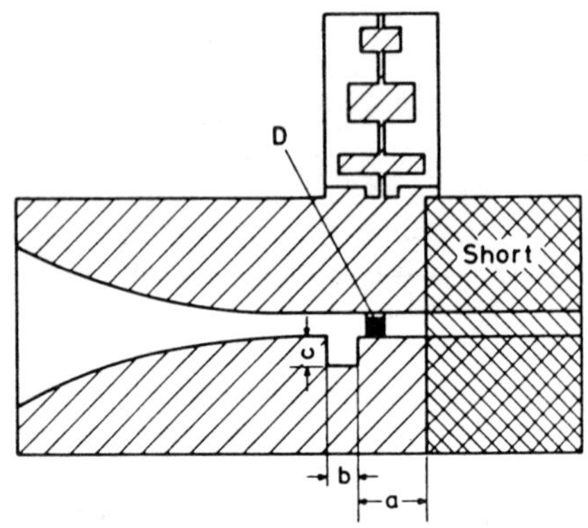

Fig. 3. Slot pattern of a reflection phase modulator with p-i-n diode D HPND-4050 of Hewlett Packard. 180-deg phase shift: $a = 3.85$ mm, $b = 0.2$ mm, $c = 1.0$ mm. 90-deg phase shift: $a = 2.9$ mm, $b = 0.2$ mm, $c = 1.5$ mm.

The slot pattern of the reflection phase modulator is shown in Fig. 3. The p-i-n diode has directly been soldered across the fin-line slot. The matching network consists of a short-circuited transmission-line section behind, and a series inductance in front of the diode. The inductance has been realized by a short section of a fin line with a large wave impedance. The short has been implemented by a metal coating on the back side of the substrate. The bias filter network is a five-section microstrip low-pass filter with a cutoff frequency of 2.6 GHz which has been integrated on a fin-line substrate.

The modulator is tuned by properly adjusting the geometrical parameters *a*, *b*, and *c*. This is performed in the following way. The distance *a* is adjusted by using a movable short circuit plunger. Once *a* has been found the plunger is replaced by short-circuiting the slot on the back side of the substrate with a coating of silver. The effective short-circuit plane has been analytically determined by using the fin-line equivalent introduced in [12]. The width *b* of the notch has been made equal to the width of the fin-line slot (0.2 mm). We have found experimentally that *a* and *c* are sufficient for tuning the modulator. The parameter *c* has been adjusted by using conductive coating as is often done in the adjustment of microstrip circuits. By this, both a 180-deg and 90-deg phase modulator have been realized. Phase shift ϕ and reflection coefficients ρ_f in forward and ρ_r in reverse state have been plotted versus frequency in Fig. 4. The modulator bandwidth can be seen to exceed 26 percent; the losses are less than 0.5 dB. These results are even slightly better for the 90-deg modulator. The bandwidth is appreciably larger than has yet been reported. Furthermore, the modulator turned out to be moderately insensitive against an exchange of the p-i-n diode. In this case it is sufficient to slightly adjust the geometrical parameters *a* and *c* or, alternatively, to insert one to two dielectric tuning screws, in order to restore the performance curves of Fig. 4. By these means, it is always possible to match the circuit to every p-i-n diode. In fact, we had to alter only the distance *a* of the short from the notch or to introduce a tuning screw between diode and short for all the p-i-n diodes we tried. (We used diodes from hp, alpha, and NEC with equal breakdown voltages and identical cases.) The penetration depth of the tuning screw was so small that the circuit loss was not increased with measurement accuracy.

The 180-deg phase modulator has successfully been scaled to twice the operating frequency. To this end, all dimensions have been reduced by a factor of 2 with the substrate thickness held constant. A final fine tuning of *a*

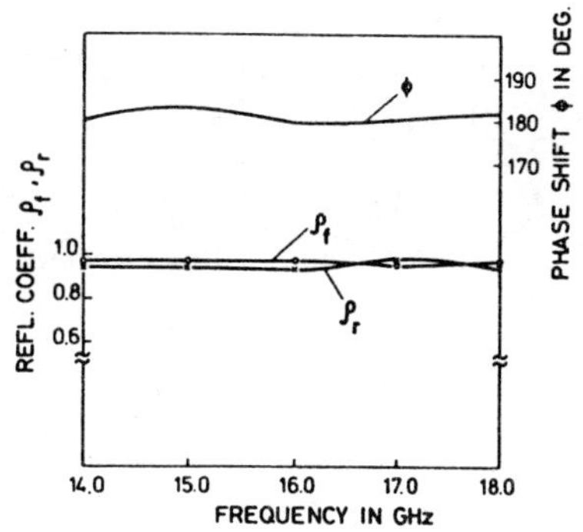

(a)

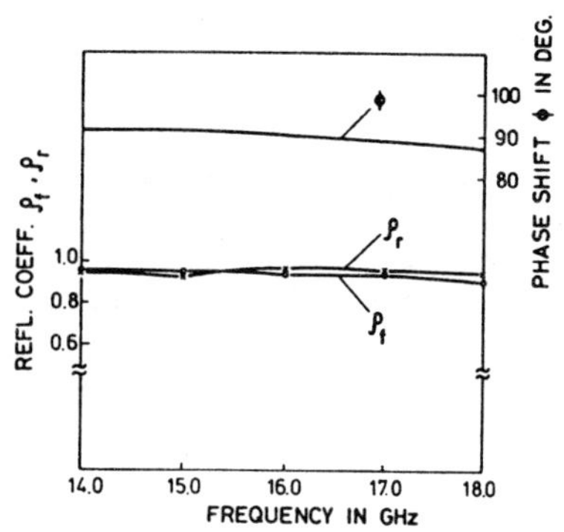

(b)

Fig. 4. Reflection coefficients ρ_f (forward bias) and ρ_r (reverse bias), and phase shift ϕ versus frequency. (a) 180-deg modulator. (b) 90-deg modulator.

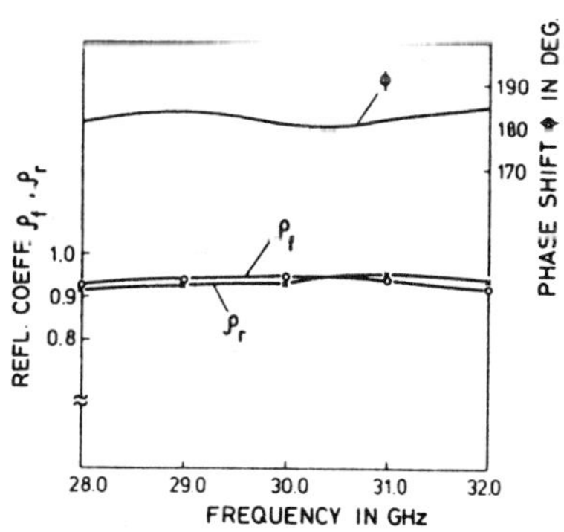

Fig. 5. Reflection coefficients ρ_f, ρ_r and phase shift ϕ of a 180-deg modulator.

and c is then sufficient to obtain the performance curves of Fig. 5. The insertion loss is slightly increased by about 0.2 dB, while the modulator bandwidth could be preserved.

III. 3-dB Hybrid

Our aim is to develop a quadriphase modulator by cascading a 180-deg modulator with a 90-deg modulator. In order to avoid nonreciprocal devices, the input and output waves of each modulator are separated in a 3-dB

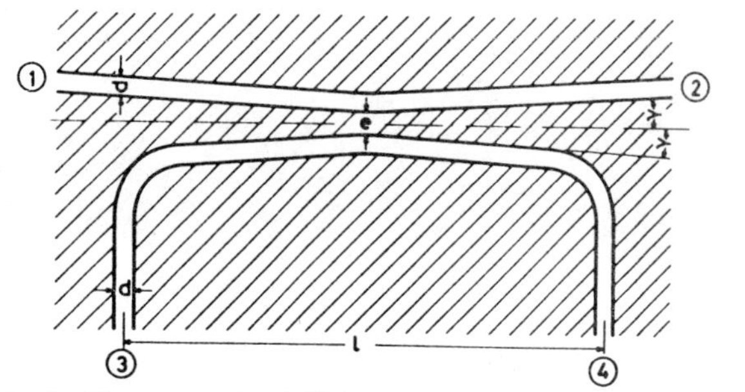

Fig. 6. Slot pattern of a 3-dB hybrid. Optimized dimensions $d=0.2$ mm, $e=1.0$ mm, $l=39.6$ mm, and $\gamma=5$ deg.

hybrid, as has been shown in [2]. Either modulator then consists of a 3-dB hybrid and of two identical reflection-type phase modulators with p-i-n diode which terminate two ports of the hybrid.

A simple configuration for a 3-dB hybrid in fin-line technique has not been reported until now. The directional coupler of [8] or the magic-T circuit of [9] could indeed be utilized to implement a hybrid-coupled phase modulator; however, they do not lead to a simple mechanical construction of the integrated device. A practical solution of the problem is the fourport, whose slot pattern on the dielectric substrate is sketched in Fig. 6. The slot width has been chosen to 0.2 mm in order to fit the dimensions of the reflection modulators.

A disadvantage of the proposed hybrid circuit is that a theoretical analysis has not yet been performed. One has to rely upon an experimental optimization. The main design parameters are the slot width d, the length l of the coupling region, the narrowest distance between the two slots e, and the angle γ. If properly designed, the fourport equally splits the power between ports 2 and 4 with 1 the input and 3 the isolated port. An extensive experimental investigation yielded the following results.

1) The coupling from port 1 to port 4 can be increased by increasing d and l and by decreasing e. Because d has been determined by the slot width of the modulators, the coupling between the two fin lines is adjusted by varying both the distance between the two slots and the length l of the coupling section.
2) The directivity is largely influenced by tapering the coupling strength along the waveguide axis, i.e., by varying the angle γ.
3) This, on the other hand, influences the effective coupling length and thus the coupling, too, as is, for instance, known from hybrids realized in inverted strip dielectric waveguide technique [10].

Broad-band performance of the 3-dB hybrid can already be achieved with a coupling section, which is 2–3 wavelengths long. The optimum design parameters are given in the caption of Fig. 6 for the case of a midband frequency of 15 GHz. The performance of the 3-dB hybrid is shown in Fig. 7. Both the forward coupling and the

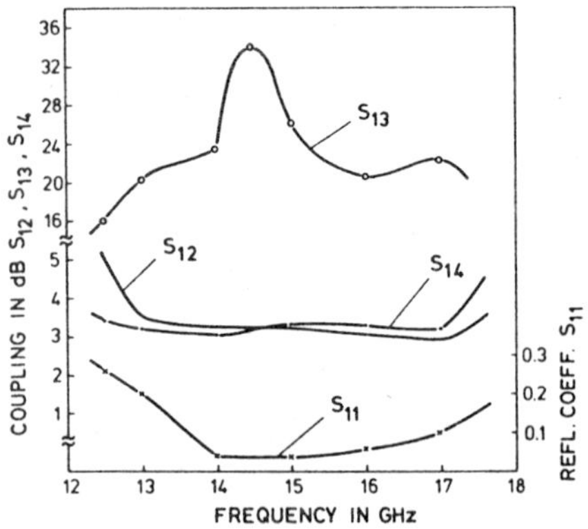

Fig. 7. Scattering coefficients of the hybrid circuit of Fig. 6 versus frequency.

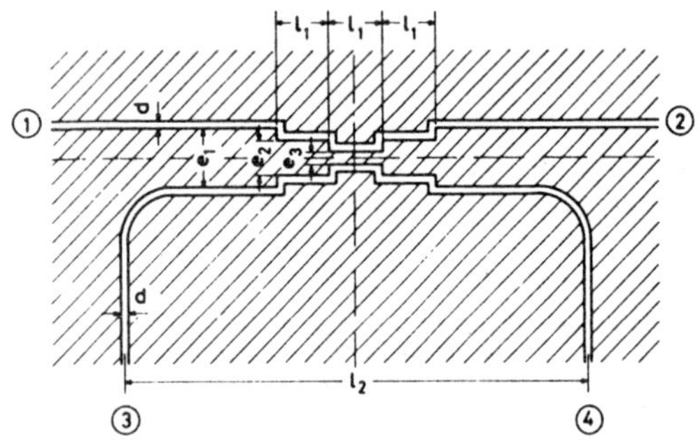

Fig. 8. Slot pattern of a 3-dB hybrid. Optimized dimensions $d=0.2$ mm, $e_1=5.0$ mm, $e_2=3.0$ mm, $e_3=1.0$ mm, $l_1=4.4$ mm, and $l_2=39.6$ mm.

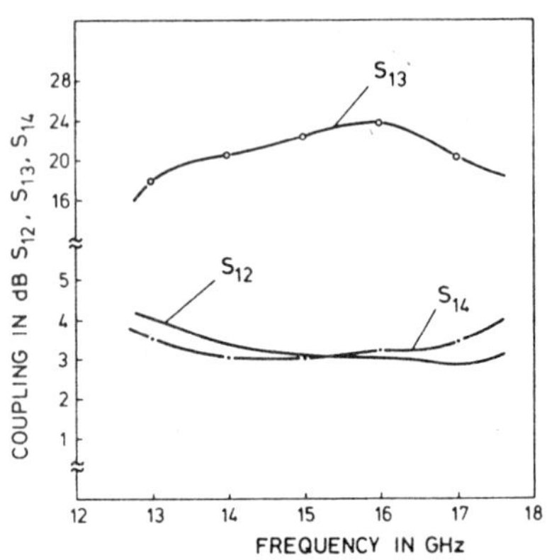

Fig. 9. Scattering coefficients of the hybrid circuit of Fig. 8 versus

directivity are flat over a frequency band from 13–17 GHz, the insertion loss amounts to 0.3 dB. The directivity is better than 20 dB. It can be increased to more than 30 dB in a fractional bandwidth of about 10 percent.

Another configuration of a 3-dB hybrid with a stepped coupling strength distribution is shown in Fig. 8. This structure is similar to known microstrip realizations. Its performance is more limited in bandwidth than that of the hybrid sketched in Fig. 6, as can be seen from the experimental results in Fig. 9. Its remarkable features are, however, a very small insertion loss of 0.1–0.2 dB in a frequency range from 15–16 GHz.

IV. Hybrid-Coupled Quadriphase Modulator

The complete quadriphase modulator consists of two hybrids whose ports 2 and 4 are both terminated either in 180-deg reflection phase modulators or in 90-deg devices. The hybrids are then cascaded to form a quadriphase modulator. The compound circuit is sketched in Fig. 10.

The 3-dB hybrid and the reflection phase modulators can individually be matched and then be coupled together due to the high directivity of the hybrid. The tuning elements of the modulator have to be changed only slightly afterwards. No corrections were necessary when the 180-deg phase shifter and the 90-deg phase shifter were cascaded. The input VSWR of the compound modulator is less than 1.3, so that a Gunn oscillator on a fin line could be directly coupled to the modulator input. The bandwidth of the quadriphase modulator is 15 percent, its overall insertion loss 1.7 dB at an operating frequency of 15 GHz. (Nearly the same performance was achieved at 30 GHz except for the insertion loss, which degraded to 2.2 dB).

The circuit of the quadriphase modulator has finally been integrated on a common substrate. The outer dimensions of the waveguide housing are 110 mm$\times$60 mm$\times$33 mm. The overall transmission loss and the phase shift are shown versus frequency in Fig. 11. These results have been obtained with beam-leaded p-i-n diodes HPND-4050 of Hewlett-Packard. The switching times exclusively depend on the turn-on and turn-off time of the p-i-n diode, because the response time of the imbedding has been measured to be less than 0.5 ns. Switching times of 2 ns have been obtained.

V. Tolerance Analysis

It has been demonstrated that the electrical performance of an integrated quadriphase modulator can compete with that of a waveguide realization, even in the millimeter-wave region. In the following we would like to stress that the fin-line technique is even an alternative to microstrip at centimeter wavelengths. Since the electrical performance of fin-line circuits can sometimes be better than that of microstrip circuits, one has still to consider the economic point of view. A fin-line housing can be produced with low cost, provided that there are no severe tolerance limits for the dimensions of the waveguide cross section.

This question has been theoretically answered in [11]. It has been shown that the sensitivity of the electrical parameters of a fin line with respect to variations in the dimensions of the waveguide housing is much less than for a rectangular waveguide, provided that the impedance of the fin line is not larger than 100 Ω. It is typically reduced to about 20 percent due to the tight coupling of the wave to the slot between the fins. It is hence not

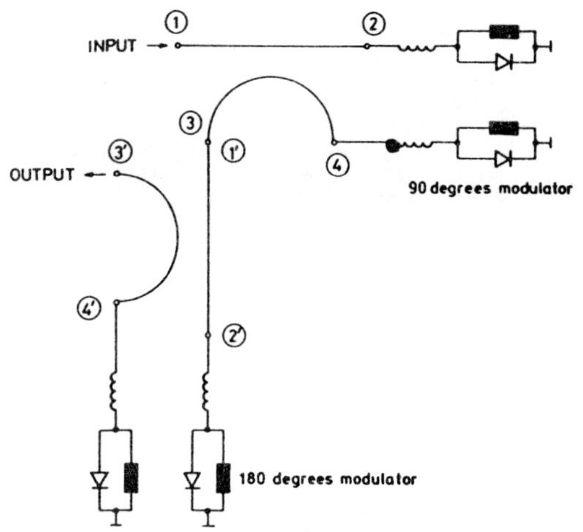

Fig. 10. Equivalent circuit of the quadriphase modulator.

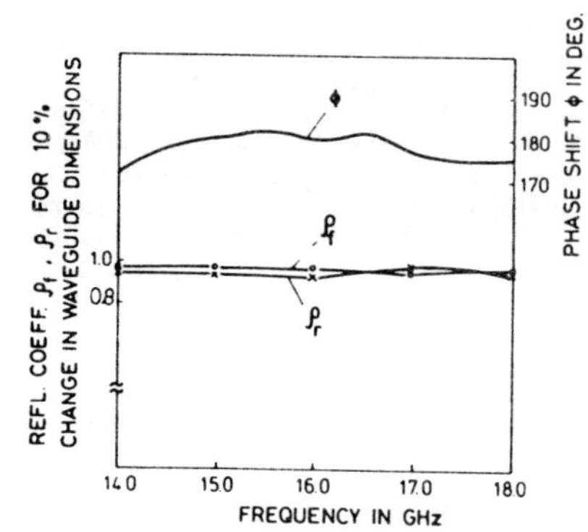

Fig. 12. Reflection coefficients ρ_f, ρ_r and phase shift ϕ of a 180-deg modulator versus frequency after varying the dimensions of the housing by 15 percent.

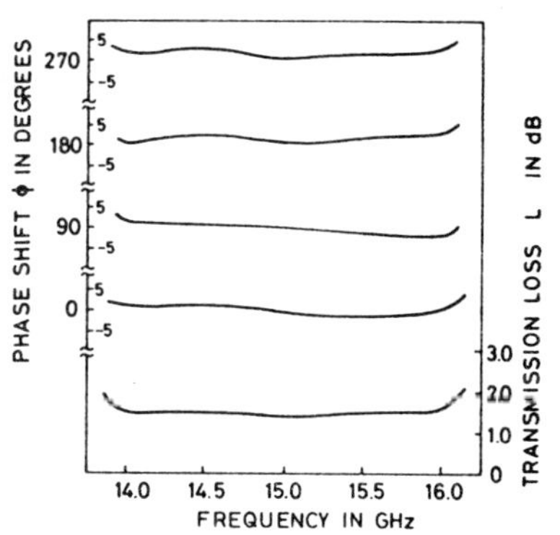

Fig. 11. Transmission loss L and phase shift ϕ of the quadriphase modulator versus frequency.

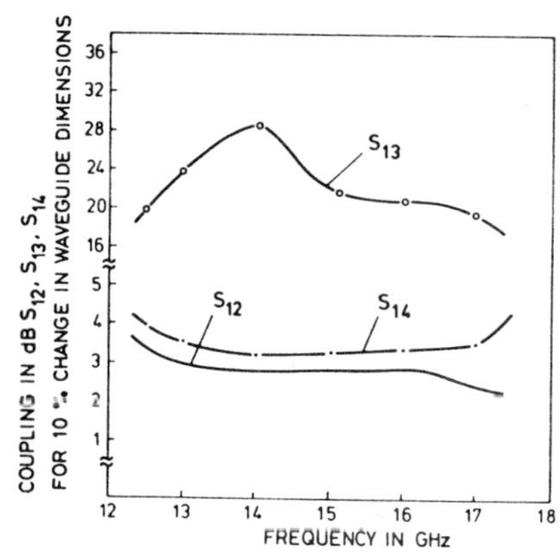

Fig. 13. Scattering coefficients of the hybrid circuit of Fig. 6 versus frequency after varying the dimensions of the housing by 15 percent.

necessary that the fin-line housing is carefully machined. This shows that fin line might be preferable to microstrip even at, say X-band frequencies, if integrated circuits with filter networks must be realized.

In order to prove the validity of these results, the dimensions of the waveguide housing (height, width, and distance of the substrate to the side walls) have been changed by 10 percent. The performance of the 180-deg modulator and of the 3-dB hybrid are shown in Figs. 12 and 13, respectively, for an unchanged slot pattern of the fin-line circuits. It can be seen that the performance deteriorates only slightly. There is no deterioration at all if the dimensions of the housing are changed by up to 7 percent.

VI. Conclusions

A Quadriphase modulator has been realized by cascading two hybrid-coupled modulators with phase shifts of 180 deg and 90 deg, respectively, in an integrated fin-line circuit. The overall insertion loss is 1.7 dB at 15 GHz, which is comparable to what has been obtained in waveguide technique, while the bandwidth is an order of magnitude larger (it amounts to 15 percent). The modulator has likewise been realized at 30 GHz. The bandwidth remained the same while the insertion loss slightly degraded to 2.2 dB. The switching times exclusively depend on the p-i-n diodes: the response time of the imbedding network has been measured to be less than 0.5 ns. Switching times of 2 ns were obtained.

References

[1] W. J. Clemetson, N. D. Kenyon, K. Kurokawa, B. Owen, and W. O. Schlosser, "An experimental mm-wave path length modulator," *Bell Syst. Tech. J.*, vol. 50, pp. 2917–2945, 1971.

[2] H. Junghans, "A Ku-band hybrid-coupled 4-phase modulator in MIC technology," in *Proc. 5th Europ. Microwave Conf.*, Hamburg, Germany, 1975, pp. 133–137.

[3] B. Glance, "A fast low-loss microstrip p-i-n phase shifter," *IEEE Trans. Microwave Theory Tech.*, vol. MTT-27, pp. 14–16, 1979.

[4] P. J. Meier, "Integrated fin-line millimeter components," *IEEE Trans. Microwave Theory Tech.*, vol. MTT-22, pp. 1209–1216, 1974.

[5] M. S. Navarro, "Modulator imbedding networks for pin-diodes," *M.I.T. RLE Quart. Progress Rep. no. 106*, pp. 23–27, July 1972.

[6] T. A. Dorschner, "Characterization of reflection phase modulators using hyperbolic geometry," in *Proc. 3rd Europ. Microwave Conf.*, Brussels, Belgium, 1973, pp. A.9.1–A.9.4.

[7] A. K. Saad and G. Begemann, "Electrical performance of fin lines of various configurations, *IEE J. Microwaves, Opt. Acoustics*, vol. 1, pp. 81–88, 1977.

[8] P. J. Meier, "Planar multiport millimeter integrated circuits," in *Proc. Int. Microwave Symp.*, San Diego, CA, 1977, pp. 385–388.

[9] G. Begemann, "An X-band balanced fin-line mixer," in *Proc. Int. Microwave Symp.*, Ottawa, Ont., Canada, 1978, pp. 24–26.

[10] R. Rudokas and T. Itoh, "Passive millimeterwave IC-components made of inverted strip dielectric waveguides," *IEEE Trans. Microwave Theory Tech.*, vol. MTT-24, pp. 978–981, 1976.

[11] E. Kpodzo, K. Schuenemann, and H. El-Hennawy, "A quadriphase modulator in fin-line technique," in *Proc. Int. Microwave Symp.*, Orlando, FL, 1979.

[12] A. K. Saad and K. Schuenemann, "A simple method for analyzing fin-line structures," *IEEE Trans. Microwave Theory Tech.*, vol. MTT-26, pp. 1002–1007, 1978.

7.2.2

MILLIMETER-WAVE QPSK MODULATOR IN FIN LINE

G.B. Gajda and C.J. Verver

Communications Research Centre
Department of Communications
Ottawa, Canada

ABSTRACT

A PIN diode QPSK modulator in unilateral fin line is described for use at millimeter-wave frequencies. A parallel configuration, consisting of two 180-degree bi-phase modulators and quadrature and in-phase hybrids, is used in order to accomodate pre-filtered data. All circuitry was implemented using unilateral fin line printed on a single side of the substrate. No stringent dimensional tolerances for either the circuit or the housing were required.

Experimental results gave an overall insertion loss of 6.5 dB, and maximum phase error of 5 degrees at the design frequency (28.4 GHz). This approach for implementing a QPSK modulator lends itself to low-cost manufacturing.

INTRODUCTION

PIN diode QPSK or quadriphase modulators realized in microstrip and waveguide have been reported extensively in the past [1]. With the increasing use of the millimeter wave bands, fin line or E-plane technology appears to be an attractive circuit medium for low-cost implementation of these components.

The QPSK modulator to be described here was implemented in unilateral fin line printed on a 0.010 inch thick RT/Duroid substrate (ε_r=2.2). The design was centred on the 28.4 GHz uplink band of an experimental earth terminal for use with the European Space Agency (ESA) Olympus satellite. The housing is of the open split-block type, and has the dimensions of WR-28 waveguide (Ka band). The circuit uses four beam-lead PIN diodes and is etched entirely on one side of the substrate. The absence of bond wires and plated via holes, along with the relatively low dimensional tolerances required for the circuit and the housing, make this approach attractive for low cost manufacturing.

The configuration of a parallel QPSK modulator is shown in Fig. 1. The circuit consists of two 180-degree bi-phase modulators fed in quadrature by the input 3 dB quadrature coupler. The two bi-phase modulated outputs are recombined in the output in-phase combiner to produce the QPSK signal. This approach is in contrast to a previously-reported QPSK modulator implemented in fin line which used 90 degree and 180 degree bi-phase modulators connected in series [2].

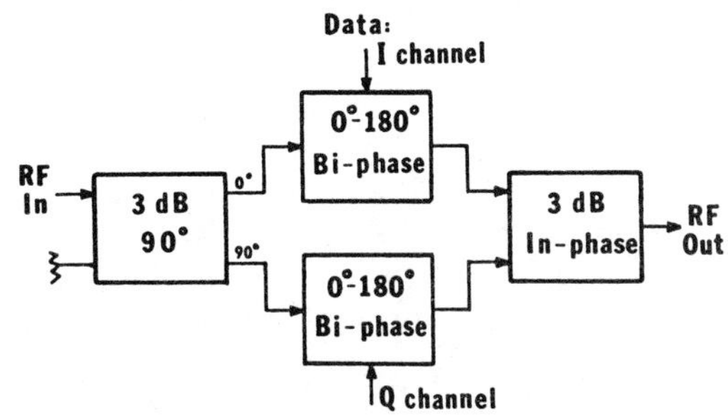

Fig. 1. Parallel QPSK Modulator.

The parallel configuration was chosen since it may be used in situations where the data is filtered prior to modulation. Under these conditions, the bi-phase modulators function as linear amplitude modulators, provided the non-linearities of the PIN diodes are compensated [3]. Maximum bit rates of 2.7 Mbs for the Olympus terminal dictate the use of a linear modulator.

The individual bi-phase modulators are of the reflection type, where a PIN diode is used to switch a quarter-wave length of line terminating each of the output ports of a 3 dB quadrature hybrid. The hybrids for the bi-phase and the input splitter are constructed using parallel-coupled-slot fin line [4]. The novel output combiner is also constructed from parallel-coupled-slot fin line and is similar to the well known Wilkinson combiner in microstrip. All individual components were designed and tested separately, followed by integration on a single substrate to produce the integrated QPSK modulator.

3 dB Quadrature Hybrid

A diagram of the coupler is shown in Fig. 2. For this type of structure, with unequal odd and even mode velocities, coupling is in the forward direction. For a good input match and isolation (i.e. from port 1 to 4), the odd and even mode impedances, Z_{oo} and Z_{oe}, should be close in value to the characteristic impedance of the four ports Z_o [5].

G.B. Gajda and C.J. Verver, *IEEE MTT-S International Microwave Symposium Digest,* 1986, pp. 233–236. Copyright © 1986 IEEE. Reprinted with permission.

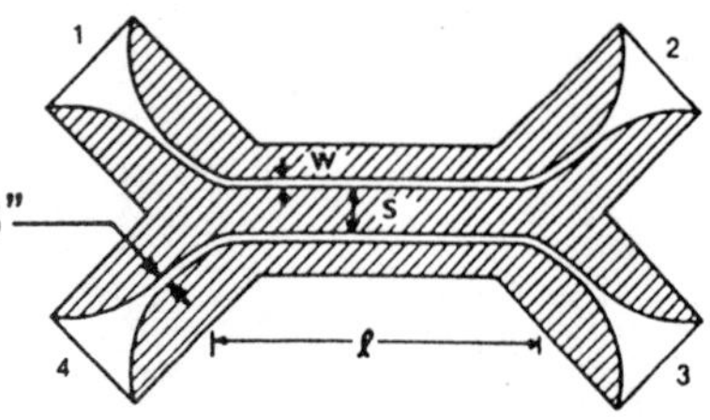

Fig. 2. Parallel-coupled-slot fin line hybrid.

By considering an even/odd mode type of analysis, the coupling coefficients of ports 2 and 3 (denoted by A_2 and A_3) are given by [6]:

$$A_2 = \frac{1}{2}(T_e + T_o), \quad A_3 = \frac{1}{2}(T_e - T_o) \tag{1}$$

$$T_{e,o} = \frac{2}{2\cos\theta_{e,o} + j\left(\frac{Z_{oe,oo}}{Z_o} + \frac{Z_o}{Z_{oe,oo}}\right)\sin\theta_{e,o}} \tag{2}$$

$$\theta_{e,o} = \frac{\omega\ell}{v_{e,o}} \tag{3}$$

where $v_{e,o}$ is the even/odd mode phase velocity and ℓ is the length of the coupler. For an equal power split and quadrature phase, the two outputs must satisfy:

$$A_3 = -jA_2 \tag{4}$$

Substituting (1) and (2) into (4), the following transcendental equation is obtained:

$$4 + \left(\frac{Z_{oe}}{Z_o} + \frac{Z_o}{Z_{oe}}\right)\left(\frac{Z_{oo}}{Z_o} + \frac{Z_o}{Z_{oo}}\right)\tan\theta_e \tan\theta_o = 0 \tag{5}$$

The required length of the coupler is found from the solution of (5). However, if the approximation $Z_{oe} \approx Z_{oo} \approx Z_o$ is used, the solution of (5) becomes:

$$\ell = \frac{\lambda_o}{4}\,\frac{v_e/c}{(1 - v_e/v_o)} \tag{6}$$

where λ_o is the free space wavelength and c is the speed of light.

A spectral domain analysis was used to calculate the even and odd mode parameters of coupled-slot fin line. The even mode has a magnetic wall symmetry (similar to coplanar waveguide), while the odd mode has an electric wall symmetry. The port impedance Z_o was calculated to be approximately 200 Ω, corresponding to a 0.015 inch slot.

Values of even and odd impedance were found to be close to Z_o if the coupled slot widths w were close to 0.015 inches, and the separation s greater than 0.050 inches. Values of v_e/c and v_o/c were computed to be in the vicinity of 0.84 and 1.0 respectively, for the afore mentioned range in values of w and s. A coupling section of nominally 0.500 inches long was used, and several couplers with different slot widths and separations were built. Initial designs with 0.015 inch slot widths gave a stronger than expected coupling at 28.4 GHz, thus the slot width was reduced and the separation increased. An acceptable coupler was achieved with parameters w=0.007 inches and s=0.063 inches. The performance of this coupler is illustrated in Fig. 3, showing an equal power split from 27 to 31 GHz. The insertion loss of 0.5 dB on the coupled outputs includes a 0.2 dB loss due to each of the 1.5 wavelength-long tapered transitions at all four ports.

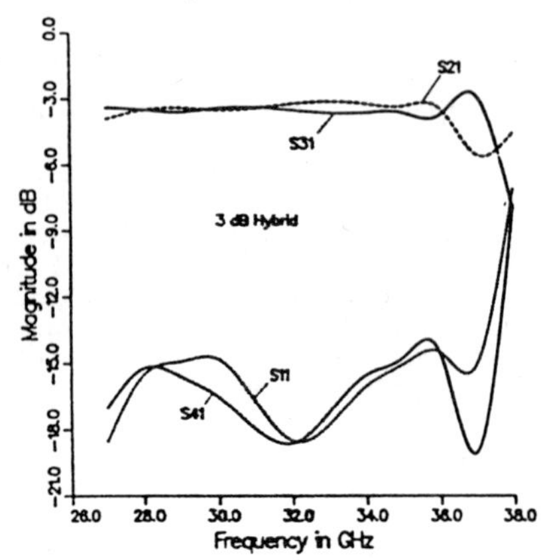

Fig. 3. Quadrature hybrid performance.

In-phase Combiner

The in-phase combiner consists of two parallel slot lines converging to a single slot line connected in series as shown in Fig. 4(a).

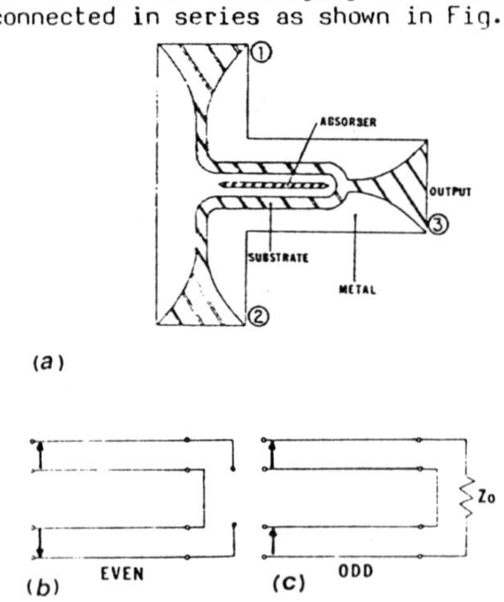

Fig. 4.(a) In-phase fin line combiner.
(b) Even mode equivalent circuit.
(c) Odd mode equivalent circuit.

The operation of this circuit may be explained by considering even and odd excitations. For an even excitation of ports 1 and 2 (analogous to the coplanar waveguide mode), the fields arrive at the load in phase. Thus no current passes through the load and it may be replaced by an open circuit (Fig. 4(b)). Since the even mode sees an open at the end of the line, it is the undesired mode and is attenuated by mounting a thin strip of absorbing material along the centre strip. If the structure is excited in the odd mode (Fig. 4(c)), the fields arrive out of phase across the load. Because of symmetry, the load may be bisected so that each line sees one half the load impedance. The effective load impedance, $Z_0/2$, may be matched to the input lines (with impedances Z_0) using the odd-mode impedance and wavelength. (i.e. Z_{oo} = .707 Z_0 and $\ell = n\lambda_{odd}/4$, n = 1, 3, ...). Since the odd mode possesses an electric wall symmetry, it is not attenuated by the absorbing strip as strongly as the even mode.

An experimental combiner was built having a 0.007 inch slot width, 0.048 inch slot separation, and a coupling length of approximately 0.25 inches. To allow for sufficient absorbtion of the even or coplanar mode, the length of the coupled slots was made equal to $3\lambda_{odd}/4$. The absorber strip was made as thin as possible in order to prevent attenuation of the odd mode field, while the optimum height of the strip was found empirically.

Experimental results shown in Fig. 5 gave an insertion loss of 1.0 dB (including 0.5 dB loss due to input/output lines and tapers), and a return loss and isolation of greater than 20 dB at the centre frequency.

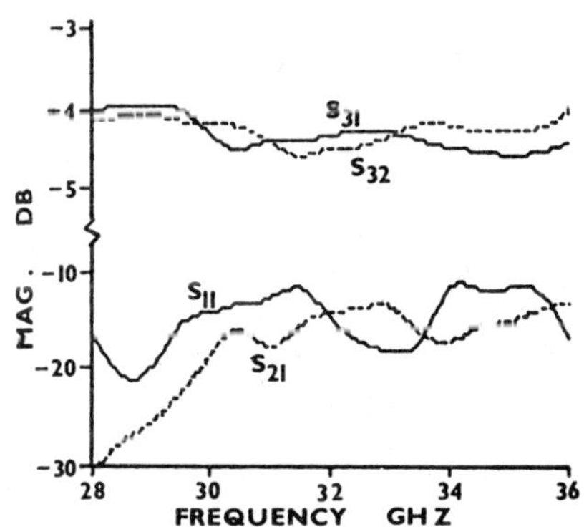

Fig. 5. Measured performance of fin line in-phase combiner.

Bi-phase Modulator

A single bi-phase modulator consists of one of the quadrature hybrids described previously, terminated with a pair of reflection phase shifters at the two coupled ports. The reflection phase shifter (Fig. 6) is implemented using a PIN diode (Alpha DSG 6474E) mounted in shunt at the input of a 90° section of short-circuited fin line. The fin line slot width was chosen to be 0.015 inches wide, and the required length of the 90-degree section was calculated using the spectral domain analysis for the guided wavelength. This length includes a correction, $\Delta\ell$, for the short-circuit end effect [7]. A narrow, half-wavelength long slot etched in the top fin functions as a DC block [4].

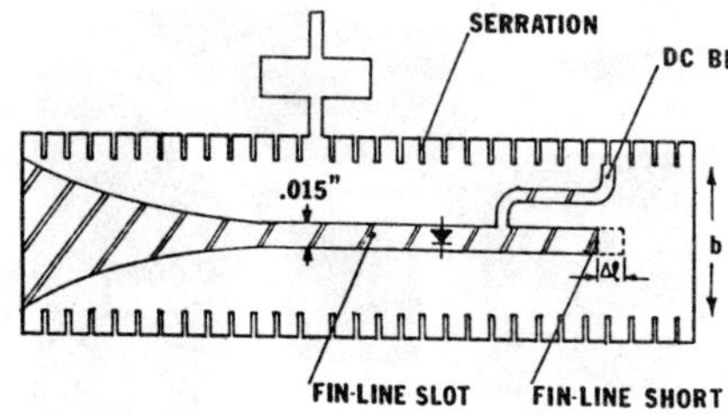

Fig. 6. Fin line reflection phase shifter test circuit.

The results of measurements of the bi-phase modulator performed from 28.0 to 30 GHz are shown in Fig. 7.

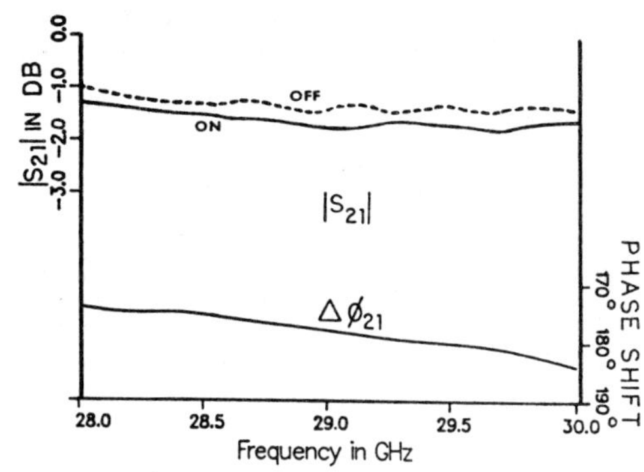

Fig. 7. Measured bi-phase modulator performnce. I_{on} = 15 mA, V_{off} = 0 volt (per diode)

The overall insertion loss through the modulator at 28.4 GHz is 1.5 dB, while the amplitude imbalance between the two diode states is about 0.25 dB. The phase shift between the two states is within 5 degrees of 180 over the 28 to 30 GHz band.

Integrated QPSK Modulator

The circuit pattern of the integrated QPSK modulator is shown in Fig. 8. Fin line tapers (1.5 wavelength long) are provided at the input and output ports, as well as the isolated port of the input coupler. Absorbant material is placed on the substrate at the isolated port to provide a matched load, while a thin strip of the material is used in the output combiner as outlined previously. The circuit occupies an area of approximately 3 square inches.

Static measurements of the integrated modulator are shown in Figs. 9 and 10. From Fig. 9, it can be seen that the maximum insertion loss through the modulator at the design frequency is 6.5 dB, with a maximum amplitude unbalance between the states of 0.5 dB. Fig. 10 shows measurements of the phase shift for the four different states of the modulator. A maximum phase error of approximately 5 degrees was obtained at the design frequency. For these measurements, the "ON" state represents a forward

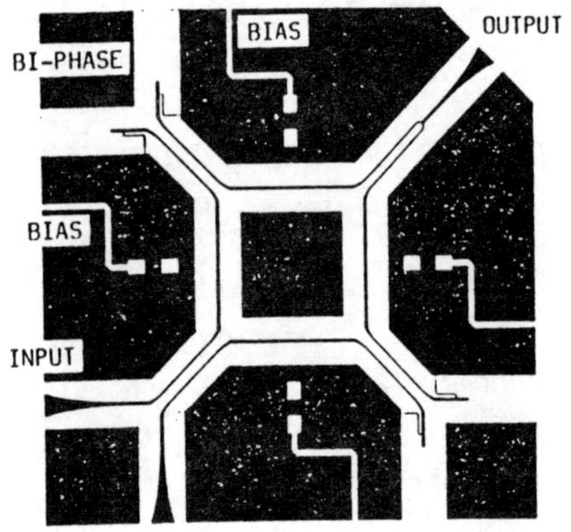

Fig. 8. Integrated QPSK modulator circuit pattern.

diode current of 15 mA and a bias voltage of approximately 1 volt, while the "OFF" state represents zero volts across the diode.

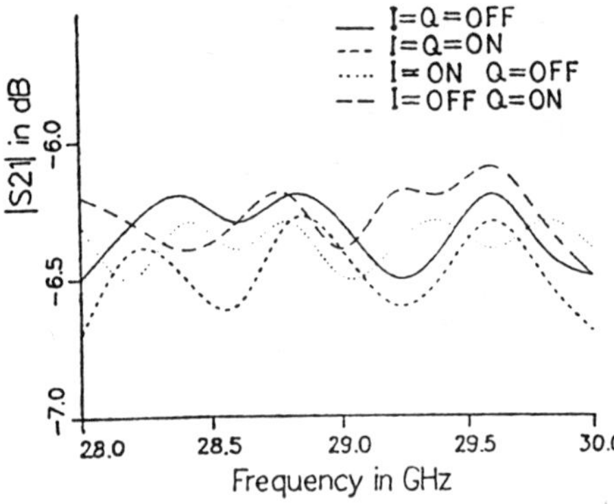

Fig. 9. QPSK modulator insertion loss.

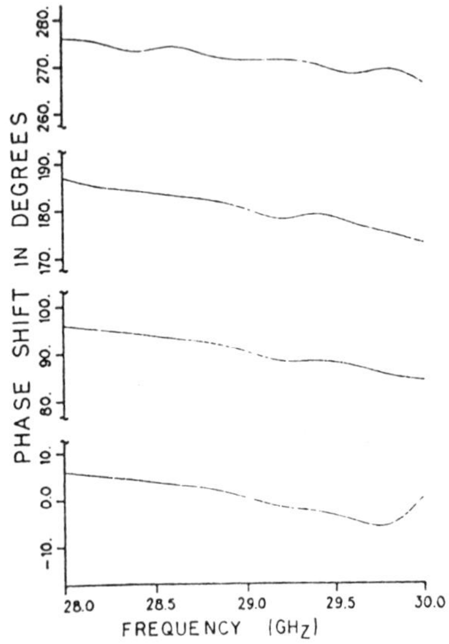

Fig. 10. QPSK modulator phase response.

CONCLUSION

The design of a parallel QPSK modulator realized totally in unilateral fin line has been presented. The design was centred at the uplink frequency of an experimental earth terminal for use with the ESA Olympus satellite (28.4 GHz). The circuit made use of two 180-degree bi-phase modulators, an input 3 dB quadrature hybrid, and an in-phase output combiner. The components were designed and tested separately, and then integrated on a single substrate. Performance results gave a 6.5 dB insertion loss with a 0.5 dB amplitude unbalance and 5 degree phase error using 4 beamlead Alpha DSG6474 PIN diodes. The circuit was etched on a single side of 0.010 inch RT/Duroid substrate using standard photo lithographic techniques. Bond wires or plated-through via holes were not required and tight fabrication tolerances were not necessary for either the circuit or the housing. This approach to millimeter-wave circuit design is attractive for low cost manufacturing.

ACKNOWLEDGEMENTS

The authors gratefully acknowledge the assistance of C. Szabo and J. Collins in the design and fabrication of the housings.

REFERENCES

[1] H. Ogawa, M. Aikawa and M. Akaike, "Integrated balanced BPSK and QPSK modulators for the Ka-band", IEEE Trans. MTT-30, pp. 227-234, March 1982.

[2] E. Kpodzo, K. Schunemann and G. Begemann, "A quadriphase fin-line modulator", IEEE Trans. MTT-28, pp. 747-752, July 1980.

[3] M. Cuhaci, G.J.P. Lo and N.S. Hitchcock, "A linearized high power microwave digital phase modulator", Proc. Int. Microwave Symp., Dallas, TX, pp. 57-59, 1982.

[4] G.B. Gajda and C.J. Verver, "Millimeter-wave bi-phase modulator in fin-line", Proceedings ELECTRONICOM '85, Toronto, Canada, 1985, pp. 578-581.

[5] J. Siegl, "Design and optimization of planar directional couplers", Proceedings of 15th European Microwave Conf., Paris, 1985, pp. 853-858.

[6] J. Reed and G.J. Wheeler, "A method of analysis of symmetrical four-port networks", IRE Trans. Microwave Theory Tech., pp. 246-252, Oct. 1956.

[7] J.B. Knorr, "Equivalent reactance of a shorting septum in a fin-line: theory and experiment", IEEE Trans. MTT-29, pp. 1196-1202, Nov. 1981.

7.2.3

AN E-PLANE BROADBAND BI-PHASE BALANCED MODULATOR FOR Ka-BAND

W Thorpe

British Telecom Research Laboratories
Martlesham Heath, Ipswich IP5 7RE, ENGLAND.

Abstract

A bi-phase p.s.k. modulator for use in 29 GHz Radio equipment has been developed that gives state of the art performance (3 dB insertion loss across 7% bandwidth and 70/100 Mbit/s data rate using beam lead PIN diodes as the switching elements). The circuit maintains the full instantaneous bandwidth without requiring any tuning or alignment during or after manufacture.

Introduction

Radio systems are increasingly using digital techniques for reasons such as robustness to interference and the ability to provide secure communications when encrypted. A commonly used form of digital modulation is phase shift keying (p.s.k.). The modulator to be described in this paper meets the specific requirement for a modulator operating in the 27.5-29.5 GHz frequency band, capable of switching r.f. powers of at least 250 mW, as needed for a local distribution millimetre-wave system under development by British Telecom[1]. The circuit is amenable to low cost fabrication techniques and is versatile in that it can be used also as an amplitude shift keyed modulator with 30 dB on-off ratio without any modification.

Balanced Modulators

The concept of the balanced circuit as a modulator has been widely reported, the most recent implementation being that by Ogawa, Aikawa and Akaike[2]. Using two anti-parallel Schottky barrier diodes as the switching elements, such circuits have severe r.f. limitations for many system applications. They offer low insertion loss at powers of a few milliwatts only, with loss increasing rapidly above 10 mW to typically around 15 dB at 250 mW. This power limitation of the circuit arises because of the use of Schottky barrier diodes as the switching elements. Directly replacing the Schottky barrier diodes with beam lead PIN diodes would allow significant increases in r.f. power, but would severely degrade the switching speed because of the inherent charge storage mechanisms associated with PIN devices; the anti-parallel connection inherent in all balanced circuits reported to date prevents adequate reverse bias being applied to overcome this speed limitation. The modulator to be described in this paper overcomes these problems by adapting the balanced configuration circuit to allow separate biasing to two PIN diodes without unduly impairing the r.f. performance. The circuit makes use of the mode transformation properties of a slotline to microstrip transition to effect a 180 degree phase change using equal length switched paths.

Circuit Configuration and Mode of Operation

The complete modulator circuit (see Fig 1) is fabricated on a Duroid 5880 substrate (t = 125 μm, Er = 2.22) incorporating printed transitions to waveguide.

It is estimated that the pair of transitions contribute 0.7 dB to the overall loss of the circuit. The beam lead PIN diodes are bonded across the emerging arms of the slotline T-junction. The circuit has separate biasing to each diode to allow fast removal of stored charge carriers by the use of a split data source and dual driver. To achieve separate biasing the annular slot-line ring is split symmetrically without introducing

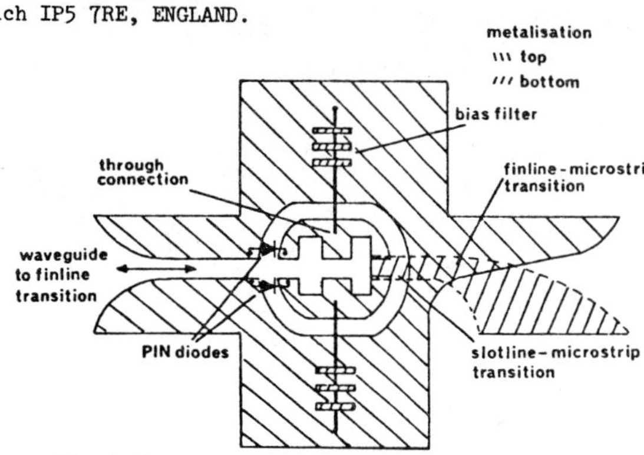

Fig 1 Circuit Layout of P.S.K. Modulator

r.f. discontinuities. This is achieved by inserting into the central 'island' of the annular slotline a double-ended quarter-wave transformer structure in slotline, forming a good open-circuit at all frequencies encountered by digital radio system modulation data streams, whilst maintaining good r.f. continuity. The circuit is the subject of a patent application.

The circuit operation can best be described by referring to Fig 2. This shows the circuit configuration with one or other PIN diode biased into forward conduction approximating a short-circuit condition, the complementary diode being reverse biased. The arrows in Fig 2 are a schematic representation of the r.f. electric field and show how the bias change results in generation of a 180 degree phase shift at the output port.

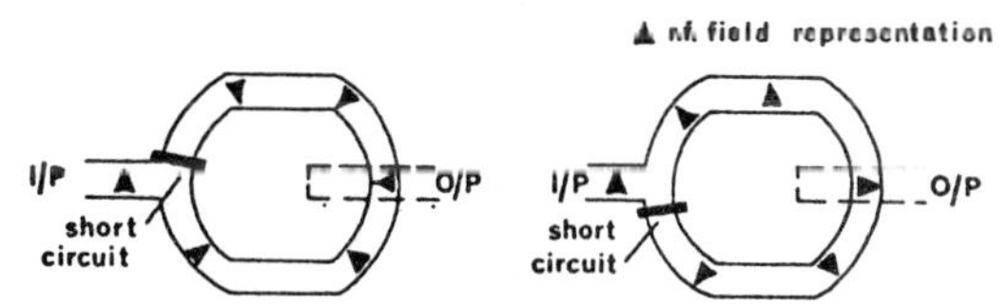

Fig 2 Principle of Circuit Operation

The circuit is potentially wideband because the phase change does not depend upon path length changes.

PIN Diode Switching

PIN diodes have two states, a low impedance forward bias state and a high impedance reverse bias state. Switching from the reverse bias state to the forward bias state is the faster of the two switching times and is dependent upon the charge build up within the diode. Charge Q is governed by the equation:

$$dQ/dt = I_f - Q/T$$

where If is the forward current
and T is the carrier lifetime of the diode.

When T is large, say more than 20 ns, then slow switching speeds result if a constant forward current If is used to inject the charge. To ensure fast switching, current spiking has in the past been used to inject a large amount of charge in a short time, and these spiking

W. Thorpe, *IEEE MTT-S International Microwave Symposium Digest*, 1983, pp. 513–515.

circuits have usually been RC networks. However, RC networks although giving fast rising edges result in slow falling edges. This limits the modulation rate of PIN diode switching. Instead of an RC spiking network a fast monostable of 10 ns duration (rising and falling edges typically 4 ns) has been used for this drive amplifier.[3] Changing the diode from its low impedance state to its high impedance state requires the stored charge to be removed, the equation governing this is:

$$dQ/dt = -Ir - Q/T$$

where Ir is the reverse current. The drive amplifier provides a high reverse voltage and a low impedance condition to remove the stored charge.

The three diode types that appeared to best meet our requirements were measured to ascertain the comparative minority carrier lifetimes under identical conditions. The measured results are shown in Fig 3 and show the GEC DC2602 preproduction devices[4] to have a shortest minority carrier lifetime. The best commercially available low capacitance device for our circuit was the Alpha 4380E diode even with its 76 ns minority carrier lifetime.

The 4001 diode has a low minority carrier lifetime but for our circuit operating at Ka-band the capacitance is too large.

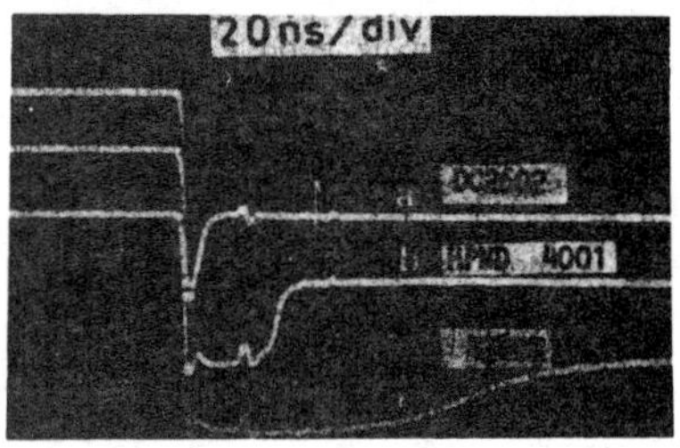

Fig 3 Measured PIN Diode Minority Carrier Lifetimes

Practical Results

Swept frequency static measurements were made of the modulator insertion loss, amplitude imbalance, and phase imbalance. Lowest insertion loss is obtained using Alpha diodes as shown in Fig 4.

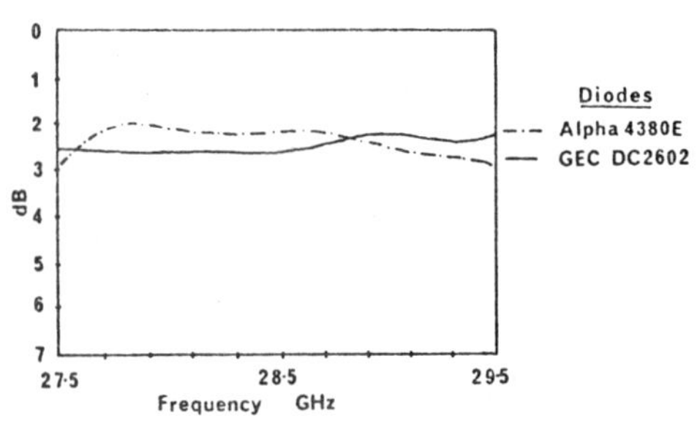

Fig 4 Modulator Insertion Loss

Amplitude imbalance remains less than 1 dB and the phase imbalance has been measured as less than 3 degrees over a 27.5-29.5 GHz band. The dynamic response was evaluated by using a pseudorandom 1023 bit sequence and measuring the demodulated waveform.

Eye patterns for a modulation rate of 70 Mbit/s are shown in Fig 5 and 6.

5 ns/div

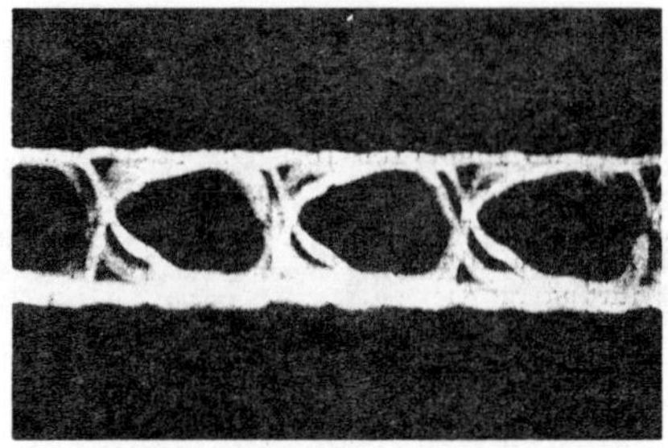

Fig 5 70 Mbit/s Eye Pattern, DC2602 Diodes

5 ns/div

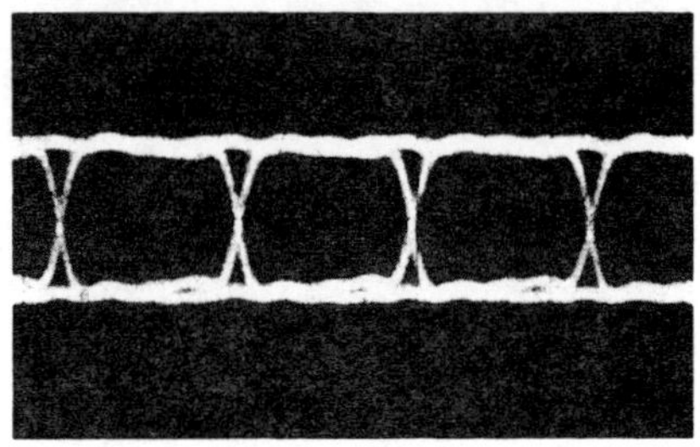

Fig 6 70 Mbit/s Eye Pattern, 4380E Diodes

A circuit using GEC preproduction diodes [3] was tested at the higher rate of 100 Mbit/s and the response is shown in Fig 7

5 ns/div

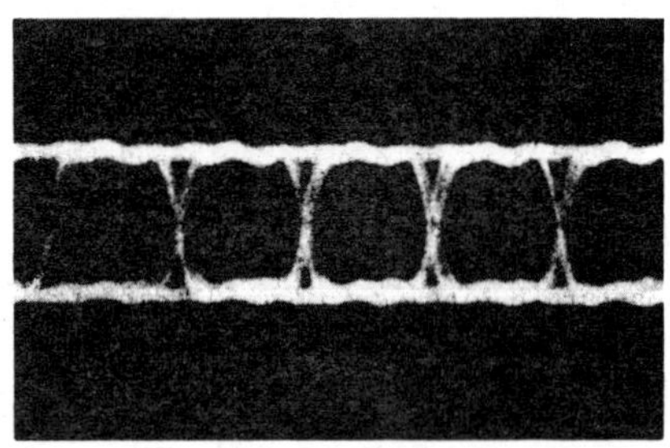

Fig 7 100 Mbit/s Eye Pattern, DC2602 Diodes

Modulators have been tested with r.f. powers up to 150 mW with no observable change in performance. Isolation between the input and output ports has been measured as greater than 25 dB over a 2 GHz bandwidth.

Amplitude Shift Keyed Modulation

A form of modulation that allows simple receiver detection systems to be used and is fairly tolerant of oscillator drift is amplitude shift keying (a.s.k.). The p.s.k. modulator described here can also be used without modification as a simple a.s.k. modulator. The principle of operation of the circuit in this mode is shown in Fig 8 illustrating that a high on-off ratio can be obtained because of the phase cancellation arising from the balanced nature of the circuit when both diodes are reverse biased. With one diode forward biased the circuit is in its on state with a low insertion loss over a wide band as for the p.s.k. modulator.

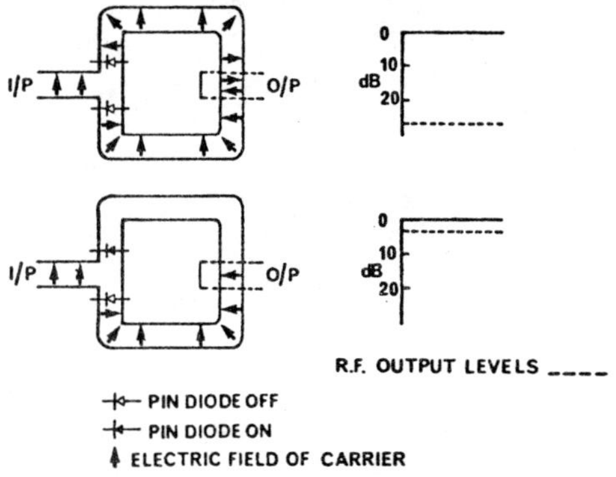

Fig 8 Principle of Operation as an A.S.K. Modulator

Amplitude measurements of the signal in the a.s.k. mode of operation have been made over the full bandwidth and show the on-off ratio to be greater than 25 dB.

Conclusions

A planar bi-phase p.s.k. modulator has been fabricated that operates over a broad bandwidth at Ka band without requiring any tuning or alignment during or after manufacture. The circuit has demonstrated switching speeds of 100 Mbit/s and shows potential for even higher data rates. The balanced nature of the circuit has provided excellent amplitude and phase balance (typically +/-0.5 and +/-3.0 degrees respectively). The circuit has been shown to be versatile in that without modification it can be operated as a simple amplitude keyed shift modulator with an on-off ratio of at least 25 dB. The balanced circuit design is simple in concept and should easily be adaptable to other frequency bands. The planar construction is amendable to low cost high volume production methods. The modulator and drive amplifier design have been incorporated in 29 GHz radio equipment being manufactured by Mullard Hazel Grove Ltd.

Acknowledgements

I would like to thank J D Gilliland of British Telecom for the drive amplifier design. Dr R N Bates of Philips Research Laboraties and I M Clarke of Mullard Hazel Grove Ltd for their assistance in production engineering, and the Director of BTRL for permission to publish this paper.

References

1 MOHAMED S, WILSON P: "29 GHz Radio Systems for Local Distribution Network", 12th European Microwave Conference Proc, Helsinki, 1982, pp 709-714.

2 OGAWA H, AIKAWA M and AKAIKE M: "Integrated Balanced BPSK and QPSK modulators for the Ka-band," IEEE Trans. MTT-30, March 1982, pp 227-234.

3 THORPE W, GILLILAND J.D, "29 GHz E-Plane Bi-Phase Modulator", ELECTRONICS LETTERS, 3rd FEBRUARY 1983 Vol 19 No 3, pp 107-109.

4 SISSON M J, BRIGGINSHAW P M et al: "Microstrip Devices for Millimetric Frequencies", 1982 IEEE MTT-S Digest, pp 212-214.

7.3.1

A FET AMPLIFIER IN FIN-LINE TECHNIQUE+

Jean L'Ecuyer*, G.B. Gajda and W.J.R. Hoefer**

Communications Research Centre
Department of Communications
Ottawa, Canada

* Presently with Apollo Microwaves, 687 Lepine Ave. Dorval, Que.
** With University of Ottawa, Ottawa, Ontario

+ A patent application based on this work is under review.

ABSTRACT

This paper describes the successful demonstration of a stable FET amplifier using fin-line technique. The circuit has been realized in integrated E-plane technology, utilizing two main fin-line ports and a microstrip/coplanar bias circuit. The amplifier was designed for 17 GHz operation using a NE67300 FET and achieved a gain of 6 dB over a bandwidth of 1 GHz.

INTRODUCTION

It can be seen from previous publications that much effort has been expended in deriving a circuit configuration suitable for mounting FETs in fin line technique. Such a mounting structure would enable the realization of FET low noise amplifiers and oscillators in fin line thus making possible completely integrated low noise receivers in fin line or E-plane technique.

Previous efforts at realizing an amplifier in E-plane technique involved employing an E-plane transition from waveguide to a FET device mounted in microstrip [1]. Absorbent material was placed at various points on the circuit as a means of stabilizing the device at frequencies below the cut-off frequency of the waveguide. Losses due to the transitions and absorbent material make this approach unattractive for use in an integrated fin line receiver.

Oscillator circuits using FETs in fin line were published by Jacob and Ansorge [2] and Meinel [3]. These circuits utilized either a cross-shaped or T-shaped arrangement of unilateral fin lines with the device mounted at the junction of the slots. Owing to the unpredictable coupling between the lines connecting the input and output terminals of the device, the designs of these oscillators were largely empirical. In addition, no provision for low frequency stabilization of the FET device was provided.

This paper presents a novel circuit configuration for mounting a FET in fin line. It allows the design of an unconditionally stable amplifier without the use of lossy waveguide-to-microstrip transitions or absorbent material. Other circuit types such as FET oscillators or mixers may also take advantage of the novel mounting structure proposed here.

The RF Circuit

The waveguide housing is a split block type consisting of two parallel rectangular waveguides sharing a common broad wall (septum). The waveguides are isolated from each other, except for a small window in which the transistor is mounted (see Fig. 1). The input of the amplifier begins with a conventional rectangular waveguide that tapers into a single fin, unilateral fin line, to which the gate of the transistor is connected (see Fig. 2). The transistor is installed in an opening in the septum, and the drain lead is bonded to an identical fin in the output waveguide. Fin line short circuits terminate the input and output lines a quarter wavelength beyond the FET terminals, thus presenting an open circuit to the transistor. The source pads of the FET chips are bonded to the printed circuit.

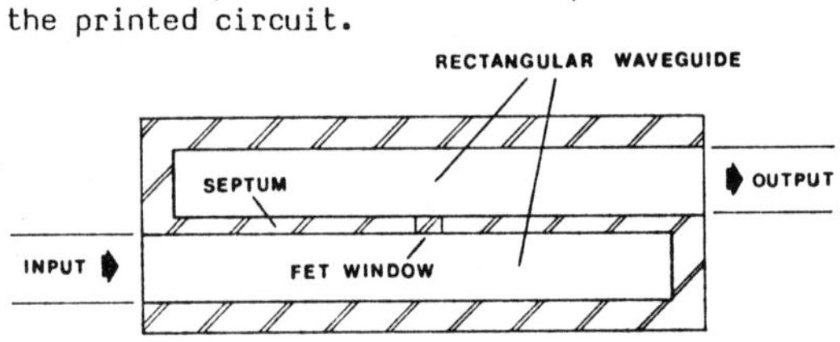

FIGURE 1 SPLIT BLOCK HOUSING

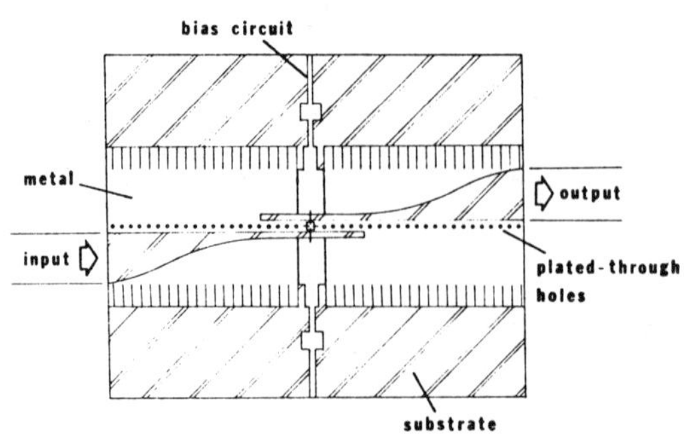

FIGURE 2 METALIZATION PATTERN OF FIN LINE AMPLIFIER

J. l'Ecuyer, G.B. Gajda, and W.J.R. Hoefer, *IEEE MTT-S International Microwave Symposium Digest*, 1986, pp. 287–290.

The circuit is printed on a 0.010 inch thick RT/Duroid substrate (ϵ_r = 2.22). In order to prevent any RF coupling between the two waveguide structures through the substrate, a series of plated-through holes is used as a shield along the common broad wall junction. The isolation between the input and output ports, without transistor, was increased to greater than 30 dB with the plated-through holes (Fig. 3).

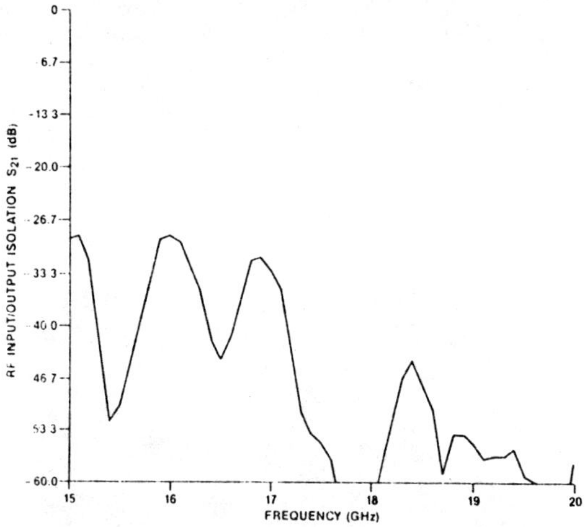

Fig. 3. RF input-to-output isolation of circuit without transistor.

The Bias Circuit

A special bias circuit was designed to present a good match to the transistor at lower frequencies where the FET is potentially unstable and the fin line structure is cut off. In addition, the bias circuit was designed to be transparent to the RF. The bias line begins at the FET terminals, with a coplanar waveguide made from two series slots in the fin (see Fig. 4). Each slot is one half wavelength long at RF and runs from the broadwall to the edge of the fin in order to reflect a short in series with the fin line slot.

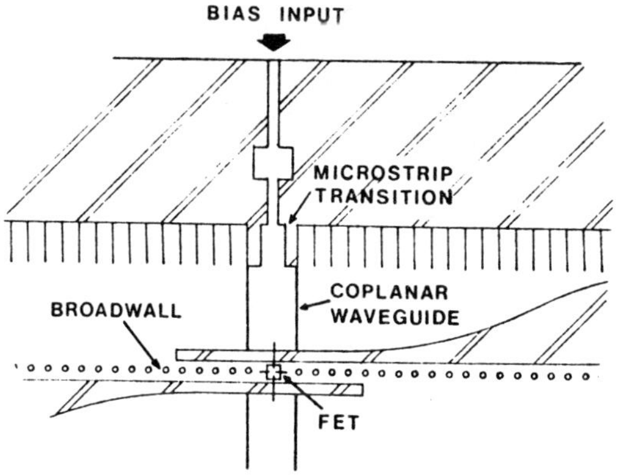

FIGURE 4 DETAILS OF BIAS CIRCUIT

At the outer broadwall plane, the coplanar line makes a transition to microstrip line that continues on to a matched load (50 Ω) through a bias tee. A filter is used to reduce RF leakage through the microstrip. Since the bias circuit is comprised of microstrip and coplanar waveguide, a low (50 Ω) impedance can be presented to the FET terminals at all frequencies from D.C. to the cut-off of the filter.

A plot of measured RF leakage, from one of the waveguide ports through the bias circuit is presented in Fig. 5. From this graph, it may be seen that the leakage is least around 15.2 GHz and begins to increase for higher frequencies. At the design frequency the leakage is approximately -27 dB. This was considered a reasonable value for our experiment.

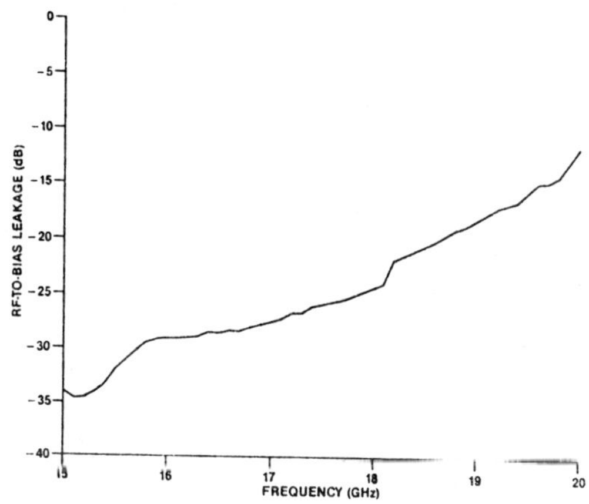

Fig. 5. Leakage from waveguide port to bias port.

For the FET device selected for this work (NE67300), the most critical potentially-unstable region is situated in the neighborhood of 4 to 5 GHz as shown in Fig. 6. Thus, the bias circuit was designed to have a low S_{11} in that critical frequency region. To verify the performance of the bias circuit, the FET was replaced by a 50 Ω chip resistor and the reflection coefficient was measured at the microstrip bias input line. The circuit showed a better than required match for all frequencies especially from 3.5 to 5.6 GHz, where S_{11} was less than -18 dB (see Fig. 7.).

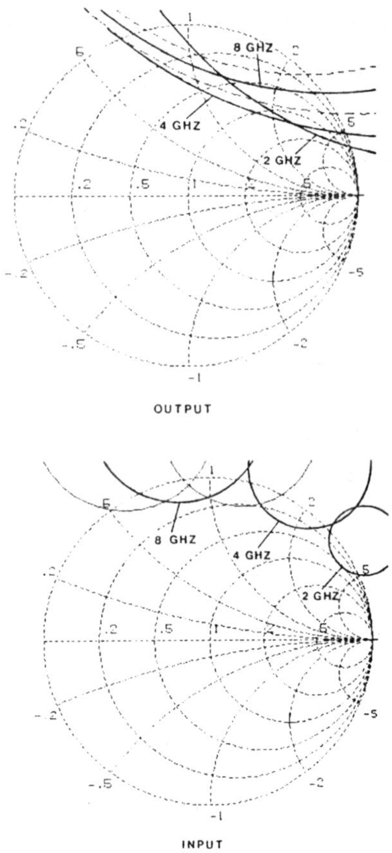

FIGURE 6 NE67300 STABILITY CIRCLES

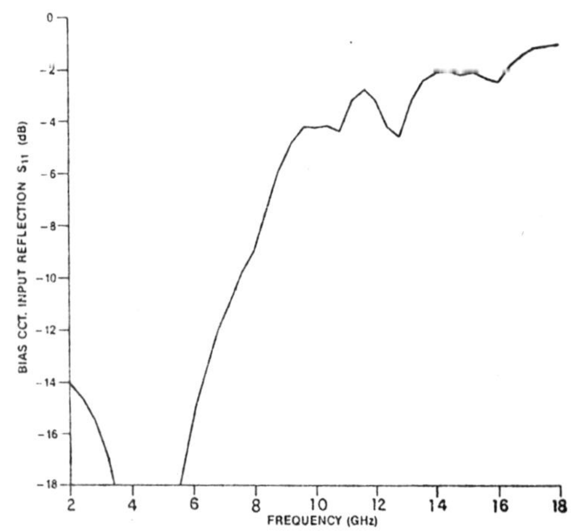

Fig. 7. Input reflection coefficient of bias circuit.

Amplifier Performance

A photograph of the complete amplifier, showing the circuit substrate and housing, is displayed in Fig. 8. The transistor was matched using short-circuited series stubs at both input and output ports. The stubs were designed using data given in [4] and were tuned using silver paint to adjust the length. Tuning was performed directly during measurement through an opening or window in the sidewall of the housing situated above the transistor. After tuning, the window was closed using a brass cover. Since most of the fields are concentrated in the slot of the fin line, the opening had no significant effect on the response of the circuit and tuning was performed easily.

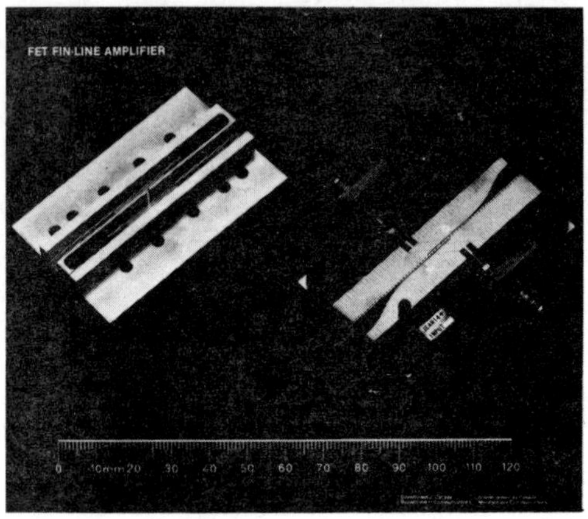

Fig. 8. Photograph of Fin Line FET Amplifier.

The final amplifier displayed unconditional stability over all frequencies. The gain was greater than 6 dB over a 1 GHz bandwidth centered at 16.8 GHz (see Fig. 8). The measured noise figure was 3.5 dB at that frequency.

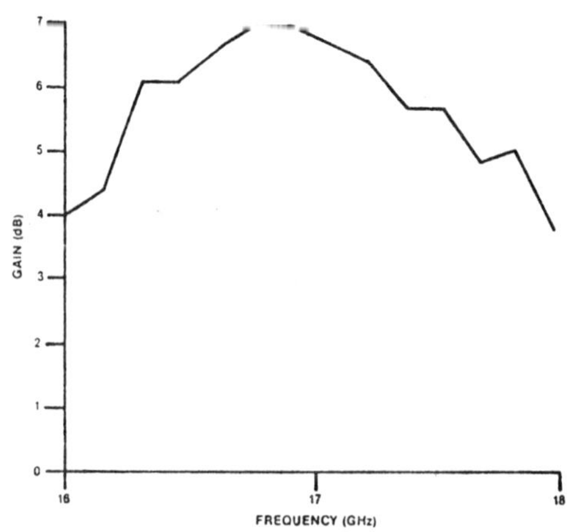

Fig. 9. Measured gain of fin line amplifier.

CONCLUSION

To this date, an unconditionally stable FET amplifier mounted in fin line has not been reported in the open literature. This paper presents a novel approach to achieving stable amplification using FETs in fin line. Stabilization is achieved by use of a unique coplanar/microstrip bias circuit while RF performance is maintained by decoupling the input and output fin lines using a combination of novel approaches including plated-through holes.

A useful amplifier having 6 dB gain and 3.5 dB noise figure over a 1 GHz bandwidth at 17 GHz was designed. This result will encourage further effort in this area to produce multi-stage amplifiers and complete integrated receivers in fin line using FETs. Also, other components such as FET oscillators and mixers may take advantage of the stabilizing and isolating properties of this circuit configuration.

ACKNOWLEDGEMENT

The authors are grateful to P. Deegan for assistance in mechanical design and construction of the housing. Also the helpful discussions and technical assistance of B. Clarke and G. Boucher are gratefully acknowledged.

REFERENCES

[1] H. Ebner, J. Opfer and E.G. Schweppe, "A New Integrated Waveguide Transistor Mount", Conference Proceedings 13th EMC, Nurnberg, West Germany, 1983.

[2] A. Jacob, C. Ansorge, "Stablized Fin-line FET Oscillators" Conference Proceedings 13th EMC Nurnberg, West Germany, 1983.

[3] H. Meinel, "A 30 GHz FET Oscillator using Fin Line Circuitry", Conference Proceedings 11th EMC, Amsterdam, 1981.

[4] Miles Burton, W.J.R. Hoefer, "An Improved Model for Short- and Open-Circuited Series Stubs in Fin Line", IEEE MTT-Symposium Digest, 1984.

[5] A. Beyer, I. Wolff, "Fin Line Taper Design Made Easy", IEEE MTT-Symposium Digest, p. 493, 1985.

7.4.1

Design and performance of microwave oscillators in integrated fin-line technique

Reinhard Knöchel

Indexing terms: *Microwave integrated circuits, Microwave oscillators*

Abstract: Guidelines are outlined for the design of oscillators for various applications, which are realised in integrated fin-line technique. The characteristic features are derived and compared to one another. The results are then applied to cavity-stabilised oscillators, injection-locked amplifiers, and varactor-tuned oscillators, which have all been realised in fin-line technique. Stabilisation by an integrated transmission cavity is shown to be the best configuration for fin-line. For injection-locked amplifiers the stable locking range has been considerably improved by reactance compensation. Finally, it is shown how a varactor diode should be integrated in order to achieve wideband electronic tunability. Performance data of realised (Gunn or Impatt) oscillators are given for each case of application.

1 Introduction

Since there is increasing activity in the millimetre-wave region directed toward the developemnt of low-cost mass-production components such as communication systems, collision-avoidance radar, and environmental-monitoring equipment, several transmission media allowing simple batch production have been proposed. The conventional microstrip proved to be disadvantageous beyond 30 GHz owing to radiation losses spurious coupling and tolerance problems, which degrade the performance of microstrip components with increasing frequency. Hence different planar structures and dielectric waveguides were considered. Among these, integrated fin-line seems to be the most promising.[1-5] It consists of a planar dielectric substrate carrying a metallisation, mounted symmetrically in the E-plane of a standard rectangular waveguide. Thus, integrated fin line combines the benefits of rectangular waveguides and planar circuits. It provides low loss and compatibility with solid-state components, even at high frequencies.

Several papers have recently been published which deal with theoretical investigations of the fin line.[6-13] Their main results are explicit expressions for wave impedance, cutoff frequency, wavelength, and effective dielectric constant of the fundamental mode, and a simple design theory for some types of filters. The state-of-the-art of the fin line theory is, however, far from allowing computer-aided design of more complex circuits. Experimental investigations deal with a description of various filter circuits, *p-i-n*-diode attenuators, single-ended and balanced mixers, and oscillators.[14-18] Owing to the lack of an effecient design theory these components have, so far, been developed by trial and error. Moreover, fin line oscillators take a special place among them: their development appears to be only in a preliminary state.

It is the aim of this paper to give a thorough investigation of the design and performance of oscillator circuits, which are compatible with fin line. Special attention is paid to the diode coupling structure, which must be shaped in a prescribed manner to obtain high efficiency, and the loaded Q-factor of the circuit. From these investigations guidelines will be outlined for the design of three classes of oscillator circuits: cavity-stabilised oscillators, injection-locked amplifiers (i.l.o.), and voltage-tuned oscillators (v.t.o.).

Paper T344 M, received 9th January 1979

Mr. Knöchel is with the Institut für Hochfrequenztechnik der Technischen Universität, Postfach 3329, D-3300 Braunschweig, West Germany

2 Diode-coupling structure

At present, the only solid-state elements providing negative resistance at mm-wave frequencies are the Gunn and impatt diode. These active devices cannot be coupled directly to the fin line slot, because the lowest impedance level obtained with this kind of waveguide is about 100 Ω.[6] The diodes must, however, be operated with a load of the order of 10 Ω or less, if maximum power output is to be delivered. For high-power diodes in particular, the impedance level has to be low. For that reason, the diode coupling structure of Fig. 1, which has been proposed in References 16 and 17, will not be adequate in many cases, because it suffers from too high an impedance level. Furthermore, there is no second degree of freedom in order to satisfy the complex oscillation equation. Hence, a special coupling network is needed to match the diode for the desired output power with low circuit losses. Four possible structures are shown in Fig. 2, where only the side views of the circuits are outlined. A good heat-sink for the diode is achieved by mounting it in

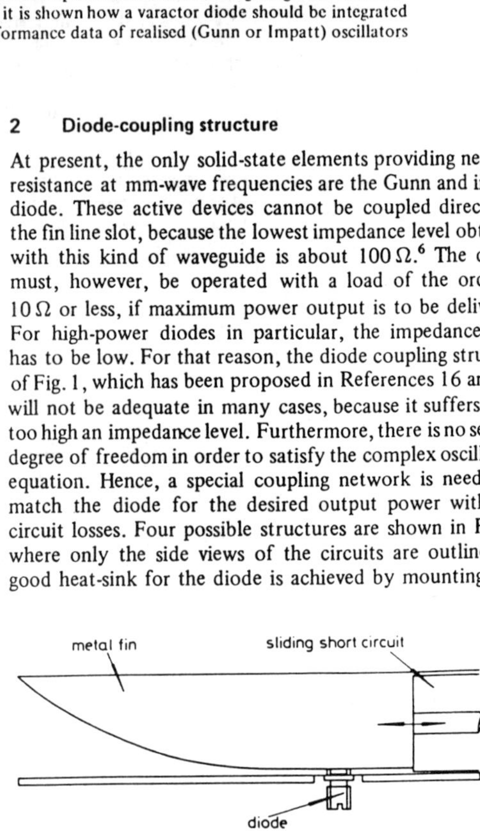

Fig. 1 *Simple fin-line oscillator*

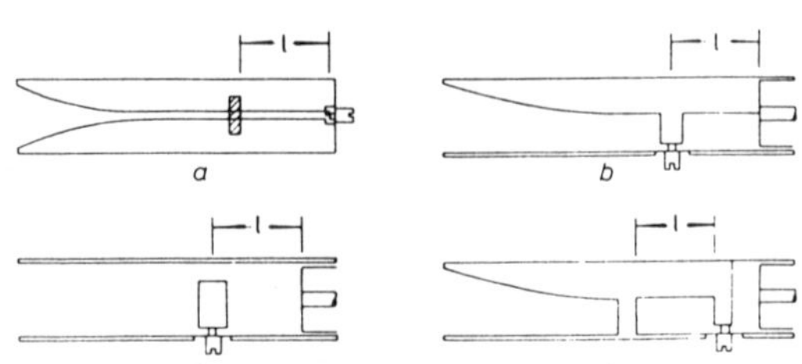

Fig. 2 *Various diode mounts*

a Transmission-cavity mount
b Conductor-strip mount
c Antenna mount
d Another transmission-cavity mount

the ground- or end-plate of the housing waveguide. Each diode mount consists of a resonant circuit and an impedance transforming element.

In the mount of Fig. 2*a*, which has been proposed in Reference 16, the diode is fixed in the end plate of the waveguide. At a distance l, about half a wavelength from the diode, the fin line slot is bridged by a discontinuity. The bridging element acts as a transformer, and the intermediate line to the diode forms the resonant loop.

The conductor-strip coupling of Fig. 2*b*, and the antenna coupling (Fig. 2*c*) show essentially the same features: A resonant cavity of length l to the right of the diode, which is mounted in the ground plate of the waveguide. Impedance transformation is accomplished through a conductor strip or antenna, respectively.

Fig. 2*d* depicts a structure where the active device is coupled by a metallic strip to one end of a resonator of length l, whereas the output waveguide is coupled to the other end. The short-circuited end plate of the housing waveguide to the right of the diode is tuned so that it presents a high impedance, as seen from the terminals of the active device. This mount and that of Fig. 2*a* have in common that the output power is extracted through the resonant circuit. They can be described by the equivalent circuit of Fig. 3*a*. The mounts of Figs. 2*b* and *c* may be modelled by the equivalent circuit of Fig. 3*b*. In these equivalent circuits R_L denotes the load resistance and $-R_D$ the resistance of the active device at maximum generated power. The various n_K are the transmission ratios of the coupling transformers. The resonant cavity is described by its resonant frequency ω_0, unloaded Q-factor Q_0 and internal loss resistance R.

The circuit of Fig. 3*a* is, of course, more versatile than that of Fig. 3*b*. It allows optimisation of either the circuit efficiency or the loaded Q-factor by adjusting n_2, while the active element is matched to the load by means of n_1. The configuration of Fig. 3*b*, on the other hand, exhibits high efficiency as long as the resonant circuit is of high unloaded Q and has small R. It has, however, one degree of freedom less than the other circuit. This difference in principle between the possible structures of Fig. 2 is completed by a practical one: the circuit of Fig. 2*a* has indeed two impedance transformers n_1 and n_2, it is, however, not obvious how to vary the coupling n_1, because the diode is directly connected to the metal fins. Hence this transformer does not provide a real degree of freedom, and a poor circuit efficiency may result. This could lead to the conclusion that the circuit of Fig. 2*d* is the best realisation of a

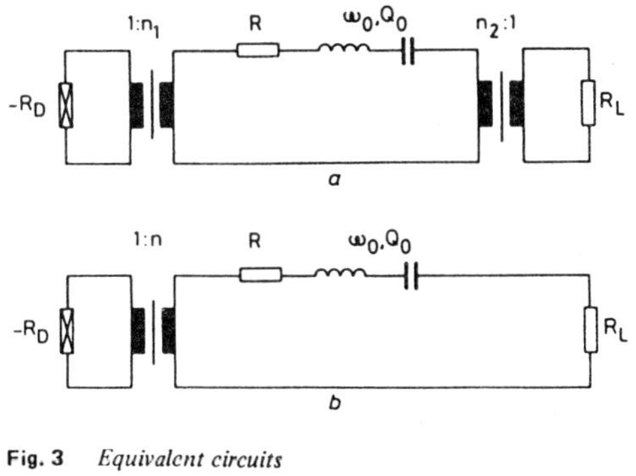

Fig. 3 *Equivalent circuits*

a Of transmission cavity mounts
b Of conductor stirp and antenna mount

fin line oscillator. There are, however, also advantages of the structures of Figs. 2*b* and 2*c*. Both circuits can easily be tuned mechanically by varying the position of the short-circuit plunger.

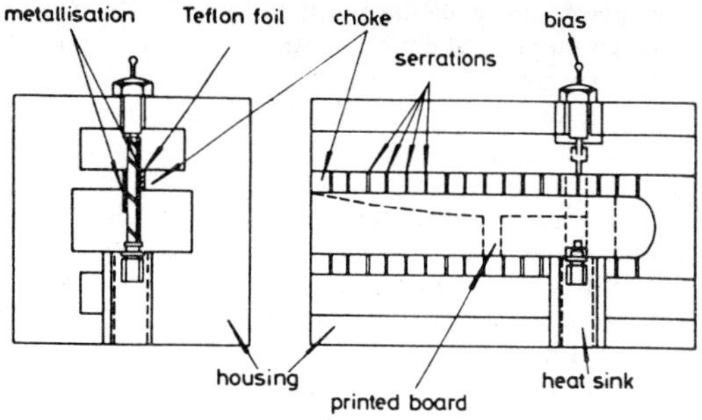

Fig. 4 *Oscillator cross-section*

3 Experimental performance of the diode mounts

The experimental investigations have been conducted at 15 GHz to enable network analyser measurements, but scaling the circuits to higher frequencies is straightforward. We have used commercially available packaged active devices. Fig. 4 shows the cross-section of the basic oscillator structure. The diode is placed in a copper heat sink, and threaded to the housing of the planar printed circuit, which is split in two parts along the E-plane of a milled-in standard rectangular waveguide. The dielectric substrate of the printed board is cut out where the hat of the diode enters. The electrical connection between the metallisation of the planar circuit and the diode is provided by a thin gold-plated ribbon of flexible metal adjacent to the hat of the device, or simply by soldering the hat to the metallisation of the circuit.

Biasing the diode is accomplished through the metallisation of the printed board. This is insulated from the metal housing by a thin Teflon foil. Alternatively, stripline filters may be used at a place, where they do not influence the coupling of the active device.

To protect the circuit against radiation loss through the longitudinally split housing, the choke construction of Reference 1 has been used. To this end the walls of the milled-in waveguide are chosen to be a quarter wavelength thick in the dielectric medium. In addition, the choke is serrated in short segments according to Reference 19, which prevents the flow of currents in a longitudinal direction. Otherwise, resonances are supported which lead to increased circuit power losses of 1 dB and more. This loss is even more pronounced in waveguide sections with a standing wave.

The experimental investigation includes the coupling structures of Figs. 2*b*–*d*. The resonant cavity of each configuration is implemented such that the full height of the rectangular waveguide is used and no metal fin projects into the inside in this section. It has been found that this gives the highest unloaded Q of the cavity and thus the best circuit efficiency of the oscillator. A short-circuit plunger has sometimes been used as an end plate of the housing waveguide, but this short circuit may be integrated on the printed board in the final construction. The integrated short-circuit is formed by defining a metal pattern on the printed board, which divides the terminating section of waveguide in two; both sections are below cutoff and are thus inductive and the actual short-circuit plane lies somewhat behind the mechanical one.

Employing a Gunn diode as an active device, all the mounts show almost the same electrical features in that the rated power can be obtained from the oscillator by properly choosing the physical dimensions of the coupling structures. The tuning characteristic of an oscillator with the mount of Fig. 2*b* is shown in Fig. 5 as an example. The curves are valid for an integrated short circuit which has subsequently been moved away from the diode mount. The oscillation frequency can be tuned over a relative bandwidth of 15%, with an output power variation of less than 1 dB over 10% bandwidth. The external Q of this circuit has been determined by a locking experiment to be about $Q_{ex} \simeq 150$. The tuning performance of such oscillators is comparable with the standard post-coupled waveguide Gunn oscillators. Allowing a metal fin to penetrate into the resonant cavity expands the tuning range of this oscillator. At the same time, however, efficiency and frequency stability are also reduced, due to a degradation of the loaded Q.

The antenna mount (Fig. 2*c*) leads to a somewhat more narrowband performance. In order to bias the diode, a thin wire was used to connect the diode coupling structure and the bias filter. The wire, however, turned out to influence the impedance level at the active device, and had to be properly positioned. Hence, the antenna coupling structure is more suitable for unbiased semiconductor devices. The mount of Fig. 2*d* requires a more complicated tuning procedure, although it allows a greater degree of freedom in choosing low impedance levels.

The mounts of Figs. 2*b* and 2*d* have also been operated with an impatt diode. It is well known that a crucial point for this kind of oscillator is the build up of bias oscillations, especially if there is shunt capacitance at the bias port of the diode. In the case of fin line oscillators, this shunt capacitance, which is due to the input capacitance of the r.f. choke, is higher than for the coaxial chokes which are normally used. For that reason, special care had to be taken to damp out any bias oscillations. To this end standard techniques have been applied.[20]

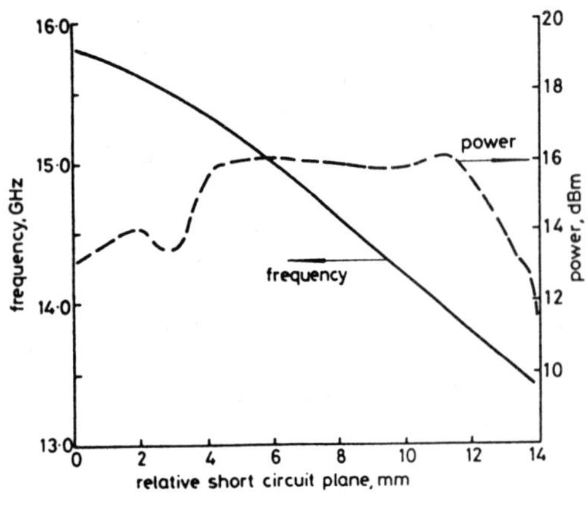

Fig. 5 *Performance of the conductor strip mount*

4 Integrated-cavity stabilized oscillators

In many cases, a better frequency stability is required than may be obtained from single tuned oscillators. To this end the oscillator structures just descibed may be coupled to high-Q cavities, which are integrated on the printed board together with the diode mount. The structures realised are outlined in Fig. 6. We have to distinguish between three different types of stabilisation,[21] which employ a reaction (*a*), reflection (*b*), or transmission (*c*) cavity.

The reaction cavity is formed by a resonant slot acting as a bandstop filter (Fig. 7*a*). Natural frequency, coupling coefficient and unloaded Q-factor depend on the geometri-

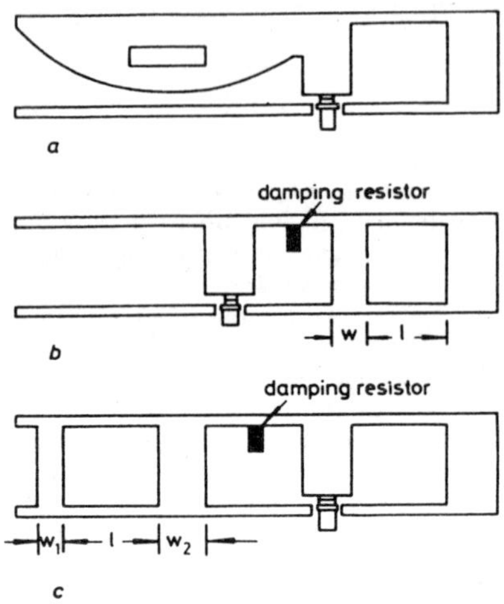

Fig. 6 *Various stabilised oscillators*

a With reaction cavity
b With reflection cavity
c With transmission cavity

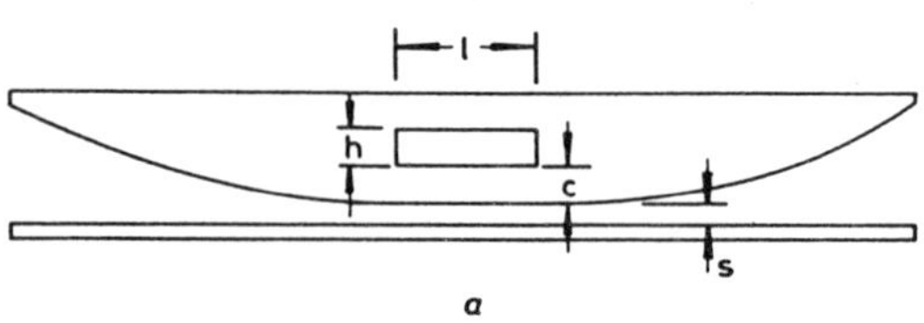

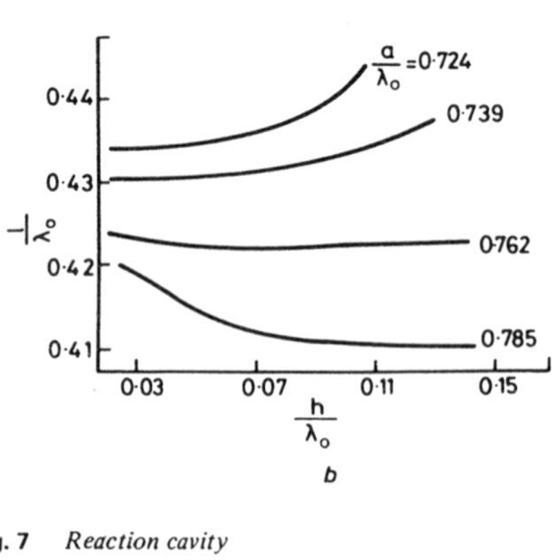

Fig. 7 *Reaction cavity*

a Circuit pattern
b Physical dimensions
a = broad dimension of housing waveguide
λ_0 = free space wavelength

cal dimensions. The unloaded Q increases with increasing slot height h. The resonant frequency is essentially given by the slot length l. It depends weakly on h in a manner which is depicted in Fig. 7*b*. These results have been found experimentally. The coupling coefficient of the reaction cavity turns out to be rather tight in practice, because it depends mainly on the slot height h but only weakly on its distance c from the fin line slot. The parameter h cannot, however, be reduced sufficiently to decrease the coupling, because this would also appreciably decrease the unloaded Q. For a constant fin line slot height $s/\lambda_0 = 0{\cdot}05$ and constant $c/\lambda_0 = 0{\cdot}1$, the coupling coefficient (which is the resistance at resonance, normalised to the wave impedance of the line) dropped by 300% when h/λ_0 was changed from 0·15 to 0·04. At the same time the unloaded Q decreased from $Q_0 = 1500$ to about 700. When h/λ_0 was held constant and c/λ_0 changed from 0·1 to 0·05, this increased the coupling coefficient by 25%, and affected the unloaded Q only weakly.

The reflection cavity (Fig. 6*b*) is mounted in shunt with the diode port, whereas the transmission cavity (Fig. 6*c*) is connected in series with the load. Both cavities are essentially formed by the same planar structure (resonant slot of length l) which is coupled by inductive strips to the output waveguides. The coupling coefficients and natural frequencies have again been determined experimentally.

As is well known, the intermediate line between a reflection or a transmission cavity and the diode mount must contain a damping resistor in order to ensure a single-mode operation, when the natural frequency of the cavity is varied. As a consequence, an additional reflection cavity means no improvement of the electrical performance of a fin line oscillator, as can be seen by comparing the circuits of Figs. 2*b* and 6*b*. Hence, an oscillator with an integrated fin line cavity of the reflection type is of no practical importance.

Furthermore, it has to be noted that stabilisation of an oscillator by an integrated cavity is advantageous only for the diode mounts of Figs. 2*b* and 2*c*. In these cases the additional cavity provides a means for an independent choice of the loaded Q-factor of the oscillator assembly. This additional degree of freedom is, however, an inherent feature of the diode mount of Fig. 2*d*. It is hence not necessary to use cavity stabilisation in connection with the latter mount.

Integrated-cavity stabilised oscillators with Gunn elements have hence been constructed only with the diode mount of Fig. 2*b*. The load impedance as seen from the diode has been measured with a network analyser. It is shown for a reaction-cavity-stabilised oscillator in Fig. 8 (solid line). The loop in the locus curve denotes the resonance effect of the stabilising cavity. Single-mode operation can be seen to be achieved, because the impedance locus curve of the active device (approximately sketched by the broken line) intersects the load impedance locus curve only once. The impedance locus curve of the transmission-cavity-stabilised oscillator looks approximately the same, but can be shaped with one more degree of freedom, because input and output coupling are sperately adjustable. Both reaction- and transmission-cavity-stabilised oscillators could be improved in f.m. noise performance by about 20 dB with a sacrifice of 6 dB in power. An absolute value of f.m. noise, which can be achieved, is 0·1 Hz/$\sqrt{\text{Hz}}$ effective frequency deviation at a modulation frequency of 100 kHz. Comparable results for oscillators which are stabilised by a high-Q TE_{01n}-mode circular waveguide cavity are 30 dB reduction in f.m. noise at the expense of typically 1 to 2 dB in power. This is of course due to the high unloaded Q of these cavities, which are more than two to three times an order of magnitude larger.

As has already been explained, the coupling of the reaction cavity cannot be made much looser, in order to increase the power output. Hence this kind of stabilisation is appropriate only if a comparatively extreme noise reduction is required. The transmission cavity, however, can be adopted with much more flexibility. In addition, it exhibits the feature of noise reduction at high modulation frequencies[22] contrary to the reaction cavity. Therefore, it should be used exclusively for integrated-cavity stabilisation of oscillators.

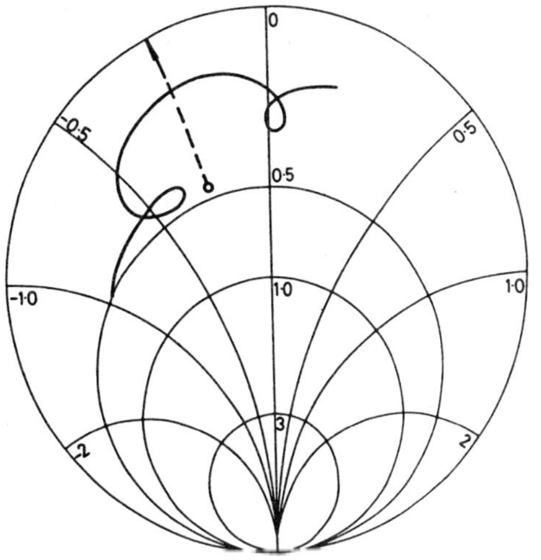

Fig. 8 *Impedance locus curve of reation-cavity-stabilised oscillator*

5 Broadband locking circuit

Since we can get high frequency stability only at the sacrifice of a significant portion of the output power, there is need for an amplifying module. The basic oscillator structure in the injection-locked mode may serve for this purpose. But, in general, the locking range of the circuit will not be large enough, if the same module is to be used in different frequency channels. It is hence desirable to increase the locking range of the oscillator. This can be achieved by using reactance-compensation techniques, as has been proposed in Reference 23. The method is to lower the Q-factor of the passive circuitry by adding another (the complementary) resonance circuit. Equivalent circuit and fin line realisations are shown in Figs. 9*a* and 9*b*, respectively. If the frequency of the locking signal deviates from the free-running frequency, one resonant loop stores capacitive and the other inductive energy, which can nearly cancel one another provided that the coupling of the circuits is correct. Analysing this compensation circuit, one finds that there is an optimum coupling with respect to a maximum locking range, which depends on the magnitude of the injected power or on the locking gain.[24] The compensation loop can be designed in order to obtain a linear phase relationship between the input and the output signal (constant time delay), or constant output amplitude (good a.m. compression), or a compromise between these features as well.

Such a broadband locking circuit has been implemented as depicted in Fig. 9*b*. The diode-structure of Fig. 2*b* has been used. The compensating resonator is provided by a capacitive stub in parallel with the output waveguide. It is tuned together with the length of the intermediate line between the stub and the diode coupling network. The coupling between the various loops has been adjusted by monitoring the load impedance as seen from the diode port with a network analyser. The locking range has been optimised for locking gains of 20 dB and 10 dB, respectively. It was broadened by a factor of 3·5 at 20 dB and doubled at 10 dB, which is roughly in accordance with the theoretical results of Reference 24. Absolute values are 1·7% relative bandwidth at 20 dB and 4·5% at 10 dB for a Gunn oscillator with a centre frequency of 14·5 GHz. These results are very similar to what can be obtained with oscillators in rectangular waveguide techniques. The amplitude and phase characteristics of the compensated injection-locked oscillator are shown in Fig. 10 for a power gain of 20 dB.

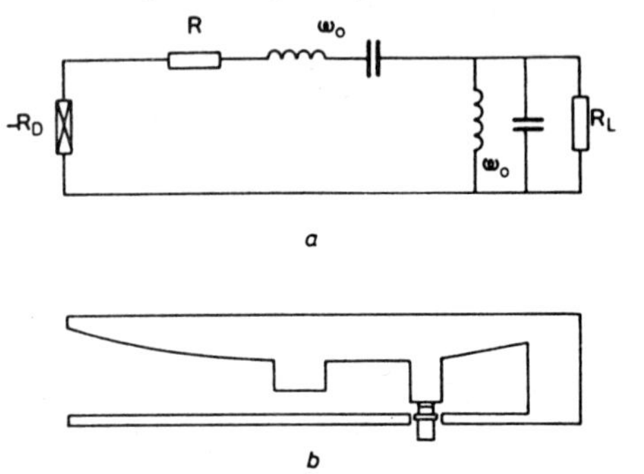

Fig. 9 *Broadband-locking circuit*
a Equivalent circuit
b Circuit pattern

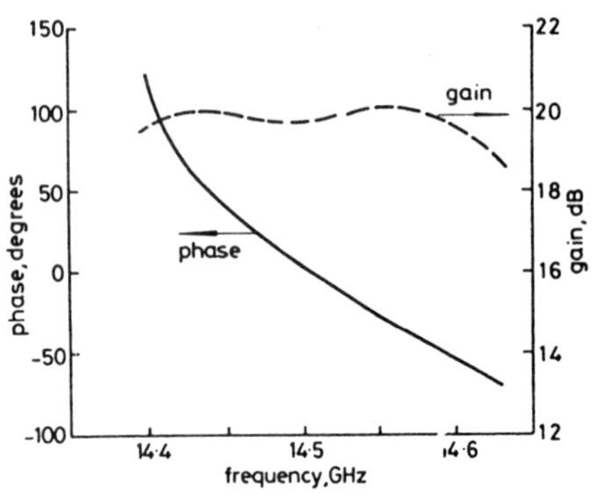

Fig. 10 *Performance of broadband-locking circuit*

6 Varactor-tuned oscillator (v.t.o.)

A severe disadvantage of the integrated circuits described here is the lack of an inexpensive possibility of mechanically tuning the frequency. Therefore, it is very important to provide electrical tuning. Then active frequency stabilisation can be easily accomplished employing well-known techniques such as automatic phase control.

Electrical tuning capability is most easily introduced by coupling a varactor diode to the basic oscillator structure. In a final application (especially at millimetre wavelengths) diode chips will be used because of their small parasitics and their low price. We have, however, used a varactor diode in an l.i.d. package owing to its easy mechanical handling. It is mounted in a separate coupling structure between the active device and the waveguide load, at a length *l* away from the diode mount. The equivalent circuit is sketched in Fig. 11*a* and the fin line slot pattern in Fig. 11*b*. We have used the diode mount of Fig. 2*b*. The length *l* has been optimised experimentally with respect to a large tuning range. The degree of coupling of the varactor diode (i.e. the power loss in this diode) is adjusted by choosing the physical dimensions of the varactor coupling strip. The l.i.d. package is soldered directly to this strip with an In-Ag alloy. The position of the varactor package must be adjusted carefully, because it has a strong influence on the performance of the oscillator.

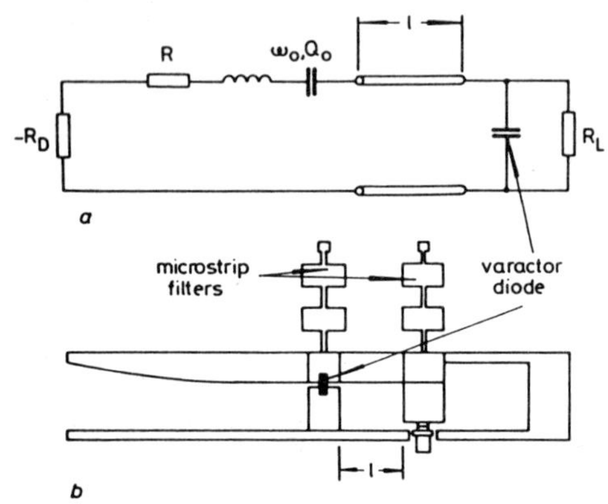

Fig. 11 *Varactor-tuned oscillator*
a Equivalent circuit
b Circuit pattern

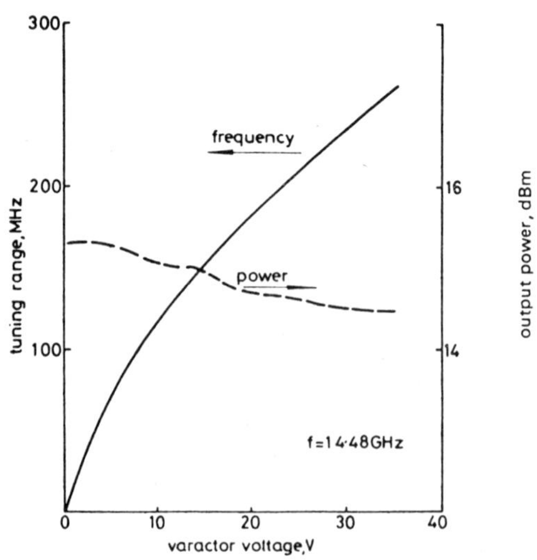

Fig. 12 *Performance of varactor-tuned oscillator*

The active device (Gunn diode) and the varactor diode have to be biased separately. For that reason, stripline filters are used, as is depicted in Fig. 11*b*. The design of these filters is simple, because the electromagnetic fields both in stripline and in fin line are found to be decoupled to a high degree in the arrangement shown. A typical example for a tuning characteristic of the oscillator structure of Fig. 11*b* is shown in Fig. 12. It corresponds to a case where the power dissipated in the varactor diode is approximately 0·5 dB, which means the output power of the oscillator degrades by this value, when the varactor diode is coupled and the free-running frequency readjusted. The tuning range is 260 MHz when the varactor bias varies between 0 and 34 V, with a power reduction of less than 1 dB over the tuning range. Again, the results are comparable to the familiar two-post v.t.o. structure in rectangular waveguide. However, as a difference, the integrated structure gives the possibility to use non encapsulated varactor diodes with low parasitics.

7 Conclusions

Various oscillator circuits have been developed in integrated fin line technique. There are several structures for coupling the active device to the passive circuitry. These diode mounts have been investigated with respect to circuit efficiency and obtainable loaded *Q*-factor. A comparison has shown that an active device should be coupled to the waveguide by a planar pattern which resembles the post-coupling structure familiar from conventional rectangular waveguide (Fig. 2*b*), by an antenna coupling (Fig. 2*c*), or through a transmission cavity, where care has to be taken that the coupling at both the diode and the load port can be adjusted separately (Fig. 2*d*). To arrive at high circuit efficiency together with a high-loaded *Q*-factor, the unloaded *Q*-factor of the frequency determining resonator should be made as high as possible. That means that no metallic fin should penetrate into the housing waveguide across the resonant length. The performance of the fin line oscillator circuits is then similar to that of conventional rectangular-waveguide oscillators.

The coupling structure utilising a transmission cavity gives considerable freedom since the oscillator loaded *Q*, which is a measure of frequency stability, can be traded with output power. The other coupling structures do not show this feature. Frequency stability (loaded oscillator *Q*) may, however, be increased for these patterns, too, by coupling them to an external integrated cavity. The possible features of such an assembly have been discussed.

The fin line oscillators can, of course, be operated as injection-locked amplifiers. Reactance compensation has been used to increase the locking range. This could be realised by integrating an additional capacitive stub on the printed board at a special distance from the diode mount. Thus, the locking range has been increased by a factor of 3·5 at 20 dB locking gain and doubled at 10 dB gain, respectively.

Finally, electronic tunability has been provided by coupling a varactor diode to the integrated oscillator. A structure is shown, which allows one to adjust the power loss in the varactor to vary the tuning range. Varactor tuning is possible with the diode mounts recommended above. The experimental performance of a varactor-tuned Gunn oscillator with medium tuning range and low varactor loss has been shown as an example.

8 Acknowledgement

The author is indebted to the Deutsche Forschungsgemeinschaft for financial support.

9 References

1 MEIER, P.J.: 'Two new integrated-circuit media with special advantages at millimeter wavelengths'. IEEE-MTT Symposium digest, Arlington Hights, Ill., USA, 1972, pp. 221–223

2 MEIER, P.J.: 'Equivalent relative permittivity and unloaded Q-factor of integrated finline', *Electron. Lett.*, 1973, **9**, pp. 162–163

3 MEIER, P.J.: 'Integrated fin-line millimeter components'. IEEE-MTT G Symposium digest, Atlanta, Georgia, USA, 1974, pp. 195–197

4 MEIER, P.J.: 'Integrated fin-line millimeter components', *IEEE Trans.* 1974, **MTT-22**, pp.1209–1216

5 MEIER, P.J.: 'Integrated fin-line, a versatile and proven millimeter transmission line', *Microwave J.* 1976, **19**, pp. 24–25

6 HOFMANN, H.: 'Fin-line dispersion', *Electron. Lett.* 1976, **12**, pp. 428–429

7 SAAD, A.M.K., and BEGEMANN, G.: 'Electrical performance of fin-lines of various configurations', *IEE J. Microwaves, Opt. & Acoust.* 1977, **1**, pp. 81–88

8 HOFMANN, H.: 'Calculation of quasi-planar lines for mm-wave application'. IEEE MTT-Symposium digest, San Diego, California, USA, 1977, pp. 381–383

9 HOFMANN, H.: 'Dispersion of planar waveguides for millimeter-wave application', *Arch. Elektron. & Uebertragungstech.*, 1977 **31**, pp. 40–44

10 HOEFER, W.J.R.: 'Fin-line design made easy'. IEEE-MTT Symposium digest, Ottawa, Canada, 1978, p. 471

11 SAAD, A.M.K., and SCHÜNEMANN, K.: 'A simple method for analyzing fin-line structures'. IEEE MTT Symposium digest, Ottawa, Canada, 1978, pp. 405–407

12 SAAD, A.M.K., and SCHÜNEMANN, K.: 'A first order design theory for integrated fin-line circuits'. Proceedings of the 6th Microcoll., Budapest, 1978, pp. III–7/46.1–III–7/46.4

13 SAAD, A.M.K., and SCHÜNEMANN, K.: 'Field description for generalized (multi-slot) fin-line structures'. Proceedings of the 8th European Microwave Conference, Paris, 1978, pp. 101–105

14 MEIER, P.J.: 'New developements with integrated fin line and related printed millimeter circuits'. IEEE-MTT Symposium digest, 1975, pp. 143–145

15 MEIER, P.J.: 'Planar multiport millimeter integrated circuits'. IEEE-MTT Symposium digest, San Diego, California, 1977, pp. 385–387

16 HOFMANN, H., MEINEL, H., and ADELSECK, B.: 'New integrated mm-wave components using fin-lines'. IEEE-MTT Symposium digest, Ottawa, Canada, 1978, pp. 21–23

17 COHEN, L.D., and MEIER, P.J.: 'Advances in E-plane printed millimeter-wave circuits'. IEEE-MTT Symposium digest, Ottawa, Canada, 1978, pp. 27–29

18 BEGEMANN, G.: 'An X-band balanced fin-line mixer" IEEE-MTT Symposium digest, Ottawa, Canada, 1978, pp. 24–26

19 TOMIYASU, K., and BOLUS, J.J.: 'Characteristics of a new serrated choke', *IRE Trans.* 1956, **MTT-4**, pp. 33–36

20 BRACKETT, C.A.: 'The elimination of tuning-induced burnout and bias-circuit oscillations in IMPATT oscillators', *Bell Syst. Tech. J.*, 1973, **52**, pp. 271–306

21 KNÖCHEL, R., SCHÜNEMANN, K., and BÜCHS, J.D.: 'Theory and performance of cavity-stabilised microwave oscillators', *IEE J. Microwaves, Opt. & Acous.* 1977, **1**, pp. 143–165

22 SCHIEK, B., and SCHÜNEMANN, K.,: 'Noise of negative resistance oscillators at high modulation frequencies', *IEEE Trans.*. 1972, **MTT-20**, pp. 635–641

23 AITCHISON, C.S., DAVIS, R., and GIBSON, P.J.: 'A simple diode parametric amplifier design for use at S, C, and X-band', *ibid.*, 1967, **MTT-15**, pp. 22–31

24 KNÖCHEL, R., and SCHÜNEMANN, K.: 'Reactance compensation techniques for injection locked amplifiers', *Arch. Elektron. & Uebertragungstech.*, 1979, **33**, pp. 49–56

7.4.2

K. Solbach, B. Adelseck, F. Sicking*)

Fin-line Varactor-Tuned Gunn-Oscillators for 35 and 60 GHz

During the last years some fin-line Gunn-oscillator designs have been represented [1 – 3]. The grating design [1], Fig. 1(a), relies on the stop-band impedance transformation feature of periodic waveguide perturbations (grating) to match the low source impedance of semiconductor active devices. In the grating oscillator the frequency of oscillation mainly is determined by the stop-band frequency while the impedance level depends on the number and magnitude of the perturbations. Since long grating structures have to be employed the design does not achieve optimum circuit efficiency.

An other design [2], Fig. 1(b), virtually translates the well-known metal waveguide post-structure into planar circuitry. While the 3-dimensional post-structure in metal waveguide features a rather wide range of impedance transformation, the planar structure is rather limited in this respect and thus is less flexible than pure metal waveguide circuits. On the other hand there is a lack of design data concerning the planar structures so that this alternative has found little application so far.

In this work a very compact fin-line oscillator circuit of simple design is presented that combines high circuit efficiency with easy implementation of a tuning-varactor.

The Basic Oscillator Circuit for 35 GHz

In Fig. 2 the basic oscillator circuit for 35 GHz is shown. The standard packaged Gunn-element is soldered to a fin-line section of length l_g which, together with the fin-line stub of length l_s, establishes the transformation circuit needed to match the source impedance of the Gunn-element. The variable short-circuit behind the Gunn-element may be used to tune the oscillator within a few percent in frequency.

The fin-line used in the Ka-band (26.5 – 40 GHz) experiments was 0.25 mm thick RT/Duroid ($\epsilon_r = 2.3$) with 0.3 mm slot widths with RG – 96 metal waveguide mount. The effective permittivity of the fin-line mode thus was $\epsilon_{eff} \approx 1.1$ and the line impedance $Z_L \approx 200\ \Omega$ [4].

In Fig. 3 the impedance transformation path for the oscillator circuit in Fig. 1 is plotted in the Smith-chart. Discontinuity effects due to the fin-line series junction and the truncation of the fin-line in the plane of the Gunn-element have been included in the plot; it has been found experimentally that the series junction mainly causes small shifts in the reference planes of the junctions on the order of 0.1 mm. The Gunn-element mounting discontinuity may be described by a shift of the terminal reference plane of about .4 mm, thus reducing the length l_g. The approximate terminal impedance of the Gunn-element was taken from [3].

The stub is operated below its $\lambda/4$ – resonance in this plot. An alternative operating point above resonance is indicated in brackets (2 – 2'). It can be seen that the two line sections l_g and l_s result in a $\lambda/2$ – resonator, terminated

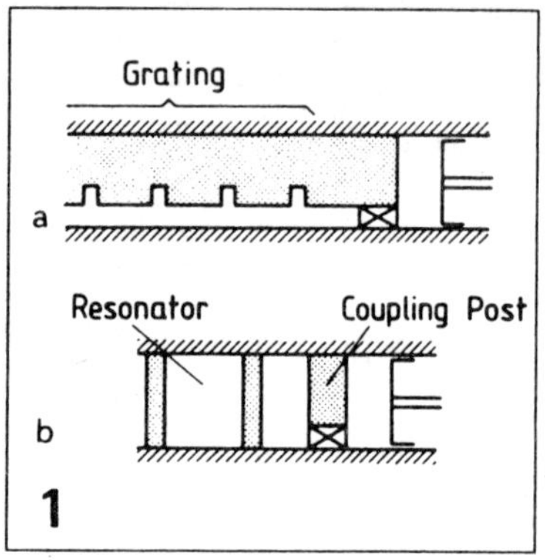

Fig. 1: Fin-line oscillator circuits due to Hofmann [1] (a) and Knöchel [2] (b)

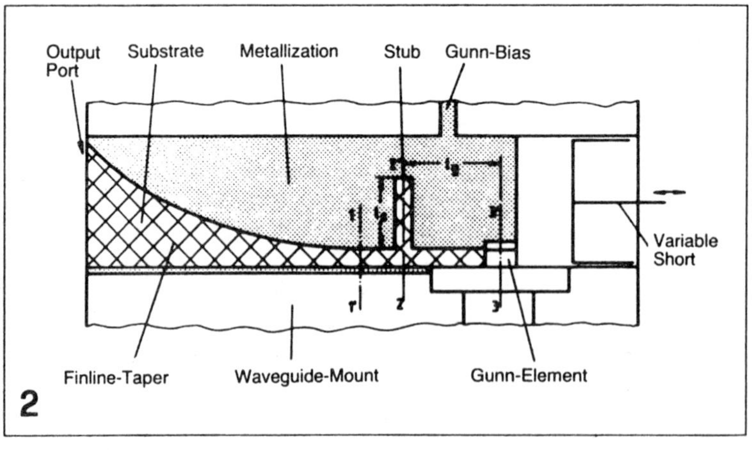

Fig. 2: The basic fin-line Gunn-oscillator structure

*) AEG-Telefunken, Geschäftsbereich Hochfrequenztechnik, A1 E32, D-7900 Ulm, FRG

K. Solbach, B. Adelseck, and F. Sicking, *Mikrowellen Magazin*, Vol. 9, No. 1, 1983, pp. 64–66.

at one end in the Gunn-element and at the other end in a short-circuit. The resonator is tapped near its voltage maximum to match the relatively high line impedance.

It is clear that to change the frequency of oscillation both line lengths must be changed with the ratio l_g/l_s kept constant; for an adjustment of the load impedance match only the ratio l_g/l_s needs to be varied. From this it follows that the adjustment of this specific oscillator circuit is not as simple as it is the case for the grating oscillator. On the other hand this configuration has higher circuit efficiency due to its shorter length. Using Gunn-elements MA 47177 (150 mW minimum power devices) 200 mW were produced at 33 GHz using the described circuit. In an oscillator designed for 35 GHz 130 mW CW were produced and 500 mW peak power were obtained in a pulsed mode of operation (U = 10 V, puls length ≥ 10 ns, PRF = 200 kHz).

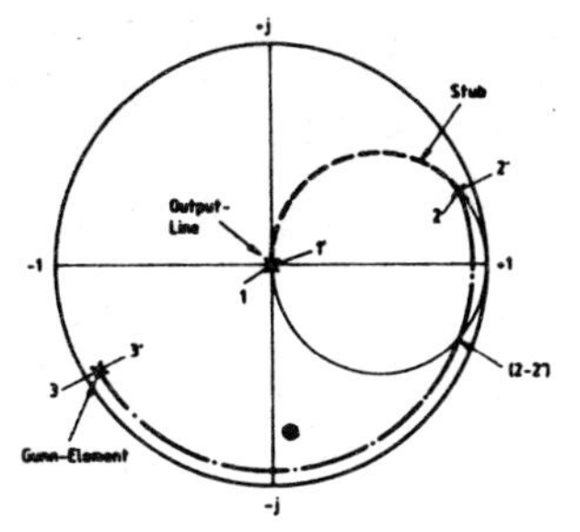

3

Fig. 3: The impedance transformation path for the basic oscillator circuit

Fig. 4: The varactor-tuned fin-line stub

Fig. 5: The measured tuning characteristic of a varactor-tuned Gunn-oscillator

Varactor-Tuning

One major advantage of the design presented here is that the oscillator may be varactor-tuned by a simple tuning the stub. In Fig. 4 the circuit used for varactor-tuning is shown.

The stub short-circuit termination in this configuration is formed by a microstrip resonator printed on the back of the fin-line substrate in order to achieve a d.c. separation of the varactor and the Gunn-element. The varactor chip MA 45064 (C_{j4} = 0.4 pF) is soldered to one side of the stub and a 200 μm wide bond-ribbon is used to connect the varactor-anode to the other side of the stub. The resultant series circuit of ribbon-inductance and varactor-capacitance is dominated by the ribbon contribution and is thus inductive on the order of j 50 Ω. Due to this impedance the stub was detuned to a higher frequency by about 4 GHz. The detuning-effect was corrected by adding a small piece of dielectric material (ε_r = 6) on the front-section of the stub.

Using a medium-power Gunn-element MA 47172 (50 mW minimum power device) an electronic tuning range of 700 MHz was achieved, Fig. 5. It was found that even with zero bias no varactor current was produced, indicating that the coupling of the varactor to the oscillator circuit was still rather low. With foreward bias up to the onset of rectification current 200 MHz additional tuning was achieved without degradation of the 35 to 40 mW output power level. The external Q of the oscillator circuit, measured by load-pulling, was found at about 140. This is essentially the same figure as has been measured for the oscillator without varactor, confirming the assumption of a low coupling of the varactor. Modified versions of the presented oscillator circuit have yielded tuning ranges in excess of 1 GHz with power variations from 15 mW to 40 mW.

During the experiments it has been observed that the adjustment of the soldered varactor-chip and the connecting bond wire was badly reproducible since both the position of the varactor relative to the stub termination and the effective length and the curvature of the wire influence the coupling to the fin-line stub. In order to improve the reproducibility of the varactor circuit beamlead varactor devices would be helpful; it is expected that in the near future such devices will become commercially available.

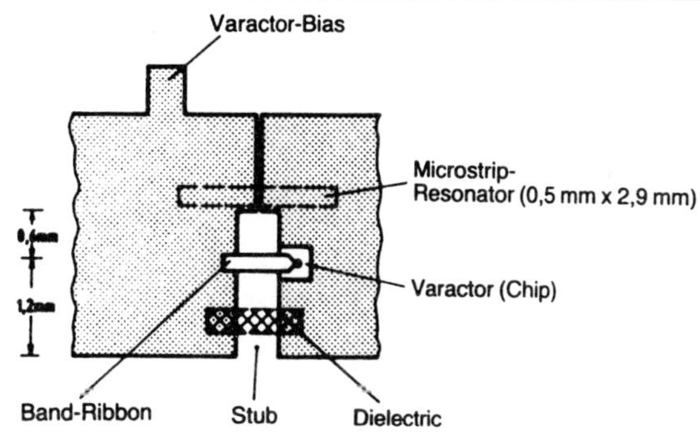

4

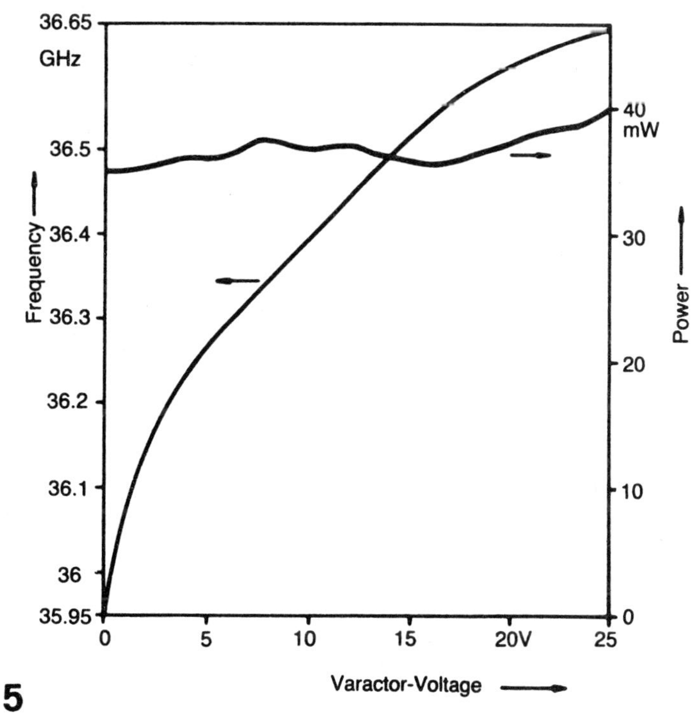

5

The 60 GHz Oscillator Circuit

The design of the 60 GHz Gunn-oscillator starts with an investigation of the source impe-

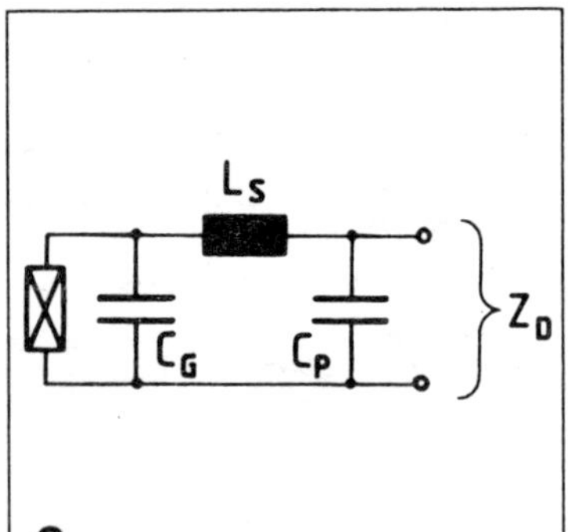

6

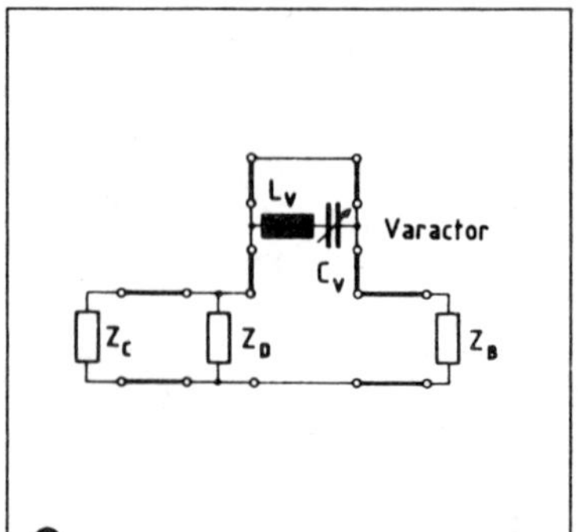

9

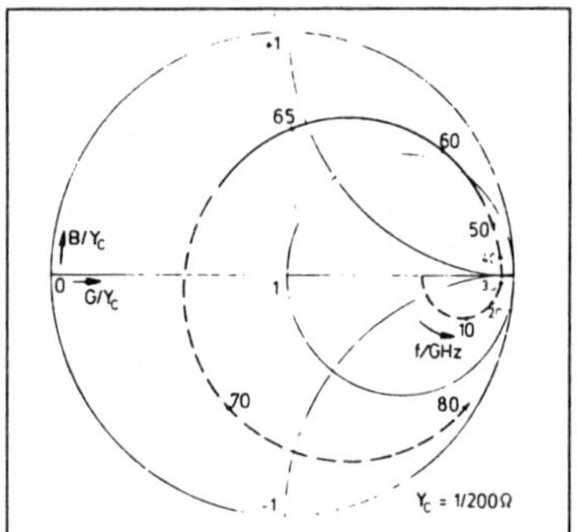

7

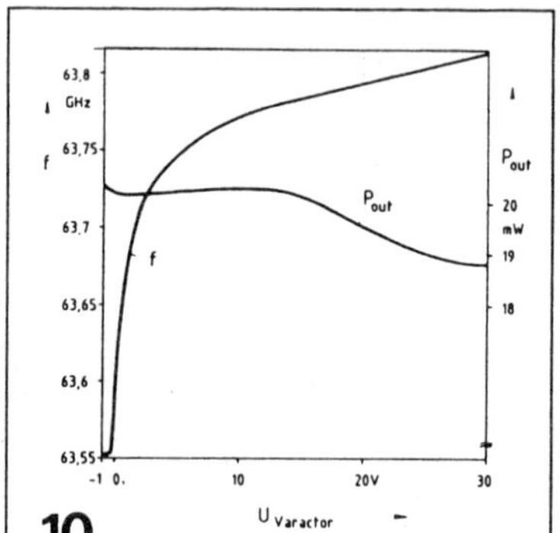

10

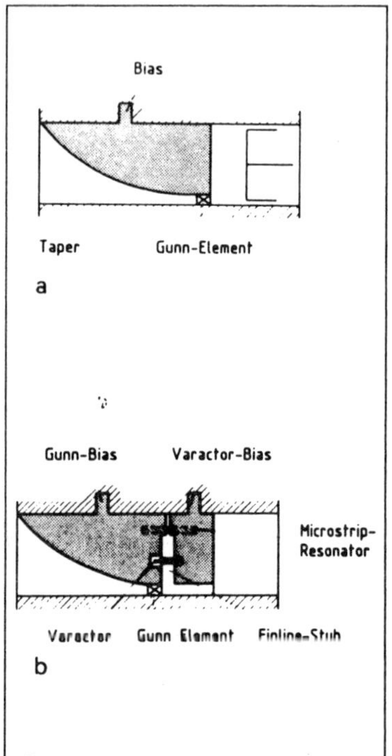

8

Fig. 6: Equivalent-circuit representation of Gunn-element

Fig. 7: Negative conjugate complex source admittance of Gunn-element as a function of the frequency

Fig. 8: Basic 60 GHz Gunn-oscillator circuit (a) and modified circuit incorporating tuning varactor (b)

Fig. 9: Equivalent-circuit representation of varactor-tuned Gunn-oscillator

Fig. 10: Tuning characteristic of varactor-tuned Gunn-oscillator

dance of the gunn-element to be employed in the circuit. Bischoff [3] has measured the equivalent circuit elements of the Varian VS E922 OS4 Gunn-elements, Fig. 6. The negative resistance is about 50 Ω while the parasitics of chip and package yield $C_g \approx 0.28$ pF, $L_s \approx 0.08$ nH and $C_p \approx 0.09$ pF. In an oscillator-circuit, the impedance connected in parallel to the Gunn-element has a) to compensate the imaginary parts and b) to match the negative resistance as transformed by the parasitic impedance components. Fig. 7 shows a plot of the complex admittance (normalized to 200 Ω) needed in a parallel-circuit oscillator. It is seen that for a frequency near 60 GHz a shunt resistance of about 200 Ω is required together with a reactive shunt component. An oscillator design satisfying these conditions has been first described by Cohen and Meier [5] and is shown in Fig. 8a. The fin-line characteristic impedance is roughly 200 Ω (slot-width 0.3 mm on RT/Duroid 0.25 mm substrate) and a short circuit backing the Gunn-element provides the necessary reactive termination. In order to tune the oscillator electronically a fin-line stub as described for the 35 GHz oscillator is used as the reactive termination in stead, Fig. 8b.

The equivalent-circuit of the new oscillator design is shown in Fig. 9. The active element is loaded with the real impedance Z_c (characteristic impedance of fin-line) to the left and with the stub input impedance connected in series to the terminating impedance of the waveguide Z_b to the right. The variable short in the right-hand waveguide section (Z_b) only slightly modifies the center frequency and output power of the oscillator since the effective length of the stub is about one quarter wavelength (high input-impedance). The resultant impedance of the varactor is highly inductive due to the high bond-ribbon inductance (on the order of j 100 Ω).

With respect to the reduction of this parasitic component in order to achieve wide tuning capability and with respect to the low reproducibility of the circuit at 60 GHz it is even more desirable to replace the chip-varactor by a beam-lead device.

The tuning characteristic of the varactor-tuned oscillator is plotted in Fig. 9. 400 MHz tuning range has been achieved with power variations of less than 2 mW.

Conclusions

Two fin-line varactor-tunable oscillator circuits have been presented featuring high efficiency, wide tuning range and simple design. One of the main advantages of fin-line, namely good reproducibility of circuits was not fully proven in the presented oscillator circuit mainly due to the inavailability of beam-lead varactor diodes. To a lesser extent the large tolerances in commercial Gunn-element characteristics and in the discontinuity effects of the soldered contact to the fin-line have an influence on the reproducibility of the oscillator circuit.

References

[1] Hofmann, H.: mm-wave gunn-oscillator with distributed feedback fin-line circuit. Proceedings of the 1980 IEEE MTT-S International Microwave Conference, pp. 59 – 61

[2] Knöchel, R.: Design and performance of microwave oscillators in integrated fin-line technique. Microwave Optics Acoustics 3, 1979, pp. 115 – 120

[3] Bischoff, M.: Ein Verfahren zur Optimierung von Millimeterwellen-Gunnoszillatoren aus dem Ersatzschaltbild. Archiv für Nachrichtentechnik der ntz (ntz Archiv) 1979, H. 7, pp. 161 – 164

[4] Schmidt, L.-P.: A comprehensive analysis of quasiplanar waveguides for millimeter-wave application. Proceedings 11th EuMC 1981, Amsterdam, pp. 315 – 320

[5] Cohen, L.D., Meier, P.J.: Advances in E-plane printed millimeter wave circuits. Proceedings of the 1978 IEEE MTT-S International Microwave Symposium, pp. 27 – 29

Kennziffer 49

7.4.3

A 30 GHz FET-Oscillator using Fin-Line Circuitry

H. Meinel

Abstract: The rapid progress in the semiconductor field has rendered possible the utilization of GaAs Field Effect Transistors for oscillator applications in the mm-wave range. Recently quasi planar line structures, like fin-line, have been investigated. Based on this know-how, a 30 GHz fin-line FET-Oscillator has been developed. 3 mW of output power were achieved with a corresponding DC-RF efficiency of nearly 4 %.

Introduction: The rapid progress in the semiconductor field has rendered possible the utilization of GaAs Field Effect Transistors for oscillator applications in the mm-wave range. Transit frequencies higher than 100 GHz have been reached today /1/. Operating frequencies as high as nearly 70 GHz have been reported only recently /2/. Being three-pole devices however, FETs are hardly suited for metal waveguide applications, the commonly used mm-wave transmission medium.

Quasi-planar line struktures, like e.g. fin-line are very well suited for such three pole devices, especially for unpackaged chip type FETs. Recently a number of papers have been published, dealing with mm-wave integrated circuits, based on fin-line technique /3,4/; design data are available e.g. in /5/. Based on this know-how, a 30 GHz fin-line FET-Oscillator has been developed.

Dipl.-Ing. Holger Meinel, c/o AEG-TELEFUNKEN, Anlagentechnik, Geschäftsbereich Hochfrequenztechnik, Grundlagenentwicklung

A1 E32, Sedanstraße 10, Postfach 1730, D-7900 Ulm/Donau, West Germany

Oscillator design: The fundamental set-up of the fin-line FET-oscillator is shown in Fig.1.

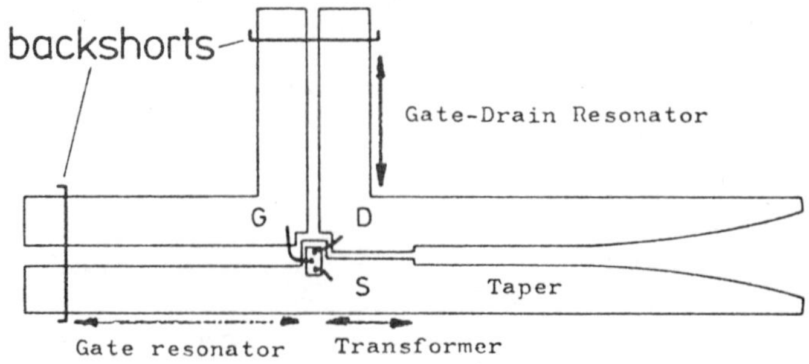

Figure 1 Basic set-up of a fin-line FET-oscillator

The FET-chip or packaged device is mounted inside a T-type fin-line series junction. The three terminals of the device are wire-bonded or soldered to three different fins respectively, thus DC isolation and seperate biasing of the three different terminals is achieved. An enlarged section of the assembled semiconductor is shown in Fig. 2.

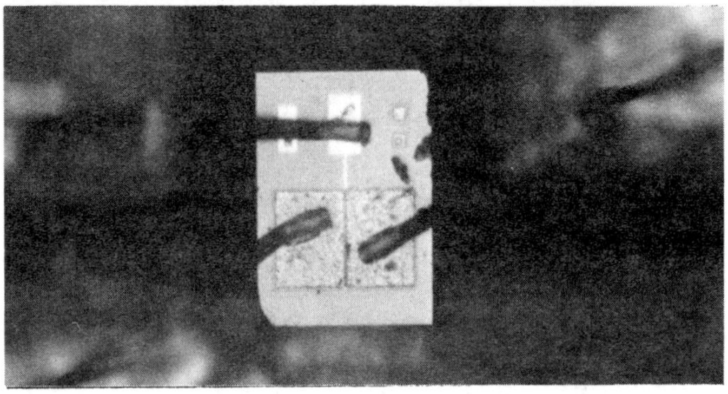

Figure 2 FET chip, bonded into the planar circuit

The oscillator operates in the common-source mode. The frequency of operation mainly depends on the length of the gate resonator (Fig. 1, left), which is adjustable by means of a fin-line backshort, which has been developed especially for this purpose. Oscillator feedback is accomplished by the Gate/Drain resonator (Fig. 1, above). Employing another fin-line backshort, the resonator length can be adjusted for maximum power output. Output matching is accomplished, using a quarter-wave transformer inserted between the FET and the output taper section (Fig. 1, right). The fin-line circuit is printed on RT-Duriod type 5880 with a thickness of 254 µm. Using a split block fin-line housing, the completely assembled circuit was mounted, as described earlier in /6/.

Fin-Line Backshort: The fin-line backshort, developed as an analog to the common metal waveguide high-low design, contains three LOW- and two HIGH-impedance quarter-wave sections. However, apart from the slot, being introduced in order to fit the fin-line configuration, Teflon instead of air is used as dielectric material in the HIGH impedance sections. In contrast to hardcoated anodized aluminum backshorts, this Teflon-isolated backshort does not destroy the fin-line metallization.

Measured results: Two different types of semiconductors have been used for the FET-oscillator described above. A packaged device, the NEC-type NE 388-06 and an open chip type device, placed at our disposal by the research laboratory of Laboratoires D'Electronique et de Physique Appliquee, France. The achieved results are listed in table 1.

Device Type	f_{op}/GHz	f_T/GHz	P_{out}/mW	η
NE 388 06	29.88	90	0.38	1.4
LEP-SAMPLE	29.88	100	2.95	3.97

Table 1 Oscillator data

Taking into account the NEC data sheet and interpolating for 30 GHz, the maximum achievable output level for the NE 388 amounts to 1.2 mW, without taking into account any losses. The measurements carried out have shown, that tuning and matching of the packaged device is by far more difficult; thus only .38 mW could be achieved. The maximum output power with the unpackaged LEP device is about 3 mW, with a corresponding DC-RF efficiency of nearly 4 %. Taking into account, that recently developed Schottky-barrier mixer-diodes only have the need for an LO-power of less then 250 μw /7/ the described oscillator will be able to fit future system requirements.

The tuning behavior of this type of FET-oscillator is given in Fig. 3 und 4.

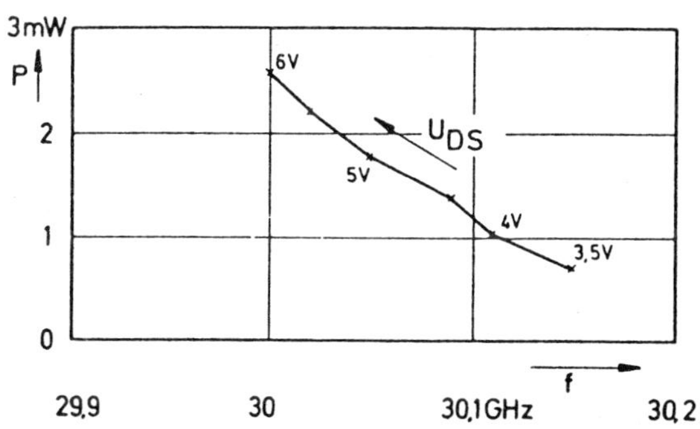

Figure 3 Fin-line FET-oscillator electronic tunability

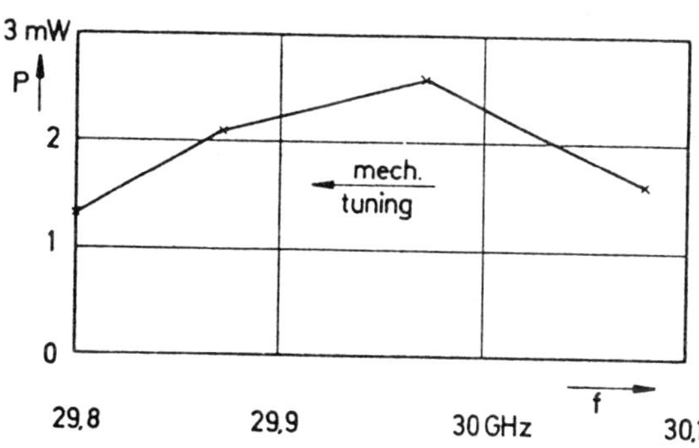

Figure 4 Fin-line FET-oscillator mechanical tunability

Output power is plotted as a function of the operating frequency with the Drain-Source voltage and the mechanical length of the gate-resonator respectively as the parameter; all other parameters were kept unchanged.

Conclusion: It has been demonstrated that it is possible to build a 30 GHz FET oscillator using fin-line technique. The efficiency of nearly 4 % is a promising result with respect to future mm-wave application. Optimizing the channel parameters of the employed semiconductor devices might lead to even better values.

Acknowledgements: The author wants to thank M. P. Baudet and and M. D. Boccon-Gibod of LEP, France for preparing and supplying the investigated chip devices, respectively. This work has been supported by the German Government (Grant No. T/RF31/91043/91335)

References

/1/ Tohyama, H., Mizuno, H., 23 GHz GaAs MES FET Reflection-Type Amplifier, IEEE MTT-27, No. 5, May 1979, 408 - 411

/2/ Schellenberg, J. M.,Yamasaki, H., Maki, D. W., A 69 GHz FET oscillator, IEEE-MTT-S, Los Angeles, Symp.Dig.

/3/ Meinel, H., Adelseck, B., Callsen, H., A survey of planar integrated mm-wave components, 2nd Military Microwave Conf., London 1980, Symp.Dig., 82 - 87

/4/ Bates, R. N., Coleman, M. D.,Nightingale, S. J., Davies, R., E Planes Drop Millimeter costs, MSN, Dec. 80, Vol.10, No. 12, 74 - 80

/5/ Schmidt, L. P., Itoh, T., Hofmann, H., Characteristics of unilateral fin-line structures with arbitrarily located slots IEEE MTT-S 1980, Wash.D.C. Symp. Dig. 255 - 257

/6/ Adelseck, B., Barth, H., Hofmann, H., Meinel H., Rembold, B., Advances in mm-wave components and Systems, AGARD-Conf. on mm- and submm-wave propagation and circuits, Munich 1978, AGARD-Proc. No. 245, 25-1-25-17

/7/ Schroth, H., High reliable low noise mm-wave mixers with whisker-contacted honey comb diodes, IEEE MTT-S 1981, Los Angeles, Symp. Dig.

7.4.4

ADVANCES IN E-PLANE PRINTED MILLIMETER-WAVE CIRCUITS*

L. D. Cohen and P. J. Meier
AIL, a division of Cutler-Hammer
Melville, N.Y. 11746

* The work reported has been partially funded under Government Contract No. DAAB07-77-C-0118.

L.D. Cohen and P.J. Meier, *IEEE MTT-S International Microwave Symposium Digest*, 1978, pp. 27–29.

I. INTRODUCTION

It has been shown that multiport components can be constructed at millimeter wavelengths by suspending printed circuits in the E-plane of rectangular waveguide (references 1 and 2). Such circuits feature standard-flange ports, printed-circuit economy, wide single-mode bandwidth, and compatibility with packaged or hybrid IC devices. The following paragraphs summarize recent progress with E-plane circuits, including wideband four-port coupling elements and one-port, mechanically tunable oscillators in the 5-mm band employing GaAs and InP Gunn devices.

II. FOUR-PORT COUPLING ELEMENTS

It has been demonstrated that directional coupling can be obtained at millimeter wavelengths with an array of printed capacitive elements, mounted in the E-plane common to a pair of parallel waveguides (reference 2). Such couplers feature printed-circuit reproducibility, housings with axially uniform coupling regions, and compatibility with other E-plane circuits such as filters and mixers. The following paragraphs will describe E-plane coupling elements with enhanced bandwidth capability.

Figure 1 shows the structure of the E-plane four-port coupling network. Also illustrated are three types of printed coupling elements: capacitive probe, L-C probe, and loop. Since the coupling of the capacitive probe increases monotonically with frequency, the bandwidth capability of this simple element is limited. The series inductance associated with the L-C probe can reverse the capacitive coupling trend and, for certain configurations, provide a relatively constant coupling across a wide band. Frequency compensation can also be obtained with the loop--a coupling element which has been extensively applied to nonplanar components at lower frequencies (reference 3). Since the loop dimensions can be adjusted to independently control the electric and magnetic coupling, highly directional coupling is feasible with one element. However, the printed-circuit approach is well suited to multielement couplers which are inherently directional. Thus the shape of the printed loop can be optimized for coupling flatness, without regard to single-element directivity.

Figure 2 shows the measured coupling versus frequency for representative E-plane coupling elements. The coupling characteristics for compensated elements and simple capacitive probes are compared for the case of tight (-8 dB) and loose (-23 dB) coupling. Across the band of 33 to 40 GHz the L-C probe provides a coupling of -8.5 ± 0.3 dB whereas the capacitive-probe coupling is -7.7 ± 1.5 dB. The L-C probe has also been investigated for the loosely coupled case, however, superior flatness has been demonstrated for the printed loop. Figure 2 shows the measured coupling is -22.5 ± 0.5 dB for the loop versus -22.7 ± 2.2 dB for a comparable capacitive probe. The plotted coupling for the loop is in the forward direction (port 1 to port 3) which is relevant to multiple-element arrays. The backward loop coupling is approximately 6 dB weaker across the band.

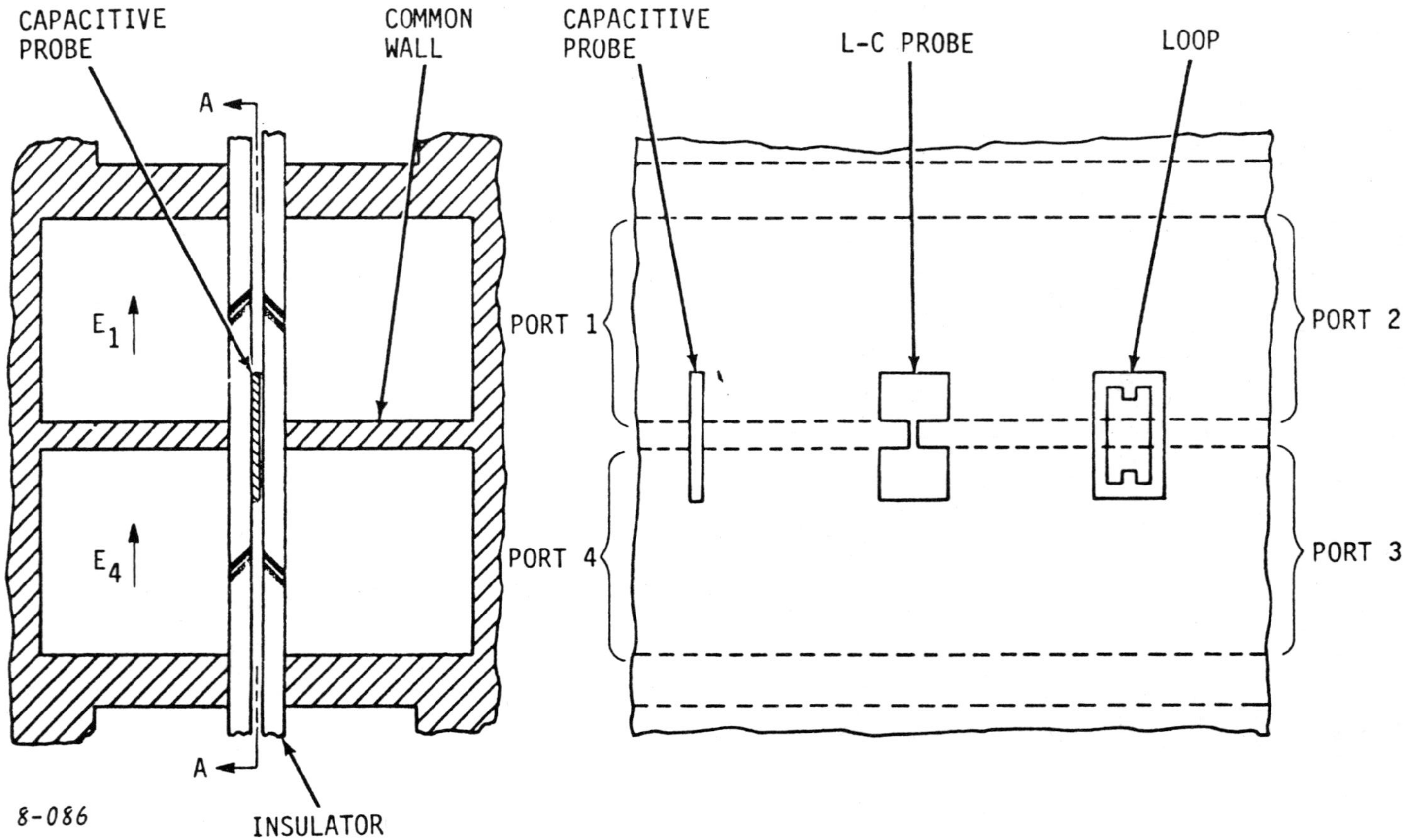

Figure 1. E-Plane Four-Port Coupler

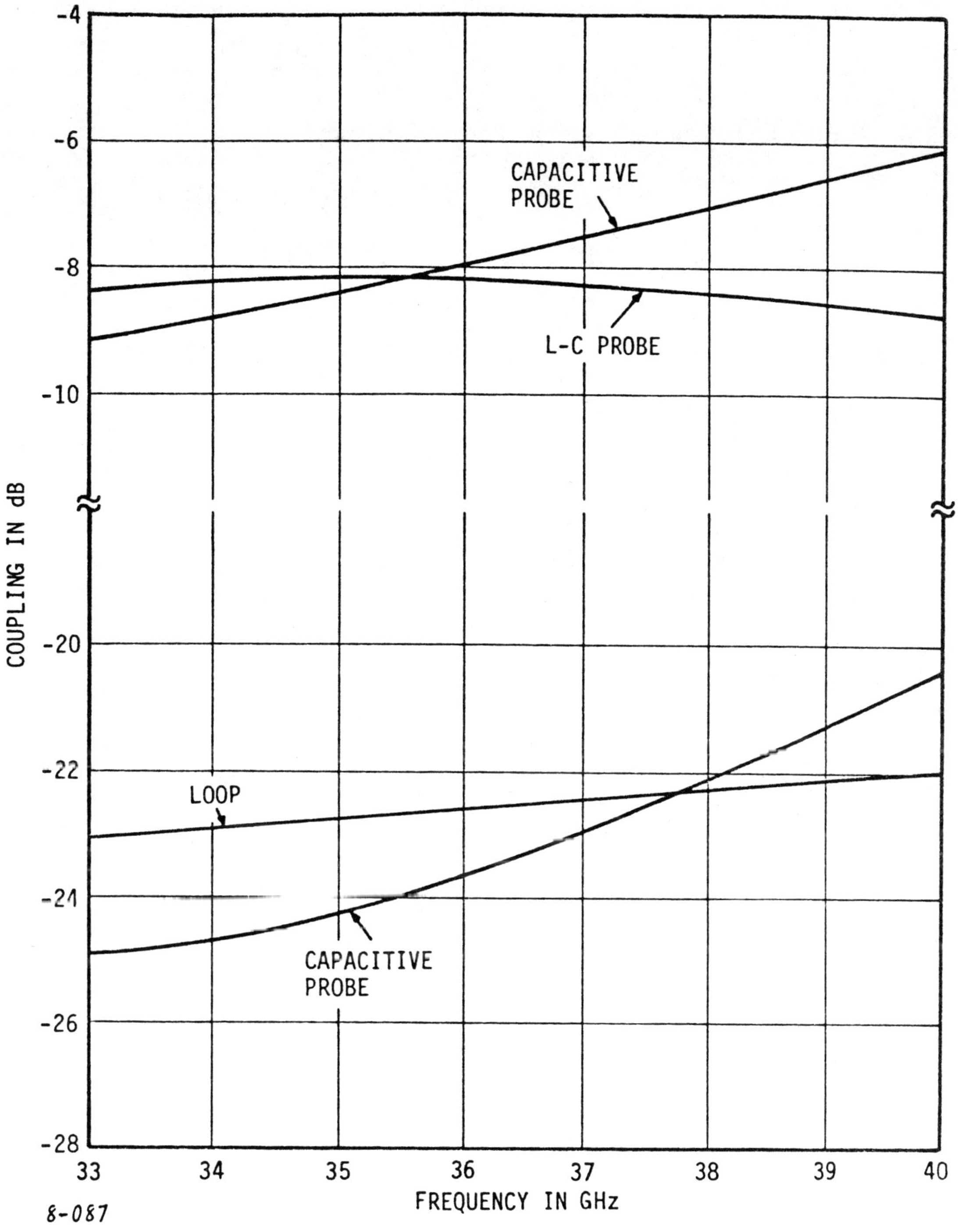

Figure 2. Coupling Versus Frequency

The characterization and optimization of E-plane coupling elements is important to the development of a wide variety of E-plane wideband components including couplers, balanced mixers, and multiplexers.

III. FIN-LINE GUNN OSCILLATORS

Single-ridge fin-line (reference 4) is an E-plane transmission line which has proven to be useful in the construction of Gunn oscillators at millimeter wavelengths (reference 5). The printed fin can be easily biased and tapered to provide wideband single-mode coupling between a packaged Gunn device and a standard waveguide port. The following paragraphs report recent progress with tunable fin-line GaAs and InP Gunn oscillators in the 5-mm band.

Figures 3 and 4 show the external and internal features of the mechanically tunable oscillators. The housing is split along the E-plane to accommodate two 0.005-inch boards (Duroid 5880) which are aligned by dielectric dowels. One of the boards contains the printed circuit, whereas the other board is unclad to insulate the tapered fin from the housing. This allows dc bias to be applied (via the printed pattern and a foil bond) to a packaged Gunn device. The Gunn device is threaded into a removable copper block which is bolted to the housing to provide a heat sink and the dc return. The tapered fin, which couples the Gunn device to standard WR-12 waveguide, is RF grounded to the upper wall of the housing by a serrated choke pattern within the broad wall. The length of the choke region (dimension L in Figure 4) is chosen to be a quarter wavelength in the dielectric medium. The serrations

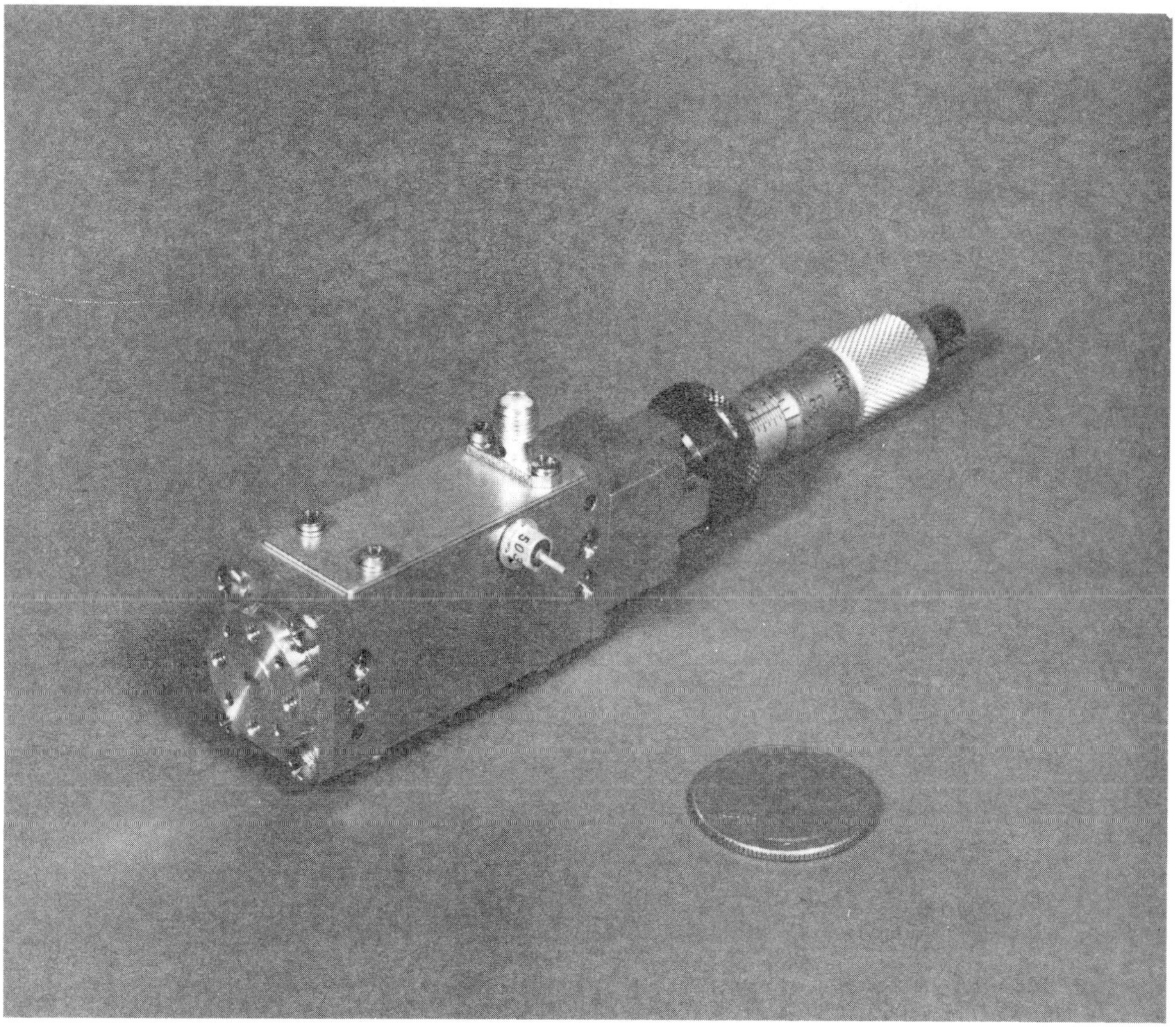

Figure 3. Mechanically Tunable Gunn Oscillator

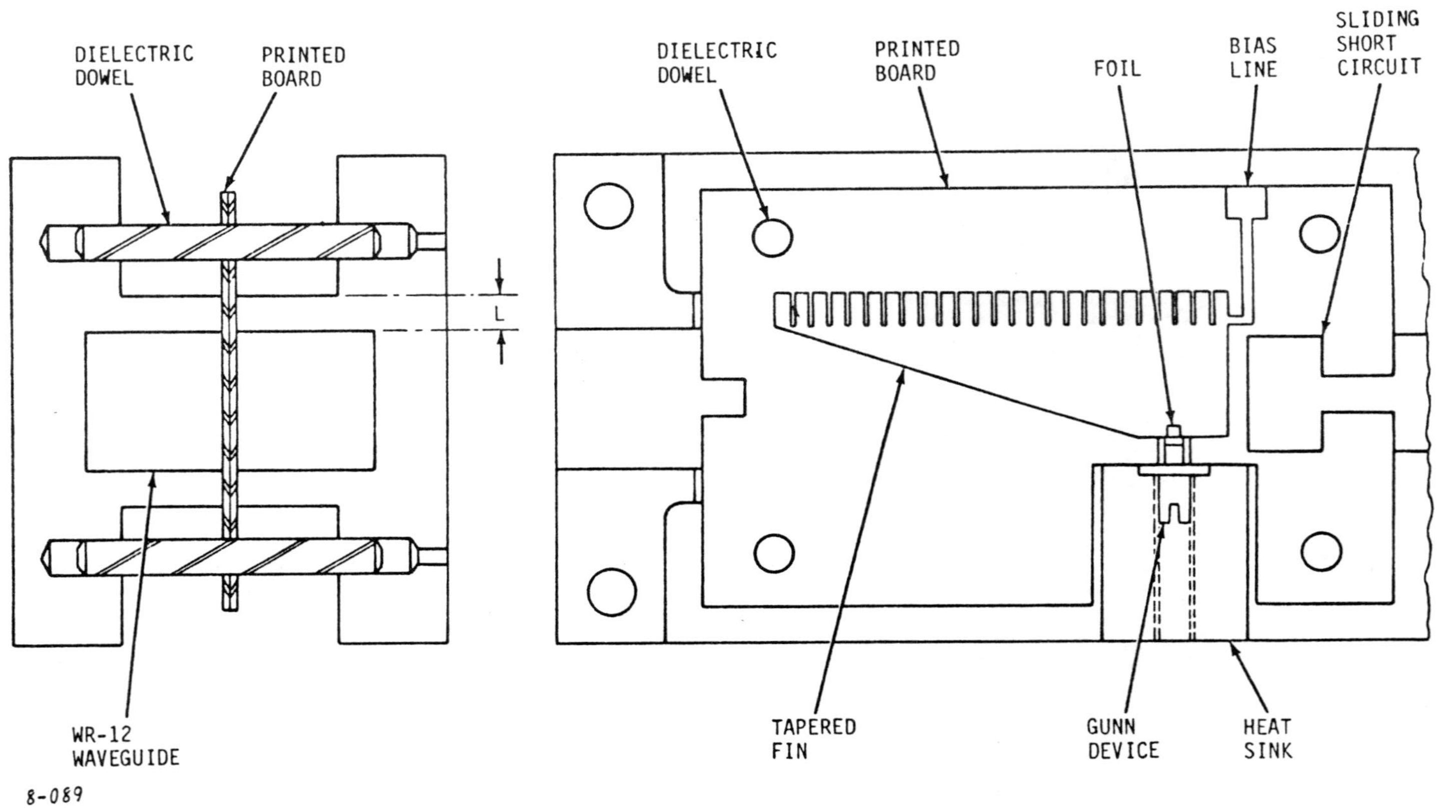

Figure 4. Oscillator Cross Section

prevent longitudinal current in the choke and thereby ensure single-mode propagation. The Gunn device is reactively terminated by a sliding noncontacting short circuit, which is mechanically driven by a resetable micrometer head with a nonrotating spindle.

The oscillator has been tested with commercially available GaAs Gunn devices (Varian VSE-9220) and Government-furnished InP devices (see acknowledgments). Measurements were performed of frequency and power output as a function of the spacing between the device and the short-circuit termination. Power output was measured with a fixed-tuned thermistor mount (Hitachi F2512) which was calibrated against a dry calorimeter (PRD 666) across the band of interest. Measurements are referenced at the output of a commercially available isolator having a forward loss of 0.4 to 0.5 dB.

Figure 5 shows the measured power output of the GaAs Gunn oscillator as a function of the short-circuit micrometer dial. As the short is moved 0.14 inch, the frequency varies smoothly across an 8-GHz band centered at 58.75 GHz and the CW power output is typically 16 dBm (40 mW) at the output of the isolator. Correcting for isolator loss, the power output of the oscillator is 18 dBm (63 mW) at the center of the operating band. The measured pushing factor is -220 MHz/volt.

Figure 6 summarizes the oscillator performance with an InP Gunn device. As the short is moved 0.09 inch, the frequency varies smoothly across a 6-GHz band centered at 66.8 GHz and the power at the isolator output is 15.1 dBm ± 1.2 dB. Correcting for the isolator loss, the maximum output of the InP Gunn oscillator is 46 mW.

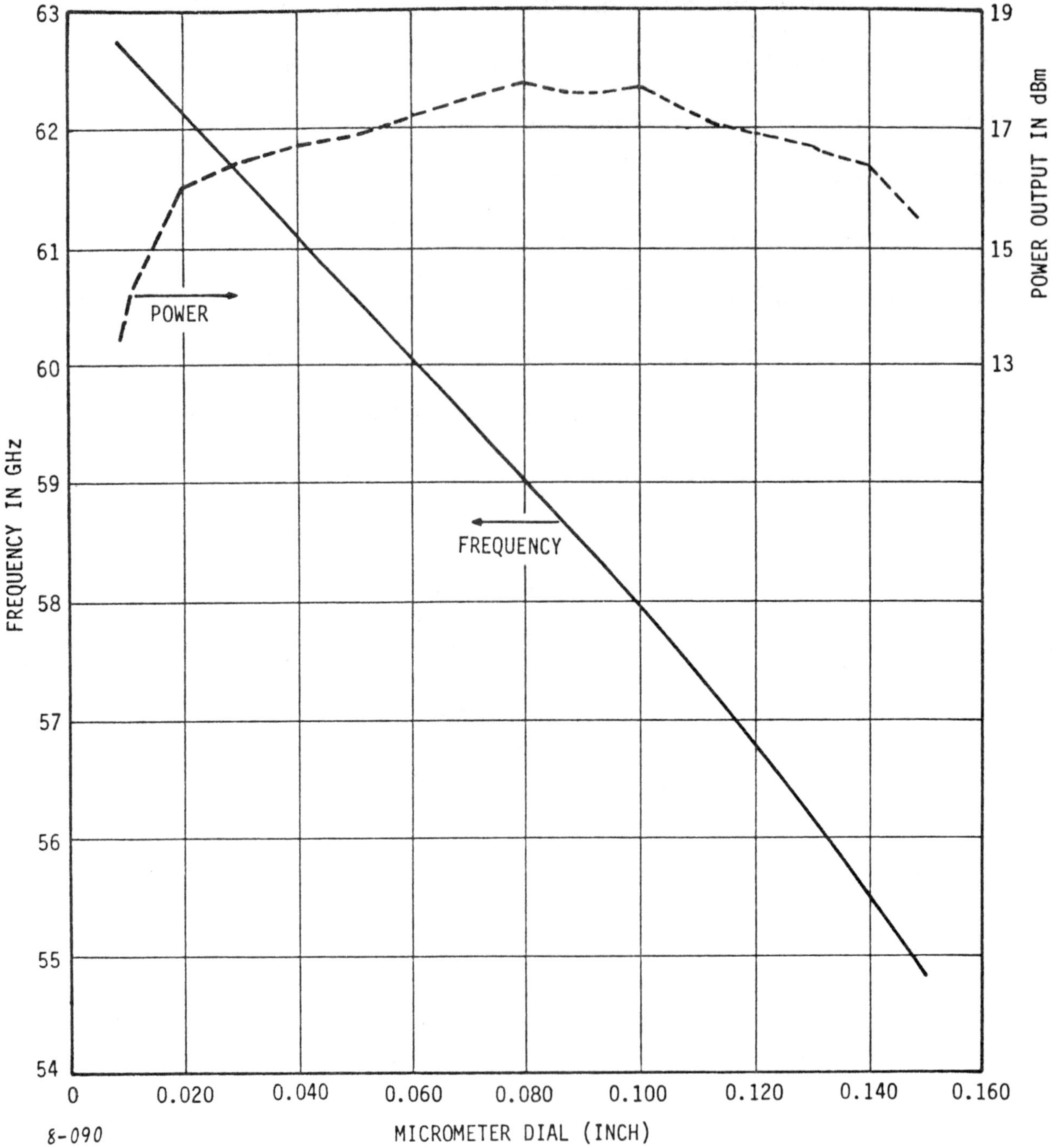

Figure 5. Oscillator Performance with GaAs Gunn Diode

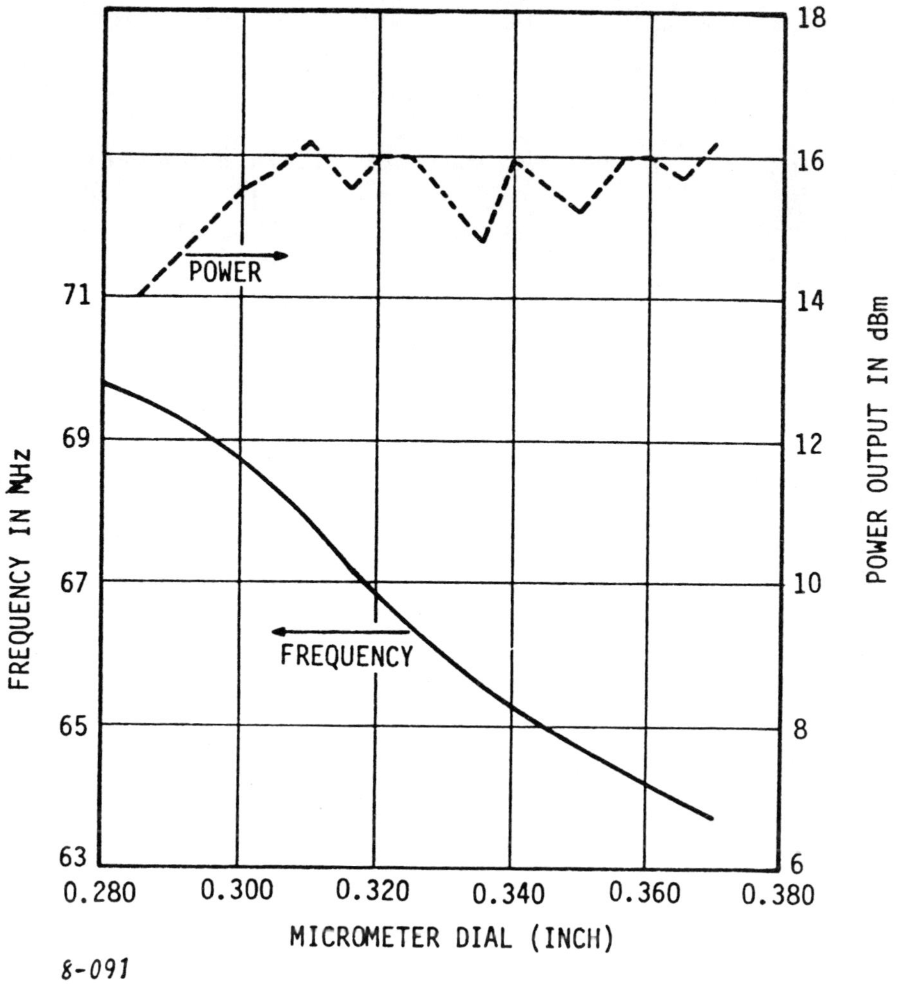

Figure 6. Oscillator Performance with InP Gunn Diode

Work is now in progress to further optimize the mechanically tunable oscillators and add varactor-tuned FM capability. The fin-line pattern will be modified to accommodate an independently biased varactor, mounted a half wave from the Gunn device. Such oscillators will be applicable to transmitters as well as widely tunable local oscillators.

IV. CONCLUSIONS

Recent work with printed E-plane millimeter-wave circuits has been reported. Improved bandwidth has been demonstrated with four-port coupling elements which are applicable to printed millimeter-wave couplers, balanced mixers, and multiplexers. Fin-line GaAs and InP Gunn oscillators have also been described which demonstrate advances in terms of operating frequency and mechanical tunability. Further progress in fin-line oscillator development is expected as improved InP Gunn devices become available and circuit refinements (including varactor tuning) are added. Multiport E-plane circuits are applicable to a variety of millimeter systems and feature standard waveguide interfaces, printed-circuit reproducibility, wide single-mode bandwidth and compatibility with packaged and hybrid IC devices.

V. ACKNOWLEDGEMENTS

The four-port coupler work was sponsored by AIL under the direction of M. T. Lebenbaum, K. S. Packard, and J. J. Whelehan with technical consultation provided by J. J. Taub. The oscillator development was sponsored by ERADCOM under the direction of I. Kullback, R. Christian, and E. Fliegler. The InP devices were developed by Varian Associates and furnished to this program through the courtesy of Dr. H. Jacobs of ERADCOM.

VI. REFERENCES

1. Meier P.J., "Integrated Fin-Line Millimeter Components," IEEE Trans, vol MTT-22, p 1209-1216, December 1974.
2. Meier P. J., "Planar Multiport Millimeter Integrated Circuits," Digest of 1977 IEEE International Microwave Symposium, p 385-388, June 1977.
3. Early H. C., "A Wide-Band Directional Coupler for Wave Guide," Proc IRE, p 883-886, November 1946.
4. Meier P. J., "New Developments with Integrated Fin-Line and Related Printed Millimeter Circuits," Digest of 1975 IEEE International Microwave Symposium, p 143-145, May 1975.
5. Fleri D. A., Taub J. J., Whelehan J. J., and Wolczok J. M., "Low Cost X-band MIC Parametric Amplifier," Digest of 1974 IEEE International Microwave Symposium, p 210-212, June 1974.

7.5.1

F. Arndt, J. Bornemann, D. Grauerholz & R. Vahldieck
Microwave Department, University of Bremen, Germany

Material and Circuit Evaluation for Millimetre-Wave Applications: A Broadband Fin-Line Mixer for Millimetre Waves*

Abstract A low-cost broadband fin-line mixer has been designed for 28 to 40 GHz operation. RT/duroid is employed as substrate material and beam-lead PIN diodes (HP 5082-2716) are used. The said mixer is well suited for millimetre-wave applications, the measured conversion loss being about 10 dB over the whole frequency band.

Résumé Un mélangeur économique à ligne nageoire à large bande a été conçu pour fonctionner entre 28 et 40 GHz. On a utilisé un substrat en RT/duroid ainsi que des diodes PIN à connexions en poutre de type HP5082–2716. Avec une perte par conversion d'environ 10 dB sur l'ensemble de la bande de fréquence, ce mélangeur est tout indiqué pour les circuits en ondes millimétriques.

* This paper, provided via ERNO Raumfahrttechnik GmbH, is based on work conducted under ESA/ESTEC Contract No. 4044/79/NL/DG(SC). A companion paper titled 'High-Q Fin-Line Filters for Millimetre Waves' was published in ESA Journal No. 1, 1981. See also page 127 of this present issue.

Introduction

It has already been demonstrated that wide-band, low-loss integrated circuits can be constructed for millimetre wavelengths using fin-line techniques[1,2]. Because of its advantages in terms of wide-band single-mode bandwidth, low insertion loss and production economy, as well as its compatibility with hybrid IC devices and waveguide instrumentation, this fin-line circuit is also well suited for use in receiver front ends or in broadband measuring equipment as a mixer unit.

Unlike the more complicated mixer structures reported in References 3–7, which utilise a variety of transmission lines integrated onto the substrate, the mixer described here was developed specifically for low cost as well as ease of manufacture and procurement, whilst still meeting the need for very broad band operation. Based on a single-ended mixer structure[8], the design incorporates a number of other low-cost circuits: an exponentially tapered unilateral fin-line circuit combined with a simple series capacitance adaptation to a sort of waveguide coplanar section, an inexpensive beam-lead diode (HP 5082-2716), and a simple wire, periodic-loaded low-pass filter.

The fin-line millimetre-wave mixer has a conversion loss of about 10 dB over the whole frequency band of 28–40 GHz.

Design

Figure 1 shows the fin-line mixer in section. The principal components are the tapered fin-line on RT/duroid material within a WR-28 waveguide housing (7.112 mm × 3.556 mm), the variable short-circuit at $\lambda_{fin}/4$, the beam-lead PIN diode, the IF low-pass circuit, and the LO input (shown in Fig. 4).

Figure 2a shows the approximate equivalent circuit of the mixer (without the fin-line taper) at midband frequency. The equivalent circuit of the diode, together with its characteristic data, is shown in Figure 2b (the HP 5082-2716 diode data have been converted to the chosen frequency range of 28–40 GHz).

The first design problem is to adapt the RF source to the diode input impedance. Two circuits are used: a fin-line taper and a series capacitance C_s (Fig. 2a). The fin-line taper consists of an exponentially tapered section (cf. Fig. 1). The slot width, chosen to be about 60 μm, corresponds to a characteristic impedance of about 100 Ω for the unilaterial fin-line.

The series capacitance C_s was realised by experimental optimisation using a silver blot on a photo lac layer (Fig. 3). Apart from the problem of adaption of the RF signal, C_s has to be small because it shunts the IF signal.

The LO circuit is coupled to the mixer by a waveguide flange beside the mixer housing (cf. the photograph, Fig. 4, of the designed mixer). A waveguide resonator is formed by means of a step in the height of the waveguide flange. The measured insertion loss was about 13.5 dB at 8–11 mW LO input and 400 μA direct diode current. It has to be pointed out that no optimum separation of the RF from the LO signal was sought. For this purpose, this simple waveguide circuit must be replaced by a more complicated structure (e.g. irises).

The IF signal is extracted from the diode mount through a standard, inexpensive low-pass filter (Fig. 5), which is designed according to Reference 9. The IF insertion loss is about 10 dB (direct diode current 400 μA, alternating IF current about 2.8 mA). The IF signal and the diode bias are separated by an HP bias-tee.

The measured conversion loss, shown in Figure 6, is about 10 dB throughout the 28–40 GHz frequency range.

Conclusion

A simple low-cost, fin-line broadband mixer has been designed for millimetre-wave applications. The technique presented allows one to use standard low-cost, beam-lead diodes. The measured conversion loss is about 10 dB.

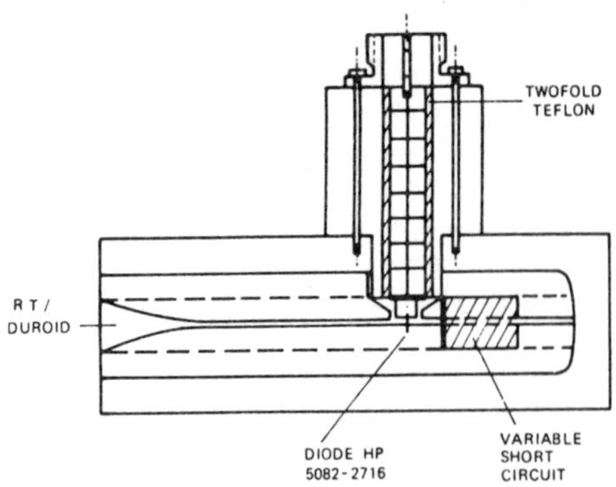

Figure 1. The new fin-line mixer for millimetre-wave (28 40 GHz) operation

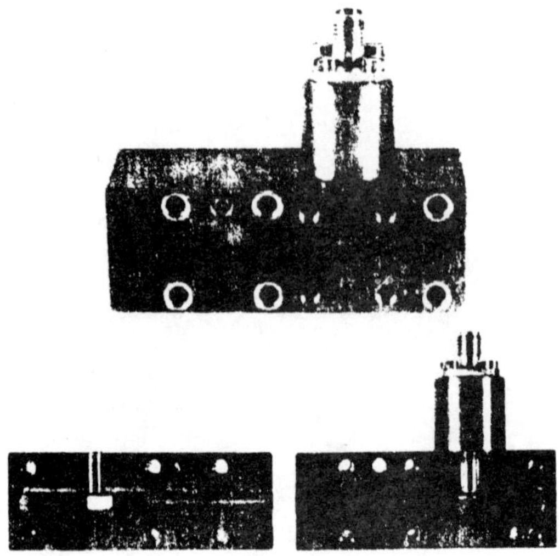

Figure 4. Photographs of the new fin-line mixer (for 28-40 GHz)

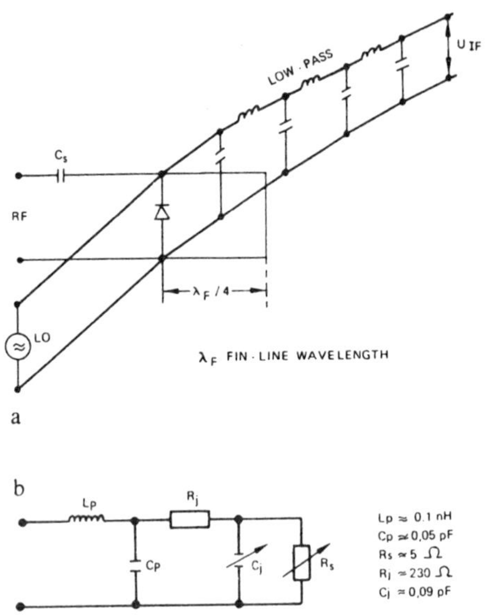

Figure 2. Approximate equivalent circuit
(a) Mixer circuit
(b) Diode (HP 5082-2716)

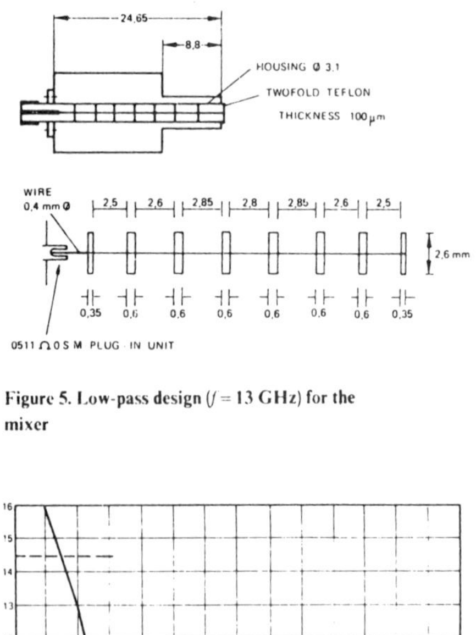

Figure 5. Low-pass design (f = 13 GHz) for the mixer

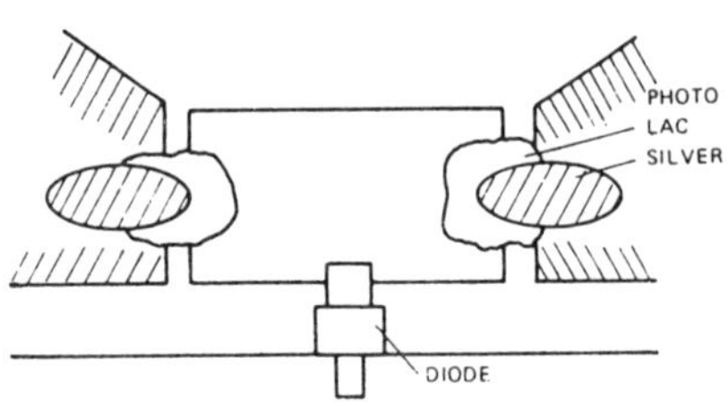

Figure 3. Realisation of the series capacitance C_s at the waveguide coplanar section in the rear of the diode (cf. Fig. 1)

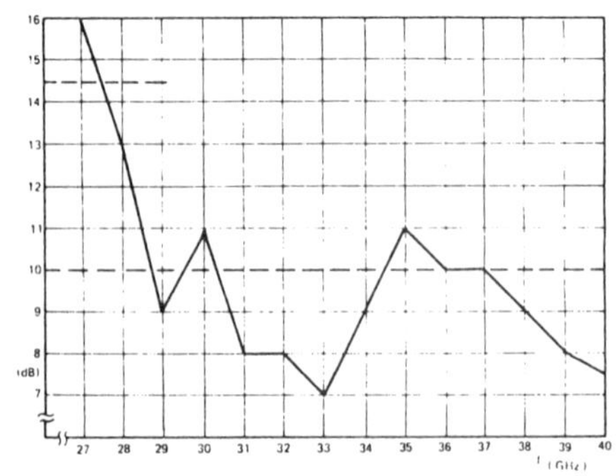

Figure 6. Measured conversion loss for the fin-line mixer as a function of operating frequency

References

1. Meier P J 1978, Millimeter integrated circuits suspended in the E-plane of rectangular waveguide, *IEEE Trans.* **MTT-26**, pp. 726–733
2. Arndt F, Bornemann J, Grauerholz D & Vahldieck R 1981, Material and Circuit Evaluation for Millimetre-Wave Applications: High-Q Fin-Line Filters for Millimetre Waves, *ESA J.* **5**, No. 1, pp. 33–42
3. Reindel J 1980, Printed WG circuits trim component costs, *Microwaves*, October 1980, pp. 60–63
4. Begemann G 1978, An X-band Balanced Fin-Line Mixer, *IEEE Trans.* **MTT-26**, pp. 1007–1011
5. Begemann G & Knöchel R 1980, KU-Band Front-End in Integrated Fin-Line Technique, *Archiv Electron. Ubertragungstechn.*, **AEU-34**, pp. 179–180
6. Bates R N & Coleman M D 1979, Millimeter Wave Fin-line Balanced Mixer, *Proc. 9th European Microwave Conference*, Brighton, pp. 721–725
7. Meier J P 1979, 4 and 5 mm Band Printed Circuits Balanced Mixers, *Microwave J.* August 1979, pp. 63–68
8. Kohl W & Olbrich G 1979, Quarzgenaue Frequenzmessung im Frequenzband 26.5–40 GHz, *Microwellen-Magazin*, **Heft 1/79**, pp. 16–18
9. Matthaei G L, Young L & Jones E M T 1964, *Microwave Filters, Impedance Matching Networks, and Coupling Structures*, McGraw Hill Book Co., New York

Manuscript received 9 March 1981.

7.5.2

Technical Feature

Wideband mm-Wave Mixers For EW Applications

D. W. Ball and L. Q. Bui
Hughes Aircraft Company
Electron Dynamics Division
Torrance, California

Introduction

In parallel with the development of millimeter-wave radar weapon and communication systems, a need for millimeter-wave surveillance receivers have been increasing. Conceptual systems call for receivers capable of covering hundreds of GH_z of bandwidth. In order to meet this requirement in a manner which will be feasible from the standpoint of production costs, mixers with broadband coverage must be developed. This requires broadbandwidth capabilities at both RF and IF bands. Consequently, low parasitic, high performance mixer diodes, wide instantaneous bandwidth mixer circuits, and multi-octave low-noise IF amplifiers are needed.

This article describes the design and development of broadband millimeter-wave mixers suitable for these EW applications. Specifically covered are diode characteristics, circuit design considerations, and performance measurements over waveguide bandwidths.

The mixer design presented here yields wide instantaneous band coverage at the RF port, with selective band constraints applied to the local oscillator and input port. The IF output is band limited via low-pass and band-pass diplexing at the local oscillator/mixer diode junction. Inherent LO to RF isolation is achieved by proper use of balanced and single-ended transmission lines, along with well-matched mixer diodes.

The mixer design takes advantage of low-cost printed circuit fabrication technology which is particularly suited to batch processing and facilitates the use of low parasitic beamlead diodes. Simple circuit mounting via "split block" housing precludes the need for multiple complex machined parts, thus further reducing fabrication costs as compared to the traditional waveguide millimeter-wave mixer with whisker contacted "honeycomb" Schottky diodes.

Fig. 1 Planar hybrid balanced mixer incorporating beam lead diodes.

It is worth noting that this mixer design approach leads to fully integrated multi-functional down-converters which can readily include a local oscillator, low noise amplifier, antenna, appropriate preselection and post-selection filters, or other integrated circuitry which complement the down-converter assembly. Integrated receivers using monolithic or millimeter-wave integrated circuitry are under particular scrutiny due to their inherent small size and low production cost. This benefit offers considerable advantages when mechanizing multi-channel receivers for applications where space is a premium.

EW Mixer Design Considerations

The EW mixer specifications require exceptional solutions to opposing performance criteria.

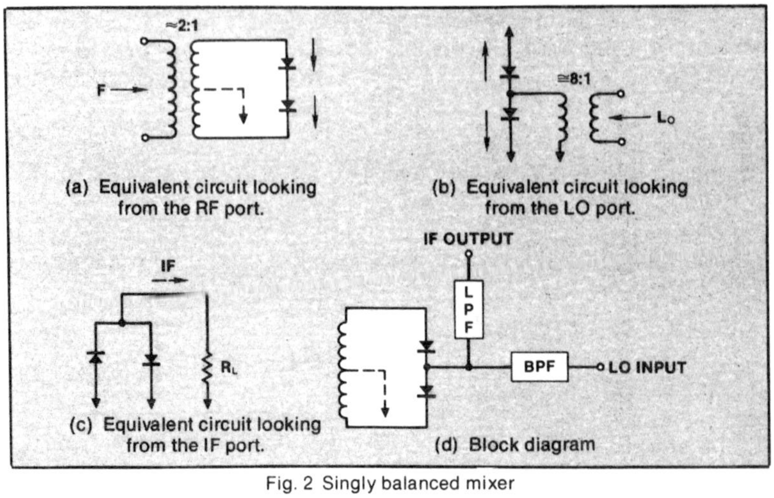

Fig. 2 Singly balanced mixer

D.W. Ball and L.Q. Bui, *Microwave Journal*, Vol. 25, No. 6, June 1982, pp. 65–76.

Typical receiver trade-offs are geared around the following requirements:

- High spurious-free dynamic range
- Flat, low-noise figure response
- Full waveguide band or greater frequency coverage
- Small size and low cost

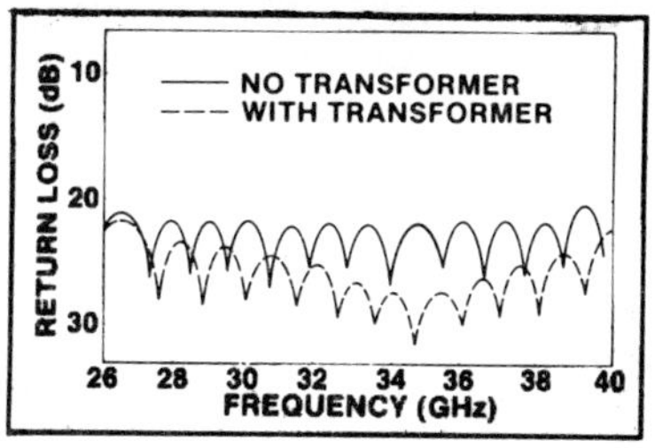

Fig. 3 Return loss comparison of matched vs. unmatched interface of the waveguide to slot transition.

Generally, EW receivers are required to process multiple emitters in a relatively dense spectral environment. Intermodulation products, single tone or multitone, generated internally or externally to the receiver are the limiters of dynamic range. They can result in false emitter detection and effectively desensitize the receiver.

To the first order, the degree of linearity of the receiver is defined by the mixer/IF amplifier single-tone and two-tone intermodulation intercept point. The higher the intercept point the higher the spurious free dynamic range. (It should also be noted that some intermods are level independant and can only be reduced by proper choice of local oscillator frequency and appropriate pre-select filtering). The value of the intercept point for the mixer is determined by a number of conditions under the control of the designer, they are:

- Mixer type (single-ended, balanced, double balanced)
- Available local oscillator power
- Mixer diode choice
- RF/LO/IF network matching circuitry.

The effects of each option noted above must be taken into consideration in terms of impedance levels and parasitics to insure wideband flat mixer response. Contrasted with dynamic range, the mixer must also provide a low noise figure to insure the probability of intercept of distant emitters. Mixer noise figure is a strong function of the quality of the converting diode(s) and its associated parasitic elements.

For millimeter-wave EW applications the diode design criteria requires that special attention be paid to the reduction of the junction Capacitance (C_j) and series inductance (L_s), while maintaining a reasonably low value of series resistance (R_s). The diode is designed to maintain a high barrier potential via appropriately selected Schottky metal so as to allow strong local oscillator drive for high intercepts without sacrificing bandwidth or noise figure.

A number of diode configurations are available. Whisker contacted diodes have shown the lowest spot noise figures reported to date[1]. The beamlead diode using mesa-etch technology is fast catching up[2,3]. Advanced planar diode technology using proton bombardment[4] designed to yield a truely planar device is being developed with the ultimate goal of growing arrays in-situ, so that, stray parasitic inductances are virtually eliminated. It should be noted that, contrary to popular beliefs, whisker contacted mixers do withstand the severe environmental conditions of space and military qualifications tests since the whiskers mass is extremely small. The main drawback associated with the whisker contacted diode is that the series inductance and post-mount capacitance, as viewed from the RF and IF port severely limit the bandwidth performance of the mixer. In addition, the assembly of whisker contacted mixers is labor intensive thus costly. For EW applications, IF frequencies in some cases extend to Ka-band and above. The beamlead diode junction capacitance typically runs 2 to 3 times that of the 2 micron dot whisker contacted device. However, the planar diode, when mounted in an appropriate structure yields a comparatively small series inductance. This results in the junction capacitance to be essentially de-embedded, thus it can be theoretically cancelled using external matching elements.

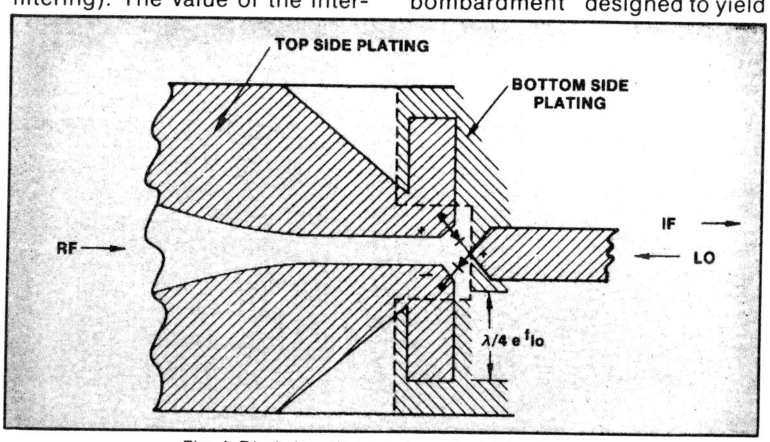

Fig. 4 Diode location on planar hybrid mixer.

Although planar circuit designs are inherently broadband, it is mandatory that high efficiency waveguide to other media transitions be developed. It is important that a number of variations be available to accommodate special mechanical configuration needs. A few key transitions are listed below:

- Waveguide to dielectrically loaded crossbar
- Waveguide to slot transition
- Probe-coupled transition
- Waveguide to microstrip transition

Since planar circuit dielectric and conductor losses can be appreciable at millimeter-waves, the mixer designer must selectively distribute these losses so as to maximize the key mixer performance parameters. This is done in conjunction with maintaining a simple low-cost housing structure. Ideally, the converting diodes are mounted as close to the RF port as possible so as to minimize path losses which add directly to noise figure. Losses due to LO injection are not as critical, but must be attended to since LO power at millimeter-wave frequencies comes at a premium. For this reason, the planar circuit mixer reported here is hybrid by nature, in that various propagation media are used to optimize performance.

Fig. 5a Prototype local oscillator band pass filter.

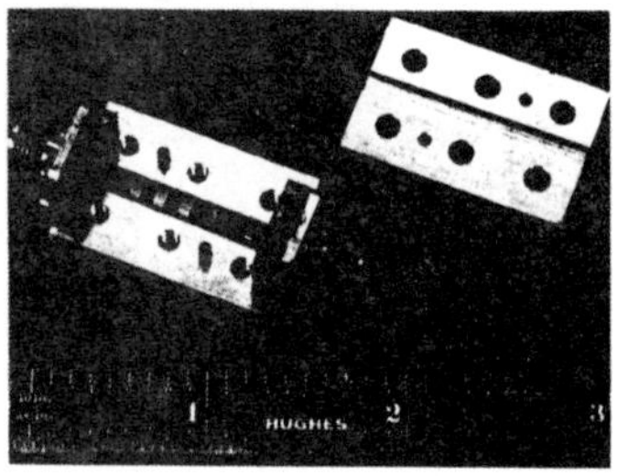

Fig. 5b Prototype IF low pass filter.

Hybrid Mixer Design

This mixer has been developed for use at K and Ka-band. The mixer is singly balanced and uses beamlead diodes mounted on a low dielectric constant substrate material.

A photograph of the mixer is shown in Figure 1. A block diagram of the mixer and its circuit elements are shown in Figure 2. The "hybrid" mixer is so called because three propagation media are used. The RF input uses a wave-guide to slotline transition centrally located in the waveguide to yield a wideband balanced line transformer which is terminated by a matched diode pair. The transition uses a cosine taper whose length is 2λ at the center frequency of Ka-band. Figure 2a shows the simplified equivalent circuit as seen from the RF port. The slot impedance was chosen to be 200 ohms which is consistent with two diodes in series at a nominal pump power of +10 dBm. The cosine taper was evaluated for electrical lengths of 1, 2 and 3 wavelengths over the full waveguide bandwidth. It is interesting to note that the return loss for all these transitions got progressively better, but not dramatically, with added length, starting at 18 dB for the 1λ case and limiting to 23 dB for the 3λ case. The 23 dB limit is attributed to the waveguide-to dielectrically loaded waveguide interface. Figure 3 shows this indeed was the case, where a matching transformer was added at the interface and provided a band-limited improvement of the return loss.

Unlike the RF port, the local oscillator is injected into the diode pair using a single-ended transmission line. Figure 2b shows the equivalent circuit as viewed from that node. The problem here is to provide a good LO return so that the diodes are well biased. This is accomplished by adding open shut stubs at on the balanced input as shown in Figure 4. The lengths of the stubs may be trimmed to optimize the LO match by observing the return loss on a reflectometer test setup.

The LO is transitioned from waveguide to microstrip and then to suspended microstripline to accommodate a low-loss, 3 pole, 0.1 dB ripple, Tchebychev filter, using end-coupled resonators. This filter type is particularly suited for fixed LO mixers. The structure facilitates fringing capacitance analysis, thus simplifies the design procedure. Figure 5 shows a photograph of the prototype filter. Figure 6 illustrates the filter response, indicating good in-band loss characteristics (<1 dB α30 GHz).

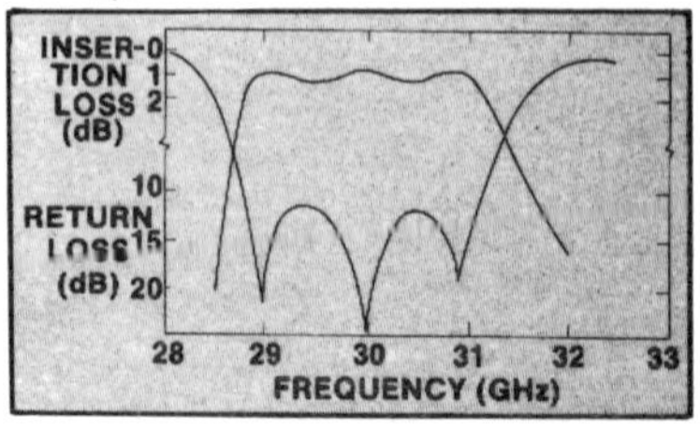

Fig. 6 Prototype filter frequency response.

The waveguide to microstrip transition uses two overlapping cosine tapers by way of two-sided substrate etching. This circuit provides a simple in-line technique for converting to the microstrip propagation mode and is very broadband. As with the waveguide to slot transition, this circuit was evaluated at electrical lengths ranging from 1 to 3 wavelengths. Test results showed that 2 wavelengths yielded 18 dB return loss, which was adequate for this application. This transition is particularly sensitive to, symetrical location in the H-plane of the wave-guide for tapper lengths of 1 and 2 wavelengths. This is attributed to extraneous modes being generated due to the assimetry in the transition region. It was noted that the sensitivity lessened as the transition was lengthened.

The IF frequency is picked off using a five-section low-pass filter designed in microstripline. Distributed lines were used to approximate lumped elements. The series elements (1's) used high impedance lines (130 0hms), whereas the shunt elements (C's) were approximated by using open shunt stubs whose lengths where chosen to be quarter-lamda at the LO frequency. The impedance of the stubs were commensurate with the net capacitance required for each unit element of the filter. The total network analysis includes the parasitic capacitance of the diodes (Figure 2c). As with LO filter, the low pass structure was optimized for flat response to 18 GHz with high insertion loss in the 26 to 30 GHz range. Since the LO filter and IF filter are in parallel, the problem reduces to a simple diplexer and should be analyzed together. This procedure insures against extraneous loading of one filter with the other.

Fabrication

As seen in Figure 1, the housing is a simple split block design, which was fabricated using a numerically controlled end-mill. Note that the mixer consists of six parts, including the diodes, which is consistant with low cost fabrication techniques.

The two-sided printed circuit was layed out and the rubylith cut using a desk top calculator (HP9825) and plotter (HP9872B). The plotter pen was modified so that a special blade could be inserted, thus enabling the use of standard graphics subroutines to modify or cut new rubyliths. The fabrication feature allowed a number of quick iterations and variations to be tried and evaluated in a short period of time.

The substrate material used was 10 mil thick Duroid® with half-ounce copper plating. Line widths and gaps were limited to 4 mils and taken into account early in the design stage so that processing yield may be maximized.

The beamlead diodes were soldered to the substrate using Indium® solder whose liquid state was 140° C.

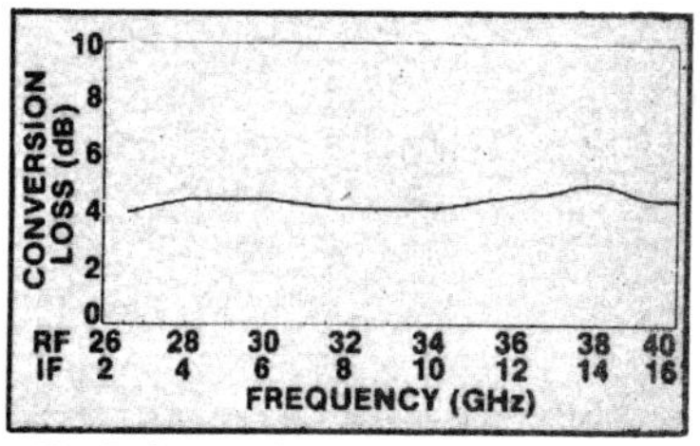

Fig. 7 Conversion loss of hybrid mixer.

Hybrid Mixer Test Results

The mixer was evaluated for RF coverage from 26 GHz to 40 GHz at a fixed LO of 24 GHz LO drive was +10 dBm. The IF band was evaluated to 16 GHz although good response is expected to extend to 18 GHz. Figure 7 shows the wideband response of the mixer at both RF and IF ports. The conversion loss ranged from 4 to 5.2 dB over the 14 GHz IF bandwidth. Subsequent experiments resulted in conversion losses as low as 4.5 dB over narrow bands (3-5 GHz) by incorporating a quarter-wave

TABLE I
HYBRID MIXER SPECIFICATIONS

RF Band	26 - 40 GHz
LO Frequency	24 GHz fixed
IF Band	2 - 16 GHz
Conversion Loss	4.6 dB ±.5 dB
Input VSWR	
RF Port	2.0:1 max
LO Port	2.0:1 max
Output VSWR	
IF Port	2.0:1 max
LO Power	+10 dBm max
3rd Order Intercept	+20 dBm

length matching section appropriately located in the slot transition preceding the diode pair. Table 1 summarizes the mixer specifications.

Conclusions

The hybird mixer described in this paper offers a considerable performance improvement over whisker-contacted mixers in terms of instantaneous bandwidth at the RF and IF ports. A major result of this mixer development is that cost savings are incurred by parts reduction and improved fabrication procedures. As a result, the mixer can be used in multichannel receivers, yet maintain overall systems cost at a reasonable level. The mixer topology lends itself to intergration with other planar structures that results in further reduction in both overall size and production cost.

Acknowledgements

The authors wish to thank Ms. Y. Weyand for fabrication of the mixer. ■

REFERENCES

1 Begemann, G., *"An X-Band Balanced Finline Mixer"*, Presented at the 1978 IEEE MTT-S International Microwave Symposium.

2. Cohen, S., *"Parallel-Coupled Transmission Line Resonator Filters"*, IEE Trans, Vol MTT-11, pp 223-231, 1958

3. Reindel, J., *"Mixers Simplify Millimeter Surveillance"*, MSN, Vol 10, pp 71-74, August 1980

4. Cardiosmenos, A. G., *"Planar Devices Make Production Practical"*, MSN, Vol 9, pp 46-56, May, 1979

7.5.3

A NEW TYPE OF E-PLANE UPCONVERTER FOR SPACE APPLICATIONS

Alexandra P. Delamer*

ABSTRACT

Published work on MIC mixers has dealt with downconverter applications [1-3]. This paper describes a low conversion loss upconverter which has been designed to operate over a wide bandwidth with a high level of intermodulation suppression in a space environment. The design and measured performance of an E-plane upconverter for frequency conversion from 4 to 20 GHz is described. The mixer uses beam-lead diodes and is realized in a single-balanced configuration on a soft substrate. The circuit is clamped in the E-plane of a waveguide housing.

The conversion loss of the mixer is 6.5 ± 1.1 dB over a 700 MHz RF bandwidth for LO power levels of +11 dBm to +15 dBm, and more than 40 dB suppression of 3rd order intermodulation products is achieved. Thermal and vibration test data show that this configuration is rugged and thus has potential for application in space and other hostile environments.

INTRODUCTION

Satellite communications are moving to higher frequencies as the lower frequency ranges become saturated, and the 30/20 GHz band is now being exploited for commercial use. Since on-board processing is carried out at baseband frequencies, down/up converters are required for frequency conversion [4-6]. Advances in planar circuit and solid-state device technologies over the last decade have led to the development of E-plane mixers which offer advantages in size, weight and production cost since a complete circuit can be printed on a single substrate using photolithographic techniques.

DESIGN AND MEASURED PERFORMANCE

The electrical specifications and performance achieved are summarized in Table 1. Conversion loss of 6.5 ± 1.1 dB is achieved over the RF band for LO power levels of +11 dBm to +15 dBm with more than 40 dB suppression of 3rd order intermodulation products. Measured data are shown in Figure 1.

---

* COM DEV Ltd., 155 Sheldon Drive, Cambridge, Ontario, Canada, N1R 7H6.

Table 1: Upconverter Performance Summary

PARAMETER	SPECIFICATION	ACHIEVED PERFORMANCE
IF 4 GHz ± 350 MHz Input Level, dBm	0	0
LO 16 GHz Input Level, dBm	+13 ± 2	+13 ± 2
RF 20 GHz ± 350 MHz Conversion Loss, dB	15 (max)	<7.7*
Image Suppression, dB	40 (min)	>50
Higher Order Product Suppression, dB	50 (min)	>50*
3rd Order IMP Suppression for Two-Tone Input at -3 dBm per tone, dB	>40 w.r.t. fundamental RF output	44†
LO to IF Isolation, dB	TBD	40
LO to RF Isolation, dB	TBD	33††
1 dB Compression Point, dBm (L.O. @ + 13 dBm)	TBD	6.7

* with ridge waveguide filter on RF port
† 20 MHz separation of RF tones
†† no filter on RF port

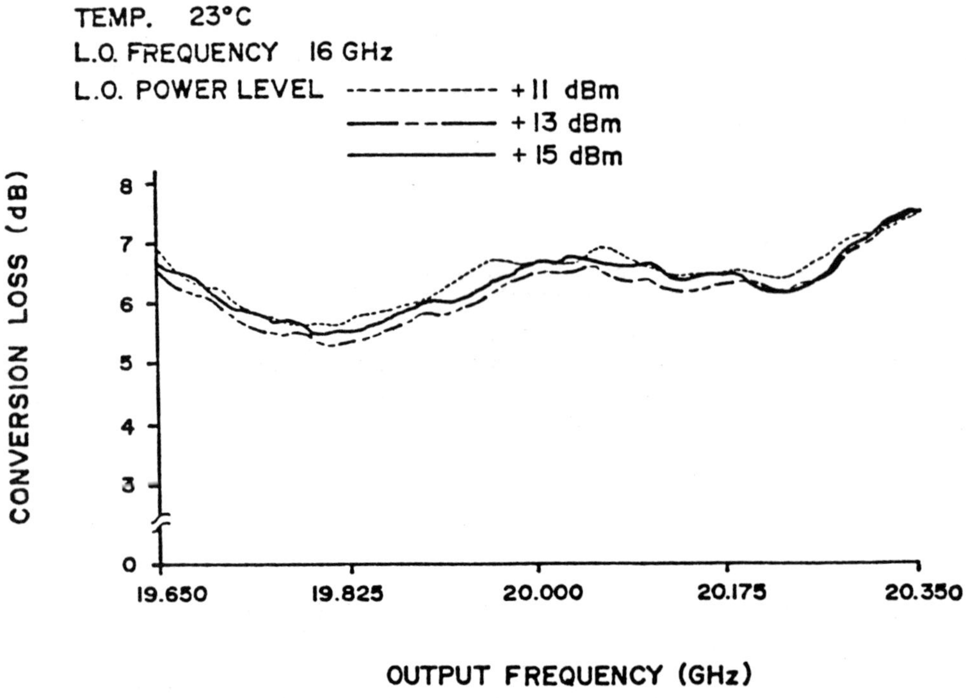

Figure 1 : Upconverter Swept Conversion Loss

The upconverter is implemented as a single-balanced mixer, and is fabricated on a soft substrate (RT/Duroid 5880). A combination of E-plane printed circuit technologies is used, exploiting the advantages of each to achieve different objectives in the circuit. Figure 2 shows the mixer configuration:

- a suspended stripline probe into waveguide was developed to meet the specification of LO input from waveguide.
- microstrip is used for the IF/LO diplexer on the unbalanced input side of the mixer.
- coplanar line completes the input impedance match and provides an accessible ground plane for mounting GaAs Schottky barrier beam-lead diodes.
- the coplanar line/fin-line balun acts as a 180 degree hybrid and gives a minimum of 20 dB isolation between input and output ports.
- fin-line is used for the balanced RF port and a fin-line taper provides a low-loss transition to the output waveguide.

Mode suppression slots [1], a quarter wavelength at the IF/LO band centre, are used to suppress any spurious signals due to non-symmetric modes. The gap in the coplanar line is made approximately equal to the length of the diode active area to minimize diode parasitics. Beam-lead diodes (Marconi DC1338) are parallel-gap welded to the substrate.

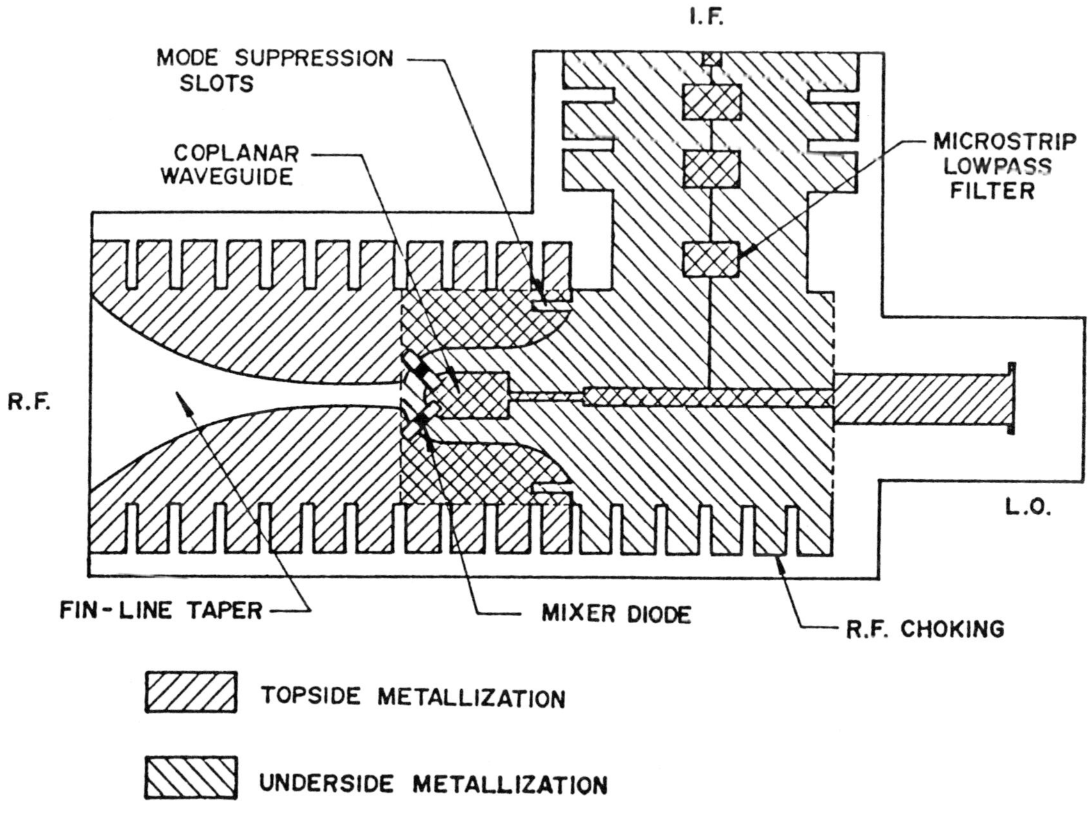

Figure 2 : E-Plane Upconverter Configuration

At the LO port the single frequency match required is accomplished more efficiently using a co-axial or printed probe, rather than a printed taper which is long and therefore lossy at this frequency. A printed suspended stripline probe for the microstrip/WR62 transition has been developed. The probe was designed to provide a match to the radiation resistance of the guide at the LO frequency, and empirically optimized to obtain the desired match (≥20 dB over 400 MHz).

A bandpass filter is required on the RF port to suppress higher order products exiting the mixer. The performance of two different types of filter was compared:

a) an E-plane (finline) bandpass filter etched on a dielectric substrate; and
b) a quasi-bandpass evanescent mode ridge waveguide filter.

Advantages offered by the E-plane filter are low-cost fabrication and ease of integration, since it can be fabricated on the same substrate as the mixer circuit. However, the E-plane filter has a narrow stopband and above 30 GHz exhibits essentially high-pass behaviour. The stopband may be widened by reducing the waveguide 'a' dimension in the filter section [7], but in practice the passband return loss does not meet the requirements (≥20 dB) for this application. The ridge filter is more costly to produce, but provides >50 dB suppression of all higher order products at the RF port. Thus, there is a trade-off between cost and stopband attenuation.

The E-plane configuration may be used at higher frequencies, and the design is directly applicable to upconverter requirements for space applications at frequencies from 30 to 100 GHz.

QUALIFICATION STATUS FOR SPACE APPLICATIONS

The mixer housing is a completely enclosed structure, Figure 3, and a large clamping area is provided for the substrate due to the presence of RF choking slots as depicted in Figure 2. The structure is therefore mechanically rugged and the substrate is shielded from radiation losses. Vibration tests of the upconverter at an rms acceleration of 20.76 g (i.e. equivalent to flight qualification levels for the U.S. domestic satellites), were successfully completed. Thermal cycling of the mixer was carried out over the range -20 to +50°C, and conversion loss variation was less than 1 dB over this temperature range.

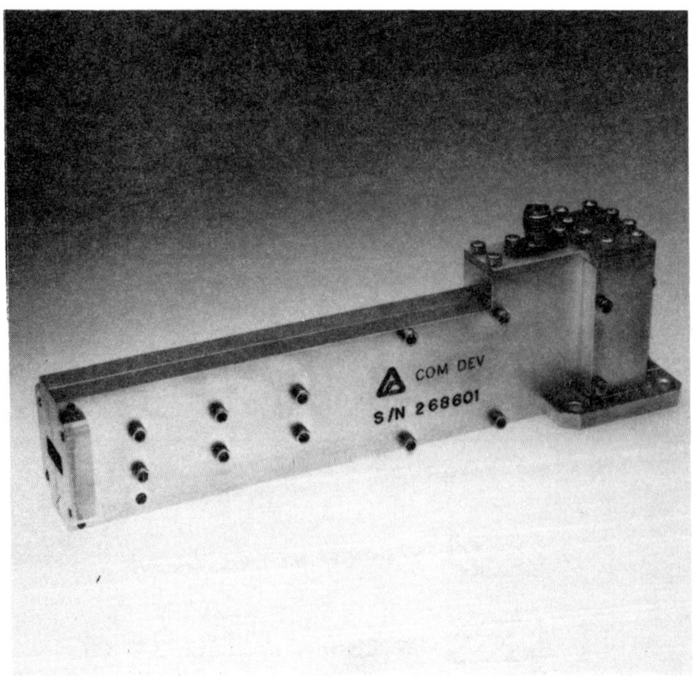

Figure 3 : Photograph of Unit

CONCLUSIONS

A single-balanced E-plane upconverter using beam-lead diodes has been developed. Good conversion loss and intermodulation product suppression have been achieved. The unit has been successfully vibrated to qualification levels of U.S. domestic satellites. Measured performance, over an extreme temperature range, shows only small variations in the performance of the upconverter.

ACKNOWLEDGEMENT

I am grateful to Mr. W. Thorpe for guidance and encouragement throughout. This work was funded by the Canadian Department of Communications, and I wish to thank the scientific authority, Mr. C. Verver. The finline filter design was obtained from Dr. R. Valhdieck and Dr. W. J. R. Hoefer. The probe development and testing was done by R. Hildesheim, and circuits fabricated and assembled by R. Buffett, R. Hunter and C. Barz.

REFERENCES

[1] R. N. Bates and M. D. Coleman, "Millimetre Wave Fin-Line Balanced Mixers", Proc. 9th European Microwave Conference, 1979, pp 721-725.

[2] P. J. Meier, "Printed-Circuit Balanced Mixer for the 4- and 5-mm Bands", IEEE MTT-S Proc. 1979, pp 84-86.

[3] T. Berceli, V. Geleji and Mrs. T. Kolumban, "Optimum Design of Single Sideband Fin-Line Mixers with Filters", Proc. 13th European Microwave Conference, 1983, pp 573-578.

[4] C.L. Cuccia et al, "Optimization of Input Sensitivity and Radiated Power in Millimeter Wave Communication Satellites", AIAA 7th Communications Satellite Systems Conference 1978, pp 231-240.

[5] F.J. Solman et al, "The Ka-Band Systems of the Lincoln Experimental Satellites LES-8 and LES-9", AIAA 7th Communications Satellite Systems Conference, 1978, pp 208-215.

[6] S. Kitazume and H. Ogawa, "14/4 GHz Receiver, Upconverter and Solid State Power Amplifier for Intelsat VI", AIAA 10th Communications Satellite Systems Conference, 1984, pp 421-427.

[7] R. Valhdieck and W.J.R. Hoefer, "Finline and Metal Insert Filters with Improved Passband Separation, and Increased Stopband Attenuation", IEEE Trans. Microwave Theory & Techniques, Vol. MTT-33, Dec. 1985, pp 1333-1339.

7.5.4

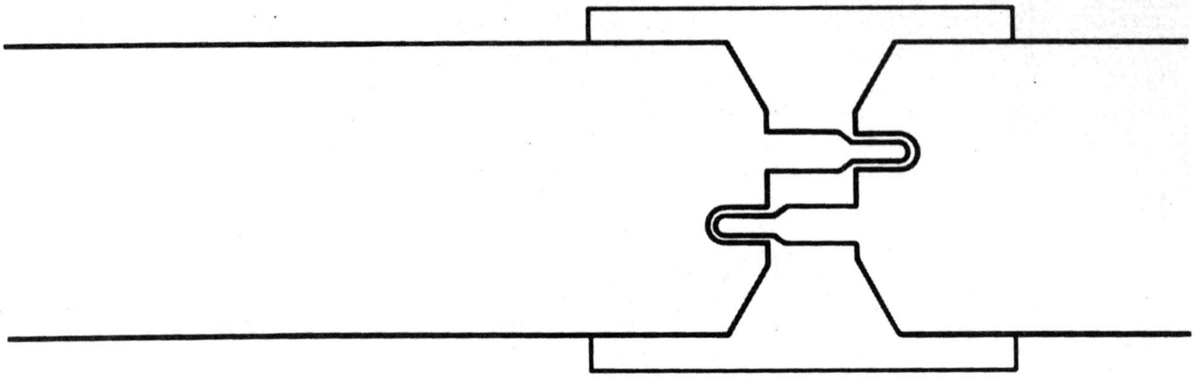

Wide-Band Subharmonically Pumped *W*-Band Mixer in Single-Ridge Fin-Line

PAUL J. MEIER, SENIOR MEMBER, IEEE, JOSEPH A. CALVIELLO, SENIOR MEMBER, IEEE, AND PAUL R. BIE, MEMBER, IEEE

***Abstract* — A subharmonic mixer is described that has an instantaneous bandwidth of 11 to 14 GHz centered near 95 GHz. A wide bandwidth is achieved by the close integration of a low-capacitance diode mount, printed circuit matching elements, and simple yet effective filters which are uniquely suited to realization in a single-ridge fin-line. The mixer also has a two-terminal shunt mount that will accept two conventional beam-lead diodes or a single dual-junction device. With the latter, a minimum conversion loss of 8.5 dB has been achieved with a drive level of only 4 dBm at 45 GHz.**

I. Introduction

Subharmonically pumped mixers have been constructed in various transmission media including stripline [1], microstrip, waveguide [2], and double-ridged fin-line [3]. Although excellent results have been obtained at millimeter wavelengths [4], tunable waveguide back-shorts have so far limited the instantaneous bandwidth. The existing shunt mounts accommodate two diodes rather than the latest dual-junction antiparallel devices [5], [6]. Since these mounts are susceptible to asymmetric bonding resulting in degradation [7], a two-terminal mount would be ultimately preferable. A wide-band fin-line mixer with a two-terminal shunt mount and unique RF/LO filters is described.

II. Circuit Description

Fig. 1 shows the construction features of the wide-band single-ridge fin-line mixer. The fin-line is printed on 5-mil Duroid which is suspended in the E-plane of a rectangular

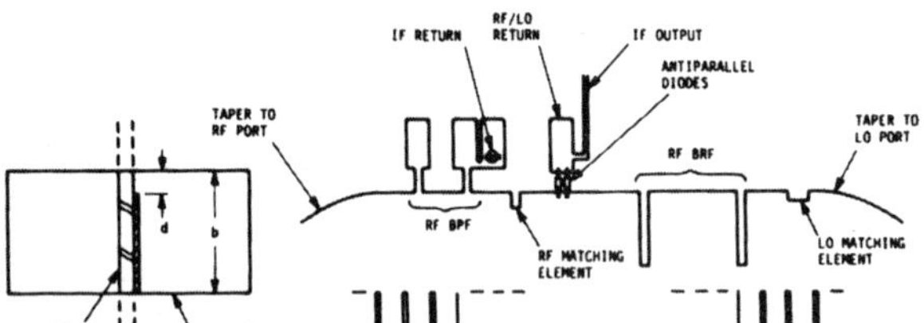

Fig. 1. Subharmonic mixer in single-ridge fin-line.

housing. The housing, in the indicated region, has the same inner dimensions as WR-10 waveguide. The gap (d) between the edge of the fin and the upper housing wall is chosen to be small enough to propagate the subharmonic local oscillator (LO), yet large enough to accommodate the antiparallel diodes. By selecting a gap ratio (d/b) equal to 0.3, the cutoff frequency of the fin-line is placed at 37 GHz which is well below the intended LO band.

The RF signal enters from the left and passes through a cosine taper, matching the fin-line to a standard WR-10 port. The signal then passes through a single-section bandpass filter (BPF). The filter is formed by two inductive strips each joining the edge of the fin to a ground return. Grounding in the RF and LO bands is achieved with low-impedance stubs within the choke region [8] of the housing. Grounding in the IF band is provided by a feedthrough that links the illustrated pattern to a rear-surface metalization. After exiting from the BPF the signal passes through the RF matching element and reaches the antiparallel diodes which are reactively terminated by the

Manuscript received March 15, 1982; revised July 15, 1982. This work was supported in part by NOSC under Contract no. N00123-79-C-1879.

The authors are with Eaton Corp., AIL Division, Melville, NY 11747.

P.J. Meier, J.A. Calviello, and P.R. Bie, *IEEE Transactions on Microwave Theory and Techniques*, Vol. MTT-30, No. 12, Dec. 1982, pp. 2184–2189.

RF band-reject filter (BRF). The diodes are mounted between the edge of the fin and a low-impedance stub (RF/LO ground return). The IF leaves the low-capacitance diode mount through a microstrip line and an SMA connector (not shown).

The LO enters the circuit from the right in Fig. 1. The LO-matching elements include a notch in the fin-line, a cosine taper, and a step in the width of the housing. The latter allows the subharmonic pump to propagate without the aid of the fin-line, and provides a direct interface with a standard WR-19 port. After emerging from the matching elements the LO passes through the RF band-reject filter and reaches the diode mount. It is then reactively terminated by the RF bandpass filter.

The fin-line filters, which diplex the RF and LO, are key components of the subject mixer.

III. RF Bandpass Filter

The fin-line bandpass filter was analyzed and optimized by computer-aided techniques, based on the equivalent circuit shown in Fig. 2. The circuit includes the normalized shunt susceptances (b_3 and b_5) separated by a line length (l_4) that controls the resonant frequency. The circuit also includes input and output lines (l_2 and l_6) that model the dissipative loss in the cosine tapers. All the transmission lines are assigned a normalized characteristic impedance of unity. (The tapers are assumed to be reflectionless.)

To utilize the equivalent circuit, information is required on the fin-line properties including the equivalent dielectric constant (k_e), cutoff wavelength (λ_c), and unloaded Q. Because this information was unavailable for the preferred single-ridge configuration, appropriate measurements were conducted. A sliding-short technique [8] provided measurements of the guide wavelength (λ_g) at widely spaced frequencies. The fin-line properties were then found from

$$\lambda_c = \sqrt{\frac{\lambda_2^2 - \lambda_1^2}{(\lambda_1/\lambda_{g1})^2 - (\lambda_2/\lambda_{g2})^2}} \qquad (1)$$

$$k_e = (\lambda_1/\lambda_{g1})^2 + (\lambda_1/\lambda_c)^2 \qquad (2)$$

where λ_{g1} is the measured guide wavelength at the free-space wavelength λ_1 and λ_{g2} is the guide wavelength at λ_2. For the preferred configuration ($d/b = 0.3$ in a WR-10 housing), k_e was found to be 1.22 and λ_c was determined to be 0.292 in. Based on other measurements of insertion loss the unloaded Q was estimated to be approximately 500.

The final parameter needed to analyze the circuit of Fig. 2 is the normalized shunt susceptance. This parameter was determined by embedding a single inductive strip between matched transitions and measuring the insertion loss across the bands of interest (40 to 60 GHz and 80 to 110 GHz). At each frequency, the normalized susceptance was calculated from

$$b_n = 2\sqrt{\log^{-1}(L/10) - 1} \qquad (3)$$

where L is the measured loss (less the transition loss) in

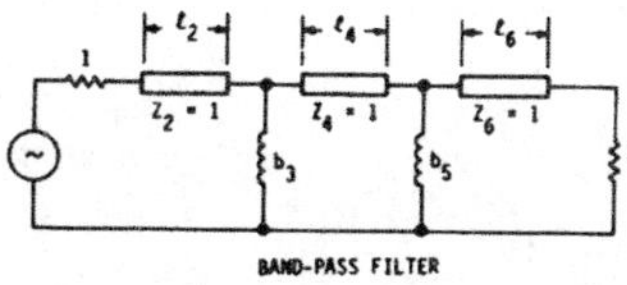

Fig. 2. Equivalent circuit of fin-line bandpass filter.

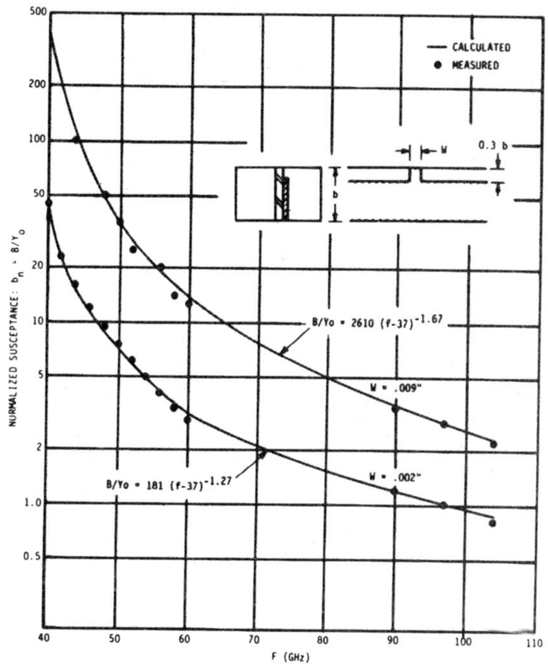

Fig. 3. Susceptance of inductive strips.

decibels. After investigating a variety of analytical expressions, it was demonstrated that the susceptance could be accurately modeled by

$$b_n = k(f - f_c)^n \qquad (4)$$

where f and f_c are the operating and cutoff frequencies, respectively. The constants k and n were determined from a log/log plot of b versus $f - f_c$.

Fig. 3 shows how the susceptance varies with frequency for typical inductive strips in single-ridge fin-line. Good agreement between (4) and the measurements has been obtained for the values of k and n appropriate to a particular strip width w. Because of the exponential variation, the susceptance is an order of magnitude larger in the LO band than it is in the RF band. This condition is highly favorable to the design of the RF bandpass filter. The exponential variation assures that high rejection in the LO band and a wide RF bandwidth can be simultaneously obtained with a simple one-section filter.

Through a computer-aided analysis of the circuit of Fig. 2, it was demonstrated that adequate rejection could be obtained in the LO band with a small (2-mil) strip width. The preliminary analysis also predicted a suitable value for the resonator length l_4. Based on the assumption that the electrical and mechanical values of l_4 were equal, a bread-

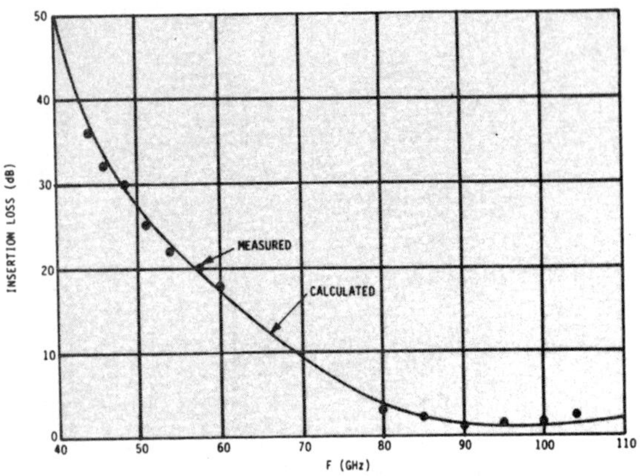

Fig. 4. Response of bandpass filter.

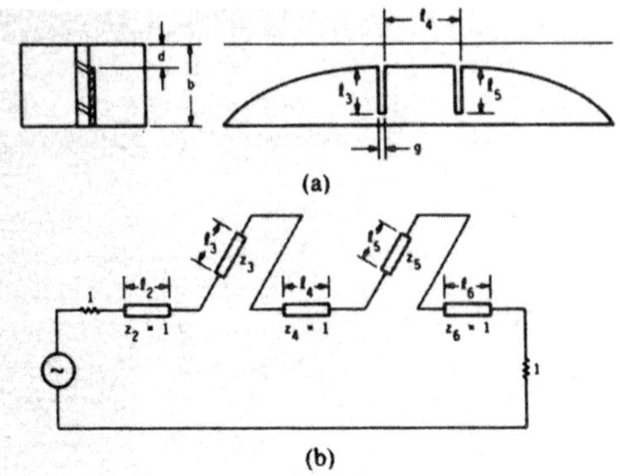

Fig. 5. RF band-reject filter. (a) Structure. (b) Equivalent Circuit.

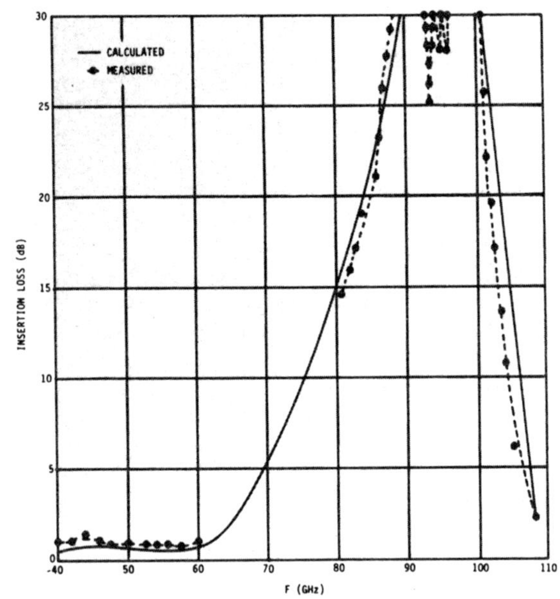

Fig. 6. Response of two-slot band-reject filter.

board model of the RF bandpass filter was constructed and tested. The measurements showed that the passband was centered at a frequency 7 percent higher than that desired. Using this information the design was refined to include an offset between l_4 and the centerline spacing of the inductive strips.

Fig. 4 shows the measured performance of the revised RF filter together with the calculated response. The measured rejection is better than 25 dB in the LO band of 42 to 51 GHz and the insertion loss is 1.4 ± 0.4 dB across the RF band of 86 to 102 GHz. The loss, measured from flange to flange in an oversized fixture, includes superfluous line lengths which were eliminated in the subharmonic mixer. The RF input loss in the final mixer design is estimated to be 0.5 dB.

It should be emphasized that the favorable asymmetric response of Fig. 4 was obtained with a single-section fin-line filter. To duplicate this performance with quasi-lumped elements in a TEM medium, at least six elements would be required [9].

IV. RF Band-Reject Filter

A second filter is required in order to complete the RF/LO diplexer. Although this filter could be a bandpass filter tuned to the subharmonic LO frequency, it would have an unsuitable susceptance variation. (In Fig. 3, the high susceptance in the passband would result in an impractically narrow bandwidth, and the low susceptance in the RF band would afford little rejection.) Consequently, a band-reject filter was chosen to block the RF from the LO port.

Fig. 5 shows the structure and the equivalent circuit of the RF band-reject filter. The filter consists of two notches that are etched into the fin-line. It has been shown that such a structure can be modeled by shortcircuited stubs that appear in a series with the main line [10]. The circuit contains the series lines (l_3 and l_5) which are each a quarter-wave long at the center of the rejection band. The stubs are spaced along the main line by l_4, which is the length that controls the match in the desired passband, as well as the center frequency of the second passband. (The latter falls above the rejection band at a frequency where l_4 is approximately a half wavelength.) The equivalent circuit includes input and output lines (l_2 and l_6) to model dissipative loss in the cosine-taper transitions to standard waveguide ports.

The paper design of the filter was started with this equivalent circuit and the aforementioned fin-line parameters. Based on earlier work [10], plausible magnitudes were assigned to z_3 and z_5, and the remaining variables (l_3 through l_5) were optimized by computer-aided calculations. After obtaining a satisfactory response, a breadboard filter was constructed and tested. The model was then refined to obtain the best fit with the measurements. Good agreement was obtained with

$z_3 = z_5 = 0.95$

l_4 = physical spacing between the slots

$l_3 = l_5$: 17 percent longer than the physical slot length.

With these parameters the computer model was utilized to optimize the filter performance for the intended application. Calculations showed that l_4 should be a half wavelength just above the stopband. This prevents a spurious response within the RF band and yet keeps l_4 as large as possible, thereby obtaining a favorable (quarterwave) spacing of the slots near the LO band. Fig. 6 compares the measured and calculated response of the optimized band-reject filter. The halfwave response has been fixed at 108 GHz which provides ample rejection at the upper edge of the RF band. Across the RF band of 86 to 102 GHz, the measured and calculated rejection is greater than 20 dB. In the LO band the measured insertion loss, from flange to flange, is typically 1 dB. Since the measurement includes superfluous line lengths and a pair of WR-19/WR-10 transitions, the LO input loss in the final mixer is estimated to be 0.5 dB.

V. Mixer Integration and Performance

The previously described filters were next integrated with antiparallel diodes and other circuit elements to form a mixer similar to that shown in Fig. 1. In the first phase of integration, however, the circuit did not include RF- or LO-matching elements. In keeping with the wide-band objective, low-parasitic beam-lead devices [5] were chosen for the antiparallel devices. Separate diodes were utilized in the preliminary breadboard [11]. They were later replaced by a dual-junction device, to be described.

The RF bandpass filter was located at a quarter LO wavelength from the diodes to provide a wide-band backshort termination. The band-reject filter, which appears as an open-circuit to the RF, was located as close to the diodes as possible. The LO- and RF-matching elements were then added, respectively, and optimized by an iterative technique.

Fig. 7 shows the breadboard mixer, with a duplicate of the circuit board which is suspended in the E-plane of the housing. As in recent designs [12], the board is held in a recessed portion of the housing and is aligned by pins external to the waveguide channel. The overall assembly measures 1.1 by 1.5 by 2.8 in.

With separate beam-lead diodes, the measured conversion loss was 12 ± 2 dB across a 12-GHz band centered at 95.5 GHz. These results were obtained with a fixed LO at 45 GHz at an input level of 16 dBm [11]. The moderate conversion loss and the high drive requirement are attributed to unequal bonding inductance [7] and the finite separation between the diodes.

The performance of the fin-line mixer was improved by replacing the separate diodes with an advanced dual-junction device [5]. This device, whose structure is shown in Fig. 8, has the following advantages:

1) The junctions are formed at almost the same location on a given wafer, thereby promoting uniformity and providing a better lumped-element circuit approximation.
2) The circulating current within the antiparallel loop is not affected by bonds external to the device.
3) Fewer bonding operations are required and the alignment during bonding is less critical.
4) The n and n^+ layers can be etched away beneath the junctions, thereby reducing the parasitic package capacitance without sacrificing mechanical strength. (Each end of the device has a wide ohmic contact.)

Fig. 7. Fin-line subharmonic mixer.

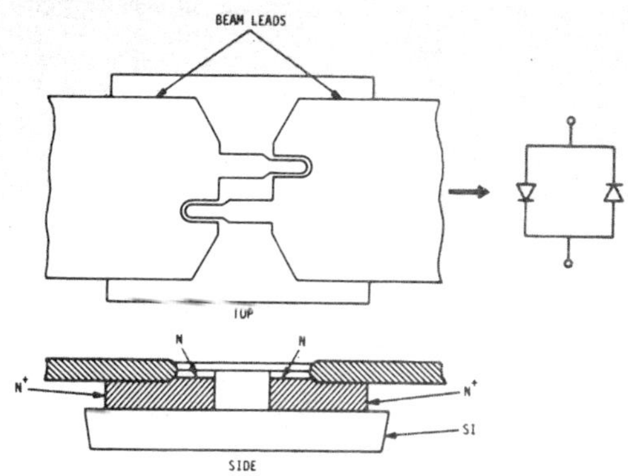

Fig. 8. Beam-lead dual-junction device structure.

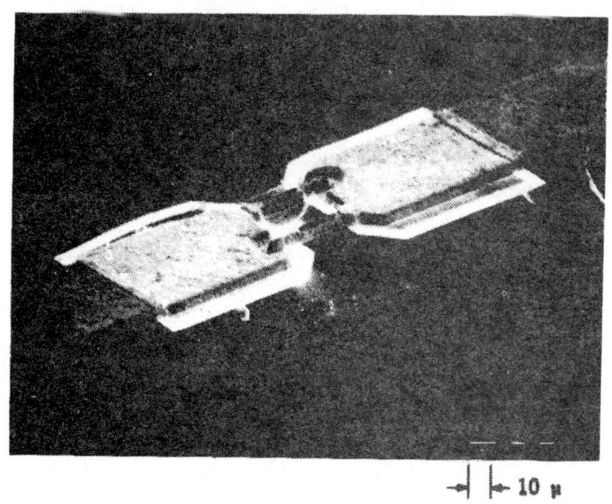

Fig. 9. SEM photograph of dual-junction device.

Fig. 9 is an SEM photograph of the dual-junction device. The device has a package capacitance of 10 fF, which is at least three times lower than that achievable with a pair of

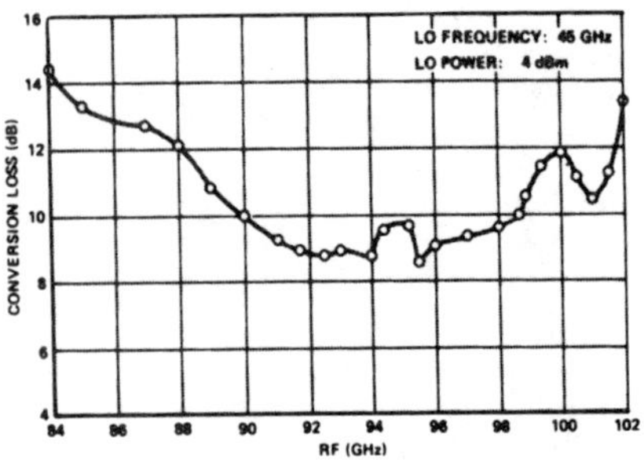

Fig. 10. Conversion loss of fin-line mixer with dual-junction device.

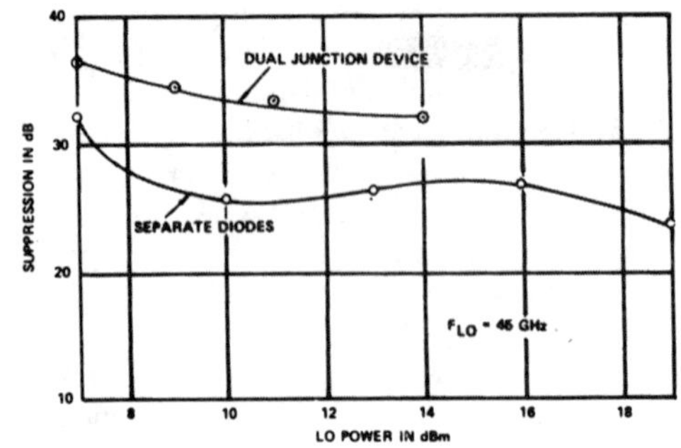

Fig. 12. Suppression of LO second harmonic.

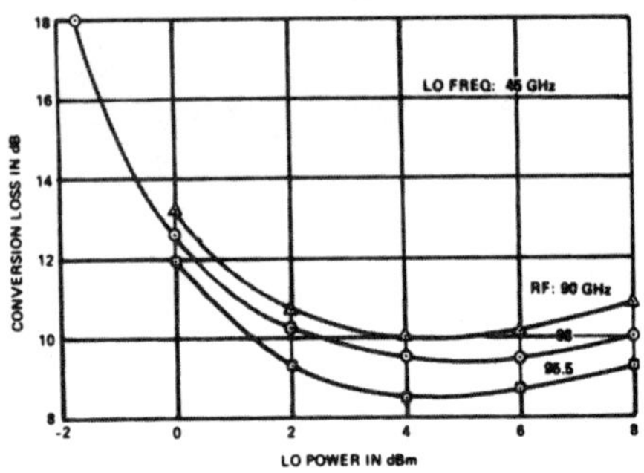

Fig. 11. Conversion loss versus LO power (dual-junction device).

the best available beam-lead diodes. Each junction has a zero-bias capacitance of 4 fF and a series resistance of 8 to 10 Ω, resulting in a minimum zero-bias cutoff frequency of 4 THz. The measured ideality factor is in the range of 1.05 to 1.07.

Fig. 10 shows the measured conversion loss of the fin-line mixer with a dual-junction device. To assure the measurement accuracy, the RF power was measured with a wideband thermistor, calibrated against a wet calorimeter. The minimum conversion loss is 8.5 dB at 95.5 GHz and the 3- and 4-dB bandwidths are 11 and 14 GHz, respectively. As shown in Fig. 11, the optimum LO drive level is 4 to 5 dBm. This drive requirement is unusually low for a pair of self-biased high-barrier junctions. This effect is attributed to efficient LO coupling, reduced diode parasitics, and favorable input/output impedances at moderate drive levels.

The useful LO bandwidth of the existing design is approximately 3 GHz. Although a wider bandwidth could be achieved by moving the LO-matching element closer to the diodes, further design iterations would be necessary to simultaneously optimize the RF and LO match.

Because the subharmonic LO is cutoff at the WR-10 RF port, the LO/RF isolation is extremely high. Of concern in some applications, however, is the radiation from the RF port of the virtual LO, which is generated at the second harmonic of the LO. Although such even-order products can be suppressed by over 30 dB at centimeter wavelengths [2], unequal diode parasitics generally limit this suppression at millimeter wavelengths. Fig 12 shows the measured suppression of the second LO harmonic for the fin-line mixer. The output level at the RF port, relative to the input level at the LO port, is plotted versus the LO input power. At normal drive levels, the suppression is typically 25 dB for separate diodes and 35 dB for the dual-junction device.

VI. Conclusion

A subharmonically pumped mixer has been developed with a useful instantaneous bandwidth of 11 to 14 GHz centered near 95 GHz. The wide bandwidth is achieved by closely integrating a low-capacitance diode mount, printed-circuit matching elements, and simple yet effective filters which are uniquely suited to realization in single-ridge fin-line.

The mixer includes a single-section fin-line filter which performs as well as a six-element filter in a TEM medium such as stripline. Also included is a unique shunt mount for conventional or dual-junction devices.

Although a wide bandwidth has been achieved with separate antiparallel devices, superior performance has been demonstrated with a dual-junction beam-lead device. The lower parasitics and better balance afforded by the dual-junction device result in lower conversion loss (8.5 versus 10 dB, minimum), reduced LO power requirements (4–5 versus 16 dBm), and better suppression of the LO second harmonic (35 versus 25 dB).

Because the performance of the subject mixer is comparable to that obtained with commercial fundamental mixers, the new design should be applicable to a wide range of advanced systems. The mixer is well suited to low-cost receivers—particularly where a simple yet stable (phase-locked) LO is required. The approach is also applicable to those RF bands (above 120 GHz) where fundamental low-noise solid-state oscillators are not currently available.

Acknowledgment

The work reported was partially funded by NOSC under the direction of J. E. Reindel. The work was performed at Eaton Corporation AIL Division under the direction of B. J. Peyton and J. J. Whelehan. The diodes were developed in our Central Research Laboratory under the direction of J. J. Taub. Technical assistance was provided by A. Cooley, A. Kunze, J. Pieper, and A. Rees.

References

[1] M. V. Schneider and W. W. Snell, "Harmonically pumped stripline down-converter," *IEEE Trans. Microwave Theory Tech.*, vol. MTT-23, pp. 271–275, Mar. 1975.

[2] M. Cohn, J. E. Degenford, and B. A. Newman, "Harmonic mixing with an antiparallel diode pair," *IEEE Trans. Microwave Theory Tech.*, vol. MTT-23, pp. 667–673, Aug. 1975.

[3] G. Begemann, "A subharmonically pumped fin-line mixer for satellite TV receiver applications," in *Dig. IEEE 1981 MTT-S Symp.*, June 1981, pp. 454–456.

[4] E. R. Carlson, M. V. Schneider, and T. T. McMaster, "Subharmonically pumped millimeter-wave mixers," *IEEE Trans. Microwave Theory Tech.*, vol. MTT-26, pp. 706–715, Oct. 1978.

[5] J. A. Calviello, S. Nussbaum, and P. R. Bie, "High-performance GaAs beam-lead mixer diodes for millimeter and submillimeter applications," in *Dig. IEEE Int. Electron Devices* Meet., Dec. 1981, pp. 692–695.

[6] G. F. Anderson *et al.*, "GaAs beam-lead antiparallel diode for millimeter wave subharmonic mixers," in *Dig. IEEE Int. Electron Devices Meet.*, Dec. 1981, pp. 688–691.

[7] R. G. Hicks and P. J. Khan, "Analysis of balanced subharmonically pumped mixers with unsymmetrical diodes," in *Dig. IEEE 1981 MTT-S Symp.*, June 1981, pp. 457–459.

[8] P. J. Meier, "Integrated fin-line millimeter components," *IEEE Trans. Microwave Theory Tech.*, vol. MTT-22, pp. 1209–1216, Dec. 1974.

[9] G. L. Matthaei, L. Young, and E. M. T. Jones, *Microwave Filters, Impedance Matching Networks and Coupling Structures.* New York: McGraw-Hill, 1964, pp. 83–103.

[10] H. Hofmann, "MM-wave Gunn oscillator with distributed feedback fin-line circuit," in *Dig. IEEE 1980 MTT-S Symp.*, May 1980, pp. 59–61.

[11] P. J. Meier, "Wide band subharmonically pumped *W*-band mixer in single-ridge fin-line," in *Dig. IEEE 1982 MTT-S Symp.*, June 1982, pp. 201–203.

[12] P. J. Meier, "Millimeter integrated circuits suspended in the *E*-plane of rectangular waveguide," *IEEE Trans. Microwave Theory Tech.*, vol. MTT-26, pp. 726–733, Oct. 1978.

Applied Electronics Division on programs resulting in the development of a high-power solid-state switch, an *X*-band MIC sweeping receiver, and a multibeam quasi-optical millimeter receiver. He is currently a Staff Consultant in the Receiver Systems & Technology Department engaged in the development of millimeter-wave hybrid and monolithic integrated circuits. Mr. Meier holds the U.S. Patent for Integrated Fin-Line and has published over 30 papers concerning microwave and millimeter-wave components.

Mr. Meier is a member of Eta Kappa Nu and is a Past-Chairman of the New York/Long Island Chapter of the MTT Committee.

7.6.1

Radar

Integrated Ka-Band Radar Increases Target Resolution

Advances in mm-wave integration techniques yield higher frequency radars able to detect targets in conditions where other sensors fail.

By Wolfgang Menzel, Heinrich Callsen, Dr. Klaus Solbach, and Holger Meinel

The performance possibilities offered by millimeter-wave systems are urgently needed in military applications. Microwave sensors at frequencies below 20 GHz, used for searching for small target vehicles in a battlefield or air targets at low elevation angles, often lack sufficient target resolution. Infra-red and electro-optical sensors are limited by adverse weather conditions such as fog, clouds, haze, dust, and smoke which amount to "nature's chaff." Bridging the gap between microwave and IR/EO sensors, mm-wave systems provide several advantages, especially for military radar applications:[1, 2]

- Reduced antenna size. For example, an antenna beamwidth of ~1.5 degrees (40 dB gain) can be achieved with a parabolic dish of only 320mm diameter at 35 GHz and 120mm at 94 GHz.
- High spatial resolution and low minimum detection distance using very short pulses.
- Possibility of increased IF bandwidth.
- Increased radar cross section, even of non-metallic objects.
- Increased Doppler frequencies.
- Smaller size and lighter weight.

Also, mm-waves offer better penetration capabilities under adverse atmospheric conditions when compared with IR/EO systems.

The frequency-dependent atmospheric attenuation of mm-waves can be used effectively for several applications, such as covert communication or low probability of intercept (LPI) radar systems.[3, 4] Because of these advantages, mm-wave systems are finding greater applications. Typical examples are missile and projectile guidance, fuzing, smart bomb sensors, ground mapping, handheld communication and, as a consequence, EW receivers.[5]

The high resolution, 200 mm parabolic dish radar module demonstrates the application of planar integration techniques for mm-wave circuits.

Fin-Line Circuits Reduce Costs

The successful development of mm-wave systems depends on the availability of efficient solid state devices and low cost production techniques for both the devices and for entire modules. Considerable advances in both areas have been made recently.[6, 7] Progress in mm-wave integration techniques, along with design and performancce of a Ka-band pulse radar, will be described here. Figure 1 shows a block diagram of the radar module. The electronic circuits—pulse

Dr. Wolfgang Menzel, Heinrich Callsen, Dr. Klaus Solbach, and Holger Meinel represent AEG-Telefunken, West Germany.

W. Menzel, H. Callsen, K. Solbach, and H. Meinel, *Defense Electronics*, Oct. 1982, pp. 95–99.

center (including pulse oscillator bias), IF amplifier, IF filter, video detector, video amplifier with limiter, and the signal processing unit—are state of the art and tailored to meet the small dimensions of the complete unit. Thus, the accent of this contribution is set on the mm-wave part of the radar, marked by a frame in Figure 1.

Millimeter-wave circuits based on standard metal waveguide require the production of fairly complicated three-dimensional waveguide structures with very tight tolerances, achieved in most cases by milling techniques. Thus, these circuits become expensive and are not well suited for mass production. For these reasons, an integrated fin-line circuit was chosen for the mm-wave front-end of the radar described here. Fin-line (Figure 2) is a kind of shielded slot line; or from another point of view, a double-ridge waveguide with very thin ridges printed on a dielectric substrate. Due to the double-ridge waveguide structure, fin-line gives the basis for very components.[5] Nearly all circuit functions are performed using metal structures printed on the substrate. In this way, the requirements for tight tolerances are shifted to the printed substrates.

The production of such planar circuits is a state of the art function. Improved tolerances for mm-wave applications do not create severe difficulties, allowing for reproducible and cost effective mass production. And the waveguide mounts containing the planar circuits are far less complex and allow much greater dimensional tolerances than pure waveguide circuits, and may be produced by casting or similar techniques.

Furthermore, fin-line exhibits a number of additional advantages such as easy implementation of semiconductors, especially beamlead devices; flexible design with the aid of a computer-controlled design and layout procedure; and the possibility of integrating several components on a single substrate, where fin-line can be combined with other types of planar transmission lines, such as microstrip or coplanar lines.

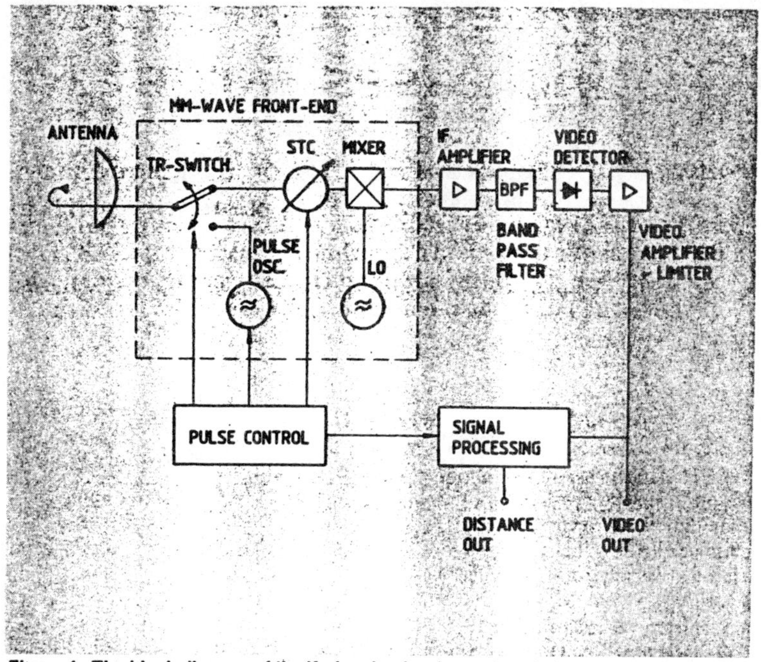

Figure 1. The block diagram of the Ka-band radar shows the relative positions of the mm-wave front end, pulse center, IF amplifier and filer, video subsystems, and signal processing unit.

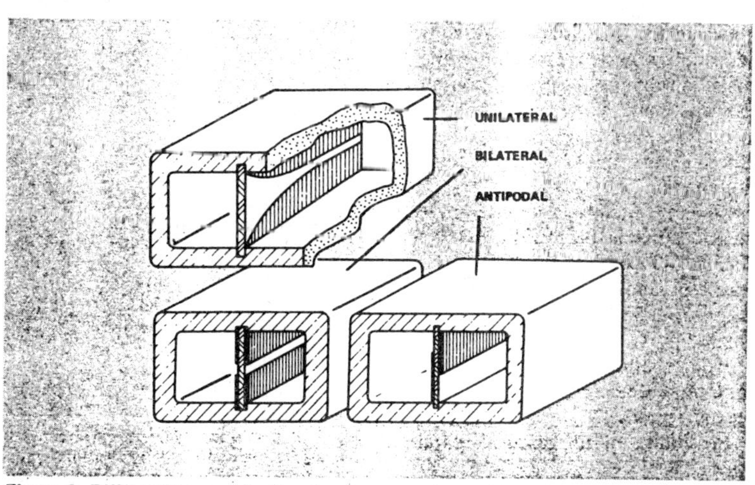

Figure 2. Different types of fin-line structures are shown. Many of the radar's circuits are achieved by using metal structures printed on substrate to ease integration of subcomponents.

The Mm-Wave Circuit

In Figure 3, the basic assembly of the planar fin-line circuit is shown.[8] The pulse oscillator, PIN-diode TR-switch, PIN-diode STC (Sensitivity Time Control), balanced mixer and local oscillator are combined on a single glass: reinforced PTFE substrate (RT-Duroid). A taper from fin-line to standard metal waveguide enables connection to a suitable antenna. The substrate is then clamped in an H-shaped waveguide mount. All the components are placed close together, thus minimizing line lossess. Compared to similar circuits composed of individual components, the number of waveguide connections, and in the case of fin-line components, the number of transitions from standard waveguide to fin-line, is significantly reduced

The Fin-line Gunn-oscillators used

in the system consist of a Gunn element mounted in the broad wall of the waveguide mount. The Gunn element is connected to an asymmetrical fin-line with one or more series stubs that determine the frequency of operation and the impedance match of the Gunn element.[9] A backshort terminates the oscillator at the rear, allowing a final adjustment of the operation frequency, and in the case of the pulse oscillator, of the pulse form. Due to the different modes of operation of the two oscillators, different configurations have been chosen for optimum performance.

Five PIN-diodes are placed parallel to the fin-line slot, and one across a series stub. In the first case and without bias, the fin-line is loaded only with the low capacitance of the PIN-diode, thus resulting in a very high shunt impedance compared to the characteristic impedance of the line. With forward bias, the low series resistance and the lead inductance of the PIN-diode combine almost short-circuiting the fin-line.

In the second case, the series stub is designed so that in the case of high attenuation, a parallel resonant circuit is formed by the stub and the diode parasitics. In the case of low attenuation, a low impedance results in series to the line, and a DC separation is introduced by the stub.[10]

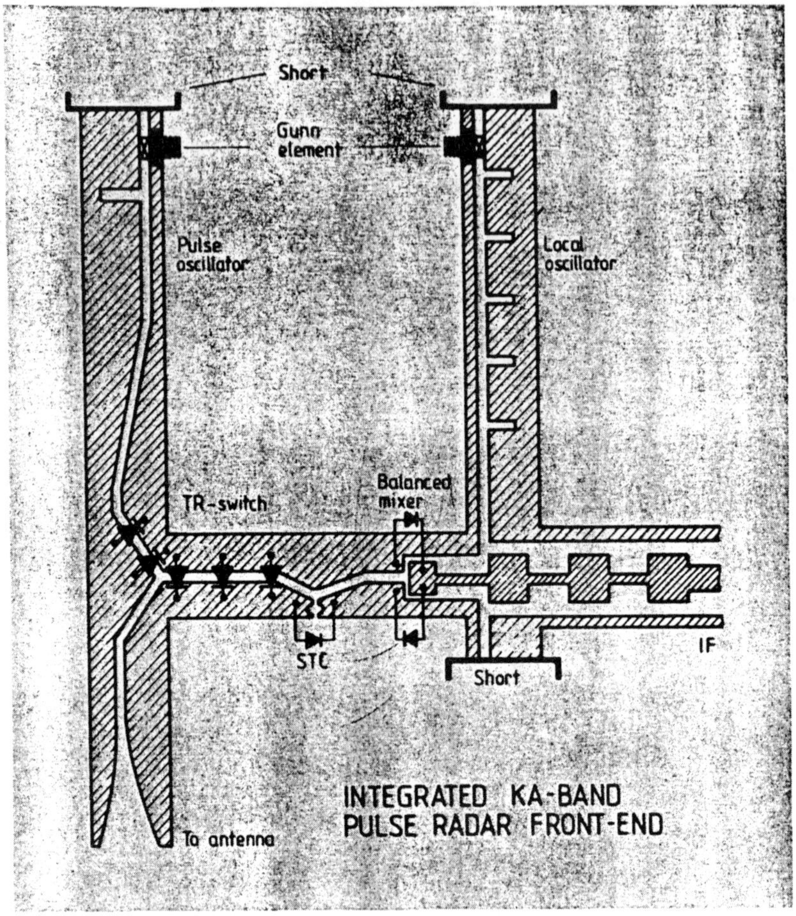

Figure 3. The principal layout of the mm-wave fin-line circuitry, showing the components as they are positioned in the H-shaped waveguide mount.

The balanced mixer is built by the combination of fin-line and coplanar line, thus forming a very broadband 180-degree hybrid. The local oscillator power is fed to the diodes via a transition from a non-symmetrical fin-line to a coplanar line, working in a way similar to a "probe-type" transition from standard waveguide to coaxial line. For proper operation, a backshort is placed a quarter-wavelength apart from the transition. This short also allows a final adjustment to the LO input impedance.

The signal path consists of a symmetrical unilateral fin-line. Schottky-barrier beam-led diodes are placed at the junction between fin-line and coplanar line. Reduced dimensions of the waveguide mount cross section

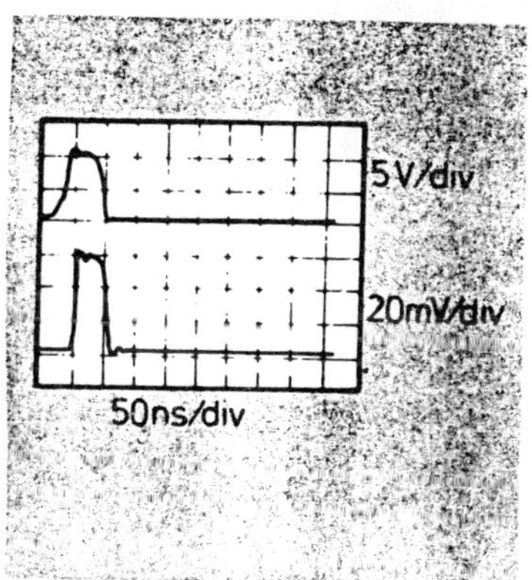

Figure 4. The graph shows the pulse forms of the Gunn bias (top) and of the detected RF signal (bottom), adjusted for a pulse width of 50 ns.

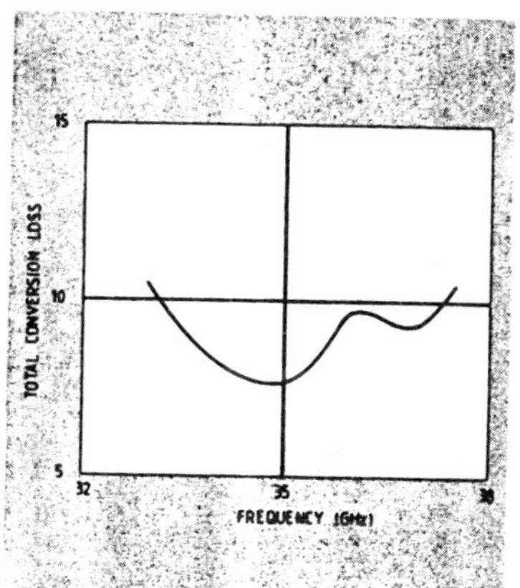

Figure 5. The graph shows the conversion loss of the RF front-end (including the losses of the PIN-diode switches), plotted against frequency. The minimum loss of 7.5 dB is shown at the lowest point of the graph.

around part of the coplanar line yield a mode filter providing the desired quasi-TEM field configuration for the LO power, as well as a reactive termination for the signal path (slot mode) behind the diodes. A microstrip IF low-pass filter is connected to the coplanar line. Line transformers are inserted in the LO and signal path to match the diode impedences.[11]

Performance Data

The radar transmitter operates at 35.55 GHz. The transmitting pulse oscillator delivers a peak power of 0.5 W measured at the output flange. The pulse width is adjustable between 5 and 100 ns, and with the shortest pulse width, a minimum detection range of a few meters is possible. The PRF is chosen at 200 kHz. Figure 4 shows the pulses of the Gunn bias (top) and of the detected RF signal (bottom) adjusted for a pulse width of 50 ns. A maximum bias voltage of 10 V was applied to the Gunn element. The signal path of the TR-switch and the STC attenuator protects the mixer during the transmit mode. The isolation between transmitter and receivers amounts to about 38 dB.

In Figure 5, the conversion loss of the receiver is plotted versus frequency. This curve includes the complete circuit, including the mixer, the transition waveguide to fin-line, the TR-switch, and STC. A minimum conversion loss of 7.5 dB is found. For simplicity of the circuit, no image filter is incorporated in the receiver. So a DSB noise figure (including IF amplifier) of 8 dB is reached, resulting in an SSB system noise figure of about 11 dB. The Gunn oscillators are frequency stabilized to some extent by thermostating their temperature at about 50 degrees centigrade. Under these conditions, IF bandwidths of 100 to 200 MHz are necessary. With these data and an antenna gain of 40 dB, a signal-to-noise ratio of 10 dB and a radar cross section of the target of 1 m^2, a maximum detection range of about 400 m is achieved. Further improvements are possible by adding an image filter and by integrating the received pulses.

Thus, a very compact but high-resolution, lightweight and potentially low cost radar module has been achieved. This special unit was designed for civil applications, where distances from a few meters up to a few hundred meters have to be measured, and not only metallic but dielectric targets are present that would be difficult to detect at lower frequencies.

The radar shows the feasibility and the great potential for planar integration techniques for mm-wave circuits. Based on these experiences, the design of similar systems with improved performance and higher frequencies[12] for military applications is possible. Successful shock acceleration measurements up to a few thousand Gs and temperature tests from −30 to 70 degrees centigrade have proven the ruggedness of fin-line circuits. ■

References

1. Witse, J.C., "Millimetre Wave Technology and Application," *Microwave Journal*, Vol. 22, August 1979, pp. 39–42.
2. Sundaram, G.S., "Millimetre Waves: The Much-awaited Technological Breakthrough?" *Int. Def. Rev.*, 2, 1979, pp. 271–277.
3. Plattner, A., and Meinel, H., "A 60 GHz Hand-Held LPI-Transceiver," *Symp. Dig.*, 12th EUMC, Helsinki, 1982.
4. Rembold, B., Wipplch, H.G., Dischoff, M., and Frank, W.F.X., "A 60 GHz Collision Warning Sensor for Helicopters," *Symp.-Dig.*, 3rd Military Microwave Conf., MM82, London.
5. Menzel, W., Meinel, H., Rembold, B., and Schmidt, L.P., "MM-wave Planar Integrated Components for Frequencies up to 110 GHz," *MSN*, Vol. 12, No. 10, Oct. 1982.
6. Kuno, H.J., "Solid-state MM-wave Power Sources and Combiners," *Microwave Journal*, Vol. 24, June 1981, pp. 21–34.
7. Adelseck, B. et al, "A Survey of Planar Integrated MM-wave Components," *The Radio and Electronic Engineer*, Vol. 52, No. 1, 1982, pp. 46–50.
8. Menzel, W., Callsen, H., Adelseck, B., and Sicking, F., "An Integrated Ka-band Receiver Front-end," *Proceedings of the 11th European Micro. Conf.*, Amsterdam 1981, pp. 361–364.
9. Hofmann, H., "MM-wave Gunn-oscillator with Distributed Feedback Fin-line Circuit," *IEEE Symp. Digest MTT-S*, Washington, D.C., 1980, pp. 59–61.
10. Meinel, H., Callsen, H., "Fin-line PIN-diode Attenuators and Switches for the 94 GHz range," *Electronics Letters*, Vol. 18, 1982, pp. 541–542.
11. Menzel, W., Callsen, H., "94 GHz Balanced Fin-line Mixer," *Electronics Letters*, Vol. 18, 1982, pp. 5–6.
12. Menzel, W., Callsen, H., "Integrated Fin-line Components for Radar and Radiometer Applications, *Symp. Digest*. 3rd Military Microwave Conf., MM82, London 1982.

7.6.2

DESIGN

A dual-channel receiver front-end for a 94-GHz radar with dual-polarization capabilities is described, which consists of a polarizer, circulator, filter, STC PIN-diode switches, balanced mixers, and a power divider. Except for the polarizer, the front-end is built using E-plane techniques.

Compact 94-GHz Dual-Polarization Radar Receiver Realized Using Different Integration Techniques

By
Wolfgang Menzel,
AEG-Telefunken

Millimeter-wave radar systems operating at 94 GHz provide high resolution with rather small antenna diameters, while still maintaining acceptable atmospheric attenuation and far better penetration into obscurants such as clouds, fog, and dust, compared to infrared. At the same time, millimeter-wave techniques have reached a status such that sophisticated applications have become possible.[1]

Additionally, planar integrated circuits (for example, finline, microstrip) have allowed the building of high-performance, rugged, but relatively low-cost mm-wave circuits, even up to 140 GHz.[2-6] Circuit elements requiring high dimensional tolerances are replaced by planar structures deposited on a dielectric substrate where high precision is no problem. Compared to microstrip, finline circuits exhibit lower losses, larger component bandwidth, and far better compatibility with waveguide circuits.

Specifically, this discussion concentrates on a dual channel receiver

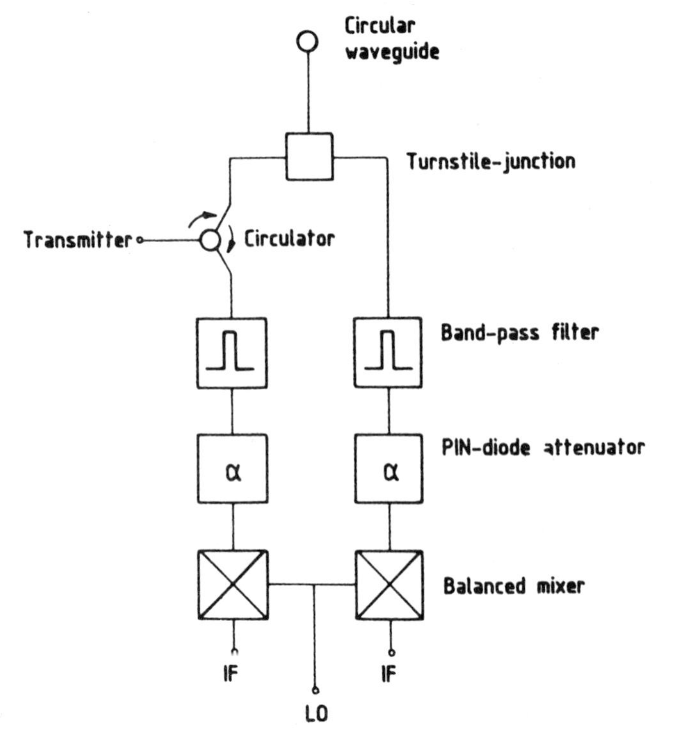

1. *In this dual-channel receiver front-end* *for a coherent 94-GHz radar, a turnstile coupler connects the two receiver channels with the circular input/output waveguide coupled to the antenna feed.*

Dr. Wolfgang Menzel is with the New Technologies section of AEG-Telefunken Anlagentechnik, AG, Sedanstrasse 10, D-7900 Ulm (Donau); Tel.: (0731) 192-0.

DESIGN

2. **Disassembled, the receiver block** and front-end mount reveal the turnstile junction, which includes matched 90-degree E-plane bends.

front-end for a coherent 94-GHz radar with dual-polarization capabilities. A block diagram of the front-end is given in Figure 1. A turnstile coupler connects the two receiver channels with the circular input/output waveguide coupled to the antenna feed. Each channel corresponds to one sense of the circular polarization in the antenna feed. In one of the receiver input channels, a circulator is incorporated as the transmit-receive diplexer. In this configuration, one polarization is transmitted, whereas co- and crosspolarization are received.

In a later version of this radar, both senses of the circular polarization are required for the transmitted signal, thus another circulator has to be added in the second channel and an SPDT switch to select the desired polarization. Following this, both channels contain bandpass filters, PIN-diode STC attenuators, and balanced mixers.

DESIGN

Finally, a power divider provides the mixers with the LO power.

Components

To achieve a compact and rugged front-end, different techniques have been combined—a standard metal waveguide circuit for the polarization diplexer and different E-plane techniques for the remainder of the circuit. In this way, the front-end (except for the turnstile junction) can be integrated in a single split-block mount. Figure 2 shows the disassembled receiver block and the complete front-end mount. The turnstile junction consists of two parts and includes matched 90-degree E-plane bends for compatibility with the remaining configuration. As the transmit/receive diplexer, an E-plane circulator was chosen. The two filters also use metal inserts in the E-plane. The PIN-diode switches, the balanced mixers, and the LO power divider are built employing finline circuitry and are integrated on a single substrate.

Turnstile Junction

This junction transforms the two circular polarized waves of the circular input waveguide into rectangular waveguide modes. Each rectangular waveguide is coupled to one sense of the circular polarization in the circular waveguide.

The dimensions of the 94-GHz version were scaled down from lower-frequency models.[7] A final adjustment can be done by moving the cylindrical post up and down in the junction center. In Figure 4, the transmission properties of the 94-GHz version of this unit are plotted. The insertion loss S_{14} includes a transition from rectangular to circular waveguide and a quarterwave-plate polarizer from linear to circular polarization. Thus,

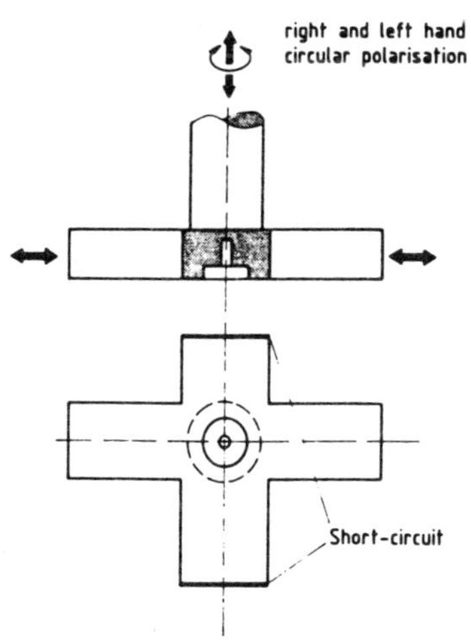

3. **At the turnstile coupler,** each rectangular waveguide is coupled to one sense of the circular polarization in the circular waveguide.

the losses of the turnstile junction are below 0.5 dB.

Circulator

In contrast to the H-plane circulators normally employed a flat ferrite disk is used in the E-plane circulator instead of a rod. The ferrite disk is fitted exactly into a symmetric E-plane waveguide junction.[8] The center frequency of the circulator is adjusted by varying the height of the ferrite disk; for 94 GHz, this value amounts to about 0.25 mm.

Magnetic bias is provided by a small rare-earth permanent magnet mounted in the waveguide block. The gap between the permanent magnet and the ferrite disk is about 0.2 mm, producing magnetic field bias on the order of 2 to 3 kGauss. The measured insertion loss characteristics of the circulator junction are plotted in Figure 5. It is seen that the insertion loss in the through-pass is about 0.5 dB, and the 20-dB-isolation bandwidth is about 1.2 GHz.

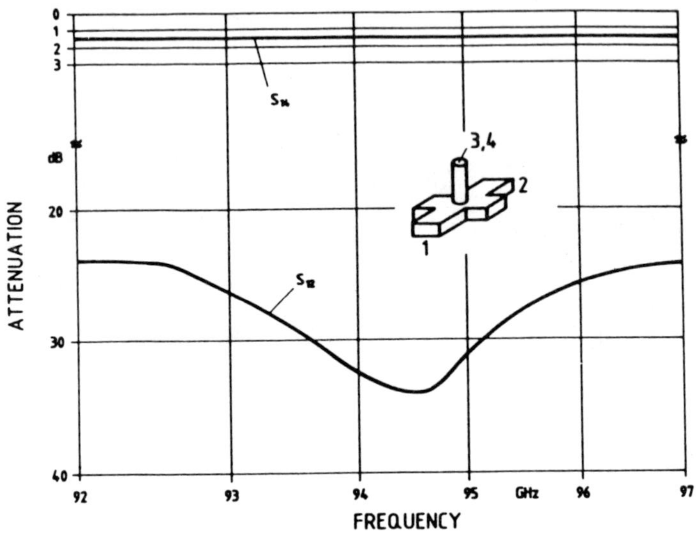

***4. A plot of the transmission properties** of the 94-GHz version of the unit shows insertion loss S_{14}, which includes a transition from rectangular to circular waveguide and a quarterwave-plate polarizer from linear to circular polarization.*

Band-pass Filters

For the rejection of image frequency, metal insert bandpass filters are used. The filter structures are etched from 50-μm-thick copper foil, gold plated, and inserted in the E-plane of the waveguide. The filter dimensions were designed by the University of Bremen using field theoretical computation procedures.[9]

A typical filter characteristic is plotted in Figure 7. The minimum insertion loss amounts to 1.6 dB, and the 3-dB bandwidth is 1100 MHz. Etching the filter substrates, the center frequency can be met within about 300 MHz. The filter characteristics are relatively good. If necessary, the substrates with the best performance are selected.

Finline Circuit

PIN-diode switches, balanced mixers, and the LO power divider are integrated on a single planar substrate using a finline technique. For the STC attenuator, two PIN diodes are placed across the finline. Ideally, with forward bias, these diodes short-circuit the finline, whereas with or without negative bias voltage, they do not affect the line.

In reality, however, the parallel capacitance of the diodes leads to

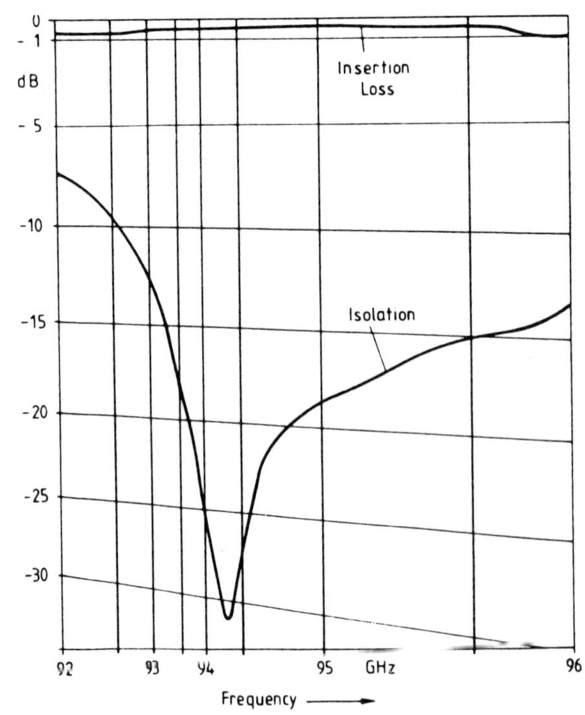

***5. In the measured insertion-loss characteristics** of the circular junction, it can be seen that the insertion loss in the through-pass is about 0.5 dB, and the 20-dB isolation bandwidth is about 1.2 GHz.*

DESIGN

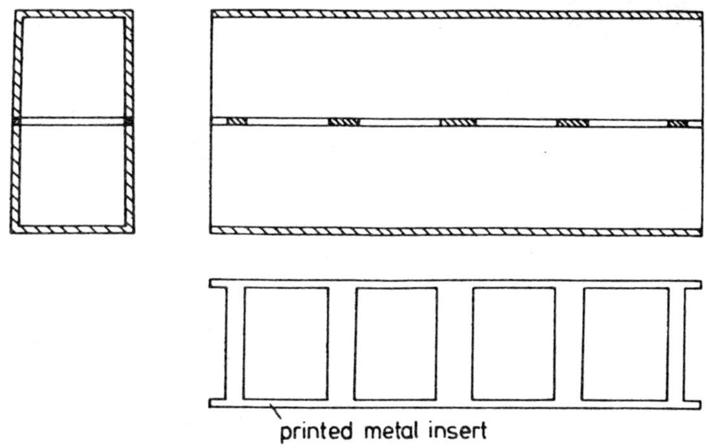

6. The filter structures are etched from 50-μm-thick copper foil—gold-plated—and are inserted in the E-plane of the waveguide.

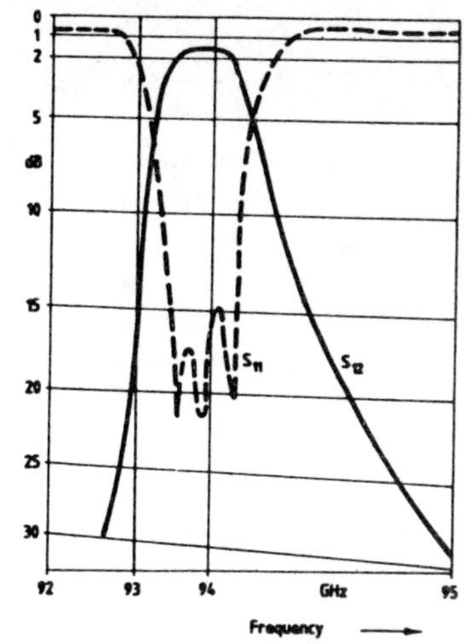

7. In this typical filter characteristic, minimum insertion loss amounts to 1.6 dB, and the 3-dB bandwidth is 1100 MHz.

finite insertion loss, and series resistance and inductance as well as higher-order mode coupling limit the maximum attenuation. By properly spacing the two diodes—about one-quarter of a wavelength apart—the diode capacitances compensate to some extent, and the maximum attenuation is approximately doubled compared to a one-diode attenuator. As a discrete component, such an attenuator exhibits an insertion loss of about 1.5 dB, and a maximum attenuation between 25 and 30 dB covering a complete waveguide band.[10]

In the integrated circuit, the insertion loss is reduced due to the lack of tapers from waveguide to finline. The finline balanced mixers are based on the combination of finline and coplanar line to give a broadband 180-degree hybrid. The signal is fed to the diodes via a symmetrical finline, whereas the LO power enters via a transition from finline to coplanar line. This transistion works in a similar way to that of a probe-type transition from waveguide to coaxial line. For proper operation, a short is placed at the transition, allowing some tuning of the LO input impedance. An IF microstrip lowpass filter is connected to the coplanar line. A reduced waveguide cross-section in the first part of the coplanar line forms a mode filter allowing only the quasi-TEM mode of the LO to propagate to the diodes.

Additionally, the RF path (slot mode) is terminated by this waveguide channel with a short about a quarter wavelength away from the diodes.

As each balanced mixer requires an LO power of 12 to 13 dBm for self-bias operation, bias is applied to the mixer diodes. Figure 9 gives the conversion loss of the two receiver channels, with 5 mW LO power for each mixer and about 2 mA bias. These data include the PIN-diode switches, but not the circulator and filter. The converson loss amounts to about 8.5 dB, approximately 7.5 dB from the mixer and 1 dB from the PIN-diode attenuator. An excellent balance is found between the two channels. At IF frequencies of 1.3 GHz, slight resonances occur due to the bias circuit; as they do not interfere with the actual IF. However, no measure was taken for their suppression. The maximum attenuation of the PIN-diode attenuators is greater than 28 dB.

Final Results

Finally, circulator, filters and planar substrate were assembled in the front-end mount and tested as a whole. In Figure 10, the overall conversion loss of this arrangement is plotted versus frequency, exhibiting a minimum value of about 11 dB. This loss is composed of the 8.5 dB

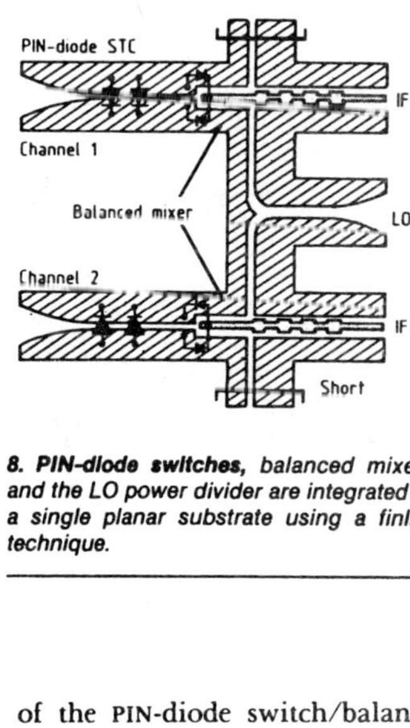

8. PIN-diode switches, balanced mixers, and the LO power divider are integrated on a single planar substrate using a finline technique.

of the PIN-diode switch/balanced mixer configuration, 1.6 dB from the filter, 0.5 dB from the circulator, and a few tenths of a dB from additional line losses and the interference of mixer and filter mismatch.

This interference is also responsible for the slight asymmetry of the curve in Figure 10. Due to the filter, a response to the image frequency is suppressed by 25 to 30 dB. A higher suppression would have been possible with a modified filter, but only at the expense of higher filter losses.

The results presented here show the possibility of realizing compact mm-wave front-ends employing dif-

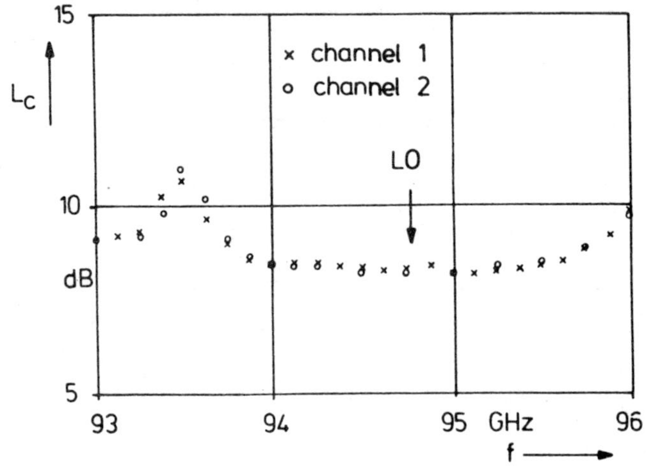

9. The conversion loss *of the two receiver channels, with 5 mW LO power for each mixer and 2 mA bias includes the PIN-diode switches, but not the circulator and filter.*

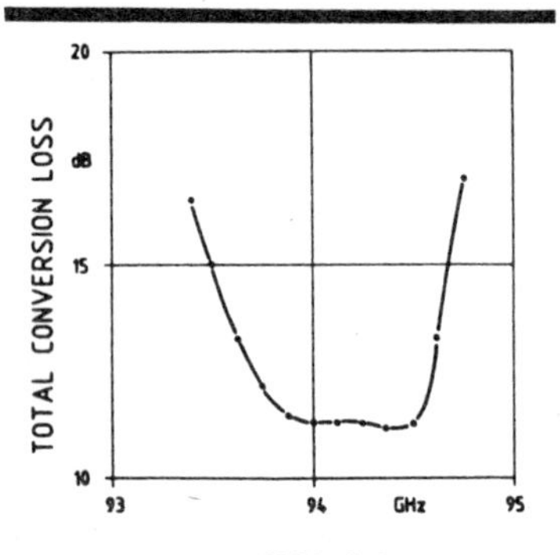

10. The overall conversion loss *of the assembled front-end is plotted versus frequency, exhibiting a minimum value of approximately 11 dB.*

ferent integration techniques. The front-end mount has outer dimensions of 76 x 32 x 20 mm. This is not, however, the final limit of miniaturization. Based on the experience with this first front-end, further reductions in size and weight are possible. Furthermore, dual-polarization operation for the transmitter will be included by adding a second circulator and a finline SPDT switch[10] to select the polarization of the transmitted signal. ■

References

1. J.C. Wiltse, "Millimeter-wave Technology & Applications," *Microwave Journal*, Aug. 1979, pp 30–42.
2. R.E. Scarman, et al, "Millimetric Microstrip Components and Their Application to Integrated Front-ends," *Military Microwave Conf.*, London, 1982, pp. 92–102.
3. W. Menzel, et al, "Planner Integrated Components Perform Up to 110 GHz," Microwave Systems News, Nov. 1982, pp. 92-102.
4. W. Menzel and H. Callsen, "Integrated Fin-line Components and Subsystems At 60 and 94 GHz," *IEEE Trans. on MTT-31* 1983, pp. 142–146.
5. W. Menzel, "A 140-GHz Balanced Mixer for Finline Integrated Circuits," *13th European Microwave Conf.*, Nurnberg, Germany, pp. 179–182.
6. H. Callsen, "Fin-line PIN-diode Attenuator and Switch for the 140 GHz Range," *8th International Conf. on Infrared and Millimeter Waves*, Miami, Florida, Dec. 1983.
7. P.A. Crandell, "A Turnstile Polarizer for Rain Cancellation," *IRE Trans. on MTT-3*, 1955, pp. 10–15.
8. K. Solbach, "E-plane Circulators 30 through 150 GHz for Integrated Mm-wave Circuits," *13th European Microwave Conf.*, Nurnberg, 1983.
9. J. Bornemann, et al, "Optimized Low-insertion loss Millimeter-wave Finline and Metal Insert Filter," *The Radio and Electronic Engineer*, Vol. 52, No. 11/12, 1982, pp. 513–521.
10. H. Meinel and H. Callsen, "Finline PIN-diode Attenuators and Switches for 94 GHz Range," *Electron Letters*, Vol. 18, 1982, 541–542.